SHUTTLEWORTH COLLEGE
LIBRARY

Shuttleworth
100445

Ecology 2

Ecology 2

Paul Colinvaux
Center of Tropical Paleoecology
Smithsonian Tropical Research Institute
Balboa, Republic of Panama

John Wiley & Sons, Inc.
New York Chichester Brisbane Toronto Singapore

ACQUISITIONS EDITOR / Sally Cheney
PRODUCTION MANAGER / Pam Kennedy
MARKETING MANAGER / Catherine Faduska
PRODUCTION SUPERVISOR / Sandra Russell
DESIGNER / Kevin Murphy
MANUFACTURING MANAGER / Andrea Price
COPY EDITING SUPERVISOR / Elizabeth Swain
PHOTO RESEARCH DIRECTOR / Stella Kupferberg
PHOTO RESEARCHER / Jennifer Atkins
PHOTO RESEARCH ASSISTANT / Lisa Passmore
ILLUSTRATION COORDINATOR / Anna Melhorn/NETWORK GRAPHICS
COVER PHOTO / Chuck Place/THE IMAGE BANK

This book was set in 10/12 (ITC) Bookman Light by Waldman Graphics, Inc. and printed and bound by R. R. Donnelley (Crawfordsville). The cover was printed by The Lehigh Press, Inc.

Recognizing the importance of preserving what has been written, it is a policy of John Wiley & Sons, Inc. to have books of enduring value published in the United States printed on acid-free paper, and we exert our best efforts to that end.

Library of Congress in Publication Data:

ISBN 0-471-55860-5

Printed in the United States of America

10 9 8 7 6 5 4 3 2 1

Preface

I like writing. Among the most treasured happenings of the quarter century I spent as a professor were the times, five of them now, when extracts from my scientific writings were used in readings meant for English classes. I see no reason why a textbook on so rich a subject as ecology should not give pleasure in the reading. To be comprehensible is not to lack rigor. I have tried to be comprehensible.

Ecology2 is actually the third incarnation of my ecology text, the first appearing in 1973 (under the title *Introduction to Ecology*) and the second in 1986. I had much new material to add to the present edition: several chapters are entirely new, and several more overlap only partly with their predecessors. This meant that the old material needed pruning, for I was determined that the book of 1993 should be no longer than that of 1986. The best way to prune is to write more concisely.

Where old material has been retained, I reviewed every paragraph, often rewriting entirely as I went. Usually I found myself recasting the entire chapter. I reordered chapters, with the aim of making the four parts of the book independent entities, each with its own beginning and ending. An extra benefit of this is that the parts of the book (Energy, Diversity, Community, System) can be shuffled more easily, so that the grander subdivisions of ecology can be read and taught in any order.

Every chapter preview has been entirely rewritten. The new versions are not summaries of what follows, nor do they refer to all the issues to be taken up. Instead they are meant as diminutive essays that tell of the more important or interesting ideas and phenomena discussed in the chapter.

In the new organization, climate is disentangled from biogeography and brought toward the front of the book, rounding out a study of the consequence of energy flows for life on earth that is the subject of the new Part I. The larger change in substance is in the treatment of population biology and speciation, the subject of the new Part II (Diversity). I have recast the ancient debate on the causes of the balance of nature in the ways suggested by the work of Robert May and the others working with difference equations and approximations to chaos. I persist in my habit of keeping math to a minimum, intending the book to be read by the literate as well as the numerate.

I review contemporary ideas of cascades in food webs in a new chapter, and I have attempted a summary of the ecological properties of oceans in another. The ecological basis for some of the ideas of environmentalism will be found in a number of chapters, in the discussions of diversity and population regulation, and in most of the chapters of Part IV (System). My new chapter on the human impact (Chapter 29) is not a tale of pollution, such as can be found in environmental science books, but rather looks at humans, and the human condition, as they are consequences of choices made by natural selection in evolving populations of hominids long ago. The consequences of our ancient traits are with us yet.

I stress the findings of my own speciality, paleoecology, without apology. Understanding the diversity of life on earth requires that we master the time-scales on which evolution happens. The end of the last ice-age was roughly 40 tree life-times ago, and nearly all contemporary plants and animals have passed most of their existence as species under conditions of the ice-age earth, rather than in the odd climates of short interglacials like the present. The time-scales of evolutionary ecology are those of millennia, or tens of millennia, more than they are those of generation change. Thus a student of evolutionary ecology needs a paleoecological perspective as much as does a student of global change. I have used paleoecological data throughout the book, not just in the review chapter called "The Ecologist's Time Machine."

The new introductory chapter, "The Science of the Universe" echoes one of G. E. Hutchinson's more grandiloquent phrases, but I use the title as an excuse for writing a preview of the whole book, and, of course, of ecology as I know it. Thence the organization is a version of the "levels of integration" approach, though the first emphasis is on energy as much as on the individual. The partitioning of energy between the quick and the dead, as it were, and the necessary adaptations of individuals to the results of that partitioning, give structure to life.

The behavior, and diversity, of species populations make up the second part, the whole naturally falling together as an inquiry into the causes of diversity. Building and persistence of communities is the natural next part, and one in which Pleistocene time-scales are crucially important. Accordingly, I begin Part III with a review of paleoecological methods, though it is a shorter review than in the 1986 book. The final part on systems is an analysis of the biospheric process, ending with a discussion of the origin and control of the earth's peculiar atmosphere.

I have kept the number of references to a minimum. Textbook writers face a dreadful temptation to pile in references; 'tis so easily done, and the result is a swelling bibliography adequate for the most rigor-minded graduate student. The references I have left in the text were meant to pass the test: Is this necessary to guide the student to the primary studies, to back up what I say, or to help the captivated student who wants to read more or find the latest work?

THE BOOK AS A TEACHING TOOL

There must be nearly as many ways of teaching ecology as there are teachers of the subject. An "integration" approach, working from individual to ecosystem, as pioneered by E. P. Odum in his classic books of the 1950s, remains one of the more popular, and *Ecology2* follows roughly this format. If time is pressing, or one is impatient to get to the big questions of diversity and community, I recommend deleting chapters 4–6 on individual energetics from course requirements. Chapter 12 on the origin of sex is also removable, and I admit, with reluctance, that the Chapter 17 review of paleoecology could be omitted in many treatments. And when time is running out at the end of a course, soil, lakes, and oceans can go easily enough. With this stratagem of judicious deletion, *Ecology2* can fit the Procrustean requirements of most courses set to the "levels of integration" approach.

Because each of the four parts of the book is designed to stand on its own, they can easily be reshuffled. The introductory chapter will work for any reordering of the four parts. A common alternative to the "levels of integration" approach is to follow deductive logic from the complex to the simple, starting with ecosystems and ending with populations. I have frequently done this myself, so that my writing reflects this possible need. I suggest the following:

For ecosystem to population approach

Introduction

Part I: Energy

Part IV: System

Part III: Community

Part II: Diversity

Postscript

The parts also lend themselves to an early emphasis on populations or ethology as follows:

For population or behavioral approach

Introduction

Part II: Diversity

Postscript

Part III: Community

Part I: Energy

Part IV: System

In Tables 1–3, I suggest reading outlines for shorter courses organized from alternative population, behavioral, or environmental points of view. Because the chapters are written with the idea that each should be an independent essay, these various organizations should let the subjects flow well enough.

TABLE 1
Abridgement with Population Emphasis

Introduction
Part II
  Chapters 8–11
  Chapters 13–16
Part III
  Chapter 17 pages 338–347
  Chapter 18
  Chapters 20–22
Postscript: The Human Impact
Part I
  Chapter 2
  Chapter 3, pages 30–36, 45–49
  Chapter 7
Part IV
  Chapter 23
  Chapter 28

TABLE 2
Abridgement with Behavioral Ecology Emphasis

Introduction
Part II
  Chapters 8–16
Postscript: The Human Impact
Part III
  Chapter 18
  Chapters 20–22
  Chapter 29
Part I
  Chapter 2
  Chapter 3, pages 30–36, 45–49
  Chapter 7
Part IV
  Chapter 23
  Chapter 28

TABLE 3
Abridgement with Environmental Emphasis

Introduction
Part I
  Chapter 2
  Chapter 3, pages 30-49
  Chapter 7
Part IV
  Chapters 23-28
Part II
  Chapters 8-11
  Chapters 14-16
Part III
  Chapter 17, pages 338-347, 360-363
  Chapter 18
  Chapters 20-22
Postscript: The Human Impact

THE ECOLOGIST'S SEVENTH SENSE

Ecology is the most privileged of sciences. We who follow it savor the life of a naturalist while using the methods of chemistry or the philosophy of mathematics. Above all we see the workings of natural selection in the form and function of every living thing. Long ago, Darwin said admiringly of mathematicians that they seemed to have a sixth sense. But ecologists have a seventh sense, even more precious and coming from their gut understanding of natural selection. The shape of a tree as a great light diffuser, the singing of birds in the morning, the puny size of a male black widow spider beside his fearsome mate, the flash of a firefly and the croaking of a frog: —everything noted by the ecologist as naturalist in a country walk seems right and good and proper before the seventh sense that understands natural selection.

I wrote most of this book in a cottage at Woods Hole, with the waters of Buzzards Bay just visible through the trees and the library of the Marine Biological Laboratory a comfortable walk away. I conclude it looking out over the Bay of Panama, with the soft growling of waves beneath my window and the lights of freighters entering the Panama Canal to my right. Over to the left the English pirate Morgan anchored 300 years ago before storming ashore in one of the most ruthless robberies in history. In the evening, greater ships than Morgan's, with jet engines throttled for their final approach, bank over his old anchorage before landing at the airport near the ruins that are all he left of a city. In my mind's eye I can see Morgan's ship there still; what if his ruffians could have imagined those wonderful crafts that would float above that anchorage on their way to land just three centuries later? But those three centuries are nothing in ecological time. The cormorants and pelicans, and the lone osprey, fish the same waters now as then. Real time is measured in millennia, and tens of millennia, and hundreds of millennia. It is into this real time that our seventh sense grasp of natural selection most easily fits.

Acknowledgments

I have a huge debt to The U.S. National Science Foundation, which has funded my research almost continuously for more than 30 years. The NSF projects grant system, with its forbidding hazards of peer review, is the protector of freedom in science. No better system of seeking out the great questions of science, or of rewarding individual initiative in science has been found by any country. I offer my heartfelt thanks to those who maintain the NSF: its staff, its reviewers, and its defenders in Congress.

This book was written after twenty-five years of teaching ecology. Inevitably I have borrowed from the minds of more than two generations of ecologists, my own and my teachers' generation. Most important to the formation of my thinking have been D. A. Livingstone, E. S. Deevey, and G. E. Hutchinson; my gratitude to the happenstance that placed me in such intellectual descent is not easily expressed. D. A. Livingstone not only guided my thesis long ago but reviewed my earlier texts as well as this one. R. H. Whittaker served as role model and counsellor when I wrote my first text of 1973. Other teachers of my formative years in ecology whose thoughts became embedded in my own were P. H. Klopfer, K. Schmidt-Nielsen, H. J. Oosting, and H. G. Andrewartha. I have a special debt of gratitude to Michael C. Miller who has been my essential companion at remote field camps in the Andes, Amazon, Siberia, and Russian Arctic for the last 15 years and from whom I have learned much. In recent years I have been privileged to work with, and to learn from, N. A. Shilo of the Russian Academy of Sciences and his research teams in Moscow and Magadan led by N. Patyk-Kara and T. Lozhkin. For intellectual companionship and wisdom I thank my research associates Kam-biu Liu, M. B. Bush, W. Eisner, and W. N. Mode and my graduate students T. Ager, E. Asanza, M. Bergstrom, P. D. Boersma, N. Carter, P. R. Colbaugh, I. Frost, D. Goodman, D. A. Greegor, D. Maxwell, L. Parrish, P. De Oliveira, M. Riedinger, F. Sarmiento, M. Scheutzow, J. Shackleton, M. Steinitz-Kannan, C. Vasquez, and J. Weiler. The coral reef biologist L. Hillis is my companion in ecology as in life, and her imprint is on all my work.

This book has benefitted from trenchant reviews from Robert Stavn, Daniel Livingstone, Robert Kalinsky, Stuart Fisher, Kathy Ann Miller, Peggy Wilzbach, Mark

Brinson, Laszlo Szijj, Julia Krebs, S. E. White, Donald Moll, and Brian C. McCarthy. Reviews of earlier versions of this text and of my other ecological books were important in the genesis of this text, including those by J. Webster, D. C. Coleman, and D. Johnston, A. J. Brook, N. Stanton, J. A. MacMahon, F. B. Golley, C. A. S. Hall, H. S. Horn, and P. D. Moore. People who have contributed on the road to this ecology text in ways ranging from field companion through critic and colleague to an illuminating conversation include M. Acosta-Solis, M. G. Barbour, E. S. Barghorn, B. Barnett, W. D. Billings, J. Blackwelder, L. C. Bliss, S. Bolotin, R. Bonnefille, I. Brodniewicz, W. S. Broecker, J. L. Brooks, J. Brown, L. Brown, J. B. Calhoun, R. A. Carpenter, R. G. Cates, P. Chesson, J. O. Connell, J. S. Creager, M. Cunningham, E. J. Cushing, M. B. Davis, D. Deneck, T. deVries, J. Doherty, J. F. Downhower, G. M. Dunnett, W. T. Edmondson, I. Eible-Eibesfeldt, M. Ewing, R. S. R. Fitter, R. E. Flint, M. Florin, D. Frey, A. S. Gaunt, Z. M. Gliwicz, H. Godwin, R. P. Goldthwaite, T. Goreau, C. E. Goulden, P. R. Grant, A. T. Grove, T. Grubb, J. P. Hailman, B. Hajek, M. P. Harris, J. Hatch, C. J. Heusser, H. Higuchi, D. M. Hopkins, S. P. Hubbel, H. W. Hunt, J. Jackson, P. L. Johnson, D. E. Johnston, L. Krissek, W. S. L. Laughlin, E. P. Leopold, W. M. Lewis, M. Lieberman, O. L. Loucks, R. H. MacArthur, P. S. Martin, R. M. May, R. McIntosh, J. H. Mercer, R. D. Mitchell, H. Nichols, H. C. Noltimier, W. J. O'Brien, K. Olson, G. H. Orians, F. Ortiz, R. Patrick, R. Perry, T. J. Peterle, G. M. Peterson, K. G. Porter, F. W. Preston, R. Pulliam, C. Racine, J. L. Richardson, J. E. Richey, M. R. Rutter, E. K. Schofield, K. Schmidt-Koenig, J. Sedell, W. M. Shields, A. Sih, C. Smith, F. E. Smith, G. Sprugel, H. and T. Steinitz, J. R. Strickler, M. Stuiver, J. C. F. Tedrow, D. Tilman, M. Tsukada, F. C. Ugolini, J. Vagvolgyi, L. Van Valen, A. L. Washburn, M. Watanabe, W. A. Watts, D. S. Webb, T. Webb, F. H. West, G. W. Wharton, T. C. R. White, D. R. Whitehead, T. G. Whitham, D. T. Wicklow, R. G. Wiegert, I. L. Wiggins, E. O. Wilson, W. C. Wimsatt, and K. Winter. M. Paton Villar brought order to my references. This latest version of my ecological writings was made possible by the vigorous promotion of the enterprise by my editor S. Cheney. I offer her my special thanks, as well as to S. Kuppferberg and the production staff at John Wiley who go about the making of books with an informal enthusiasm.

Contents

Chapter 1

Introduction: The Science of the Universe

Ecology is the study of how the world works, "the science of the universe" as G. Evelyn Hutchinson, one of the greatest of ecologists, once called it. Most of us working ecologists narrow this definition somewhat, restricting our studies to the planet we live on, and of course its energy supplies. This still leaves us plenty to do.

Our generation knows "ecology" to grapple with pressing questions of public concern: greenhouse warming, pollution, habitat destruction, spreading deserts, feeding the hungry, saving rain forests. This "ecology" fits well in the definition of "how the world works." The public affairs side of ecology thus overlaps geography, climatology, and even economics, in studying the systems that run the earth.

But at its core, ecology is a biological discipline. Ecologists concern themselves first with life on earth, asking how the living part of the earth system works. They see a biosphere in which life is driven by the energy of the sun. Green plants are staked out under that sun in every part of the earth that is moist, warm, or nutrient-rich enough to let them live, using solar energy to drive a chemical synthesis of carbohydrate fuels. The rest of life (with the exception of remarkable chemists among the bacteria) persists as a tax on the fuel hoards that the green plants make. Animals eat plants; fungi and bacteria rot plant corpses. All give off heat, so that the energy once trapped from the sun by green plants is radiated back to outer space. Elegant in its simplicity is this ecologist's overview of how the world works.

But the details are not simple at all. Consider a meadow or old pasture; a place of moisture, warmth, and good soil where the green plants are indeed staked out under the sun as expected. In keeping with the simplicity of the model, all are synthesizing the same carbohydrate fuel using essentially the same complex chemistry.

But the meadow plants are of many different species, all apparently doing the same essential thing. Why this redundancy? Why dozens of plant designs? Why not just one perfect pasture plant? Or consider the tropical rain forest, now in the news for its breath-taking diversity. A botanist recently identified 300 species of good-sized tree growing in a hectare (10,000 m^2 or 2½ acres) of Amazon rain forest, although the total number of trees on that hectare was only 600 (Gentry, 1988). Not the kind of thing a commercial forester would aim for.

Variety is the outstanding characteristic of life; variety that comes as variations on common themes. At present counting, the world holds about half a million species of green plant, all of them getting their energy supplies in the same way, through the common biochemistry of photosynthesis. The redundancy seems absurd. And it illustrates how the intellectual tasks of ecologists are of a different order from those of cellular and molecular biologists who study life at a chemical level. The chemistry is awesome, and the skills needed to understand it perhaps more awesome still. But the chemistry is all the same, so that when you have mastered the chemistry of photosynthesis for one green plant you know pretty well how they all work. Remaining is the basic ecological question of diversity: "Why are there so many different kinds of plants, all wielding the same chemistry to perform the same synthesis?"

The diversity of animals is far greater even than the diversity of plants. Ecologists love to tell the tale of the small talk in an English railway carriage early in this century, when an eminent churchman found himself shut in for the journey with the celebrated evolutionary scholar, J. B. S. Haldane. Apparently desperate to find common ground for conversation, the churchman asked Haldane what, as an evolutionist, he considered to be a principal characteristic of the Creator. Haldane answered, "An inordinate fondness for beetles."

Present estimates of the numbers of species still living run anywhere from 3 million to 30 million, most of them small and most of them insects. Some of these are common, whereas others are seemingly always rare. All can be seen to be capable of prodigious feats of reproduction, and yet the common stay common and the rare stay rare, with not even the most rapid breeders displacing the others from the systems in which they live. Numbers are controlled; diversity is maintained.

The world works the way it does because immense arrays of different animals and plants work in unison, achieving that elegant simplicity of fueling the biosphere from the sun, then radiating back the energy as heat. An ecological imperative is to understand the causes and maintenance of variety, of diversity, of the very complexity that underlies this seemingly simple system.

And because life is diverse, the different species must live together. Living is communal living, and resources are shared. The environment for the animals and plants of a place is set as much by the other animals and plants as by the physical conditions for life. A second imperative of ecology is to discover the laws of communal living.

Because ecology studies the causes of diversity, the limits to number, and the ways species live together it is at its core the most evolutionary of subjects. "Why are there so many species?" and "How have they come to live together?" are questions that must be addressed to the process of natural selection. Ecology is the study of how life was fashioned by natural selection.

The word "ecology" is well over a hundred years old, being coined by the

German biologist Ernst Haeckel in 1866 to express his idea of the varied animals of a place evolving to live together, perforce to share an earthly home and its resources. Thus the Greek word **oikos**, for "homestead," combined with **logos**, for "wisdom," to yield **œkology**, now rendered **ecology**, to mean the study of the household of nature. Charles Elton of Oxford defined the modern subject of ecology nearly 70 years ago as "the study of animals (and plants) in relation to habit and habitat" (Elton, 1927).

Economics, the dismal science, takes its name from that same Greek word for household. When Darwin first wrote of the "struggle for existence," he actually borrowed the language of economics to talk of the "economy of nature." Economists might now gain wisdom from ecologists by following Elton and defining their subject as the study of people in relation to habit and habitat.

NATURAL SELECTION AS AN OBSERVABLE PROCESS

Natural selection is a necessary consequence of three essential properties of living things, properties so obvious that they must have been common knowledge as long as thinking humans have existed:

1. All animals and plants procreate with such vigor that the average number of offspring is more than is needed to replace the parents.
2. Individuals differ in ways that can be shown to give them different chances of survival.
3. Many differences between individuals (traits) can be shown to be hereditable.

Because too many young are always produced, many do not survive; they perish without issue, as a genealogist would say. Because the young are all different, their chances of survival are not equal and the winnowing of the generations is selective. And because the differences are heritable, the effects of winnowing the generations are cumulative. This is natural selection.

Natural selection is the differential reproduction and survival of individuals carrying alternative inherited traits.

Natural selection works by destruction; it kills individuals or it stops individuals from breeding. Change comes about because the culling process of natural selection hits some varieties harder than others. It is this endless sifting out of arrays of chance or contrived variety that ensures that existing species are suited to the environment in which they live.

Why They Have so Many Babies

An inevitable consequence of the process of natural selection is that all individuals of all living species must breed to the uttermost. Success is measured by offspring thrust into the next generation through the meshes of natural selection's net. This means that the more eggs or young an individual makes, or the more effort she puts into care of her young, the more the chance there is of some surviving.

Thus a positive feed-back drives the reproductive behavior of all organisms. Because too many young are born or raised, many must die without themselves entering the breeding population. But because the deaths are selective deaths, individual parents are most successful at leaving surviving offspring if they make many and various young. Therefore all parents must make too many young.

THE MEANING OF "FITNESS"

The success of any parent can be measured as the number of her offspring that avoid early death and successfully enter the breeding population of the next generation. We call this number the individual's **fitness**. Fitness is the number of offspring thrust by a parent into the breeding population. Or, more formally:

Fitness is the individual's relative contribution of progeny to the population.

This meaning of the word "fitness" is not the meaning of everyday speech. Ecological fitness is described as a number. If the Joneses raise four children, all of whom grow up to marry and have children of their own, then Mr. and Mrs. Jones each have a fitness of 4 (or 2 if the context suggests allowing for the fact that each parent has only a half interest in each child).

Charles Darwin first introduced the concept of "fitness" when he wrote of natural selection as promoting "the survival of the fittest." He did not, by "fit," mean the good athletes, the healthy, or those who were clever fighters. Being "fit" in Darwin's usage simply meant escaping removal by untimely death or from failing to reproduce more successfully than others did. This usage is criticized by some as being tautologous, because it merely says that the survivors (the fit) are those who survive. The modern definition of **fitness** avoids this trap by defining fitness as relative success in parenthood. This yields a powerful tool for studying the causes of diversity because all organisms now living do so because of the relative success (fitness) of long ancestral lines of parents.

It is our working hypothesis that all adaptations, behaviors, or structures of organisms promote fitness, letting an ecologist examine every property of an animal or plant with this in mind. The two components of fitness are survival and reproduction. Because fitness can be measured, it is possible to experiment, or to compare fitness of organisms in different natural circumstances, in order to investigate the adaptive significance of any particular trait.

Geneticists define fitness in terms of gene frequencies, instead of numbers of offspring, and give definitions like *fitness is a measure of the relative change in the frequency of an allele owing to selection* (Valentine and Campbell, 1975). But the concept is essentially the same; success is still measured by parental copies thrust into the next generation. The geneticist's definition, however, recognizes that parts of the parental blue-print pass between generations through relatives as well as through direct reproduction. Your brothers and sisters (siblings), or your cousins and their children, pass on some portion of your own genes.

The geneticist's definition lets fitness be measured as the number of copies of a gene that appears in the next generation, regardless of which organism carries them, the resulting measure being called **inclusive fitness**. This allows the parallel concept of **kin selection** as an explanation of what would seem to be otherwise altruistic acts. Any action by an individual that helps the relatives of that individual survive and reproduce necessarily gives some reward in inclusive fitness because those relatives carry copies of that individual's genes. The closer the relationship, the higher the probability of sharing genes. Hamilton (1964) showed that an act of helping relatives at cost to the helper (apparent altruism) can be selected for if the gain in inclusive fitness resulting from relatedness outweighs the apparent costs in fitness of the act to the altruist itself. Thus a sterile worker bee gets her fitness through the reproduction of her mother, the Queen, whom she feeds.

EVERY SPECIES HAS ITS NICHE

The vast variety of species represents an equal variety of life-styles. These may be subtly different. All the plants in a meadow, for instance, are indeed doing the "same thing" to the extent that all synthesize carbohydrate from carbon dioxide and water using common chemistry. But the different species live different lives. They have different niches: some are prone, some erect; some flower early, some flower late; some are annuals, some live for years; all have different defenses against the herbivores of the place. The complexity of the meadow is the interlocking of the different lives; the fitting together of different niches.

Niche is a word used by ecologists to describe a species in functional terms. Consider the niche of an imaginary species of wolf spider, an animal of the woodlands that gets its living by hunting smaller animals on the forest floor. The spider is a superb hunter, pouncing with dreadful fury on the smaller and weaker animals that are her prey. She has those skills as a heritage of endless generations when hunting success led to successful motherhood. But as she hunts, her eight eyes must ever be on the watch for the movement that warns of the terror that flies by day, the robin, the chickadee, or the blackbird. A dead spider makes no babies. Skills as a hunter, and skills at avoiding hunters, are important parameters of this imaginary niche. But the spider must also be programmed to do the right thing when it rains, or when winter comes. She must be able to recognize the smaller male of her species when he does his recognition dance, and so delay eating him until his semen is satisfactorily transferred. She must be able to carry her egg cocoon; perhaps also to let the baby spiders ride with her for a time as some species do. And the baby spiders must obey, accurately, the behavioral programs that guide the growing spider to food and shelter on that dangerous journey from the egg to a place in the breeding population of the next generation.

Thus the profession of "wolf-spidering," the complex vocation of my imaginary species. A way of life like this is generally what an ecologist means by "niche." Every species has its niche, its very special place in the grand scheme of things. The concept gives a first understanding of why the individuals of a species are so similar: only those who do everything right live to breed, and to breed successfully. Deviant behavior or structure has no future. Natural selection winnows out from the baby throng all but the few who perform perfectly in all the many parameters of the niche. Thus species appear to breed "true," despite the variety appearing in the offspring of every generation.

Why Don't They Overrun the Earth?

The concept of niche also allows an intuitive understanding of limits to populations. The niche of an animal or plant may be likened to a profession. For success, all the resources of that profession must be available. My imaginary wolf-spider, for instance, requires a woodland floor, in the right kind of climate, with the right kind of prey, no unfamiliar predators, the right kind of baby-spider food, suitable places of shelter, and so on. This combination of circumstances will certainly be limited, at once setting a limit to the number of wolf spider niches, and hence to the number of wolf spiders that can exist. Thus 'niche' sets "number," or at least an upper limit to number.

Yet a general theory of population control is not easily come by. Common sense, fortified by powerful mathematical argument, suggests that we look to the effects of crowding as the numbers rise. More animals or plants growing up to require the same resources should mean more death, or less reproduction, or both, until numbers are fitted to resources. Ecologists call these postulated effects of crowding "density dependent;" the denser the crowd, the greater the inhibition. Yet field ecologists often have great difficulty in discovering the crowds.

Populations of small animals often do not seem to be crowded at all. The numbers rise until winter comes, or some other catastrophe strikes them down. Next year, or after the drought or whatever, numbers take off again. The expected dense crowding never seems to come. Control (if that is what it is) is independent of density.

Predators seem as uncertain as density in controlling numbers. Some predators, particularly insects, can be so devastatingly efficient that prey animals live as scattered refugees. But many other predators encounter prey with powerful defenses so that the controlling effects of predation in these animals is minimal.

Concocting general theories of population control in these circumstances has been difficult, yet many have tried. Entertaining polemics have resulted, as proponents of density-dependent or density-independent control did battle in the literature. The debate has died down now, as ecologists have adopted a case-by-case approach, seeking to find what factors are limiting the numbers of a particular organism at a particular time. It remains true to say that the effects of crowding will certainly control numbers if no other process intervenes to prevent the crowds from forming.

THE TIME-SCALES OF ECOLOGY

Species may be very old. An expert in fossil beetles finds virtual constancy in the forms of large carabid ground beetles from the Miocene to the present, a span of more than 10 million years (Coope, 1979). Recent work suggests that trees of the genus *Platanus* (plane trees, known as sycamores in North America) lived in the early Tertiary in wet bottom lands, just as they do today, implying virtually constant niches for 60 million years; leaves of the sumac *Rhus stellariaefolia* can be recognized to species in a 40 million-year-old deposit (Fortey, 1991). These are the record examples. Yet many of the genera, and perhaps species too, of known forest trees have left fossils in Tertiary deposits several million years old (Figure 1.1).

Probably most animal species have much shorter spans, coming and going while species of forest tree persist. Several elephant species, like mammoths and mastodons, came and went in the million or two years of the Pleistocene ice ages, for instance; and swarms of endemic fish species turn up in lakes that may be only tens of thousands of years old. Our own species, *Homo sapiens*, may not much predate the onset of the last ice-age 110,000 years ago, although attempts to date our origins more precisely with molecular clocks have not yet given reliable results (Vigilant et al., 1991; Templeton, 1992).

But even newcomers like you and me have persisted while immense climatic changes have reordered life on earth. The last ice-age ended just 10,000 years ago, say 40 tree lifetimes away from us. For most of the 100,000 years before the final

FIGURE 1.1
***Platanus*, ancient genus of the Tertiary Epoch.** Plane trees, known as sycamores in North America, are closely similar to plane trees of the Eocene some 50 million years ago. The causes of species diversity thus often lie far in the geological past and the existence of species sometimes spans huge climatic changes.

melting of the ice, the earth remained in ice-age mode. Ice of mountain thickness covered Scandinavia and Northern Europe, down to the valley of the Thames. An even larger ice sheet in North America buried all of Canada and the northern parts of the United States to well south of the line of the Great Lakes. Shallow seas were drained because of the huge volumes of water locked up in the ice.

The ice sheets themselves were great reflecting mirrors, sending solar energy back into outer space. Every hectare of the earth's surface had a different climate then; some climates were hostile to life, others were merely cooler: many, as in Africa, were drier. Some of the drained sea bottoms might have had climates that would be thought ideal by modern humans, but those places are all drowned now.

All the important species of the contemporary earth lived through that last ice age, and many of them lived through the long procession of ice ages that preceded the last. For most species, ice ages are normality, because each lasted far longer than the brief intervening warm times like the present interglacial period. Thus it is a mistake to think of ice ages as times hostile to life as we know it. Ice-age climates gave good living to all the forest trees we know, and most other living things as well. But the pattern of life on earth was certainly different in the ice-age.

The message to ecologists of this ice-age history is that all the communities (as opposed to the species) of the contemporary earth, even the forests, are likely to be young—when "young" means less than 10,000 years old. This is indeed young when set against the age of the important species of the earth, from our own scant 100,000–200,000 years to the million-year spans of forest trees.

Figure 1.2 is a reconstruction of the changing ranges of hickory trees and pines in eastern North America since the ice-age, based on radiocarbon-dated fossil evidence (see Chapter 17). The two species advanced to their modern ranges at different rates and by different routes. The pine advance has scarcely ended, and others of the American trees may still be on the move (Davis, 1986; Webb, 1986).

To understand the make-up of the larger communities we must grasp this time-span: it took 10,000 years to build the eastern American forests. Nor can we have any sense that the job will ever be completed, because the climate will shift

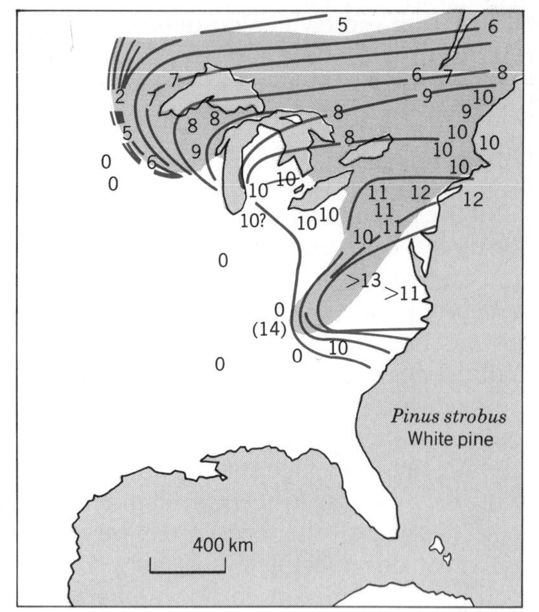

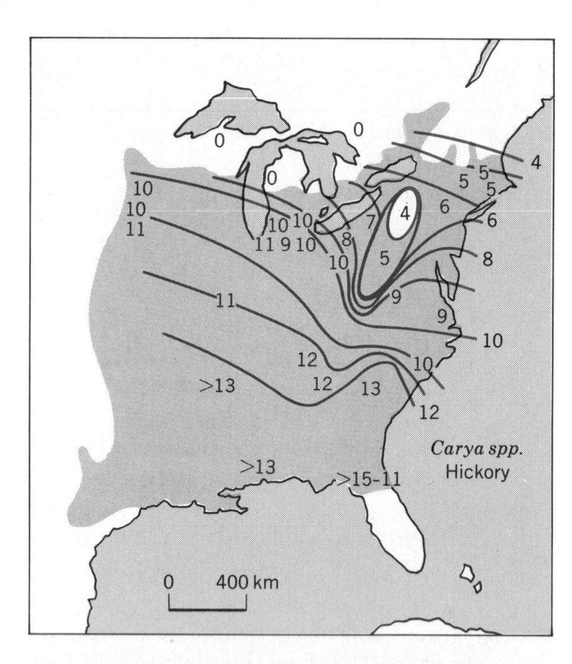

FIGURE 1.2
The spread of hickory and pine in eastern North America since the last retreat of continental ice. The lines (isopleths) show the edge of distributions at thousand-year intervals as shown by pollen preserved in sediments. Shaded areas are modern distributions. This use of the ecologist's time machine (see Chapter 17) shows that the complex forests of eastern North America are of recent construction, despite the fact that the component trees are ancient (M. Davis, 1982).

again before it is done (Figure 1.3). The present interglacial period is nearly over, and the next ice-age is about to begin (within a few thousand years). The trees will go on reshuffling into new combinations as their descendants settle down again to the ice-age earth known to their ancestors. This process, and this time-scale, influences every regional community from the Amazon forests to the arctic tundras.

In the Blinking of an Eye

Locally, communities can be put together at rates more easily understood: mere decades and centuries for forest. The process sometimes suggests a natural order in the building of communities. Community building on this time-scale is called **ecological succession**.

When a farm is abandoned in Europe or temperate America, a sequence of plant invasions begins. The year after the farmer quits, the fields are choked with annual weeds, but by the second or third year the first weeds are nearly all gone, to be replaced by perennials, plants like michaelmas daisies or goldenrod. This is a great wild-life habitat for bird watchers and shooters; but preserving this habitat would be hard for it is about to change. The fields of perennial herbs are invaded by brambles, shrubs, or small trees adapted to full sunlight, like the American white pine. The shrubs choke out the herbs and the small trees become big trees, casting shade. Years are needed before the former

FIGURE 1.3
Young forest: old species
The deciduous forest of eastern North America is complex, with many ancient tree species that have inhabited the continent since before the ice ages began, but the actual mixture of trees that make up the modern forest community is no more than a few thousand years old. The forest community is an ephemeral accommodation between species set by the peculiarity of recent climates and historical chance.

fields can crudely be described as woodland, though less than a human lifetime. Even then the process of community building is still far from over.

The crude woodland continues to change as seedlings of the real forest trees begin to appear in the shade now cast by the pioneer trees. Slowly the first trees die out and the newcomers assert their place in the canopy. Now the pace of change is slower, requiring more than a human lifetime to complete, but the succession seems well set to end with a forest like that which held the land before it was farmed, or at least like local woodlands left untouched by farming. The **ecological succession** will reach its **climax**.

Ecological successions were the first phenomena studied intensively by the founding ecologists at the beginning of this century. Their conclusions influence ecology still. Outstanding among their observations were that the process of community building through succession appeared to be ordered, directional, and leading to the most complex community appropriate to the locality.

The orderliness of an ecological succession can indeed be remarkable, for the plants of each community in the series usually appear in the right order and after the right lapse of years. A good local botanist could walk into an old field and tell nearly to the year when the farmer quit, merely by looking at the mix of plants. Moreover, it seemed that complex communities could not be built without this rigamarole of successional replacements' being gone through first. Some early ecologists saw in all this more than a hint of a mysterious community-organizing force in nature.

Modern research has stripped the mysticism from our understanding of successions (see Chapter 20). Orders of arrival are set by relative dispersal powers. Replacements of trees depend on individual adaptations to light or shade. Change never ends, meaning that climax never comes. In favorable circumstances, seedlings of forest trees sometimes actually arrive early, get a jump on the competition, and deny the whole successional process. Any succession we like to examine can be shown to result from the scramble of colonization as individuals in every species do their own thing. No organizing power is needed. Yet, from hyperbole about succession, ecology acquired philosophies about the holistic ordering of nature that are with us still, particularly in "green" political movements.

At a more mundane level, change on successional time scales illustrates the normal conditions of communities we study. Communities are always changing in response to disturbance, past or present. Farmers are not the only disturbers of communities. Climatic change, the weather, fires, and environmental accident disturb all the rest. It is a good working hypothesis to predict that climax never comes and that every community examined is under some stress of colonization.

THE COMING OF THE ECOSYSTEM

The early ecologists learned that all communities changed in space as well as in time through succession. Communities had no edges, but graded one into another across landscapes. The only real boundaries to communities found by field ecologists were physical boundaries: the edge of a stream, the strike of a rock formation, a change in slope. Practical ecologists, therefore, found themselves studying communities defined by habitat.

As ecology came to its first maturity in the years between the two world wars, the proper communal unit to study became a matter of prime debate. Hard experience counseled that the unit must be set by habitat, but the ethos of ecology replied that interaction between species was what mattered, not mere physical factor control. Ecologists staged a compromise. They would study both together and call the result the ecosystem.

"Ecosystem" linked most of the grander ideas of ecology. Into the space chosen to be studied as an ecosystem streamed the energy of sunlight. The massed ranks of plants made their carbohydrate fuels, with one chemistry but in many shapes. Animals ate the plants, degrading stored energy from the plant fuels back to heat, starting the food chains destined to end with the top predators. Communities were built and numbers were controlled within the ecosystem. Above all the habitat was maintained, as the restless patterns of life and death within the community served to cycle and recycle the resources of living. With the invention of the term "ecosystem" a new paradigm had entered science.

THE MAINTENANCE OF THE BIOSPHERE

The biosphere is the combined ecosystems of the whole earth treated as one: all living plants, all animals, and all processes of decomposition together with the air, soil, and waters in which life persists. The same processes apply in this giant ecosystem as in the very smallest.

Throughout the biosphere, every property and action of a living thing has been, and is, subject to natural selection. Individual advantage, measured as fitness, is the only validity of all that biospheric activity. Every organism is fashioned to reproduce to the utmost, and every impact on the environment, whether beneficial or not, is a mere by-product of that reproductive drive.

Throughout the biosphere, the vast arrays of plants, animals, and tiny protists have always performed the same few essential functions of life. But the scale of the biosphere is so large that the consequences of life process have been cumulative. Actual physical conditions for life on earth have been utterly changed by these combined activities.

Little doubt now remains that the mixture of free oxygen and nitrogen in our air is a by-product of the activities of plants and protists working together on a global scale. Likewise, the curious chemistry of the oceans seems in part to be regulated by the life of the seas (see Chapter 27). And the soils that sustain forests and farms would not exist without the life that covers them.

The mightiest of these effects have been achieved slowly, and mostly long ago when the array of species was quite different. For photosynthesis to free enough oxygen to flood the air with it, for instance, required the burial of almost unimaginable tonnages of reduced carbon compounds (see Chapter 28). This took many hundreds of millions of years, during which the banded iron formations were deposited more than half a billion years ago. Paleoecologists reconstruct the vital workings of that all-important past in the manner of a geochemical detective story, more thrilling in its way even than the lives and deaths of dinosaurs a mere 200 million years back.

But the combined activities of biospheric life can be detected still. The massed pumping of water by forest trees (transpiration) modifies local habitats and local climates. When the forest is as large as the Amazon, the effects on climate may be more than local. Perhaps more interesting still is the role of life on the carbon dioxide concentration of the atmosphere. Year-round measures of carbon dioxide in the Hawaiian air show a pulsing rhythm on an annual cycle (Figure 1.4). In the northern summer, plants of the great

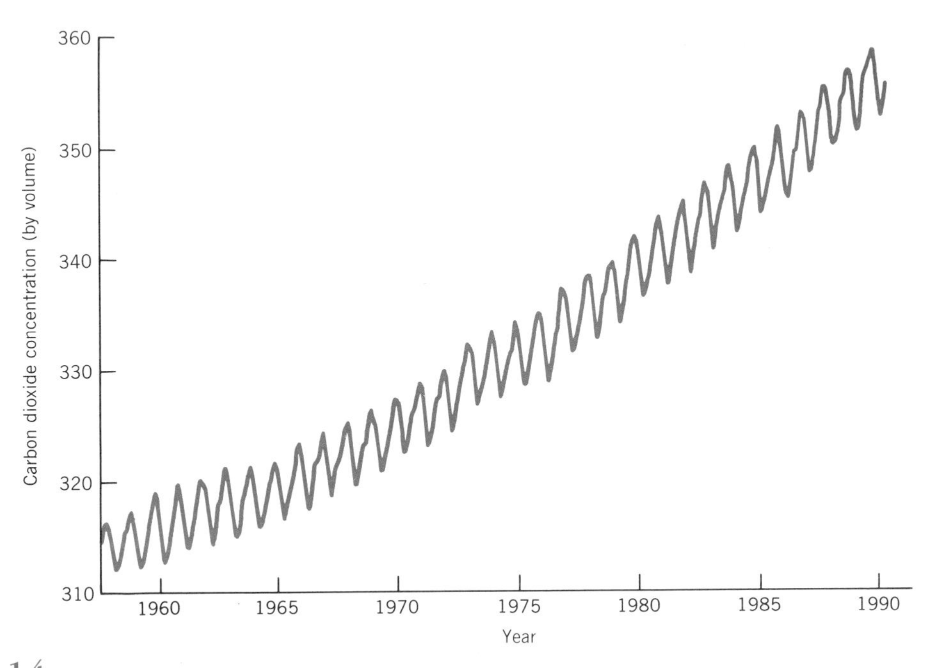

FIGURE 1.4
Biological rhythm in the atmosphere.
Concentration of the comparatively rare gas carbon dioxide fluctuates annually, rising in the northern winter with respiration in leafless landscapes, falling in the northern summer as photosynthesis draws down the atmospheric supply. The measurements are from Hawaii, where the lack of local industry allows measures of unpolluted Pacific Ocean air. The global carbon pulse registers events in the northern hemisphere because most land, and hence most plant production is in the north. The sinister upward trend in the data is the best evidence that CO_2 concentration is rising as a result of burning fossil fuels and clearing forests. (Data from L. Machta, NOAA Air Resources Laboratories, 1992).

northern continents fix carbon in concert, drawing down the carbon dioxide stocks of the atmosphere in the process. In the northern winter, the massed sun-driven factories of the plants are dormant, though life goes on. Now the stored carbon fuels from summer are drawn down by life in the mass, returning carbon dioxide to the air through respiration. In the mixed air above Hawaii this seasonal building and withdrawal of carbon reserves shows up like the beating pulse of the living earth.

THE ECOLOGY OF HUMANS

Mass action by people can change environments. This is now so well known as to be trite. We abolish forests by clear-cutting, drain marshes, impound rivers, dig up grasslands, pour concrete, and pump water from ground to desert. Through most of recorded history all these doings have won praise, riches, and power for those who directed them. Ancient Romans, from Republic to Empire, curried favor by draining lakes to make farmland (read about it in Suetonius, Livy and the rest). People who followed the frontier in the American West were proud to "tame" the prairie with the plow. City people live happily when a great river is made safe by "channelization"; walk in the springtime through splendid capital cities along the so-called embankments of the Thames, Seine, Danube, or Moscva rivers, and know that it is good to be human.

And yet the aggregate of all our changes becomes alarming. Can we go on clear-cutting the last forests until none are left at all? Surely if we do, our descendants will curse us forever. The joys of being a civilized human depend on harnessing energy for habitat control, travel, and manufacture, but an extraordinary consequence of using the most available source of energy (buried production of ancient ecosystems, known as coal or oil) is that we enrich the air with carbon dioxide, even beyond the annual enrichment of the northern winter (Figure 1.2). The excess carbon dioxide absorbs heat and will be certain to raise the temperature of the earth, a fact that we ecologists debated more than a quarter of a century ago but which has only now penetrated the minds of those set in political authority over us.

I once walked down the main street of Magadan with T. Lohzkin, talking, as ecologists will, about the state of the earth. Magadan is a city in Siberia, once the administrative capital of the **gulag** camp system, with an appalling winter, surely one of the world's harshest dwelling places. My companion lives and works there. Of the greenhouse effect, he turned to me and said, "Paul, anything that warms this place is all right with me." The decisions to curb human conduct are political decisions, and it is not always clear where the human interest lies. But the science of ecology can at least tell us why things are happening to us, without which knowledge even the advent of wise politicians could not help.

PART ONE

ENERGY

Preview to Chapter 2

Thalarctos maritimus: Energy from the sea.

THE energy flow model so important to modern ecological thinking arrived late to the discipline, the first explicit statement being in 1942 during the Second World War. First expression of the model was prompted by the need to explain the rarity of large predatory animals relative to the great abundance of small animals. This familiar observation of the pattern of life on earth had been formalized by the observations of Charles Elton, who had described the result in 1927 as the "pyramid of numbers." That predators should be discretely and significantly larger than their prey in many circumstances had been satisfactorily explained by Elton's principle of food size, being the requirement that a predator should usually be powerful enough to overwhelm prey easily. But the great reduction in numbers of animals high on food chains could not be explained until it was understood that food energy was attenuated at every link in a food chain. The energy flow model is consistent with the second law of thermodynamics, yet energy degradation in ecosystems is both more rapid than required by the second law and distorted by adaptations of animals and plants that tend to prevent energy transfers. For these reasons, physical models of ecosystem function based on thermodynamic theory have been less than fruitful.

Chapter 2

The Energy Flow Paradigm

The flowing energy model of how the world works was first developed to explain some of the more familiar of the observations of naturalists. These were that large, predatory animals were always rare but that herbivores, particularly small herbivores, were much more abundant. Plants were the commonest organisms of all, even big plants like trees. Moreover, all the animals and plants of a place were linked together with ties of eating and being eaten, what fishery biologists early in this century came to call food chains.

A **food chain** is a chain of eating and being eaten that connects large and carnivorous animals to their ultimate plant food. An example from the sea might be

phytoplankton → copepod → herring → seal

though in most ocean systems a series of little fish being eaten by ever larger fish is possible, an idea now sufficiently familiar to be a favorite subject of cartoonists (Figure 2.1).

On land a likely food chain might be:

pine trees → aphids → spiders → titmice → hawks

Food chains are written thus, with the arrows pointing the way they do to show that food passes from animal to animal up the chain from the plants through the various animals to the top predator of the system.

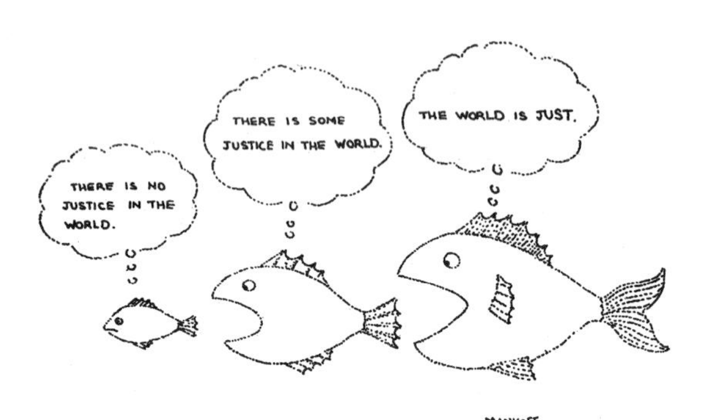

FIGURE 2.1
Not quite what Elton meant.
Elton's principle of food size structuring communities along food chains is now so familiar as to be a subject for popular cartoonists.

This essential structure of natural communities was described formally in a 1927 textbook by a young self-styled animal ecologist, Charles Elton, with far-reaching consequences. Elton went expeditioning in the arctic islands of the Spitzbergen Archipelago, a place of tundra, no trees, and highly visible animals, from polar bears to birds and insects. Elton set himself to working out who fed on whom.

The conspicuous arctic foxes fed on birds—the resident ptarmigan, migrants

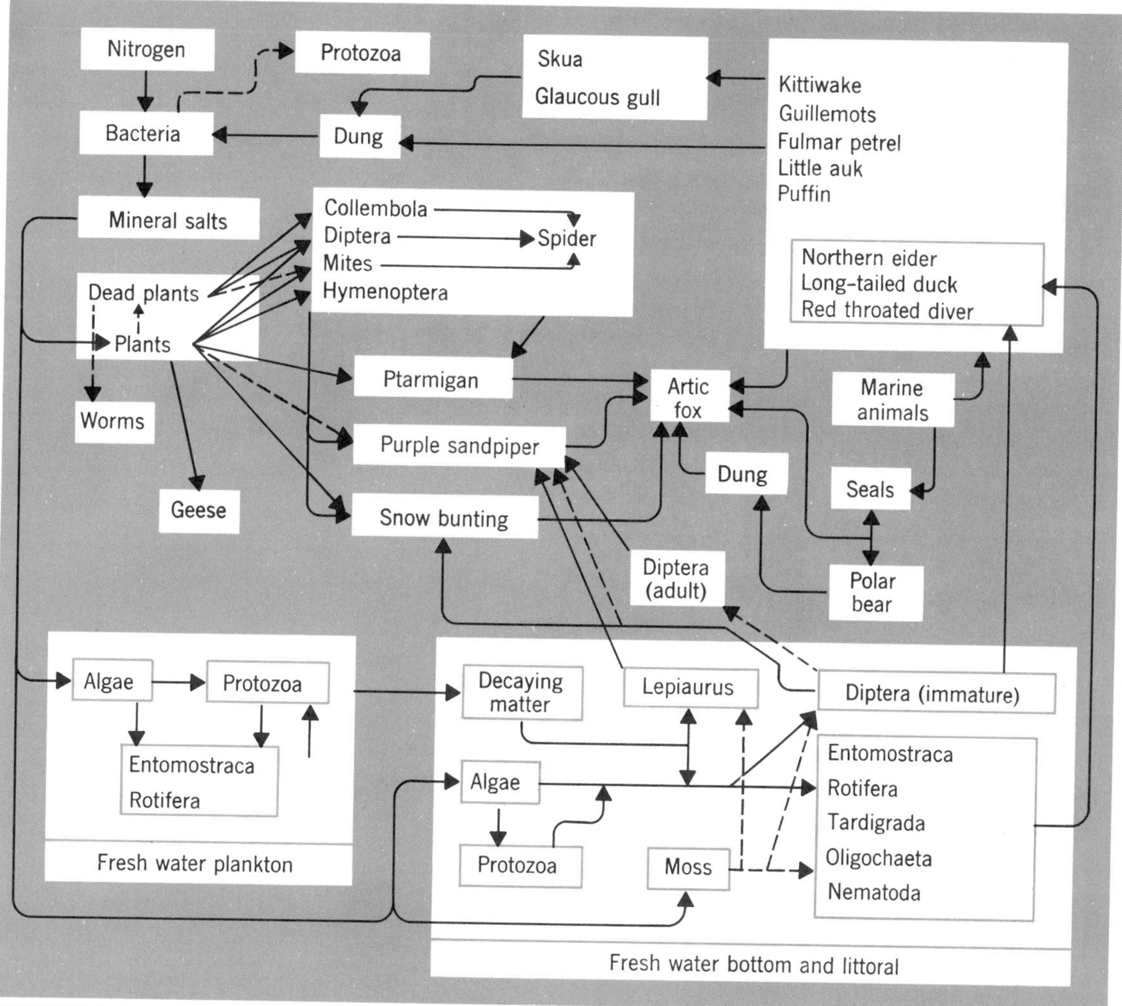

FIGURE 2.2
The Bear Island food cycle that Elton found.
Elton worked out the details of this food cycle by seeing in the mind's eye the movement of nitrogen through the community. He was able to confirm some of the pathways by direct observation, and others (shown as dashed lines) he could only infer. Since Elton's time more food webs have been worked out in much greater detail than Elton could attempt, allowing comparisons to be made in search of general principles of community organization (see Chapter 22). (From Summerhayes and Elton, 1923.)

like sandpipers and buntings, and various sea birds. Thus the foxes evidently got their living both from the tundra and from the plants of the sea via different food chains. In winter the sea was particularly important because the foxes survived by eating polar bear dung and the remains of seals that the polar bears killed. Elton and his party filled in as many details as possible, revealing complex interconnectedness among all the animals they were able to observe (Figure 2.2).

Many food chains ran through the Spitzbergen community in what has come to be called a **food web**. Elton realized that the organizing principle of this and all animal communities was the common need for food. But as yet there was no talk of energetics; the days of counting calories were yet to come. Instead Elton talked of a **food cycle**.

To Elton, food in his tundra community was typified by indestructible nutrients, particularly the combined nitrogen that is nearly always scarce in arctic landscapes. Elton postulated that much nitrogenous fertilizer came from the sea via the excrement of sea birds, and that this nitrogen was then cycled up the food chains and back again. Figure 2.2 is thus really what we now would call a recycling diagram, with nitrogen the scarce resource for which the whole undertaking is thought to exist.

Elton's second major observation, however, was more potent. It was his formal recognition of what people have always known, that big animals are less numerous than small animals. A polar bear was a rare sight, foxes would be encountered every now and then, the place was thick with birds, and insects came in clouds. Big, fierce animals were rare on Spitzbergen, as they are everywhere else. Elton played with graphs and called the result the **pyramid of numbers**, sometimes now called the Eltonian pyramid.

When animal size is plotted against number the result is a graph of this general form:

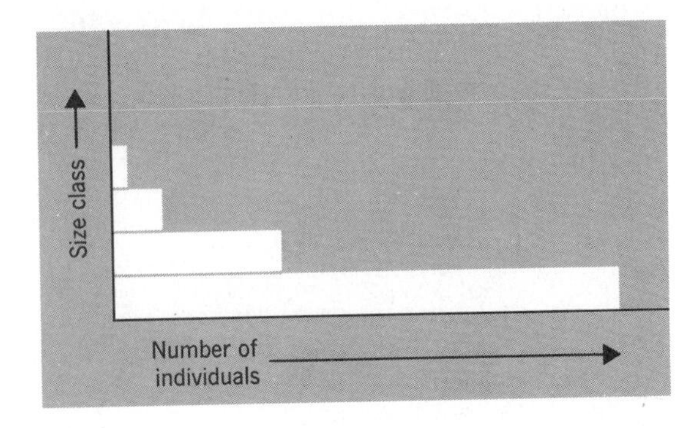

Shifting the *x* axis to the middle of the figure gives the following result:

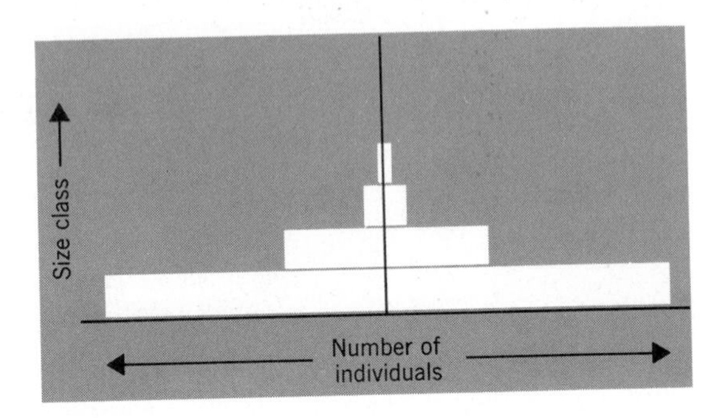

Each layer on the pyramid represents kinds of animals living at parallel levels on food chains: all herbivores on one level, all primary carnivores on the next level, all secondary carnivores on the next level, and so on. These levels are now called **trophic levels** (after the Greek word for nursing, in the sense of a mother suckling her young, because all the animals of one level operate on a common feeding plan).

A pyramid of numbers can be drawn for any community from real numbers by anyone with the patience to do an adequate census. Figure 2.3 shows a pyramid of numbers of arthropods on a tropical forest floor, a result probably typical of most terrestrial sites (see also Chapter 22). Elton

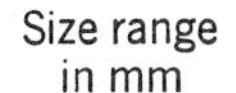

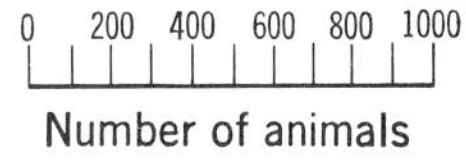

FIGURE 2.3
Eltonian pyramid of numbers on the floor of a forest in Panama.
Williams (1941) collected all the animals in small samples of the litter on the floor of the forest, counted the individuals in his catches, and sorted them into size fractions. The smallest and most numerous animals were Collembola (spring tails) and mites, both of which are herbivores or scavengers feeding in the litter. The larger, rarer animals, such as ground beetles and spiders, were carnivores.

himself (1927) described the possibilities for water, saying, "In a small pond, the numbers of protozoa may run into millions, those of *Daphnia* and *Cyclops* into hundreds of thousands, while there will be far fewer beetle larvae, and only a very few small fish."

Elton's strange pyramidal graph was seen to describe fundamental relationships in nature. Predators are generally bigger than their prey, and much less common. Food chains hold communities together, tying the powerful flesh eater to the plant providers as firmly as the herbivores are tied. It was true that not all communal pyramids would look as neat as the ones Elton described; predatory parasites, for instance, are smaller than their prey. Such exceptions have their own stories to tell. But the ubiquity of the structure obviously required explanation. Why does life have this pyramidal ordering?

A first oddity of Elton's result was the steps on his pyramid. These steps show that animals of successive trophic levels are discretely different in size. This is not at all intuitively obvious because common sense wants to say that animals are graduated into any possible size. But the Eltonian pyramidal graph states that animals of each trophic level are expected to be discretely larger than the animals of the trophic layer below. There should be no intermediate sizes.

Elton explained these steps easily enough with his **principle of food size**. Typical predators had to be much bigger than their prey in order to overwhelm it safely. Many predators in fact are relatively so large that they can stuff prey whole into their mouths, or at least engulf it with a few vast bites. Chickadees, titmice, and robins swallow their writhing invertebrate prey alive and whole; great white sharks require a minimum of bites.

Whenever predators fit in with Elton's model of being large enough and powerful enough to overwhelm their prey easily, we can expect discretely different sizes in animals of successive trophic levels. This actually reveals problems for the life-history strategies of many animals; how to grow to the proper size. Birds and mammals overcome this problem by nursing the young through the difficult sizes. Ectothermic animals, like reptiles, have other solutions, like switching prey after long periods of dormancy, effectively jumping from one trophic level to another (see Chapter 6).

Some predators do hunt in a non-Eltonian way, as do wolves when they take ungulate prey larger than themselves. This does little to distort the general structure revealed by the pyramidal graph, particularly because wolves do much of their feeding on rodents in a satisfactorily Eltonian manner. The steps in the sides of an Eltonian pyramid, therefore, are real. They demonstrate fundamental constraints on the sizes of predatory animals set by their ways of life.

Unfortunately, general biology textbooks usually show some version of a figure with sloping sides as the pyramid of numbers. This may take the form of Figure 2.4, the first of its kind introduced by the limnologist Juday (1940), though the cone-shaped drawing is often embellished with pictures of big cats snarling at animals lower down, or even with a human on top brandishing a spear. Nonecologists apparently are reluctant to accept that animals of discrete communities come in discrete sizes, but they do. Pyramidal diagrams that omit the steps are in error for most real, functioning communities and they obscure fundamental truths that Elton saw.

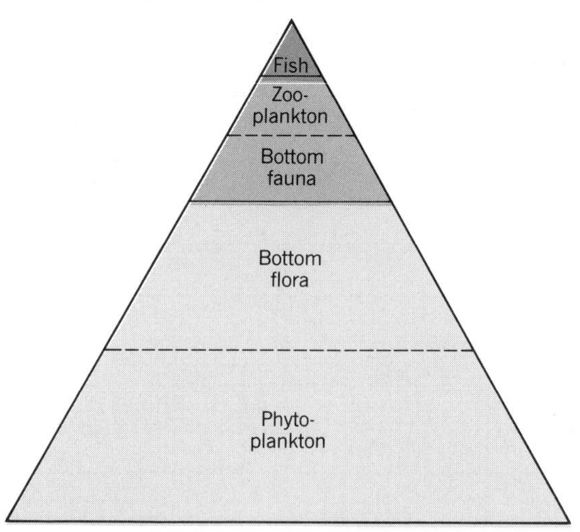

FIGURE 2.4
Cone-shaped version of the pyramid of numbers.
This figure, taken from Juday (1940), is probably the first pyramid figure with smooth sloping sides to be published, and it is now often imitated. Elton's original vision was of a pyramid with stepped sides, looking in profile like the stepped pyramids of Egypt. In a lake system like that studied by Juday it is probably more realistic to separate the organisms of the different trophic levels into discrete size fractions so that the steps on the pyramid show. This is because the animals of the different levels are in the main large enough to engulf their food whole.

So why are big, fierce animals rare, as the Elton diagrams show? The answer we now give depends on energetics, but ecologists of the 1920s and 1930s were not trained to think in terms of energetics. Elton tried a different explanation, which in fact does not work. He suggested that small animals reproduced more quickly than large animals, but lived shorter lives. This rapid turnover of small animals resulted in large numbers of individuals. Slowly reproducing large animals (the big, fierce beasts on top), however, could not maintain such large populations because they lived long and reproduced slowly.

The essence of Elton's suggestion is that population size is a function of reproduction rate. But this cannot be, except for short intervals of population growth before limits are reached. As I suggested in Chapter 1, at all other times niche, or some critical resource, or perhaps a devastating predator promoting refugee status sets number. Rate of reproduction has nothing to do with the numbers that can live, only with the speed with which these limiting numbers are reached. Elton's explanation, therefore, could not apply in most circumstances.

The Eltonian pyramid also cannot be explained away as the result of mere geometry. It might, for instance, seem that small animals are numerous merely as a result of cutting the cake into smaller slices in an equation that says lots of little animals equal one big one. This is not the answer; weigh all the little ones and they will weigh more than the combined weight of the big ones.

For 15 years Elton's pyramid remained unexplained, though it was prominently displayed in the leading textbook of the times. Then Raymond Lindeman (1942) introduced the modern explanation based on energy flow that he and G. E Hutchinson had come upon independently. Lindeman and Hutchinson thought of food as calories that could be used up, not as nutrients that cycled. With this view, the food chains crossing the trophic levels had a different meaning.

Small animals low on the pyramid are the food of larger animals higher up so that energy, as food calories, enters the pyramid at the bottom. In maintaining life, energy is constantly being dissipated as heat. This means that the usable food energy flux received by each successive trophic level is less than that received by the trophic level below (Figure 2.5).

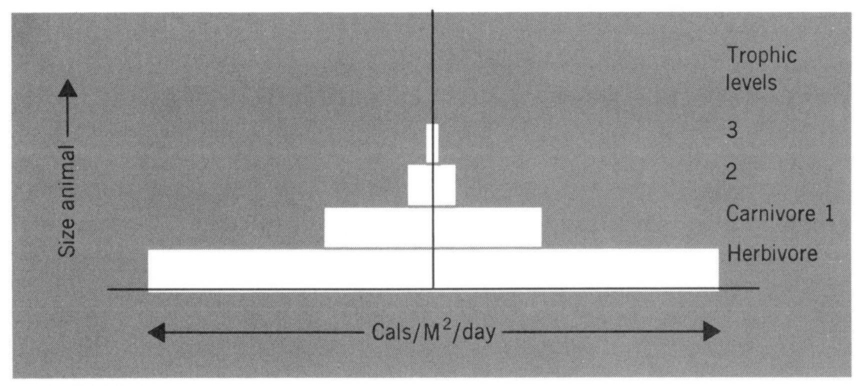

FIGURE 2.5
Energy pyramid.
The figure illustrates the concept of energy degradation at each transformation between trophic levels that was introduced to ecology by Lindeman and Hutchinson.

Even if the animals at all levels of the pyramid were of the same size, those high on food chains would of necessity be less numerous than those below. Since, because of the principle of food size, those high on the chains tend to be large, the energy available for body-building and maintenance is sufficient for even fewer of them.

Plants can be both big and common (forest trees) because they "feed" directly on the sun. The plant trophic level has more energy at its disposal with which to make bodies and do work than any other trophic level. Herbivores come next, having at their disposal all the energy stored as plant biomass that they can actually manage to eat. Because most herbivores are insects on land and plankton in the sea they are also tiny; tiny plus huge potential energy supplies equals very large numbers. When elephants eat vegetation, or baleen whales filter out plankton, they are both huge and common, yielding truncated Eltonian pyramids with only two or three levels.

For carnivores, energy supplies are always bleak because they must feed on those herbivores they can catch, taking only a tax of the energy that was available as herbivore flesh before the prey used most of it in the activity of living. For predators of predators, possible energy supplies become exponentially bleaker. Great white sharks are not numerous. A potential predator of a great white shark would have to be so large and terrible, and so rare, that it has never evolved.

Thus was Elton's trenchant observation of the pyramid of numbers finally explained. The essential relationships between predator, prey, and plant were defined by the ever-diminishing flow of energy as it makes its way up food chains. In a sense, the organization of communities was a function of the second law of thermodynamics.

ECOLOGY AND THE LAWS OF THERMODYNAMICS

The **first law of thermodynamics** states that when energy is converted from one form into another energy is neither gained nor lost. This first law is also called the law of conservation of energy and is, of course, strictly true only if matter is considered a form of energy.

The **second law of thermodynamics** states that every energy transformation results in a reduction of the free energy of the system. In the language of physicists, this law states that all energy transformations result in an increase in entropy, requiring that only a fraction of the energy conserved under the first law is available to do useful work within the system.

The rarity of the large and fierce demonstrated by Elton's pyramids is in part a necessary consequence of this second law of thermodynamics. Energy to fuel the animal community is limited to that flowing into the plant trophic level through photosynthesis. But for herbivores to have use of this energy requires an energy transformation, from plant carbohydrate to animal carbohydrate. Because this energy transformation cannot be 100% efficient (second law), the animals must have less energy, and must therefore be rarer, than the plants they feed on. And so it goes, up the food chains, through the successive energy transformations of each trophic level. Energy gets less and less; animals get rarer and rarer.

Invoking the second law invites some heady ideas into ecology. If commonness and rarity are set by the second law of thermodynamics, might it not be possible to make purely physical and mathematical models of living systems, based on the laws of thermodynamics alone? Early attempts were not encouraging (Slobodkin, 1962). The main difficulty is that the second law is far too simplistic a model of what is actually happening in living communities. The wastage of energy between trophic levels is far greater than could be accounted for by the inefficiency of physical energy transformations alone. Animals and plants burn calories, they don't just transform them. Furthermore, natural selection has seen to it that animals are equipped to avoid capture and that plants are adapted to be hard to eat (woody tissue, allelochemics; see Chapter 15). Organisms resist the free transfer of their hard-won calories, making the second law of little actual relevance.

EMPIRICAL MODELS OF COMMUNITY ENERGETICS

Lindeman and Hutchinson introduced their energy flow model of communities just seven years after the coining of the word "ecosystem" in 1935 (Chapter 1; Tansley, 1935). For the ensuing generation of ecologists, the two concepts merged. The unit of study should be the ecosystem, with convenient boundaries set by the in-

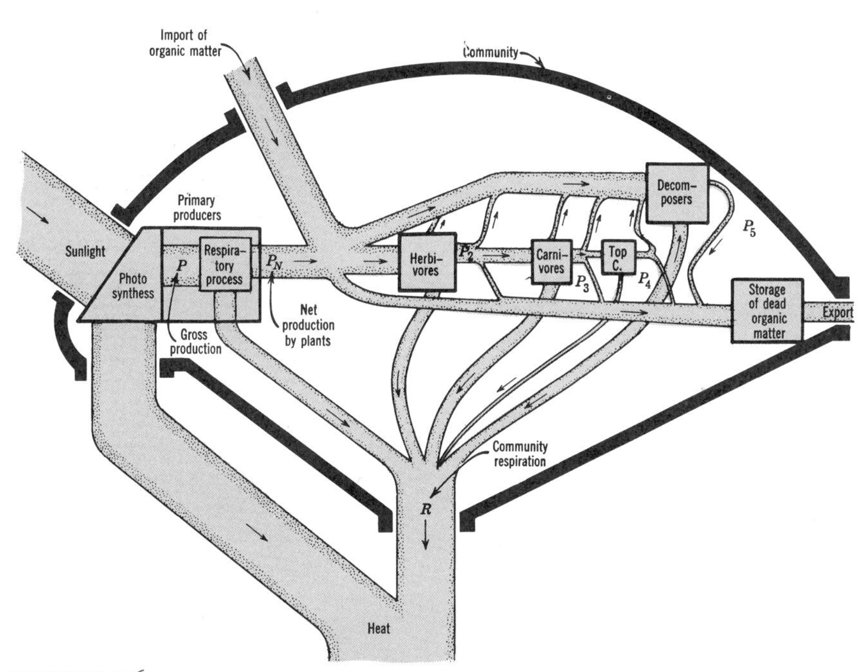

FIGURE 2.6
The hydraulic analogy of H. T. Odum.
In this analogy the energy cascade is imagined as being channeled through pipes whose thickness is proportional to the rates of energy flow. A prism placed at the entrance (or some hydraulic equivalent of a prism) deflects most of the sunlight from the community to represent that proportion of incident light not used in photosynthesis. From then on the degradation of energy at each trophic level is shown by pipes running to the heat outlet. (After Odum, 1956.)

vestigator. But the model to guide the collection and interpretation of data should be the energy flow paradigm, as life in the defined system was driven by the solar energy falling on it. The well-known hydraulic analogy diagram of H. T. Odum illustrates this approach (Figure 2.6).

With this model in place, the task then became the measuring of energy flows in real populations—calorie-counting on a grand scale. This can be difficult. Not only must calories present as sensible biomass be measured, but so must the calories that are constantly being burned. Our limited successes with this undertaking are described in the next two chapters. Meanwhile, interesting results were obtained by simply measuring biomasses as proxy data for energy.

Standing Crop

The most practicable measure in field ecology is of mass present at the time of a visit, or the standing crop. *Standing crop is the biomass present at unit time in unit area.* Standing crop is usually measured in grams dry weight per square meter (gm^{-2}), or similar convenient units. Dry weight can be converted to grams of carbon or calories readily enough when the appropriate conversion factors are worked out.

"Standing crop" must not be confused with the agricultural term crop from which it is derived. When ecologists talk of the standing crop of a cornfield they mean not just the grain but the leaves, stems, and roots as well. Standing crops can be measured for every trophic level, thus being used for animals or bacteria as well as for plants.

Figure 2.7 shows some collected standing crop data from the period when ecologists first set out to measure the energetics

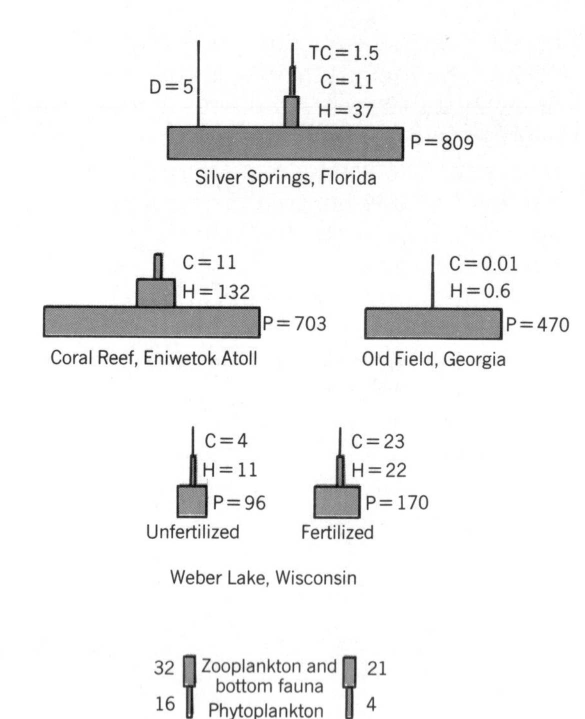

FIGURE 2.7
Pyramids of biomass from various communities. In the terrestrial and freshwater ecosystems the pyramids of biomass are the shape we would expect the pyramids of energy to be if we had sufficient data to draw them: large energy flows low on the food chains are reflected in large biomass. In the two samples of productive ocean over shallow continental shelves, however, the biomass pyramids are inverted. This has been shown to be caused by rapid turnover of biomass in the producer trophic level, which occurs because the plants of this system are microscopic and short-lived. Pyramids are drawn approximately to scale, and figures represent grams biomass per square meter. P = Producers, H = herbivores, C = carnivores, TC = top carnivores, and D = decomposers. (Data from Odum (1971) from various sources.)

of natural communities. In an old field, a coral reef, a lake, and a clear-water stream the pyramids of biomass appear similar to a pyramid of numbers, with orders-of-magnitude reductions in mass with each higher trophic level. This reflects the degradation of energy up the food chains as expected. But in two fertile patches of the oceans, Long Island Sound and the English Channel (bottom of diagram), the pyramids of biomass are inverted. How can a great mass of grazing zooplankters possibly be supported by a much smaller mass of tiny planktonic plants?

A hypothesis to explain this inversion of Eltonian prediction is easily offered: the mass of phytoplanktonic plants is a poor predictor of the energy flowing through the plant trophic level. The tiny plants are short-lived and burn up their photosynthesis-derived energy instead of storing it. The longer-living, and larger, zooplanktonic animals do store energy. Thus the sensible biomass in the animal trophic level is larger than that in the plant trophic level, even though more energy flows through the plants (Figure 2.8).

The most beautiful test of this hypothesis would be from direct measure of the energy flowing through each trophic level. Those data we do not have. But it has been possible to calculate the flux of dry matter through each trophic level, working from growth rates of the more important organisms. If the hypothesis is correct, then the dry matter flux should be greater in the phytoplankton than in the zooplankton, reflecting the greater energy passing through. Figure 2.9 gives Harvey's (1950) estimates for dry matter flux through the English Channel food chains. The daily flux of dry matter through the phytoplankton trophic level turns out to be larger than that through the animal trophic levels as predicted.

FIGURE 2.8
Standing crop in water and onshore.
The plant standing crop of the open water is the mass of phytoplankton living there at the time of sampling, just as the standing crop of the surrounding forest is the mass of spruce trees, trunks, leaves, roots and all.

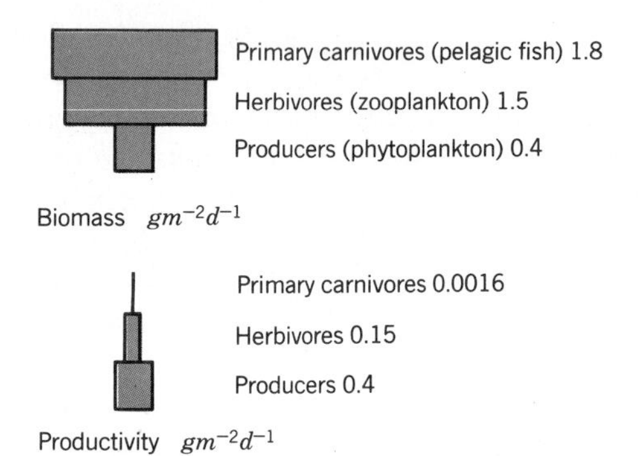

FIGURE 2.9
Pyramid of dry matter flow in the English Channel.
Biomass produced by the producers (phytoplankton) per day is greater than the biomass produced per day by the consumers of the English Channel ecosystem. This large flux of biomass requires a large flux of energy so these data provide a satisfactory explanation for the anomalous inverted pyramids of standing crop biomass found in shallow productive marine systems. (Data from Harvey, 1950.)

HOME RANGE, ENERGETICS, AND SIZE

As animals become larger and rarer, they also have larger home ranges. Elton recognized this in his original work and elaborated on the idea later when he talked of the "inverse pyramid of habitats" (Elton, 1966). Like the original observation of the pyramid itself, this is a necessary consequence of the quest for energy. Large animals need a large energy flux, and they must hunt or crop a large area to get it. Whereas plants can support themselves on the energy flowing directly onto the spot where they stand, animals that receive their energy after one or two inefficient energy transfers up food chains from the plants require the pickings of a much larger area. Rarity and being spaced out are connected states, both necessary when energy is in short supply high on food chains.

Wide spacing will also be forced on animals with specialized diets, again ultimately for energetic reasons. Widely spaced food energy implies a low flux per unit area, thus enforcing comparative rarity. This is nicely revealed by McNab (1963), who compiled data for home ranges of mammals ranging in size from mouse to moose and with various eating habits (Figure 2.10). A strong logarithmic relation-

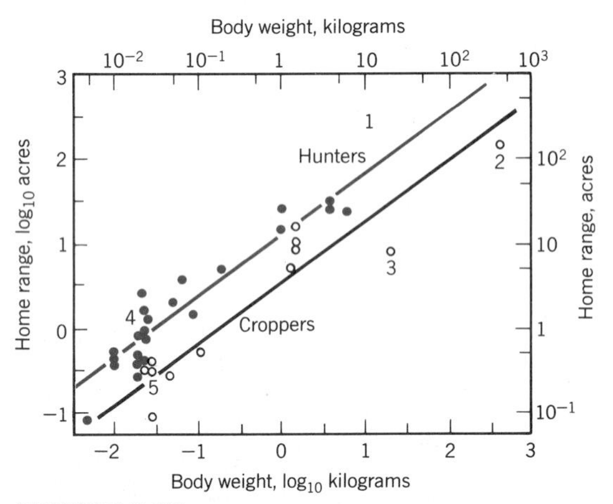

FIGURE 2.10
Relationship of size of home range to weight of body in mammals.
There is a significant relationship between the size of the animal and the area over which it must range to get its food. Herbivorous browsing animals like moose or deer need about four times less land than carnivores of the same size, a relationship that is easy to understand on energetic grounds. Of particular interest, however, is that herbivores or omnivores that seek out discrete particles of food like seeds need about the same area as true predaceous animals of their own size. McNab (1963) refers to both the true predators and those that seek out patchily distributed plant food with the common term "hunters." These are distinguished from the true browsing or grazing animals, which are "croppers." Open circles = croppers, solid circles = hunters, 1 = raccoon (*Procyon lotor*), 2 = moose (*Alces alces*), 3 = beaver (*Castor canadensis*), 4 = deer mouse (*Peromyscus maniculatus*). (From McNab, 1963.)

ship between range and body weight appears as expected: larger animals need larger home ranges. But the data also separate animals into two groups depending on whether they seek out isolated food or not.

The range–weight relationship of carnivores in McNab's sample (top line, Figure 2.9) is shared by vegetarian animals that eat scattered, high-calorie foods like seeds. Both kinds of animals are "hunters," whether true carnivorous hunters or those that "hunted" seeds or fruits. The energy flux per unit area from both kinds of hunting was low, with the result that animals with both habits lived well spread out. Much less space, about a fourth, was needed for animals that browsed or grazed on vegetation, animals that McNab called croppers.

OPTIMAL FORAGING THEORY

These successes of the energy flow model in accounting for community structures allow the general hypothesis that energy supplies are crucial to the lives of individuals. This is, of course, little more than a statement of the obvious. If individual organisms have been fashioned by natural selection to leave the most possible surviving offspring (fitness), then clearly all individuals must be magnificent harvesters of the energy needed for survival and reproduction. But explicit statement of the hypothesis lets us follow the route of formal prediction about individual behavior, followed by at least tentative testing.

If animals have been programmed by natural selection to maximize their energy intakes, as the hypothesis predicts, then they should allocate time spent feeding so as to maximize their energy returns. This statement can be refined for those foraging animals that must seek out food as discrete items, the hunters in McNab's model, because foragers with a choice of food items ought to be able to concentrate on those items that yield the best return. Formal statement of this principle is called "optimal foraging theory" (MacArthur and Pianka, 1966; MacArthur, 1972).

The simplest model is of a forager offered a choice between large, scarce, hard-to-catch prey rich in calories or abundant, small prey of low caloric value. Should it hunt the one or the other? Optimal foraging theory states that the forager should be able to direct its efforts to which of the two gives it the best return in calories for the time spent hunting and handling the prey. One elegant test of the model used shore crabs eating mussels in experimental tanks (Elner and Hughes, 1978).

Shore crabs (*Carcinus maenas*) can eat even the largest specimens of the mussel (the bivalve *Mytilus edulis*), but it takes the crabs significant time to open the tough shells and eat the contents. Small mussels are devoured quickly, but the return in calories is of course smaller. By watching crabs eat mussels of different sizes and weighing the mussels, it is possible to plot a curve of energy return per unit time when the crabs eat mussels of different sizes. The curve is roughly bell-shaped, showing that the best return for the crabs comes from eating mussels of intermediate size (Figure 2.11). When feeding the crabs to prepare this curve, the crabs had no choice but to eat the size mussel given them.

Optimal foraging theory now predicts that if the crabs do have a choice they should be able to concentrate on mussels of intermediate size, because only in that way could they maximize their energy returns per unit time. The histograms in Figure 2.10 show the results of feeding trials in which the crabs were placed in tanks with mussels of all sizes. They did indeed

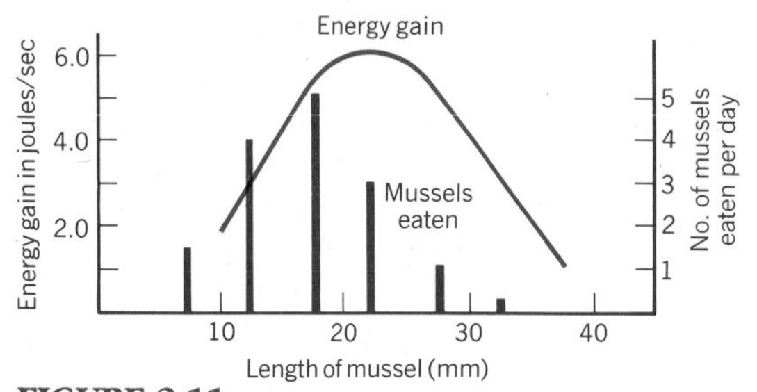

FIGURE 2.11

Optimal foraging by shore crabs.

Large (6.0–6.5 cm) shore crabs (*Carcinus maenas*) break up mussels (*Mytilus edulis*) with their claws and eat the soft parts. When offered an unlimited choice of mussels of different sizes, the crabs concentrate on mussels of intermediate size (bars). This choice is as predicted if the crabs are constrained to act in ways that yield the largest energy returns per unit time spent feeding, as shown by the match to the apex of the solid curve of energy yield per second of handling time, a curve derived from watching crabs handle mussels of different sizes. (From Elner and Hughes, 1978.)

concentrate on middling-sized mussels as predicted.

Optimal foraging theory has now become a subdiscipline of behavioral ecology, because it can guide observations on the foraging behavior of animals. Many studies show that animals as different as back-swimming insects (water-boatmen, Notonectidae), starfish, and birds of all kinds are able to direct their foraging behavior in ways that do in fact maximize their energy intake as predicted by the model (Krebs, 1978). At the very least, these studies encourage our belief that the successes of animals, and hence their relative abundance, is critically dependent on the local flux of energy.

CONCLUSION: THE ENERGY FLOW PARADIGM

The energy flow model gives a wholly satisfying explanation of the most basic structures of natural communities. Plants are abundant and carpet the earth because their energy supplies come directly from the sun. All animal life exists as a tax on the fuel reserves of plants, so that total animal biomass is but a tithe of the biomass of plants. Large predatory animals must always be rare. The model has powerful implications for human activities, showing that beyond some critical density human populations must be vegetarian. The eating of meat, or even more the eating of large fish like tuna that hunt at the ends of long food chains, are luxuries of less-crowded human populations.

Preview to Chapter 3

Multilayered light trap. *Quercus robur* in the open.

THE proportion of solar energy actually transduced into the energy of carbohydrate fuels by green plants is astonishingly low, generally about 2%. This compares with about 35% for the efficiency with which incoming photons can be converted in the process of photosynthesis itself. Only part of this discrepancy can be accounted for by allowing for light wasted on moribund tissue, or because half the energy of sunlight is in wavelengths that cannot be used by photosynthesis. Rapidly growing young plants in ideal conditions achieve maximum conversion efficiencies of only about 8%, still far below the 35% of photosynthesis itself. A general ecological explanation for low ecological efficiency is that plants are adapted to maximize energy production over a lifetime, during which incident solar energy fluctuates repeatedly between darkness and overheating intensities, in an ambience of fluctuating resources and chronically low carbon dioxide. Light absorption capacity might yield optimum returns if arranged to work best at intermediate light intensities. C4 plants are provided with a special costly but effective carbon uptake system in addition to the mechanism found in all other plants, which lets them continue to take up carbon rapidly in brighter light than can the more normal C3 plants. With the CAM variant of this system, a cactus can take on carbon at night, then pass through the day as a sealed capsule, carrying on photosynthesis behind closed stomates using stored carbon as raw material. Disadvantages of C4 photosynthesis must be large because the C4 mechanism is found only in plants of stressed habitats, though it is apparently widely available throughout the families of vascular plants. The structure of a tree with scattered leaves through which light diffuses is explained as an adaptation to spread light of comparatively low intensity to as many leaves as possible. In bright light, many layers of leaves should be optimal to expose the maximum leaf area to adequate light. Shade-adapted trees should have leaves compressed to a monolayer to allow all the leaves adequate light. The model successfully predicts the relative layering in the canopies of trees at all stages in an ecological succession, from open field to closed canopy woodland. Plants at the bottom of the sea all encounter reduced light and are equipped with dense concentrations of pigment, or red and brown auxiliary pigments, that ensure virtually complete absorption of available light.

Chapter 3

Ecological Efficiency and Plant Design

The plant trophic level is fueled by the sun. The efficiency with which the plants of a community harvest energy is the efficiency with which energy is transformed into the 6-carbon sugar, glucose, by photosynthesis on a field scale. Plants should be highly efficient at this, because the fitness of each individual plant must be a function of the solar energy it can harvest as 6-carbon fuel. The chemistry of photosynthesis is essentially the same in all green plants (cyanobacteria are different), and probably has been the same for more than a thousand million years. Natural selection, therefore, has had plenty of time to cull less efficient versions.

The rate at which plants synthesize glucose can be arrived at by various indirect measures, but most easily in field crops where the plants grow in synchrony. The method is nicely illustrated by the first attempt, which not only had the benefit of making as many simplifying assumptions as possible but also yielded a result that later work has shown to be approximately correct (Transeau, 1926).

Assume that one acre of good agricultural land in the American Midwest supports 10,000 corn (*Zea*) plants at harvest and that they should take 100 days from sprouting to harvest. Transeau then proceeded as follows:

Total dry weight of 10,000 corn plants (roots, stems, leaves, and fruits)	6000 kg (dry corn)
Total ash content of 10,000 corn plants (minerals from soil left after burning)	322 kg (ash)
Therefore, total organic content per acre	5678 kg (dry carbohydrate)

Average organic matter contains 44.58% of carbon, therefore carbon per acre	2675 kg (carbon)
Converting to glucose, 10,000 corn plants produced on one acre in 100 days	6678 kg (glucose)

This figure of 6678 kg glucose represents the standing crop of corn at harvest described as a mass of glucose. The farmer's crop, or yield, of grain would of course be much lower. But the standing crop represents only a portion of the glucose that had originally been made, that portion, in fact, that was left after the plants had used glucose fuel in the business of living and growing for 100 days. The glucose fuel consumed could be measured as the carbon dioxide respired by the field of corn. Transeau had his own measures of corn-plant respiration on which to draw. He had kept corn plants in dark chambers through which he passed continuous streams of air. He collected the carbon dioxide at the outlet in an alkali solution and estimated the quantity evolved in unit time by titration. Measurements made on typical corn plants of various ages gave him an average figure for respiration of 1% of the mass of each plant per day, which enabled him to complete his calculation as follows:

Since the crop at the end of the season weighed 6000 kg, the average dry weight for the season was	3000 kg (dry corn)
Average respiration was 1% of this, which	= 30 kg (dry corn)
Therefore, the total CO_2 released in 100 days is 30 times 100	= 3000 kg (CO_2)
Carbon equivalent of 3000 kg CO_2	= 818 kg (carbon)
Glucose equivalent of 818 kg carbon	= 2045 kg (glucose)
Gross primary production of glucose equals net primary production plus respiration equals 6687 kg plus 2045 kg	= 8732 kg (glucose)
But the energy required to produce 1 kg glucose is 3760 kg cal (a figure found by bomb calorimetry).	
Therefore, total energy consumed in photosynthesis of one acre of corn in 100 days equals 8732 times 3760 equals approximately	33,000,000 kg cal
Energy received by one acre of Illinois in 100 days equals	2,043,000,000 kg cal

Therefore, efficiency of photosynthesis equals

$$\frac{33 \times 10^6}{2043 \times 10^6} \times 100 = 1.6\%$$

This result of 1.6% efficiency at culling energy seems a poor result after a billion years of selection for efficiency. And yet the calculation is correct. Numerous measurements since Transeau's time have been made on field crops and wild vegetation, with direct and detailed measures of growth and by measuring respiration with more sophisticated methods than were available to Transeau (see Chapter 24). Some of the results are given in Table 3.1, data that suggest that 2% efficiency is about the norm in the better habitats and climates. The low efficiency of plant trophic levels at harvesting sunlight for the ecosystem is a demonstrated fact (Figure 3.1).

This result is all the more shocking in view of the high efficiency of the photosyn-

TABLE 3.1
Ecological (Lindeman) Efficiencies of the Plant Trophic Level
Transeau's and Gaastra's data as discussed in text. Remainder rearranged from Odum (1971).

	Kcal/m²/day	Efficiency (%)
Transeau's cornfield	33	1.6
Gaastra's sugar beet field	—	2.2
Sugarcane	74	1.8
Water hyacinths	20-40	1.5
Tropical forest plantation	28	0.7
Microscopic alga culture on pilot scale	72	3.0
Sewage pond on seven-day turnover	144	2.8
Tropical rain forest	131	3.5
Coral reefs	39-151	2.4
Tropical marine meadows	20-144	2.0
Galveston Bay, Texas (fertilized by wastes)	80-232	2.5
Silver Springs, Florida (vegetated bottom)	70	2.7
Subtropical blue water (open sea)	2.9	0.09
Hot deserts	0.4	0.05
Arctic tundra	1.8	0.08

FIGURE 3.1
***Zea mais:* As inefficient as the old prairie or forest it displaced.**
A corn field converts about 2% of incident solar energy into reduced carbon fuel, a performance about as good as wild vegetation. Even productive ecosystems are powered with only a small fraction of the energy theoretically available to them.

thetic chemistry itself. A biochemist's calculation of the efficiency of the whole process of photosynthesis, from the absorption of photons in photosystem II until the completion of a molecule of glucose, goes like this: The total free energy supplied in the photosynthesis of glucose is

$$6CO_2 + 12H_2O \rightarrow C_6H_{12}O_6 + 6O_2 + 6H_2O$$

This can be compared with the energy put into the synthesis by estimating the number of electrons that pass through the pathway in the synthesis of glucose, this number being 24. But each electron is energized twice (in photosystem II followed by photosystem I), requiring a total of 48 photons of light per electron. The longest wavelength these photons are likely to have is 700 μm (Figure 3.2), which gives a minimum of 41 kcal/mol per photon, yielding an energy input to all 24 electrons of 41 × 48 = 1968 kcal. Then the biochemical efficiency of photosynthesis equals:

$$\frac{\text{increase in free energy}}{\text{energy of photons consumed}} = \frac{686}{1968} = 35\%$$

Thus photosynthesis, as a biochemical process, is splendidly efficient: 35% of usable sunlight entering a reaction site is converted to the potential energy of glucose. Apparently natural selection had indeed been alive and well all the billion-plus years available to perfect the process. So why is the efficiency of plants on a field scale only about 2%, even in good conditions?

The first glaring difference between the two measures is that a biochemist is only concerned with the efficiency with which energy actually introduced into the system is transformed: 48 photons are used, so let us forget all the rest that might have been absorbed into a photosystem but were not. Let us even forget those photons that did enter the photosystem, and which excited electrons that then dropped back to lower energy states without doing useful work. Only the used energy will be counted. But calculating the ecological efficiency of a plant requires that all these portions of the total energy flux be included.

When photosynthesis is used by an individual plant to yield the resources it will need to gain maximum fitness, the test must be: How much of the energy theoretically available to the plant was actually trapped and stored as glucose? The numerator of the ecological efficiency equation might be the same as the biochemist's, but the denominator, including as it must all the photons that got away, will be much larger.

So part of the wide gap between the biochemical efficiency of photosynthesis by a corn plant (35%) and the ecological efficiency of a field of corn (1.6%) is represented by the photons that got away from the corn plants. It is easy to see where many of these escaped. In the first weeks of the corn's hundred days, bare ground lay between the seedlings and most photons went to heat soil. In the last weeks of the hundred days many incoming photons struck moribund tissue of rustling yellow corn plants at the end of their lives, and these photons did no more than warm the corn. Losses of photons to unproductive surfaces must apply in wild vegetation as well as to crops, as growing seasons begin

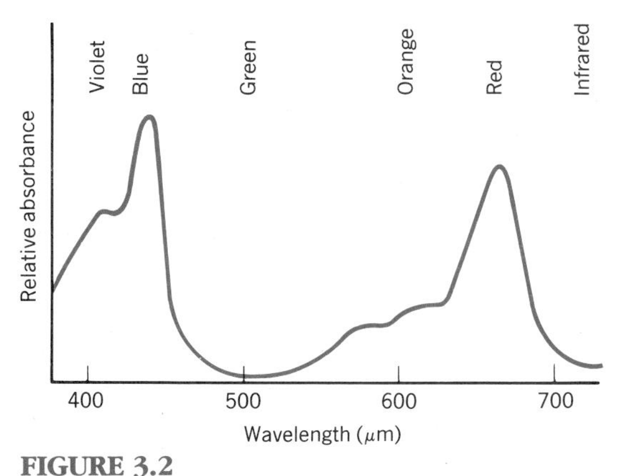

FIGURE 3.2
Absorption spectrum of chlorophyll *a*.
Only light with wavelengths between 0.38 and 0.78 μm have quantum energies that cause electron transitions when absorbed. These are both the wavelengths of visible light and the wavelengths in which most solar energy reaches the earth's surface (Figure. 3.2). Chlorophyll *a* absorbs strongly in these wavelengths. The "green gap," where chlorophyll does not absorb strongly, is plugged by secondary pigments like *carotenoids*, *xanthophylls*, and *phycobillins*. Electron excitations induced in these pigments by absorbed quanta are passed to chlorophyll *a* by resonance.

and end or as individual leaves are young or old.

A second source of lost photons results because only about half the energy of incoming solar radiation is in wavelengths sufficiently energetic to effect the electron shifts necessary for photosynthesis. Figure 3.3 shows the solar spectrum reaching the surface of the earth. Only the visible part of the spectrum can be used in photosynthesis (vision no doubt requires the same high-energy photons needed for chemical synthesis). Plants have adapted to this reality by having green pigments that absorb light strongly in the most useful part of the spectrum (see Fig. 3.1). Ecological measures include all sunlight, usable and unusable. The biochemists include only that actually used. To be fair to the plants, therefore, we should multiply their typical ecological efficiency by 2, making them effectively 4% efficient.

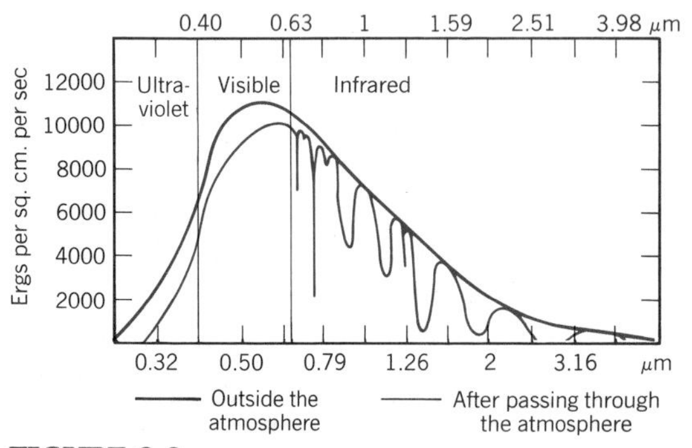

FIGURE 3.3

The solar spectrum above and below the atmosphere.

Although only about 20% of the incoming energy is degraded in the atmosphere, selective absorption by atmospheric gases (particularly CO_2 and H_2O) presents plants with an irregular array of wavelengths. Photosynthesis needs energy of wavelengths roughly comparable to that of visible light. The energy flux is actually greatest (peak of the curve) in the visible wavelengths, so the narrow range of wavelengths used in photosynthesis actually includes about half the total energy flux. The data are from measurements made in clear weather from a mountain observatory in Australia by C. W. Allen. (From Lamb, 1972.)

Photons that get away, and photons of the wrong wavelength, however, are still quite inadequate to explain the huge discrepancy between the low ecological efficiency of field vegetation and the high efficiency of the synthesis itself, even though they do narrow the gap. Vigorous, healthy, young, growing plants or intact leaves turn out to transform light falling directly upon them with only about 8% efficiency at light intensities adjusted to yield highest possible productivity. Individual plants at the peak of their form, not just whole vegetation, are relatively inefficient.

Possible causes for the low ecological efficiency of plants are many. Plants operate in a real world of many constraints and fluctuating ambient conditions. All plants are exposed to a huge range of different light intensities, from the glimmers of dawn and eve to the brilliance of the noonday sun; their designs must yield the best return for a lifetime of dawns and dusks, not just for an hour of bright light. And all plants are resource-limited, whether the scarce resource is water, nutrients, or the essential raw material of photosynthesis—carbon dioxide.

LIGHT AND DARK REACTIONS: TWO SYSTEMS WITH POTENTIAL LIMITS

Photosynthesis is a process of many stages but they may be divided into two groups, collectively known as the light reactions and the dark reactions. The **light reactions** essentially transform quanta of light (photons) into forms of energy usable by a plant to power chemical synthesis, and the **dark reactions** work the synthesis itself.

The essentials of the light reactions are the dissociation of water into molecular oxygen, hydrogen ions, and electrons, followed by the excitation of the captured electrons by light. Molecular oxygen is vented to the atmosphere, and energetic molecules (ATP and the like) are left to power the actual glucose synthesis in the Calvin–Benson cycles. This synthesis following the light reactions could be performed as well in the dark as the light, provided ATP were to be available, hence the term "dark reactions." Obviously, the light and dark reactions might be limited by different environmental factors, to both sets of which plant designs must be accommodated as well as possible.

THE BONNER LIGHT-ABSORPTION HYPOTHESIS

In the morning, in the evening, under overcast skies, or in the shade, leaves receive only dim light. Probably most leaves, most of the time, are thus dimly lit, requiring that they be adapted to operate efficiently at low light intensities. Bonner (1962) identified what may be an important adaptation to this end. All plants have what appears to be a large excess of chlorophyll molecules for the number of synthesis sites we can identify. In bright light much of this absorbing chlorophyll is largely redundant, as more than enough photons are trapped at any instant. But coupling a reaction site to this apparent surplus of chlorophyll should pay off in dim light, because it would ensure that sufficient photons were fed simultaneously into each production site to give electrons their necessary double kicks through photosystems II and I.

A leaf with surplus chlorophyll molecules will inevitably be inefficient in bright light, as the photons striking them take no part in photosynthesis. To the extent, therefore, that leaves are indeed engineered to perform well in dim light, their comparatively poor performance as energy converters in bright light is understandable.

THE RESOURCE LIMITS HYPOTHESIS

Rates of photosynthesis for small plants or leaves can be measured in experimental systems by the rates of oxygen evolved or carbon dioxide absorbed. Figure 3.4 shows data for photosynthesis measured as CO_2 absorbed by a preparation of a sugar beet leaf at various light intensities. A curve of photosynthetic efficiency is superimposed on the curve of photosynthetic rate. Efficiency is high in dim light—nearly 18%—but this falls to 8% at high light intensities. Increasing the intensity of light still further yields no increase in photosynthesis, and the system is said to be "light saturated."

Data sets like that of Figure 3.3 for sugar beets are typical of all plants, from crops to algae. Wassink (1959) got closely comparable results for a culture of the green alga *Chlorella*, for instance: 20% efficient in the dimmest light, 8% when light saturation was reached, and even lower efficiencies at still higher light intensities. Thus algae are no more and no less efficient than higher plants.

Even at very low light intensities, when photosynthesis is a linear function of light intensity, the efficiency of energy conversion revealed by these laboratory measures is still only half the biochemist's efficiency of the process itself (18% versus 35%). This is probably a measure of the photons that escape capture, being absorbed by other leaf structures or reflected, or because some of the light was the wrong wavelength. Given the appar-

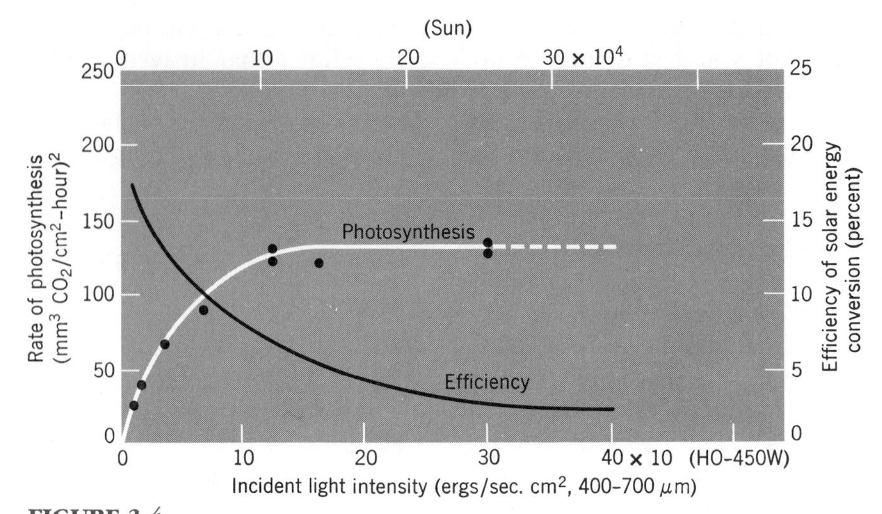

FIGURE 3.4

Rate and efficiency of photosynthesis as a function of light intensity.
The data are for preparation of a sugar beet leaf. The rate of photosynthesis would not be raised for light intensities much above 10 ergs per second per square centimeter, and accordingly the efficiency of the process progressively falls. It is important to note, however, that the yield of energy to the plant was highest at high light intensities even though the efficiency was low. Note also that this sugar beet leaf in dim light was quite as efficient as a green alga culture in equally dim light. It is untrue to claim that algae are more efficient than other plants. (From Gaastra, 1958.)

ently extraordinary difficulty of sweeping incoming radiation clean of all useful photons, an efficiency of 20% must seem an extraordinary achievement. Only in brighter light does the process work less well.

High efficiency in dim light but low in bright light seems broadly consistent with the Bonner hypothesis of leaves supplied with excess chloroplasts for operating in dim light: high efficiency (about 18%) is attained until photons absorbed by all available chlorophyll molecules are in use, after which the rate of photosynthesis cannot be increased further. The rate of photosynthesis then remains constant, even as light intensity continues to rise, and the ratio of the two (efficiency) falls progressively. But the same results can be explained even more simply by a hypothesis of resource limits of a kind well known in classical economics.

A resource limit appears particularly likely because the carbon needed for glucose synthesis is available only as the rare atmospheric gas, carbon dioxide, present in the ambient air at a concentration of only 0.03% by volume. When light is dim, on this hypothesis, the rate of photosynthesis is so low that carbon extracted by pumping even this diffuse gas should be ample. But it is reasonable to expect that the rate of carbon uptake would become critical at high rates of photosynthesis. Thus plots of photosynthetic rate against light intensity can be explained as well by a resource limits hypothesis as by a dim-light adaptation hypothesis.

The hypothesis of CO_2 limitation predicts that light saturation would be delayed if extra CO_2 were provided to the system, a prediction that is confirmed by experiment (Figure 3.5). A rich supply of CO_2 increases the rate of photosynthesis in full light.

The **carbon dioxide hypothesis** is of respectable antiquity, having first been put forward when F. F. Blackman demonstrated results like those of Figure 3.3 back in 1905, and postulated that something other than light, that is, something working in the dark, limited photosynthesis in bright light. This was the crucial inference that let Blackman conclude that photosynthesis had two components, which he called the light and dark reactions.

If Blackman was correct in concluding that the dark reactions were resource-limited, then the most likely ultimate limiting resource for land plants must often be carbon dioxide. Plant design and photosynthetic performance, however, must also be limited in other ways, such as by shortage of nutrients or water and fluctuating temperature. Ecologists do not accept a general hypothesis of resource limits by any one factor. Nevertheless, such alternative photosynthetic pathways as C4 or CAM systems, as well as many physical structures of plants, seem particularly directed to maximizing carbon dioxide uptake in different circumstances.

C4 PHOTOSYNTHESIS

Among the adaptations fielded by plants in their gathering of energy are alternative ways of extracting carbon dioxide from the air. Some, called C4 plants, operate an extra CO_2-fixing system, in effect having an

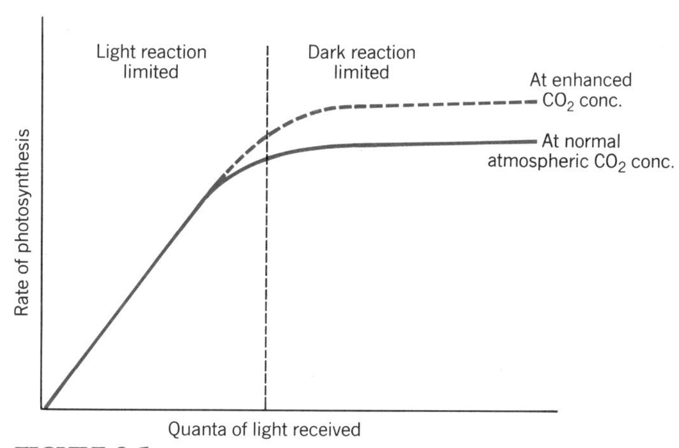

FIGURE 3.5

Effect of CO_2 concentration on photosynthesis.

Increasing the concentration of CO_2 in the experimental vessel raises the rate at which photosynthesis becomes light-saturated, thus suggesting that the dark reaction limit can be set by carbon dioxide concentration. Similar results have been obtained for field crops provided with enhanced CO_2 under plastic tents.

extra carbon pump. These are the C4 plants.

Carbon uptake in most plants (called C3 plants) is straightforward. The gas diffuses through the open stomates to the wet membrane directly under the surface of the leaf, where it is taken up by the enzyme RuBP carboxylase. The enzyme then passes the CO_2 directly into the synthetic mechanism of the Calvin–Benson cycle. This was discovered in the classic work of Calvin's team with isotopically labeled carbon dioxide ($^{14}CO_2$). The labeled carbon passed rapidly through RuBP and into the 3-carbon phosphoglycerate (PGA).

The carbon pump of the C4 system was discovered when the Calvin approach was applied to samples of sugarcane (Hatch and Slack, 1970). Labeled carbon appeared first in several 4-carbon compounds instead of in 3-carbon PGA. Sugarcane is a C4 plant instead of a C3.

The essentials of the C4 system are shown in Figure 3.6. The enzyme PEP carboxylase fixes CO_2 as it enters the leaf through the stomates. Carbon dioxide is added to the PEP (phosphoenolpyruvate), converting it from a 3-carbon molecule into a 4-carbon molecule. Oxaloacetic acid, malic acid, and aspartic acid (all 4-

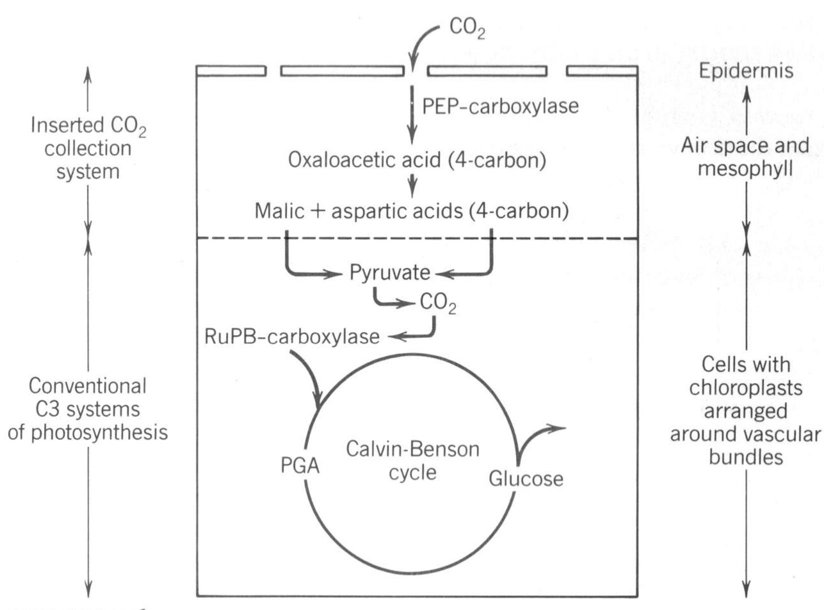

FIGURE 3.6
C4 photosynthetic pathway.
C4 plants have the same system for light collection and glucose synthesis as C3 plants, but the chloroplasts where synthesis is performed are placed deep within the leaf around the vascular bundles. Carbon dioxide is collected from air spaces inside the stomates by an enzyme, PEP carboxylase, that passes the CO_2 to a series of 4-carbon acids. These are transported across cells of the mesophyll to the cells containing chloroplasts, where the CO_2 is regenerated. In some environmental circumstances the energetic costs of this extra CO_2 transfer can be met by extra glucose production from more efficient use of light.

carbon) are then produced. These are transported across the cells of the leaf mesophyll to the interior, where cells green with chloroplasts await them. Once arrived at a production site, the 4-carbon acids are decomposed to yield their CO_2, and their residues are returned across the mesophyll for more. The regenerated CO_2 is then fed into a Calvin–Benson synthesis cycle via RuBP carboxylase in the usual way, the synthesis being powered by photons that have penetrated the transparent mesophyll to reach these production sites deep in the leaf.

Possessing a C4-carbon transport system clearly incurs costs that most plants do not have to pay, because extra work is done in manipulating CO_2 and 4-carbon acids through the mesophyll. This work must detract from calories that the plant can use to win fitness. Compensating advantages must, therefore, exist. These seem to be of two kinds: better scavenging of CO_2 and prevention of photorespiration. Both these advantages reflect on unsatisfactory properties of RuBP carboxylase when used, as by a C3 plant, to take CO_2 directly from the air.

The enzyme RuBP carboxylase has the unfortunate quality that, in the presence of free oxygen, it catalyzes a reaction that combines free oxygen with RuBP to split it into one molecule of PGA and some useless by-products that are oxidized back to CO_2. The enzyme thus undoes much of its own good work. On the one hand, the enzyme adds carbon to RuBP but on the other, it oxidizes carbon already held in RuBP and vents it back as CO_2. This curious behavior of RuBP carboxylase is called **photorespiration** (because CO_2 is respired in the presence of light). In light and in the presence of free oxygen, this always happens. In a C4 plant this unfortunate side effect of RuBP carboxylase is blocked, because the sites of activity of the enzyme are removed from the air to deep within the leaf, where they are isolated from ambient free oxygen.

More exciting for an ecologist is the observation that C4 plants can scavenge CO_2 from the air more rapidly and completely than can C3 plants. In the special circumstance of hot dry conditions, C4 plants appear to take up carbon dioxide with so little restraint that it can be argued that photosynthesis in these plants cannot be carbon-limited. However, the restriction of the C4 system to particular life forms and habitats shows that the system must have serious drawbacks

The rate at which CO_2 is taken from air by a C3 plant with RuBP carboxylase falls rapidly at concentrations much below the atmospheric norm of 300 parts per million, (ppm; 300 ppm = 0.03%) and ceases entirely at about 50 ppm (Figure 3.7). PEP carboxylase of a C4 plant, however, continues to take up CO_2 at a high rate down to concentrations so close to zero that they are hard to measure.

Experiments to demonstrate the apparent superiority of C4 photosynthesis are comparatively easy because it often happens that different, closely related species (sympatric species, see Chapter 9) occur in pairs, one using the C3 pathway and one using the C4. Both species can then be raised in a greenhouse in identical conditions and their carbon uptake measured in experimental chambers. Figure 3.6 describes the results of this procedure when applied to two species of desert saltbush (*Atriplex*), suggesting that the C4 plants were more "efficient."

Yet, to an ecologist, the idea that "efficient" plants and "nonefficient" plants could coexist is an idea so unlikely as to verge on the impossible. The efficient plants should win more resources, leave more offspring, and drive the nonefficient plants to rapid extinction. The only logical

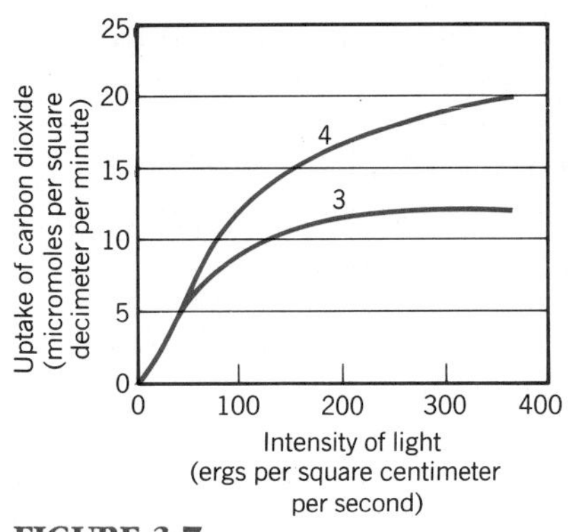

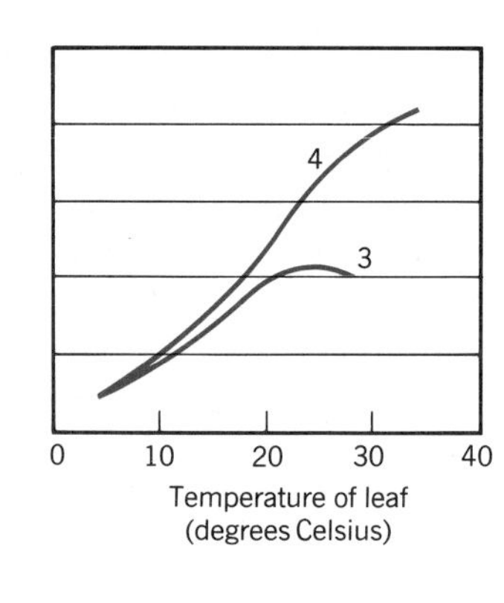

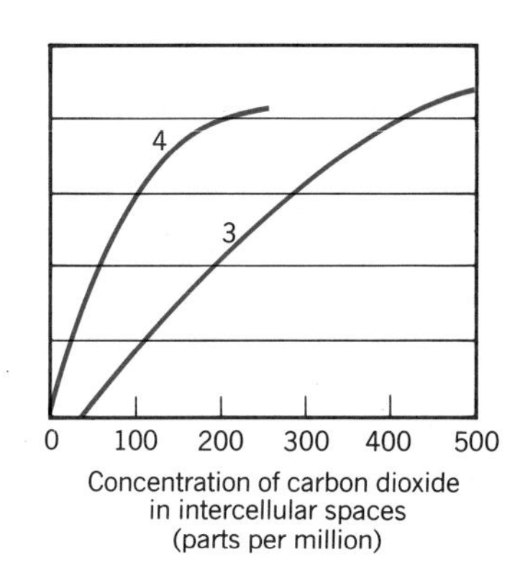

FIGURE 3.7
Uptake of CO_2 in C3 and C4 plants.
The data are for greenhouse-grown plants of two species of the same genus, one C3 and the other C4, grown under identical conditions. For both, the rate of CO_2 uptake depends on light energy, temperature, and CO_2 concentration, but the rate for the C4 species is always higher. The data, however, describe only the rate of CO_2 uptake of plants for the experimental runs, and this is not necessarily proportional to the rate of glucose synthesis either during the run or over the life of the plant. Furthermore, a calculation of relative efficiency requires an estimate of the extra cost of the C4 operation. The plants are two species of desert saltbush *Atriplex patula* (C3) and *Atriplex rosea* (C4). (After Björkman and Berry, 1973.)

way in which this coexistence can be conceived is that the process of replacement is still in progress. Convincing evidence shows, however, that no such replacement is happening. The evidence comes from studies of the spread and antiquity of the C4 pathway.

Whether a plant is C3 or C4 can be determined from dead herbarium material by using a mass spectrometer. The two enzymes RuBP carboxylase and PEP carboxylase discriminate differently against the rare stable isotope of carbon, ^{13}C, so that a mass spectrographic measure of the $^{13}C/^{12}C$ ratio on dead material reveals which pathway the living plant used. This method has shown that the C4 mechanism is possessed by more than 100 genera in ten families of flowering plants (Burris and Black, 1976). The method has also shown that different species within the same genus possess one or the other, as in the *Atriplex* pair of Figure 3.6. Clearly, the C4 pathway cannot be a new evolutionary invention that is even now changing the balance of life on earth but rather it is an alternative that has long been available.

In fact, all plants possess PEP carboxylase and the related chemistry of a C4 plant. What they lack is the leaf structure that makes C4 photosynthesis possible. The leaf geometry of a C4 plant, called the **Kranz syndrome**, is illustrated in Figure 3.8. This provides the essential separation of the chloroplasts from the epidermis that allows PEP carboxylase and 4-carbon acids to serve their functions as carbon transporters to the buried chloroplasts. Because this structure occurs so widely in so many unrelated genera, only simple genetic changes can be needed to bring it about, possibly only a change at one locus.

These arguments make it clear that C4 photosynthesis is, as it were, an option always open to natural selection. Whether the individuals of a species possess a C3

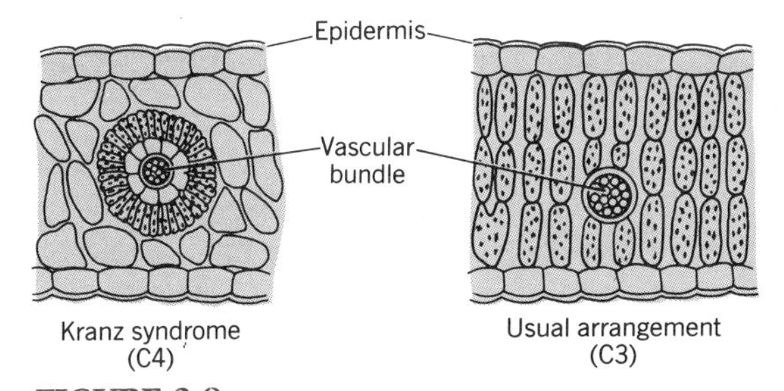

FIGURE 3.8
Leaf structure in C4 and C3 plants.
The Kranz syndrome of C4 plants groups cells containing chloroplasts clear of the epidermis and the ambient air. In the leaf structure of the more usual C3 plants, cells containing chloroplasts are distributed throughout the leaf.

system or a C4 system must depend on the suitability of one system or the other to the local circumstance. Neither can be more efficient throughout the life of an individual or in all circumstances, because then only one system would be found. And the C3 system is the most common: it follows that in most circumstances the C3 system must give the better return despite the advantage that C4 seems to give in carbon uptake.

When C4 Photosynthesis Gives the Advantage

The circumstances in which the minority of plants that do use the C4 pathway have an advantage are suggested by the fact that they are found in hot, dry places, or places with special water stresses. All plants of more mesic habitats and all trees use C3. The only plant of cool temperate regions known to use C4 is the salt marsh plant *Spartina*, which may be thought to be under water stress because of the strong salt solutions in which it lives (Long et al., 1975).

Plants of hot, dry places must open their stomates, exposing wet tissue to the air, when they take in CO_2. This suggests that the C4 pathway works to reduce water loss in conditions of drought. The PEP carboxylase can scavenge air within the leaf very thoroughly, allowing the stomates to be closed for longer periods, and the air spaces between the productive cells and the epidermis offered by the Kranz syndrome can collect and store the CO_2 of respiration, thus allowing it to be recycled behind closed stomates. This water conservation aspect of the C4 pathway appears in its most refined form in CAM plants, described in the next section.

The recent discovery that carbon dioxide of the air is much reduced in ice-ages suggests that C4 plants may have a particular advantage in glacial times. The present preponderance of C3 plants thus may be exaggerated in the ephemeral interglacial warmth in which we live (see Chapters 7, 17, and 28).

The C4 Variant Called CAM

CAM stands for crassulacean acid metabolism, so named because it was first discovered in desert succulent plants of the family Crassulaceae, though it is used by cacti and other desert succulents as well (Figure 3.9). These plants operate a C4 pathway of a sort, but without a Kranze syndrome leaf. Green photosynthetic tissue is directly under the stomates as in a C3 plant, though well supplied with PEP. Their most elegant adaptation is an ability to work in shifts, a night shift and a day shift. At night they open their stomates, collect CO_2 on PEP carboxylase, and make the array of C4 acids. A desert plant can do all this at night with little loss of water because gas collection is powered by ATP and can be done in the dark. The next stages of production require light, but they can be put off until daylight (Winter, 1985).

FIGURE 3.9
***Sempervivum tectorum* CAM plant using air by night and sun by day.**
The plant dubbed by gardeners as "hens-and-chickens" is typical of the family Crassulaceae in using Crassulacean Acid Metabolism or CAM. In arid times the plants open stomates at night and store carbon in 4-carbon acids, like C4 plants. By day the leaves are sealed off from the air and photosynthesis proceeds using the carbon reservoir of the night's accumulation of 4-carbon acids.

In the daytime, CAM plants leave their stomates shut, taking in no more CO_2. Instead they use the CO_2 stored overnight in 4-carbon acids and run their Calvin–Benson cycles in the abundant desert light but behind closed stomates. It is pleasing to think of a large cactus under the desert sun splitting water to combine hydrogen with CO_2 in the synthesis of glucose and using the sun as power but otherwise almost entirely insulated from desert stress, a capsule journeying through the light-filled day toward the next night when it will take on supplies again.

The most elegant desert variants of all are perhaps plants that can use CAM or C3 photosynthesis, or both, as occasion requires: so-called **inducible CAM plants**. An example is the salt-tolerant (halophytic) plant *Mesembryanthemum crystallinum* of Israel and California. In the wet season they grow quickly using C3 photosynthesis. In the dry season they switch to CAM and reproduce, opening their stomates only at night. Winter (1992) kept experimental plants in CO_2-free air at night, letting them have access to CO_2 only in daylight. They set only 1/10th the number of seeds of plants allowed CO_2 night and day.

C3 or C4: Practical Implications

The Hatch–Slack pathway of C4 and CAM plants is used only in hot places, dry places, or both. Building a Kranz syndrome leaf and meeting the costs of a PEP carboxylase pump, therefore, only offer a commensurate reward of fitness in these places. Where water is abundant or the temperature moderate, the C3 pathway is more efficient.
I offer the couplet,

In different circumstance it pays
To gather gas in different ways.

Much research has recently gone into breeding C4 strains of crop plants, many of which are C3. The aim is to make them more efficient and thus increase yields. The breeding program will probably succeed, since the genetic changes needed to produce a Kranz syndrome leaf seem to be few. Whether the yields will indeed be better with the C4 varieties must depend on the conditions in which the plants grow. The C3 mechanism of the wild ancestors of the crops must have been the most efficient system for the niches and habitats of those ancestors. The C4 varieties will be better only if grown where the C4 is known to be best: in relatively hot, dry places. In fact, agriculture does create mini-deserts in which crops have to grow: for instance, bare fields under the scorching sun of summer. It may be, therefore, that C4 wheat will yield more than the present C3 wheat on vast people-made prairies.

LIGHT DIFFUSION BY TREES

A tree is an array of solar panels spread on stalks. The panels are green leaves. They are usually small, say, five centimeters or less at the widest continuous part (Figure 3.10). When people build solar panels to heat a house or power a spaceship, they make them in large continuous sheets perhaps several meters across. But trees make their solar panels small, while spreading them to the sky in thick mats of overlapping ranks. Presumably the green solar panels work better in this broken array. But why?

A convincing explanation is that a tree is an arrangement of leaves stacked in broken layers through which light streams to increase photosynthesis by exposing the maximum leaf area to lowered light intensities where photosynthesis is most efficient. What follows comes from the analysis of Horn (1971).

A large, intact sheet-like leaf spread out under the noonday sun would be light-saturated, converting energy with low efficiency. Let the photosynthesis of a hypothetical tree equipped with a single huge leaf be one unit (Figure 3.11*a*). If instead of an intact leaf the tree was equipped with a perforated solar cell that was half holes, it would achieve only ½ unit of photosynthesis. But then it would be possible to

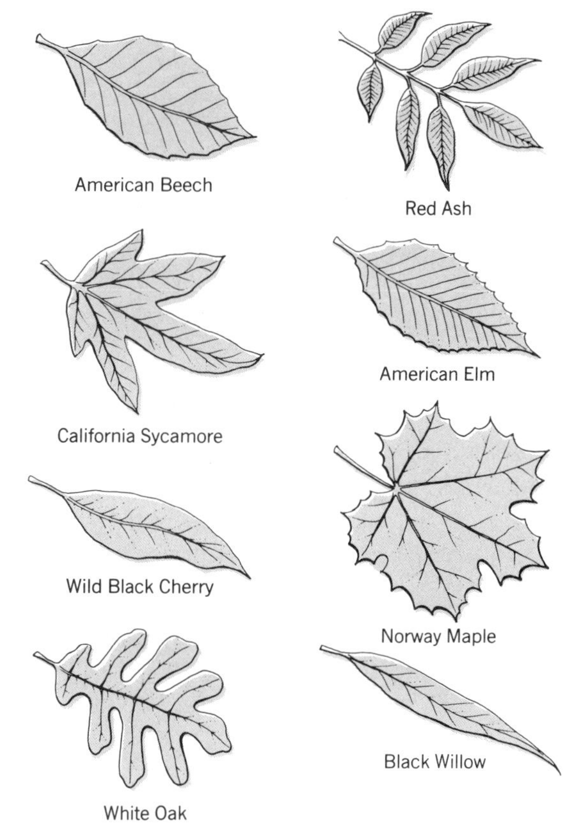

FIGURE 3.10
Solar panels used by trees of the temperate forests.
Trees do not make solar panels in large sheets but instead use small leaves, often of irregular design. The use of small leaves instead of large solar panels results partly from mechanical and heat budget considerations (Chapter 5), but a small leaf design also allows the diffusion of light through layers of leaves, thus increasing the effective area for CO_2 uptake.

spread out a second perforated sheet below the first to receive the diffused light (Figure 3.10*b*). If the light on the second layer were still sufficiently intense for maximum photosynthesis, then the second perforated sheet would also achieve $\frac{1}{2}$ unit of photosynthesis, and the two perforated sheets would have the same total produc-

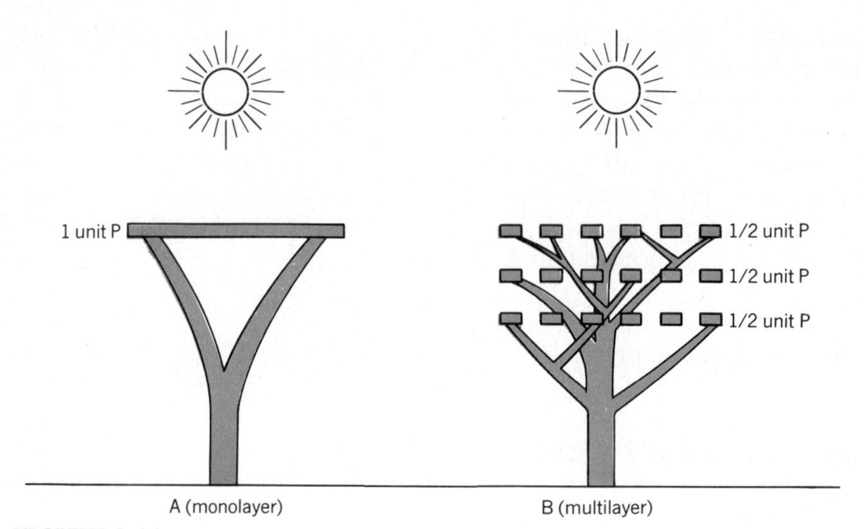

FIGURE 3.11
The advantage of a perforated leaf system.
Tree *a* has a single huge leaf that traps all incident sunlight. Photosynthesis is 1, a rate set by CO_2 uptake and light saturation. Tree *b* has a perforated leaf structure in layers such that each layer is half leaf and half holes. The leaf area in each layer can achieve maximum photosynthesis in the diffused light because each has a private CO_2 supply. Each layer of tree *b* achieves ½ unit of photosynthesis and the whole tree *b* achieves one and a half times as much photosynthesis as tree *a*.

tion as the one intact sheet. Despite this, light would still stream through the second perforated sheet, making possible yet a third punctured solar cell spread in the light beneath, as in Figure 3.10*b*.

If the bottom perforated solar cell in a stack of three worked as well as the top two, the three-layered device would outperform the single intact sheet by three to two. More layers would increase the advantage of the stacked design still further, provided light intensities were high enough to allow this. This is the essence of Horn's explanation for the smallness of leaves. They serve to produce the effect of perforated sheets stacked one above another. The payoff is increased photosynthesis in bright light.

However, a stack of perforated solar collectors would work only if the lower layers could be kept clear of the shadows of those above. Shadows can be avoided provided the distance between the layers is large enough. Because of the earth–sun distance and the diameter of the sun, a leaf casts a shadow for about 100 leaf diameters (Figure 3.12). If a leaf is 1 cm across, therefore, the second layer would only have to be 1 m below the first to escape shadows. Leaves of conventional size would all be suited to layer arrangements with layers up to 2 to 3 m apart—well within the branching distance of most trees.

Larger leaves can be accommodated by irregular shapes that reduce their effective

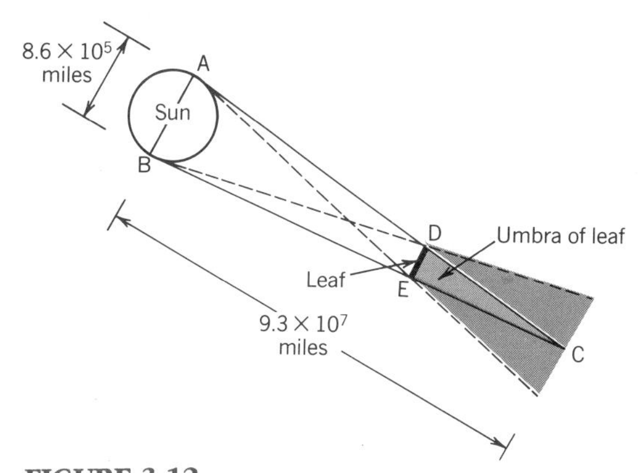

FIGURE 3.12
Shadow cast by a leaf.
The triangle ABC is similar to the triangle DEC. Hence (length of umbra)/(distance from tip of umbra to sun) = DE/AB. Therefore length of umbra = $(9.3 \times 10^7)(DE)(8.6 \times 10^5)$ = 108 DE. This means that a leaf casts a shadow to 108 diameters and that light farther away than 108 diameters is evenly diffused. Leaves with solid widths of 2 cm need to be in layers about 2 m apart for the lower leaves to escape the shadow of those above. (Redrawn from Horn, 1973.)

diameters. Here is an explanation of the deeply notched margins of so many leaves: the notches are devices to reduce effective leaf diameter in bright light, thus turning a layer into the optimum perforated sheet (Figure 3.9).

However, real trees do not live under a sun that is stationary at the zenith as in Figure 3.10, but are illuminated from different directions at different hours of the day. Real leaf arrangements must be adapted to catching light from changing angles. Thus real trees should not be built in discrete, horizontal layers but only to tend toward them.

Horn's model thus describes the typical great tree in the open as a huge light-diffusing and carbon-gathering device. Evenly diffused light adequate for maximum photosynthetic rate is spread over the largest possible leaf area. The trick is to light as many leaves as possible.

Testable Predictions of the Hypothesis

Horn describes the ideal great tree in the open as a "**multilayer.**" But many trees in the real world are adapted to grow in the shade of others, where the light supply is not sufficient for photosynthesis to be light-saturated. For a tree in so gloomy a place, a multilayer design would be a mistake, and the tree might do best with a single dense solar collector that lets no light pass. The design of shade trees thus should be close to that of a **monolayer.**

The hypothesis thus predicts that, although trees living in bright light should have a multilayer design, shade-adapted species should be without layers. This could be achieved with huge, nonperforated leaves, designs that are in fact found in understory plants of the tropical rain forest, like the house plants *Monstera* and rubber plants. That these are adapted to low-light conditions is shown by their survival in dimly lighted office corridors. More interesting for a test of the hypothesis is that shade trees with small leaves should have their leaves arranged close together in a nonoverlapping array, thus making them functional monolayers.

Detailed predictions of the hypothesis include the following:

1. Trees growing in the open should have their leaves arranged in depth, probably in a random array (multilayer design).
2. Understory trees growing in dense shade should have larger leaves, or their leaves should be arranged nonrandomly (to avoid overlap) and within a short vertical distance of one another (monolayer design).
3. A single branch of a tree adapted to growth in the open (multilayer)

should cast less shade than a single branch of a tree adapted to grow in the shade (monolayer).

4. Trees growing in the open should have small leaves of irregular shape.
5. Understory trees and shrubs should have leaves of regular shape, and these should tend to be larger than leaves of trees growing in the open.
6. Total leaf area will be larger than the ground area covered in trees growing in the open, but nearly equal to the ground area covered for trees growing in shade.

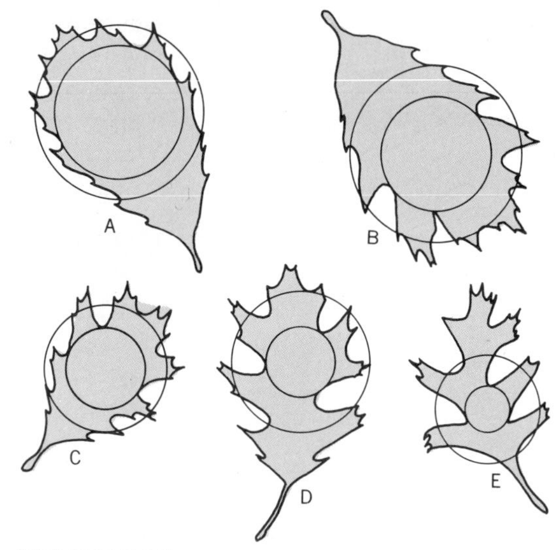

FIGURE 3.13
Oak leaves in shade and sun.
The leaves are all from black oak (*Quercus nigra*). (*a*) Leaf of a seedling. (*b*) Leaf from a shaded branch near the ground. (*c–e*) Leaves from progressively higher on the tree. The small circle is the largest circle that can be inscribed in each leaf. The relative sizes of the two circles show how well lobing adapts the leaf to its multilayered role of letting half the light pass through the layer of which it is a member. (From Horn, 1971.)

Predictions 4 and 5 receive qualitative support from standard observations of botanists. Large-leaved plants grow on the gloomy floors of tropical rain forests. Common emergent canopy trees of the tropical forests, on the other hand, are of the family Leguminosae, with their typical finely divided pinnate leaves. Familiar temperate trees that colonize open fields are ashes (*Fraxinus*) and maples (*Acer*), with their characteristic leaves of thin or irregular shape. Figure 3.13 shows how the leaves of an oak tree have only small teeth on the margin when born by a sapling growing in the shade of larger trees, whereas they are deeply indented when exposed to the direct rays of the sun in the canopy of a mature oak tree.

These qualitative observations on leaf shape are more anecdote than hard evidence, yet they are consistent with the hypothesis. Much stronger support comes from Horn's demonstration that the relative layering of trees, and relative leaf area, can be measured and shown to correlate with their relative tolerance to shade (predictions 3 and 6).

To measure both the relative number of layers and relative leaf area, Horn worked backward by measuring the light actually penetrating different parts of a tree. A single branch of a true monolayer, for instance, should allow virtually no light to penetrate, but a single branch of any multilayer should allow much light through (prediction 3).

Horn used a specially calibrated light meter to compare the light coming through a branch with the light in the open, the difference giving him a measure of relative leaf density (p) of a branch. At the same time he measured the proportion of light coming through the whole tree (u). From the three measures—light in the open, light through a branch, and light through the whole tree—he was able to calculate both relative number of layers and relative area:

$$u = (1 - p)^n$$

where n is the relative number of layers. Solving for n we get

$$n = \log(u)/\log(1 - p) \quad (3.1)$$

Table 3.2 gives an example of Horn's results. Trees are ranked according to their known order of appearance in ecological succession. Pioneer trees like birch and sassafras enter old fields, growing as saplings in full sunlight. Accordingly, they should have the properties of extreme multilayers. Beeches and maples at the end of succession (the "climax community" of succession theorists) must grow as saplings under the shade of others. Accordingly, they should approach the monolayer state. Horn's data show that these predictions are upheld.

Horn's hypothesis, therefore, generates predictions that fit the facts of tree life very well. Trees spread their leaves in a dappled pattern against the sky as an essential strategy in the quest for energy with which to win fitness. Tree design is essentially an adaptation made necessary because photosynthesis is limited in full daylight.

Inclined Planes and Tiny Leaves on Plants of Short Stature

If light strikes a leaf at an oblique angle it will be spread out, and the intensity of the radiation received on unit area will be less

TABLE 3.2
Tree Layering and Shade Tolerance
The tree species are arranged in the conventional order in which they appear in secondary succession. The order also represents a ranking by shade tolerance known to foresters. Horn's calculations of relative number of layers show that there is a relative shift from multilayer to monolayer designs as succession proceeds as predicted. The ratio of total leaf area to ground covered also is reduced as succession proceeds. (From Horn, 1971.)

Species	Number Measured	Percentage Light/ Branch	Percentage Light/ Tree	Number of Layers ± SE	Leaf Area/ Ground Area
		Early succession			
Gray birch	10	44	3.6	4.3 ± 0.4	2.4
Bigtooth aspen	6	45	6.9	3.8 ± 0.5	2.1
White pine	13	25	0.8	3.8 ± 0.4	2.9
Sassafras	3	14	0.8	2.7 ± 0.7	2.4
		Mid-succession (on moist soil)			
Ash	10	26	3.0	2.7 ± 0.2	2.0
Blackgum	7	15	1.4	2.6 ± 0.5	2.2
Red maple	21	20	1.8	2.7 ± 0.2	2.2
Tuliptree	6	17	2.3	2.2 ± 0.2	1.8
		Mid-succession (late on dry soil)			
Red oak	19	23	2.6	2.7 ± 0.2	2.1
Shagbark hickory	12	18	1.4	2.7 ± 0.2	2.2
Flowering dogwood	13	5	2.1	1.4 ± 0.1	1.3
		Late succession			
Sugar maple	8	9	1.2	1.9 ± 0.1	1.7
American beech	16	6	1.5	1.5 ± 0.1	1.4
Eastern hemlock	13	8	2.1	1.6 ± 0.1	1.4

than if the leaf is square to the sun. The light received by a horizontal leaf of length AB (Figure 3.14) would be spread over the length BC of an inclined leaf, and the relationship between the two is the familiar Pythagorean one. Two effects of this relationship accrue to the inclined leaf: it absorbs less heat on unit area of surface, and it spreads the light to more photosynthetic tissue at reduced intensity. Inclined leaves are possessed by many plants, most notably grasses, cereals, and the like of prairie landscapes.

Adapting the Horn hypothesis to lance-shaped leaves of grasses suggests that fitness is served by spreading light at reduced intensity to additional leaf area.

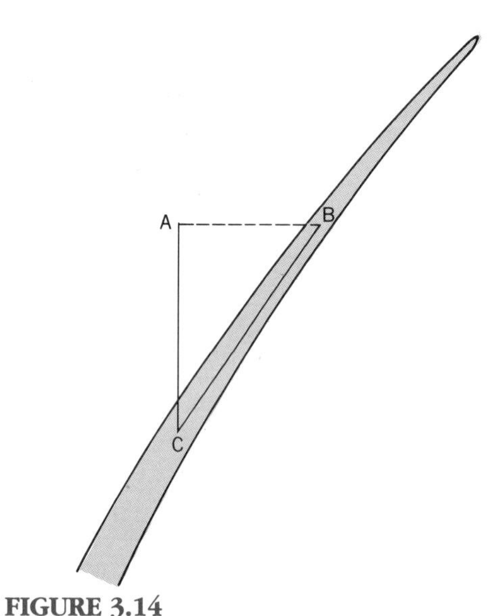

FIGURE 3.14
Light on an inclined leaf.
Light intensity per unit area is reduced by the oblique presentation of the leaf to the sun. This increases the number of chloroplasts and CO_2-collecting stomates that can be brought into production in bright light.

This is a solution to the excess radiant energy problem available for plants that are too short for the geometry of multilayers to be possible.

Since incident heat load is reduced on inclined planes, it is likely that plants with long narrow leaves lose less water by transpiration than would plants with horizontal broad leaves, which should be adaptive in dry places. The apparent adaptive advantage of the inclined leaf, therefore, offers an acceptable general explanation for the existence of grassy prairies in drier regions where trees do not grow. The shapes of the dominant prairie plants maximize production while conserving water.

Many papers in the agricultural literature discuss how the attitudes and shapes of crop leaves affect net primary production (Loomis et al., 1971). Rice plants, for instance, have long, narrow leaves typical of the grass family Gramineae, although some cultivated varieties hold their leaves in different attitudes. In one study it was shown that a variety with drooping leaves produced less dry matter in unit time than another variety with erect leaves. This would be expected if the leaves had to operate in bright sunlight as crop plants do.

Herbs such as cereal do not have the opportunities a tree has for stacking leaves over several meters. They may, however, achieve the same effects by keeping their leaves small or narrow, because it is the width of the leaf that determines how far apart they must be if they are to avoid being in each other's shade. It has been shown (Nichiporovich, 1961) that the occlusion of skylight by leaves is given by the relationship

$$Y = 2 \tan^{-1} (w/2d) \qquad (3.2)$$

where Y is the angle of skylight occlusion,

w is the width of a leaf, and d is the distance from a shaded leaf.

When $w/2d$ is large, shading of the lower leaf is complete. Shading is reduced as w is reduced. Most cereals seem to have leaf structures such that

$$d > 2w(\Upsilon < 28° \qquad (3.3)$$

which appears to yield an optimum spread of light.

Like a tree, therefore, a grass plant is shaped to spread light to as many chloroplasts as can be operated at the maximum rate. The erect posture helps by spreading the noonday sun over as much area of each leaf as possible. At the same time the leaves are narrow to minimize their shadows and let the diffuse light pass down to power the next narrow leaf on line.

AT THE BOTTOM OF THE SEA: PLANTS IN WATER-FILTERED LIGHT

Benthic marine plants (plants that grow anchored to the bottom of the sea) have photosynthetic problems quite different from those of land plants. They receive their carbon in solution, principally from the bicarbonate ion, and this carbon source appears to be ample for any likely rate of photosynthesis. Benthic plants are also much less affected than floating planktonic plants by the low concentration of dissolved nutrients, which reduces productivity in the open sea to the level of hot deserts on land (see Chapter 24). Anchoring itself is the plant's answer to nutrient shortage because it allows water continually to flush past it, providing a natural conveyor belt that brings unlimited fresh supplies of nutrients. Thus anchored plants on the sea floor seldom lack for raw materials or essential nutrients. Their problem is attenuated light.

Light is rapidly absorbed by water, and its energy is degraded to heat. Figure 3.15 shows how rapidly light is absorbed by even clear ocean water, the red end of the spectrum being lost with particular speed. In turbid coastal waters, absorption by suspended sediments and colloids may occlude light ten times more quickly still. Except when very near the surface, therefore, efficient photosynthesis in the sea requires special attention to the absorption of light by the pigment array.

The large anchored plants of the sea are of different colors—green, red, brown, and blue-green. On the face of it, these different colors might be thought of as primary adaptations to the spectral composition of light filtered through seawater. In the shallows, light is little changed from what passes through the air, and we find green algae like the sea lettuce *Ulva* and the thin,

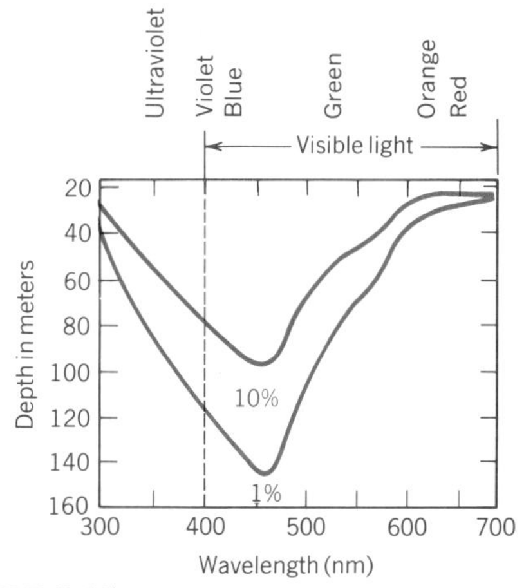

FIGURE 3.15
Light extinction by clear water.
Depths at which 10% and 1% of normal sunlight penetrate into the clearest seawater. Red light is absorbed rapidly, but even the blue end of the visible spectrum is mostly extinguished by 140 m. The deeper a plant lives, the more it is dependent on a narrow band of wavelengths. (Modified from Jerlov, 1951.)

short ribbons of green *Enteromorpha* that make rocks in estuaries slippery. But deeper in the sea the longer wavelengths are selectively absorbed, and we find brown and red algae. So attractive is this line of reasoning that it was long asserted that the colors of seaweeds were correlated with depth, first the greens, then the browns, then the reds. But it is now known that this scheme is in error (Ramus et al., 1976, Ramus, 1980). For instance, the shallowest algae of coral reefs are the red calcareous members of the genus *Lithothamnion*, which present their partially submerged mass to both the equatorial sun and the fury of the waves. And green algae of the genus *Halimeda* are now known to flourish down to depths of at least 140 meters.

One data set long cited to support the postulate that the colors of seaweeds were correlated with depth is the distribution of common and conspicuous plants in and below the intertidal region along northern temperate shores. Along a typical temperate coastline, like that of Massachusetts or southern England, there seems to be a rough zoning of the colors of seaweeds with depth. Near the high-tide line are the green sea lettuces (*Ulva*) and *Enteromorpha*. Occupying the bulk of the intertidal region are brown seaweeds like *Fucus* and *Ascophyllum* and the kelps like *Laminaria*. Below the low-water mark, red seaweeds with delicate fern-like fronds become numerous. This zonation shows green algae exposed to most light, brown algae in the intertidal, and red algae only at depth. But the implied correlation with light intensity is spurious.

Direct evidence that factors other than pigments suited to different light set the depths of both brown and red algae even in the classical coastal zoning of the North Atlantic came from a study comparing distributions between Massachusetts and the Bay of Fundy in eastern Canada and Labrador: the same species of both red and brown algae were found at different depths at the three sites, there being a general shallowing of the range of all three to the north (Hillis, 1966). If a simple hypothesis of limiting factors explains these distributions, temperature works better than light (Figure 3.16).

The truth is that no simple correlation exists between the colors of seaweeds and their depth—browns, reds, greens, and blue-greens can be found throughout the lighted layer of the sea where photosynthesis is possible, the so-called **euphotic zone**. Why, then, is there such an array of colored photosynthetic pigments in sea plants, whereas almost all plants on land are of the same color, green? The answer probably lies in a combination of the dimness of light in the sea and the variety of ancient lineages of marine plants.

Green, brown, red, and blue-green algae are so different in structure that taxonomists place them in separate divisions of the plant kingdom. The zoological equivalent of the botanical division is the phylum. This means that taxonomists consider that green and red algae are as distant from each other as are worms and birds, or starfish and insects. All the green plants of the land, with the exception of the mosses, belong to a single division, Tracheophyta. Each group of algae has inherited a pigment pattern from remote ancestors, and its color cannot be changed. An interesting question for an ecologist then becomes, "How have each of these different colors been adapted to life in the dim light of the sea?"

When light is limiting, the problem for a plant is to absorb as much of it as possible. Ideally, the plant ought to be black, though this requirement can be relaxed since

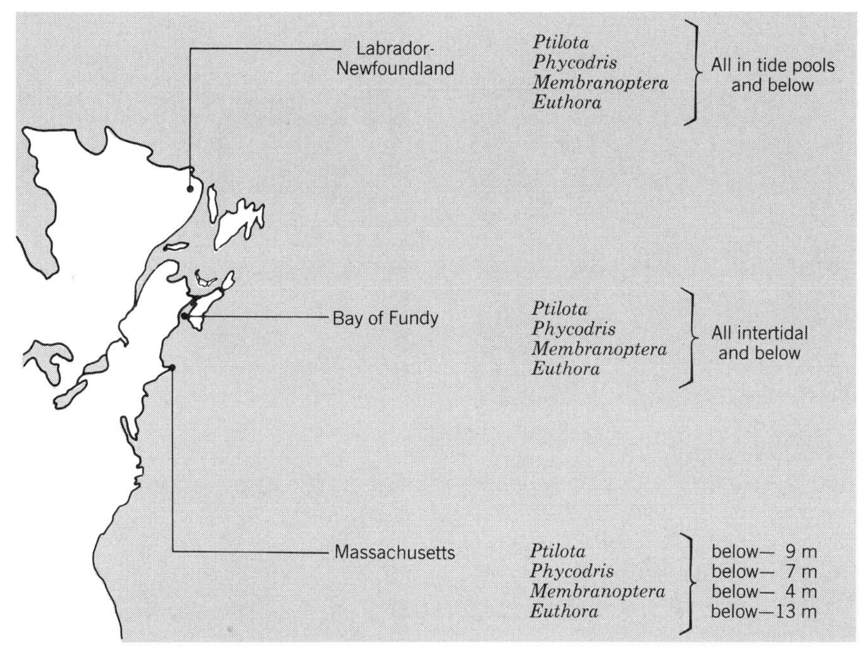

FIGURE 3.16
Depths of red algae in northeastern America.
The four genera of red algae are found at progressively shallower depths as one goes from south to north. It is likely that water temperature (or some correlate of temperature) controls the depth distribution of these red algae more than light does. Massachusetts data are in depths below mean tide level. (From data of Hillis, 1966.)

some colored wavelengths are useless for photosynthesis. But blackness can be approached with almost any highly absorbent pigment if this is deployed in sufficient density. Red and brown algae are dark colored—it is possible to say that they do not have the luxury of being able to reflect in the green as land plants do. Therefore redness and brownness can be thought of as adaptations to dim light at depths as in the classical depth–color hypothesis. But green algae can solve this problem with a close array of chloroplasts, each absorbing slightly in the green.

Species of the reef-building alga *Halimeda* grow from depths exposed at low tide down to at least 140 meters, but their close relative *Penicillus*, the merman's shaving brush, is apparently confined to shallow water (Hillis, 1980; Figures 3.17 and 3.18). The green pigment array, therefore, can be adapted to any depth, and questions about actual distribution become questions of species niche and strategy rather than questions that can be answered in terms of light as a limiting factor. The fact that sea grasses have entered only the shallows of the sea may likewise have little to do with their green pigments (Figure 3.19).

Colored algae do have elegant solutions to the absorption problem. They have **accessory pigments** arranged around their chlorophyll *a* molecules, and these pigments absorb many of the wavelengths that would be reflected by chlorophyll. When a plant has a battery of these pig-

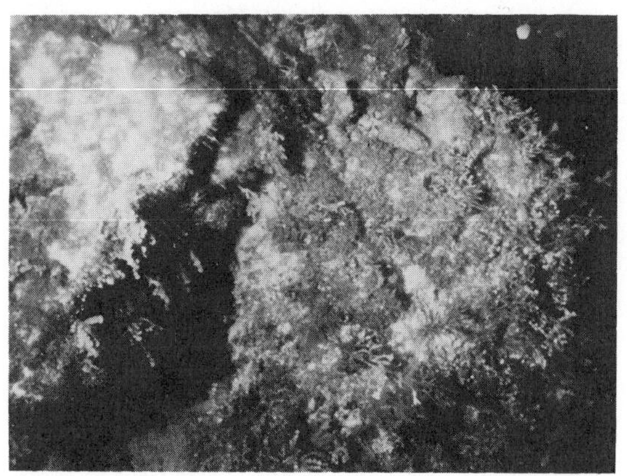

FIGURE 3.17
Green algae in the deep sea.
This community of *Halimeda* was photographed from a submersible at about 140 m depth on the outer face of the coral atoll of Enewetak. Light at this depth is but a trace of 1% of sunlight and yet these green algae achieve extensive cover. *Halimeda* contributes much of the carbonate mass to coral reefs.

ments as well as chlorophyll *a*, it is able to absorb virtually all wavelengths that deliver photons energetic enough for photosynthesis. Figure 3.20 illustrates one such array of pigments (collectively called phycobilisomes). These form a series with overlapping absorption spectra. Energy can be passed from one to another down a chain by a process of resonance, a process analogous to striking the first tuning fork in a series, which in turn causes other tuning forks to resonate in sequence. At each transfer the wavelengths of induced resonance are larger and less energetic, but energy can still be passed down the chain to chlorophyll *a* with an efficiency of more than 80%.

Possession of **phycobilisomes** gives brown and red algae their colors: they reflect red and orange light when viewed in bright light. At the depths where many of them grow they may reflect very little light at all.

FIGURE 3.18
***Penicillus:* The merman's shaving brush.**
A green calcareous alga of shallow water in the tropical Atlantic Ocean. The plants grow spread out on sandy bottoms, reproducing asexually by rhizoids through the sand or by sexual episodes that result in the death of the adult plant (see Chapter 10).

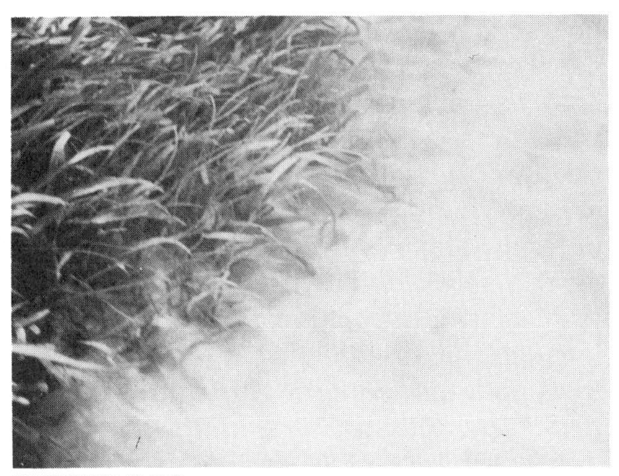

FIGURE 3.19
Tropical sea grasses.
Stand of sea grass (*Thalassia*) in about 1.5 m of water on a sand flat behind the fringing reef, north shore of Jamaica. Alone of the angiosperms, the sea grasses have penetrated the sea but are confined to very shallow water.

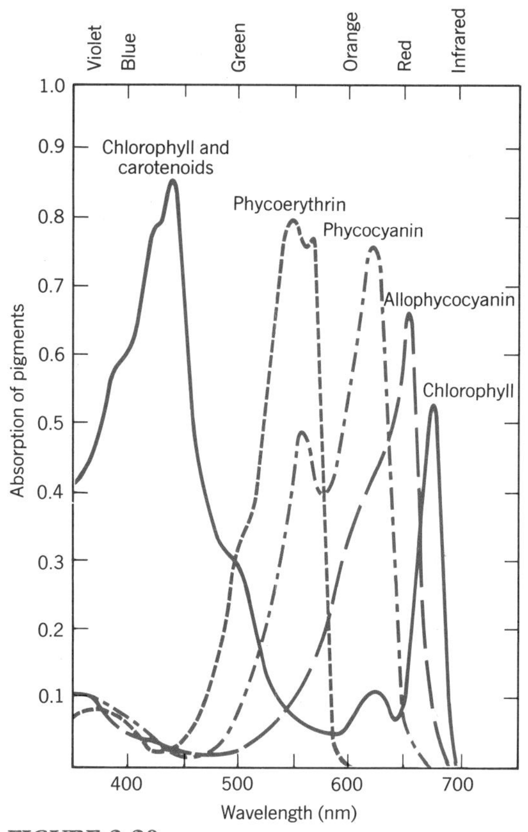

FIGURE 3.20
Light absorption by accessory pigments of seaweeds.
The chlorophyll *a* and carotenoid system absorb the blue and red ends of the visible spectrum as in land plants, but brown and red algae also possess pigments that absorb light in the wavelengths reflected by chlorophyll *a*. Energy is passed to chlorophyll *a* from these pigments by resonance. (Modified from Gantt, 1975.)

SUMMARY: THE ECOLOGICAL LIMITS OF PHOTOSYNTHESIS

In dim light plants are superb operators, both on land and in the sea. Laboratory measurements show that any green plant, from alga to beet, converts dim light to the energy of reduced carbon with an efficiency of about 20%; a fine performance at photon scavenging, granted that the conversion efficiency of photons actually used in synthesis is 36%. But stronger ecological constraints apply in bright light. Even in seemingly ideal conditions for a plant, most green plants (C3 plants and all algae) apparently never do better than 8%, and in the brightest sunlight the phenomenon of light saturation means they do considerably less well. No general explanation for this comparatively low efficiency in bright light is yet accepted by most ecologists.

The problem is particularly perplexing because C4 plants can be shown to continue to take up carbon as a function of light intensity in quite bright light (Figure 3.6), an observation which is frequently used to deny that the carbon supply is limiting to photosynthesis in any terrestrial plants. Against this conclusion, however, is the fact that C4 plants growing in their natural hot, dry habitats are not highly productive plants. Dry grasslands of hot climates and deserts, the natural homes of C4 plants, are far from being highly productive places. Productivity in ecosystems where C4 plants are common probably is limited by water shortage, and these special plants with their Kranz syndrome leaves are thus specially adapted to boost rates of carbon uptake while minimizing water loss.

The highly productive ecosystems of the earth are mostly, if not entirely, fueled by C3 plants. C4 plants do not invade the communities of these productive systems, suggesting strongly that they are competitively inferior to C3 plants, except in the hot dry places in which they live. Every living tree is a C3 plant.

The argument that plants are always engineered to operate best in dim light because that is what their leaves receive most of the time (the Bonner hypothesis, page 37) is powerful; and it gives a plausible explanation of light saturation as resulting when production capacity is overloaded. But the costs of lost energy surely must be high, suggesting that natural selection should have equipped plants of the open sunlight, like pioneer trees, with a different ratio of absorbing chlorophyll molecules to production sites, if this would yield greater net energy returns. The fact that C3 leaf structures do seem to be arranged to work best in dim light itself suggests that converting the much greater flux of energy available in full sunlight cannot be done for some other reason.

Thus the hypothesis that photosynthesis of plants in highly productive places can be limited by the carbon supply cannot be entirely dismissed. The layered and separated leaves of trees are understood best as parts of a light-diffusing system that spreads light at optimum intensity for C3 photosynthesis to as large a total leaf area as possible. But this same separation of leaves also results in increased surface for the uptake of carbon dioxide.

What is clear is that the comparatively low ecological efficiency of plants results from environmental circumstance. With a perfect photon flux in an environment conducive to a plant's well-being, plant chemistry can achieve astonishing energy conversion efficiencies of 36%. But the photon flux is never optimum, ranging from darkness to rates that would cause death by overheating, were plants not equipped with cooling systems designed to dissipate much of the incident energy (see Chapter 5). Plants must transduce energy from this variety of inputs in a way that results in the most reproduction over a lifetime. And they must do so when the essential raw material, carbon dioxide, is available only in low concentrations and when vital supplies of water and nutrients themselves fluctuate, over lifetimes or between generations.

The arrangements different plants use to bring their splendidly efficient production sites on line are thus always trade-offs between designs for different energy intensities, varying water supplies, possible scarcity of nutrients, and fluctuating temperatures, all against a background of chronically low carbon dioxide concentration. In real plant lifetimes, the average ecological efficiency is less than 2%.

Preview to Chapter 4

Canis lupus: **Ecological efficiency 1.4%.**

AN imaginary super-herbivore ought to be able to eat a plant's sugars as soon as the plant makes them, so that none are wasted. A comparable super-carnivore ought to eat its prey as fast as its prey did its own feeding. The ecological (Lindeman) efficiencies of animals are measures of how close they come to these impossible ideals, which are ratios of gross productivities. Lindeman tried to calculate these efficiencies for the different trophic levels in lakes by reconstructing biomasses of different populations from standing crops and turnover times (time taken to replace the biomass of a trophic level with entirely new organic carbon), then adding respiration. This method cannot work, and Lindeman's results were in large error by being too big. His results gave rise to the often cited rule of thumb that Lindeman efficiencies are about 10%. Better estimates can be calculated in theory from measures of yields of corpses and respiration. Measures of the efficiency of water fleas and hydra in culture by this method gave results close to the 10% rule of thumb, but measures for wolves gave results of about 1%. Actual Lindeman efficiencies in nature are probably much less than 10% usually, which fully explains the shape of Eltonian pyramids.

Chapter 4

The Ecological Efficiency of Animals

An individual animal may be thought of as a device programmed by a base-pair sequence of DNA to collect reduced-carbon fuel and to process as much of this fuel as possible into offspring. All animals should be highly efficient at this, since the stakes are fitness or nonfitness, survival or oblivion.

Yet the drive to energetic efficiency meets powerful constraints. Plants and prey animals have adaptations that work to prevent them from being eaten. Other organisms may compete for the food. Always food is taken against resistance. The efficiency of energy transfer, therefore, depends on fundamental relationships between animals. They are common or rare, and arranged in pyramids along food chains, according to the efficiencies with which they turn plants or other animals into food.

One commanding statistic must always be remembered when considering energetic constraints on animal life. It is that 98% of solar energy goes to heat; only 2% is stored in living tissue through photosynthesis. All animals combined have at their disposal only that fraction of 2% which has not already been sent to the sink of heat by the activities of the plants themselves.

Some of the common terms used in community energetics are given in the accompanying box.

THE TRANSFER OF ENERGY BETWEEN TROPHIC LEVELS

The efficiency of all the plant eaters combined (the whole herbivore trophic level) is a measure of how much of the potential energy of plants they convert to their own use.

$$E_{\mathrm{h}} = \frac{\lambda_n \text{ (herbivores)}}{\lambda_{n-1} \text{ (plants)}} \times 100 \quad (4.1)$$

Terms Used in Community Energetics

Gross production is the energy represented by the biomass produced, together with the energy that went into the work of producing it.

Net production is the energy represented by the biomass produced.

Respiration is the difference between gross production and net production.

(Respiration of course refers directly to exhaled carbon dioxide and is a term I suppose ultimately derived from medicine. The ecological adoption of the term is quite reasonable, however, because the difference between gross production and net production does appear as exhaled carbon dioxide, the inevitable consequence of burning carbohydrate fuel.)

Producers are the plants, the base trophic level that is responsible for transforming solar energy into the potential form of fixed carbon compounds.

Consumers are the representatives of all other trophic levels and include herbivores, carnivores, omnivores, and parasites. **Primary consumers** are the herbivores (plant eaters); **secondary consumers** are the *primary carnivores* (flesh eaters); **tertiary consumers** are the *secondary carnivores*, and so on. Animals of the highest link in any food chain are often called **top carnivores**. A **parasite chain** typically runs from a small animal (flea) to an even smaller animal (protozoan parasite of a flea) to an even smaller animal still (bacterial disease of the protozoan). The animals of these parasite chains may be directly assigned to appropriate trophic levels along with the other consumers.

Decomposers are organisms that feed on corpses or dead organic matter. **Saprobe** and **saprophage** (corpse-eating) are terms that are sometimes used instead of "decomposer." A **decomposer chain** or a **saprobe chain** might run from a large organism (fungus) to smaller organisms (protozoa and bacteria).

Productivity may be qualified as primary (energy fixed by green plants) or secondary (energy flowing into any level other than that of primary green plants).

where E_h is the ecological efficiency of herbivores, λ_n is energy flowing into a trophic level in unit time, and λ_{n-1} is energy flowing into the next lower trophic level, using symbols introduced by Lindeman (1942).

The efficiency of the primary carnivore trophic level is

$$E_c1 = \frac{\lambda_{n+1}(\text{primary carnivores})}{\lambda_n(\text{herbivores})} \times 100 \qquad (4.2)$$

where E_{c1} is the ecological efficiency of primary carnivores.

Equations 4.1 and 4.2 both use ratios of **gross productivity**. The ecological efficiency of the herbivore trophic level, for instance, is biomass and respiration of herbivores divided by biomass and respiration of plants. The ratios are strictly comparable to that used in calculating the ecological efficiency of plants, where the ratio was biomass and respiration of plants divided by total solar energy received (see Chapter 3).

These equations describe the fact that the theoretical upper limit of energy available to a trophic level is all the energy flowing into the trophic level below. This includes even the energy that is degraded (used) in the lower trophic level and thus is not available to be passed on. The formulation assumes that a super-herbivore would be able to get all the plant production, even before the plants respire, and a super-carnivore would be able to eat all the starch reserve of the prey without the prey's using any of those reserves for running about. Only animals able to achieve these impossible ideals as food gatherers could be 100% efficient. It will be recalled that the low ecological efficiency of 2% for plants was largely accounted for by the fact that most of the solar energy theoretically available never entered active photosynthetic tissue at all. The efficiency of animal trophic levels is usually low for similar reasons—most of the energy is degraded in the trophic level below as respiration or goes to decomposers, and thus can never be transferred as food.

Transferring energy from one trophic level to another involves many processes, the efficiency of which varies (Figure 4.1). The efficiency with which plant or prey organisms maintain biomass per unit of respiration determines how much net production is available as food. The efficiency of cropping or hunting determines what portion of this net production is taken. The efficiency of digestion determines how much of the energy eaten goes into net production, and hence how much becomes potential food for the next trophic level. Ecologists define, and seek to measure, these many efficiencies. The ecological efficiency of energy transfer between whole trophic levels depends on the cumulative effects of these other efficiencies (Table 4.1).

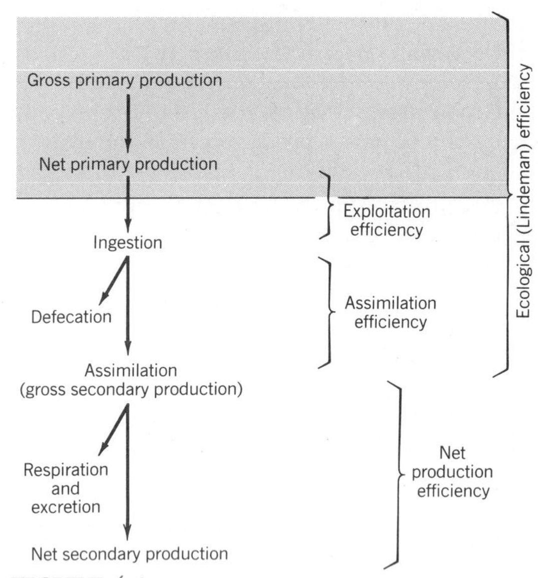

FIGURE 4.1
Relationships of efficiencies of energy transfer. The various efficiencies are defined in Table 4.1. All these efficiencies refer to the rates at which animals collect fuel to use to win fitness. All may be expected to be maximized by natural selection. Differences in calculated efficiencies must therefore reveal the local circumstances in which the animals live.

Lindeman (1942) made the first attempt to measure transfer efficiencies between whole animal trophic levels as part of the seminal study that introduced the energy flow paradigm into ecology (see Chapter 2). He not only put forward the hypothesis that Elton's pyramids of numbers reflected energy reductions between trophic levels, but he set out to measure the energy reductions themselves. The necessary measures were to be the ratios of gross productivity for whole animal trophic levels in nature, for which reason "ecological efficiency" is often called "**Lindeman efficiency.**" But the terms ecological efficiency, transfer efficiency, Lindeman efficiency, and even the "**assimilation efficiency**" of plant physiologists, are synonymous.

Lindeman's results from this first attempt at measurement were in error by some large but undetermined amount, yet they still influence ecological judgment. The difficulties he encountered were so forbidding as to discourage many imitators. It is necessary to know that Lindeman tried to measure the efficiency named for him, and to understand why his method would not work, to understand some of the limits of our present knowledge.

TABLE 4.1
Efficiencies of Energy Transfer
The ecological (Lindeman) efficiencies of energy transfer between trophic levels are the fundamental efficiencies that set the energy flux to the next trophic levels. All other efficiencies result from processes that also affect Lindeman efficiency. All efficiencies, therefore, are related to the fundamental Lindeman efficiency.

$$\text{Ecological efficiency of plants (Lindeman and Transeau efficiencies)} = \frac{\text{Rate of assimilation (photosynthesis) by plants}}{\text{Solar flux}} \times 100$$

$$\text{Lindeman efficiency of herbivores} = \frac{\text{Rate of assimilation by herbivores}}{\text{Gross productivity of plants}} \times 100$$

$$\text{Lindeman efficiency of primary carnivores} = \frac{\text{Rate of assimilation of primary carnivores}}{\text{Rate of assimilation of herbivores}} \times 100$$

$$\text{Assimilation efficiency} = \frac{\text{Food assimilated (digestible energy)}}{\text{Food ingested}} \times 100$$

$$\text{Exploitation efficiency} = \frac{\text{Food ingested}}{\text{Available net production}} \times 100$$

$$\text{Net production efficiency} = \frac{\text{Net production (growth and reproduction)}}{\text{Food assimilation}} \times 100$$

LINDEMAN'S FLAWED CALCULATIONS

Lindeman tried to calculate his efficiencies from data on standing crops in a manner analogous to that of Transeau when he calculated the efficiency of a field of corn from data on standing crop and respiration (see Chapter 3). Lindeman studied Cedar Bog Lake in Minnesota for five years, measuring standing crops of all the more important animals and plants of each trophic level and year by year (Lindeman, 1941). The problem then was how to calculate the energy input for each trophic level over a typical year, and to do this from the standing crop measurements alone. All the animals and plants of the system were respiring, and this must be measured as for Transeau's corn. But all were reproducing and dying as well, some "turning over" rapidly and some slowly. Lindeman had to allow for the missing individuals as well as for the living in a system that was in constant flux. Lindeman tried to account for all these replacements of individuals, and for the energy that was dissipated in all this flux, by multiplying each standing crop by an appropriate "**turnover time**": one week for algae, two weeks for zooplankton, a whole year for rooted aquatic plants. These were the times it took to replace the biomass of those particular organisms of a trophic level with entirely new organic matter imported from the trophic level below; essentially one lifetime. To this turnover time Lindeman had to add a figure for respiration to the reconstituted standing crops.

The respiration data that Lindeman used were of course collected to allow for the contributions of all the animals and plants that had occupied a given trophic level during the year. But this measure inevitably included the energy dissipated in the activity of turning over as well as that used for finding food, and turnover had been allowed for already. Lindeman's energy input was, therefore, too high by some large but unknown factor. He then compounded the error by adding estimates for animals taken by predators and animals that died other deaths and went to the decomposers. The energy represented by these bodies had already been allowed for in the turnover calculation and was now appearing in the equation twice.

Lindeman had wanted to solve the following equation:

$$\text{Productivity} = (\text{yield of dry matter in calories}) + (\text{calories respired})$$

but what he had actually done was to say:

$$\text{Productivity} = \left(\begin{array}{c}\text{standing crop in calories} \times \\ \text{turnover time}\end{array}\right) + \left(\begin{array}{c}\text{calories} \\ \text{respired}\end{array}\right) + \left(\begin{array}{c}\text{calories lost} \\ \text{to mortality}\end{array}\right)$$

This second equation has both mortality and respiration appearing twice, once disguised as turnover and once in their proper places (Slobodkin, 1962).

Lindeman's results, together with a more recent aquatic study using comparable logic, are given in Table 4.2. They look plausible. The highest efficiency of any animal trophic level is given as about 20%, and a conservative figure of about 10% looks to be about the norm. For this reason ecologists advising such organizations as the United Nations Educational, Scientific and Cultural Organization, the Food and Agriculture Organization of the United Nations, and other organizations concerned with the energy resources of the earth usable as food for people have suggested that 10% conversion should be accepted as a rule of thumb. Seldom in the history of nations can so many decisions have been made on the basis of so obviously suspect data.

But Lindeman's estimates were quite good enough for his primary purpose, which was to explain Eltonian pyramids. The plant trophic level was but 2% efficient, as was already known in 1942 from the work of Transeau and those who had come after him. Lindeman's results suggested that a top carnivore in even a short, three-link food chain got 10% of 10% of the 2% of the energy that plants got from the sun. With longer food chains the 10% of 10% hymn sung on. Thus the rarity of the large and fierce was satisfactorily explained.

But Lindeman's results were actually in error, as are all other estimates based on

TABLE 4.2
Lindeman Efficiencies Calculated from Standing Crops and Turnover
These estimates are in error, being too high. They are often quoted, however, and they lead to the ecological rule of thumb that energy is usually transferred between trophic levels with an efficiency of about 10%. Real ecological (Lindeman) efficiencies may usually be significantly lower than 10%. (From Lindeman, 1942, and H. T. Odum, 1957.)

	Cedar Bog Lake (%)	Lake Mendota (%)	Silver Springs, Florida (%)
Primary consumers (herbivores)	13.3	8.7	16
Secondary consumers (1° carnivores)	22.3	5.5	11
Tertiary consumers (2° carnivores)	No data	13.0	6

standing crops and turnover times. They err in the direction of being too high, perhaps by a large amount, because the measures of turnover and respiration overlap. At the very least, the rule of thumb of 10% efficiency had best be taken as an upper limit without strong reasons for thinking otherwise. For top carnivores on land, for instance, the 10% figure may in some circumstances be nearly ten times too large (see below).

Ecological Efficiency in Laboratory Cultures

Lawrence Slobodkin demonstrated the uncertainty that lay in Lindeman's calculations in 1962, just 20 years after the calculations were published. By then the 10% rule of thumb for transfer efficiencies had penetrated all the way to agricultural economics and world affairs. Lindeman's work itself, of course, had come 17 years after the publication of the Eltonian pyramid he had so trenchantly explained. I am tempted to comment on the inadequate resources that society puts into ecology. Be that as it may, Slobodkin (1962) offered a new set of measurements in pursuit of Lindeman's goal.

Slobodkin argued that measurement of standing crops in real ecosystems could never be used as a basis for ecological efficiency measures because the mix-up between respiration and turnover could never be resolved. The only practical hope was work with systems where respiration and the yield of corpses (mortality) could be measured separately and independently. The energy that flows into a trophic level has one of only two possible fates: either it is degraded within the trophic level, in which event it appears as respiration, or it is passed out of the trophic level in a dead animal body. No other fates for the energy were possible. It follows that, in a trophic level,

$$\text{Respiration} + \text{mortality} = \text{gross production}$$

This was the equation that Transeau had solved for corn plants when he measured their mass at harvest and added a figure for lifetime respiration (see Chapter 3). To apply similar logic to animal communities meant finding a simple ecosystem that provided the animal equivalent of a field of corn for both the predator and prey trophic levels. Slobodkin's answer was to contrive the requisite communities in the laboratory.

In the laboratory, stable ecosystems with few species can be designed in which populations are both constant and food-limited. Slobodkin (1962) and his students used water fleas of the genus *Daphnia* and the freshwater coelenterate *Hydra*, cultures of which could be kept in small laboratory jars provided with manipulated supplies of food (Richman, 1958; Armstrong, 1960; Slobodkin, 1962). *Daphnia* and *Hydra* populations provided two alternative predator trophic levels. The food in both systems was green flagellates, and the supply was manipulated until a roughly constant yield of growth and mortality for the predators was obtained. Micro-bomb calorimeters were designed to give accurate estimates of the calorific value of *Daphnia* and *Hydra* corpses.

Results from Slobodkin's laboratory were remarkably close to Lindeman's results, flawed though the latter undoubtedly were. The *Hydra* population trophic level had an ecological efficiency of 7% and the *Daphnia* population trophic level 13%. The use of the 10% rule of thumb for transfer efficiencies was quite consistent with these results.

Ecological Efficiency of the Wolf-Moose System

Colinvaux and Barnett (1979) applied Slobodkin's method to a population of wolves feeding on moose (Figures 4.2 and 4.3). The wolves and moose lived on an island in Lake Superior. The Isle Royale is about 100 km long, but for animals as big as moose and wolves this is a small place. In theory, both wolves and moose can travel off the island by swimming or by walking across the winter ice, but over the study period there was little exchange with the mainland. The experimental container called Isle Royale, therefore, was effectively sealed. Studies over many years showed how many wolves and moose lived on the island, how long they lived, and that both populations had been stable for a number of years (Mech, 1966; see Chapter 14). This means that the prime necessities for reasonable calculations—simplicity and stable populations—were met.

FIGURE 4.2
***Canis lupus*, 1.3% efficient.**
The ecological (Lindeman) efficiency of a pack of wolves feeding on moose is about 1.3%, or about one-tenth of that of water fleas in captivity. This low efficiency is probably typical of large carnivores of high metabolic rate that hunt prey equipped with defenses.

Another rare circumstance favored the investigation in that it was known both what the wolves ate and the cause of all moose mortality. The wolf diet on Isle Royale over the study period was almost entirely moose, certainly 70 to 80% of the diet, anyway. And the only fate of a moose was to be eaten by wolves. The wolves killed some young moose when they could catch them, and they killed all sick and old moose. Thus there was on Isle Royale a trophic level of primary carnivores (wolves) almost entirely subsisting on a single-species trophic level of herbivores (moose). Other carnivores of Isle Royale, like birds and small mammals feeding on rodents, could be expected to have little impact on the wolf–moose system.

Two quantities now had to be calculated for both the moose and wolf populations: yield of corpses (in calories) and respiration. Mortality data were available from published life tables of both moose and wolves; in other words the yields of moose and wolf corpses had already been published. Converting these data to calories required consulting the agricultural literature and the assumption that cows and dogs of appropriate size could be substituted for moose and wolves.

Respiration of wolves and moose was calculated from the established function relating body weight to respiration of homeotherms (Kleiber, 1961). About 20 wolves on Isle Royale subsisted on 600 moose, which looks like a reasonable relationship for two layers of an Eltonian

FIGURE 4.3
***Alces americana:* Large prey makes inefficient predators.**
Because large prey are not easily killed by predators, they can be taken with only low ecological efficiency.

pyramid. The ecological (Lindeman) efficiency of wolves turned out to be only 1.3% (Table 4.3).

Wolves on Isle Royale are evidently much less efficient than *Daphnia* (13%) or *Hydra* (7%) in laboratory tanks. This should not be a cause for surprise. For one thing the laboratory animals were fed simple food in ideal circumstances: seasons did not change in the tanks, and the measurements were made over a time span when conditions for life were optimum. But the wolves had to hunt through the changing seasons of a whole year. More-

TABLE 4.3
Trophic Level Data for Moose and Wolves
Published data for moose and wolves on Isle Royale are used to calculate the ecological (Lindeman) efficiency of wolves. (Reworked from Colinvaux and Barnett, 1979.)

	Moose	Wolves
Population	600	20
Yield of young corpses (calves: pups) per year	52	5
Yield of adults per year	111	1.57
Total corpse yield per year	123,321,825 kcal	600,439 kcal
Annual respiration	910,349,091 kcal	12,786,382 kcal
Energy flux into trophic level	1,033,670,916 kcal	13,386,821 kcal

$$\text{Ecological (Lindeman) efficiency of wolves} = \frac{13{,}386{,}821}{1{,}033{,}670{,}916} \times 100 = 1.3\%$$

over, the wolves tackled prey larger than themselves, prey that, apart from calves, they could pull down only when the moose was old or sick.

It is worth noting from Table 4.3 how critical the measurement of respiration is to the calculations. Most of the energy ingested into each trophic level reappears as respiration, not as mortality. Measures of respiration are absolutely critical in calculating transfer efficiencies. Large changes in the respiration estimates of wolves and moose, however, would be needed to bring the estimate of ecological efficiency to the 10% rule of thumb.

FIGURE 4.4
***Dipodomys merriami:* Efficient harvester of desert seeds.**
A population of kangaroo rats harvested 90% of the total net production of their principal food plant and nearly 7% of net primary productivity of the ecosystem in which they lived.

Ecological Efficiency Calculated from an Energy Budget

The energy flowing into an animal population can in theory be measured from a detailed energy budget of the animals. It is necessary to know the respiration rate of individuals of all ages of animals at all times in their lives, young and old, resting and active. It is also necessary to know growth and reproduction rates. But if all these measurements can be made, the total flux of energy into the population can be arrived at without any need to monitor what the animals eat. If the animals are herbivores, the energy flux in the plant trophic level can be approximated by harvesting with plant respiration added as Transeau did for corn, or by more complex sampling (see Chapter 24). This provides all that is necessary to calculate the ecological (Lindeman) efficiency of the herbivores.

Exhaustive studies undertaken to produce an energy budget for kangaroo rats lets this approach be used (Soholt 1973; Chew and Chew 1970). The kangaroo rat *Dipodomys merriami* is a desert animal (Figure 4.4). It rests by day underground in a nest and comes to the surface at night for a few hours to feed. The energy budget, therefore, includes the energy needed for maintenance when the animal is resting in its burrow, the energy of maintenance when the animal is in the open at night, and the energy used to support its nighttime activity. These data have to be known for animals of various ages and for the ambient temperatures prevailing at different times of the year. Then

$$E = E_m + E_g \tag{4.3}$$

where

E = energy flowing into the population
E_m = maintenance energy
E_g = growth energy

Chew and Chew and Soholt measured carbon dioxide flux from captive animals in various conditions and measured growth rates by repeatedly weighing both captive and wild animals. They then proceeded to calculate an energy budget for kangaroo rats as follows:

$$E_m = E_r + E_a \tag{4.4}$$

where

E_r = maintenance energy of resting animal

E_a = maintenance energy of active animal

$$E_r = (T)(S_c)(E_{rn} \cdot t_n + E_{rs}) \cdot t_s \tag{4.5}$$

where

T = time interval of measurement

S_c = standing crop

E_{rn} = maintenance energy of animal resting in nest

E_{rs} = maintenance energy of animal resting at surface

t_n, t_s = time animal spends in rest and on surface

$$E_a = (T)(I_a)(t_s)(S_c) \tag{4.6}$$

where

I_a = estimated rate of energy expenditure for activity

$$E_g = \frac{NP_1 + NP_2 + NP_3}{GE_n} \tag{4.7}$$

where

NP_1 = production of biomass in prenatal growth

NP_2 = production of biomass while nursing

NP_3 = production of biomass after weaning

GE_n = net efficiency of growth

$$GE_n = \frac{NP}{E_g} \tag{4.8}$$

where

NP = net secondary production

To solve for the growth efficiency GE_n[3] we must know the growth energy E_g[3], but this cannot be found without knowing GE_n itself (equation 4.9); hence an apparent impasse. This difficulty was avoided in a later study by Soholt (1973) that used data for growth efficiency GE_n reported in the literature for white rats, which suggested that the correct figure for kangaroo rats would be about 50%. By measuring all the other parameters needed for solving equations 4.5 through 4.9 and applying the resulting energy flux (E) to his standing crop data, Soholt calculated that 85.5 megacalories flowed through the kangaroo rat population per hectare per year.

Thus we know the numerator of the efficiency equation for a population of kangaroo rats in the California desert. Energy budgets of this detail are not available for many wild populations for obvious reasons. To proceed further and extract a measure of ecological efficiency out of the data, it is necessary to assume (though to know would be better) that the kangaroo rat is the only serious herbivore feeding on the wild vegetation of the region. Because of the simplicity of the local desert ecosystem, this may not be too risky an assumption.

The habitat of the kangaroo rats studied by Soholt (1973) was mostly vegetated with annual plants, although there were some perennial bushes. This meant that rough estimates of net primary productivity could be made by measuring the standing crop of plants at the end of the season. Using Soholt's data it is possible to arrive at an approximation of Lindeman efficiency as follows:

(a) net primary productivity = 1400 megacalories per hectare per year
(b) productivity of plant species found in kangaroo rat stomachs (available productivity) = 900 megacalories per hectare per year
(c) assume respiration is one third of net productivity (approximation of Transeau's finding)
(d) gross available productivity = 1200 megacalories per hectare per year (900 + ⅓)
(e) approximate ecological (Lindeman) efficiency of population of kangaroo rats is

$$\frac{(85.5 \times 100)}{1200} = 7.13\%.$$

The wild kangaroo rat results thus puts them in the range of Slobodkin's laboratory animals, or the 10% rule of thumb, rather than in the calculated range for wolves.

SUMMARY: ECOLOGICAL EFFICIENCIES OF ANIMAL TROPHIC LEVELS

The first thing that must be said is that we do not have a satisfactory calculation of ecological efficiency for a whole animal trophic level of even a moderately complex natural ecosystem. Quite a number of standing crop studies of marine systems are available from measurements made on shipboard, but these necessarily encounter Lindeman's old difficulty of separating turnover from respiration.

Adding to the measurements described previously, a single estimate for domestic sheep grazing a pasture yields the meager crop of results listed in Table 4.4.

Of the three field examples in Table 4.4, the kangaroo rats suggest relatively high efficiency (7%), whereas wolves and sheep are low (1%). Both of the herbivore measures include significant uncertainties, since for neither is it certain that the population of herbivores or their food plants are at a steady state. Indeed, the very high exploitation of the seeds on one particular plant by the rats is reason for caution. The rats ate 95% of the seeds of their principal food plant *Erodium* and 90% of the plant's

TABLE 4.4
Ecological (Lindeman) Efficiencies
*These results are free from the dangers inherent in using standing crops and turnover times and are offered as generally conservative figures. (*Daphnia *and* Hydra *from Slobodkin, 1962; kangaroo rat calculated from data of Chew, 1974, and Soholt, 1973; sheep from Perkins, 1978; wolf from Colinvaux and Barnett, 1979.)*

	Ecological (Lindeman) Efficiency (%)
Plankton feeders in culture	
Daphnia	13
Hydra	7
Wild herbivores	
Kangaroo rat	<7.3
Domestic herbivores	
Sheep	0.96
Wild carnivores	
Wolf	1.3

total net production. This suggests that the system was sampled at a time when rats may have been overly numerous. If the rats were overexploiting their food, then the calculated efficiency is too high. The estimate for sheep is likely to have been influenced by stocking decisions in an agricultural system, and should perhaps be viewed with even more caution.

All that can really be said about transfer efficiencies in animal trophic levels is that they are certainly low enough to validate the Lindeman–Hutchinson energy flow paradigm. Pyramids of numbers are fully explained. And it might be prudent to expect ecological (Lindeman) efficiencies to be significantly lower than the 10% rule of thumb for many natural ecosystems.

A More Tractable Efficiency for Comparing Animal Performance

Two parameters of wild populations that can be measured with reasonable ease are net production and food eaten. Net production can be measured for almost anything with a suitable census program repeated at intervals (see Chapter 24). Food eaten can nearly always be measured with sufficient ingenuity using marked food, controlling its supply or monitoring its reduction. It is then possible to calculate the efficiency with which an animal harvests the visible production. This efficiency is called the **exploitation efficiency**.

$$\text{Exploitation efficiency} = \frac{\text{Energy ingested}}{\text{Net production of food species}}$$

For an individual herbivore species population this becomes

$$\text{Exploitation efficiency} = \frac{\text{Food ingested by herbivore population}}{\text{Available net production of food plant}}$$

and for a whole herbivore trophic level this becomes

$$\text{Exploitation efficiency} = \frac{\text{Food ingested by herbivores}}{\text{Net primary productivity}}$$

Table 4.5 lists exploitation efficiencies for different animals. These data reveal at a glance the relative importance of animals in different lifestyles in different ecosystems. As a cause of biospheric process, zooplankters in the sea appear to be more interesting than mammals in the Arizona desert.

But a special word of caution about exploitation efficiency is warranted concerning the state of the populations at the time of measurement. Blatant overgrazing, for instance, will result in a very high exploitation efficiency, but this is likely to be a short-lived phenomenon.

CONCLUDING NOTE

All animals covet large supplies of food energy. This must be so, because food is the ultimate resource that leads to fitness. It is not so self-evident that the fuel of life should be used efficiently, at least not without reflecting on what the word "efficient" means. A racing car is efficient at going fast, though poor on fuel conservation. An economy car is efficient in going far on a tankful of fuel. Both the hare and the tortoise strategies certainly are found in the use of fuel by animals.

If the animals of a given trophic level are prodigal in their use of fuel, this might be because prodigality is the route to fitness, just as prodigality in use of fuel lets a formula-one car win its race. But a prodigal trophic level must have a small standing

TABLE 4.5
Exploitation Efficiencies of Herbivores in Different Systems
Exploitation efficiencies suggest the importance of the animals to their respective ecosystems. The data need to be treated with care, however. The lemming exploitation efficiency must refer to the local population concentration of a lemming high (see Chapter 13), and lemmings may be rare at some times and in some places when their exploitation efficiency would be correspondingly low. In general, exploitation efficiencies refer to the performance of animal populations at particular times and places. (From Chew, 1974, and Sobolt, 1973.)

Consumer Category	System	Exploitation Efficiency
Mammals	Desert, Arizona	2% total NPP (above ground) 5.5% available NPP 86% seed production
Kangaroo rat	Desert, California	10.7% available NPP 95% available seed product
Mammals	Oak-pine forest, Poland	0.7% available NPP
Mammals	Talga forest, Alaska	13.5% available NPP
Leaf feeders	Deciduous forest	3-8% of foliage
Soil-litter fauna	Deciduous forest	4-8% annual litter 1% litter
Insects, except ants	Old field, herbland	1% total NPP
Ungulates	Grassland, Africa	28-60% total NPP
Domestic animals	Rangeland	30-45% for maximum sustained yield
Lemming	Tundra monocots	93% NPP
Zooplankton	English Channel	98.5% phytoplankton
Seed predators	General	10-90% predispersal seed crop

NPP = net primary productivity.

crop, requiring that the next higher trophic level will have only a meager flux of food energy available to it. I suggest that moose are prodigal with the use of energy (the animals keep warm in a Canadian winter), and the evidence is that most of the calories flowing through the moose population appear as respiration. The wolf trophic level thus has very little chance of turning in a high transfer efficiency.

All animals also must be made to hang on to the energy they have, in that they are programmed by natural selection to avoid being eaten. A good defense makes it hard for a predator to be efficient at transferring the prey's energy to its own use.

Thus measures of animal energetics and energy efficiencies are likely to tell significant things about the animals themselves but need not have generality. The one grand generality that is possible is that most energy entering any trophic level is spent (degraded to heat) within that trophic level. Ecological efficiencies, therefore, can never be high. And large predatory animals must always be rare.

Preview to Chapter 5

Cereus giganteus: **Built to control temperature.**

ALL terrestrial organisms must be adapted to manipulate gains and losses of heat so that body temperature is maintained within the limits required by body chemistry. The massive importance of manipulating heat depends on the fact that >98% of incoming radiation goes to heat, not into tissue via photosynthesis. Where the gradient between core and ambient temperatures is large, as in warm-blooded endothermic animals, control of surface temperature with insulation or countercurrent heat exchangers is critical. Control of heating is also imperative for plants whose way of life requires that they be staked out under the sun. Leaves in moist forests are water-cooled, a mechanism that can regulate operating temperature within a few degrees of an optimum biochemical temperature. In deserts water cannot be spared for cooling, requiring that plants be designed to minimize heat loading. An organ pipe cactus exposes the minimum surface to the noonday sun and does away with leaves all together. The needle leaves of conifers let the flow of air past them be nearly laminar, as a result of which leaf temperature tracks air temperature closely. Needles thus can be air-cooled in a semidesert environment where many pines live. More importantly, coniferous trees can be air-warmed on cold, cloudless nights, which is probably the reason that evergreen conifers grow in northern forests. In cold arctic or alpine regions, trees are at a disadvantage because they reach up into cold, moving air, whereas ground-hugging plants of arctic and alpine tundras live in a thin layer of ground air warmed by radiations from the ground.

Chapter 5

Heat Budgets and Life Forms

A general conclusion from studies of energetic efficiency was that more than 98% of incident solar energy is dissipated in physical systems of the biosphere. Plants and animals have the remaining 2% to support life and gain fitness. Individual living things, therefore, cannot control their environment but must adapt to physical realities of habitats made wet or dry, hot or cold, windy or calm, by the power of sunlight wielded through physical cycles.

ORGANISMS HAVE HEAT BUDGET PROBLEMS

All organisms must balance their heat budgets. They produce heat as they release energy to do the work of living, and they gain heat from the sun. It follows, therefore, that they must lose heat at an equal rate or they would soon heat up to the point where their proteins were precipitated, causing death. This is true even of a plant, perhaps especially for a plant because it is designed as a solar trap. Plants, therefore, must "thermoregulate" even as all animals do. That warm-blooded animals thermoregulate is obvious; ecologists call them "**endothermic**." But cold-blooded animals must regulate temperature also; they do so by such methods as moving in and out of the sun, and ecologists call them "**ectothermic**" (see Chapter 6). Plants thermoregulate to within remarkable tolerances, though I know of no special name given to the processes they use. Forest trees and many other plants typically are water-cooled. Where surplus water is not available for cooling, plants must be shaped to reduce heat input or to lose heat by radiation or convection.

An animal standing under the sun (Figure 5.1) receives energy from many sources: directly from the sun, considered as a point source; by the radiation of skylight, considered as a hemispherical bowl above the animal; by reflection from its surroundings; and by infrared radiations from the ground, surrounding objects, and

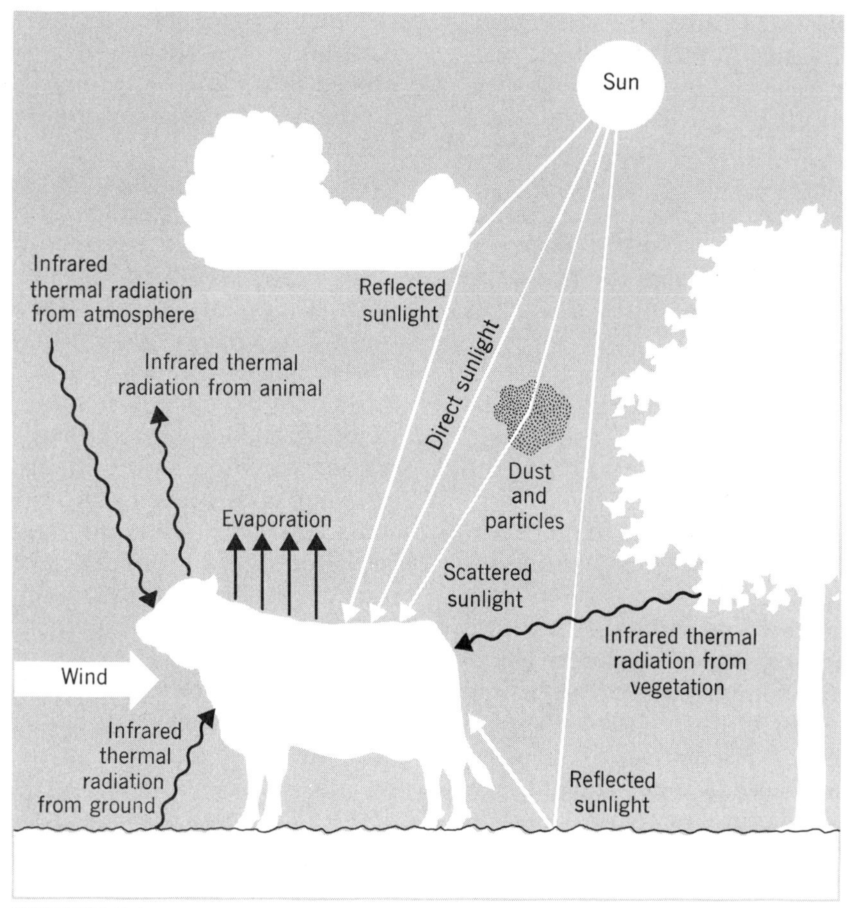

FIGURE 5.1
Energy exchange between an animal and the environment. (After Gates 1968*a*.)

the clouds. If the air is hotter than the animal, it may also receive heat by convection, though convection is usually a source of heat loss in real environments. And an animal receives much energy from its food, called **metabolic energy**. On the other side of the heat budget, the animal loses energy by blackbody radiation, by convection, perhaps a little by conduction, and by evaporation of water at the respiratory or skin surfaces. A heat budget for an animal may be drawn as follows:

$$\begin{pmatrix}\text{Metabolic}\\\text{energy}\end{pmatrix} + \begin{pmatrix}\text{energy absorbed}\\\text{from direct}\\\text{sunlight}\end{pmatrix} + \begin{pmatrix}\text{energy absorbed}\\\text{from}\\\text{skylight}\end{pmatrix} + \begin{pmatrix}\text{energy absorbed}\\\text{from radiations}\\\text{coming from ground}\end{pmatrix} + \begin{pmatrix}\text{energy absorbed}\\\text{from radiations}\\\text{coming from clouds}\end{pmatrix}$$
$$= \begin{pmatrix}\text{energy lost}\\\text{by blackbody}\\\text{radiation}\end{pmatrix} + \begin{pmatrix}\text{energy lost}\\\text{by}\\\text{convection}\end{pmatrix} + \begin{pmatrix}\text{energy lost}\\\text{by}\\\text{evaporation}\end{pmatrix} + \begin{pmatrix}\text{energy lost}\\\text{by}\\\text{conduction}\end{pmatrix}$$

Animals can act to alter their heat budgets by manipulating both inputs and outputs. Inputs are strongly influenced by the environments in which the animals live, yet animals can regulate them by behavior, by moving to sun or shade, for instance, or by spending time underground. Heat inputs from metabolism, however, are less under the control of the organism since they are a function of the resources that the animal uses.

Energy outputs can be manipulated in a number of ways. Heat loss by blackbody radiation is proportional to the fourth power of the absolute temperature of the body surface, so the animal can reduce this loss by insulating itself with fat or fur, a stratagem that keeps the body surface at a lower temperature than the body core. Alternatively, both the radiation loss and the convection loss can be enhanced by arranging to have a large surface area heated to nearly the temperature of the body core. This is done by animals native to deserts and other hot places by means of intensely vascular tissues that can be used as radiators. The large ears of desert lagomorphs (rabbits and hares) and of elephants can function in this way, and it is likely that the spiny fin of the Mesozoic *Dimetrodon* also worked as a heat regulator by radiating heat away to outer space when the animal faced the sun or absorbing heat when the animal turned at right angles to the sun (Figure 5.2).

This analysis explains why many animals of hot places have large extremities; they are needed to radiate heat. That extremities in warm latitudes are larger than extremities in the cold north in some groups of animals has long been known. Nineteenth-century biogeography recognized this phenomenon as **Allen's rule**. But the superficial explanation offered for Allen's so-called rule was that large extremities were harmful in cold climates. On the principle of frostbitten fingers, perhaps they are. But where extremities are indeed larger in the tropics, the cause is probably the use of extremities as potential radiators.

All heat budgets may be powerfully affected by the evaporation of water. Since the latent heat of the evaporation of water is high (about 590 kcal/g) water loss represents a potent source of heat loss. Humans function in deserts by evaporating water (sweat) from all parts of the body surface, a practice that makes us splendidly adapted to sunbelt living as long as we have unlimited water to drink, but less adapted otherwise. Dogs and other animals cope with temporary heat overloads by panting. Yet evaporating water to balance heat budgets in hot places is not something that is equally open to all animals, the utility of the stratagem being critically dependent on body size. Figure 5.3 shows how the water demanded to balance heat budgets of animals exposed to hot desert conditions would be much greater for small species than for large if evaporation were the principal method of keeping cool.

Plants, which are essentially solar traps unable to hide from the noonday sun, often make particular use of evaporating water to balance their heat budgets. They do this with their transpiration streams, a mechanism so effective that we think of plants as typically being "cool." Yet water shortage in dry heat or in cold wind may drive plants to other expedients that lead to the appearance of deserts or tree lines.

It will be evident that any one kind of animal or plant has a limited number of ways to manipulate its surface temperature or its water use in order to balance its heat budget at a temperature satisfactory for life processes. Animals with good insulation of fur or feathers and an ability to maintain their core temperature by releas-

FIGURE 5.2

Animal heat loss by induced radiation.

Maintaining a large surface at close to core temperature will result in a net loss of heat by radiation if surrounding bodies are at lower temperatures. The best surrounding "body" for this purpose is outer space, at which a desert jack rabbit (*a*) directs its large ears while sitting at the mouth of a burrow. It is likely that the back "fin" of *Dimetrodon* likewise allowed loss of body heat by radiation (*b*). *Dimetrodon* could also have gained heat rapidly by turning its fin at right angles to the sun (*c*).

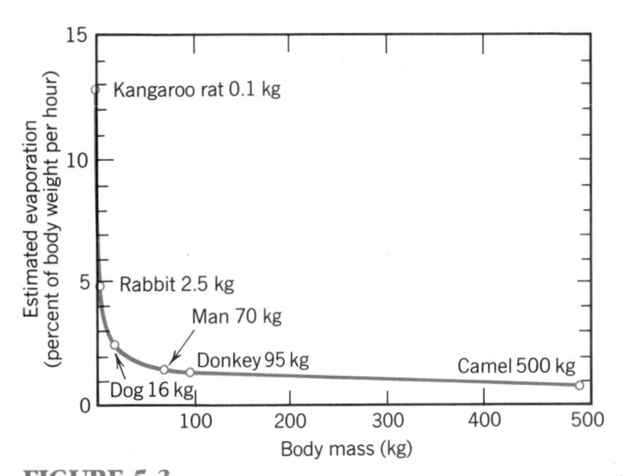

FIGURE 5.3
Evaporation requirements for animals exposed to hot desert conditions.
The curve is calculated on the assumption that heat load is proportional to body surface. Small animals have a body surface that is exponentially greater than that of large animals. Animals like kangaroo rats of deserts must avoid the desert sun, since they are without the necessary large flux of water. (From Schmidt-Nielsen, 1975.)

ing metabolic heat can be well-suited to life in cold places. Relatively large animals that do not maintain core temperatures by metabolic heat, like lizards, fail to inhabit the Arctic, presumably because they cannot survive the inevitable freezing. Yet insects of many kinds do very well in the Arctic, possibly because they seldom maintain an adult existence through the winter, or because they synthesize antifreeze solutions that preserve them from freezing but in a state of torpor.

The potential advantage of large size for life in the arctic gave rise to an old dictum of biogeography called **Bergman's rule**. This said that in warm-blooded animals, individuals living in cold places tend to be larger than close relatives living in warm places. This "rule" is a gross oversimplification of reality, as can be seen from the patent observation that many very large animals live in the tropics. Bergman's rule is mainly of historical interest for showing that nineteenth-century biologists began to think in terms of ecological energetics. One selection pressure on warm-blooded animals of cold regions, among the many determining size, is that energy conservation is favored by large size.

ANIMAL HEAT BUDGETS

To make a heat budget for an animal we can start with the simplifying assumption that animals are cylindrical and with negligible appendages; not really too desperate an oversimplification considering the shape of a deer (Gates, 1968*a,b*). The cylindrical animal then consists of a body core, an insulating sleeve of fat bound by skin, and perhaps a sleeve of fur or feathers separating the skin from the radiating surface of the animal, as shown in Figure 5.4.

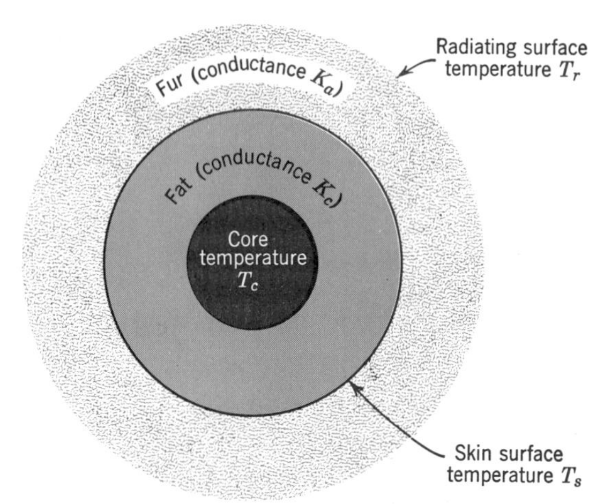

FIGURE 5.4
Heat balance cross section of idealized animal.

The following symbols are used in the equations below.

M = metabolic energy as cal cm^{-2} min^{-1} generated at skin surface
E_1 = energy lost by sweating cal cm^{-2} min^{-1}
E_2 = energy lost by evaporation at respiratory surface cal cm^{-2} min^{-1}
K_a = conductance of insulation (fur or feathers)
K_c = conductance of body fat
T_c = temperature of body core
T_s = temperature of body surface
Q = radiation absorbed by body surface h_c is the convection coefficient
T_a = air temperature
ϵ = emissivity of the skin surface δ is the Stefan–Boltzmann constant for radiation
T_r = temperature of effective radiating surface at outside of insulator

The energy conducted from the body core to the skin surface is

$$\begin{pmatrix}\text{Metabolic energy}\\ \text{generated at}\\ \text{skin surface}\end{pmatrix} = \begin{pmatrix}\text{energy passed}\\ \text{by simple}\\ \text{conduction}\end{pmatrix} + \begin{pmatrix}\text{energy lost}\\ \text{at respiratory}\\ \text{surface}\end{pmatrix}$$

$$M = K_c(T_c - T_s) + E_2 \tag{5.1}$$

If the animal has no fur or feathers the energy budget of the external surface is

$$\begin{pmatrix}\text{Energy passed}\\ \text{by simple}\\ \text{conduction}\end{pmatrix} + \begin{pmatrix}\text{radiations}\\ \text{absorbed from}\\ \text{environment}\end{pmatrix}$$

$$K_c(T_c - T_s) + Q$$

$$= \begin{pmatrix}\text{radiant energy}\\ \text{lost as blackbody}\end{pmatrix} + \begin{pmatrix}\text{convective}\\ \text{heat}\\ \text{loss}\end{pmatrix} + \begin{pmatrix}\text{energy lost}\\ \text{from skin}\\ \text{evaporation}\end{pmatrix}$$

$$= \epsilon\delta T_s^4 + h_c(T_s - T_a) + E_1 \tag{5.2}$$

Eliminating $K_c(T_c - T_s)$ from equation 5.2 we get

$$M + Q = \epsilon\delta T_s^4 + h_c(T_s - T_a) + E_1 + E_2 \tag{5.3}$$

If the animal has fur or feathers, the energy budget at the radiating surface is given by

$$\begin{pmatrix}\text{Energy reaching}\\ \text{skin surface}\\ \text{by conduction}\end{pmatrix} = \begin{pmatrix}\text{energy lost}\\ \text{from skin}\\ \text{evaporation}\end{pmatrix} + \begin{pmatrix}\text{energy conducted}\\ \text{across insulating}\\ \text{layer}\end{pmatrix}$$

$$K_c(T_c - T_s) = E_1 + K_a(T_s - T_r) \tag{5.4}$$

and the energy budget for the radiating surface is given by

$$\begin{pmatrix}\text{Energy conducted}\\ \text{across}\\ \text{insulating layer}\end{pmatrix} + \begin{pmatrix}\text{radiations}\\ \text{absorbed from}\\ \text{environment}\end{pmatrix}$$

$$K_a(T_s - T_r) + Q$$

$$= \begin{pmatrix}\text{radiant energy}\\ \text{lost as blackbody}\end{pmatrix} + \begin{pmatrix}\text{convective}\\ \text{heat loss}\end{pmatrix}$$

$$= \epsilon\delta T_r^4 + h_c(T_r - T_a) \tag{5.5}$$

This last equation describes how the environmental energy flow may affect the heat budget of the animal, but it is not easy to put numbers against various terms in the equation. Consider the difficulty of measuring K_a or ϵ for a real animal, for instance. But this difficulty may be avoided as follows: By eliminating $K_c(T_c - T_s)$ from equations 5.1 and 5.4 we get

$$M - E_2 - E_1 = K_a(T_s - T_r) \tag{5.6}$$

Solving for T_s, and substituting back into equation 5.1, we get

$$T_c - T_r = \frac{M - E_2 - E_1}{K_a} + \frac{M - E_2}{K_c} \tag{5.7}$$

Now

$$\frac{M - E_2}{K_c}$$

is the temperature difference between the core and the skin surface, and

$$\frac{M - E_2 - E_1}{K_a}$$

is the temperature difference between the skin and the radiant surface. Equation 5.7 therefore shows that temperature differences between the various layers of an animal will always be proportional to the temperature difference between the core and the radiant surface. This means that measures of temperature of the core and the surface are all that are needed to infer the ability of an animal to regulate temperature.

This apparently rather self-evident result of Gates (1968*a,b*) in fact gives formal understanding to some well-known properties of animals. Mammals and birds commonly maintain larger differences between core temperatures and surface temperatures than do reptiles and fish. It is not so much that one group is warm-blooded and the other cold-blooded that matters as that the former can maintain a larger temperature difference between core and surface. Warm-blooded animals can withstand greater variations in the energy flowing through their environments. This gives fresh insight into the relative successes of the two engineering designs: the cold-blooded poikilothermous designs of fish and reptiles are suited to oceans and equatorial regions because these are places where variance in environment energy flux is least.

Manipulating Surface Temperatures with Countercurrents

Maintaining steep temperature gradients between the core and the periphery can certainly be partly achieved with thick insulation, but more active systems of heat-gradient maintenance are even better. The active principle is that of the so-called **countercurrent systems**. Cool and warm liquids flowing in opposite directions work as a heat exchanger. The system as it applies to the fin of a marine mammal swimming in arctic waters will serve as an illustration. Because of countercurrent flow,

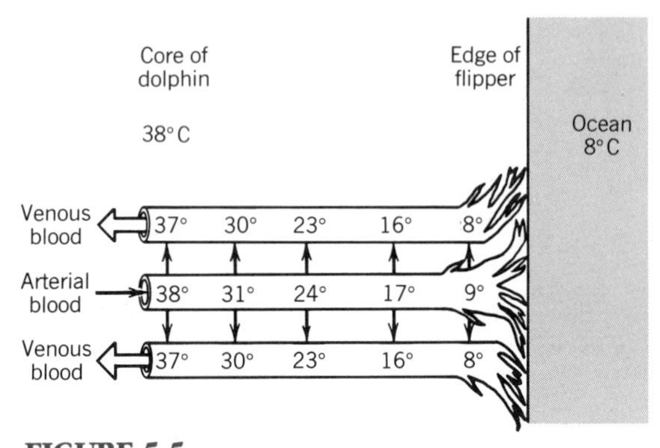

FIGURE 5.5
A countercurrent flow heat exchanger.
In the flipper of a dolphin, each artery that goes toward the skin is wrapped in a bundle of veins carrying blood away from the skin. This arrangement means that the outgoing arterial blood is always bathed in venous blood of slightly lower temperature. Arterial blood is progressively cooled as it approaches the cool surface of the fin, and venous blood is progressively warmed as it approaches the region of the core temperature. Small arrows from artery to veins in the figure show that a constant temperature gradient is maintained. This heat exchanger can cool arterial blood, or warm venous blood, to within 1°C of the ambient temperature. The result is that there need be very little heat loss from flippers despite the absence of insulation in their thin cross section.

the fin is kept close to the temperature of seawater, a condition in which heat loss is minimal.

When hot and cold liquids flow in opposite directions down narrow pipes (capillaries) laid side by side there will always be a small temperature difference between the adjacent flows (Figure 5.5). Arteries and veins in a porpoise flipper are arranged such that bundles made up of veins are wrapped around arteries. The outgoing arterial blood is progressively cooled, and the returning venous blood is warmed. Instead of losing heat to the environment, the heat is "recycled" as much as is thermodynamically possible within the flipper. In practice, the heat loss of an operating porpoise flipper in cold water may be small (Schmidt-Nielsen, 1975). This manipulation of countercurrents as heat exchangers was an essential adaptation for the penetration of cold oceans by warm-blooded swimming mammals (Figure 5.6).

Some fast-moving fish also use heat exchange countercurrents at the periphery. High-speed sharks and tuna do this (Schmidt-Nielsen, 1972). The system allows the animals to maintain temperatures in the swimming muscles well above that of the surrounding water, providing them with a modified endothermy. Warm muscles give these predatory fish the ability to maintain rapid movement in pursuit of food, thus letting them function as do endothermic predators on land. Sharks, therefore, have some of the energetic adaptations of mammalian predators. Countercurrent heat exchangers are particularly necessary for these fish because they are without significant insulation and would not be able to maintain

FIGURE 5.6
Triumph of countercurrent systems.
Dolphins and other cetaceans conserve heat and distill salt water using countercurrent flow devices. By these physiological adaptations the sea was opened to air-breathing endotherms.

high core temperatures without a system for cooling blood before it reaches the periphery.

Temperature at the surface is also critical for desert animals that must minimize the loss of water from their breathing. Countercurrent heat exchangers then work to cool the exhaled air, effectively letting its moisture condense within the animal. Exhaled air of desert rodents is dried by being cooled in this way before it is released from the nostrils.

A countercurrent heat exchanger is arranged in the convoluted nasal passages so that incoming air cools the outgoing air, an arrangement comparable to the opposed blood vessels in the flippers of porpoises. As the outgoing air cools it loses water, which condenses in the nose or is absorbed by the incoming air as this gains temperature. Evaporation within the nasal passages can actually cool the exhaled air below the temperature of the outside air, again without loss of water to the animal. An easily detectable consequence of this mechanism is that the animal has a cool nose. But the important consequence detected by natural selection is that the animal does not lose water from its cold nose while breathing.

This adaptation is also one of the many that equip camels, the most celebrated of desert animals, to exploit the carbon resources of desert lands. They cool and dry their exhaled air, they manipulate the temperature at the periphery with fur on the back to keep the sun out and bare skin on the belly to lose heat in their own shadow, they concentrate urine, they tolerate desiccation, and they store energy in a fatty hump; the whole making a package of adaptations that allows a warm blooded animal to survive through the leaner times of desert living (Figure 5.7).

Countercurrent heat exchangers probably serve widely, alongside insulation, to maintain that critical temperature gradient between core and periphery necessary to prevent ruinous heat and water losses in all kinds of animals that maintain high core temperatures.

HEAT BUDGETS AND TEMPERATURE REGULATION BY PLANTS

Plants live in habitats that display every form of variance in energy flux, from polar regions to the equator. They must cope with these different degrees of variance almost entirely by manipulating their shapes or the flow of water.

Figure 5.8 illustrates the energy exchange between a plant and its environment. A plant receives all its energy directly from radiation or convection from the contemporary environment, except when it is metabolizing the reserves of previous years, as when a potato sprouts from a tuber. A small fraction of the incoming solar radiation is transformed into reduced carbon compounds (see Chapter 3), and the rest is converted to heat. The plant must balance this income of heat by loss—through radiation, convection, and transpiration—of an equal amount.

The active parts of plants are the leaves, and these are so thin that the exterior temperature must approximate the core temperature. Insulation, therefore, is of little importance to a plant's heat budget. Leaf behavior also matters little, though there are possibilities, as when heated leaves droop through wilting or turgor changes, events that move the flat surface so that it no longer faces the sun. The adaptation of plants to different temperature regimens is, however, mostly restricted to adopting appropriate shape and controlling the

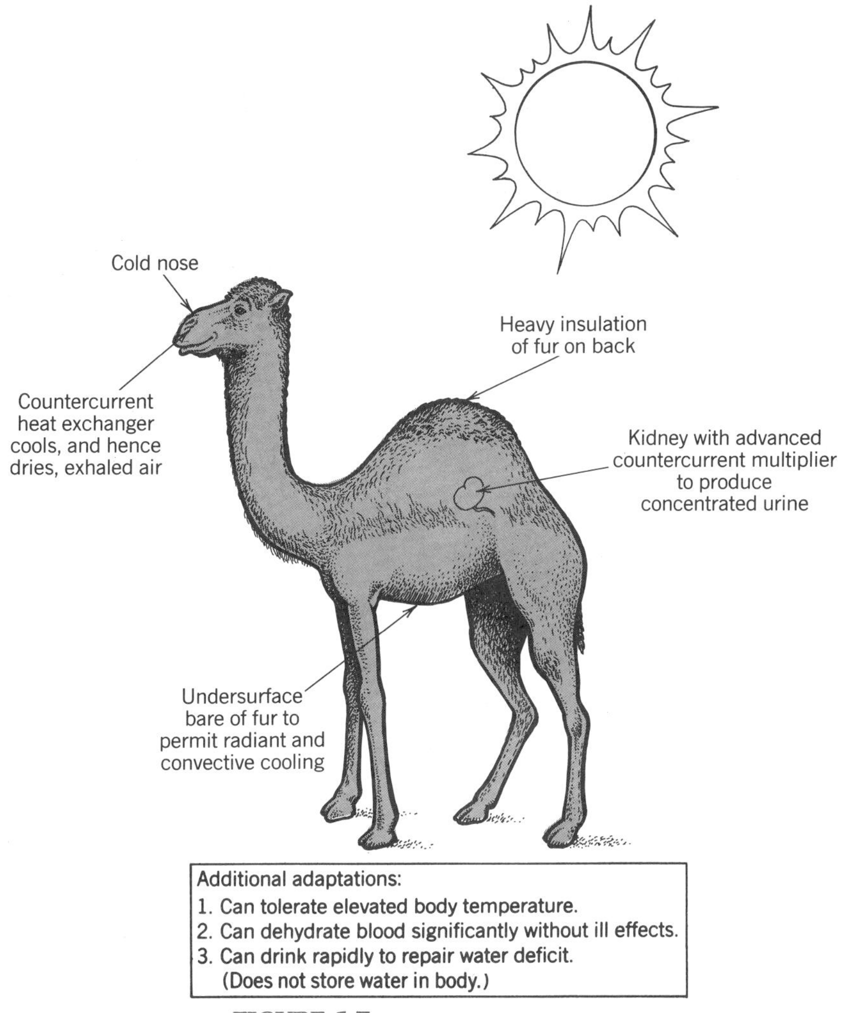

FIGURE 5.7
Adaptations of camels to desert life.

transpiration flow by manipulating the stomates.

Gates (1965) studied leaf and ambient temperatures of monkey flowers (genus *Mimulus*) in Nevada. He found that *Mimulus lewisii* populations growing at 10,600 feet in the Nevada range on a day of still air at 19°C had leaf temperatures of between 25 and 28°C. At 1300 feet on the same mountain, the air temperature was as high as 37°C, but the leaf temperature of the local monkey flowers (*Mimulus cardinalis*) was between 30 and 35°C, or very close to the temperature of those in cool air near the mountaintop. The plants in cool air manipulated their heat budget so that they were warmer than their surroundings; the plants in hot air manipulated theirs so that they were cooler. The two species are not radically different in design, thus re-

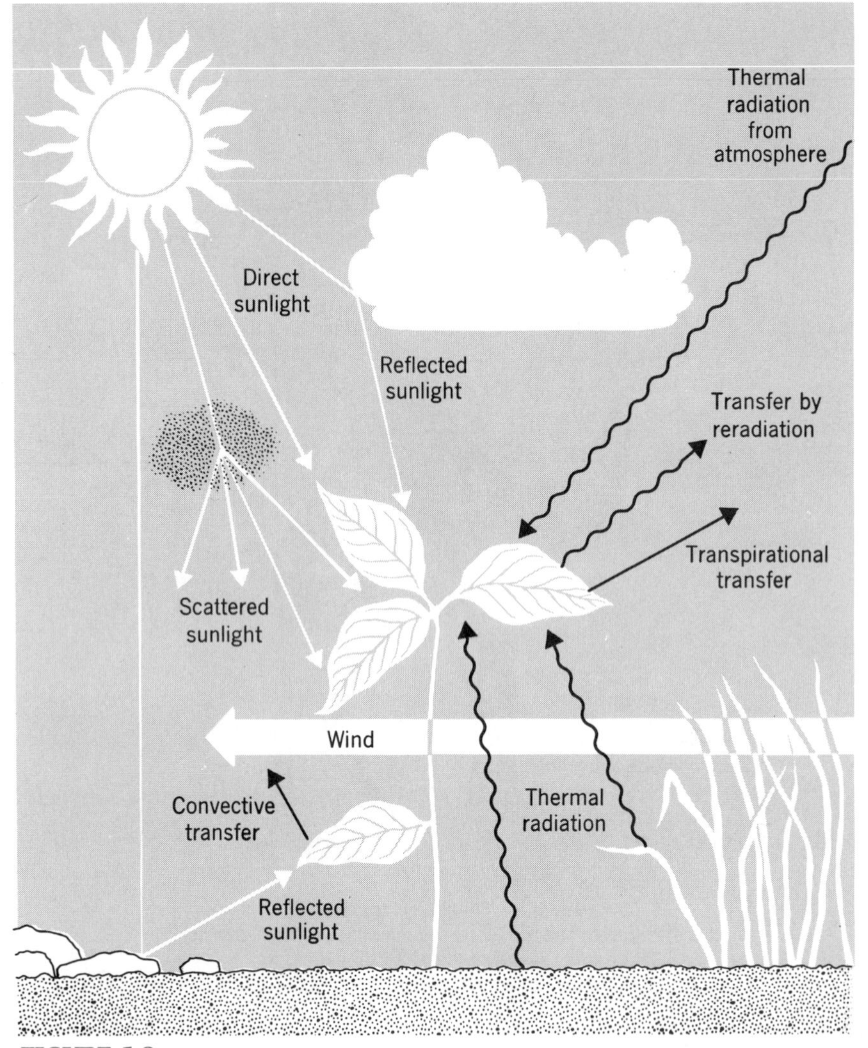

FIGURE 5.8
Energy exchange between a plant and its environment.
For an energy budget in the short term, a small part of absorbed solar radiation is sequestered as the energy of reduced carbon, but the remainder must be balanced by radiative and connective losses. (After Gates, 1968*b*.)

quiring that their feats of temperature regulation be done by manipulating the flow of water from the leaves by opening or closing stomates.

It seems likely that *M. cardinalis* of the hotter elevations in Nevada is limited to habitats with an adequate supply of water, since these organisms, like *Homo sapiens*, solve the problem of a too-hot habitat by evaporating water. The niches of *Mimulus* and *Homo* include the use of water, and water becomes a limiting factor of their environment so important that isohyets (lines joining places of equal precipitation) on maps might well be found to run parallel to the borders of their species distributions.

Mimulus plants can get a living by photosynthesis in hot places using broad, thin leaves, if the local species is adapted to pump large quantities of water through the leaves. But if water is not available, the flat, thin leaf cannot be kept cool, and plants with this structure cannot live. Plants of hot places that are also very dry are made without conventional leaves, resulting in the barrel or organ-pipe designs of the family Cactaceae (the true cacti) or their old-world analogs in the Euphorbiaceae. (Figure 5.9).

The Energetics of Plant Shape

The design of a cactus is known to be suited to conservation of water: the surface-to-volume ratio is small, the surface is covered with a thick cuticle that is impermeable to water, and the stomates are few and tend to be sunk in pits. Furthermore, these plants use the C4 system of photosynthesis, or even resort to CAM metabolism so that they can synthesize sugar from carbon stored overnight, working behind closed stomates when supplied with the energy of the sun (see Chapter 3). These adaptations obviously make good sense for life in a dry place, but they leave the plant with a large heat budget problem that cannot be solved by the usual expedient of evaporating water. A solution in structural engineering is used instead.

Figure 5.10 offers an analysis of heat budget control in an organ pipe cactus of the type of saguaro (*Carnegiea gigantea*) of the Sonoran desert of Arizona. The tall, narrow shape presents the smallest area to the noonday sun and the largest area to the oblique sun of early morning and late

FIGURE 5.9
Shapes to balance heat budgets in deserts. Reduced leaves and vertical stance minimize heat loading from the noonday sun. The cacti of the new world (upper photo) and the Euphorbiaceae of the old world are ecological equivalents given the same shapes by natural selection to minimize heat stress in hot dry places. The extended ears of the animals in the lower photograph are working as heat-shedding radiators.

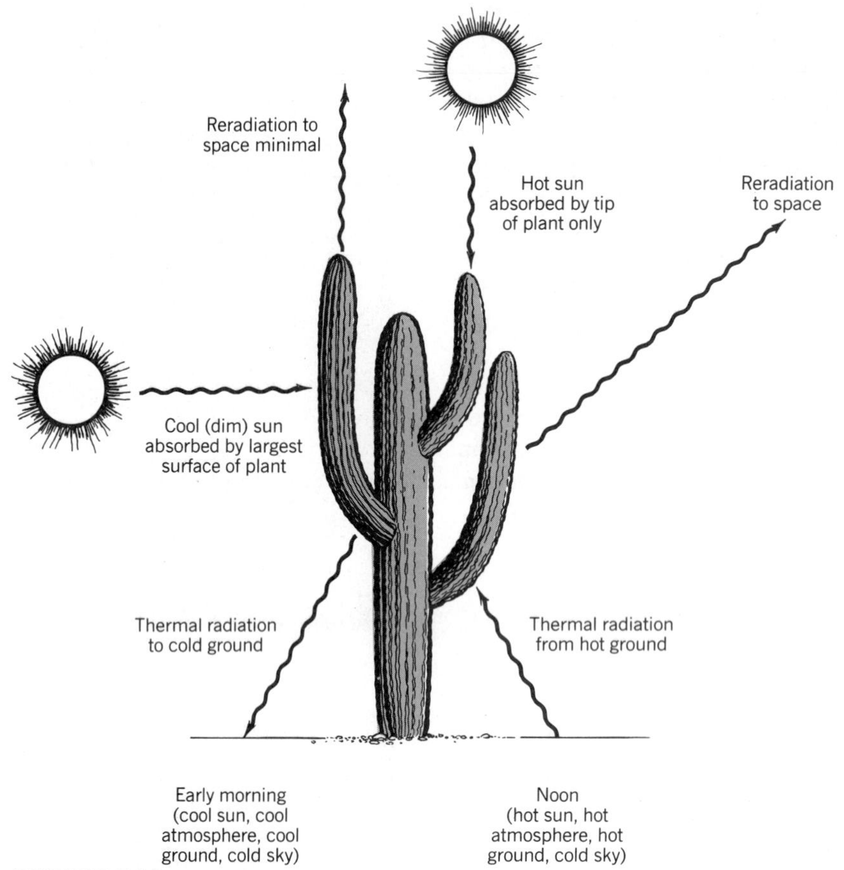

FIGURE 5.10
Heat budget control by a saguaro cactus.
At high noon only the point of the plant faces the sun and the largest possible surface faces the sky (which is cold since it is without clouds). In early morning and late evening the sides of the cactus face the sun to allow maximum absorption, permitting maximum photosynthesis and keeping the plant warm enough for efficient operation.

evening. It seems a good working hypothesis, therefore, that the shape of a cactus not only reflects conservation of water but also is suited to minimize heat load under the hottest sun. Cactus shapes are among the more extreme examples of plant engineering in deserts, but the essential themes appear in other taxa also—notably reduced leaves, reduced horizontal shapes, and green, photosynthetic stems. These structural designs are approximated in plants of different ancestry (Figure 5.11). They reflect the fact that in hot deserts heat budgets must be balanced without resort to evaporating water. The result is the range of life forms found in hot deserts.

In other hot places, however, water is abundant and the typical life form is that found in the tropical rain forest. Trees are evergreen and their leaves are flat sheets. The geometry of these trees and their leaves can be explained as maximizing photosynthesis by diffusing light so that the largest possible number of leaves are exposed to light at the optimum intensity for photosynthesis (see Chapter 3). The heat budget problem is solved by pumping water and has little effect on engineering design.

In places that are hot only part of the year, alternative structural designs exist. In the middle latitudes the basic leaf and tree plan of the tropical forest is maintained, but the trees drop their leaves in winter. In the more northerly region of the boreal forest, leaves are needle-shaped and are kept year-round. Some hint of an explanation of the transition from deciduous broad-leaved trees to evergreen needle-leaved forms in the north comes from Gates's (1965) analysis of heat balance in the two kinds of leaves, for the two shapes have quite different convective properties.

Heat gain or loss by convection depends on the relative temperatures of the leaves and the surrounding air, on the speed of air flow, and, most importantly, on the intimacy of contact between the flowing air and the leaf surface. As air flows over a surface, a thin film of almost stationary air remains at the junction, called the **boundary layer**. This boundary layer turns out to be much thinner for a needle-shaped leaf than for a flat leaf, implying that convection would produce more rapid temperature changes in needle-shaped leaves. Gates was able to test this with wind tunnel experiments, using silver replicas of leaves. Electrodes fastened to the silver leaves recorded rates of heat gain or loss when the leaves were exposed to warm or cold air moving past them. A pine needle gained or lost heat many times more rapidly than a poplar leaf. A practical consequence of this is that needle shapes of leaves tend to force a passive balancing of the heat budget toward the temperature of the surrounding air.

In the boreal forest, then, trees are designed in a way that keeps their operating parts close to air temperature without further manipulation by the plant. In cool summers of boreal latitudes, possession of needles keeps plants cool with a minimum of transpiration. But the real significance of this design is probably found by night and in winter, because then the leaves will be kept relatively warm at air temperature despite the tendency for them to cool much below this by radiation to the cold blackbody of space. Even in a northern winter, the surface of a spruce tree is warm compared to the black cloudless sky above it, and so must be losing heat by radiation according to the fourth power of its absolute temperature. Rapid convection from the surrounding air balances this heat loss, perhaps keeping the leaf several degrees warmer than possible for a flat leaf.

Since needle leaves tend to adopt, by convection, the temperature of the air around them, the design is suited to warm deserts as well as to the dry cold of the north temperate belt. This probably explains why many pines exist in semiarid regions, and others invade old fields in the early years of secondary succession, where conditions are often locally semiarid (bare fields, direct sun) (Figure 5.11).

Understanding the implications of having needle-shaped leaves, therefore, allows plausible hypotheses to explain their distribution. Plants with this design strategy live in places that may be seasonally hot and dry, because high heat loss by convection reduces the demand for transpired water. More interestingly, the great latitude spread of the boreal forest reflects the suitability of the design both for preventing leaf temperature from falling below air

(*a*)

(*b*)

(*c*)

FIGURE 5.11
Needle-leaved trees: when cold or dry.
Needle leaves lose or gain heat by convection many times faster than flat leaves. Radiation in bundles of needles is more to each other than to the sky. These combined properties result in higher leaf temperatures in cold times than would be possible for flat leaves, suggesting an advantage in the north. Needles also need less cooling by transpiration under a hot sun, suggesting that they are adaptive in hot, dry places or in dry northern summers. (*a*) A pine in a hot, semidesert place. (*b*) A pine in an old field succession exposed to relative heat and drought. (*c*) Trees of the boreal forest.

temperature in cold times and for conserving water in dry times. Presumably the needle-leaved design may pay a price in photosynthetic efficiency, since flat leaves are the rule whenever there are no constraints of temperature or water.

Remaining is the problem of the Pacific Northwest of the United States and Canada, where cool rain forests of coniferous trees achieve the highest standing biomass of any ecosystem on earth. Their environment is neither dry nor particularly cold; so why the needle-leaf design? The answer may be connected to the fact that the larger part of the annual production in this ecosystem is during the mild winters (M. A. Witzbach, personal communication). Or it may be that water cooling is impracticable in cool saturated air, thus giving an edge to air-cooled structures.

In the broad belt of deciduous forest, between the evergreen rain forest of the wet tropics and the evergreen needle-leaved forest of the cold temperate north, a different strategy apparently is optimum. Summers in this belt have much in common with the climate of wet tropics—they are warm, moist, and cloudy. Trees using flat leaves produce very well in these summers, but their leaves would be colder in winter than needles would. Apparently it pays in mid-latitudes to behave like a tropical tree in summer and simply close down production for the winter months.

Further north, the summers are so constrained (shorter, drier, or both) that a strategy of making the most of all seasons with needle leaves may well be optimum. Evergreen conifers, like all evergreens, do shed leaves (about a third annually), but in the boreal forest they do not do this synchronously or with the seasons. They hold their air-warmed, air-cooled needles available for production on good days at all times.

Forest formations, therefore, like desert plant forms, can be understood as the result of mechanical adaptation to make the most of local heat and water. The result is the most basic fact of biogeography: evergreen forest at the equator, deciduous forest at mid-latitudes, evergreen conifers in the north, odd-ball plants in deserts—all a result of engineering to manipulate different regimes of heat (see Chapter 18).

The Setting of Tree Lines

The disjunction at a tree line is one of the most striking of natural phenomena: to one side the forest, to the other an open space, the transition abrupt enough for easy mapping (Figure 5.12). On high mountains the tree line can be sharp indeed, a progression from trees to no trees within a span of a few meters. In the Arctic the tree line is a more gradual affair and follows a progressive shortening of trees across tens of kilometers. The line is broken by long fingers of trees stretching up river valleys and depends on the direction in which the slope faces. Yet the change is clear enough to pose teasing questions: Why should there be an edge to the forest? Why should there be no trees in the Arctic?

The puzzle of the absence of trees in the Arctic is not simple. Clearly, cold weather has something to do with the phenomenon but it has never been enough to make statements like "the arctic climate is too harsh for trees" because the arctic ground is often completely carpeted with herbs that survive the harshness well enough, and woody plants of low stature live in the Arctic. Tree distribution does correlate roughly with permanently frozen ground (permafrost), but the synchrony of tree line and permafrost is far from complete. And yet an analysis of the heat budget problem of a hypothetical arctic tree yields a plausible explanation.

A tree in the far north in the few weeks of summer must balance its heat budget when the heat input is from low-intensity sunlight supplemented by radiations from the ground and neighbors. Set against these inputs are the unavoidable losses by

FIGURE 5.12
Tree line in Idaho.
Many continental tree lines look like this as trees finger their way up valleys. This Idaho scene can be duplicated in the arctic (*see* Chapter 14).

radiation to the sky, by transpiration (some of which will always be necessary to provide the leaves with a flux of moisture), and by convection to cold air. The arctic air is cold most of the time, and usually in motion, so it follows that the leaves of an arctic tree will always be relatively cold. This means low productivity, a serious matter for a plant with only a short growing season and the metabolic expenses of a large body to be met.

This poor heat budget outlook for an arctic tree may be compared with the prospects of an arctic herb or low shrub like a prostrate willow. Their leaves are close to the ground, very often within the layer of air that is almost stationary as winds pass over it. Still, air between ground-level leaves will be warmed by radiation and conduction from the plant leaves as they themselves are warmed by the sun. A dwarf plant, therefore, will be able to maintain higher leaf temperatures in the Arctic than is possible for a tree. Dwarfs also have smaller maintenance costs because their bodies are smaller. A low life form, therefore, can be expected to yield a positive photosynthetic balance in cold northern places where a tree cannot. Thus, one simple answer to the question, "Why are there no trees in the Arctic?" is, "because a heat budget cannot be balanced at a satisfactory temperature for a tall plant in cold polar air."

A more complicated answer to the tree-line problem must take into account the effects of cold on the water supply, together with such matters as length of growing season on the energy budget. These complications have been studied most thoroughly for alpine tree lines.

A common pattern at alpine tree lines is a solid block of trees approaching to a definite line beyond which there are essentially no trees (Figure 5.13). Inside the block of forest, the environment of each tree must be modified by the presence of its neighbors. Outside the block, or at its edge in the front rank, trees do not have the shelter of neighbors. Tree seedlings exposed beyond the tree line fail to gain weight. This makes them peculiarly vulnerable to damage by frost after drying.

FIGURE 5.13
Natural tree line in the Alps.
The trees are pines (*Pinus cembra*). Although a few trees stand out at the edge, the forest approaches its limit as a solid block.

Conifer needles can withstand frost as long as they do not dry out, but if they become desiccated they are easily killed. In a strong frost the ground itself is likely to be frozen, particularly if the alpine snow cover is light. Trees are then denied access to water, yet the needles, and even twigs, must remain turgid. Resistance to intense cold by coniferous trees, therefore, is a function of the ability of their needles to retain water when the ground is so frozen that fresh supplies of water cannot be taken up. A well-formed needle, with a thick cuticle right to the tip, retains water well and survives long periods of freezing temperatures, but a needle with a thin cuticle does not. The ability of needles to survive frost is thus a function of the previous growing season, the length and warmth of which determine how well-formed the needles are at the onset of winter (Table 5.1; Tranquillini, 1979).

This work on alpine trees shows that physiologically induced drought in winter

TABLE 5.1
The Liability of Immature Spruce Needles to Winter Desiccation.
The needles of Picea engelmannii *were sampled at tree line. By February the water content of needles that started the winter without being matured had seriously decreased. Winter desiccation is a crucial factor in setting tree lines. (From Wardle, 1968.)*

	Mature Needles	Immature Needles
Shoot length (cm)	0.07–2.7	0.4–1.2
Needle length (mm)	12–16	4–11
Needle spatial density (no. per cm shoot)	29–44	50–72
Water content (% dry wt.)		
17 Nov.	134	152
22 Dec.	125	130
3 Feb.	126	51

can be a cause of tree death, but that this death is linked to the length of the preceding summer. Net production during the summer, however, is partly determined by the temperature at which a tree balances its heat budget (as discussed above), and this is a function of summer temperature. Low-growing plants high in the mountains or in polar regions produce more in summer and make leaves better able to resist desiccation in the physiological drought of a cold winter. This combined explanation of why trees do not grow in cold seasonal places applies alike to the tops of high mountains and to the Arctic (Figure 5.14).

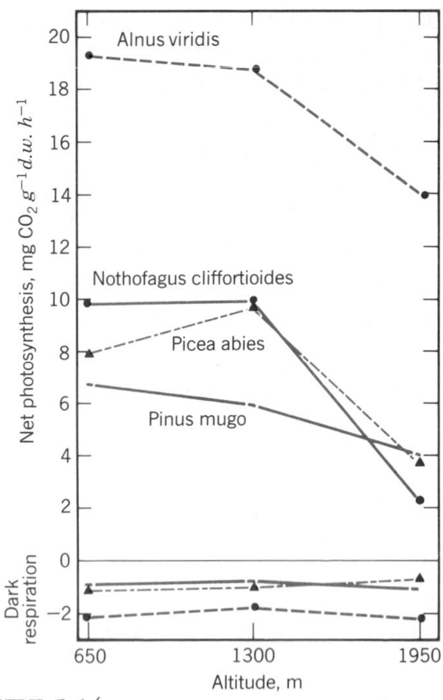

FIGURE 5.14
Effect of altitude on net photosynthesis and dark respiration of seedlings.
The four species were grown in pots at the three altitudes and net photosynthesis and respiration were measured as gas exchanged under controlled conditions. (From Benecke, 1972.)

Balancing Heat Budgets Defines the Structures of Ecosystems

The familiar plant geography of the earth is set by the response of plants to different energy regimes. In the warm and wet tropics, plants can dissipate their enormous incident heat loads by evaporating water. Leaves are flat, many, and stacked layer upon layer to diffuse incident light to as many of them as possible. The result is the tropical rain forest.

In the desert tropics a water-cooled lifestyle is impossible. If leaves are present at all, they are needle shaped structures that maintain the temperature of the ambient air by convection. But many desert plants lack leaves entirely, being shaped like pillars. The pillar design reduces the amount of heat absorbed while serving as a radiator.

Temperate latitudes with tropical summers interspersed with winters allow water cooling in summer. Winters are met by dropping leaves entirely. Further north, the optimal solution to maximizing productivity appears to be an evergreen strategy that includes preventing loss of tissue by freezing. Needle leaves are the answer. The temperature of needles remains close to that of ambient air in both summer and winter. In the very cold climates of arctic and alpine regions, tall structures are elevated to air masses too cold for rapid growth even in summer.

A final observation is the commonsensical one that temperature regimes fluctuate over most of the earth. As a result, many real ecosystems include plants with different solutions to the energy problem. Between the boreal forest of Canada and the eastern deciduous forest of the United States, for instance, is a broad land where assorted broad-leaved and coniferous trees grow in mixed communities, and where some conifers, like tamarack (*Larix*) are deciduous. An ecologist with a time machine would probably see that this broad belt was in continual flux of invasions from north and south as opportunities of weather opened the community to competitive invasion by first one form and then the other.

Preview to Chapter 6

Amblyrhynchos cristatus: Ectotherm using solar heating.

COLD-BLOODED, or poikilothermic, animals like reptiles or fish are best called ectothermic animals. They regulate temperature when they are active to within 5°C of the temperature of homeotherms, but they use external heat sources to do so. This lets them conserve metabolic energy and go for long periods without food. A prime drawback is that bouts of strenuous activity must be short, since their metabolism rapidly causes oxygen debts in the muscles. Warm-blooded homeothermic animals are best called endotherms, because they regulate temperature by using metabolic heat. They have the big disadvantage of large energy costs for metabolism, but they are able to be active at night and in cold places. Endotherms can maintain high rates of activity for prolonged periods, letting them run down ectothermic prey or escape from ectothermic predators. They can also migrate long distances, letting them tap seasonal fluxes of resources denied to ectotherms. Insects get the best of both systems. Insects are ectotherms, with all the advantages of energy conservation and ability to go without food for long periods that this implies, but they also are capable of prolonged exertion like endotherms. Insects avoid the oxygen debts that build up in the muscles of reptiles by piping air directly to muscles in the system of tracheae. Locomotion is an important charge to the energy budget. Flying and swimming require less energy for the weight moved than does running. The style of locomotion on land, however, whether hopping or running, using two legs or four, makes no difference to the energy cost. Details of style are functions of evolutionary accident, rather than conditions that can be explained by energetics.

Chapter 6

The Energetics of Lifestyle

Animals have long been divided into warm-blooded (birds and mammals) and cold-blooded (the rest, but reptiles are particularly called to mind). The terms **homeothermic** (keeps the same temperature) and **poikilothermic** (the body temperature varies) commonly replace "warm" and "cold" bloodedness. This makes the point that the important difference is not temperature alone but the way in which one group can maintain a constant temperature independent of the environment whereas the other cannot. Most fish (poikilotherms) must let their body temperature be close to that of the water around them, but whales stay warm even in cold water.

Talking of homeothermy or poikilothermy in land animals can be misleading, however, because active animals of either design regulate their body temperatures within fairly narrow limits, to within a few degrees of 38°C. Poikilothermic reptiles do let their body temperature change when they are active or when they rest, so that at night, for instance, their temperature may be close to that of the air. But hibernating mammals let their temperature fall also. Hummingbirds of the high Andes Mountains let their tiny bodies cool at night, thus slipping into a state of torpor, a stratagem that saves energy for them just as it does for the reptiles called poikilotherms (Calder and Booster, 1973). Clearly the terms "homeothermic" and "poikilothermic" do not offer an unambiguous classification.

The terms **endothermic** and **ectothermic** are more revealing for ecologists, since they direct attention to the energy source used to reach the temperatures at which the animal is active. Endothermic animals use energy ingested as food to heat their bodies. Ectothermic animals tend to use solar power for their heating.

Ectothermy is the more ancient plan, and it is still used by most living species. In one sense, therefore, ectothermy can be called the most "successful" plan, which is a good observation with which to start a discussion. It is endothermy (in the guise of homeothermy) that is often thought to be the "better" or "more successful." **Endotherms** regulate their temperatures precisely, with advantages in running body chemistry that are easy to enumerate. Also, we humans are endotherms, a circumstance encouraging us to think that endothermy must be superior. But if number of species, or number of individuals among the followers of a plan, is a measure of success, then there can be no doubt that ectothermy is the better way. What, then, are the properties of ectothermy that lead to this success? It is useful to consider first ectothermy in the lives of tetrapod (four-legged) vertebrates, the reptiles and amphibians, with which tetrapod vertebrate endotherms (mammals and birds) can be compared directly.

BENEFITS AND COSTS OF ECTOTHERMY

Ectothermy is one way of manipulating a heat budget and regulating body temperature. In its simplest form, the hot animal seeks the shade and the cold animal the sun. This is a cheap way of meeting the costs of temperature regulation. Few food calories are used by a reptile to regulate its temperature. The resting metabolic rate of a reptile is low compared to that of a mammal, and its production efficiency (the ratio of calories ingested to calories used for growth and reproduction) is high. Because energy is used for reproduction rather than for keeping warm, a clear payoff in fitness should follow.

Pough (1980) examined production efficiencies for 16 species of small mammals and birds and eight species of reptiles and amphibians of comparable size and found that the efficiency of the endotherms was 1.4% whereas that of the ectotherms had an average of 43.6%. Ectotherms, then, seem to be remarkable machines for making much biomass or many offspring from the ingested energy available. Pough (1982) calls ectotherms "**low-energy systems.**"

When food is scarce, an ectotherm can relapse into torpor. It then uses very little energy. This ability to switch off as an energy consumer for long periods lets ectotherms exploit episodic or periodic resources in ways that are very difficult for endotherms. The lizard *Sauromalus obesus*, for instance, relies on desert plants for both calories and water. When the plants dry up, the lizard crawls into a rock crevice (where it can stay cool) and remains there for up to 8 months (Figure 6.1). When rain makes the plants grow again, all the lizard has to do is work its way into the sun, warm up to a nice operating temperature near 38°C, and start feeding again (Nagy, 1972). A similar ability to be active and grow when food suddenly becomes available is shown by the many snakes that can do well eating bird eggs, though these are available for only short periods.

An even more striking result of the property of eating when there is food and relapsing to torpor when there is none is the ability to switch from food to food as animals grow. For ectotherms this opens the possibility of life histories that include larvae or small young that feed on food quite different from that of adults. Table 6.1 lists the changing diet of a snake with age, showing a series of food preferences possible only because the snake can make do with nothing between the various crops of small frogs. This property of ectothermy is used to even greater advantage by many fish and invertebrates where dispersing young feed on different food in different places at different times.

Switching food by ectotherms also offers a ready explanation of the puzzling steps in the sides of Elton's pyramids of num-

FIGURE 6.1
***Phrynosoma cornatum:* design for desert heat.**
The horned toad of Texas and thereabouts is a desert lizard able to sit out weeks without food in cool torpor, hidden in rocks or holes, and can answer the call of eventual food by working out into the sun to get back up to standard animal operating temperature.

bers (Chapter 2). The growing ectotherm changes trophic level by switching prey after a period of dormancy or after dispersing to a fresh habitat. When the ectotherm resumes feeding, the ecosystem around it has changed or the animal is in another ecosystem physically separated from the first, as is common in the oceans.

TABLE 6.1
Changes in Diet of Growing Garter Snakes (*Thamnophis sirtalis*)
The food of the snake changes as it grows. Snakes can endure long periods of time between supplies of food of different sizes because the energy cost of their maintenance in a state of torpor is low. (Data from Fitch, 1965.)

Length (cm snout to vent)	Prey	Mass of Prey as % Mass of Snake
20–29	Earthworms	3.8
30–39	Small frogs	5.6
40–49	Large frogs	94
≥50	Mice	50

Yet other possibilities of the low-energy system are small size and long shape. Resting metabolism increases at about the 0.75 power of the body mass, which means that energy at rest goes up very steeply as animals get smaller. This is not very serious for ectotherms because resting metabolism is always low, since at rest the animals are cold. But for endotherms to be small involves a heavy energy cost (Figure 6.2). Mice, shrews, and chickadees are the smallest endotherms, and yet they are huge by the standards of most ectotherms. Even most lizards and amphibians are

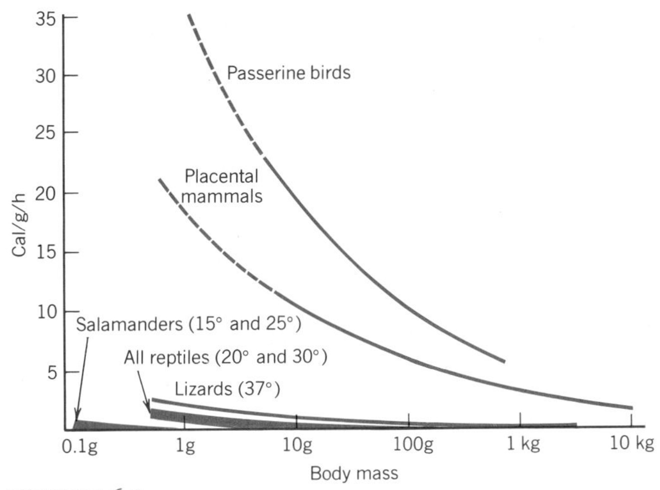

FIGURE 6.2

Resting metabolism as a function of body size and temperature.

Ectothermic animals can be small because their energy consumption for maintenance is low. A bird weighing 1 gram would use about 30 times the energy of a reptile of the same weight. Not surprisingly, 1-gram reptiles exist but 1-gram birds do not. (From Pough, 1983.)

much smaller than chickadees and may be only one tenth the mass. Ectothermy, therefore, offers all the advantages of small size (ability to hide, short growth time to reproduction, possibly larger food resources). It is one of the keys to the success of insects.

Ectothermy also allows the long, thin shape of a snake that would impose a ruinously expensive energy cost for an endotherm because of the large surface area over which heat would be lost. The peculiar hunting methods of snakes must be successful by any test of what "success" means because there has been a massive radiation of snakes since they first appeared in the Cretaceous period, and they have occupied all habitats except the very coldest (Figure 6.3). The serpentine shape must be intimately linked with these hunting methods and their success, yet this life form and lifestyle are forbidden to endotherms because of the energy cost. Even the somewhat elongated bodies of weasels apparently require twice the maintenance costs of a normal mammal of comparable size (Brown and Lasiewski, 1972).

Finally, ectothermy allows short bursts of violent activity (flight or fight) at minimum cost. It does this by using anaerobic metabolism to support the activity (Figure 6.4). Anaerobic metabolism is itself an expensive way of releasing energy, because the lactate molecules stored must be resynthesized to glycogen after the period of intense metabolism is over (Figure 6.5). But the cheaper way of releasing energy for a violent burst of activity (at least in large animals; insects have solved the problem) requires the high basic metabolism of an endotherm, a cost that is quite out of proportion to the small expense of synthesizing a little glycogen from lactate whenever an intense burst of energy is needed.

FIGURE 6.3
***Crotalus horridus:* thin through ectothermy.**
The long thin shape of snakes makes them superb hunters of the warm night. Even more than deadly venom, this thin shape of a rattle snake serves its hunting role, a shape possible for an ectotherm, but ruinously expensive in lost heat for an endotherm.

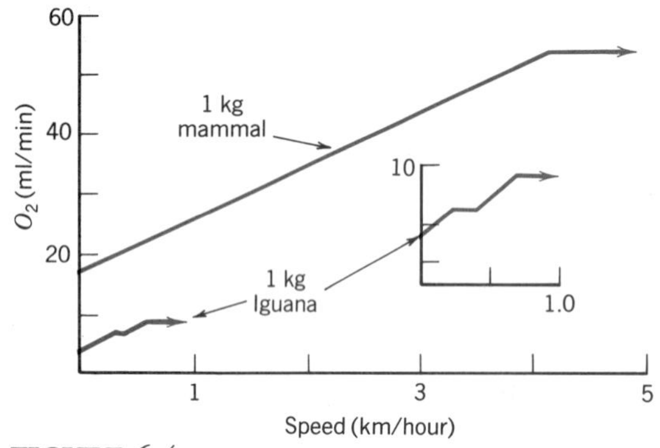

FIGURE 6.4
Oxygen consumption and speed in endotherms and ectotherms.
The data for a green iguana are from measurements on captive animals. The curve for a mammal of the same size is calculated from generalized mammalian data. Mammals attain five times the oxygen consumption of reptiles. (From Bennett and Ruben, 1979.)

One little noticed pattern among reptiles and amphibians is that they occupy the size classes between insects and birds or mammals, letting them extend into niches denied to endotherms by the energy cost and denied to insects by such factors as gas exchange or the excess musculature that would be required to operate huge exoskeletons on land. Some useful properties of ectothermy are summarized in Table 6.2.

BENEFITS AND COSTS OF ENDOTHERMY

Endotherms can do things that ectotherms cannot. Their burst of energy, for

FIGURE 6.5
***Varanus komodoenis:* low-energy carnivore.**
Ectotherms can deliver short bursts of very high activity, a pattern suited to large ambush predators in places with sufficient sun to warm their bodies for the spurt. Crocodiles are the classic example on the contemporary earth, but the Komodo dragon shows what was once a common life style.

instance, can last longer. An endotherm uses body chemistry that releases energy from ATP by continual oxidation, a property attested to by the fact that the oxygen consumption of a mammal under hard exercise may be five times that of a lizard being similarly driven. The lizard quickly builds an oxygen debt and has to reduce activity for a long while for lactate to be resynthesized to glycogen, but a mammal can keep going with only a modest and manageable oxygen debt. Hard-running time for a lizard is never more than a few minutes, but mammals can produce a marathon runner.

Endotherms can penetrate the shade and hunt by night, undertakings that are virtually denied to large ectotherms in colder habitats (Figure 6.6). The advantage of this to endotherms is more time to forage. The endothermic animal uses fuel rapidly to stay warm and keep going, but it has the compensation of having more time in which to collect its fuel. It can properly be called a "**high-energy system**" (Pough 1982).

TABLE 6.2
Special Properties of Ectothermy
Ectothermy represents a strategy for heating the body into an activity temperature range *suited to foraging as cheaply as possible. Extra advantages accrue because the animal easily relapses into torpor at other times. Table compiled from the analysis of Pough (1982).*

1. Resting metabolic rate low
2. Production efficiency high
3. Adverse periods easily passed in torpor
4. Maximum advantage taken of episodic gluts of food
5. Remaining motionless while cryptically colored; a cheap defense against predators
6. Short burst of rapid activity (flight or fight) possible by anaerobic means
7. Small size possible
8. Growth from small to large facilitated
9. Serpentine or elongated shape possible

FIGURE 6.6
Endothermic hunter of the night.
The high-energy system of endothermy is inefficient in its use of fuel energy but makes possible activity when ectotherms are sluggish.

Clearly, a high-energy system does not let an animal be an efficient producer of growth or young if "efficiency" is taken to mean the ratio of energy absorbed to energy put into growth and reproduction (**production efficiency**). Like racing cars, endothermic animals do not use fuel "efficiently" but use large quantities of energy for speed and other activity, and yet many animals gain fitness by this "inefficient" behavior.

A cost that must have made endothermy difficult to introduce by natural selection in the first place is the constraint it places on methods of reproduction. Young cannot grow by switching food and position in food chains as they become larger, as do ectotherm young. This is because endotherm young cannot drop out of the system into a dormant state while waiting for a supply of the next food item to become available. In practice this means that nearly all endotherms must be fed by their parents until they are fully grown and able to eat adult food. Mammals nurse and care for their young throughout the juvenile days, months, or even years. Altricial birds, those with helpless offspring like familiar songbirds, feed their young in nests until they are fully grown. Precocial game birds, like pheasants or domestic fowl, start with young big enough to hunt small adult food from the day of hatching, and even they are tended by parents through most of their growth.

Endotherms make their contribution to the steps in Elton's pyramids by the reproductive habit of raising the young to full adult size before it is released into the ecosystem. The principle of food size determines the size of the animal relative to its prey, and the method of rearing young ensures that no half-sized juvenile individual enters the ecosystem.

This reproductive necessity of endotherms makes a contribution to the debate over the possibility of dinosaurs having been warm-blooded. If any dinosaurs were true endotherms, then it is hard to see how they could have failed to have had systems of parental care for the young to grow to the large adult sizes. The discovery of what look like fossil nests with young dinosaurs in them is thus of interest for more than just emotional reasons (Horner, 1982).

SUMMARY: ENDOTHERMY AND ECTOTHERMY COMPARED

Ectothermy and endothermy are two quite different strategies for gaining fitness. Each has properties that the other lacks. Both systems allow regulation of body temperature to within narrow limits while active, and tetrapods with either system operate at near 38°C, a temperature apparently required by their common body chemistry. Endothermic tetrapods certainly keep close to this biochemically optimum temperature of 38°C, usually to within about a degree, but even lizards control temperature to the point of not allowing their bodies to fluctuate by more than 5°C when active.

Endotherms can forage more easily in the shade, the cold, and the dark; they can also indulge in protracted chases, either as pursuer or as fugitive. But ectotherms can feed in good times, pass the bad times in torpor, and spread through habitats and food webs as they grow by switching from food to food. They can also escape, or mount an ambush, as a torpid, semianimate, camouflaged object hidden in the background. The difference between the two strategies is based in energetics, so they are well described by Pough (1982) as low- and high-energy systems. They represent quite different possibilities for obtaining a living. When the small endothermic tetrapods, which would one day give rise to mammals and birds, first arose, the number of species possible on the earth was multiplied by all the possibilities of foraging at night and of the more relentless and persevering kinds of hunting.

NATURAL HISTORY OF ENDOTHERMY AND ECTOTHERMY

That the inefficiency of squandering energy as heat and movement can bring remarkable returns in fitness is shown nicely by the natural history of hummingbirds. A hummingbird is tiny, has a large surface area, and is an active forager so that it loses heat rapidly. It must meet its high energy costs from continued high-energy intakes, requiring in fact a diet of sugar solution laced with amino acids and other high-nutrition ingredients. And yet hummingbirds are both abundant and highly diverse throughout the New World tropics. This appears to be because the hummingbird lifestyle can be used by plants to give them fitness by spreading pollen. Plants with nectaries well supplied with nectar to meet the high energy demands of hummingbirds are at a selective advantage over plants that do not.

A parallel process in Africa, starting with different bird stocks, has led to another array of related species of high-energy systems called sunbirds. They are somewhat larger than hummingbirds and drink their nectar while perching as much as possible, but they are in all other respects very much like hummingbirds (Figure 6.7; Wolf and Hainsworth, 1982). Sunbirds and hummingbirds succeed because other living things can gain fitness by supplying them with energy for their profligate habits.

FIGURE 6.7

Nectar feeders: the ultimate high-energy systems.

Both the hummingbirds of the new world and the sunbirds of the old world support high activity in tiny bodies on copious supplies of high-energy food (nectar). Plants support this profligate energy addiction of the birds by using them as "flying penises" for the plants' own reproduction (*see* Chapter 15).

A life of fishing in a tropical river is apparently well suited to a low-energy approach like that of crocodilians. The riverbank is a handy place to raise their temperature under the sun, the ambush and dash of a reptilian carnivore appears to be a good fishing technique, and the river provides food in many sizes, most of it small, of which the changing appetite of a growing ectotherm can take good advantage. Possibly the one necessary adaptation for a tetrapod ectotherm to thrive as a riverine fisher is a defense against endotherm predators hunting the banks for animals recovering from torpor in the sun (Figure 6.8). Large crocodiles clearly have the defenses needed to support their low-energy foraging system. They have remained unchanged as the perfection of fishers of tropical rivers for a hundred million years.

In theory, the crocodilian, ectotherm, low-energy system must often have advantages for foraging in shallows close to ocean shores also, but the strategy does not seem common outside tropical rivers. Reflections on the Galapagos marine iguana may suggest why this is. Marine iguanas (*Amblyrhynchus cristatus*) crop seaweeds just below the high-tide line, penetrating water of cold upwellings to do so. The animals are black and well adapted

FIGURE 6.8
***Crocodylus niloticus:* Low energy predator of water with a bank for warming.**
For a large predator in tropical rivers ectothermy has not been bettered: safe on the bank behind armor and weapons while heating up, and fast in the water.

to rapid absorption of the equatorial sun when lying on shore, but they are known only on the Galapagos Islands. No other herbivorous ectothermic tetrapod forages at sea anywhere in the world. A working hypothesis to explain the success of the strategy in the Galapagos and nowhere else is that there are no mammalian predators on the Galapagos Islands, and no avian predators able to attack adults. If there were, an iguana still torpid from its cold ocean swim, or huddled with its fellows on a cold, cloudless night, would be easy prey. But most of the warm coasts of the earth are in reach of endothermic predators. Thus, as admirably efficient, and even elegant, as the lifestyle of the marine iguana is, there is no fitness to be gained by using its energetic logic on most coastal environments of the contemporary earth

OXYGEN TRANSPORT AND THE SUCCESS OF INSECTS

The most numerous practitioners of ectothermy on land are insects and arachnids (spiders and allies). These reap all the advantages of small size, episodes of torpor, young that can switch food, and the rest that are always available with ecto-

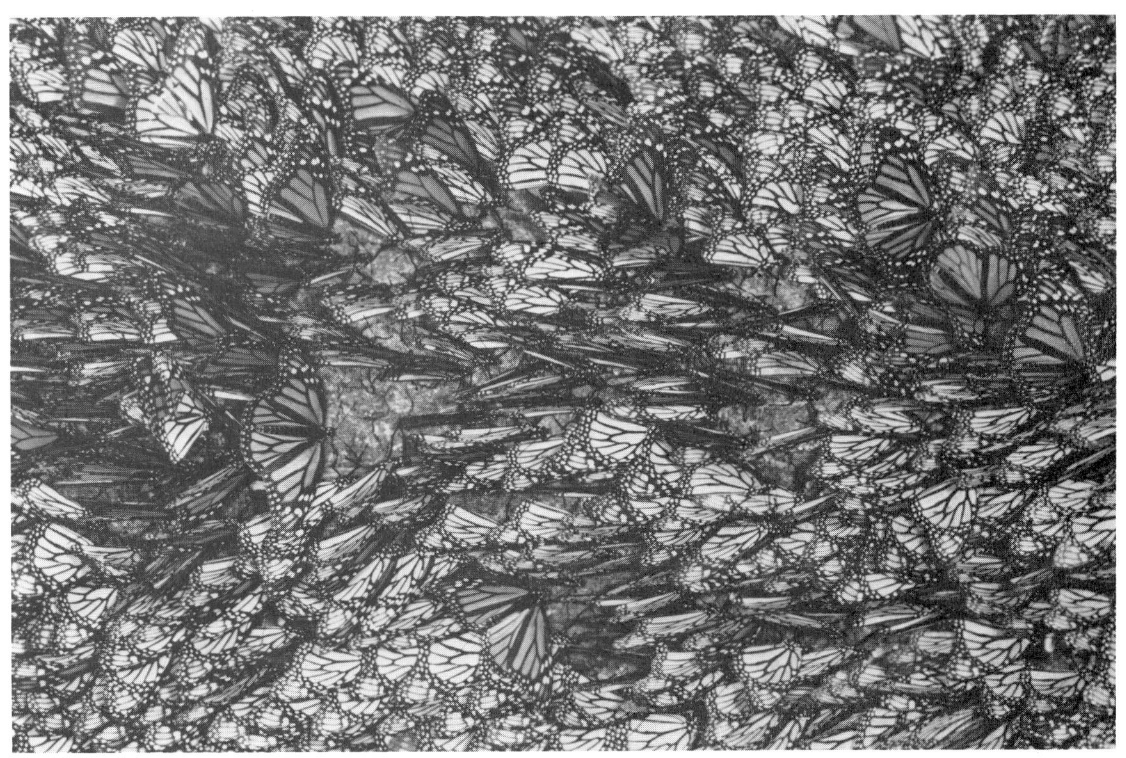

FIGURE 6.9

***Danaus plexippus:* long distance travel by ectotherms.**

Monarch butterflies overwintering in Mexico. They have flown south from all over the American Northwest, having accomplished journeys comparable to those of migrating birds. Like the sustained flight of bees, this is possible because of the air supplied directly to their muscles in the system of tracheae.

thermy. Yet they do so without paying the cost of limiting the duration of intense activity. The secret of this success appears to be their unique system of gas exchange with the air.

Insects deliver oxygen to the muscles directly through a system of branching tubes and tubelets called **tracheae** and **tracheoles**. Insects do not have to transport oxygen in solution at all, nor do they have to return carbon dioxide in solution to an evacuation point like the tetrapod lung. Air flows, partly passively and partly through pumping motions of body movements, through the short tubes directly to where it is needed. In a sense, each tissue does its own breathing. This makes possible tiny ectotherms that can sustain high activity, essentially until their energy reserves run out. Butterflies and dragonflies can fly nonstop for hundreds of kilometers, and heavy bees hum and hover at speed all day long under the sun (Figures 6.9 and 6.10). Perhaps more remarkably still, moths fly on cold nights, and the hours of darkness are fit for the activity of insects, though it may be a time of torpor for most ectothermic tetrapods.

FIGURE 6.10
***Apis mellifera:* hardworker with oxygen pump.** A bee flies on when an ectothermic vertebrate collapses because air pumped directly to muscles in systems of tiny tubes allows muscle metabolism to continue without the long periods of rest needed by vertebrate ectotherms.

THE ENERGETICS OF TRANSPORT

Apart from maintenance, the major cost to be met by the energy budget of an animal is movement. The animal must move to where the food is, to shelter, or for purposes of flight or fight. Natural selection arguments require that movement be efficient, and yet many kinds of locomotion are in use. There are different patterns of flight, different gaits on land, running or hopping, two legs, four legs, or six.

In Figure 6.11 the costs of swimming, running, and flying are compared for animals of different sizes and for various machines. The animal data are derived from measurements of metabolic rates of experimental animals in motion as follows:

Pi = metabolic rate = power output

V = velocity

W = weight = mass × gravity = the force to be overcome

Then

$$W \times V = \text{Power}$$

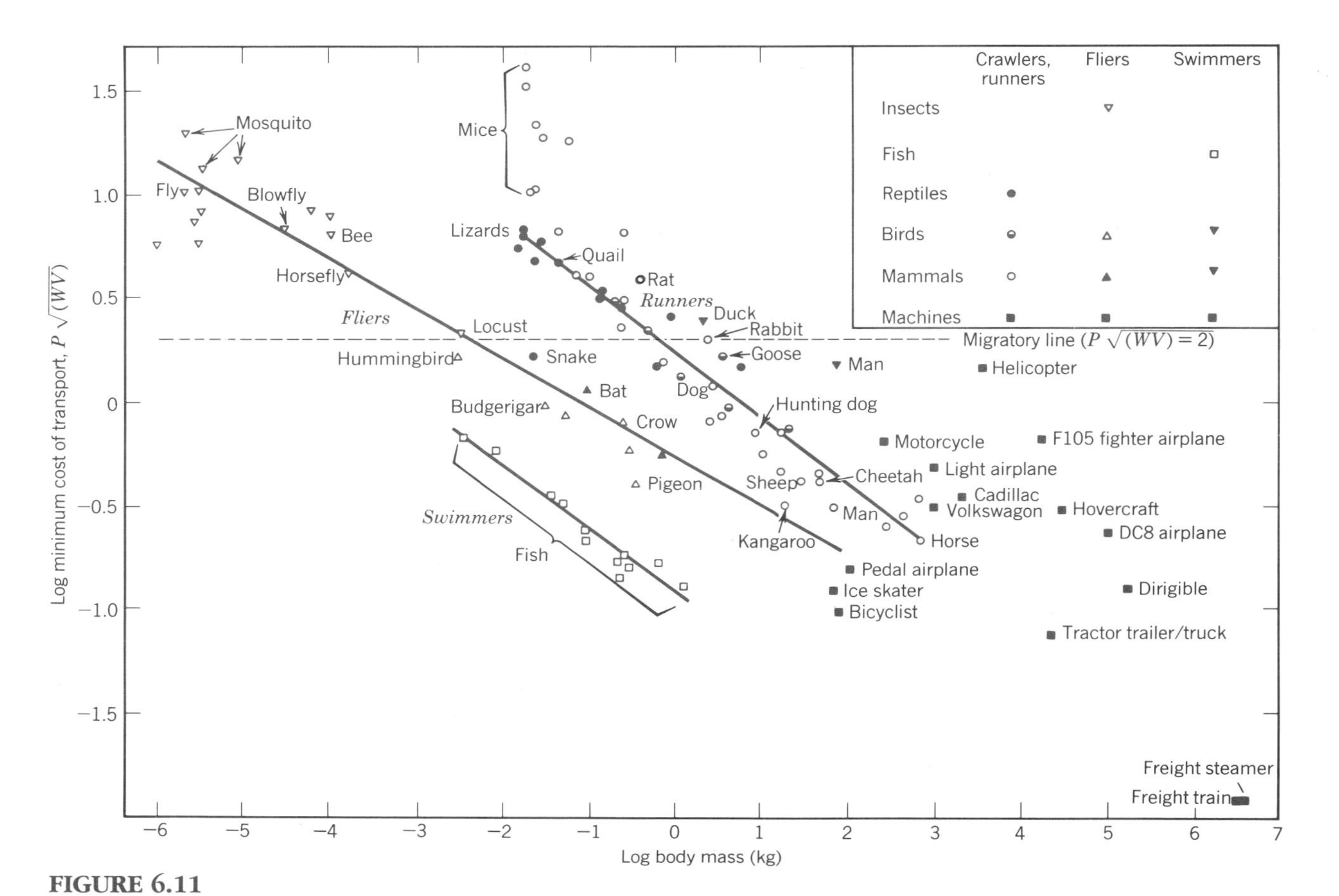

FIGURE 6.11

Energy costs of swimming, flying, and running.

The data suggest that mass to be moved is the decisive constraint once the organism is committed to a particular gait. (From Tucker, 1975.)

and

$$\text{Cost of transport} = Pi/WV$$

By measuring all movement only in terms of metabolic rate, weight, and velocity, this function (the transport cost) has the advantage of allowing the transport costs of animals of different sizes to be compared. In Figure 6.11 these transport costs are plotted against body masses on logarithmic scales. Swimming and flying use less energy than running, but otherwise the mass to be moved seems the determinant function (Tucker, 1975; Schmidt-Nielsen, 1972).

These data reveal some apparent puzzles. Why are flying and swimming less costly than running? A flying animal has to support its weight in a diffuse fluid medium by expending energy, and a swimmer has to force its way through a dense viscid medium. By contrast, a runner encounters negligible friction from the air and an amount so small from contact with the ground that it is within the margin of error of our measurements of metabolic rate. Why, then, is running so costly in energy? The answer must lie in the energy spent in extending muscles and levers, and in absorbing changes in angular momentum as heavy limbs continually change their direction of motion. Birds must pay these costs too, but their greater speed spreads the cost (Figure 6.12). Greater speed in

FIGURE 6.12
***Chen hyperborea:* energy from the front-runner.**
The vee formations of snowgeese may result as following individuals ride the waves of turbulence generated by those in front.

proportion to muscular movement is also the reason why cycling is more efficient than running. For a swimmer designed on proper hydrodynamic lines, these costs of muscle flexure and angular momentum are much reduced, and a correspondingly greater part of the cost goes to overcoming friction. Swimmers that work by thrashing limbs about, like ducks and people, face both kinds of cost and are inefficient.

Whether to Run or Leap: The Choice of History

Common sense suggests different levels of efficiency within any class of locomotion depending on such things as numbers of limbs and gait, yet these turn out to be small. For instance, there has been speculation in the literature that the hominid line met special energetic costs when the change from four-legged locomotion to bipedalism was first made. This has been tested by Taylor and Rowntree (1973), who trained chimpanzees to work treadmills both with two legs and with four. No difference could be detected in the metabolic rate of the chimpanzees, whichever way they worked the treadmill.

Of greater interest is the evidence of kangaroos. In Australia kangaroos are the only large grazing mammals, taking the place occupied by ungulate grazers on all other continents. Kangaroos hop. It is not certain if hopping gives kangaroos an energetic advantage or disadvantage. In Figure 6.11 they are shown as being slightly more efficient than quadruped ungulates, but the difference is small when the difficulty of measuring metabolic rate in a running kangaroo is considered. But why should the principal grazing endothermic tetrapod consumer on one continent hop while its equivalents on all other continents run on all fours? Figure 6.13 The answer seems to have nothing to do with energetics.

FIGURE 6.13
***Macropus giganteus:* fast by bound and ricochet.**
Hopping by kangaroos is as efficient and fast as running by deer. Which method is used is contingent on evolutionary history which leaves structures suited to one method or the other.

Marshall (1974) traces the evolution of the kangaroo foot and compares it with the feet of *artiodactyles* (two-toed ungulates like cows) and *perissodactyles* (one-toed ungulates like horses). From the original five-toed plan, cows run on toes III and IV, whereas horses run on toe III alone (Figure 6.14). But the ancestors of kangaroos had toe III fused to toe II, so that it was off center and not available to serve as an ungulate hoof. This left toe IV to be the principal prop of kangaroo motion when its ancestors took to a grazing life on the plains. The attachment of toe IV to the complex of bones in the ankle is profoundly different from the attachment of the other toes and results in a different positioning of stress and balance. Marshall argues that this anchoring of toe IV allowed the ancestral animal to gather speed by a process of spring and ricochet but not by the weight transfers required of ordinary running. Kanga-

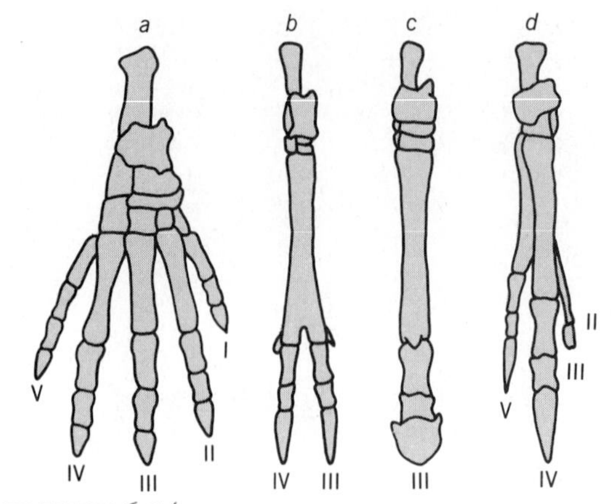

FIGURE 6.14

Ungulate and kangaroo feet.

Artiodactyles run on toes III and IV and perissodactyles on toe III alone. But kangaroos run on toe IV alone, since toe III was fused to toe II in the ancestral stock. (*a*) Hypothetical placental ancestor; (*b*) artiodactyle; (*c*) perissodactyle; (*d*) kangaroo. (From Marshall, 1974.)

roos hop, therefore, not because hopping is more or less efficient than running but because this was the best solution to the speed problem allowed by their anatomy. Grazing herbivores need speed if they are to achieve fitness in a world stocked with predators. Making a fast herbivore from the Australian marsupial line required that the animal hop.

Wondering why a kangaroo hops is like wondering why a spider has eight legs but an insect only six. These arrangements resulted from selection in distant ancestral lines to adapt animals to the niches then open to them. Subsequent adapting of these arrangements to other niches was always done starting with the engineering plan at hand. Natural selection cannot go back and start over.

What is interesting to an ecologist is that the way of life dictates the tasks that the structure must perform. A tasty grazer had best run fast: it matters not whether it hops or sprints. Intense activity requires that the body be held near 38°C; it matters not whether this regulation is achieved by ectothermy or endothermy. Really prolonged activity requires aerobic metabolism, whether achieved through endothermy and an efficient blood-vascular system or by the direct piping of air.

Preview to Chapter 7

Cumulonimbus: Heat engine to drive a storm.

THE massive flux of solar energy falls unequally on different parts of the earth. The equator receives intense radiation in 12-hour doses whereas polar regions have much less intense radiation for six months at a time. The equator is hot; the poles are cold. The earth is thus forced to become a massive heat engine, doing work to disperse heat from equator to poles in winds and ocean currents. Winds and currents are impeded in their circulating paths by the dynamics of the system, with winds descending at about 30°N and S latitude to form a line of deserts, twisting by the changes of angular momentum of Coriolis effects to yield trade winds and westerlies. The result is the familiar patchwork of earthly climates. This is the patchwork to which all life has had to adapt. But the patches are not constant, particularly on the glacial–interglacial time scales spanned by the existence of species, or from changes in concentrations of greenhouse gases. And most climates suffer havoc in ecologically instantaneous time as vast concentrations of energy launch storms and hurricanes against ecosystems.

Chapter 7

Climate

Living parts of ecosystems subvert 2% or less of incoming solar energy, as the ecological efficiency of vegetation is kept low by the carbon, water, temperature, or nutrient limits to photosynthesis. But all the incoming energy drives climate, both the 98% or more of energy the vegetation cannot get and the tiny 2% ration of the plants, which is destined to be degraded to heat in ecosystems eventually.

WINDS ON A SPINNING EARTH

The earth is a spinning sphere, the equator of which is just a few degrees away from being parallel with the plane of the earth's orbit around the sun (Figure 7.1). Because the earth is round, incident radiation at high latitudes (i.e., nearer the poles) is spread over a large area, rather as light is spread over a large area on an inclined leaf of grass (see Chapter 3). This means that the intensity of irradiance per unit area is always less at high latitudes than it is at the equator. High latitudes, therefore, get less heat than low latitudes, this being the primary inequality in energy supply that imposes a patchwork of climates on the earth.

The second inequality is in the distribution of day and night. All parts of the earth have exactly the same amount of night and day in the course of a year—six months of each. But at the equator light comes in alternations of a 12-hour day and a 12-hour night whereas at the poles a six-month day is followed by a six-month night; low latitudes therefore have more constant heat. That low latitudes have more heat, and more constant heat, than high latitudes means that low latitudes always are warmer than the polar regions. This results in a mass transfer of heat from equator to poles in fluid flows, both of air and water. A significant part of the heat transfer is by ocean currents, because the specific heat of water is so high. But movements in the fluid air are more rapid, giving rise to the spectacular properties of climate.

Most of the energy of sunlight penetrates the atmosphere and strikes the surface of the solid or liquid earth, where the

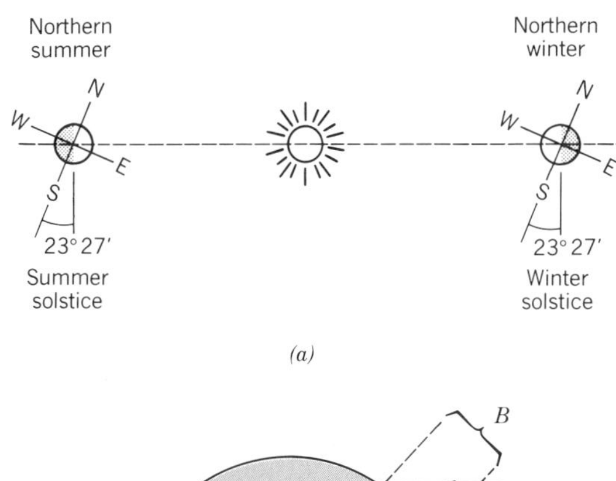

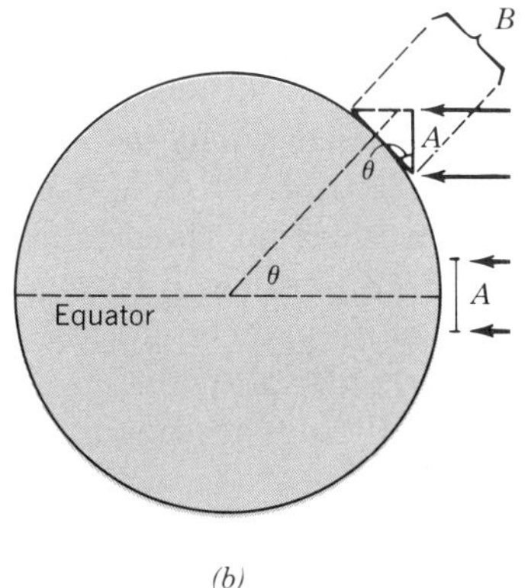

FIGURE 7.1
(*a*) Relative positions of earth and sun at summer and winter solstice. The sun is shown as effectively a point source. The plane of the earth's rotation is 23°27′ from being parallel with the plane of the orbit. (*b*) The effect of latitude on insolation received per unit area at the equinox (when the sun is over the equator). The sun's rays are assumed to be parallel. At high latitudes the same radiation received by area A is spread over the larger area B such that $B = A/\cos\theta$ where θ is the latitude in degrees.

high-energy wavelengths of light are absorbed and reemitted as radiant heat. The primary heating of the atmosphere, therefore, is from below, from the ground and ocean surfaces. The warmed air rises; as it rises it expands; as it expands it cools.

Cooling of rising air as it expands is the reason why mountaintops are cold. The cooling results because packets of expanding air must do work to push aside other packets of air that are pushing back. Energy to do this work can come only from the air itself, and it loses temperature accordingly. Since no external source of energy is involved, the cooling is called **adiabatic** (from the Greek meaning "impassable"). The cooling is a simple function of altitude, and the **adiabatic lapse rate** in dry air is approximately 10°C per 1000 meters of elevation. In moist air the lapse rate is less because heat is gained as water condenses, and the adiabatic lapse rate in moist air is about 6°C per kilometer.

The major wind systems of the earth result from the fact that large masses of air around the equator are forced to rise by bottom heating, causing air to rush in from high latitudes to fill the pending void. This of course lowers atmospheric pressure in high latitudes, which is balanced by descending air. Winds represent the working of a heat-distributing engine. Heat is applied at the equator, the air moves up and then poleward, after which it descends and returns equatorward along the earth's surface. But this simple system is modified in two ways: by some properties of scale and by Coriolis effects.

On the real earth the poleward-moving masses of upper air cannot penetrate to more than about 30°N and S latitude. This seems to be due to statistical properties of moving air masses, or what the wise mathematical ecologist Robert MacArthur (1972) called "no good reason." The actual pattern that results is given in Figure 7.2. The atmosphere is thicker at the equator than at the poles and a second circulation cell is inserted between 30° and 60°N and S.

The second modification comes because the earth revolves. An object at the equator spins at about 24,000 miles a day. Farther north or south the object would be traveling more slowly, as the distance to be traveled for one revolution of the earth is less. If you straddled a pole for 24 hours, each

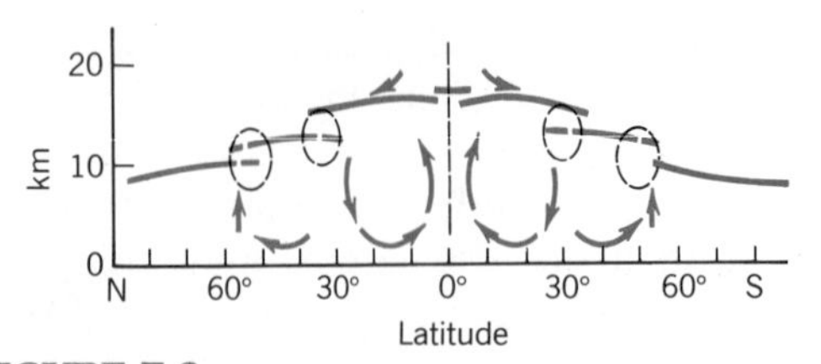

FIGURE 7.2
Profile of the atmosphere.
The dashed circles are tracks of the jet streams. (From Flohn, 1969.)

foot would have moved perhaps 1 meter in that time. A more interesting latitude for climate dynamics might be 45°N (Minneapolis, Montreal, Bordeaux, Milan) where the surface velocity is about 17,000 miles per day. A parcel of air moving due north from the equator starts with a powerful west to east momentum from its initial 24,000 mile-per-day kick, and as it travels into regions where the surface moves more slowly it will veer in the direction of its superior momentum, to the right. This diversion of winds through their own angular momentum is called the **Coriolis effect** or Coriolis force, named after seventeenth-century French mathematician G. G. Coriolis.

A parcel of air traveling in the opposite direction, southward, in the northern hemisphere is being thrust into air masses that are moving west to east faster than itself, with the result that it will drag its feet, conserving its slower angular momentum and be deflected west. So southward-traveling air in the northern hemisphere will be deflected to the right, just as is northward-traveling air. In the southern hemisphere the logic is the same but the directions are reversed. **Coriolis effect** *is the deflection of moving air or water to the right in the northern hemisphere and to the left in the southern hemisphere.*

At tropical latitudes, the surface air that rushes to fill the equatorial void from the north is deflected to the right and becomes the **northeast trade wind**. Somewhere near the equator it must meet similar air coming from the south that was deflected left—the **southeast trade wind**. But less land is present to slow air with assorted barriers in the southern hemisphere, where most of the time the winds pass over oceans. Accordingly the trades do not, on the average, meet at the geographical equator as you would expect, but somewhat to the north. The convergence, called the **intertropical convergence zone** (ITCZ), is displaced north of the equator, this being particularly visible over the Pacific Ocean. Heated air rises most rapidly along the ITCZ, causing high cumulus clouds and the heavy rains typical of Caribbean islands and Hawaii.

Events at latitudes higher than 30° depend on the unexpected sinking of the upper air that closes the circle started at the equator (Figure 7.2). Impelled by the pressure of sinking air from above, winds from 30 to 40° latitude set off poleward, but they are misdirected by strong Coriolis effects; again to the right in the northern hemisphere and to the left in the southern hemisphere. The result is westerly winds in both hemispheres at about 40° latitude (Figure 7.3). In the southern hemisphere, latitude 40° has virtually no land to slow the winds, the result being the **roaring forties**, unceasing winds of high velocity that track the fortieth parallel, round and round the earth.

Upper atmosphere winds along both fortieth parallels are also westerlies, the **jet streams** (Figure 7.2). The initial poleward movement in the upper air comes about because the atmosphere is piled up high at the equator but is thinner farther north. Atmospheric pressure is higher at low latitudes, therefore, and upper air flows poleward to where pressure is lower. Again, Coriolis effects take charge of this motion,

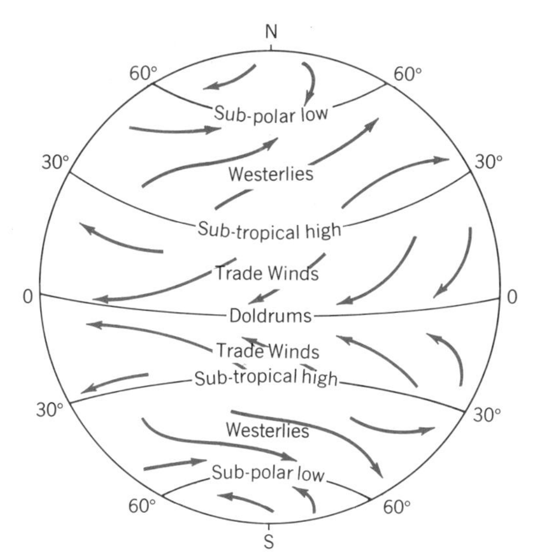

FIGURE 7.3
Winds at sea level.
Compare with Figure 7.2. The doldrums at the Intertropical Convergence Zone (ITCZ) are shown at the geographical equator as they would be on an ideal earth. The distribution of land and sea actually displaces the circulation system a few degrees to the north. (From Lamb, 1972.)

diverting it eastward in both hemispheres and leading to westerly jet streams. Placing an airliner in the jet stream when crossing the Atlantic from America to Europe cuts an hour or more off the flight time.

At the poles are the cold regions where the lack of sufficient irradiance has its own direct consequence for climate. One pole is covered with a floating ice pack and the other by glacial ice. Both sheets of ice act as reflecting mirrors, increasing the **albedo** (reflectivity) and exacerbating the already low heating of the air. Dense cold air sits over the poles, therefore, flexing with the seasons and flowing out as east winds into subpolar regions, the cold east winds that northern Europe knows (Figure 7.3).

THE CIRCULATION OF THE OCEANS

The world oceans are stirred by the winds. Of prime importance is the drive provided by the trade winds that thrust great masses of water before them, piling it up against continental dikes downwind. What this implies is neatly illustrated by the circulation of the North Atlantic, which begins as water is piled up in the shallow Caribbean Sea, where it is heated by the tropical sun. So effective is this piling up of water by the northeast trades that mean sea level is a meter or two higher on the Atlantic side of the Isthmus of Panama than on the Pacific side.

The high water level of the Caribbean Sea makes an ecological hazard of plans to dig a sea-level canal through the Isthmus of Panama. Water would flow through the canal from the Atlantic to the Pacific Ocean, inevitably dragging with it animals and plants from one ocean to invade the quite different communities of the other. The present Panama Canal avoids this hazard because ships are lifted in a series of locks over a low mountain divide. Furthermore, part of the isthmus is crossed in a freshwater lake, thus making more certain the barrier between the oceans (Figure 7.4). Ecologists shudder at the thought of a big ditch joining unique communities that have been separated at least since the early Pliocene, 4 million years ago.

The water piled up in the Caribbean escapes northward along the American coast as the **Gulf Stream**. But moving water is turned by Coriolis effects, just like moving air. The Gulf Stream veers right toward northern Europe and provides a warm-water bath for the island of Britain, which accordingly has a much warmer climate than places in Canada well to the south of it. The Gulf Stream goes on veering right

FIGURE 7.4
Panama Canal: barrier between the seas.
Because the present Panama Canal raises ships over the continental divide in a series of locks, and because much of the crossing is through a freshwater lake, life of the Atlantic and Pacific Oceans is still kept separate. A sea level canal, if one were to be built, would unite the biota of the two oceans with unpredictable effects (Rubinoff, 1968).

to complete a gyre until the water is taken once more by the trades (Figure 7.5).

In the southern hemisphere, gyres in the ocean basins rotate in the opposite direction, counterclockwise. In these oceans, where the land barriers are smaller, the correlation of currents with winds is shown most clearly by the way ocean currents follow the roaring forties in a perpetual circling of the globe.

One of the consequences for land climates of these circling currents is, of course, the warming of places like Britain. But even more interesting is what happens when currents turn back toward the equator along the lee of a continent. Consider the circulation of the North Pacific (Figure 7.5). The warm current flowing north and veering right is the **Japan Current**, which bathes southern Alaska and British Columbia with water that is comparatively warm, which may even be warmer than the adjacent land during much of the year. These coasts accordingly have fog caused in just the same way as the steam from a warm bath. But then the ocean currents veer right again and down the coast of California, where the Coriolis effect turns them away from the coast and out to sea. What then happens is that water must be drawn

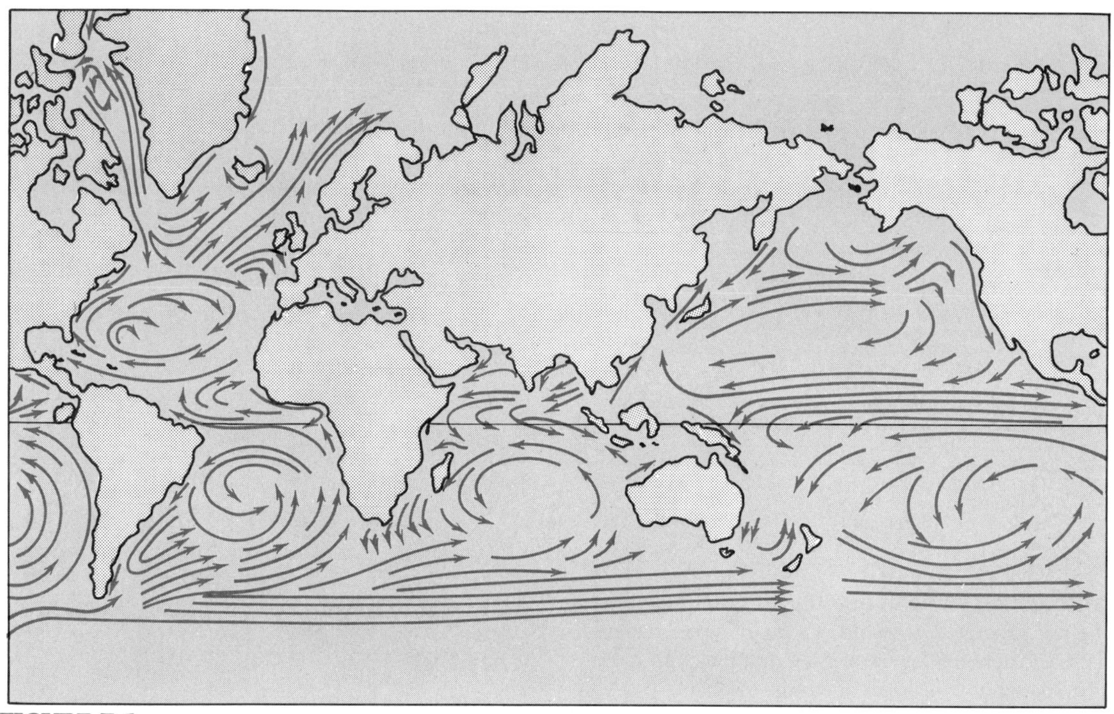

FIGURE 7.5
Surface ocean currents.
Winds and the Coriolis force move surface water in gyres around the ocean basins, clockwise in the northern hemisphere and counterclockwise in the southern hemisphere. Whether a current is approaching or departing a coast is decisive for the local climate. (From MacArthur and Connell, 1966.)

up from below to take the place of that dragged away. This is cold water taken from the dark depths of the ocean. Cold bottom water surfacing in this way is called an **upwelling**. It promotes fisheries because the water is nutrient-rich. But upwellings make coastal regions dry, because the upwelled sea is usually colder than the land so that air cannot transfer moisture from the sea.

WHY AND WHERE IT RAINS

Rain falls when moist air is cooled. This will happen when high irradiance of the sea along the ITCZ raises moist air to great heights, resulting first in the condensation that causes cumulus clouds, then in rain. But air is also raised, and causes rain, when winds are deflected by mountain ranges. Figure 7.6 shows the progress of events. Air arriving from a sea journey rises up a mountain, cooling as it goes with the adiabatic lapse rate (6–10°C/km, depending on water content) until water condenses. A layer of cloud or rain forms high against the mountainside. The air continues to rise, but now cools only at the moist air lapse rate of 6°C/km, crosses the mountains, and begins to descend. On the descent, air warms (because it is being compressed) at 10°C/km. This warming, descending air will take up moisture

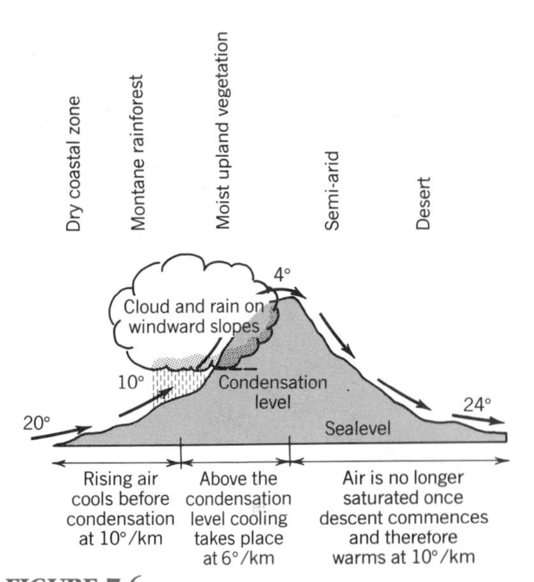

FIGURE 7.6
Precipitation along mountain ranges.
The rain shadow downwind of a mountain results because the descending air, already dry, picks up moisture. (Based on a diagram of Flohn, 1969.)

FIGURE 7.7
Under descending air: Africa 30° north.
Great deserts lie at thirty degrees latitude where the mechanics of atmospheric circulation cause air to descend.

rather than drop it, giving rise to a **rain shadow** on the lee side of a mountain.

Descending air should always absorb moisture rather than drop it. Air at 30° latitude, both north and south of the equator, is sinking, "for no good reason." This means that land along and above the thirtieth parallels should be without rain (Figure 7.2). Major deserts like the Sahara, the Grand Gobi and the American Southwest lie in this belt (Figure 7.7). Arizona is not only close to this latitude but is in the rain shadow of the Sierras.

The necessary ingredient for precipitation is rising air, but air can be prevented from rising over large geographic regions by a condition known as an **inversion**. An inversion is warm air floating over cold, a relationship that is physically stable (we say "inverted" because the usual pattern is cold air over warm, since the atmosphere is heated from below). The commonest form of inversion results when winds blow across cold ocean water, because then the moving air is cooled from the bottom instead of being heated. The cold winds blow on, bottom-heavy, with little turbulence. They are actually comparatively dry, even though crossing the ocean, because of the low temperature of the surface wind. Land under such an inversion can be desert.

The Galapagos Islands are desert because they are under an inversion, even though they are at the equator. The hot air rising from heated Galapagos rocks causes no more than a thin layer of stratus lying at the inversion. If the Galapagos Islands were only a few degrees farther north they would get rain from the ITCZ, but they are instead in the path of stable, inverted air moving toward that convergence and so remain deserts. The most complete deserts known are on western coasts of continents, as in northern Chile, where cold upwellings along the shore force inversions on the local winds, after which the cold, bottom-heavy air crosses hot land from which it actually tends to remove water.

Much of the vegetated landmass of the earth consists, in fact, of continents lying north of 30°N latitude and in the path of the prevailing westerly winds. A mountain range, such as that which lines the west coast in America, puts the continental mass to the east in a rain shadow. But the American interior gets rain all the same because it is under air masses that twist to the right under the direction of Coriolis force, bringing in moist air from warm seas to the southeast. Much rain for the U.S. Midwest comes from the Gulf of Mexico, from winds corkscrewing to the right across the Great Plains, and so eastward across New England. China is watered in a similar way, lying, like eastern North America, north of the latitude where air descends for "no good reason" (Figure 7.2).

VEGETATION MAPS CLIMATE

The foregoing is a simplified account of the circulation of the atmosphere and the causes of rain (see Lamb, 1972, for a full review), but even this elementary analysis shows that the earthly climates occur in blocks. The climatic domains tend to be reflected in the vegetation, as plants are engineered by natural selection to balance heat budgets and manage water in appropriate ways (see Chapter 5). Deserts, with their characteristic radiator and convector plant shapes, lie in the southwestern areas of continents and behind mountain ranges; lands along the ITCZ are wet, warm, and covered with rain forest; the northern edge of the boreal forest is the limit of cold, dense, polar air; and so on. Because the earth spins and is heated unequally, its pattern of precipitation is as given in Figure 7.8. This is so fundamental a pattern to life on earth that the first climate map and the standard Köppen classification of climate were based on vegetation maps (Figure 7.9). Cartographers drew maps of plants and called the result "climate."

VEGETATION RESISTS

Solar energy directed at a spinning sphere in orbit around the sun thus sets the pattern of climate. Life, with its paltry ration of 2% or less of incident solar power, must adapt to local necessity set by this overwhelming flux of energy. And yet forest trees, in the aggregate, turn out to have properties that stabilize whole landscapes against the vagaries of weather. This is in part a consequence of their use of water for cooling.

The water balance of whole forests has now been measured many times, one notable example being that of a forest occupying a whole drainage, or **watershed**, in the eastern United States (see the Hubbard Brook study, Chapter 23). All the water entering the Hubbard Brook watershed as rain was measured, together with all the water leaving in drainage streams. The difference was water evaporated. But it could be shown that virtually all evaporation was in fact directed by the trees in water-cooling activities of their transpiration streams. The trees, therefore, "controlled" evaporation. Since the latent heat of evaporation is known, a simple calculation yields the energy used to drive the transpiration streams—approximately 40% of incident solar energy.

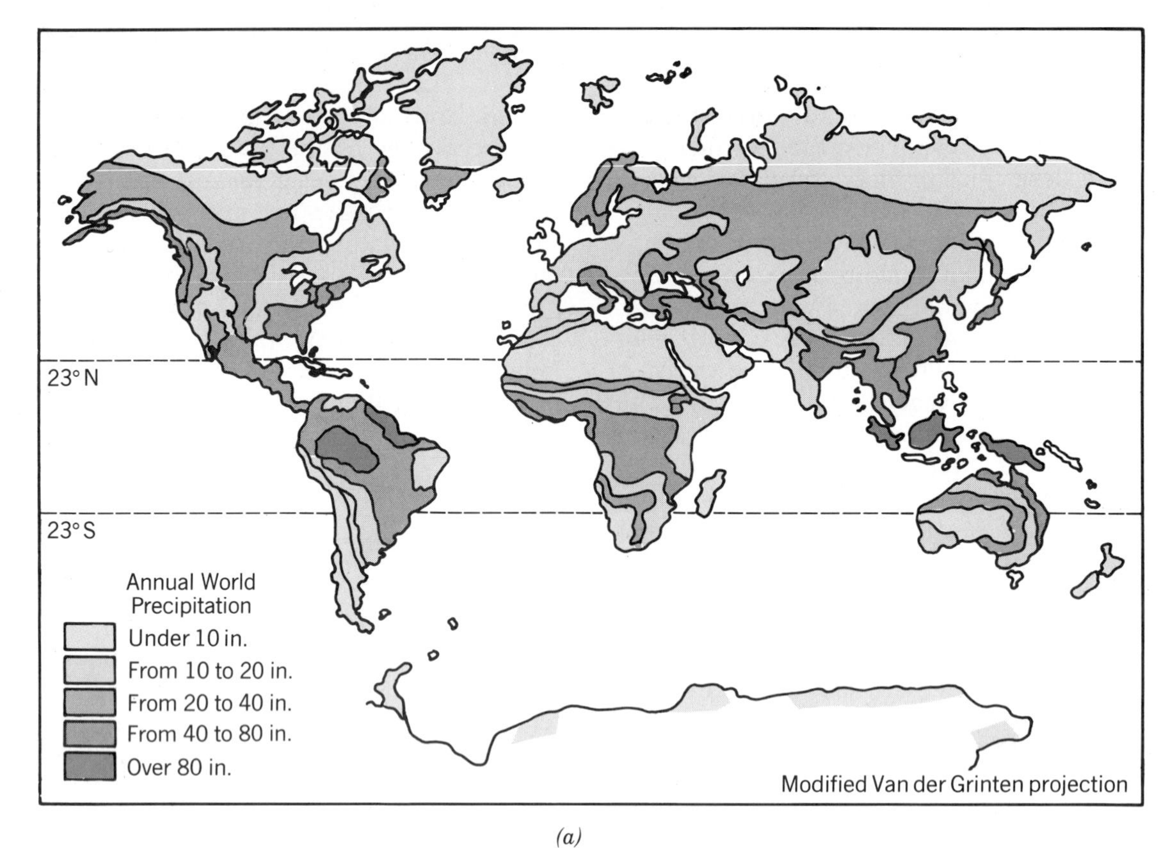

(a)

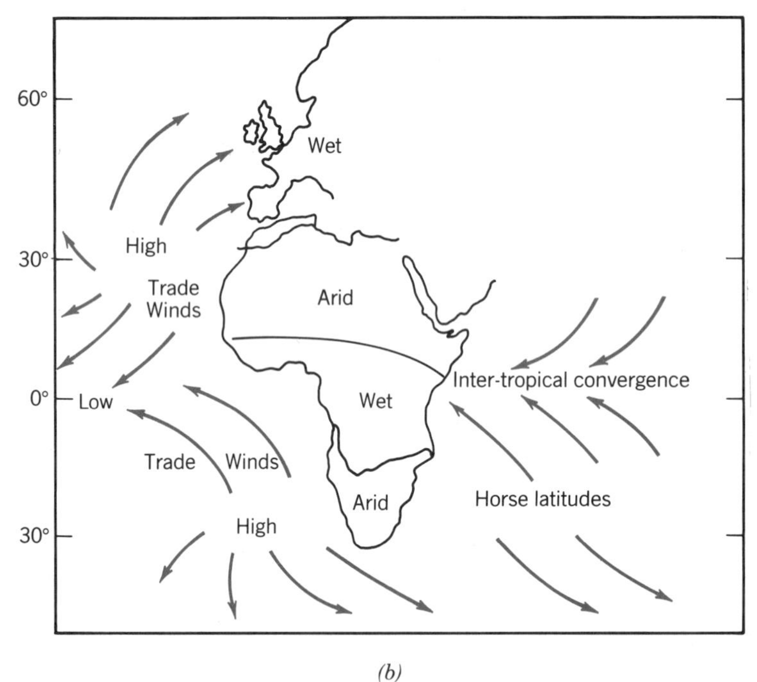

(b)

FIGURE 7.8
World precipitation.
(*a*) World map. The wettest and driest areas can easily be related to mountain ranges, the map of surface winds, and the map of ocean currents. (From MacArthur and Connell, 1966.) (*b*) Detail of causes of precipitation over Africa. (From Cox and Moore, 1980.)

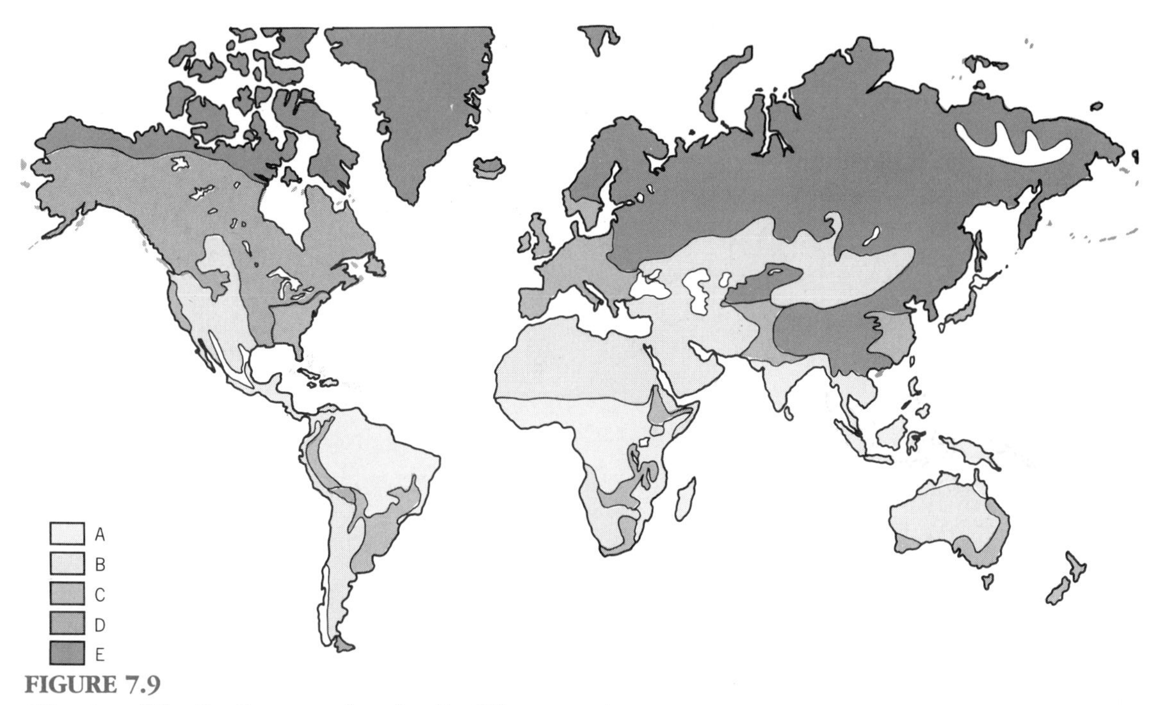

FIGURE 7.9
Climates of the Earth mapped under the Köppen system.
Köppen worked on the hypothesis that the boundaries between different kinds of vegetation were set by climate. Working in the late nineteenth century, when few actual climatic records were available, he mapped the plants and called the result climate. His climate map was actually a map of the great plant formations, what we should now call "biomes." That a useful climate map was produced in this way demonstrated that plant formation boundaries were in fact set by climate (see Chapter 18).

The deep roots of trees, the soil collecting under their shade, and the barrier mass of the trees themselves set between land and weather add physical stability to a landscape. But the trees of the forest work together as a water pump of massive power. The forest is run not just with the 1 to 2% of solar energy that the trees win by photosynthesis, but also by the 40% of energy used in the transpiration water pumps. Some part of the resistance of all vegetated landscapes to disruption by the weather must likewise be powered by solar energy used to evaporate water as well as by the much smaller flux used to synthesize carbohydrate fuel.

HAVOC

Water piled into the shallow basins of the Caribbean Sea by the trade winds can be heated top to bottom by the tropical sun. The warm sea heats and wets the air above. Let lateral movement of air and water be minimal for a critical number of days, then columns of air can rise rapidly over the warm sea. Surface winds are sucked in to feed the rising column. But the rising air is saturated with water. In rising, the air expands; in expanding, it cools; in cooling, the water vapor turns to droplets that give up their heat of condensation high above the surface of the sea.

Now the rapidly rising column is heated from near the top as well as from the bottom. Faster it rises, and faster; a violent demonstration of positive feedback. To supply the inexorable climb of wet air, yet more air is sucked in at the base as surface winds.

Inevitably the surface winds angle in from Coriolis effects. The whole structure begins to spin: a great vortex, heated at the bottom, heated at the top, sucking in surface winds, faster and faster, unstoppable unless the bottom heating is removed. At last the regional wind systems assert themselves and the whole spinning heat-engine begins to travel north and west, carried by the usually benign winds that bring Atlantic rains to eastern North America. But the mighty vortex they bring with them this time is called a hurricane (Figure 7.10).

Once over land, the hurricane loses its seed of moisture-laden warm air and must eventually collapse. Meanwhile, momen-

FIGURE 7.10
Hurricane: havoc by positive feedback.
A column of rising air is heated at the top as water condenses, and revolves through Coriolis effects. The energy that can be unleashed locally is one of the greatest rejuvenators of ecosystems.

tum drives it on. Still the warm air rises, expands, cools, and lets go its water. The forests over which it passes are waterlogged with soaking rain, putting maximum loads on branches. Then surface winds strike the soaking trees at a hundred miles an hour, at its fiercest cutting swathes through the forest. Recent pollen studies from Alabama apparently record the incidence of the most destructive hurricanes known (class five) as crossing land about once in 600 years, making them rare in the history of the United States, yet only about two tree lifetimes apart. The forests of southern Alabama have maintained themselves in a dynamic relationship with this destructive force for at least the last 5000 years (Liu et al., 1991).

A hurricane striking a forest does not work the desolation of clear-cut lumbering, but it does remove many of the older trees, making gaps in the forest and even tearing out roots to make bare ground. These gaps in canopy and soil open the forest to invaders, starting secondary successions. Hurricanes and other storms are thus the great rejuvenators of ecosystems, providing an opportunity for colonizing species to live.

CLIMATE CHANGE IN THE LIVES OF SPECIES

The climate pattern so familiar to us has been skewed from what has been the norm for the last 100,000 years, for most of which time the earth was in the ice-age mode typical of the last 2 million years. Mean annual temperatures were certainly lower on every part of the earth, but only in the range of 6°C or so lower. Some part of this cooling, it now seems certain, came because the earth actually received less solar heat in the northern hemisphere as cyclical changes in the earth's motion in space conspired to alter, ever so slightly, its seasonal attitudes to the sun. This is the so-called astronomical or **Milankovitch theory of ice ages** (Berger et al., 1984; Pielou, 1991). The cycling of the earth's motions is due to bring us back into the ice-age mode within a thousand years or so from now.

But a major part of the cooling was undoubtedly due to the ice sheets themselves, which acted as huge reflecting mirrors to send solar energy back to outer space without heating the earth. Initial cooling let snow persist in the north; the snow then reflected sunlight to cool the earth still more. Once the ice sheets were truly grown, the reflecting mirror effect (**albedo**) helped maintain the earth in the ice-age mode.

The heat engines of climate worked then as now—rising air at the equator, descending air at somewhere near 30° latitude "for no good reason" creating deserts, winds diverted by Coriolis effects, and so on. Yet it is certain that the presence of the northern glaciers skewed the resulting climatic pattern because the edge of the arctic high-pressure system was brought far south of its present position. Moreover, cooler land surfaces must have altered the temperature gradients between land and sea on which so many rainfall patterns depend, particularly monsoons.

So complex is climate that models of it must be empirical and based on actual measurements of temperature and precipitation. Thus to understand the climate of the ice age earth in which most extant species spent most of their existence it is necessary to make meteorological measurements of the climates of long ago. Most of the key measures are supplied by paleoecology, ocean temperatures from foraminifera tests, and land temperatures from fossil pollen (see Chapter 17).

Some 600 species of foraminifera are known from the oceans—tiny protists (single-celled organisms) with calcareous cases called tests (Figure 7.11). Many live in the floating plankton near the sea surface, and many of them appear to be sensitive to water temperature. The calcium carbonate tests are often well-preserved in ocean mud so that a core from the bottom of the sea records changes in species composition of the local foraminiferal fauna through time. To the extent that species really are temperature-sensitive, therefore, sediments thick with foraminiferal fossils provide a paleothermometer that can be calibrated by correlating the distributions of modern foraminifera with water masses of different temperature. Figure 7.12 shows a water temperature map for 18,000 years BP (Before Present) made by this method.

On land, plants—notably forest trees—are the most widely used paleothermometers, though beetles, lake organisms, and even rodent fossils also serve as proxies for climatic data (see Chapter 17). The tree record is deduced largely from reconstruction of ancient rains of pollen preserved in the mud of lakes and bogs. A cubic centimeter of lake mud is likely to hold on the order of 100,000 pollen grains, empty dead husks that yet retain all the ornamentation that made them distinctive in life (Figure 7.13). Not all trees oblige by

FIGURE 7.11
***Peveraplis:* foraminifera as animal thermometers in the sea.**
Foraminifera are planktonic animals of the oceans. Some live floating in surface waters, and some of these have their ranges restricted by water temperature. Their calcareous skeletons, taken from the mud of deep sea cores, can be used to infer the temperature of surface waters long ago, and hence a history of climate.

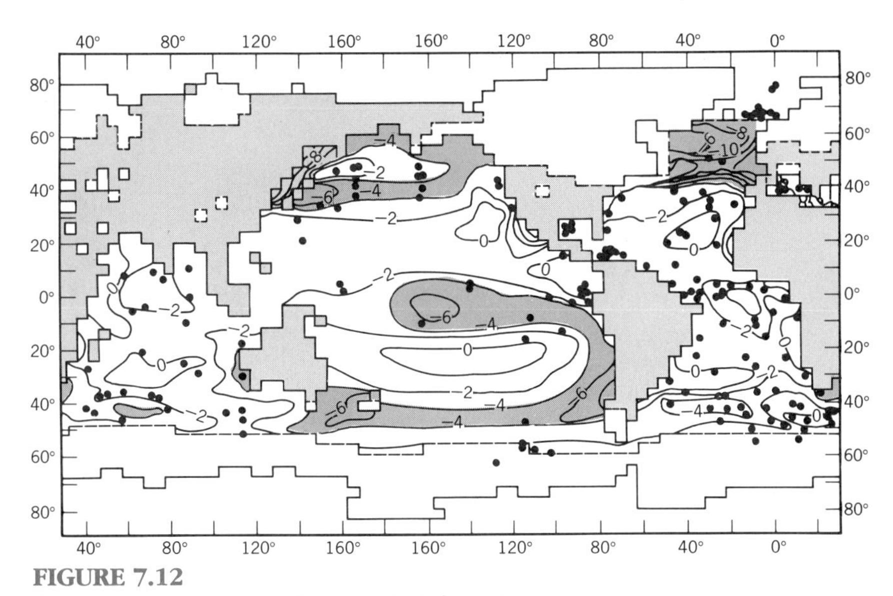

FIGURE 7.12

Ocean temperatures at the last glacial maximum.

Isotherms show temperatures as they differed from the present day. The data were reconstructed from species lists of foraminifera from deep sea cores using transfer functions correlating modern populations with temperature at the sea surface. Data sites (deep sea cores) are dots. Coldest water is shaded. Recent work suggests that tropical oceans might have been cooler than they are shown in this map (e.g. Emiliani, 1992). (From CLIMAP, 1976.)

making pollen profusely enough to send their signals to lake mud, and few of their pollen can be identified beyond the genus. Yet analysis of the ancient pollen mixtures does reveal the ranges of familiar trees in ancient times. As with the foraminifera fossils, to the extent that the trees are sensitive to temperature, pollen analysis can serve as a paleothermometer that can be calibrated against modern pollen distributions.

Reconstructing the climatic history of the ice ages is an active research area with its problems as well as its triumphs (see Chapter 17). But already the data are yielding climatic models that reveal an earth in which almost every climate was significantly different from climates of the present day (Kutzbach and Guetter, 1986). Europe, northern Asia, and North America were cold by modern standards, even when not actually under ice. Tundra parklands and open forests unlike any of the modern world covered much of this land (Huntley and Birks, 1983; Prentice et al., 1991). Africa south of the Sahara was drier than now, with the habitats for rain forest trees more limited, as monsoon rains were reduced (Livingstone, 1993). The great Sahara desert, however, was less dry, even holding lakes during part of the glacial period (Bonnefille et al., 1990). The Amazon was colder, with at least some of its forest trees living in communities that have now vanished from the earth (Bush et al., 1990).

The ending of the ice age was rapid; some data suggest that the biggest

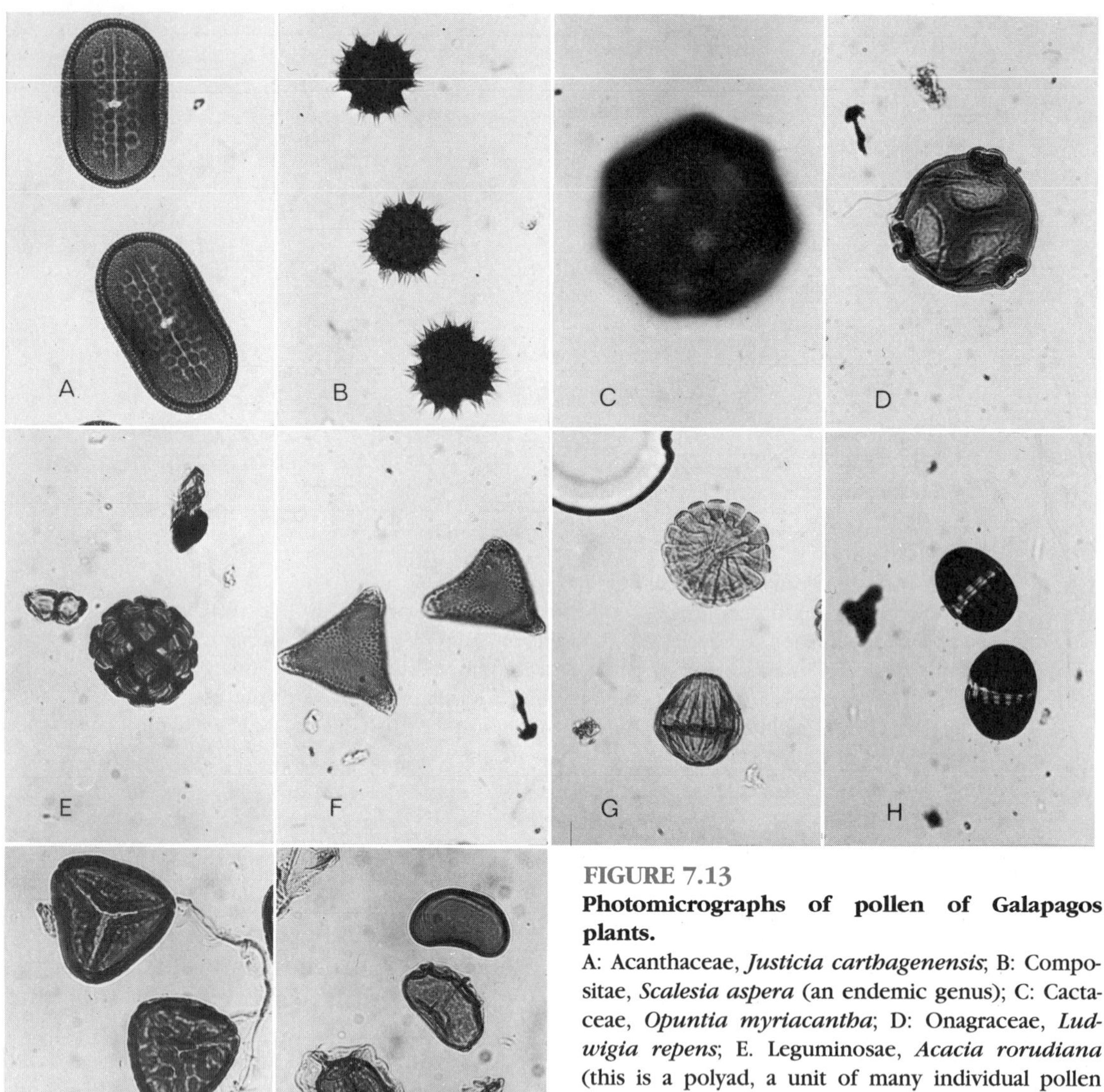

FIGURE 7.13
Photomicrographs of pollen of Galapagos plants.
A: Acanthaceae, *Justicia carthagenensis*; B: Compositae, *Scalesia aspera* (an endemic genus); C: Cactaceae, *Opuntia myriacantha*; D: Onagraceae, *Ludwigia repens*; E. Leguminosae, *Acacia rorudiana* (this is a polyad, a unit of many individual pollen grains); F: Sapindaceae, *Cardiospermum galapageium*; G: Lentibulariaceae, *Utricularia foliosa*. H: Polygalaceae, *Monnina chanduyensis*; I: Polypodiaceae, *Anogramma leptophylla* (a trilete fern spore); J: Polypodiaceae, *Asplenium serratum* (a monolete fern spore with ornate outer coat that easily separates from the smooth spore).

changes took but a few centuries to complete, though the physical melting of the residual ice took much longer (Berger et al., 1985). Ten thousand years ago virtually all the large and important species that still live on the earth had to win their fitness in this world of rapidly changing climate. Populations changed on a massive scale. Our familiar earth was built by their descendants rather quickly.

GREENHOUSE

Perhaps the most vital fact of climatology is that atmospheres are heated from below. The sun punches through the air with minimal absorption of energy to heat the land and ocean surface. The hot planet then radiates heat to and through the air. Some of the outgoing radiation is absorbed by components of the air, notably carbon dioxide, methane, and water vapor, thus warming the air. The warmed air then reradiates heat to outer space. The air is thus warmed from below and cooled from above, rather like a down quilt over a sleeping person in a cold room. A different analogy compares the atmosphere to the glass of a greenhouse: incoming sunlight punches through the glass (atmosphere), but the passage of outgoing heat is hindered by that same glass. The actual temperature of the earth surface is critically dependent on this **greenhouse effect**.

The temperature at the surface of a planet depends on the heat that can be absorbed by its atmosphere far more than it depends on the intensity of incoming radiation from the sun. Jupiter, for instance, is appallingly hot, though much more remote from the sun than earth, because its atmosphere is rich in methane, which heats far more readily than the oxygen and nitrogen of our air. Earthly temperatures are largely set by comparative traces of heat-absorbing (greenhouse) gases. Carbon dioxide, one of the important greenhouse gases, is present at only 0.03% by volume. Obviously, mean annual temperatures on earth could be very different if the concentration of heat-absorbing greenhouse gases were significantly changed.

Large changes in the percentage of carbon dioxide in the air through geologic time increasingly seem to offer a reasonable explanation of parts of the fossil record, particularly times of warmth in the Mesozoic and carbonate deposition in the oceans during the Cretaceous (Berner, 1990). But the clearest signal is from the cold of the last ice age. Actual samples of ice-age air have been retrieved as air bubbles in ancient ice from the Greenland and Antarctic ice sheets, showing that atmospheric CO_2 was down to about 0.02% by volume (Lorius et al., 1990). The greenhouse effect should have been less pronounced and the earth, therefore, colder. Colder it was, but researchers are still tackling a chicken-or-egg problem: did a lowered CO_2 greenhouse cool the earth, or did a cool earth result in lowered CO_2 (Leuenberger et al., 1992). A feedback between temperature and greenhouse gas is certainly implied, one way or the other.

Industrial societies, of course, are now enriching the CO_2 greenhouse (see Chapter 28), and the resulting changes in climate will probably be on the scale of an ice age, though with heating rather than cooling. The present is already an unusually warm time in the history of most living species, whose normal ambience has been the ice-age earth. It will be interesting to see how they respond to this new stress of warming. Some will lose fitness, some will gain fitness, some will go extinct.

PART TWO

DIVERSITY

Preview to Chapter 8

Resource competition: Nutcrackers in winter.

EXPLAINING the diversity of life on earth must begin with an understanding of why species are distinct. The ecological hypothesis for the origin of species is based on the concept of competition for resources. Individuals of any species have characteristic needs for resources based on their ways of life, or, as ecologists say, on their niches. The term "niche" can be used to describe function in a community, as in "there goes a pollinating bee," or it can be used more precisely to describe the functions of an individual species. This latter "species niche" can also be defined operationally as the set (or hypervolume) of resources needed by a member of a species population. This set of resources is the target for intraspecific competition when a confined breeding population begins to crowd. Population growth histories of experimental animals, in which eventual control is caused by density-dependent competition in this way, are s-shaped (sigmoid). The hypothesis that these histories reflect competition that changes with density can be described by the logistic equation. This logistic hypothesis is tested when it is used to predict the outcome of competition between different species in confinement. This development, provided by the Lotka–Volterra equations of competition, predicts the counterintuitive result that strong competition will lead to the total extinction of one species. The prediction has been tested in numerous laboratory experiments and shown to be true, hence this result can be stated as the principle of competitive exclusion, known in ecological shorthand as "one species: one niche." An implication of paramount importance in ecology is that species populations are kept distinct by the selective removal of individuals that tend to compete with individuals of other species. Natural selection preserves traits that allow competition to be avoided.

Chapter 8

The Competition Hypothesis

"Why are there so many different kinds of plants and animals?" must, by any accounting, be one of the most fundamentally interesting questions we can ask of the world around us. And yet we are hampered at the outset by not knowing how many species exist, even by an order of magnitude. Do 5 million or 50 million species share the earth with us? We do not know (Wilson, 1985). Worse, our ignorance is skewed. We know within tolerable limits how many bird and mammal species still exist; perhaps with birds we are within a hundred or two of the right answer. The huge ignorance is with insects and the smaller classes, making it difficult to come to an understanding of diversity that should apply to all size classes (May, 1985).

Our failure to have a decent catalog of life on earth tells us much about contemporary human society, where evidently finance and respect are not available to those who could so easily provide the missing data. But the partial catalog compiled so far does show that a complete catalog is possible. Life really is organized into species that can unambiguously be recognized for what they are. On the way to asking about limits to numbers of species, therefore, are questions about distinct species themselves. Why should individual animals and plants be organized into regimented populations of look-alikes and behave-alikes? This is the ecological part to the origin of species.

The ecological explanation for distinct species is that a successful way of life brooks no variance. To be deviant exacts a penalty in fitness. Traits inherited from parents are successful traits and are changed only at peril, and it is for this reason that species breed true. The result is that life on earth is organized around populations of look-alikes cast in the image of some successful ancestor, as it were, carrying trade-union cards for a narrowly defined profession in life.

This vision of species as collections of individuals with a common trade is captured by the term **niche**. Saying, "every species has its niche" not only expresses an ecologist's idea of a species as a collection of individuals with common habits reflected in common shape, but it also implies some extra perils of deviance. The individual whose design has strayed too far from the parental mold is likely to come into deleterious contact with individuals of other, not too dissimilar, species. The early California naturalist J. Grinnel, arguably the first person to use the word "niche" in its ecological sense, caught this idea when he wrote in 1904, "Two species of approximately the same food habits are not likely to remain long evenly balanced in numbers in the same region. One will crowd out the other" (Grinnel, 1904). The message is to feed as your own crowd feeds, because taking on some of the other crowd's habits means competition from both sides. Be definitely different and you have only your own crowd to worry about.

This intuitive vision of competitive life says that food needs must be distinctly different for populations to coexist, and the question of the number of species now becomes, "How is a limit set to the possible ways of life that can coexist?" But the concept rests on more than just intuition. The idea gained respectability with the apparent success of a simple mathematical model designed to predict the outcomes of competition between species. Known as the Lotka–Volterra equations, this model sought to describe competition for resource between individuals, realistically but with large simplifying assumptions. The model first described the consequences of competition in a crowd of act-alikes, and then went on to examine the real consequences of competition between crowds from two similar species of the kind that Grinnel had imagined when he said "one would crowd out the other." Ecologists came to like the predictions this model gave. It can be argued that the model is an infuriating oversimplification of real life, and yet surprisingly useful insights have been gained from its use. The Lotka–Volterra equations themselves came to the prominent attention of ecologists in the 1930s, but their logic is so simple and powerful that models based on them continue to be made (Pimm et al., 1991).

The train of logic runs as follows:

1. Individuals of a species population have closely similar needs for resources (they live in the same niche).
2. When crowded, a limit is set to a species population by competition for resources.
3. A population history leading to crowding and resource limitation can be described mathematically.
4. Pairs of related species can be found with enough overlap in their resource needs that they can potentially come into competition with each other.
5. Equations describing population growth terminated by the effects of crowding can be used to predict the outcome of competitions between species of overlapping habit.
6. Predictions by these Lotka–Volterra equations that coexistence requires separate resource needs can be tested by experimental manipulation of animal populations.
7. The exercise offers a satisfying explanation of how species can be kept distinct by natural selection.

The following account of the development of the Lotka–Volterra model provides a more formal introduction to the niche concept before introducing the mathematical descriptions of population growth and competition between species.

THE SEVERAL COMPATIBLE MEANINGS OF NICHE

Ecologists say that every species has its niche, a unique position in a food web or ecosystem, or a unique function in life, or a unique set of resources or factors needed for survival. There are several subtly different meanings behind the idea of niche, though all are linked.

Niche as Community Function (Class I Niche)

The term "niche" entered the normal usage of ecology only after Elton defined it in his 1927 textbook. Elton's use was rather special in that he referred to functions within a community when he talked of niches. He said that when he saw a badger as he went for an early morning stroll in an Oxford wood, he tried to think he was seeing a functionary of the local community, rather like seeing a man walk down a village street and saying, "There goes a plumber," or "There goes a farmer." The badger's niche in the Oxford woodland was its part in the running of the community, just as a farmer or plumber had a niche in the community of the village. Every kind of plant or animal must have its own function in the community, its "niche." Elton offered a formal definition, saying that "*niche means the animal's place in the biotic environment, its relation to food and enemies.*" Niche so defined is sometimes referred to as the **Eltonian niche**.

We follow Elton's usage whenever we talk of niches being filled, as in, "the niche of pollinating bird is filled by hummingbirds in South America but by sunbirds in Africa," or "the niche of active top carnivore filled by *Allosaurus* in a Jurassic plains ecosystem is filled by tigers in modern India." **Ecological equivalent** species fill comparable niches in ecosystems of different places.

Niche in the Species Definition (Class II Niche)

The theoretically more useful concept of niche is represented by the earlier usage of Grinnel (1904), which defines a niche more precisely as the property of an individual species population. Then a niche is a "*specific set of capabilities for extracting resources, for surviving hazard, and for competing, coupled with a corresponding set of needs*" (Colinvaux, 1982).

In this second, more precise definition, niche continues to define the functions of animals (or plants), but the function is so narrowly constrained that only individuals of a single species can occupy it. Take worm-pulling thrushes, their specific properties, and their roles in communities. On North American lawns, common worm-pullers are American robins, *Turdus migratorius*. Both the Class I (Eltonian) and the Class II (species) niche concepts can be applied to this animal. In the former, the American robin plays a role in the community as a puller of worms and food for hawks; in the latter, an American robin pulls worms and avoids hawks as part of a program working to thrust more robins into the next generation.

Only individuals of *T. migratorius* can fill the Class II niche of this species, but the community function fulfilled by American robins can in theory be filled by similar but different kinds of birds. American robins are the only common thrushes (family Turdidae) that pull worms on lawns of the eastern United States, but in northern Europe other species of thrush perform this function, in particular the European blackbird (*T. merula*) and the song thrush (*T. ericetorum*). These birds fill the worm-pulling niche (Class I) in Europe, though their species niches (Class II) are significantly different from that of the American *T. migratorius*.

Niche as a Quality of the Environment (Class III Niche)

If a niche is a property of a species it must yet be exercised in a suitable environment. Specialization for a food resource is possible only where that food resource is present. A bird with feet adapted to perching on small twigs can live only where the twigs are small, and so on. The concept of a species niche, therefore, allows the complementary concept of an environmental space in which that niche is exercised, a mirror image niche, as it were. Then "a niche is *that set of ecological conditions under which a species can exploit a source of energy effectively enough to be able to reproduce and colonize further such sets of conditions*" (MacFadyen, 1957).

Describing a niche as the package of resources needed by the animal offers the best hope of measurement and quantitative description of niches. Resources can be measured in a way that is difficult for diffuse concepts of species behavior. The niche then becomes *a "multi-dimensional hypervolume of resource axes,"* as G. E. Hutchinson (1958) described it in one of his seminal papers.

Every variable that affects a particular species can be thought of as being linear, in which event two resource axes can be expressed as x and y on a conventional two-dimensional graph. When a third variable is added (say, perch size to food size and height above ground for a kind of small bird) then a three-dimensional figure is needed to plot an outline of the niche that appears as a volume (Figure 8.1). When more dimensions than three are included, the resulting theoretical space can no longer be drawn but can still be described mathematically as a hypervolume:

$$x', x'', x''', \ldots, x^n \quad (8.1)$$

where

x', x'', etc., are niche axes.

Two classes of niche axes can be separated by their probable effects on the size of the niche hypervolume. If a species population lives without competitors or other organisms that would interfere with it, then the size of the niche should be set by physical needs and food alone. Hutchinson (1958) called the resulting niche the **fundamental niche** of the species. But wherever the resources of the environment have to be shared with other species, as usually happens, then the niche hyper-

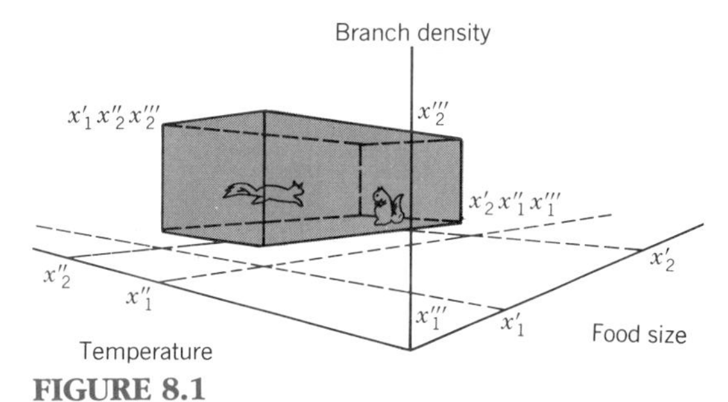

FIGURE 8.1
A three-dimensional orthogonal niche.
Three niche axes likely to be important to a squirrel are temperature, food size, and branch density. If these are represented by x', x'', and x''', the rectangular space is a niche volume (Class III niche) of only three axes. Adding more axes to make a hypervolume can be understood intuitively and mathematically although it cannot be drawn. (From Hutchinson, 1978.)

volume should be smaller. This narrower niche is the **realized niche** of the population.

THE MEANING OF COMPETITION

"Competition" is a word with a clear meaning, valid and hallowed in English usage. There is competition whenever two or more individuals or groups "strive together" (the literal meaning of the Latin roots) for something in short supply. People compete for prizes, and only one of them, or one group of those competing, can win first prize. Yet competition between individuals for niches must be subtly different from competition between long-distance runners for a trophy. In ecological competition whole lifestyles must be pitted against one another. Real success is measured a generation away in the success of offspring at reproducing themselves. Meanwhile, there must have been many interactions with neighbors and other animals and plants of the community in which it lives: perhaps competition for food energy, success at rapid dispersal to occupy an empty habitat or nesting place, better success than others at escaping predators, or the opportunity to find refuge in times of frost or drought.

Competition in nature can be difficult to detect, as might be anything the consequences of which are a generation away. And defining what we mean by competition is not straightforward. Two scientists much concerned to make their meaning clear came up with the following mass of words: "*Competition occurs whenever a valuable or necessary resource is sought together by a number of animals or plants (of the same kind or of different kinds) when that resource is in short supply; or if the resource is not in short supply, competition occurs when the animals or plants seeking that resource nevertheless harm one another in the process*" (Andrewartha and Birch, 1954).

As that tongue-twisting definition suggests, competition can be with the same kind or with different kinds. Competition with the same kind may be viewed as struggling for the opportunity of filling one of the niches vacated by death in the next generation, or **intraspecific competition**. Competition with different kinds is **interspecific competition**. Those not familiar with Latin might care to note that *intra* means "within," as "on the inside," and *inter* means "between." They sound similar to Anglo-Saxon ears, but their meanings are opposite.

When, as in this chapter and the next, the concept of competition is used to understand how species populations are kept distinct, it is satisfactory to think of competition as a zero-sum game, with clear winners and losers. This is **contest competition**, as distinct from what Nicholson (1954) called **scramble competition** (Varley et al., 1973). Scramble competition describes what happens when maggots are crowded on a piece of meat; the growing maggots all eat meat even though only a few will survive to pupation. Meanwhile, a large part of the meat energy is lost to the population in the maggots that must die. This is not a zero-sum contest, in that the eventual victors end up with only a small portion of the energy that was once available. As Lomnicki (1988) points out, real population histories will be much altered by whether competition is a straight contest for resources or whether resources are wasted in a scramble. Crowded populations will grow more slowly when competition is of the scramble kind because of wasted resources.

THE LOGISTIC MODEL OF POPULATION GROWTH

Every beginning biology textbook has a graph of population growth that is s-shaped:

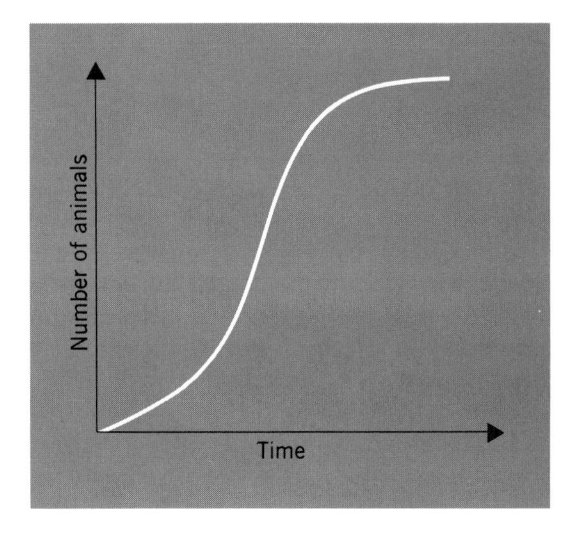

Various captions might accompany the figure, but the impression is often given that this is a generalized curve for the growth of a typical population. This is misleading. What these s-shaped (usually called "sigmoid," being the Greek for s-shaped) curves actually describe is population growth as observed in a number of experimental systems where populations of single species are kept under strict confinement. No such complete histories have been seen in the wild, though two histories of bird invasions are known that come close (see Chapter 10). The real significance of sigmoid population histories to ecology is that, being simple, they let us make plausible hypotheses to explain them. Moreover, since they have been observed in experimental systems, tests for the hypotheses are possible. A typical sigmoid population history in the laboratory is illustrated with the cultures of *Paramecium* kept by the Russian ecologist G. F. Gause[1] (1934).

Paramecia are protozoans that feed on bacteria and other small particles. Gause found that he could make a standard medium of oatmeal that served as excellent food for paramecia and one that he could add to their water in measured amounts. In a typical experiment, he placed 40 individuals of *Paramecium caudatum* in small tubes containing 10 cc of water and a few drops of his oatmeal medium. Each day thereafter he would take out a subsample with a pipette to count the paramecia present, which gave him an estimate of the total population. He would also spin the tubes in a centrifuge, which drove the animals to the bottom and allowed him to pour off the old water with its unused suspended food in order to replace it with fresh. There was thus a daily constant but limited input of food to the system. Under these conditions, the paramecia reproduced quickly at first, their numbers increasing exponentially until the water was cloudy with them. But when there were about 4000 animals in each tube the rate of growth leveled off, and the population stopped growing. The top of the sigmoid trace had now been reached. The population remained at the same level indefinitely, as long as the daily food ration was provided.

Similar sigmoid population histories result in laboratory cultures of yeasts, molds, bacteria, various protozoa, and even insects, such as fruit flies in banana mash and *Tribolium* beetles in flour. All that is required is an animal of simple wants, a way of confining them, and an energy source. Attempts to contrive sigmoid

[1]Pronounced "Gauzer." This is not only phonetically correct, but avoids confusion with the mathematician Gauss, whose work is also often cited by ecologists.

population histories in the laboratory for more complicated animals with more complicated wants do not work. Mice, for instance, reproduce well enough in large pens until they are crowded, after which they suffer mass death or failure to breed, even when given plenty of food, water, and bedding (Calhoun, 1962). This history is far from being sigmoid.

It is natural to suggest that sigmoid growth histories reflect changing intensities of competition. A few animals in a container with plenty of food should not suffer any competition. **Fecundity**, or *the rate at which individuals produce offspring*, should be high. Virtually unlimited food ensures rapid growth and thus a short "generation time." The result of these reproductive properties is that the whole population grows rapidly with, as we say, a high **intrinsic rate of increase**. Soon there should be many animals in the container, all reproducing rapidly. Population growth would be exponential or geometric. But then the animals would become crowded and competition would be important to life in the container. Fecundity, survival, or both should drop, and the death rate should rise. Population growth would slow until it ceased altogether, with that degree of crowding where births exactly balance deaths.

This hypothesis of a population regulated by crowding to an equilibrium number is basic to the prevailing ecological model of species distinctness. It is developed by writing it down in mathematical form as follows:

$$R = rN\,(1 - N/K) \qquad (8.2)$$

or

$$\frac{dN}{dt} = rN\,(1 - N/K) \qquad (8.3)$$

where

- R is the population growth rate (best defined as the differential dN/dt)
- r is the intrinsic rate of increase
- N is the number of animals present at time t
- K is the number of animals able to live in the container at population equilibrium

Equation (8.3) is called the logistic equation. It has been of interest to biologists since its properties were first pointed out by P. F. Verhulst in the middle of the last century (Hutchinson, 1978). It neatly describes the hypothesis that population growth in a closed container is first damped, then regulated, by intraspecific competition.

A logistic equation is, of course, a gross oversimplification of the events that must take place during population growth. Being a differential equation, it ought fairly to be applied only to populations where growth is continuous; a condition probably met by Gause's paramecium cultures, but certainly not by temperate insects in the wild or deer. For studies of population growth in real animals, difference equations are required instead (see Chapter 10). Modern theoretical work shows that both logistic equations incorporating lag terms and difference equations describe population histories that are far more complex than the equilibrium around a single stable point of the logistic (May, 1986; see Chapter 10). Yet the crass simplicity of the logistic hypothesis nevertheless revealed properties of species populations that have proved valid. Nor is it necessary to be able to manipulate the equations in order to understand the splendor of the theorem that results.

Crucial to the logistic statement is the

concept of **carrying capacity** (K). K is a constant, a property of the container, and is expressed as a number—the maximum number of individuals that can persist under the conditions specified. The logistic hypothesis states that as the population number N approaches the saturation number K, then rN becomes zero and population growth ceases.

The logistic equation in its differential form, therefore, is merely a mathematical description of a hypothesis. The equation states that the rate of population growth first rises then falls, yielding a parabolic plot:

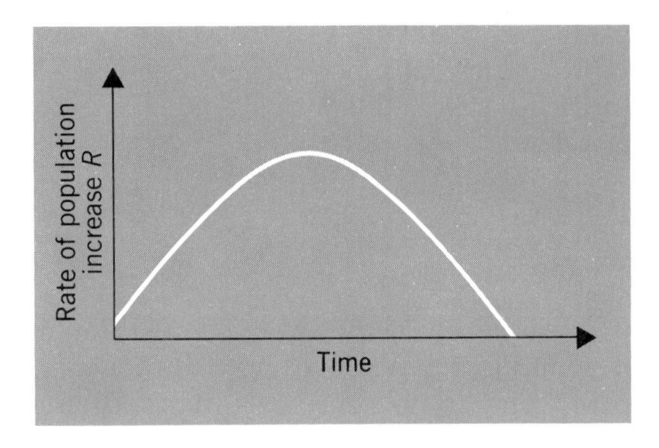

When the equation is integrated to yield changing numbers, it is then of the form:

$$N_{(t)} = N_{(0)}e^{rt} \frac{K}{K - N_{(0)}(l - e^{rt})} \quad (8.4)$$

and a plot of equation 8.4 would be sigmoid.

Saying that competition controls experimental populations when they are crowded and writing a logistic equation are two ways of stating the same hypothesis. A third way is to say that populations are brought under **density-dependent control**. Competition should always be dependent on density, because the denser the crowd, the more the competition. But many other processes that might not automatically be called competition also act in a density-dependent way. Cannibalism might occur in crowds, or crowded animals might spoil food even though they did not actually compete for it, or the chance of finding a refuge might be less in a crowd (Chapman, 1928; Park et al., 1965; Figure 8.2). So the logistic hypothesis is also a hypothesis of control by density-dependent factors.

A logistic equation can be written to fit any growth history that is truly sigmoid. Conversely, any animal populations growing when confined with limited food, the individuals of which are capable of interfering with each other in density-dependent ways, will describe a logistic growth history. The fact of writing an equation describing constraints in an experimental system thus does not, in itself, seem to get us very far. But it lets us model what

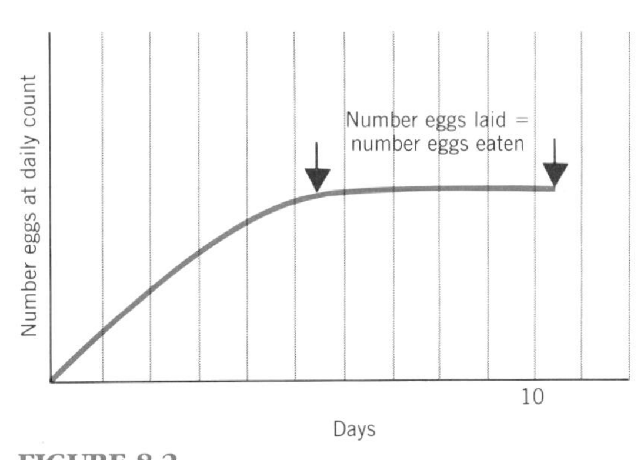

FIGURE 8.2
Density-dependent cannibalism.
Flour beetles (*Tribolium confusum*) were kept in dishes of flour, through which they burrow and on which they feed. Investigators changed the eggs every five days to prevent hatching and kept the adult numbers constant. The only population events possible were egg laying and egg cannibalism. At equilibrium, egg numbers were stabilized by density-dependent cannibalism. (From Boyce 1946.)

should happen in more complicated circumstances, then to test the resulting theoretical predictions. The logistic hypothesis can be manipulated to predict the results of interspecific competition between species populations with overlapping resource requirements. This is the exercise lying at the heart of the ecological model of species.

THE LOTKA–VOLTERRA MODEL OF INTERSPECIFIC COMPETITION

If populations of two species compete for a resource it must follow that the carrying capacity of container or habitat for each is reduced from what it would have been if one species were in sole occupancy. Intuitively, it might seem obvious that making different species populations compete should force the number of each down at equilibrium so that coexistence with much-reduced populations would be the result. Intuition, however, turns out to be wrong. The mathematical model predicts that significant competition in a confined space always ends with the elimination of one species and the total victory of the other.

The necessary manipulation of the logistic equation was performed by two mathematicians working independently in the early part of this century, Vittorio Volterra in Italy (Chapman, 1931) and Alfred Lotka (1925) at Johns Hopkins University. The resulting equations are known as the Lotka–Volterra equations.

For two competing species 1 and 2, the population growth curve of species 1 living alone is

$$\frac{dN_1}{dt} = r_1N_1\left(1 - \frac{N_1}{K_1}\right)$$

and the population growth curve of species 2 living alone is

$$\frac{dN_2}{dt} = r_2N_2\left(1 - \frac{N_2}{K_2}\right)$$

When two populations are grown together, neither one's growth rate will be influenced at first by the presence of a few individuals of the other species because there is still lots of space and food. But when the animals are numerous, such that N_1 approaches K_1 or N_2 approaches K_2, the effects of competition should be significant. Competition must be dependent on numbers and the strength with which each individual is able to compete. The growth curve of a population of species 1 in the presence of species 2 is, therefore, given by

$$\frac{dN_1}{dt} = r_1N_1\left(1 - \frac{N_1}{K_1} - \frac{\alpha N_2}{K_1}\right)$$

where α is the coefficient of competition of population 2. If population 2 does not compete, contrary to our expectations, then α will be zero, αN_2 will also be zero, and the history of population 1 will be quite unaffected. But if there is competition, then αN_2 will be positive, will combine with N_1, and will reduce the term

$$\left(1 - \frac{N_1}{K_1} - \frac{\alpha N_2}{K_1}\right)$$

to zero before the saturation value of population 1 is reached. Assigning the coefficient β to represent the competition of the other species, we may write a pair of equations as follows:

$$\frac{dN_1}{dt} = r_1N_1\left(1 - \frac{N_1}{K_1} - \frac{\alpha N_2}{K_1}\right)$$

$$\frac{dN_2}{dt} = r_2N_2\left(1 - \frac{N_2}{K_2} - \frac{\beta N_1}{K_2}\right)$$

The outcome of the competition between the species would be revealed when the system finally came to equilibrium. Then

$$\frac{dN_1}{dt} = \frac{dN_2}{dt} = 0$$

and

$$r_1N_1\left(1 - \frac{N_1}{K_1} - \frac{\alpha N_2}{K_1}\right) = r_2N_2\left(1 - \frac{N_2}{K_2} - \frac{\beta N_1}{K_2}\right) = 0$$

and

$$\begin{cases} \dfrac{dN_1}{dt} = r_1N_1\left(1 - \dfrac{N_1}{K_1} - \dfrac{\alpha N_2}{K_1}\right) \\ \dfrac{dN_2}{dt} = r_2N_2\left(1 - \dfrac{N_2}{K_2} - \dfrac{\beta N_1}{K_2}\right) \end{cases}$$

The pair of equations cannot be solved simultaneously by analytical techniques. Mathophobe ecologists, therefore, can be excused from trying. But it is possible to show that a population equilibrium is not possible with strong competition unless the population of one of the competing pair becomes zero. If the competition coefficients α or β are large, one population will inevitably be eliminated. The Lotka–Volterra equations, therefore, predict that the outcome of prolonged competition between species with significantly overlapping requirements is extinction of the one and total victory of the other.

This conclusion suggests why species are distinct in nature. Existing species populations are descendants of individuals that have survived competitions because they avoided them. Speciation is, in part, a process of avoiding competition by promoting difference.

EXPERIMENTAL TESTS OF THE LOGISTIC HYPOTHESIS

The logistic hypothesis predicts that:

1. Strongly competing species cannot coexist indefinitely, and
2. Where populations of different species do share a resource their competition is muted and weak.

These predictions can be tested by laboratory experiment. Gause's classic work provides the basic tests of the hypothesis. Gause (1934) explored the Lotka–Volterra competition equations and their implications with a series of experiments with populations of simple organisms, notably yeast plants and protozoans. It can be argued that his work was not designed primarily as a formal test of the logistic hypothesis, since Gause was engaged in a general experimental investigation into as many aspects as possible of struggles for existence in experimental populations. His work, however, together with that of other experimentalists, does provide formal tests of the hypothesis.

Gause grew various species of *Paramecium* on his oatmeal medium in centrifuge tubes as described previously. Many species of *Paramecium* did thrive on this medium and attained population equilibria after a sigmoid growth history. The centrifuge tube habitats were structurally simple and provided only one kind of food so that animals as similar as different species of *Paramecium* should be expected to have to compete if introduced into the same tube. When an inoculum of both *Paramecium caudatum* and *P. aurelia* was introduced into the same tube, population histories resulted as shown in Figure 8.3. *P. caudatum* is a relatively large and slow-growing species, whereas *P. aurelia* is smaller and reproduces more quickly. In

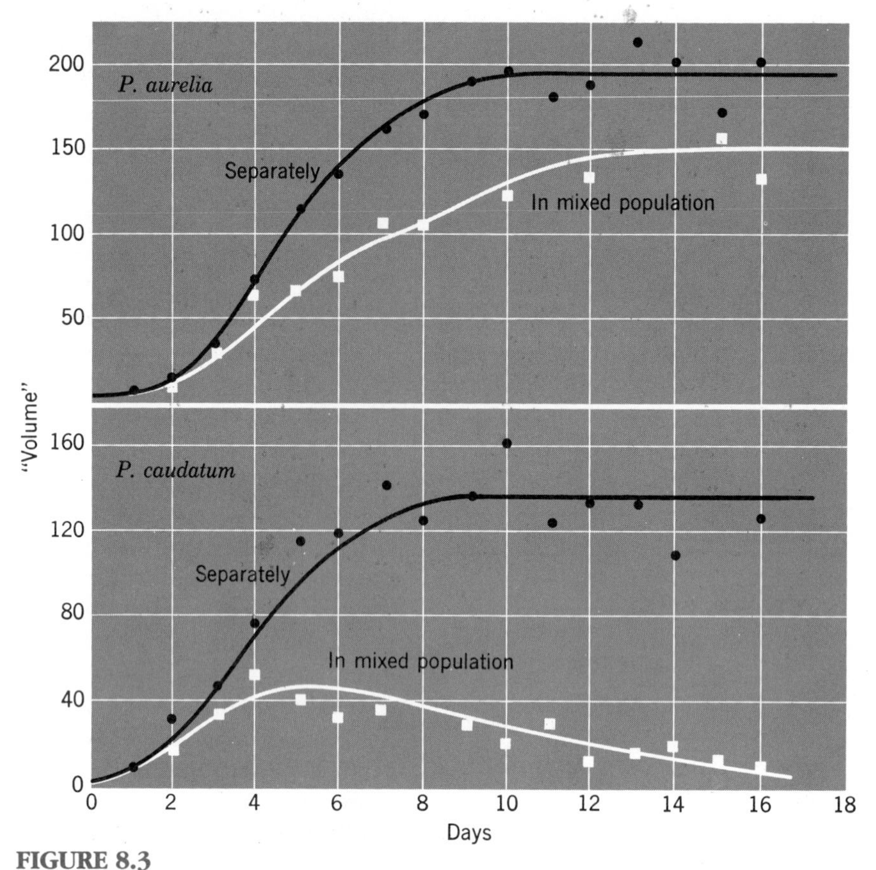

FIGURE 8.3

Competition between species of *Paramecium*.

Example of data used by Gause to confirm the Lotka-Volterra prediction that closely related species could not coexist when forced to share the same niche. Under the conditions of this experiment (daily changes of water and constant inputs of food), *Paramecium aurelia* always persisted, while *P. caudatum* died out. Gause was able to alter conditions so that *P. caudatum* could inevitably win the competition instead. Notice that the size of each population is expressed in volumes, a device used by Gause to eliminate the effect of different sizes of the two species from his graphs. (After Gause, 1934.)

combined culture, the population of the rapidly reproducing *P. aurelia* rose more quickly, and continued to grow even as the population of *P. caudatum* first leveled off and then declined to extinction. The prediction of the Lotka–Volterra competition equations is fulfilled by this experiment, the inevitable victory by the stronger suggesting model 1 or 2.

The niche requirements of *P. aurelia* appear to have been more closely met in centrifuge tubes of oatmeal medium changed daily than were those of the vanquished *P. caudatum*. There should, however, be different circumstances in which the outcome of the experiment could be reversed. Gause tried a simple change in experimental procedure. Instead of daily changing the medium in which the paramecia lived, he left the old medium in place and added a daily dose of food concentrate. When this was done, the outcome of the competition was reversed. It is thought that the new conditions favored *P. caudatum* because the changed experimental conditions allowed chemicals secreted by the paramecia to accumulate in the water. Many protozoa do secrete substances that inhibit the growth of other protozoa (ectocrine substances). These substances might serve a species of *Paramecium* as a competitive mechanism. It is at least a good working hypothesis that the ability of the slow-growing *P. caudatum* to outcompete the fast-growing *P. aurelia* in this second series of experiments was due to its relative immunity to the ectocrines that collected in the centrifuge tubes. The important aspect of this second series of experiments, however, is the continued demonstration that only one of a pair of strongly competing species continued to exist.

Gause made many other experimental tests of the Lotka–Volterra predictions, using yeasts as well as protozoa. His systems were always simple: small closed containers with food energy supplied in one simple form. The possibilities of different ways of life in these containers were deliberately few so that the animals would compete. There was little chance of the animals' avoiding the competition by living different lives, as by occupying different niches. And then one of Gause's *Paramecium* experiments had an unexpected outcome; both populations persisted.

When *P. aurelia* and *P. bursaria* were grown together, neither became extinct, but the population of each leveled off at about half the number it would attain when alone (Figure 8.4). Gause (1936) soon saw why this was so. When the two species were coexisting they were separated in space, *P. aurelia* being in pure culture at the top of the tubes and *P. bursaria* concentrated at the very bottom of the tubes, with a narrow zone of overlap in the middle where both species could be found. Furthermore, Gause discovered adaptations of the two species that provided an understanding of the different lifestyles they were able to adopt in their unnatural habitat of centrifuge tubes, one at the top and one at the bottom. Food was concentrated at the bottom of the tubes as bacteria in the culture medium settled out, suggesting that the bottom of the tube would be a good place for a bacteria-eating protozoan to live. However, this high density of bacteria drew heavily on the reserve of dissolved oxygen in the water so that the bottom of the tube tended to become anoxic; food was available but lack of oxygen tended to deny this food to an aerobic animal like most paramecia. *P. aurelia* suffocated in the bottom water and was denied the food resource.

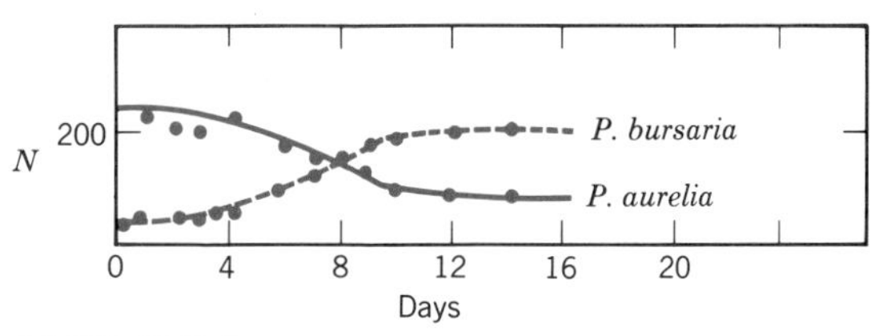

FIGURE 8.4
Competition between *Paramecium aurelia* and *P. bursaria*.
An inoculum of *P. bursaria* is introduced into a *P. aurelia* culture at the start of the experiment.

But *P. bursaria* had a private oxygen supply. It is a "green" *Paramecium* possessing symbiotic algae called *Zoochlorella*. In the light, these *Zoochlorella* performed photosynthesis, produced oxygen, and let their carriers, members of the *P. bursaria* population, forage in the anoxic bottom water (Figure 8.5). Without this adaptation, the *P. aurelia* population

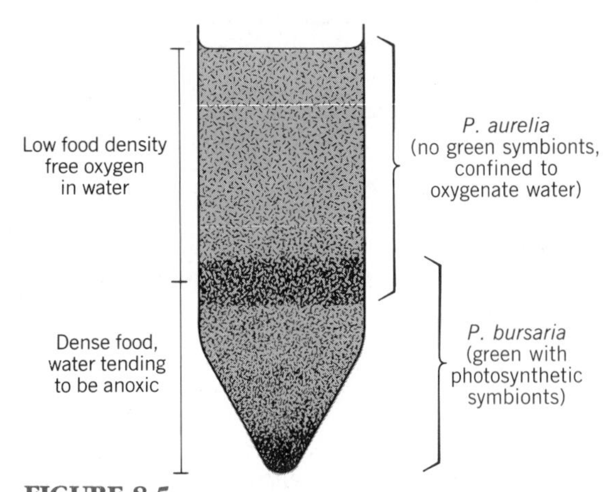

FIGURE 8.5
Coexistence of *Paramecium aurelia* and *P. bursaria*.
The green *P. bursaria* can survive in anoxic water where food is concentrated but *P. aurelia* cannot. The two species coexist with minor competition in the zone of overlap.

was forced to rely on the bacterial food in the upper water, where it apparently had a competitive advantage. The *P. aurelia* and *P. bursaria* system, therefore, was an example of model 4 of the Lotka–Volterra formulation, that which allows coexistence if competition is substantially avoided.

Gause would now conclude that the Lotka–Volterra predictions were verified, so their predictions could be stated as a working principle. It is now known as the Gause principle, or, using the language of Hardin (1960), the **principle of competitive exclusion**, and may be stated as follows: "*Stable populations of two or more species cannot continuously occupy the same niche.*" Or, more simply, as "*One species: one niche.*"

Others have followed Gause's example, choosing their favorite animals and matching species against species. Provided the systems were simple enough, the results have always been as expected: one population of strongly competing pairs has always died out. Figure 8.6 shows the results of competition between flour beetles in dishes of flour. It is known that cannibalism, particularly of eggs, is the most restrictive density-dependent pressure regulating population size when flour beetles are forced to pass their days in the unnatural habitat of miller's flour in a dish. Crombie's (1946) work showed that flour beetles of different genera could be launched against each other in dishes of flour to the utter elimination of one kind by the other (Figure 8.6*a*). Crombie was able to show that *Tribolium* was the more

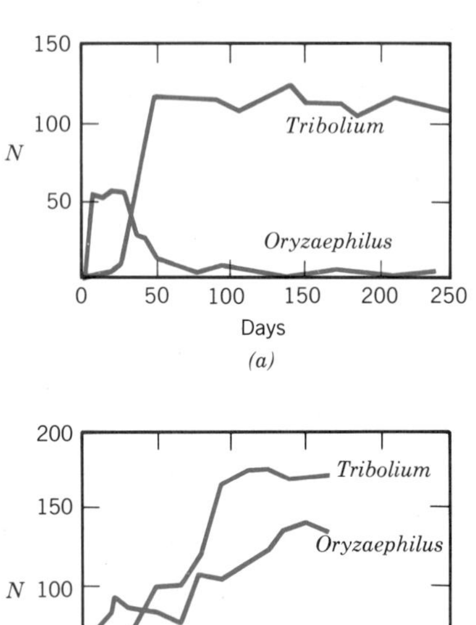

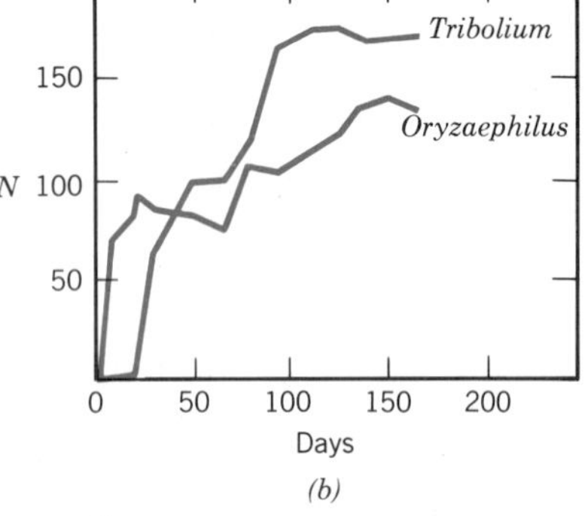

FIGURE 8.6
Competition between flour beetles of different genera.
When cultured in dishes containing nothing but flour, *Oryzaephilus* was always eliminated (*a*). When short lengths of capillary tubing gave *Oryzaephilus* safe sites for pupation, the populations coexisted (*b*). (From Crombie, 1946.)

voracious of the two, having a larger appetite for eggs and being particularly likely to eat the pupae of *Oryzaephilus*. *Oryzaephilus*, however, could not eat *Tribolium* pupae. This meant that *Tribolium* did more damage to *Oryzaephilus* as it blundered through that unnatural medium of flour than its population received from *Oryzaephilus* in return. *Tribolium* triumphed in the competition and eventually occupied the flour dishes alone, in pure culture.

This competition between the flour beetles shows how necessary is a complicated definition of competition like that of Andrewartha and Birch given earlier, which stressed that competition may sometimes involve physical harm. Food is not in short supply for the beetles, and neither is living space. One species eliminates another merely because one accidentally harms the other as it carries on its normal feeding process in unusual circumstances. Here is no competition for a single prize, yet the way in which the animals interact means that one population leaves surviving offspring whereas the other does not. Natural selection recognizes this competition even though the meaning has been strained somewhat from classical English usage.

Understanding why *Tribolium* could eliminate *Oryzaephilus* from a dish of flour through an appetite that took in pupae and eggs as readily as its normal herbivorous diet let Crombie design systems in which predation by *Tribolium* would be less possible. He mixed short lengths of capillary tubing in the flour that could be used by *Oryzaephilus* as hiding places. *Oryzaephilus* then pupated in the tubes where the tunneling *Tribolium* did not come across them and, in these flour and tubing mixtures, both species coexisted (Figure 8.6*b*).

A feature common to all these experiments is that there are likely to be surprises both as to which potential competitor succeeds and to what curious circumstance may lead to the actual outcome. The Lotka–Volterra–Gause model of the struggle for existence, therefore, tells us that direct competition is always fatal for one of the protagonists, but gives us no insight into stratagems that might avoid such competition and thus be preserved by natural selection.

AVOIDANCE OF COMPETITION IS THE GOAL

The most far-reaching implication of this work is a view of species kept distinct by the necessity of avoiding competition. Individuals most different from an invading population will compete less, will have a greater chance of survival, and will gain more fitness than relatives whose needs overlap those of the invaders. Significant differences between species populations thus serve to minimize competition between species. The result over time is populations cast in the narrow mold of the parental populations and significantly different from populations of other species. It is this pattern that we note when we describe life on earth as being made up of distinct species.

Preview to Chapter 9

East African plains: Coexistence through specialized habits.

FOR the principle of competitive exclusion to hold for wild populations, interspecific competition must always be muted. This is tested by examining closely related animals of apparently similar habit who live together (are sympatric). In every test tried, whether of sympatric pairs or whole guilds of animals sharing a resource, it has been shown that resources are indeed partitioned in ways that minimize interspecific competition. This suggests that speciation serves the function of allocating resources in ways that minimize competition. In this model, the most different individuals in populations of overlapping requirements should suffer the least competition and thus should leave the most surviving offspring. Mean characters of the populations would then be displaced and kept separate. A premium would be placed on any trait that prevents cross-breeding, with its attendant danger that offspring would have intermediate traits putting them at competitive risk. This is the character displacement model. More modest character displacement has been shown to enhance local differences between at least one pair of preexisting species. Evidence that speciation is promoted when interspecific competition is low comes from the fossil record and from islands. After mass extinctions, or after first colonists reach new islands, ecological release allows rapid radiations of new species. Between times of ecological release, species change slowly as selection keeps pace with environmental change, the so-called Red Queen hypothesis.

Chapter 9

The Ecology of the Origin of Species

The competition model led to the theory of competition avoided. This was the message of the Lotka–Volterra equations and their experimental verification by Gause (see Chapter 8). Individuals of a species population are distinctly different from individuals of related species because the differences serve to minimize interspecific competition. Working ecologists were well prepared by their own field studies to accept this, because the exclusion principle of "one species: one niche" satisfactorily described species as the functional units ecologists knew them to be.

More than this, the formal statement of the exclusion principle by Gause was based on the classical scientific method of model-making and experiment. It at once suggested that formal tests of the principle with field data were possible, so that what was once the subject of musings by gifted naturalists alone became a working body of hypotheses to guide the field work of a young discipline. The first general test of the exclusion principle was already at hand—it lay in the experience of classical taxonomy.

Species are classified by shapes of their bodies, but these shapes reflect function so closely that you can usually deduce much about the niches animals occupy merely by looking at a corpse: talons mean a carnivore, hooves mean fleetness of foot, opposing thumbs mean climbing trees, and so on. The deduction of niche from shape is what paleontologists do every time they reconstruct the life of an extinct animal. To a remarkable extent, therefore, museum taxonomy had already tested the prediction of the exclusion principle that niches of different species should be quite distinct. Indeed, taxonomists may be said to have discovered the workings of the exclusion principle on their own whenever they classified organisms of similar shape by function.

Pathogenic bacteria can be classified by testing them against a host, and parasitic nematodes by the plant hosts in which they are found. This experience of classical taxonomy explains why the exclusion principle could gain ready acceptance as a working tool. A species could be thought of as a morphological expression of the animal's way of life, of its niche. Animal and plant species were unique; they reflected unique niches; one species, one niche.

THE TEST OF CLOSELY RELATED SYMPATRIC SPECIES

Formal testing of the validity of the exclusion principle comes from studies of closely related species living together, or so-called **sympatric species** (literally, *species that live in the same country*). Closely related species might be expected to have similar niches and, therefore, be in danger of competing where their ranges overlapped, which is to say where they were sympatric. The exclusion principle requires that all such sympatric relatives should nevertheless be living in ways that avoid competition.

One of the first to realize the significance of Gause's Moscow studies was a British ornithologist, David Lack, who set himself the task of testing the exclusion principle against all the pairs of closely related sympatric bird species in the British list. He could not find enough data on feeding habits to be sure about many, but for all those for which there were good data he was able to show that they fed differently (Lack, 1945). One of the nicest sets of data Lack could find was for the two species of British cormorant, the common cormorant *Phalacrocorax carbo* and the shag *P. aristotelis*. These two birds are strikingly similar (Figure 9.1). They live on the same stretches of shore, they both feed by swimming underwater after fish, and they both nest on sea cliffs. Lack used data on stomach contents collected by government agents investigating complaints against cormorants by fishermen to determine their diets.

Shags ate mostly sand eels and sprats. The common cormorants ate various things, particularly shrimps, as well as a few small flatfish but no sand eels or sprats. The food of the two species was obviously quite different, so that they avoided competition and the exclusion principle was upheld. The fisheries study had shown further how the catching of different fish was ensured, because the shags did their fishing in shallow estuaries while the common cormorants went farther out to sea. Lack was also able to show that the nesting requirements of the two birds were different even though they did nest on the same cliffs. The shags nested low among boulders or on narrow ledges, whereas the common cormorant nested on the high tops or on broad ledges. In short, these closely related birds, so similar to look at, had niches that were quite distinct. In their normal lives they were unlikely ever to come into serious competition.

Since Lack's early work, two generations of ecologists have filled the literature with case studies of how competition is muted between closely related sympatric species, many of them studies of birds but including virtually all kinds of animals (e.g., Pulliam, 1983; Pratt and Stiles, 1985; Schoener, 1974).

Probably all ecologists now have their own personal tales of sympatric species and competition avoided, whether they have taken the trouble to publish or not. One vivid in my memory is of hairy and downy woodpeckers a'woodpeckering, one after the other, on the trunk of an old apple tree outside the window of the Ohio farmhouse in which I once lived. These birds are so similar that they are trial for a non-

FIGURE 9.1
***Phalacrocorax carbo* and *P. aristotelis*: sympatric relatives without competition.**
Cormorants and shags nest on the same cliffs and fish in the offshore waters. But cormorants nest high, fish well out to sea, eat many fish but no sand eels or sprats. And shags nest low, fish inshore, and mostly subsist on sand eels and sprats.

ornithologist to tell apart (Figure 9.2). What could be more specialized than a woodpecker's habit of propping on stubby tail and feet while lambasting a tree trunk to get insects? How can it avoid competing with a neighbor who hammers the same tree? Because the hairy woodpecker (*Dryobates villosus*) is somewhat larger than the downy (*D. pubescens*), it hammers with a larger bill and it catches larger prey. Someone shot a few to check their stomach contents, and sure enough these sympatric relatives are not competing for food even though woodpeckering on the same trees.

But our personal anecdotes are really just that, anecdotes. A critic could say that naturalists had always known that different species did things differently, and so what?

MacArthur's Warblers: A Decisive Study of Subtleties

The study that finally settled the matter in many minds was MacArthur's (1958) demonstration of separation of niches by a set of bird species that even ate the same food. The birds, migratory warblers, seemed to defy the rule of competitive exclusion. MacArthur showed that they were no exception; making them, in the lawyers' phrase, the exception that proved the rule.

Migratory warblers are mostly so simi-

FIGURE 9.2
***Dryobates villosus* and *D. pubescens*: different chisels extract different grubs.**
The heavy bill of the larger, hairy woodpecker takes a different portion of the insect crop to that taken by the more delicate chisel of the downy woodpecker. Both species can go woodpeckering on the same trees without serious competition for food.

lar, even in coloring, that learning to tell all except breeding males apart is one of the trials of a bird watcher. In the eastern United States each spring a particularly frustrating assortment of them appears, flying north from their tropical winters along common flyways to their common breeding grounds in the woods of New England and eastern Canada. Five species in particular nest in Maine and Vermont. The vegetation in which they breed is spruce forest without obvious variety. The beaks of the birds are all the same size and are similar, suggesting that they can eat the same food. Investigations of stomach contents have shown that their food is, indeed, roughly the same. Thus the usual measurements of food eaten or habitat preferred revealed no significant differences among the niches of the warblers.

In his approach to this problem, MacArthur laid the foundations not only to its solution, but also to quantitative behavioral studies that have flourished since under the name **behavioral ecology** (Krebs and Davies, 1981). He measured patterns of behavior in the birds, spending many long hours in the springs of several years watching warblers. Each time he saw one he noted exactly where it was: on top of a tree, at the side of a tree, on the ground, flying around, and he started a stopwatch so that he could measure in seconds just how long the warbler spent where it was. He timed the motions of each kind of warbler, noting how long each spent hovering, running along branches, or slowly plodding, and was able to show that each species had a characteristic pattern of doing things. This was a tedious and time-con-

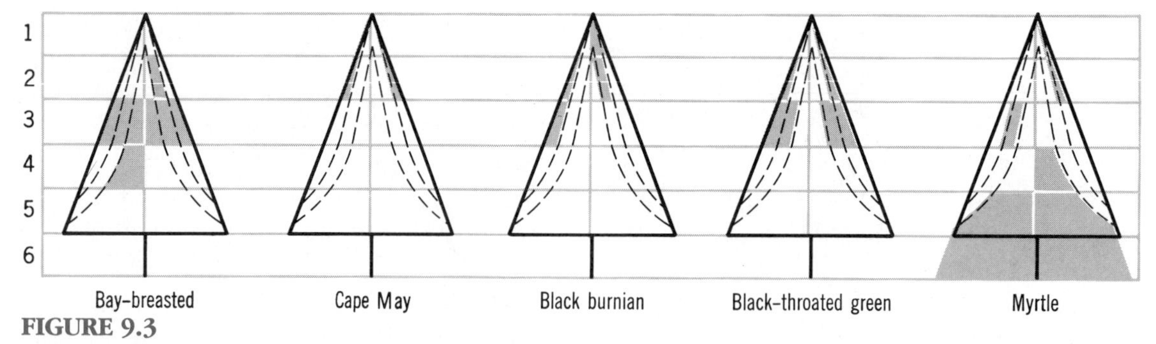

FIGURE 9.3

Separation of niches by warblers eating the same kinds of caterpillars.

The five kinds of warbler feed on bud worms on the same spruce trees. These diagrams illustrate MacArthur's study, which showed that the birds hunt in different parts of the trees so that each kind has a private crop. MacArthur divided the trees into five layers, and then considered each branch as having a top, a middle, and a bottom, which let him divide each tree into 15 compartments. A sixteenth compartment was provided by the ground underneath. MacArthur then noted in which compartment each warbler occurred, timing some of them to see how they divided their time between compartments. The stippled areas show where they spent more than 50% of their time or where they were on more than 50% of his observations. The results of timing the birds are on the left of each tree; the results of other observations are on the right. It is evident that the birds go some way to avoiding competition by hunting in different parts of the trees. Different methods of hunting further separate them. (From MacArthur, 1958.)

suming undertaking, for the little birds are hard to see in the dense spruce forest and they never remain in sight for more than a few seconds. A few summers of watching produced only a few minutes of accurately timed behavior. But this was enough.

The warblers worked in substantially different parts of the trees, each with stereotyped foraging behavior that took a specialized part of the caterpillar crop (Figure 9.3). Sharing a tree by occupying different parts was far from perfect, so that their wanderings overlapped, but behavioral traits stopped them from poaching many of each other's caterpillars even then. One was more active than others; another was more deliberate. There seemed little doubt that these different activities reflected different hunting methods. One kind of warbler got caterpillars on tops of needles, another kind got caterpillars hidden under needles, and so on. Even though a warbler might poach another warbler's space and hunt the same kind of caterpillar there, the two might still minimize competition because the hunting method of each caught a different portion of the total crop.

But MacArthur did not supply only this elegant demonstration of how competition was avoided, because he went on to use his data to calculate the coefficients of competition for sets of Lotka–Volterra equations on the assumption that the warblers compete whenever they stray into each other's part of the spruce trees (MacArthur and Levins, 1968). The assumption is reasonable because all the warblers take the same caterpillar food: time trespassing equals time competing equals a numerical estimate of a coefficient of competition. Thus the relative frequency of trespass and the relative frequencies P_{ih} and P_{jh} of the ith and jth species in the hth microhabitat measures competition relative to P_{ih}^2 and

$$\alpha_{ij} = \sum P_{ih} \cdot P_{jh} \Big/ \sum P_{ih}^2 \qquad (9.1)$$

where α is the coefficient of competition. For each of four of the five species of warbler in Figure 9.3, MacArthur knew the mean number of pairs (N_i) in the five-acre woodlot he studied, and he could calculate the competition coefficients from his stopwatch data and equation 9.1. At equilibrium the Lotka–Volterra statement of relative populations is

$$\frac{dN_i}{dt} = N_{ir_i}(K_i - N_i - \alpha_{ij} N_j) = 0 \quad (9.2)$$

Evaluating K_i from the set of values for N_i and α_{ij} now lets equation 9.2 be solved for each of the four species, and the results can be expressed in matrix notation as

$$K = A\,N \quad (9.3)$$

where K and N are the columns giving the saturation values K_i and the actual populations N_i and

$$A = \begin{matrix} \alpha_{11} & \alpha_{12} & \alpha_{13} & \alpha_{14} \\ \alpha_{21} & \alpha_{22} & \alpha_{23} & \alpha_{24} \\ \alpha_{31} & \alpha_{32} & \alpha_{33} & \alpha_{34} \\ \alpha_{41} & \alpha_{42} & \alpha_{43} & \alpha_{44} \end{matrix}$$

Results for the four warblers: myrtle, black-throated green, blackburnian, and bay-breasted, were

$$A = \begin{matrix} 1 & 0.490 & 0.480 & 0.420 \\ 0.519 & 1 & 0.959 & 0.695 \\ 0.344 & 0.654 & 1 & 0.363 \\ 0.545 & 0.854 & 0.654 & 1 \end{matrix}$$

This resulting matrix is often called a **community matrix**. Unfortunately this gives a meaning to "community" quite different from that generally meant in ecology in that it describes four sympatric warblers as a community by themselves. **Guild matrix** is a better term, for it describes relationships among the guild (see below; Hutchinson, 1978).

In both its qualitative and quantitative forms, the warbler study showed that warbler behavior did indeed minimize competition. The coefficients were low enough for coexistence to be possible. Yet residual competition was present, as must be intuitively obvious for animals feeding on the same population of prey. With their matrix MacArthur and Levins (1968) had measured the relative competitive effect of each species, making possible statements like:

"Blackburnians have a smaller competitive effect on bay-breasteds than on myrtles."

"Black-throated greens are strong competitors of bay-breasteds."

"The presence of blackburnians actually increases the chances of bay-breasteds' being able to live in the habitat (because other warblers should be less numerous if blackburnians were there)."

When avoidance of competition could be understood with this kind of precision, the principle of competitive exclusion, which predicts that interspecific competition must be minimal in coexisting populations, became established as a central working principle of ecology.

The Concept of Guild

Many species sharing a common resource in sympatry like MacArthur's warblers constitute a **guild**, a term first used by Root (1967), who borrowed it from the medieval cooperatives of that name. Members of a medieval guild worked at the same trade, so animals in an ecologist's guild exploit a resource together, though they do it in a way that minimizes niche overlap.

The concept of guild stretches to embrace the self-evident fact that groups of animals sharing a common resource can be put together from animals less related than character displacement pairs. MacArthur's five warblers, for instance, are not all placed in the same genus by ornithologists, yet it is clear that they share a common resource, the spruce bud worms, in ways that minimize competition.

The existence of guilds is predicated on the assumption that resources commonly come in classes that determine parallel ways of life, of which caterpillars, seeds, and flying insects are only some of the more obvious. A special example is provided by the insects present on the barks of trees, which invite the specialized ways of feeding of woodpeckers, nuthatches, and tree-creepers. In any one forest habitat a guild of such bark-feeding birds is possible, each member of which collects bark insects in special ways that minimizes competition with others.

One of the many contributing factors to the high diversity of birds in tropical rain forest is probably that rain forests provide opportunities for specialist guilds not possible in other forests, ant-following guilds, for instance, in which birds hunt insects fleeing the march of army ants, or the guilds of birds that feed on large succulent fruits (Terborgh, 1992).

SPECIES DEFINED IN ECOLOGICAL TERMS

Species are kept distinct by natural selection, removing all individuals that so differ from the species norm that they engage in significant interspecific competition. At the same time, all individuals must be able to compete strongly with their own kind for the resources available to the species niche. The struggle for existence must indeed be stern when the competition is intraspecific. But a consequence of competitive exclusion is relatively peaceful coexistence between species.

Traditional definitions of species stress reproduction, or genetic mechanisms, stating that members of a species can mate with their own kind but not with individuals of other species. In the ecological view, this not mating with others is merely a means to the end of maintaining the specific niche. Crossbreeds would likely be deviants unable to struggle effectively with their own kind, or likely to suffer ruinous interspecific competition with others in a guild. Thus traits preventing crossbreeding are preserved by natural selection.

The real significance of the discrete species population is that it shares a common niche while keeping interspecific competition to a minimum. I offer an ecological definition of a species: *A species is a number of related populations the members of which compete more with their own kind than with members of other species.*

An Ecological Model for the Origin

The principle of competitive exclusion not only requires species populations to be distinct, it also allows a ready explanation of how new species are forged, thus addressing directly the Darwinian mystery of the origin of species.

Consider two hypothetical, closely related populations *A* and *B* of similar but not identical food habits that have long lived in geographic isolation from each other (i.e. they were **allopatric**, living in other countries). Suppose that some accident of history has introduced both populations into the same habitat (sympatry) so that the two must try to partition their resources as in Figure 9.4. Competition between the two populations is the inevitable result. The simplest outcome is the elimination of one population according to the exclusion principle. But if the merging populations have passed some critical threshold of difference, an alternative result is possible.

Long isolation in allopatry should equip the merging populations with different, if overlapping, resource needs. In particular, individuals at the extremes of variability

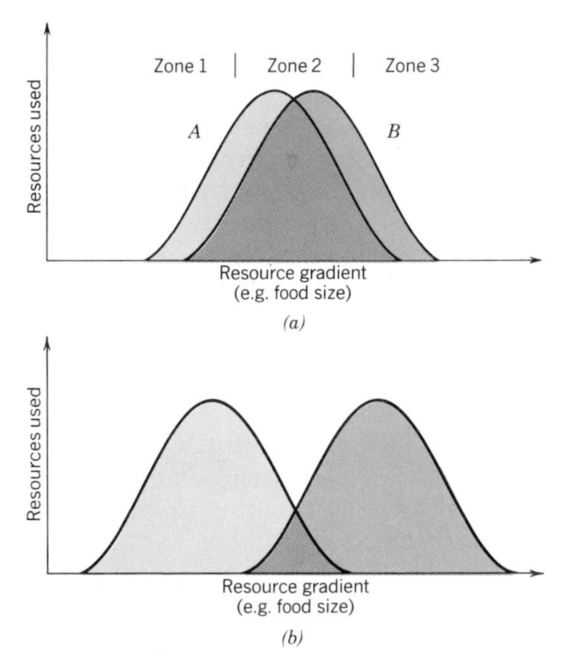

FIGURE 9.4
The concept of character displacement
Individuals of species *A* and *B* whose niche preferences bring them into competition will leave fewer surviving offspring than individuals who avoid competition. In time niche-overlap will be reduced and the spread of resource use in sympatric populations will change from a pattern like that in (*a*) to a pattern like that in (*b*).

allowed in each population should possess traits not found at all in the other population. These extreme individuals might escape interspecific competition altogether. In Figure 9.4*a* these are the individuals whose resource needs fall to the left of the figure for population *A* and to the right of the figure for population *B*. These individuals should leave more surviving offspring than the mass of individuals of both populations whose resource needs fall in the zone of overlap, zone 2.

Feeding in zone 1 results in high fitness. Feeding in zone 2 results in exclusion and low fitness. Feeding in zone 3 results in high fitness. After a few generations natural selection should see to it that genes for feeding in zone 2 become rare in the population, but that genes for extreme forms of both populations multiply. Because extreme types have strong advantages, we should actually expect even more extreme types to appear, with the result that the distribution of traits within the two populations would be as in Figure 9.4*b*.

Natural selection, through favoring individuals that suffer least interspecific competition, thus should separate the mean traits of competing populations, always with the proviso that individuals with marked differences of habit already exist in the populations. What Darwin called "**divergence of character**" is thus enhanced by selection favoring extreme types. The ecologists who first formally suggested this mechanism called it **character displacement** (Brown and Wilson, 1956).

Character displacement becomes another in the string of predictions developed from the Lotka–Volterra equations of competition. First was competitive exclusion, with its implication that interspecific competition must be minimal in all sympatric species; a natural doctrine of peaceful coexistence. Next is the satisfactory explanation of the distinctness of species, resulting because individuals too similar to foreign species suffer excess competition and are excluded. And lastly character displacement accounts for the origin of species themselves.

This model for the origin relies on initial divergence of character in isolation, in common with all preceding thoughts about the speciation process. By far the most likely method of isolation is separation by geography into far-off populations, leading to a life in allopatry during which mean characters shift under selection for local circumstance or by genetic drift. But the ecological model puts reasonable limits on how far characters must diverge by chance. Merging populations do not have to be utterly different for new species to be forged, because selection to avoid competition will displace the divergent characters rapidly once a critical threshold of mean population difference is reached. Once the presumptive species populations are coexisting in sympatry, any trait serving to prevent the production of intermediate offspring will be preserved by natural selection. Genetic isolating mechanisms thus should soon appear and the establishment of two new species should thereby be confirmed.

The Search for Character Displacement

Most species are massively ancient on a scale of human lifetimes. Our forest trees are Tertiary, some of them early Tertiary, and more than 50 million years old (see Chapter 1). Virtually all the species of macroscopic plants and animals still living probably antedate the last ice age and have been in existence for hundreds of thousands, if not millions, of years. All a contemporary biologist can hope to see is part of the predicted process of character displacement in progress. Two plausible examples of character displacement in action have been proposed: nuthatches in Asia and Darwin's finches. Both examples have been used to postulate the fitting together of overlapping niches of species already genetically isolated, not the formation of new species. Demonstrating even this modest consequence of competition has been difficult enough.

Nuthatches (family Sittidae) are small birds with long woodpecker-like bills that search for food by climbing up and down trees or rocks hunting in cracks and crannies. The physical adaptations for this way of life (long bills, short stubby tails, and big feet) give nuthatches a characteristic appearance, making them all look much alike (Figure 9.5). Strikingly similar are the common species of Greece and Turkey, *Sitta neumayer*, and of central Asia, *S. tephronota*. The nuthatches of these sites 1000 miles apart, so closely related and so

FIGURE 9.5
***Sitta carolinensis:* sympatric nuthatches need to be different.**
The shape and stance of this bird imply specialized feeding. Where sympatric, two Asian species of nuthatch are found to have bills of different length; where allopatric, traits of the two Asian nuthatch species overlap.

similar, should be occupying closely similar niches. Whatever differences exist between them may be explained in terms of their geographical isolation.

In Iran, which lies between the two distant populations, the ranges of the nuthatches overlap and they live in sympatry (Figure 9.6). But the Iranian populations are distinctly different (Vaurie, 1951). Their bills are of different sizes and one bird has a thick black stripe from eye to shoulder, whereas the other has almost lost its eye stripe. The different bill lengths suggest different feeding habits, and the different markings suggest distinctive patterns that could be used in recognition of suitable mates. These are just the sorts of differences to be expected if characters have been displaced in the zone of overlap (Brown and Wilson, 1956). Unfortunately, a closer looks at the data shows that the story is not so simple.

Grant (1975) showed that, except in the matter of the eye stripes, the Iranian populations could be fitted to gradients of types that continued into the surrounding regions where a species lived alone, the regions of *allopatry*. Bill lengths in particular were clinal, so that the different lengths in the zone of overlap could be no more than a coincidence as the big-beaked end of one cline crossed the small beaked end of the other (Figure 9.7*a*). Selection for bill length to avoid competition in Iran was thus not necessary to explain the result (which is not to say that the proposed selection for bill size did not occur, merely that it was not necessary to explain the data).

Prominence of eye stripes, however, are not clinal, their distributions being like those shown in Figure 9.7*b*. Moreover, the individuals that have small eye stripes in Iran probably pay a price for the loss of pigment around the eye. Pigment gives a very decided advantage in regions of strong sunlight as a device for reducing glare and improving clearness of vision. But eye

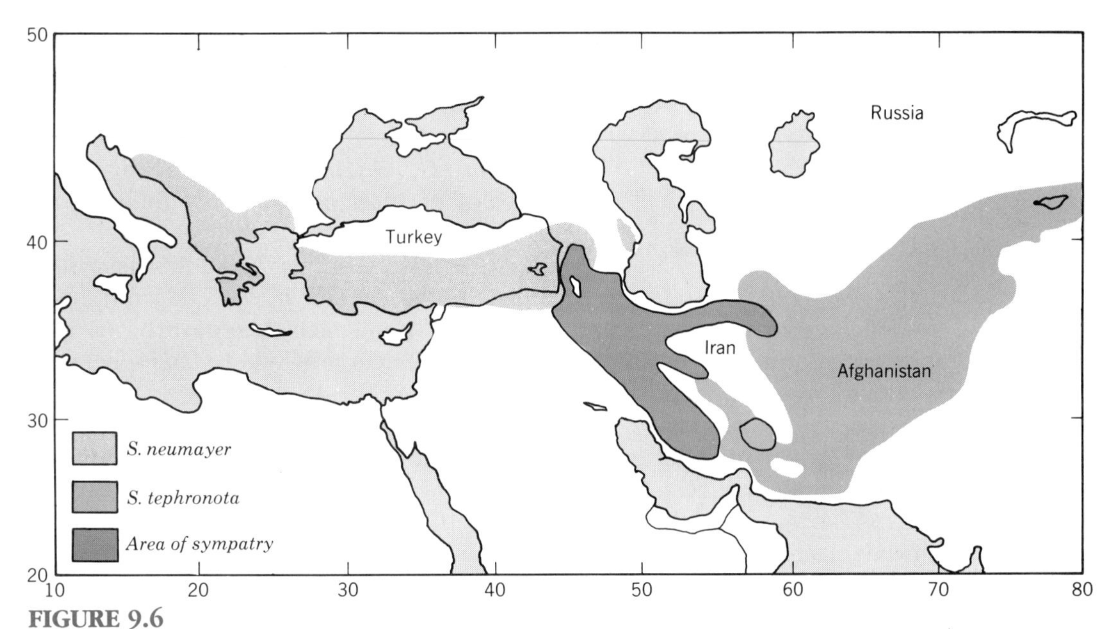

FIGURE 9.6
Nuthatch distributions and the zone of sympatry.

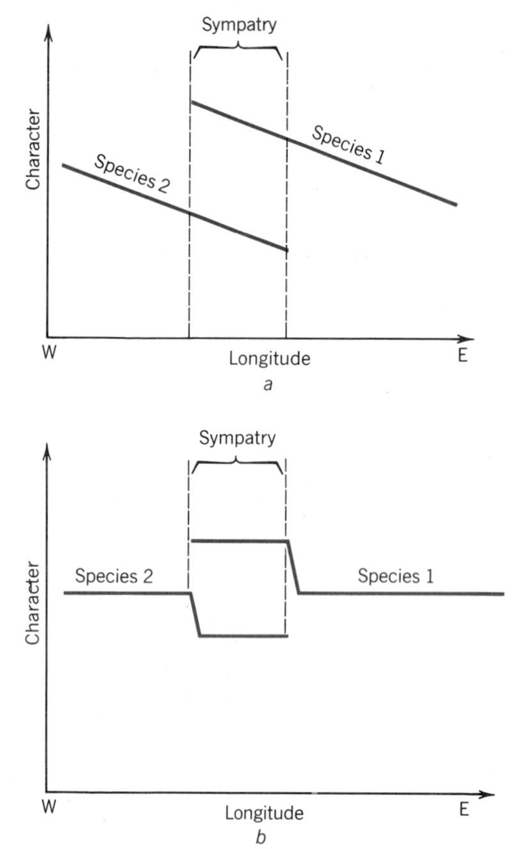

FIGURE 9.7
Alternative explanations of differences in sympatry.
If two species have characters such as size graded in space (a *cline*), then they may automatically have different characteristics in the zone of sympatry (*a*). If differences at sympatry are solely the result of character displacement, characters should be distributed in space as in (*b*).

stripes are important signals in helping the nuthatches choose mates of the right species. Grant tested wild nuthatches in the breeding season with models and stuffed skins to see what the response would be. Sexual displays, aggression, or indifference showed clearly whether the model was recognized as being of the same species or not. Grant found that the length of the eye stripe was all-important, and concluded that reproductive selection for large or small eye stripe caused the difference in the zone of sympatry.

Although, therefore, the obvious difference in bill length between the two nuthatches in the zone of sympatry could be a coincidence convenient for the birds, selection seems definitely to have displaced the characters needed for species recognition. Selection, therefore, has certainly worked to emphasize living in the separate niches and has helped to isolate the species where the ranges overlap.

But the most unequivocal example of character displacement in action is provided by Darwin's finches on the Galapagos Islands. The possibility was pointed out early (Lack, 1947), though only recently has it been possible to demonstrate that character displacement had actually occurred (Grant, 1991). Lack's original observation was that where the two species *Geospiza fulginosa* and *G. fortis* lived together, as they do on two of the larger Galapagos Islands, their beak sizes are quite separate with no overlap. But the tiny island of Daphne held only *G. fortis*, and these birds had beaks overlapping the range of sizes of both species on the larger islands. On the equally tiny Crossman Island, *G. fulginosa* lived alone, again with beaks of intermediate size (Figures 9.8, 9.9).

The pattern shown by Lack's data in Figure 9.9 is precisely what should be expected if beak size is separated by character displacement where the species live together on the larger island. In the isolated populations on the smaller islands, selection, faced with no interspecific competition, favors individuals with an intermediate size of beak. But there is a catch to this argument, because it is known that beak size in these finches is correlated with the size of seeds they eat. Natural selection tracks seed size with beak size.

FIGURE 9.8

Darwin's finch habitats of vastly different size.

The small island of Daphne is only an islet in a bay of the much larger Santa Cruz island. *Geospiza fortis* finches, the only Galapagos finches on Daphne, have smaller bills on the small island, where other finches are present. The smaller bills on Daphne of *G. fortis* have been shown to result from lack of competition rather than being dependent on seed size.

Thus the pattern shown in Figure 9.9 could potentially have nothing to do with interspecific competition between finches and everything to do with the size of seeds available on the different islands (Bowman, 1961; Grant, 1986).

The recent development has been field measurements designed to control for seed size (Schluter et al., 1985). The procedure is as follows:

1. Identify seeds available to each bill size.
2. Measure finch density for each species as a function of seed size.
3. Measure seed density of Santa Cruz, Daphne, and Crossman Islands.
4. Compute finch density per bill size expected for each of the three islands on the assumption that density of each bill size is set only by available seeds.

The expected density of each bill size can then be compared with actual densities on the three islands. Departures from the predicted densities must then be due to some factor other than seed size (Figure 9.10). On Santa Cruz, where the species live in sympatry, departures from predicted size were large. On Daphne, the isolated population of *G. fortis* had bills as predicted by seed distributions. On Crossman, the isolated population of *G. fuliginosa* had its beak size shifted to the seed peak that would have been occupied by *G. fortis* had this been present. These results are entirely consistent with the observed shifts in bill size being due to the presence or absence of competition, and they demolish the alternative explanation of the pattern is being set by the distribution of seed size.

Thus character displacement within populations of existing species has been demonstrated in at least one guild of birds. It remains plausible that a similar mechanism could be important in fixing the original characteristics of species populations, though we cannot test this by direct observation.

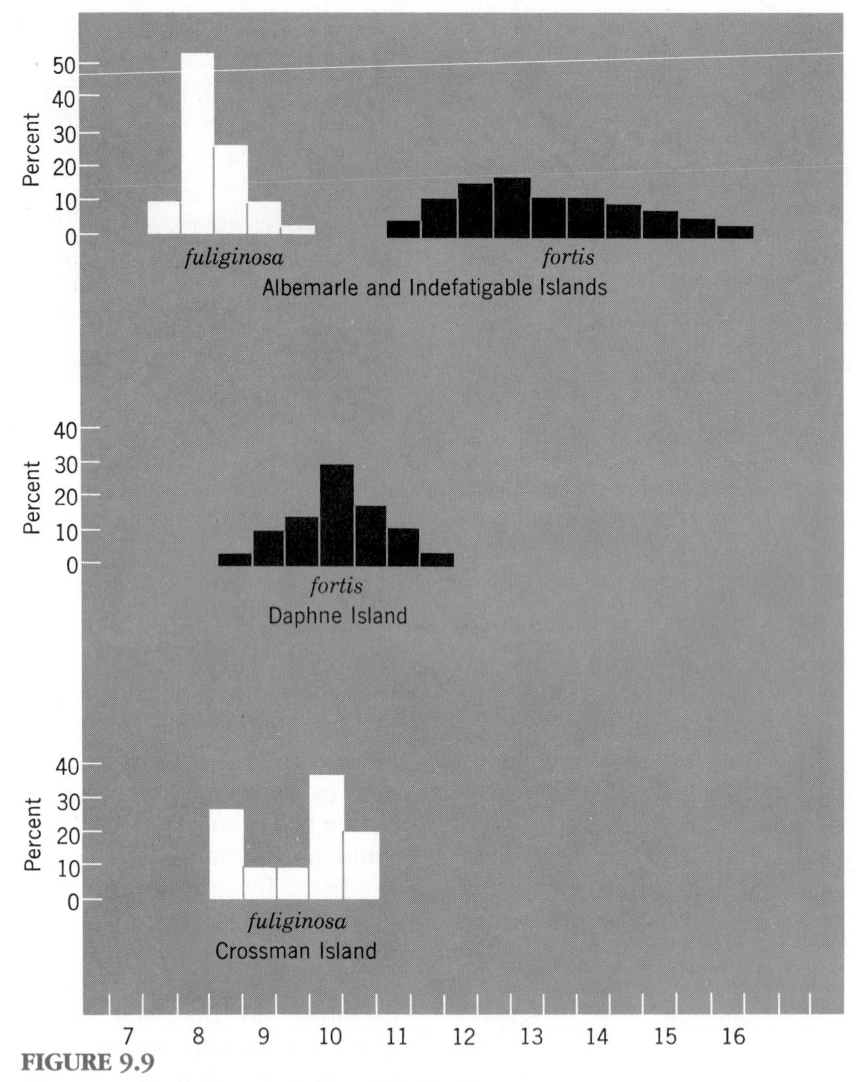

FIGURE 9.9

Character displacement in Darwin's finches.

Histograms of beak depth in *Geospiza* species (Darwin's finches). Measurements in millimeters are placed horizontally and the percentage of specimens of each size are shown vertically. If the beak depths on Daphne and Crossman Islands are indicative of optima in the absence of close competitors, the displacement on islands with competitors is most easily interpreted as character displacement. An alternative explanation is that the sizes of available seeds vary on large and small islands, and that natural selection tracks seed size with bill size. (Lack, 1947.)

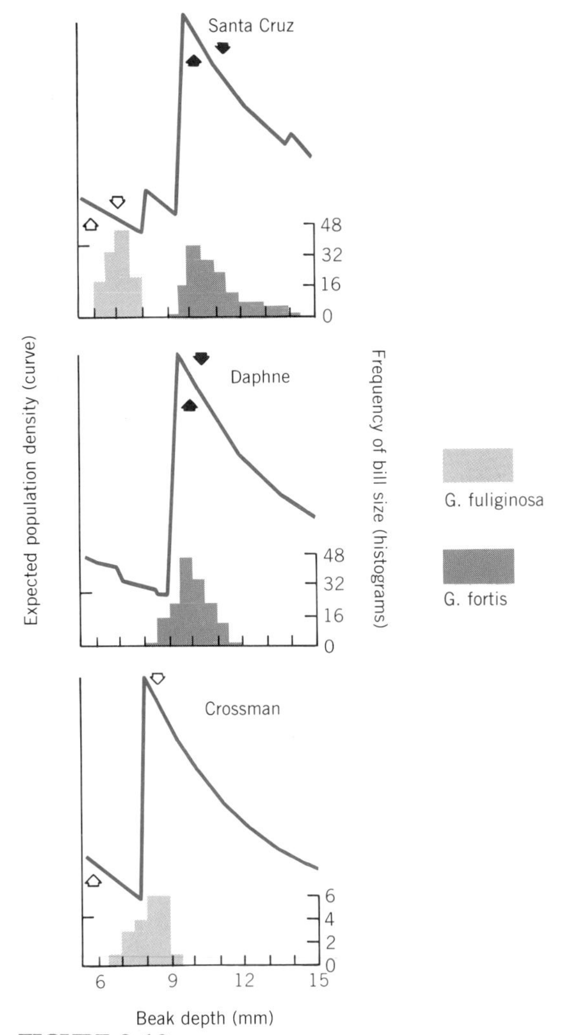

FIGURE 9.10

Character displacement in Darwin's finches confirmed.

Curves and upward arrows are predicted population densities if seed size alone determined bill size. Histograms and downward facing arrows are observed frequencies of bill size. Bill sizes in sympatry (Santa Cruz) differ more from incidence predicted by seed data than in allopatry (Daphne). On Crossman, selection has moved the allopatric population of *G. fortis* from one seed size specialization to the other, but frequency of bill size again closely coincides with one of the sizes predicted from seed data. Character displacement is confirmed in the sympatric populations. (From Grant, 1986.)

Populations Kept Apart by Competitive Exclusion

Water ferns of the genus *Azolla* on the Galapagos Islands reveal a history that suggests how ranges may be kept separate in nature by competition when differences of species in allopatry are not sufficient to permit a tenuous first coexistence of immigrants. *Azolla* grows floating on small ponds like duckweed, completely covering the water (Figure 9.11). Only one species, *A. microphylla*, is known from the islands.

The record of *Azolla* spores in drill cores of mud from a Galapagos lake shows the history of the genus on the islands (Schofield and Colinvaux, 1969). *A. microphylla* spores are abundant in all the mud deposited over the last 10,000 years. But in mud deep down in the sediments, shown by radiocarbon dating to be more than 48,000 years old, we find only spores of another species, *A. filiculoides*, a plant known from South America but not now growing anywhere in the Galapagos. Sometime before 10,000 years ago one species of *Azolla* replaced another.

A parsimonious explanation for this replacement of species invokes both the climatic changes of the last ice age and the principle of competitive exclusion. When glaciation was most extensive 18,000 years ago, the Galapagos had a drier climate even than now (Colinvaux, 1972). The lake with the *Azolla* spores dried completely, as recorded by red clay deposits from ice-age time, and drought probably drove *Azolla* populations to extinction on the islands. When the moister climate of the postglacial period began 10,000 years ago, the islands were colonized by *A. microphylla*, presumably brought from mainland South America on ducks' feet.

But we know that before the ice-age extinctions the islands were inhabited by *A. filiculoides*. If both *Azolla* species can

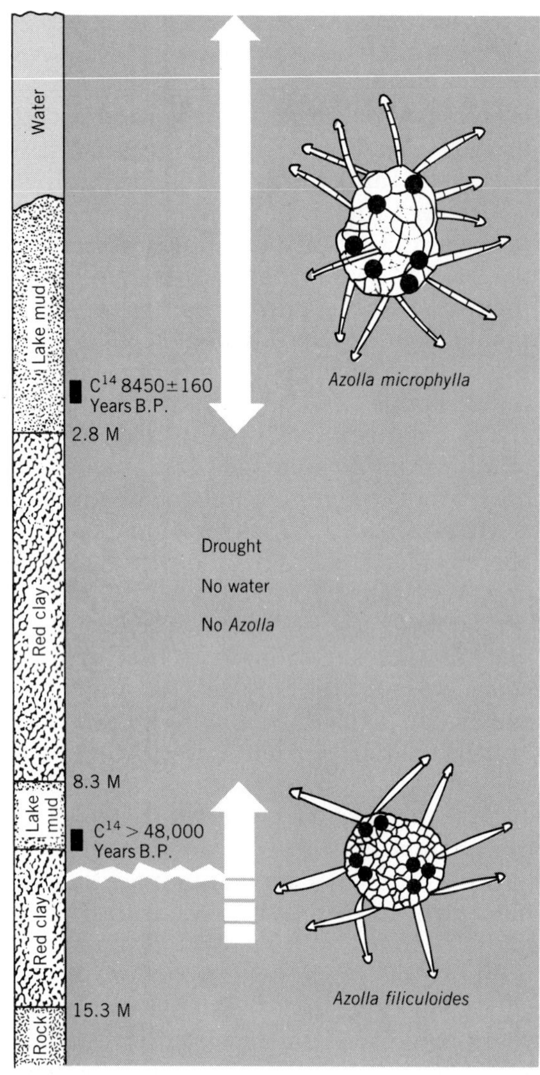

FIGURE 9.11
Water ferns: competitive exclusion.
Water ferns of the genus *Azolla* loosen their spores into the water encased in structures armed with hooks and called massulae. The shapes of these structures are different in the different species. The lake dried up for many thousands of years in its middle history, after which the new lake was occupied by a different species of *Azolla*, and the only one now known from the Galapagos Islands. This history probably illustrates competitive exclusion, the triumphant species each time being that which reached the Galapagos first by chance transport from mainland South America in each of the wet epochs. The long drought between the two invasions represented the ice-age climate of the Galapagos Islands.

reach the islands, then the presence of only one at any one time suggests that the *Azolla* niche is preempted by possession so that later immigrants are excluded.

The Ecological Hypothesis for Species Radiations

Species come and go on geological time scales. The fossil record of their passing is most imperfect, letting us see more the rhythm of passing clades than of individual species. Genera, families, or orders pass before our view in the geological column, each with its brief appearance of a few million, or tens of millions, of years. But the passing of the clades is not always a steady progression, being interrupted by mass extinctions that are followed by episodes of species creation called **radiations**.

Mass extinctions attract the human mind, with its morbid concern for catastrophe. And in truth the causes of mass extinctions provide intellectual puzzles of high quality. The trail of reasoning that led from anonymously high iridium concentrations in a layer of soil to the hypothesis of a collision between the earth and a hurtling sphere (bolide) from outer space, causing an explosion so devastating that the environmental effects eliminated the dinosaurs and much else besides, is one of the *tours de force* of science (Alvarez et al., 1980). And the quest is on for the causes of others in the sequence of mass extinctions that have defined the fossil record (Raup and Sepkoski, 1986; Stanley, 1987).

These true mass extinctions clear away species regardless of their relationships, their niches, their function, their geography, or even their numerical abundance (Jablonski, 1989). Figure 9.12 shows extinction profiles for a random sample of marine fossils from throughout the last 500 million years or so, confirming that

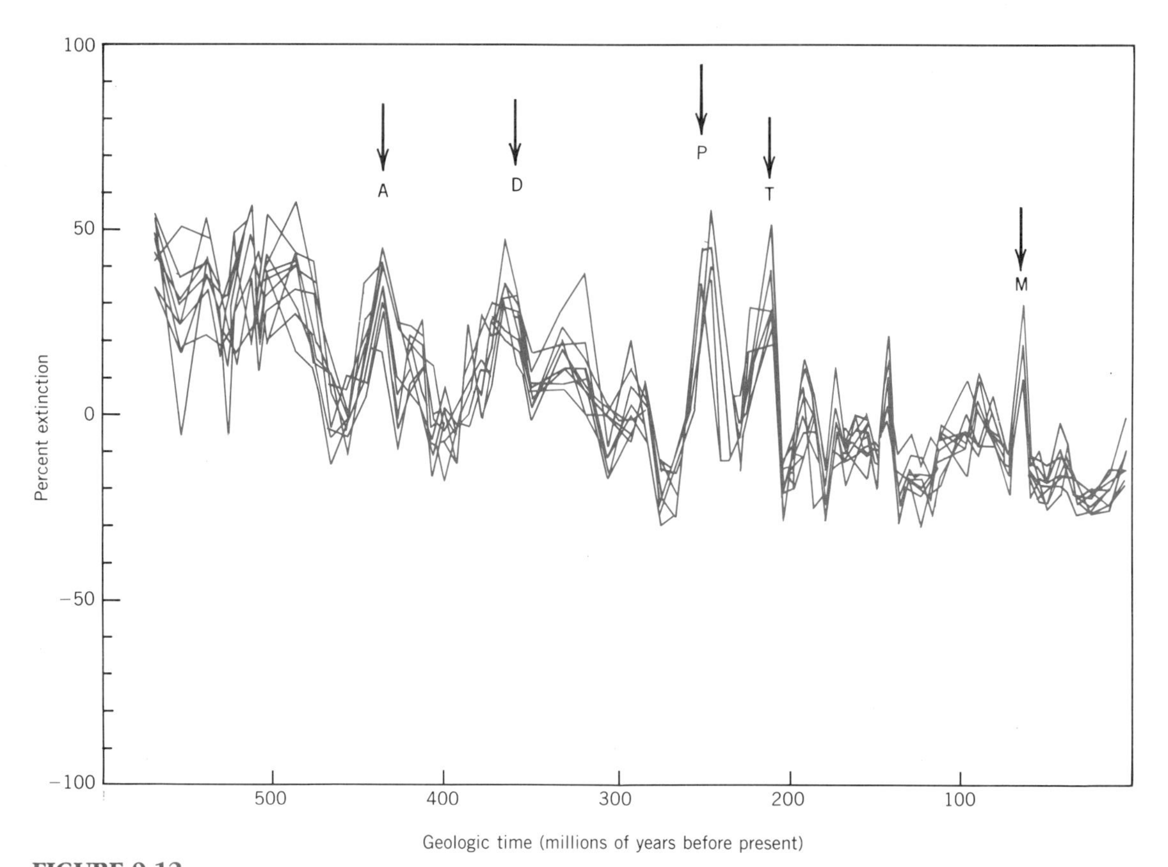

FIGURE 9.12
Mass extinctions strike at random.
Disappearance of marine taxa from ten randomly chosen groups from the fossil record appears synchronous, with peak removal of all kinds at the five great extinctions: Ashgill, Late Devonian, Late Permian, Late Triassic, and Maestrichtion. (From Raup and Boyajian 1988.)

the five largest extinctions struck all groups indiscriminately.

Each of the mass extinctions is followed by the rapid appearance of new species, as whole new clades occupy niches previously occupied by unrelated animals. A striking example is the radiation of mammals following the extinction of the dinosaurs, although early mammals appear to be as old as the dinosaurs themselves. Equally striking was the radiation of dinosaurs following the demise of the therapsids (mammal-like reptiles) in an earlier extinction (Kemp, 1982). It is now becoming clear that each of these radiations of species took place into a world in which potential competitors had already been removed (Jablonski, 1986). These were not competitive replacements; they were speciations into empty niche space.

In the long span of the fossil record, therefore, the appearance of new taxa is of two kinds. First is normality; prolonged periods when turnover of taxa is roughly constant. New species appear in established orders, families, or genera, as others are withdrawn by extinction. New inventions are exploited, as when rock boring by clams led to a radiation of boring species in the Triassic (Jablonski, 1989). But wholesale replacements of the niches of one clade by members of another do not happen.

Second are the wholesale replacements that follow mass extinctions. In these, whole new clades, representing different organizations of animals (e.g. mammals for reptiles), produce species that occupy niches once held by members of other clades. The parsimonious explanation for these radiations is that the previous extinctions removed potential competitors. This is the **ecological release hypothesis** (Jablonski, 1991).

The ecological release hypothesis is consistent with the record of speciation throughout the fossil record. In the normal, or background, condition, speciation takes place in a world in which all attempts at novelty will encounter interspecific competition and thus will be resisted. But when mass extinction is imposed from outside the system, the surviving animals, probably drawn at random from the clades existing before the event, are released from competition until niche space is again preempted for another prolonged period of competition-mediated normality.

The latest in the series of mass extinctions imposed from outside the systems is, of course, now taking place by human action. The thought that this will lead to exciting evolutionary developments following ecological release might be tempered, however, by the reflection that these brief episodes in evolutionary time yet take several million years to work themselves out.

Ecological Release into Empty Islands

The emergence of new islands from beneath the sea creates unused resources as decisively as do mass extinctions. This must be why interesting endemic species are found on remote islands. The earliest colonists live somewhat as the survivors of an extinction event must have lived. Their universe is inhabited, perhaps densely inhabited, but only by a limited array of species. After a mass extinction, the inhabitants are descendants of surviving species. On a remote island, the inhabitants are descendants of those few species able to colonize so remote a place. But the effect of life in a world of reduced interspecific competition is the same. A species radiation follows.

The radiation on islands consequent on ecological release can be swift. The time needed to produce 14 species of Galapagos finch out of one common ancestor has been estimated by electrophoretic analysis of proteins (Yang and Patton, 1981). The method assumes that selectively neutral alleles coding for proteins are substituted at a constant rate, and that this rate is known by calibration, including the use of datable fossils. This is a widely used biochemical tool. Results suggest that the oldest separation from the ancestral line could have occurred as recently as half a million years ago. However, the uncertainty of the assumptions behind the dating method make plausible older ages, back even as far as origin of the islands, dated fairly securely at about 5 million years. But the half-million years for 14 species figure shows what might be possible for birds when potential interspecific competition is minimal: fast by evolutionary standards but not giving an ornithologist much chance of watching while character displacement takes place.

Background: The Running of the Red Queen

In epochs between radiations, with interspecific competition strong and all niches filled, it is tempting to ask whether evolutionary change can happen at all. Each organism is constrained to its species niche by selection that removes all deviant offspring, particularly those that might tend to compete with neighbors. In a sense, all organisms are then perfectly adapted to the niche they are in, and therefore cannot be adapted any further by natural selection. On the one hand, selection should always be winnowing the babies, promoting change; on the other, the near impossibility of speciating and the perfection of niche seem to decree that change is impossible.

This apparent paradox is resolved by the Red Queen hypothesis (Van Valen, 1973). This hypothesis states that the environment of all organisms is always changing, either from climatic or habitat alterations or from the arrival or removal of other species. But if adaptation to the existing environment were already perfect, then all changes in the environment represent effective decay. Natural selection continually adapts organisms to meet these changes. The process is not one of perfecting an adaptation that is already effectively perfect, but of troubleshooting an adaptation for an environment that always changes for the worse. Evolution "runs hard" just to maintain perfection, like the Red Queen in Lewis Carroll's *Through the Looking Glass*, who told Alice she ran so hard to stay where she was.

A test of the Red Queen hypothesis is provided by the fossil record. If the hypothesis is correct, then the survival time of comparable species should be roughly similar on the average. But if natural selection really worked to make species better adapted all the time, then the longer a species had been exposed to natural selection the better adapted it would be and the smaller would be its chances of going extinct. The fossil record (Figure 9.13) for a number of evolutionary lines for which

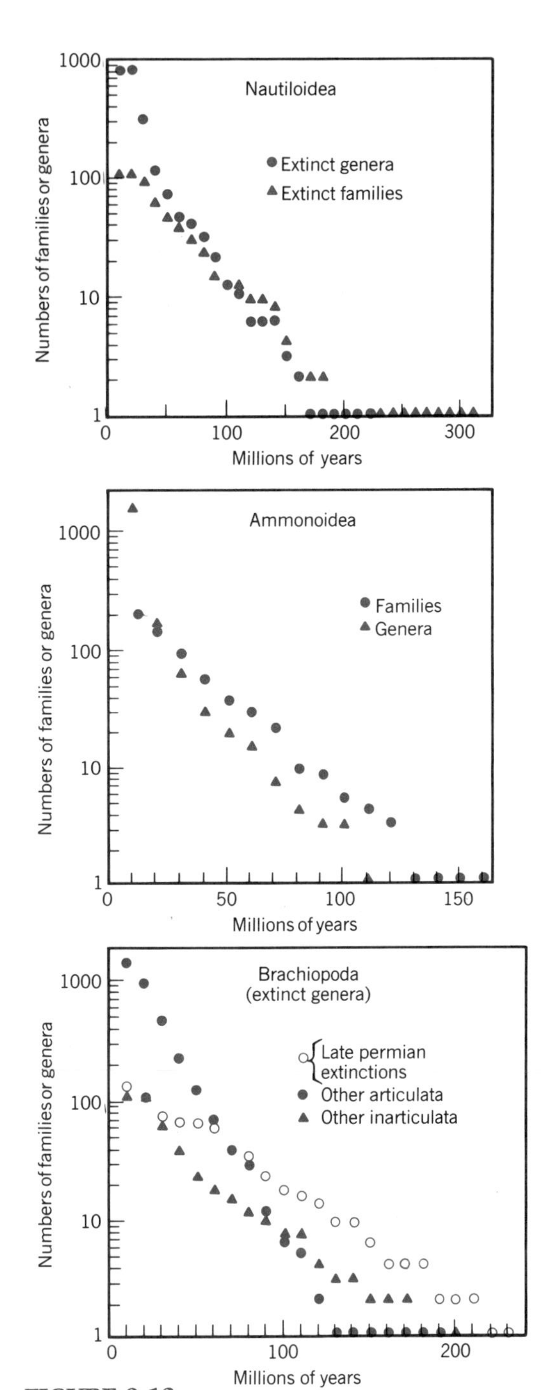

FIGURE 9.13

Test of the Red Queen hypothesis.

The plots are survival times in two groups of mollusks and one of brachiopods. The Red Queen hypothesis predicts that survival will be independent of the age of a taxon, which the data confirm. (From Van Valen, 1973.)

good data are available shows the constant extinction predicted by the Red Queen hypothesis.

Thus background normality between times of mass extinction and subsequent radiation can be understood as the result of a living system narrowly constrained by the imminence of competition. Novelties and brave experiments are forbidden, unless they are so neat that they open up an approach to resources hitherto untouched—the invention of flight, for instance. But conditions for life do change subtly for each generation, by changing population, climate, or geographic range, and natural selection winnows the offspring accordingly.

SUMMARY: A UNIVERSE OF DISTINCT SPECIES

Starting from the logic of the Lotka–Volterra equations, a satisfying ecological model for the origin of species does indeed result. Strangely, the fact that the logistic equations used are a simplistic, and far from universal, description of population growth does not affect the universality of the final theorem. Competitive exclusion would be predicted by more realistic mathematical expressions of competition under confinement, and it is this condition of competitive exclusion that is the basis of our model for the origin of species.

Central to this model is the concept of interspecific competition avoided, for to compete strongly is to risk exclusion. Selection to avoid interspecific competition in populations with initially overlapping habits is a plausible method for the separation of new species. Rapid speciation would be expected at times of ecological release following mass extinctions, which is consistent with the evidence for radiation of species at those times. Slower background speciation at other times can be understood if natural selection constantly adjusts species populations to changing circumstances. Taken together, these arguments give a convincing account of how a universe of distinct Linnaean species should be the logical outcome of a world in which natural selection operates.

Preview to Chapter 10

Sula capensis: Breeding every year, do numbers stay the same?

ECOLOGISTS have long argued about the real importance of competition and density dependence in wild populations. For some, the success of the logistic hypothesis in predicting competitive exclusion, together with the apparent constancy in number of familiar birds, were compelling reasons to conclude that almost all populations were regulated by density dependence. Other ecologists, particularly those working with insects, denied that most animals were under density-dependent control, noting that dense insect populations usually met catastrophic and density-independent death following a change in the weather. Predators like those used in biological controls also devastated prey so that prey numbers were seldom numerous enough for density-dependent effects to matter. Contemporary theoretical models of population dynamics using difference equations, however, show that the old concept that density dependence necessarily yielded equilibrium or stable populations was wrong, so that it becomes nearly impossible to deduce the system of control from data on population histories alone. When per capita intrinsic rates of increase are high, these new models of density dependence can predict periodic oscillations, or even population histories indistinguishable from chaos. These population dynamics offer parsimonious explanations of naturally oscillating populations like lemming cycles. Recent statistical analysis of 30 year histories of fluctuating insect numbers demonstrates unequivocally that the fluctuations are within bounds set by density dependence. Within these bounds the weather, random walk phenomena and the effects of high fecundity on separated generations produce the wide variability in numbers that we see. On ecological timescales populations are rather narrowly constrained as the common stay common and the rare stay rare.

Chapter 10

The Natural Regulation of Number

The common stay common and the rare stay rare. These are among the more merciful of the properties of the world around us, providing a comforting sense of normality to life. Lay people and scientists alike refer to a vague something called "the balance of nature," which ensures this comforting normality.

Control on number is inherent in the way that life is organized into species populations with fixed niches. If the niche is a narrowly defined trade or profession, then the opportunities for life will be as narrowly limited. Niche sets number, or at least an upper limit to number (see Chapter 1). The sky holds few hawks and many pigeons, and will continue to do so as long as hawks and pigeons exist.

Yet it remains true that all organisms have too many babies as an inevitable consequence of the working of natural selection. Populations, therefore, must be continuously culled to fit numbers to niches. Culling may be applied ever more strongly as individuals are crowded together, a concept called "density dependence" (see Chapter 8). This cull could work simply by raising the death rate or by interfering with breeding as competition makes resources for reproduction scarce. Or the density-dependent cull could be preempted by factors like drought or fire that kill whether individuals are crowded or not, so-called density independence. Another possibility is predation by efficient predators that keep local prey populations low. One way or another, rising tides of young are always prevented. The common stay common, the rare stay rare, and numbers are fitted to opportunities for the different niches.

In some circumstances the "balance" in nature seems elegantly done. The eighteenth century English naturalist Gilbert White remarked that, year after year, just eight pairs of swifts nested in the village of Selborne, although, as Gilbert White put it, the "eight pairs breed yearly eight pairs more." Two hundred years later, with some modest changes in the neighborhood, the breeding population of swifts at Selborne is 12 pairs (Lawton and May, 1983). Such constancy is common for breeding birds (Figure 10.1). The constancy in the number of familiar breeding birds in the north temperate springs doubtless colors the impression held in Western civilization of the fineness of nature's balance.

Yet for most species the balance of nature can be a vexingly irregular balance. Insect numbers come and go, with populations in some years a thousand times the populations of years in between. Locusts have the classic boom and bust population histories, but Australian grasshoppers (discussed in this chapter) come close, and many similar moth histories are known (Figure 10.2). Even without discovering plagues, it seems that any long census of insect numbers reveals wide fluctuations, and this is true in all latitudes, temperate, tropical, and arctic. Figure 10.3 shows a long run of counts for a Japanese moth coming to light traps. Ecologists once had a suspicion that irregular histories of this kind might have something to do with the strong seasonality of temperate climates like those of Japan and Europe, but light trap counts for tropical forest now show that tropical insect numbers fluctuate just as much (Wolda, 1992).

More striking even than large fluctuations in number are the haunting periodicities of many populations. Lemming numbers multiply every four years or so, and arctic hares every ten years (Figures 10.4, 10.5). A ten-year cycle for the larch budmoth in Switzerland had been documented for more than a hundred years, in which the population highs tend to be four orders of magnitude larger than the lows (Figure 10.6).

Thus the evidence suggests that natural populations have histories that range from extreme constancy through irregularity to wild fluctuations that look random or chaotic. And in the midst of these are the fascinating populations that oscillate. Of this mixture is the balance of nature made.

A modern synthesis of the processes leading to this whole range of population histories is at last emerging (May, 1986*a*, *b*; Chesson, 1985). The key is to understand the dynamics of resource competition as generations change, while at the same time allowing for forcing effects of environmental change. But some mighty debates had to be waged before a guild of mathematical ecologists could forge the new synthesis.

For ten or fifteen years in the 1950s and 1960s, many ecologists debated whether numbers of most organisms were under density-dependent control or not. A key year was 1954, when books were published setting out rival views of the matter, "the natural regulation of animal numbers" (Lack, 1954) and "the distribution and abundance of animals" (Andrewartha and Birch, 1954). The Lack book, drawing much of its evidence from studies of birds, set out the view that most populations most of the time were stable at a density-dependent equilibrium. In an influential paper published the same year, Nicholson (1954) also argued for density-dependent normality, though he supported his contentions with experimental data from blow flies feeding on carcass meat.

The Andrewartha and Birch book explicitly denied the contention that density-dependent equilibria were common. Instead the authors used other insect data and the experience of entomologists in strongly seasonal climates to argue that

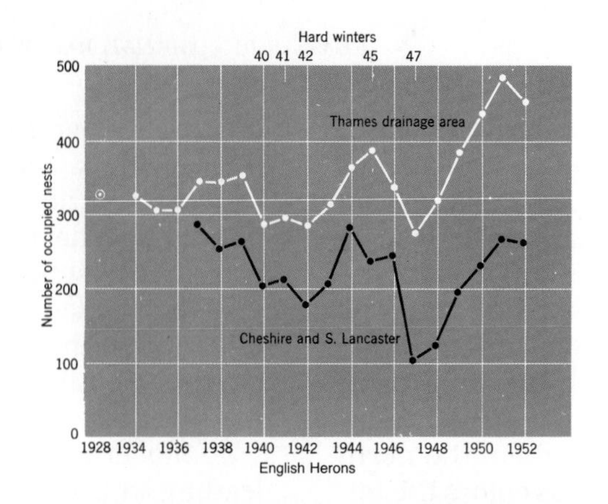

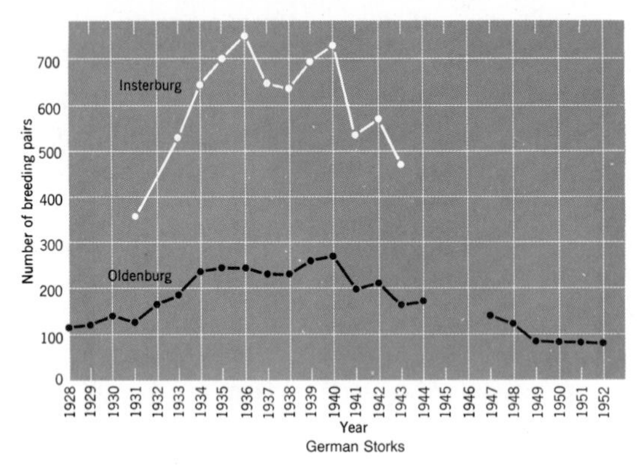

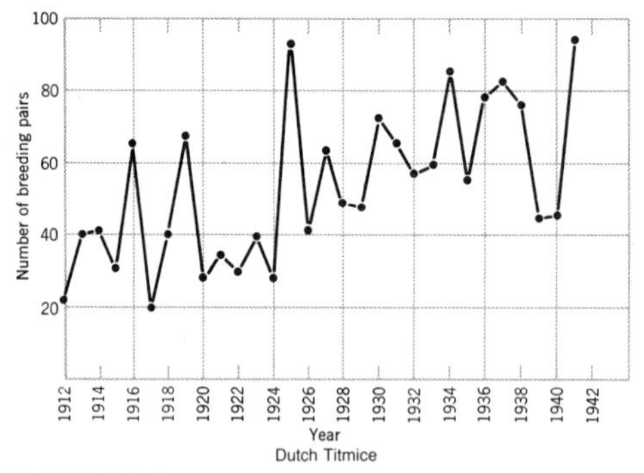

FIGURE 10.1

The classic data for constant bird numbers.

Spring census of northern temperate birds shows constant numbers from year to year, a result that was long used to support the hypothesis that density-dependent control should lead to minimal population fluctuation. Later thinking is that constancy of breeding bird numbers is a side effect of territorial behavior. In the heron data, the hard English winter of 1947 is reflected by a sparsity of breeding pairs the following spring, but the population was restored in subsequent years. (From Lack, 1954).

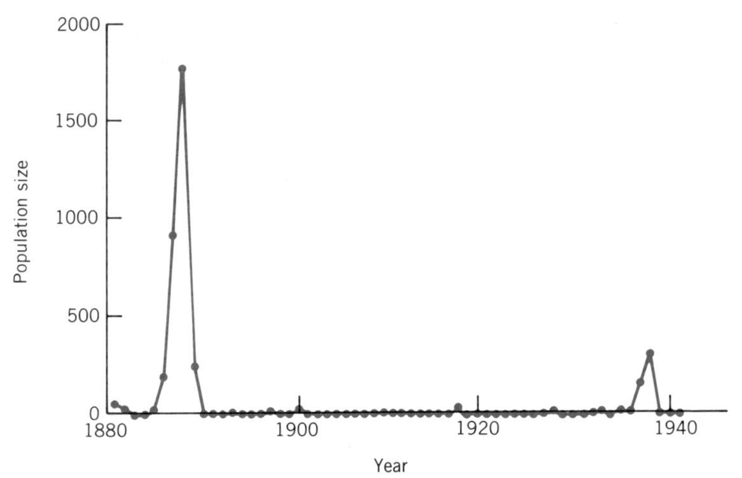

FIGURE 10.2
Sixty-year history of the large pine moth (*Dendrolimus pini*) in Germany.
(From May, 1986a.)

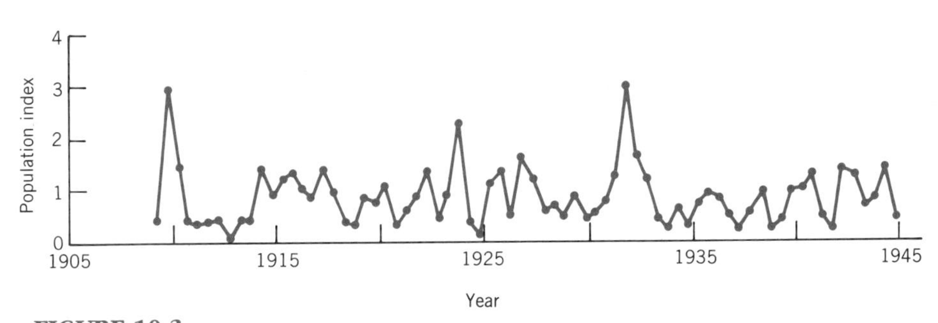

FIGURE 10.3
Forty-year record of the moth *Chilo suppressalis* caught in light traps in Japan.
(From May, 1986a.)

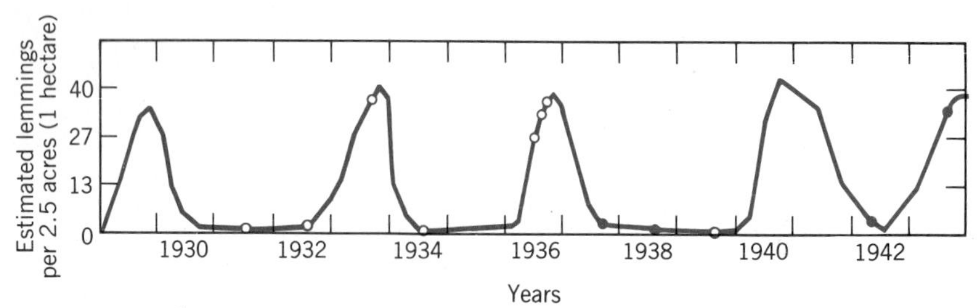

FIGURE 10.4
Lemming cycles in northern Manitoba.
Data are from field censuses by Shelford. (From Finnerty, 1980.)

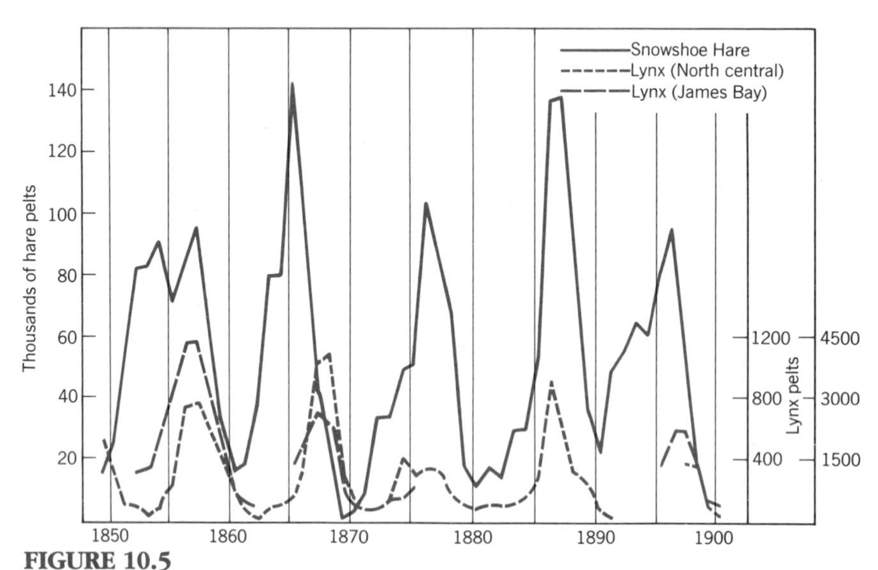

FIGURE 10.5
Relationship of hare and lynx populations in the Canadian arctic.
Data are from sales records of traders dealing in both hare (*Lepus americanus*) and lynx (*Lynx canadensis*) skins. Lynx records are from two nearby trapping regions. The data show not only that the populations are linked but that numbers of the predator follow numbers of the prey. (From Finerty, 1980.)

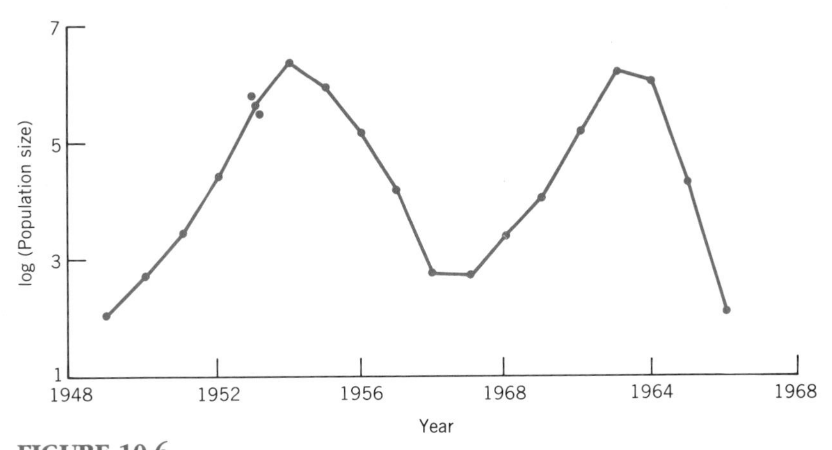

FIGURE 10.6
Twenty years from a century of oscillating population of the larch bud-moth (*Zeiraphera dinigna*) in Switzerland.
Notice the logarithmic scale; the population highs are 10,000 times the lows. (After May, 1986*b*.)

numbers in the wild were usually kept at low levels by accidents of weather or by devastating predation.

THE OLD ARGUMENT FOR DENSITY DEPENDENCE

The argument for density dependence was straightforward. Control must be density dependent or there could be no control. For numbers to be constant, the system of culling must be driven by feedback from the numbers present—by definition density dependence. So far, perhaps, so good. But the hypothesis was carried further, with the density-dependent control likened to the equilibrium final stage of a sigmoid growth history as described by a logistic equation (see Chapter 8). Numbers were to be both at equilibrium and constant.

The most beguiling intellectual support for this position appealed to the success of the Lotka–Volterra equations, and the resulting principle of competitive exclusion, in explaining the uniqueness of species (see Chapter 9). The Lotka–Volterra-Gause model predicts exclusion only when populations are at equilibrium. It seemed, therefore, reasonable to argue:

1. The existence of species is explained by a model describing interspecific competition between species populations at a density-dependent equilibrium, and
2. Therefore the populations of all species should usually be at a density-dependent equilibrium.

Elaborations of this proposition are given in Figure 10.7.

Proponents of the density-dependent viewpoint thought that this reasoning made their position intellectually almost impregnable from the start. They said that numbers in nature usually are constant from year to year because populations are

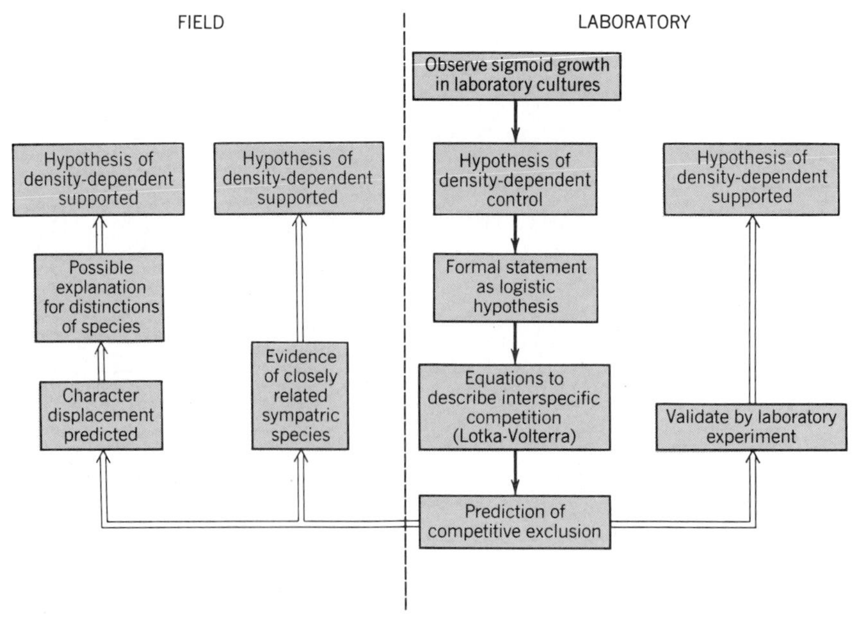

FIGURE 10.7
Arguments offered to show that natural populations are under density-dependent control.

in equilibrium such that

$$\frac{dN}{dt} = rN\left(\frac{1 - N}{K}\right) = 0 \qquad (10.1)$$

(For symbols see Chapter 8.)

Outspoken proponents of this view, like David Lack, were familiar with population census data for birds that did show remarkable constancy in number. Bird census was always for breeding pairs in the spring, because birds are easy to count on their nests. Classical results used in the argument, for herons, storks, and titmice are given in Figure 10.1. When it is reflected that breeding pairs normally fledged many chicks, it might indeed seem remarkable that only the original number turned up to breed the following year.

Particularly telling in these data sets was the way in which numbers could sometimes rebound from catastrophe to the original population level, but no further. This can be seen in the English heron data of Figure 10.1 during and after the hard winter of 1947. In that year, unusual for England, temperatures fell sufficiently to freeze tidal flats and estuaries, killing many herons. Next spring the breeding population was down by as much as a third, but numbers rapidly recovered in the next few years, to level off at the old familiar number of breeding pairs.

Histories of **irruptions** were also cited as evidence that breeding populations of birds were under precise density-dependent control. Irruptions are the sudden winter arrival of unusual species in the bird-watching belt of northern temperate latitudes. It may be crossbills (*Loxia curvirostra*) in England, snowy owls (*Nyctea scandiaca*) in Massachusetts, or grosbeaks, waxwings, or whatever. The irruptions are the result of unusual breeding success the preceding spring all across the far northern regions where the birds breed. When winter comes, intense intraspecific competition forces more migration than usual on the crowded young, and hence the sudden descent on the suburban belt. Only the usual numbers breed in the north next year, however, and the irruption is over. These histories are fully consistent with the hypothesis of numbers that are normally finely controlled. An accident of weather brings on a glut of birds, but density-dependence lets the cull that follows remove only the surplus before the next breeding season.

The Response of the Density-Independent School

Andrewartha and Birch (1954) established their alternative view of population control from empirical data from their native Australia, which seemed to show conclusively that particular insect pests lived in populations the size of which fluctuated from rarity to superabundance in ways that could not easily be associated with ideas of a density-dependent equilibrium. They went from these data to argue the position that, since some animals were not under density-dependent control, then perhaps almost no animals were.

Much of southern Australia has a Mediterranean climate, with moist winters and dry summers. Winter and spring rains flush a short-lived green pasture, and allow a crop of wheat to be grown. The winters are so mild that water rarely freezes, but the summers are hot and dry. The wettest year of a 20-year period may have more than three times the rainfall of the driest. A major agricultural pest of the region is the grasshopper *Austroicetes cruciata*. This animal has one generation a year, which passes the dry summer and cool winter as dormant eggs, a condition called **diapause**. The rising temperature of spring breaks diapause; the young nymphs hatch ready to feed, and grow to maturity on the green herbage of spring (Figure 10.8).

Despite having only one generation a year, *A. cruciata* achieves massive populations on agricultural crops. From 1935 to 1939, grasshopper swarms were present every spring, closely watched by entomologists who then could do little more than watch. But in 1940 the swarms went away, with the entomologists still closely watching. They saw the animals die. And the cause of death was starvation when the rains failed. Almost no eggs were laid to seed the next year's grasshopper populations, and it would be years before another plague appeared. This was mass death, catastrophic death, density-independent control.

The population history of the grasshopper was a direct function of weather. A run of wet springs let populations build up, but one dry year would knock them down again. The population history was like the imaginary history of Figure 10.9. Competition and negative feedback from density seemed to have nothing to do with it.

A second Australian pest, a thrips, had a similar population history set by weather, but with this animal the booms and busts were annual, although populations at the "boom" only occasionally were bad enough to damage crops. At low densities *Thrips imaginis* does little damage, but at 40 thrips to an apple flower, which happens over large areas in some years, they destroy a crop.

The thrips spend their whole lives in flowers, except when they come down to the soil to pupate (Figure 10.10). There are many generations a year, and no resting stage, or diapause. Thus the thrips must find flowers of some kind in every season of the year, the dry season as well as the

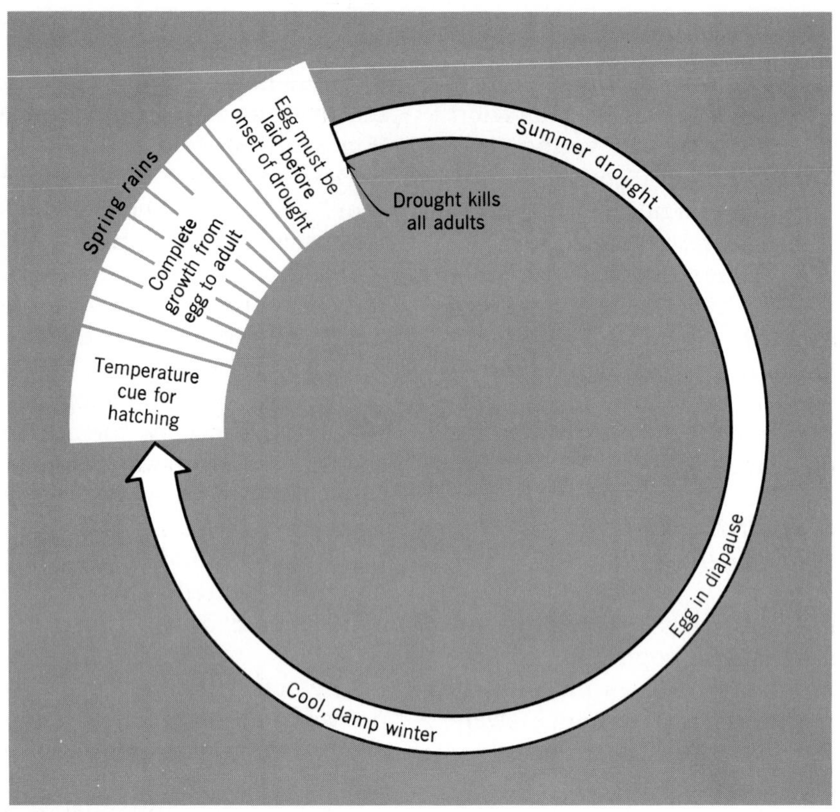

FIGURE 10.8
Life cycle of *Austroicetes cruciata*.

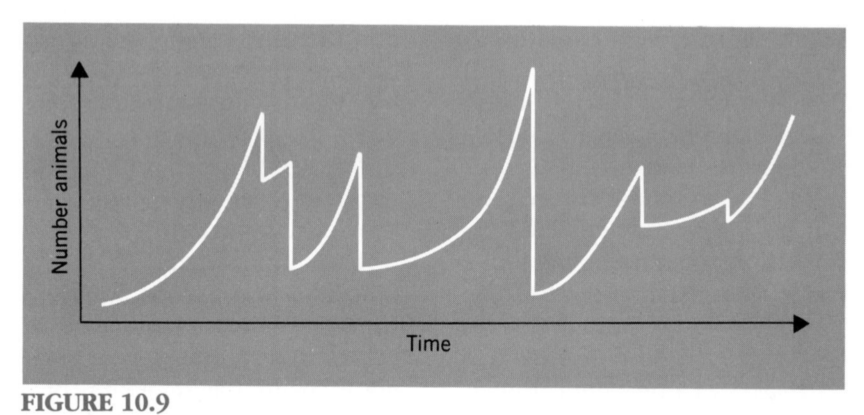

FIGURE 10.9
Hypothetical population history where a density-dependent population equilibrium is not attained.

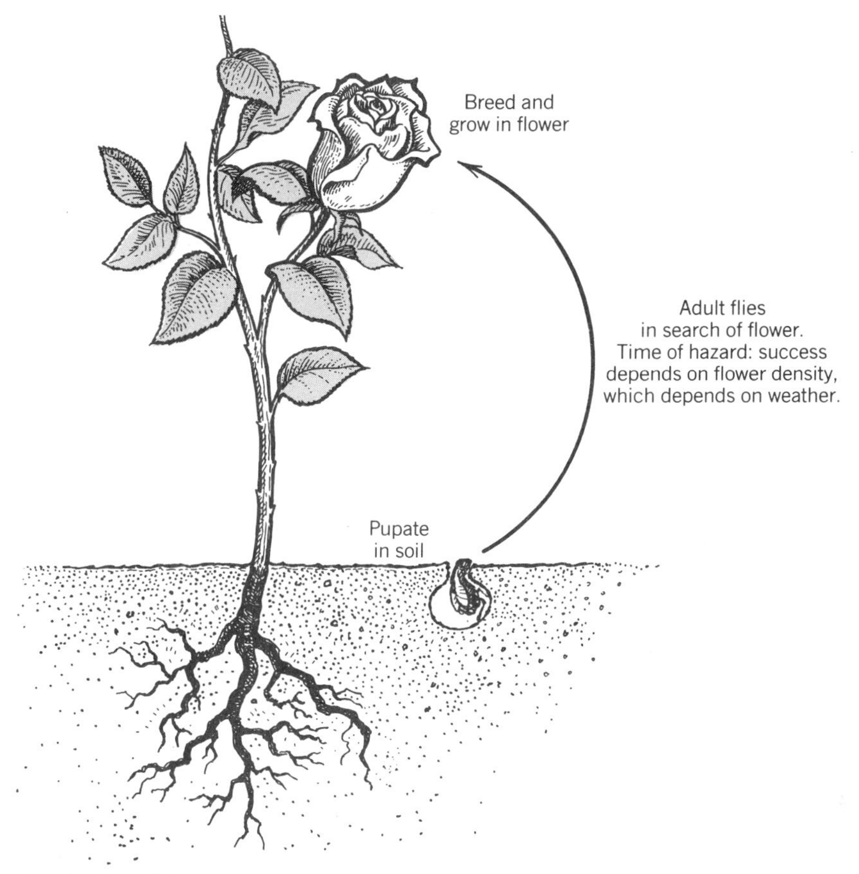

FIGURE 10.10
Life cycle of *Thrips imaginis*.

wet. In the life of every thrips is one desperate gamble—the flight of the newly emerged adult to find a flower in which to live and reproduce.

Figure 10.11 shows four years of thrips population history, recorded by counting thrips in flowers taken at random throughout the year. The numbers climbed each spring, then crashed when the rains failed and flowers became rare. Furthermore, the height of the population peak each spring, and hence whether the thrips were a pest or not, depended entirely on the weather acting through the control of flowering. This was shown quantitatively by a model correlating temperature and precipitation with outbreaks of thrips.

Andrewartha and Birch (1954) argued that at no time in the yearly cycle was there any feedback between the density of thrips and the reproductive success or survival of the animals. When rains failed at population highs, the next generation of adult thrips could find no flowers in that first critical flight, and so perished. The search for flowers was equally hazardous for the few survivors in the dry season. The tiny few that did find a flower, however, had it to themselves without competition. Thus Andrewartha and Birch argued that at no

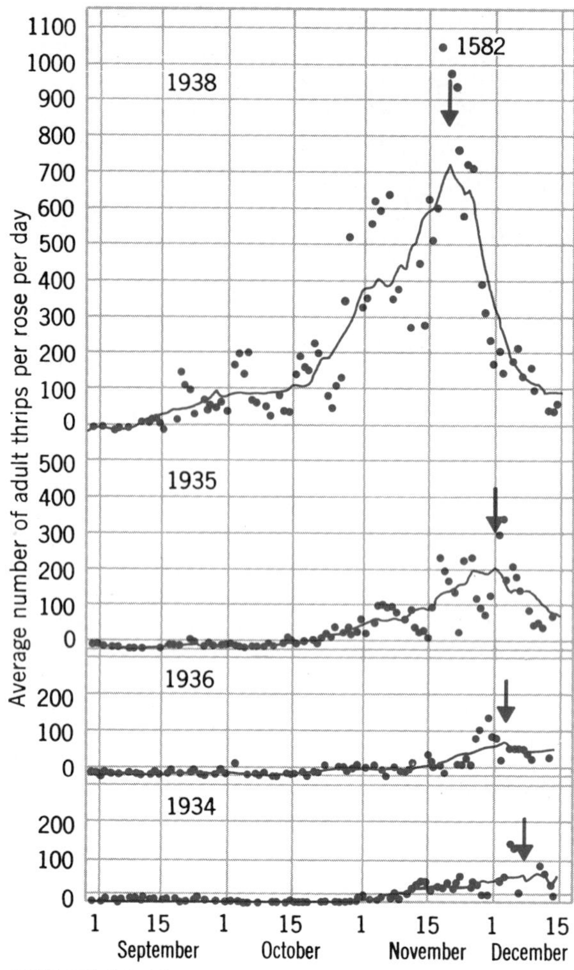

FIGURE 10.11
Annual irruption of *Thrips imaginis* in Australia. These data were offered by Andrewartha and Birch (1954) in support of their contention that insect outbreaks were not subject to density-dependent checks.

time in the population history of this animal was competition or density dependence important. Thrips' numbers were set quixotically by flowers and weather.

Booms and busts in pest numbers are the common experience of farmers and the entomologists who advise them. Often the role of weather is patently obvious, as runs of favorable days let the numbers of animals with short generation times build up. The end always comes with a population collapse as the good times end. Mass death, or mass failure to breed, is the consequence. Population highs end as in Figure 10.9.

A First Level of Resolution

Andrewartha and Birch argued that very many insects have population histories comparable to those of the grasshoppers and thrips. Population highs of these insects should rarely become sufficiently crowded for density-dependent effects to become important. Population growth was exponential but not logistic. The weather encountered by insects, whether in marginal regions of Australia, regions with cold winters or even tropical oscillations between wet and dry seasons, should rarely allow a density-dependent equilibrium to be established after a population high. Most animals are insects living in conditions like these. Therefore most animal populations are not normally under density-dependent control. The weather, not competition or niche space, controls numbers in nature.

This conclusion could not be accepted by anyone impressed by the evidence for competitive exclusion or the character displacement model for speciation. Even the separation of niches between pairs of closely related sympatric species becomes meaningless if animals never crowd to the point of competing (see Figure 8.1 and Chapter 9). Some ecologists were briefly tempted by the idea that big animals like birds and mammals were usually under density-dependent control whereas pest insects were not. This was counsel of despair, because speciation and competition arguments cannot respect taxonomic boundaries.

A more useful possibility was that the population "lows" of boom and bust insects might be at equilibrium, instead of the highs. Summer highs of thrips or plagues of grasshoppers should then have no more significance than irruptions of crossbills or snowy owls. In Figure 10.11 can be seen the suggestion that thrips populations are at equilibrium in winter—the animals irrupt from roughly similar winter populations in the spring of every year (Varley et al., 1973).

The more extreme versions of a weather-dependent life history are perhaps possible only where animals invade repeatedly from some more predictable bastion. Perhaps Andrewartha's grasshoppers always persist at the moist southern edge of their range (Figure 10.12). In the moist Australian south, local populations may well be at equilibrium and experiencing intraspecific competition. From these crowded local populations, winged migrants could be the founders of fresh northern populations after each disaster. An extreme variant on this grasshopper theme is provided by locusts, whose lives and plagues are but a magnified version of this history.

Weather does cause the populations of many short-lived animals or plants to fluctuate very widely above a baseline set by density-dependent factors or a finite supply of refuges. With this reasoning, the initial violence of the argument between density-dependent and density-independent schools could be allowed to die down. A more satisfactory bridging of the viewpoints had to await the work of mathematical theoretical ecologists of the 1980s, who were to show both that competition could yield population histories as widely fluctuating as any produced by density-independent factors (Chesson, 1985; May, 1986b). This new work was to show that animal numbers could fluctuate with large irregularities from both causes, such that constancy, or equilibrium, should be the exception rather than the rule.

Complication by Predators

Applying simple ideas of density-dependent effects to the relationship between predator and prey produces models not of constancy in equilibrium, but of violent oscillation, if not extermination. Both Lotka and Volterra adapted their equations in an attempt to predict the outcome when a

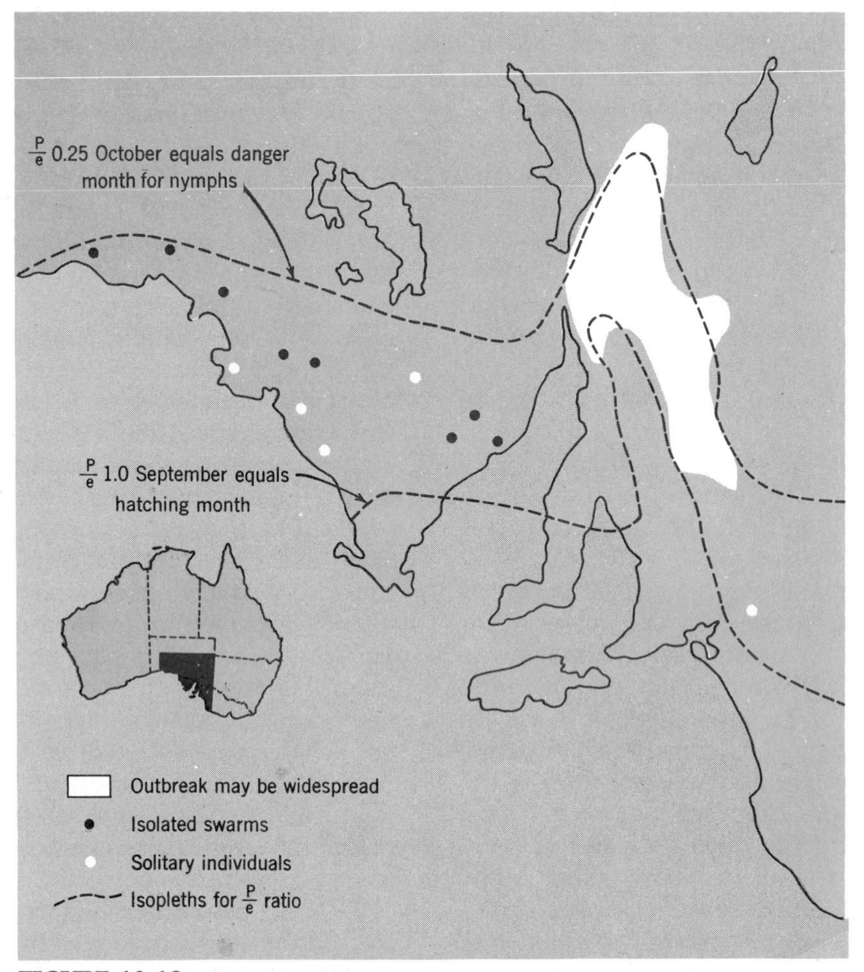

FIGURE 10.12

The area of Southern Australia where *Austroicetes cruciata* may, during a run of favorable years, maintain a dense population.

The isopleth (precipitation/evaporation ratio) lines show that the distribution parallels a climatic distribution. The northern limit may be set by the shortness of spring, but the southern limit is probably set by other animals (see text). (After Andrewartha and Birch, 1954.)

population of efficient predators meets a population of its prey in a confined system. The rate of growth of predator populations is a function of the rate of successful encounters with prey and the rate of predator deaths from all sources. The rate of growth of the prey population can be assumed to be a function of the natural rate of increase without predators, together with the rate of fatal encounters between prey and predators. Then, for the prey population (species 1),

$$\frac{dN_1}{dt} = r_1N_1 - \gamma_1N_1N_2 \qquad (10.2)$$

and for the predator population (species 2),

$$\frac{dN_2}{dt} = \gamma_2 N_1 N_2 - d_2 N_2 \qquad (10.3)$$

where

r_1 = intrinsic rate of increase of prey
N_1 = number of prey, N_2 = number of predators
γ_1 = fraction of contacts that prove fatal to prey
d_2 = rate of death of predators in absence of prey
γ_2 = rate of growth allowed predator per unit contact with prey.

Equation 10.3 relates change in the prey population to its intrinsic rate of increase, as in the logistic equation. The limit to growth, however, is seen to be set not by carrying capacity, K, but by the rate of fatal encounters with the predators.

Equation 10.3 ignores the intrinsic rate of increase of the predator population as being irrelevant. Real predators, says the equation, have their rates of growth set by the food supply, which is in turn set by the rate of fatal encounters with prey. Furthermore, no carrying capacity for the predator population is assumed because limits to predator numbers are assigned to a death rate from unknown causes, possibly old age.

The system described by these equations oscillates, with numbers of predators and prey inversely coupled. Figure 10.13 is the solution plotted as numbers against time. When predators are scarce but prey are numerous, the predators should be able to build their population quickly, inevitably reducing the population of their prey. Eventually the predators should be numerous and the prey scarce, at which time the predators should compete with each other so vigorously for food that they would suffer enormous mortality from starvation. The predation pressure on the prey would then be relaxed, and the prey population should expand, thus completing the cycle. There should always be "too many" predators or "too many" prey. Figure 10.14 plots the number of predators against the number of prey. The numbers circle a singular point, there being no stable number of either.

The idea that oscillations in related processes should be coupled is a general one with many examples in studies far removed from ecology. Economists are familiar with the idea as the theory of supply and demand, and a form of it was well and bitterly known in the American Midwest of the Depression years as the corn–hog cy-

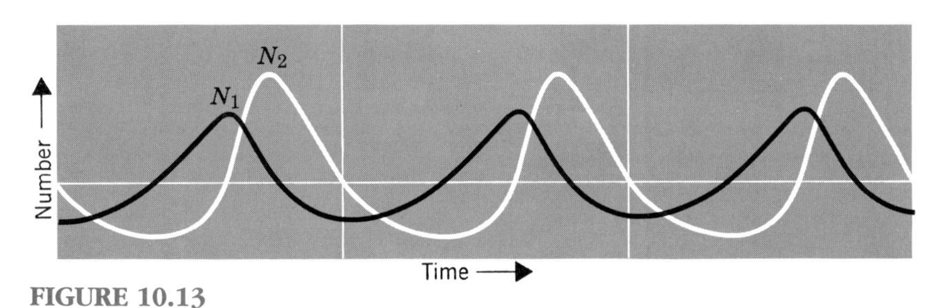

FIGURE 10.13

Coupled oscillations predicted by the Lotka–Volterra equations for predation. (After Volterra, from Chapman, 1931.)

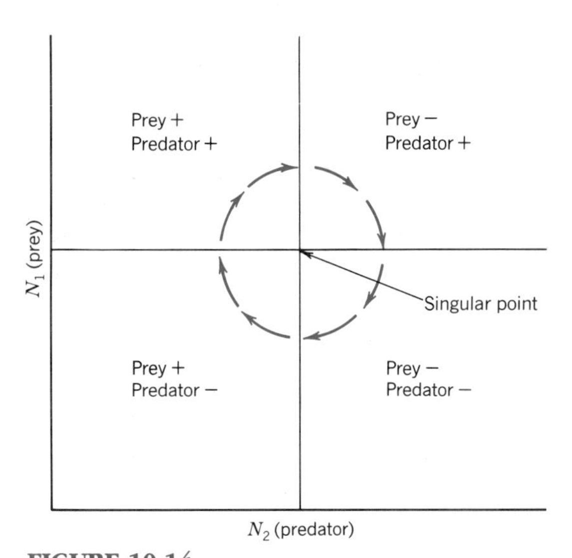

FIGURE 10.14
Population trajectories in a perfect coupled oscillation.
Numbers of both predators and prey circle a singular stable point in perpetually balanced imbalance. No such system has been demonstrated in nature.

cle. When corn sold well, farmers planned next year to keep more corn for sale even if it meant rearing fewer hogs. This resulted in a corn glut and falling prices. Many farmers responded by keeping their corn back next year to fatten hogs instead, which resulted in a glut of hogs and a shortage of corn. There would be recurrent gluts of hogs or of corn, each glut being attended by a crop of farmer bankruptcies and a shortage of one of the coupled commodities. In engineering, such coupled relationships form the foundation of servomechanism theory.

Simple laboratory tests of the Lotka–Volterra prediction of a coupled oscillation between predator and prey shows it to be wrong. The predators always exterminate their prey. Indeed, it is intuitively obvious that this should be so, because the model requires a time when predators should be massively numerous but the prey scarce—a bad time to be prey; one chicken in a run full of foxes comes to mind. The last prey should be killed, after which the predators starve to death. Gause (1934) showed that this was the outcome, using his paramecia cultures as prey and the protozoan *Didinium* as predator (Figures 10.15, 10.16).

The Lotka–Volterra equations are obviously gross oversimplifications of the relationships between predators and prey. The model is without generation times, lacks allowances for time to grow up, and has no opportunity for immigration, emigration, or refuge for either population. Remedying some of these defects, however, while still confining efficient predators with their prey, merely predicts what actually happens: annihilation of the prey

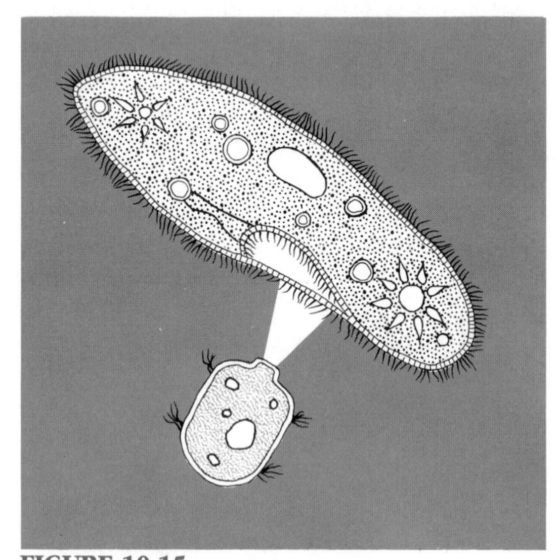

FIGURE 10.15
***Didinium nasutum* devouring *Paramecium caudatum*.**
Didinium is so effective a predator of *Paramecium* that introducing a few into a *Paramecium* culture results in the complete extermination of the paramecia. A *Didinium-Paramecium* system can be made to oscillate by adding "immigrant" paramecia at the time when the *Didinium* population is reduced through starvation.

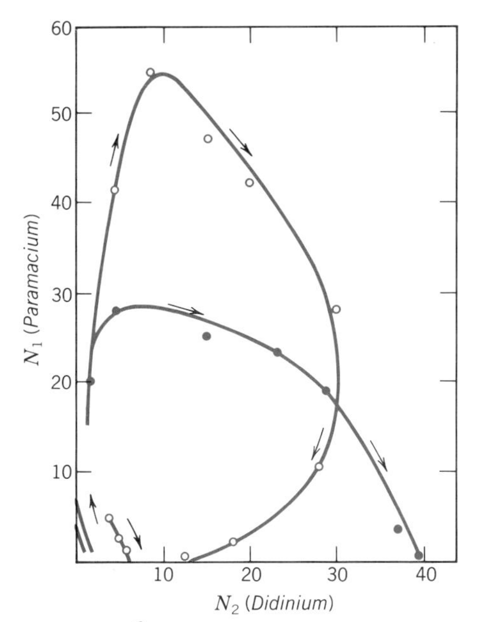

FIGURE 10.16
Trajectories of populations of *Didinium* and *Paramecium*.
Compare with Figure 10.14 for perfect coupled oscillations. (From Gause, 1934.)

followed by death by starvation for the predator (Nicholson and Bailey, 1935).

In the real world predators do not exterminate their prey, though local devastation can be common. Systems are known where prey escape devastating predators by dispersal, so that the prey live in patches that have temporarily escaped the predator while other patches have been wiped out. Two of these systems, both successful results of biological control of pests, were important to the debate in that they let in the argument that density dependence might be irrelevant to control.

Biological Control

In the late 1880s the California citrus industry faced complete ruin from a plague of scale insects, *Icerya purchasi*, inadvertently brought from Australia. In a classic of biological control, the scales were overcome by a ladybird beetle *Rodolia cardinalis*, now generally known as "the vedalia." (De Bach, 1964).

Vedalia beetles and their larvae can subsist on *Icerya*. One hundred twenty-nine live beetles were placed on the branches of one infested tree, which was then wrapped in muslin so that the beetles could not get out. After about four months a large population of the beetles was present on the tree and it was virtually clear of the scale insects, at which time the muslin cover was removed. In three more months an entire orchard was free from *Icerya*. Colonies of beetles were shipped throughout the state, and within one year the infestation of scale insects on the whole California orange crop had been controlled.

The control measure worked indefinitely. Once a tree or an orchard was effectively cleared of the *Icerya* infestation, however, this did not mean that the scale insect had become extinct. Apparently it persisted in very low numbers and in scattered populations. These were unimportant to the citrus farmers, but they were sufficient to allow a low population of the predatory vedalia to be maintained.

Ladybird beetles can fly, and their larval instars run actively. The beetle lifestyle is thus well suited to hunting down sessile prey that is scattered over a large area. *Icerya*, on the other hand, is not entirely helpless. The fluffy covering probably gives some defense against attack, particularly from the younger beetle instars, and it may be that scales can drop from the host plant when molested. By these means some individuals manage to escape local buildups of the predators until they have all flown away to seek food elsewhere. The result is a pattern of search and destroy, survive and flourish, which is maintained across the orchards of California.

Control of prickly pear cactus in Australia by caterpillars worked in a similar way. *Opuntia* cactus is not native to Australia, but escaped from gardens and then became a major nuisance on range land. Entomologists found in Arizona, the homeland of *Opuntia*, a moth whose caterpillars ate the cactus, and they released a few in Australia with gratifying results. The cactus prey was sessile, so that only a very few individuals could survive the dense populations of predatory caterpillars that could build up locally. The adult predator was active, being able to fly in search of patches of fresh prey on which to lay eggs, which its caterpillars then exterminated. The end result was low populations of the prey species pursued by low populations of predator, both of which persisted indefinitely (Figure 10.17; Dodd, 1940; Holloway, 1964).

In this relationship between predator and prey, the density of both populations certainly has effects. But for the prey at least, it could be argued that density-dependent competition was not likely to be important. Efficient predators kept prey numbers far below the carrying capacity of the food plants.

WHY THE EARTH IS GREEN

When the argument over the possibility of control by weather was at its strongest in the early 1960s an argument came to be used on both sides of the debate that turned on the observation, or claim, that the earth was green.

The "green earth argument" goes like this: The earth is carpeted green with plants, implying plenty of plants available to be eaten at the surface of the earth. But the earth holds huge numbers of herbivores. If herbivores were limited by food energy and if they competed for food, then we should expect the food supply of herbivores to be constantly used and restricted, in which event the earth should not be green but eaten piebald by the herbivores. Therefore, since the earth is green, herbivore numbers are kept down in some way to below the level at which density-dependent regulation of numbers should be important.

In first making the green earth argument, Hairston, Smith, and Slobodkin (1960) proposed that herbivore numbers are kept down by predators. In this model herbivores are harried by predators so that competition for food is not important in the life of a herbivore. The competition that matters is between predator and predator over the bodies of herbivores. Plants are the real gainers as predator stalks prey.

Hairston, Smith, and Slobodkin then developed their explanation of the green earth into an elegant overview of attractive simplicity. Because plants escaped herbivores to lead crowded lives in the green carpet, the plants must be controlled by competition for resources. And because herbivores do not eat down the plants, herbivore numbers must be set by predation. This makes the predators food—limited, since they control the herbivores. Thus the whole plant trophic level is resource-limited; the whole herbivore trophic level is controlled by predators; and the whole predator trophic level, like the plants, is resource-limited. This attractive overview was so talked about in ecology that it came to be known by the shorthand "the **HSS model**," after the initials of its authors.

But the Andrewartha and Birch school could reply that what kept the herbivore numbers down was weather. Population highs of pest insects were usually ended only by a change in the weather, never by the pests running out of food and competing for what was left. The supposed greenness of the earth, for Andrewartha and

FIGURE 10.17

Rangeland in Australia before and after the moth *Cactoblastis cactorum* was introduced.

Larvae of *Cactoblastis cactorum* feed only on species of *Opuntia*. Introduction of the moth to rangeland dominated by *Opuntia* led to rapid population growth of the moth whose larvae almost totally destroyed the cacti in every part of the range. Once the wave of caterpillars passed, both the cactus and the moth became rare, with isolated patches of cactus springing up until a wandering moth finds them and lays a clutch of eggs.

Birch, was actually due to the fact that herbivores were under density-independent control by weather. In their rival summary, some insects are controlled by weather, most animals are herbivorous insects, these herbivores eat so little that the earth is green, therefore most animals are probably controlled by weather.

Neither of these rival explanations for the greenness of the earth looks satisfying now. One important reason for the greenness of the earth, scarcely hinted at in the debates of the 1960s, is that many plants have effective means of resisting herbivore attack. Plants are inedible, or poisonous, or grow too tall, or disperse out of the reach of specialized herbivores (see Chapter 15). To the extent that plants are successful in frustrating herbivores, the earth is green (where it really is green) because herbivore numbers are kept down neither by weather nor by predators but by the plant's own defenses.

Of even more generality is an argument based on energetics. Plants are not energy-limited, in the sense that they are usually exposed to more solar insolation than can be used in photosynthesis (see Chapter 3). But plants are space-limited, and any unit of bare space in watered land may be considered as representing an unused unit of photosynthesis. This means that whenever a plant is destroyed or cut back by a herbivore, a space is cleared for another plant, or part of a plant, to grow.

An array of many species of plants under herbivore attack is in the position of a defending army provided with unlimited reserves. As fast as a gap is torn in the vegetation, so that gap can be filled by another plant, perhaps of a different species. The energy available to the herbivore trophic level is necessarily a small fraction of the energy available to the plants (see Chapter 4). So the green parts of the earth stay that color because herbivores, as a whole trophic level, cannot command an energy flux sufficient to keep land free of plants. This is why a well-grazed meadow remains green. The energy flow paradigm (see Chapter 2), not herbivores and not weather, is at the bottom of the greenness of the earth.

STABILITY, CYCLES, AND CHAOS THROUGH DENSITY DEPENDENCE

Contemporary ecology is beholden to Robert May (1986*a*, *b*) for gently showing population biologists that population histories ranging from stable equilibria through cyclic oscillations to eccentric fluctuations indistinguishable from chaos are predictable outcomes of different intensities of density dependence. May regards the old debates over density dependence or independence as over little more than semantics. And he notes that equations for extreme density dependence may yield population histories "practically indistinguishable from Andrewartha and Birchian density independence."

Consider populations that have discrete, nonoverlapping generations, as many temperate-zone insects do. Let these populations exist in a closed, homogeneous universe in which birth rates and death rates are determined by population density alone, thus providing that all changes in population growth are density-dependent. Population growth in these populations is the difference in density in generation $t + 1$, N_{t+1}, to density N_t in the preceding generation. When numbers are relatively low in the closed universe, N should increase between generations, but for high densities beyond the sustainable carrying capacity, N should decrease between generations. This relationship can be expressed by plotting density at the next generation against density at the generation before, yielding a map relating densities across generations (Figure 10.18).

The shape of the map in Figure 10.18 is a function of the intrinsic rate of increase (I): very "humpy" if r is high as in curve d, but less so for low values of r as in curve a. The no-growth condition is given for each map where the curves meet at $X_{t+1} = X_t$. Stable populations set by density dependence in these populations would, of course, remain at the constant population line, with no change in number between generations. It turns out that whether a population remains at this stable point or not depends on the steepness of the map, which is to say on the value of the intrinsic rate of increase.

May (1986*b*) explores the approach of maps of different steepness to the stable point with difference equations as follows (Moran, 1950; Ricker, 1954):

$$N_{t+1} = N_t \exp[r\,(1 - N_t/K)] \quad (10.4)$$

This equation predicts future numbers at time $t + 1$ in terms of the familiar per capita intrinsic rate of increase, r, and the carrying capacity K.

A simpler expression of this relationship is

$$X_{t+1} = a\,X_t\,(1 - X_t) \quad (10.5)$$

Population density is now a dimensionless variable, X_t, and the assumption is made that the right-hand side becomes zero if

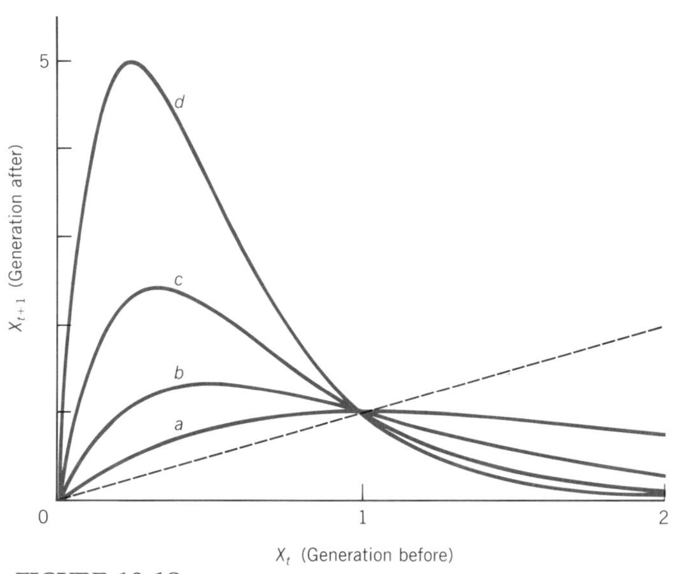

FIGURE 10.18

Maps of possible population densities as functions of density of the preceding population.

The model assumes discrete, nonoverlapping generations in a deterministic world. Populations tend to increase between generations (i.e., lie above the dotted line) at relatively low densities, but to fall (lie below the dotted line as in *a*) when at high densities. (From May, 1986*b*.)

$X_t > 1$. The parameter a corresponds to $a = 1 + r$, thus reflecting the rapidity of population growth (May and Oster, 1976).

When humps are low (intrinsic rate of increase is low), the population history is as in Figure 10.19a. The point A where the

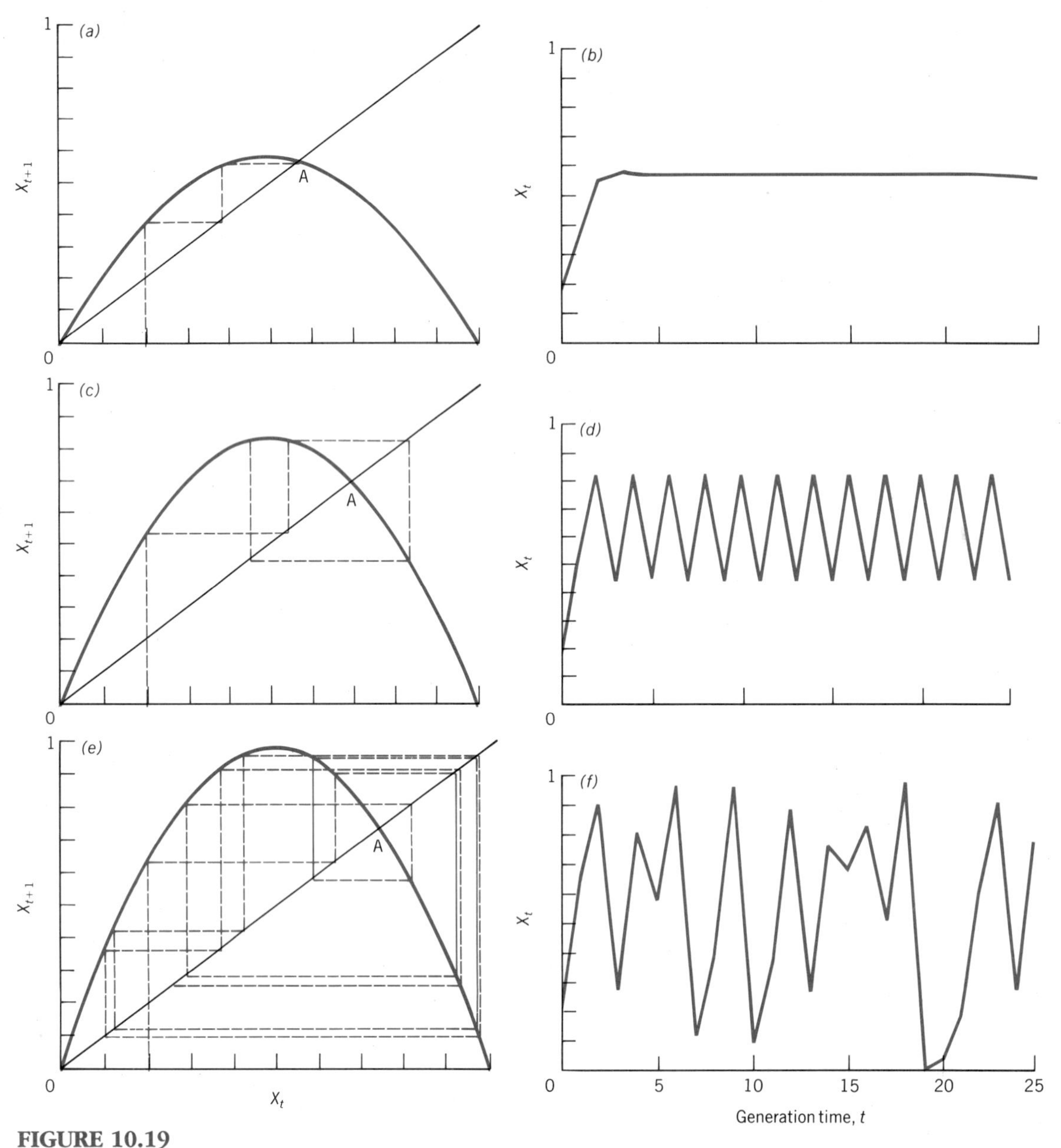

FIGURE 10.19
Maps of population change between generations and their historical consequences.
At low intrinsic rate of increase (a, b) population can be constant at a single stable point. At intermediate intrinsic rates of increase, two stable points result in an oscillating population (c, d). At high intrinsic rates of increase, numbers wander in a way indistinguishable from chaos (e, f). (From May, 1986b.)

map intersects the constant population line of no growth is the stable point. The dotted line shows the trajectory of the population as it changes with each generation. With this low-humped map, the population rapidly arrives at A, leading to a stable population equilibrium, shown in the right-hand plot (Figure 10.19*b*).

If the hump is steeper, as shown in Figure 10.19*c*, the stable point A cannot be reached, the population continually overshooting and undershooting in a stable two-point cycle. This yields an oscillating population history as in Figure 10.19*d*.

Finally, a still more humpy curve can yield a trajectory and population history like those shown in Figure 10.19*e* and *f*. This population history looks like chaos or pure randomness, but is in fact a fully deterministic consequence of density dependence. This was the history that led Robert May to remark that density dependence could yield results indistinguishable from the insect histories put forward by Andrewartha and Birch as evidence for control by weather.

Difference Equations and Oscillating Rodents

The behavior of difference equations shows that virtually all known population histories could be the result of simple density dependence acting alone. In plainer language, crowding can have unexpected and unpredictable effects on reproductive success. Population growth is not a smooth process but happens in jumps following synchronized reproductive episodes. Numbers in whole generations jump up or down together. How chaotic the resulting population history becomes depends on the rapidity with which individuals reproduce. Thus the effects of crowding, which is to say of density dependence, can yield virtually any known population history.

This is an immensely useful result because it removes the sense of mystery from populations that oscillate, produce plagues, collapse, or seem to signal chaos. These various histories should be expected as normal consequences of population dynamics in a world still under control. But this by no means implies that simple density dependence is necessarily the cause of all the population changes we see around us, merely that density dependence is always possible. Density-independent effects do exist: the environment does kill; real populations of real animals have more diversity than is allowed for in equations 10.4 and 10.5. The signals for these various effects may hide each other in the census data (Hassell, 1985).

Figure 10.20 shows the fit between May's (1976) model for a lemming population history and actual census data given previously in Figure 10.4. Many factors are known to influence lemming numbers (Finnerty, 1980). The model does no more than assume density dependence and allow a 0.72-year lag in response time, and yet it seems to explain the population history nicely. As May (1986*b*) put it, his explanation of four-year cycles as arising naturally for populations with high intrinsic growth rates in strongly seasonal environments (where regulatory effects are likely to operate with one-year lags) has been reasonably criticized by Stenseth (1985) as being "too simple to be sensible." Perhaps.

First the data. The best evidence for cyclic numbers of arctic mammals comes from the Canadian fur trade, which left us the legacy of number of furs traded every year for more than a century. Because trapping effort varied within reasonable limits, this gives a reasonable estimate of relative population size. Principal skins traded were of arctic fox (*Alopex lagopus*) and the Canada lynx (*Lynx canadensis*), though some hares were also trapped. Fox

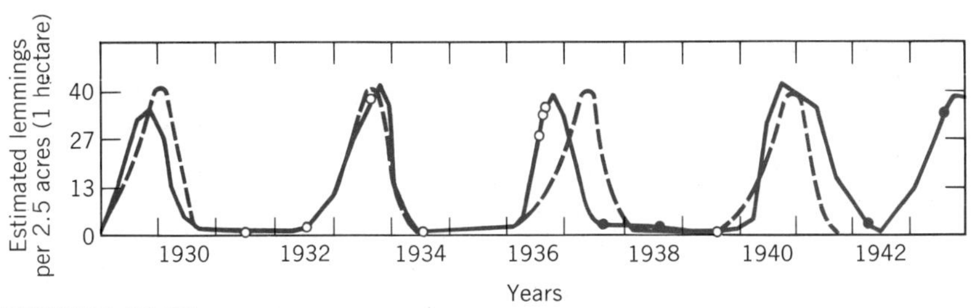

FIGURE 10.20

Lemming population history: density dependence with a 0.72-year lag in response time.

The history is consistent with the generation change model of Figure 10.19*c*, *d*. (From May, 1976.)

skins show a four-year cycle and lynx skins a ten-year cycle. It was quickly realized, and has been demonstrated, that the fox and lynx populations were acting as census agents for the rodents on which they fed. The underlying cause of boom and bust in the fur trade, therefore, was cycling oscillations in the rodent prey. Foxes fed on lemmings and other rodents with a four-year cycle; lynx fed on hares with a ten-year cycle. Because the whole fur trade experienced boom and bust, cycles must be in phase throughout the Arctic, leaving little doubt that they have been synchronized by arctic weather (Elton, 1942; Finnerty, 1980).

The cycles in the fur trade are real cycles, not just progressions of random events that seem to imitate cycles. Sometimes it does happen that a series of unconnected ups and downs can look cyclic, a happening for which ecologists have been wary ever since the mathematical possibilities inherent in random numbers were pointed out to them by Cole (1954). But the cycles of arctic populations are real (Figure 10.21).

Although predators kill many lemmings and hares, these are not 'coupled oscillations' between predator and prey of the kind conceived by the Lotka–Volterra equations in their predator–prey mode. Numbers of the predatory lynx and foxes are actually in phase with prey numbers, helpless captives, as it were, of changing supplies of resource. This is most beautifully revealed in a set of hare–lynx data based on trapping records of both animals (Figure 10.5).

Nor are rodent numbers coupled with disease organisms or with their food plants, or the result of stress in crowded populations. A determined investigator can discover disease in lemmings, or the effects of stress in dense lemming populations, or damage to tundra after a lemming high, yet years of research have not shown that mortality from any of these can generate the population cycles (Finnerty, 1980).

These studies have shown, however, that drastic culling of lemming populations by predation, disease, stress, and perhaps even malnutrition is likely from time to time, even though it can be shown that none of these alone is responsible for the cyclic behavior of the population. Is the model that leaves them out "too simple to be sensible"?

Arctic summers are six weeks long, the lemmings passing the rest of the year in tunnels under the snow, feeding on roots

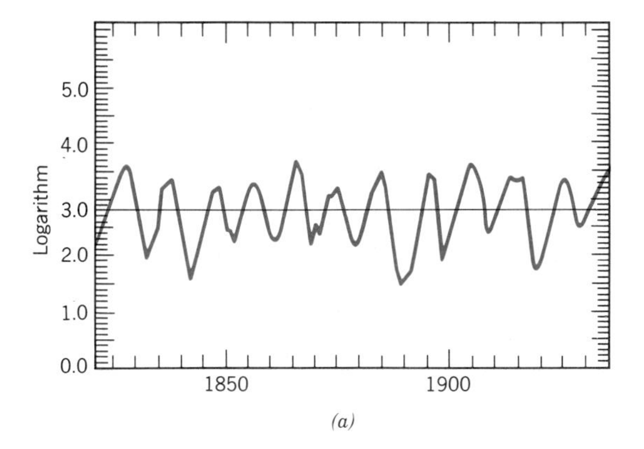

(a)

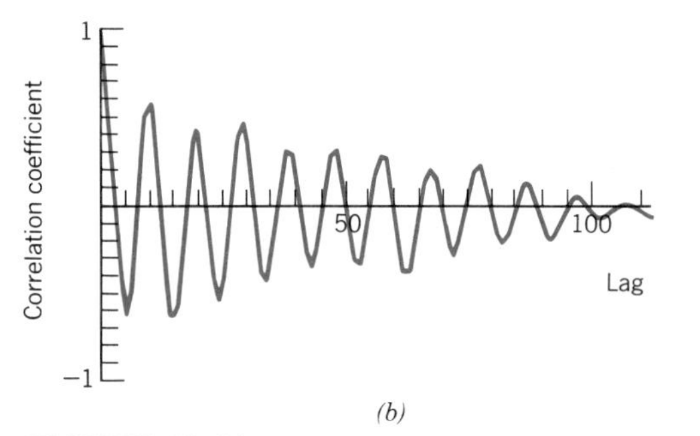

(b)

FIGURE 10.21
Cyclicity of lynx populations.
(*a*) Logarithms of numbers of lynx skins traded in the Mackenzie region. These data show strong periodicity that is shown by the autocorrelogram (*b*) to be truly cyclic. That the peaks do not follow the decades precisely in (*b*) shows that the cycle is not precisely 10 years but actually about 9.6 years. (From Finerty, 1980.)

and natural hay. No lemming in the wild has been shown to live longer than a single year, but a female lemming can raise several litters in this short time, even granted that most of the time passes in the arctic night. She can conceive when she is 27 days old and gestation time is 21 days, which means that the time between the birth of the mother and the birth of her young can be as little as seven weeks. In a good summer a female can wean two litters, each having between four and nine young. As a reproductive machine, therefore, a lemming is beautifully adapted to cope with arctic seasons, though weather does synchronize her reproduction with that of her sisters. The synchronized females should yield a high intrinsic rate of increase, fitting lemming population dynamics to the models with "humpy" maps (Figure 10.19*c*–*f*).

An apparent difficulty in accepting the model as a full and sufficient explanation of lemming cycles probably lies in accounting for population crashes. How do lemmings die? Density-dependent starvation in winter when the hay crop from overgrazed tundra lacks both nutrients and calories is one possibility (Schultz, 1969). More likely is mass failure to breed when on short commons; the adults will all die of old age within a year anyway, so that breeding failure should rapidly dispose of the crowd.

Lemming cycles are not strict in their four-year spacing, some highs being separated by six years and some by three. Sometimes a high lasts for two years. In these vagaries can be seen the print of changing weather for the breeding season, or different incidence of predation, disease, or plant growth. Thus are complexities added to lemming dynamics. But May's (1981) hypothesis for the cycles themselves might still be sensible as well as simple.

ON BIRDS AND GRASSHOPPERS

Temperate and arctic bird numbers do not oscillate like arctic rodents, though they do irrupt from time to time. Some are moderately fecund (six eggs, two clutches in some years), and all have generations synchronized by spring breeding seasons, allowing high per capita intrinsic rates of increase, suggesting humpy maps predicting oscillations. But their populations tend to be as stable as any we know (Figure 10.1).

Bird breeding behavior does turn out to evade the simple conclusions of the density-dependent model by its complexity, or perhaps sophistication would be a better word. The birds in question are all to some extent territorial (see Chapter 13), with the result that the number of birds breeding each spring is a function of the space available for breeding. Space is inflexible. It is thus not surprising that breeding numbers are constant from year to year; the census takers were actually using the birds as proxy measures for space. The dynamics of bird populations, like the populations of all territorial animals, therefore, are partly constrained by space. The stability in bird numbers used by Lack (1954) as the mainstay of his theory of density dependence was not in fact the result of simple density dependence, but a consequence of territorial behavior (see Chapter 13).

That breeding numbers of territorial birds are set by spacing does not, however, free individual birds from the rigors of density-dependent competition. Not only must they engage in contest competition for the breeding space itself, but the larger populations that follow each breeding episode are exposed to competition in the nonbreeding season. In cold winters, for instance, the food supplies of the coldest day might be a critical resource for competition as birds congregate on the few most favorable feeding grounds (Fretwell, 1972).

Hints about control systems given by insect population histories can be even more uncertain than the information given by bird numbers. If outbreaks can be predicted by modeling the weather, as apparently is true for the grasshoppers and thrips of Andrewartha and Birchian fame, then they can hardly be due to inherent dynamics of the population. In the short term at least, we witness sudden population changes that are indeed outside the control of density.

All insects with widely fluctuating or chaotic population histories are likely to be highly fecund (most lay many small eggs), requiring their populations to have large per capita intrinsic rates of increase. Most have discrete generations, since reproduction is synchronized by environmental cues. This is true even in tropical rain forests, where changes in rainfall work to synchronize both flowering and insect life cycles. As Robert May's models show, these animals can have irregular population histories like the grasshoppers, even when density dependence is the ultimate regulator.

The thrust of May's conclusions is thus that erratic population histories are quite consistent with density dependence. This conclusion has now had a striking vindication from the first comprehensive statistical study of comparatively long time-series data for insect populations. The data are numbers of moths coming to light traps and numbers of aphids caught in suction traps maintained for the last thirty years at the Rothampstead Experimental Station in England (Woiwod and Hanski, 1992). The populations are for 263 species of moth and 94 species of aphid, in all 5,715 time-series. All numbers fluctuated in the familiar, apparently erratic ways. But statistical tests for response to density when applied over the span of each time-series demonstrate the prevalence of density dependent effects even in these erratic population histories.

Thus density dependence is ubiquitous, even for insects. Over spans of only a few years numbers can change in random ways, but only within limits set by density dependence. A long enough record will always make manifest the overriding consequences of intermittent crowding or release from crowding.

Density dependence is most hidden when intrinsic rates of increase are extremely large when, as May's models show, the resulting jumps and crashes in number can be indistinguishable from chaos in the short population histories that ecologists usually have to work with. The most highly fecund of insects can appear chaotic like this but the most chaotic histories of all are likely to be in some plant populations, the fecundity of which can be prodigious (keys from a maple, cotton from a cottonwood, seeds from dandelion clocks). Are the changing numbers of weeds indistinguishable from chaos? It seems not improbable. Yet a long enough history even of weeds, say a few thousand years, should certainly reveal the workings of density dependence.

Density dependence has now been shown to be ubiquitous in all organisms, whether territorial birds, grasshoppers, or indeed weedy plants. The grasshoppers and the weeds fluctuate widely within the limits set by density dependence when viewed on the time-scale of a human lifetime. Birds fluctuate less. The difference is a function of fecundity and generation times as much as it is of weather.

REMEMBER THE TREES

The time scales of longest census are trivial: a few decades for crop pests, a century for lemmings and some insect counts. But species typically exist for millions of years, their population histories facing such forcing functions as the comings and goings of ice ages.

For forest trees a census of a few thousand years might still be inadequate, so long are generation times. Furthermore, a sequence of tree generations long enough to test hypotheses of stability, oscillations, or apparent chaos would span significant climatic change. Figure 10.22 shows the population history of hemlock trees (*Tsuga*) in the American Midwest over the last 6000 years. The history is reconstructed from percentage composition of hemlock pollen in lake sediments dated by radiocarbon, as described in Chapter 17. The period of population growth spanned all of 4000 years. At the end of this long period of population growth, 2000 years ago, pollen percentages in many lake deposits were in the same range as in samples representing the last centuries before the forests were cut down. Sampling intervals over the last 2000 years are not adequate to measure the subtleties of population dynamics near the potential equilibrium stable point, but it is doubtful if 2000 years is sufficient for more than the first pass at a potential stable point. And likely oscillations in hemlock numbers around a stable point would extend into the domain of the next ice age.

THE NATURAL REGULATION OF NUMBER

The ultimate determinate of population size is the energy supply: tigers, hawks, and great white sharks will always be rare (Figure 10.23). To the extent that Eltonian pyramids keep their shape, numbers must be controlled by a feedback between energy supply and population size. But available energy is divided among discrete species populations with specialized ways of life. For them niche sets number, at least to the extent that niche sets the stable point on a plot of numbers in one generation against numbers in the next (Figure 10.19*a*–*d*).

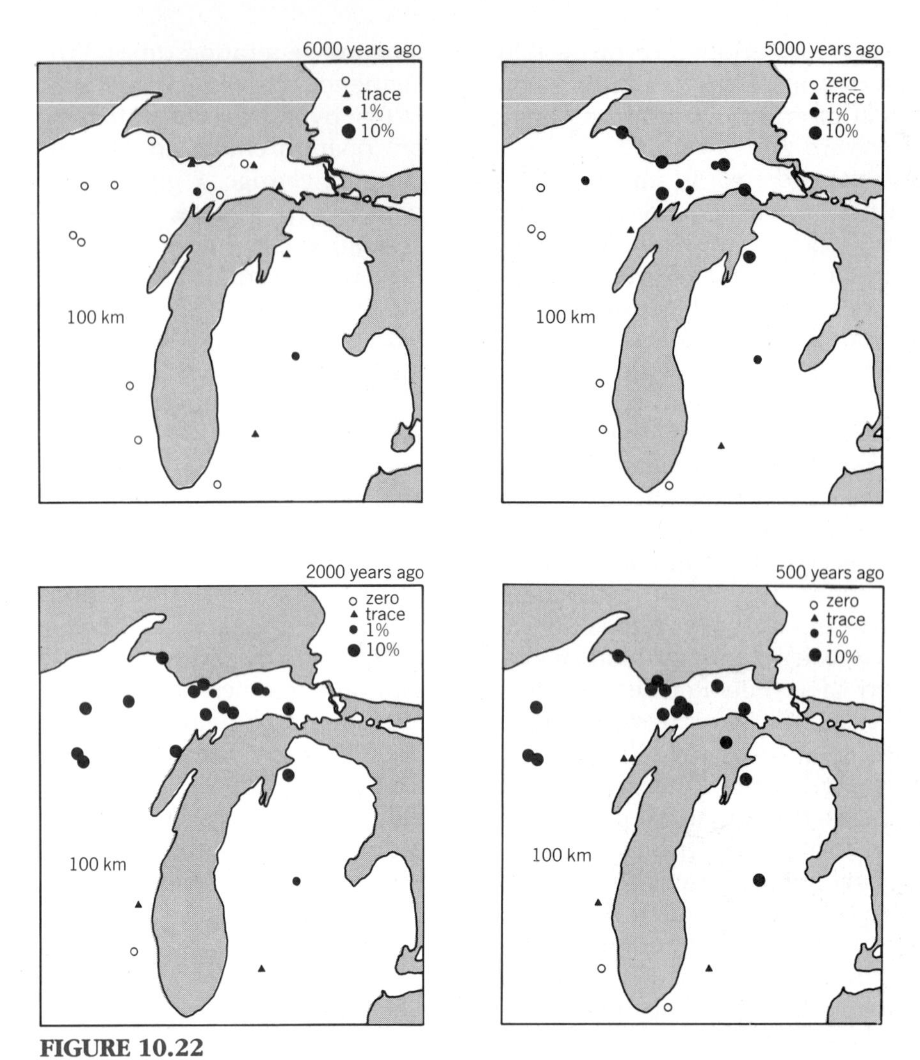

FIGURE 10.22
Migration of hemlock in the American midwest.
The data are percentage hemlock (*Tsuga canadensis*) pollen in lake sediments dated by radiocarbon. The process of invading the region west of Lake Michigan took 4000 years to complete. (From Davis, 1982.)

Numbers in all species populations must be directed by density-dependent intraspecific competition at frequent intervals in the history of the species. Whether, however, this leads to population stability, to oscillation, or to apparent chaotic population histories depends on such population parameters as the per capita intrinsic rate of increase. Even apparent chaos, however, is tied to the apparent stable point set by the species niche; probably relative abundance changes less than absolute abundance. Populations may escape the normal controls of density de-

FIGURE 10.23
***Carcharodon carcharias:* decreed rare by energetics.**
Eltonian hunters high on food chains must always have low populations because of the scant energy supplies for their way of life. Swimmers do not meet great white sharks very often.

pendence when presented with gluts of resource, as happens for insects of marginal habitats in chance good years, or as can happen with a thrips presented by farmers with a glut of food every fruiting season. These populations suffer density-independent death when the factors sustaining the resource glut, usually the weather, are changed.

Complications are introduced by the patchiness of the environment, whether caused by geographical patchiness, patchy distribution of food, or devastating predation. In patchy environments, local populations might be out of phase with each other, perhaps exchanging migrants, certainly contributing to mean densities of an area that can obscure the actual progress of population dynamics.

All population control is control by death. Perhaps the commonest form of "death" is not being born at all because the stress of competition prevents a parent from laying an egg or carrying an infant. Perhaps almost as important is the concealed form of death that comes from failure to mate. From the population point of view an individual is just as "dead" if it is a lifelong bachelor or spinster as if it were killed at sexual maturity. The following seems to be a complete list of possible forms of death:

1. Starvation (outright energy deficiency: food calories or light)
2. Malnutrition (lack of nutrients, water, or essential compounds)
3. Predation
4. Parasitic disease

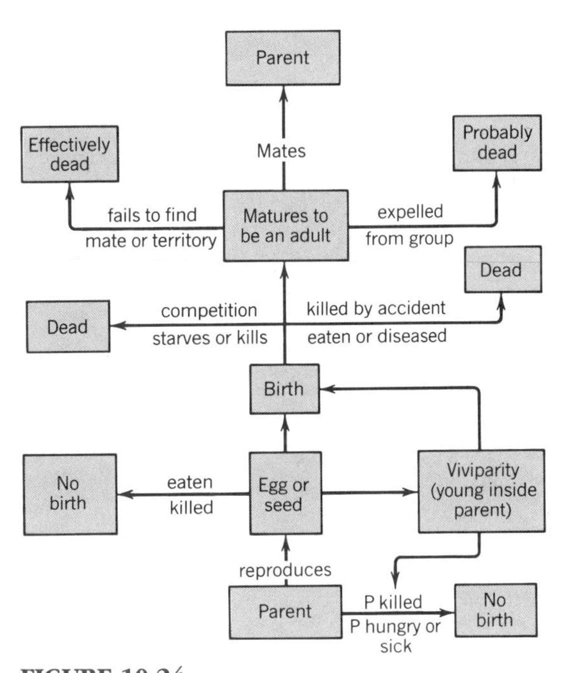

FIGURE 10.24
The fitness stakes.
Populations can be controlled at any or all of the obstacles in a parent-to-parent cycle.

5. Accident (probably weather-induced)
6. Failure to be born
7. Failure to find a mate

Obviously items 6 and 7 in this list can be debited to births instead of crediting them to deaths in the ledger, and they are so treated in life tables. But it may be more interesting to think of them as forms of death. A population can be controlled by invoking any or all of the seven ways of death.

As niche spaces fill up, the seven hostilities become ever-more pressing on individuals, with the population consequences we have seen. It is in choosing the individuals that triumph in the fitness stakes (Figure 10.24) that natural selection works, creating species by the uniformity it imposes (see Chapter 9).

Preview to Chapter 11

Ursus Horribilis **with cuddly young: Defense included in the breeding strategy.**

DIFFERENT life histories can be thought of as different strategies for gaining fitness. Life histories suited to roles as colonists include provision for rapid multiplication and rapid dispersal. These are "opportunist strategies." Colonists or weeds are also properly called "fugitive species," since they are not well-adapted to enduring competition and may respond to pressures of competition by emigrating. The opposite strategy is the life history of an "equilibrium species" that diverts calories away from high fecundity toward behavor or structures that allow persistence. A continuum of possible strategies between the opportunist and equilibrium extremes is predicted. Old field successions are explained by the strategic-continuum hypothesis as opportunist colonizers are replaced by larger plants, slower to colonize but with increasing investments in structures that allow persistence. The strategic continuum concept has become linked to the parallel concept of r *and* K *selection, although the two concepts are not the same. Islands should normally be colonized by opportunists with high intrinsic rates of increase, called "*r*-selected." Once the island is crowded with their descendants, selection should favor traits that survive competition in densities near carrying capacity, hence "*K *selection."* K *selection has been demonstrated within a few populations whose circumstances fit the model.* r *and* K *selection is a "within-species" concept, as opposed to the opportunist-equilibrium continuum, which is a "between-species" concept. Reproductive strategies can also be organized along a continuum of possibilities, from profligate small eggs to prudential large young and parental care. Small-egg strategies are suited to dispersal of many propagules and allow high fecundity, but pay a heavy cost in dead young. Large-young strategies reduce the costs of reproduction because the death rate of young is kept to a minimum. When animals with a large-young strategy raise a family or clutch, it should be regulated to an optimum size, since attempting to raise one more large youngster than resources will allow imperils the survival of the whole clutch. Extremely conservative breeders like condors are adapted to scarce or unpredictable food resources by balancing low reproductive output with long adult life. The long juvenile period and distinctive coloring of juveniles of familiar birds like seagulls are adaptations that serve breeding strategies when extra food needed for breeding is scarce or unreliable. Traits like breeding once in a lifetime (semelparity) or repeatedly (iteroparity) and having young that are helpless or young that are able to look after themselves can all be understood as alternative strategies to cope with alternative patterns of resource.*

Chapter 11

Strategies of Species Populations

Ecologists now make much use of the word "strategy." By "strategy" we mean any pattern of life history or behavior of an individual, or all members of a population, that is adapted to gaining fitness by the efficient collection or use of resources. We talk of opportunistic, fugitive, and equilibrium strategies, or alternative strategies for breeding, for surviving in deserts, and so on.

Consciousness of the importance of different strategies to the study of populations grew out of the old debate over the supposed importance of density-dependent and density-independent control of populations (see Chapter 10). Well before that debate was resolved, it was clear that more sterile parts of the argument compared population events in utterly different circumstances. Birds of the suburban belt and blow fly maggots on carcasses had in common the ability to live in stable populations at what looked like (perhaps misleadingly; see Chapter 10) the density-dependent equilibrium described by the logistic equation. Set against these animals of settled existence were insects of desert margins, like Australian grasshoppers, with their boom and bust population histories. On one side, lives of settled order with competition plain to see; on the other side the vagrants of desert places. The lives of animals at these opposite extremes were differently organized; they needed different strategies.

Since settled existence and vagrancy are opposite extremes, it should follow that compromise strategies are possible. We should expect most living things to have a strategy that is a compromise between the extremes, letting us talk of a possible continuum of strategies between extremes of settled folk like the swifts in Gilbert White's village, still settled there in the same numbers 200 years later (see Chapter 10), and opportunist weeds or insects of desert places, wiped out today, recolonizing tomorrow.

This continuum of possible strategies has gone by various names, particularly "opportunist–equilibrium" and *r* and *K*. As literal descriptors these terms are both inaccurate, and the term *r* and *K* continuum once widely used, is now rejected as a synonym for a continuum of life-history strategies by some ecologists on what appear to be good grounds (Hairston et al, 1970). The opportunist–equilibrium continuum describes a range of strategies observed among different species. The *r*- and *K*-selection concept describes processes acting within species. In this account the two are treated separately. Both are concepts that have released powerful understandings in ecology (Pianka, 1970; Boyce, 1984).

OPPORTUNIST SPECIES

In marginal habitats many life histories are clearly adapted to rapid population growth during short favorable seasons, with other adaptations letting individuals survive the hostile times that inevitably come. These are the life histories of opportunist species. Necessary to the opportunist lifestyle is the ability to disperse rapidly so that remote habitats can be colonized, or recolonized, quickly from places of refuge or chance survival. Essential to the opportunist strategy, therefore, is acceptance of a high chance of death in the lean seasons. Individuals compete to win fitness with a survival and dispersal package for their offspring where the best results come from meeting adversity with a large number of widely scattered survival units.

Archetypal of the opportunist strategy was Andrewartha's Australian grasshopper, *Austroicetes cruciata* (see Chapter 10), but comparable strategies are found in all short-lived animals of highly seasonal environments. The strategy also is adapted to exploiting ephemeral habitats of equable places and is the strategy followed by many of the familiar plants we call weeds.

Weeds are adapted to placing a propagule, usually a seed, in a chance patch of bare earth, to grow rapidly in that earth and to set seed before the benefits of a private plot of bare ground are lost as other plants crowd in. The trick of colonization may be worked by saturation of a landscape with airborne seeds, like those of dandelions (*Taraxacum*), which ride the wind beneath a pappus of fine hairs, or by leaving very resistant seeds in the soil as time capsules ready to germinate when the land is bare again (Figure 11.1). Either way, the chances for survival of any seed are slim, but the chance for fitness of the parent plant is good because each surviving seed grows in a highly favorable place freed from interference by other plants. Growth of a plant from one seed can be followed by a large seed set of its own in the next generation.

We conceive, therefore, of extreme opportunist plants or animals as being well adapted to exploit chance good times, whether these occur in time or space. Rapid reproduction, high fecundity, excellent powers of dispersal, short life, and the ability to sit out hard times in a refuge are characteristic of this lifestyle.

FUGITIVE SPECIES

Another way of looking at dwellers in marginal habitats is as fugitives from more desirable places, for one who journeys to the desert must have gone away from a more fertile place (an ecologist might say a more "mesic" place). Yet what could be the advantage of fleeing from the desirable or the permanent for the harsh or ephemeral?

FIGURE 11.1
***Taraxacum officinale:* to place a seed on bare ground.**
Dandelions are fugitives, opportunists, or called weeds. They are highly fecund and disperse their massive seed set in the wind. Reproduction is a gamble with most parachuting seeds doomed, but the tiny few that find bare ground can expect to grow rapidly with minimal competition. Compare the breeding strategy of the bears of the preview photograph, whose young are large, costly, and defended.

The answer has to be that the flight is from the other inhabitants of permanent or desirable places. This logic leads to a view that opportunist species are not so much those that take advantage of scarce and fleeting habitat as they are species on the run from competition. They make the best of what is left for them.

Hutchinson (1951) introduced the term "fugitive species" in a paper that was the first opening to the eventual concept of a continuum of strategies. He used data of copepod distributions, particularly observations of Elton published nearly a quarter of a century earlier. He wrote as David Lack and others were using the constancy of bird numbers to support their contention that numbers were regulated to the sort of stable equilibrium defined by a logistic equation (see Chapter 10). So he called his paper "Copepodology for ornithologists."

Elton (1927) had noticed that ponds associated with the Oxford municipal sewage works usually supported populations of just one species of copepod, *Eurytemora lacinulata*, but that farm or village ponds in the region had other copepods of the genus *Diaptomus.* To Elton the important difference between the sewage-works ponds and the others was in their age. Ponds in the sewage works were part of the filter system, and they were regularly drained and cleaned out. *Eurytemora* always appeared in these ponds of perpetual youth, but *Diaptomus* never did, even though *Diaptomus* was the commonest copepod in the local bodies of water. Elton suggested that *Diaptomus* was somehow inimical to *Eurytemora* so that *Eurytemora* could live only where *Diaptomus* could not get to.

A further hint of what might be happening among the copepods was given by the pattern of distribution in a river estuary near Liverpool, 200 miles to the north. The brackish waters of the estuary, like the Oxford filter beds, supported a population of *E. lacinulata* but no other copepods (Figure 11.2). Elton argued that this distribution had something to do with salinity, since one could correlate the abundance of *Eurytemora* with the reach of brackish water. But the riddle of distribution could not be solved by a hypothesis of salt toleration alone because elsewhere *Eurytemora* thrived in fresh water. Elton reasoned that this copepod must be prevented from penetrating to the fresher reaches of the river by some hostile ani-

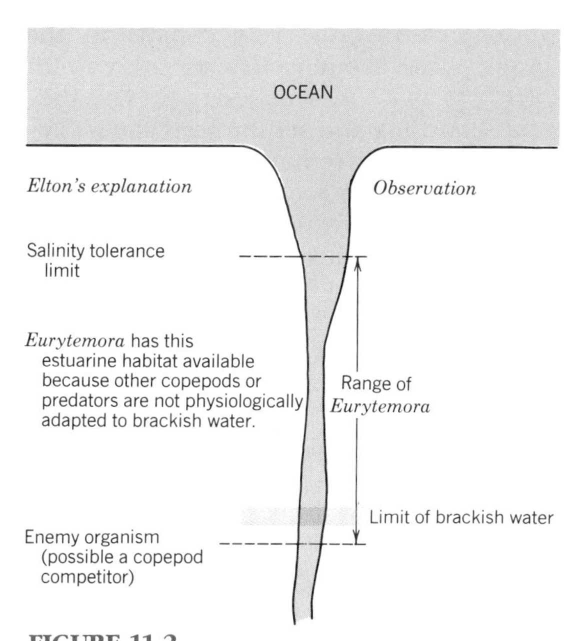

FIGURE 11.2
Copepods in an estuary.
Elton's (1927) observation of the distribution of *Eurytemora lacinulata* in an estuary in northern England.

mal, but not *Diaptomus* this time because the river was without *Diaptomus*.

Hutchinson (1951) invoked Gause's work on competitive exclusion (see Chapter 9) to suggest that the hostile animals in old ponds or fresh river water might well be formal competitors and *Eurytemora* might well be removed from those places by competitive exclusion. The animal existed only by colonizing places without close competitors. But no stable or common habitat could long be free from competitors, requiring *Eurytemora* to disperse efficiently to habitats the others could not quickly reach, there to multiply until the competition arrived. *Eurytemora* lived the life of a fugitive, and Hutchinson called it a fugitive species. Life was divided into fugitives and competitors.

But a fugitive must also be an opportunist, living in the unlikely places where the others are not found. Thus *Eurytemora* had colonized a brackish estuary as well as ephemeral freshwater ponds. Part of its stock in trade was to be a generalist.

POWERFUL COMPETITORS AND THE STRATEGIC CONTINUUM

The extreme alternative strategy to the fugitive opportunist is the strong competitor that sends the opportunist away to the desert, or to ephemeral habitats, as a fugitive. Essential to life in a stable population are adaptations that allow persistence. For want of a better word, these competitors can be called equilibrium species. Each individual of an equilibrium species should compete strongly with its own kind and with others, and be programmed to live in a crowded habitat. Dispersal is less important than persistence; recovery from adversity less important than perseverance; high fecundity of less importance than survival of young in a crowd. The result is a life history distinctly different from that of opportunist or fugitive species.

If the animal of equilibrium design is made to invest calories in mechanisms for defense or endurance, then these calories are not available for extra reproduction. All animals, of course, must invest the largest possible number of calories in reproduction, because this is the first requirement of natural selection (see Chapter 1). But the allocation of calories is a zero-sum game, so that calories spent on defense or persistence are not available for reproduction. A heavy competitor inevitably pays a direct cost in calories available for breeding, though it actually may gain higher fitness as a result of greater access to resources stemming from its success as a competitor.

Equilibrium "standfast" strategies are alternatives to opportunist strategies in ecosystems in which both strategies have possibilities. This is nicely shown by plants of the Sonoran desert of Arizona or of the coastal regions of the Galapagos Islands (Figure 11.3). Both places are difficult for plants, not only because water is scarce but because the supply of water is highly irregular. Many months with no rainfall at all may be followed by short seasons of ample or even excessive rain. Both opportunist and equilibrium strategies offer solutions to living in these desert places.

The opportunist strategy in deserts is simply that of the annual weed scattering seeds that lie on the desert soil until made to germinate by the coming of rain. The young plants grow very rapidly in the bright desert sun following rain, with plenty of space in which to grow. But they start flowering and setting seed almost immediately. They remain small, making no elaborate stem or root system packed with reserves for the future. All calories gained that are not needed for immediate metabolism are put into flowering and seed production. The result is the desert made colorful by a carpet of flowers following rain.

Yet Arizona and the Galapagos are well-supplied with large plants in their deserts: cactus, thorn bushes, dry bushes of sage (*Artemisia*) and other composite shrubs with leathery or pubescent leaves, and even small trees. These meet the problems of desert living with long lives that span many of the scarce seasons of rain. They

FIGURE 11.3
Arid land for two strategies.
A semiarid habitat on Galapagos James Island. *Bursera graveolens* trees have a long-lived equilibrium strategy, putting out leaves during rare periods of rain but being dormant through most of the year as in this photograph. Other plants are adapted to this same pattern of semiaridity with rare periods of rain by an opportunist strategy of short life, semelparity, and survival of the drought by seeds. In a wet season these opportunist would carpet the ground green.

survive the lean times, in the main, not as seeds but as stems or roots, sometimes even with leaves, all heavily protected against water loss or grazing animals. They too grow rapidly in the rains, but very many of their surplus calories go into making stems or thorns, laying down insulation, and waterproofing. There may be comparatively little energy left for reproduction, so the seed set may be less than that of an annual plant, or the plant may flower and set seeds only at very wide intervals, like the celebrated century plant.

An equilibrium strategy for desert plants surely must require perpetual intraspecific competition. Field observations are consistent with this, most notably the strikingly even spacing of many desert shrubs. Figure 11.2 shows a stand of the small tree *Bursera graveolens* on the Galapagos island of Santiago (James). The trees give the impression of a commercial orchard, so evenly spaced are they. The *Bursera* trees in the figure are without leaves, which is their normal aspect. Their primary adaptation to an equilibrium strategy in the Galapagos desert is to put out leaves when it rains and to drop those leaves again when the ground dries up. They are deciduous, but not as trees of temperate regions are deciduous, because no seasonal rhythms control the dropping of leaves.

TEST OF THE CONTINUUM HYPOTHESIS BY OLD FIELD SUCCESSION

Secondary succession in old fields (see Chapters 1 and 20) develops as annual weeds, perennial herbs, shrubs, small trees, and larger trees occupy the site in succession. The sense of deterministic building of plant communities in this process preoccupied many early ecologists, who saw a mystical process of superorganismic growth in which the weeds, herbs, shrubs, etc. were doing the community's will (see Chapter 20). But thinking about alternative strategies allowed a more mundane explanation.

Annual weeds are classical opportunists, and mature trees of the climax may be thought to be the ultimate equilibrium species with an enormous investment in structures of persistence. The intermediate species of the succession can be seen to invest ever more strongly in persistence: perennating organs, stout woody stems, increasing size. All the actors in an old field succession live within dispersal range of the field, and all will get there. But the first to arrive in significant numbers are the opportunist weeds that are able to saturate the field with their seeds. The annuals among them live rapidly and well, make their huge seed sets, and die. But in year two the perennials among the weeds take the land, growing from bastions of root stocks or buried stems. From then on the succession results from a series of contest competitions as plants allocating an ever-larger percentage of calories to organs of competition and persistence like trunks of trees take possession of the field. Thus the hypothesis of a strategic continuum provides a fully mechanistic explanation of a common phenomenon, and secondary succession can be said to be a "consequence of having species with a variety of strategies living in the same country" (Colinvaux, 1973; Drury and Nisbet, 1973).

r and K Selection

Imagine a young volcanic island, recently raised from the sea and without terrestrial animals or plants. This island will be colonized rapidly by migrants arriving from across the sea. Plants will come as windblown seeds, birds and insects on their own wings, seeds and small animals clinging to the birds, and many things, from small mammals and lizards to living plants, drifting out on floating flotsam. All these immigrants will have in common an ability to disperse. The overwhelming number of arrivals—perhaps all—will be opportunist or fugitive species well provided with mechanisms for dispersal.

Once settled on the island, the opportunist immigrants, adapted as they are to high fecundity, should achieve rapid population growth. Soon the island will be crowded with their descendants, which ought to establish population equilibria. But these animals and plants are not adapted to crowded lives; they are usually fugitives from other species that do better in crowds. On their island, weather may not strike them down and their strongly competing enemies cannot reach them. Natural selection should then work to find the most competitive varieties from the island's opportunists. This is the process now known as *K* selection.

The original migrants were highly dispersing, high fecundity individuals. This means that the per capita intrinsic rate of increase *r* of their populations was large. But they now have to live in populations holding the saturation number *K*. The symbols *r* and *K* come, of course, from the conventional writing of a logistic equation (see Chapter 8). Life at *K* means that resources have to be diverted to structures and habits that serve persistence or competition, taking away from the reproductive effort. Using the island argument, MacArthur and Wilson (1967) said that the migrants were descended from populations that were originally *r*-selected, but that the attainment of sustained high density on the island meant that they would be subjected to strong *K* selection.

Note that the terms *r* selection and *K* selection are not themselves synonymous with the terms opportunist or fugitive and equilibrium species, though they do invite comparison. The new language speaks to selection within existing species populations, not to comparison between species. This is a prime objection to using the terms *r* selection and *K* selection in place of "opportunist" and "equilibrium" species. It is easy, in principle, to modify the terms to make them equivalent: opportunists and fugitives are then called *r* strategists and equilibrium species are called *K* strategists, but the distinction is sometimes not made. It is safer to keep *r* and *K* for selection within species populations and to refer to species strategies as "opportunist" and "equilibrium."

The concept of *r* and *K* selection (as opposed to *r* and *K* strategies) is powerful since it leads to testable hypotheses. The postulated selection ought to occur rapidly. Furthermore *K* selection of an originally *r*-selected population will take place in a rapidly reproducing population with short generations, and this offers hope for experiments to show results reasonably quickly.

Ayala (1968) worked with populations of *Drosophila* in typical laboratory culture. *Drosophila* populations achieve population equilibria in standard laboratory bottles if the culture medium is maintained. The flies have high fecundity and in the wild must disperse to find discrete parcels of food. But when crowded in laboratory bottles they have no need for dispersal traits; instead they need to compete for a fixed food supply. *K* selection should result but might be slow because the amount of genetic diversity present in small laboratory populations should not offer selection a large substrate on which to work. Ayala increased the genetic diversity of his flies by irradiating them. If powerful *K* selection then worked on the irradiated flies, the survivors should be more skilled at life in crowded bottles than control populations of nonirradiated flies. This should mean that descendants of the survivors of the irradiation episode would be able to live at higher densities than flies in the control populations. Figure 11.4 gives Ayala's results. The new species equilibrium attained after irradiation allowed, as predicted, more flies to coexist. *K* selection, therefore, was demonstrated.

Twenty years and several hundred *Drosophila* generations later, a team from Ayala's laboratory demonstrated that *K* selection can be demonstrated without radiation, and that the result is indeed tradeoffs between traits needed for fitness at high population densities and at low density as predicted by the model (Mueller et al, 1991). The flies were kept at low population densities for 200 generations, then crowded for 25 generations, thus subjecting originally *r*-selected populations to *K* selection. Growth rates, and productivity of eggs and larvae, were measured. After the 25 generations of crowding, the performance of these *K* selected animals was elevated relative to controls. When the newly *K* selected animals were returned to low population densities, their performance was depressed relative to controls.

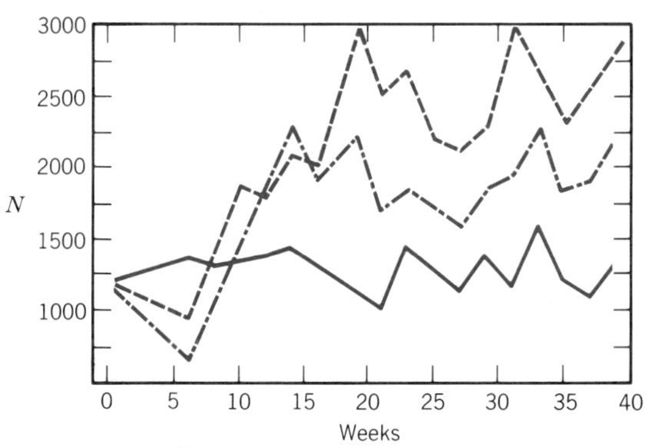

FIGURE 11.4
Experimentally induced *K* selection in *Drosophila birchii*.
Two irradiated populations (broken lines) eventually reach higher equilibrium numbers than a control population (solid line). Notice that the first effect of irradiation is a population decrease, presumably due to irradiation damage. Selection then finds individuals able to live at high densities. (From Ayala, 1968.)

Finding evidence for the *r* and *K* selection process at work in the field is more difficult. An elegant study on dandelions, however, seems convincing. The common dandelion (*Taraxacum officinale*) is an opportunist weed, but dandelions persist at high density for long periods in old grazed meadows or trampled lawns, where *K* selection might be expected. Studying selection in dandelions is helped because the plants set seeds asexually, that is, they are **apomictic**. Apart from mutations, therefore, there should be no genetic change from generation to generation. Yet a number of different dandelion genotypes coexist, each reproduced asexually. These are called **biotypes**, and they can be recognized by isozyme patterns revealed by electrophoresis. Selection in dandelions, therefore, has only a few biotypes from which to choose, and we can recognize those biotypes with a little routine laboratory work.

Solbrig and Simpson (1974) found three populations of dandelions on the University of Michigan campus growing on sites that were differently disturbed but lay within 500 meters of each other. One site was in the path of student traffic and was much trampled with bare ground showing, one site had less traffic, and the third was in the corner of a meadow, mowed annually but scarcely visited. It was reasonable to expect stronger competition between dandelions in the scarcely disturbed meadow than on the other sites and thus more *K* selection.

As expected, a mix of biotypes was present but proportions were different in the

TABLE 11.1
Percentage of Each of Four Biotypes in Three Michigan Populations of Dandelions
(From Solbrig and Simpson, 1974).

Habitat	Number in Sample	Biotype			
		A	B	C	D
1. Dry, full sun, highly disturbed	94	73	13	14	0
2. Dry, shade, medium disturbed	96	53	32	14	1
3. Wet, semishade, undisturbed	94	17	8	11	64

three sites (Table 11.1). The D biotype did well in the undisturbed meadow, for instance, though poorly in the walked-over sites. Selection, therefore, was demonstrated.[1] But was this *K* selection or selection for some unknown physical tolerance? This could be answered only through experiment. Solbrig and Simpson took 100 plants from each site and grew them together in a garden. They found that the D biotype in the garden grew to be bigger, and set less seed, than the A biotype, and that the data for the other biotypes were consistent. Therefore the biotypes did differ in the ways expected if they represented stages along an *r* and *K* continuum. Selection of biotype for the three sites was *r* and *K* selection and not just selection for physical tolerances.

r and *K* Selection on Climatic Time Scales

In the latest experiment from Ayala's laboratory, 200 nonoverlapping generations of *Drosophila* were passed in low densities before trying the effects of 25 generations of crowding (Mueller et al, 1991). A similar experiment with elephants would take a minimum of nearly 3000 years, assuming reproductive maturity at ten years followed by prolonged gestation (Wu and Botkin, 1980). For elephants in their natural habitat the experiment would take far longer. The animals reproduce intermittently in years of high rainfall only, and have young one at a time. Life span is typically 60 years, suggesting that perhaps 30 years is a realistic minimum generation replacement time. Thus 6750 years is the minimum time needed to duplicate the *Drosophila* experiment with elephants. Twice that span brings us to the time of full glaciation at the start of the great meltdown 14,000 years ago.

Climatic change in Africa over the last 14,000 years has been huge. Ice-age climates were drier, principally because of reduced monsoon rains, and evidence is mounting that mean annual temperatures might have been 6 to 9°C below those of today (Livingstone, 1993). Elephants had been subjected to that cooler, drier regime for some 80,000 years before the meltdown began 14,000 years ago, long enough for significant *K* selection to have adapted their population characteristics to living crowded in those prevailing conditions. Reproduction only in a good rainy season does seem a likely adaptation, making an elephant the animal equivalent of a *Bursera* tree on the Galapagos. The warm Holocene has been available to elephants

[1]The presence of different biotypes at the three sites could conceivably be due to founder effects, that is, the sites were by chance colonized by different biotypes in the first place and these subsequently bred true. The subsequent demonstration of phenotypic traits suited to the sites, however, makes this explanation implausible.

for 10,000 years, but significant climatic changes continued in Africa, particularly during the first 4000 years of the Holocene (Figure 11.5). We have no data on elephant densities over this time, but generation times make it certain there have been minimal opportunities for appropriate selection, whether to changed density or simply to changed habitat. The elephants we know are ice-age animals living in an unaccustomed, and hence hostile, world.

Trees appear to arrange things better than do elephants, being highly fecund and setting their massive seed crops at frequent intervals throughout their lives. This must make rapid selection possible. Thus, although the end of the last ice age is only 40 tree generations away measured in lifetimes, significant selection should be possible within the usual spans of climatic change.

The history of spruce trees in North America suggests that selection for life in different densities, as well as in different communities, has indeed taken place with glacial cycles. Pollen diagrams (see Chapter 17) from the American Midwest show spruce as prominent plants in ice-age times of what are now the prairies (Figure 11.6; COHMAP, 1988). But this was not spruce forest as we know it today; the concentration of spruce pollen in sediments, and the plant company kept by the spruce, are unlike any modern spruce forests. We call the old vegetation "spruce parkland," which is a habitat that has vanished from the contemporary earth. With the coming of the Holocene, spruce colonized the de-

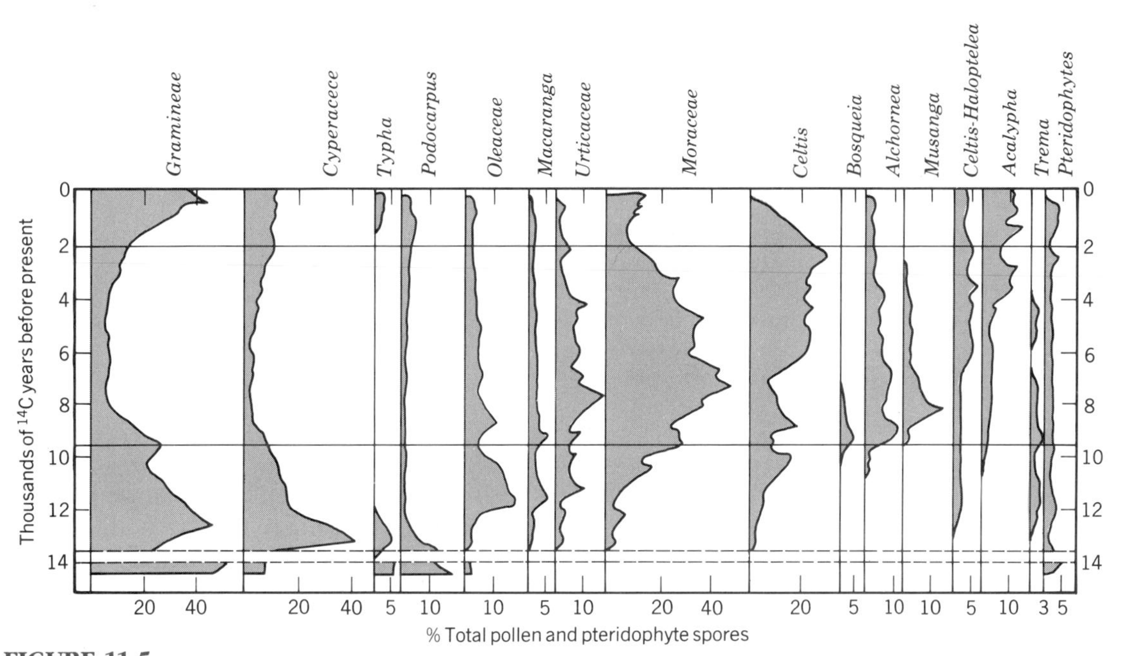

FIGURE 11.5

Pollen percentage diagram from Lake Victoria, East Africa.

The reduction in pollen of tree taxa such as Moraceae and *Celtis* before 10,000 years ago suggests that rain forest was absent from the region in ice age time. Changes over the last few thousand years may have resulted both from human activities and from changing climate. (From Kendall, 1969.)

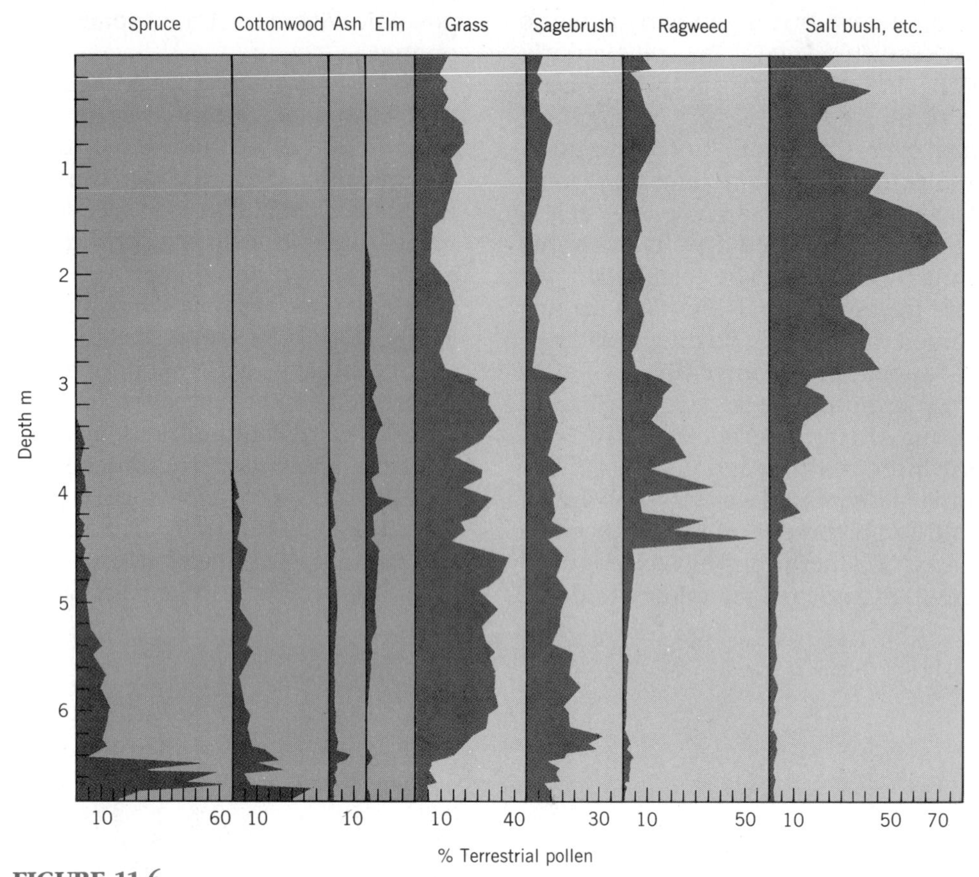

FIGURE 11.6

Spruce parkland to prairie: The adaptability of spruce.

Figure is a pollen percentage diagram from lake sediments spanning about the last 11,000 years in South Dakota. In late-glacial times, spruce and poplar grew among the sagebrush elements of what is now a prairie. Spruce must be at least as adapted to life in these parklands as it is to life in dense stands of the modern boreal forest. (See Chapter 17 for review of pollen analysis.) (Data of E. C. Grimm, from Barnosky et al., 1987.)

glaciated lands of Canada from its bastion in the Midwest parkland, finally completing the familiar Canadian spruce forests about 5000 years ago (Liu, 1991). Postulating intense K selection on seedlings throughout those 5000 years, on the r-selected populations of the old parklands seems not unreasonable.

Thus tree populations may well cope with climatic oscillations as large even as the ice ages by selective winnowing of their huge seed sets. This hypothesis predicts the existence of very varied genotypes in all tree populations. And it reminds us that all modern forests, from equatorial to subarctic, might be aberrant communities put together by winnowing and selection of the more extreme genotypes of trees.

STRATEGIC THINKING ABOUT REPRODUCTION

Animals can lay large or small eggs, or they can bear their young alive. The young can fend for themselves from the day they are hatched, or they can be looked after by their parents. Young can be defended by parents or become the prey of cannibals. Even plant seeds can be large or small, protected from seed predators or provided with lures so that animals will seek them out, produced all at once or scattered through the seasons.

The most obviously different of extreme strategies are that of laying large numbers of tiny eggs and that of raising a few large young. Hutchinson (1978) described these two strategies as "**profligate**" or "**prudential**" reproduction, but perhaps the terms "**small-egg strategy**" and "**large-young strategy**" convey the two alternatives more clearly. These opposing strategies must be explained as working to give fitness to individuals of the species that use them, though obviously in widely different ways.

Profligate animals with a small-egg strategy make small eggs, each capable of developing into a sexually mature adult. The cost of each egg is small so that many can be made for a modest investment in calories. The relative fitness of an individual becomes a function of the number of eggs laid so that natural selection will program all individuals to make as many eggs as possible. This, surely, is the simplest solution to the fitness problem: design tiny young and make many of them.

But real fitness is not simply a measure of fecundity; it is the number that survive to reproduce. Few of the tiny young hatched from small eggs in fact live long enough to breed because they die by accident, by predators, or even by cannibalistic adults who are their parents' neighbors. Calories invested in these many young that die are wasted. Therefore, though at first sight the strategy of many small eggs seems a good way of flooding the future with one's descendants, from a cost-effective point of view it can be a very poor way. A female salmon, for instance, spends a year or two foraging so that she can lay many eggs that will be eaten by other animals.

Prudential reproduction of the large-young strategy is a quite different outlay of resources that in many circumstances can bring a larger return on investment. The mother makes large young (a reptile's egg, for instance). The mother's capital of calories will not let her make so many of these large eggs, but the young will each have more chance of survival than, say, a fish fry. Compared to investing the same amount of calories in many more small eggs, making a few large eggs may mean more surviving offspring that themselves live to reproduce. It can be, therefore, that restrained fecundity can result in greater fitness.

The logical extension of the large-young strategy is to reserve some calories for defense and nurture of the young. Many birds are programmed to lay few eggs, then to stop laying, even though they have food resources to lay more. They divert these extra resources to feeding and defending their young. Animals that bear their young alive, whether mammals, viviparous reptiles and fish, or even viviparous insects, carry this process one step further. Essentially they are making their young even bigger at the time of birth by letting them start development within the mother. All these large-young strategies should be seen as ways of maximizing fitness by safeguarding the original investment. When done really well, no calories are wasted, as

FIGURE 11.7
***Phasianus colchicus:* precocial young.**
Pheasant chicks, like domestic chickens, can run and peck for food almost immediately after hatching, requiring minimal parental care. Eggs must be large enough for the chick to be well-grown at hatching, but the mother has no need to prepare for feeding young and so can invest maximum calories in a large clutch even of large eggs.

every calorie spent in reproduction goes to a young individual that eventually will itself reproduce. Contrast this idea with the certainty of mass death and mass waste that must result from broadcasting tiny eggs.

Precocial and Altricial Breeding Strategies

Obvious variants of the large-young strategy are shown by birds that have naked, helpless offspring ("nidicolous" or "altricial" birds) and those whose feathered young can forage for themselves near their mother ("nidifugous" or "precocial").[2] The nidifugous or precocial lifestyle is possible

[2]These terms are all from Latin roots. "Precocial" comes from the direct transfer of the Latin word *praecox*, meaning "ripe before its time," which is itself derived from *prae-* (before) and *coquere* (to cook). "Altricial" is derived from the Latin *altrix* (nourishment), and thus implies young that need feeding as opposed to precocial young, which can feed themselves. The Latin *nidus* means "nest," and *fugere* means "to flee," so nidifugous birds have young that leave the nest as soon as they hatch, which implies that they probably are also precocial. The Latin *colere* means "to cultivate," so a nidicolous bird is one that feeds (cultivates) its young in the nest, requiring that the young are certainly altricial.

only where food is of a kind that can be found by baby birds. Game birds like pheasant and partridge, and many water birds like ducks and shore birds, are precocial and nidifugous (Figure 11.7). The food of the chicks in these species can be found by run, swim, and peck behavior. On the other hand, typical nesting songbirds, or hawks and owls, are altricial and nidicolous because insects, seeds on bushes, and animal prey can be collected only through the adult behavior of flight (Figure 11.8). And so the food supply provides the selection pressure that drives toward either the nidifugous or the nidicolous condition. It is not surprising to find that eggs of nidifugous and precocial birds are generally larger: parental investment has gone into a large egg from which develops a more advanced young, instead of hatching a less-developed chick and then feeding it (Clutton-Brock, 1991).

A first thing to be stressed about the opposed reproductive strategies of profligate and prudential reproduction is that they are alternative ways of pursuing fitness, not alternative approaches to population problems. A small family of large young does not represent restraint on breeding when compared with some insects and fish engaged in profligate reproduction. The real difference is not in reproductive effort but in the way that resources are disbursed to promote reproduction. It can be argued, for instance, that a strategy of a few large young will actually tend to increase population faster than would a strategy of many small eggs because the large young give a better return of survivors for every investment calorie.

FIGURE 11.8
***Buteo borealis:* altricial and helpless.**
A red-tailed hawk starts life helpless and almost naked, requiring a large investment in parental care. Clutch is small, because the parents can feed no more from the meat they can supply from hunting. The young must be raised until they are themselves full-grown hunters.

CLUTCH SIZE REGULATION

Large-young strategies are of two kinds: one-at-a-time or a clutch-at-a-time. The typical ungulate mammal, such as a cow, bears one young at a time and then raises it to be self-supporting before the next reproductive episode. Fitness of the mother is a simple function of how long she can stay alive and active as a reproductive machine. But when a number of young are born synchronously, as with most mammalian carnivores and most birds, then there is the added complication of regulating the size of the clutch.

If a bird laying a clutch of eggs is programmed to lay one egg fewer than its neighbors, and if the neighbors are able to fledge a chick from their extra egg, then the bird with a small clutch will be at a selective disadvantage and its genes will disappear from the population. That much is obvious. But a bird might be at an even greater disadvantage if it were programmed to lay one more egg than its neighbors, because the resulting extra mouth to feed might endanger the survival of the whole brood. Raising a clutch of young, therefore, requires that clutch size be regulated to an optimum number.

Regulating clutch size in birds requires that the female be programmed to stop laying when the optimal number of eggs has accumulated. This could be arranged in two ways: the number of eggs could be fixed by a genetic mechanism or the bird could be equipped with a sensing device that regulated the number of eggs according to the resources available in the local habitat (Figure 11.9). Both patterns are known to exist in different species of birds.

The classical data for fluctuating clutch size are those for the European robin given by Lack (1954). Robins (*Erithacus rubecula*) are found all across Europe—from North Africa to Finland—as year-round residents (Figure 11.10). The average clutch size corresponds closely to latitude, there being more eggs in the typical nest in the northern parts of the range. This observation certainly shows that clutch size is regulated to suit environmental circumstance. One environmental condition that might favor the larger clutches of the north is the longer day in which the parents may forage, thus providing more food. It may also be that the northern spring and early summer have insects in greater abundance than are found in lower latitudes, or that there are fewer competitors for them. These arguments suggest that clutch is being adapted to the food supply available to the parents.

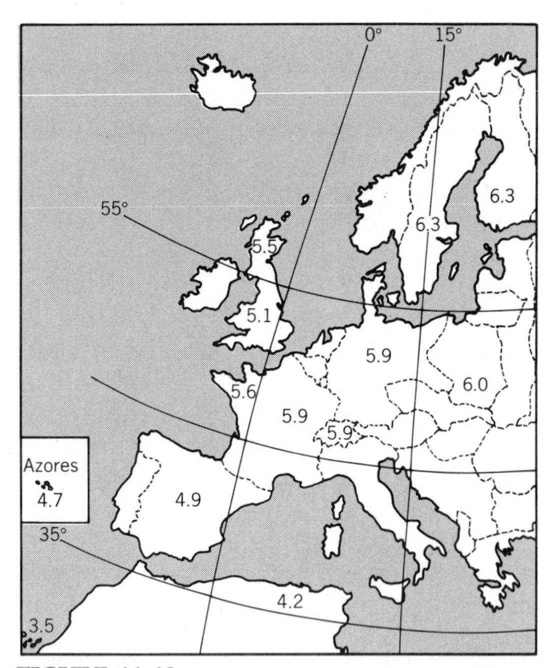

FIGURE 11.10
Average clutch size of the European Robin (*Erithacus rubecula*) along the north–south axis of its range.
(From Lack, 1954.)

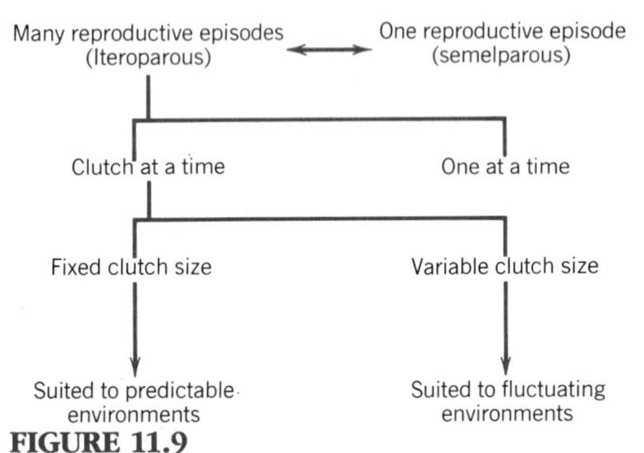

FIGURE 11.9
Variations on the strategy of producing and nurturing large young.

Another factor that may be involved in reducing optimal clutch in the southern part of the range is predation. If the warmer latitudes are home to more predators, then the parents should minimize the number of visits they make to their nest each day, a requirement that suggests a smaller clutch (Skutch, 1949; Cody, 1966). Apparently female robins are able to assess such things as available food or danger from predators, and this assessment regulates egg laying. This is no great feat because a simple hormonal response to maternal fat, or feeding rate, or number of frightening predator alarms is all that is needed.

More spectacular regulating of clutch size is shown by snowy owls (*Nyctea scandiaca*) of the Arctic. Snowy owls rear their young on the meat of lemmings and other arctic rodents whose numbers fluctuate widely from year to year (see Chapter 10). In a high lemming year on the arctic coast of Alaska, each owl nest may hold ten or so eggs, and the owls raise and fledge all the hatchlings, but in a low lemming year a nest may hold only one or two eggs, if the owls breed at all. Snowy owl physiology apparently can detect the density of lemmings: perhaps excitement at seeing how much succulent food runs about does the trick.

A large literature exists to show that clutch size in birds may be correlated with the food supply. Lack (1954, 1968) first produced persuasive summaries testifying to this, and there have been numerous studies since. We take this literature as providing sufficient test of the hypothesis that clutch should be regulated to an optimum in animals that lay large eggs or nurture their young as to raise this concept to a working principle.

REPRODUCTIVE STRATEGIES OF OPPORTUNIST AND EQUILIBRIUM SPECIES

Profligate reproduction with a small-egg strategy seems well suited to a species that must disperse rapidly and well. Classical weed plants with small, wind-blown seeds are certainly profligate reproducers: Andrewartha's Australian grasshoppers (see Chapter 10), mosquitoes whose larvae live in temporary puddles, and flies that raise young in isolated bits of rotting fruit (*Drosophila*). All these use a small-egg strategy and all are opportunists or fugitives, suggesting that profligate breeding is suited to these lifestyles.

Yet the broadcasting of small propagules is also done by some forest trees and by long-lived animals (many fish) that might well exist in near-equilibrium numbers. Trees of the climax forest that use wind as a dispersal agent have small, though not necessarily tiny, seeds. Sugar maples (*Acer saccharum*), codominants of the beech-maple forests of the American Midwest, for instance, make a large number of smallish seeds, each equipped with a wing-like structure. The sugar maples are true standfast, equilibrium plants with massive trunks, long-lived, and with the habit of setting aside energy reserves for the future. But their numerous wind-dispersed seeds give them some capabilities as colonists so that it is not impossible that they might sometimes colonize bare ground like true opportunists, thus preempting the herbs of the secondary succession (see Chapter 20). But large propagules also have their place in dispersal, as demonstrated by coconut trees, which have truly huge seeds and yet may be one of the most widely dispersed trees in the world, able to cross great oceans.

What seems true is that a small-egg strategy has particular value to many lifestyles that require wide dispersal, because dispersal is always a gamble in which many of the would-be migrants must perish. Making these migrants small is then actually prudent because the inevitable losses are minimized. The gambling analogy is particularly appropriate here. Some chips (dispersed seeds) do find a good site, effectively hit the jackpot, and produce a large return in fitness in the next generation. The strategy works in a patchy environment when the cost of dead offspring is the price of finding the right patch with the lucky ones. But this does not mean that the strategy is without value to other lifestyles (like maple trees).

Prudential reproduction, large young, and parental care, however, may also be fine adaptations to harsh or unpredictable environments. This is particularly clear for parental care, because a survey of its incidence across all taxa in which it is known shows that parental care is most highly developed where eggs or young face adverse environments, high rates of predation, parasitism, or intense interspecific competition (Clutton-Brock, 1991). That this should be so seems intuitively reasonable, for where danger is highest the cost of parental care is likely to yield the highest return in surviving offspring. But notice that the list of adversities includes intense interspecific competition. Prudential reproduction and parental care is evidently a good tactic in some instances for both opportunists facing harsh environments and strong competitors of productive places for whom harshness is intraspecific competition.

In many instances the basic breeding strategy is in fact contingent on the past history of the species. Lemmings, being mammals, have a large-young strategy with parental care. They have, however, many traits of opportunists: short lives, high relative fecundity, high growth rates, etc. Their system of parental care certainly is suited to the harsh Arctic, and to the probability that they suffer severe intraspecific competition leading to density-dependent-driven oscillations in number (see Chapter 10). But lemmings practice parental care because they are mammals, not because they live in a harsh environment.

FIXED CLUTCH OR VARIABLE CLUTCH?

Many birds have clutch sizes that are fixed, and thus cannot be altered to suite environmental circumstance or the food supply. It might be expected that a fixed clutch size would be well suited to predictable conditions and life at a species equilibrium. Most birds do in fact have fixed clutch sizes, which, given the evidence that birds tend to be under density-dependent control (see Chapter 10), seems consistent with this expectation. But a fixed clutch can also be the answer to unpredictable environments in some circumstances.

The fixed clutch size (two eggs) of the Galapagos penguin turns out to be an effective response to a highly unpredictable food supply. Galapagos penguins (*Spheniscus mendiculus*; Figure 11.11) wander the seas throughout the 1000 square kilometers or so of water that surrounds the Galapagos Archipelago, but they nest only on the western coasts of the westernmost islands (Figure 11.12). This distribution can be understood from a knowledge of ocean currents of the region because the coasts where the penguins breed are those that intercept, from time to time, the submerged Cromwell current coming from the west. When this cold current strikes a Galapagos island, cold, nutrient-rich water upwells along the shore causing a bloom of algae and thence of fish (Maxwell, 1974). This productive water provides the food on which penguins rear their young. But the arrival of an upwelling is highly unpredictable; it may happen in any month of the year and may last for days or for weeks. The penguins are equipped with a breeding strategy, involving a fixed clutch of two eggs, that lets them cope with the problem of an unpredictable food source. The main elements of this strategy, as worked out by Boersma (1977), are as follows:

1. Molt as soon as an upwelling is detected (probably using water temperature as the cue).
2. Copulate as soon as molting is complete.

FIGURE 11.11
***Spheniscus mendiculus:* parent that abandons young.**
The Galapagos penguin lives in an unpredictable environment, using inshore food supplies of intermittent upwellings to feed young. When upwellings cease during breeding, which happens more often than not, the parents abandon the young. This breeding behavior works to produce fitness because the adults are long-lived and feed safely out at sea.

3. Lay two eggs at once and start incubating.
4. Abandon the weaker chick when finding sufficient food is difficult.
5. Abandon nest and both chicks at once if the upwelling fails.
6. Live many years.

The striking thing about this breeding strategy is the readiness of the penguin mothers to abandon chicks in times of adversity. Galapagos penguins do not have human ideas of motherhood. Indeed, Boersma (1977) found that the usual outcome of a breeding effort was abandonment of *both* chicks, and it was rare that both were raised. A successful breeding effort, in penguin terms, is to raise one chick, and the most usual result is failure to rear any.

The proximal reason for a Galapagos penguin to abandon chicks is obvious, because it is done when the upwelling fails. No upwelling means little food within swimming range of the nest. If the penguin mother tried to sit out the failure of an upwelling, not only would there be no food for chicks but she herself might starve. The proper response is for her to swim away to the open sea where she can survive herself and come back to breed during another upwelling.

Two eggs rather than one also make sense in the penguin's circumstances, even though Galapagos penguins seldom

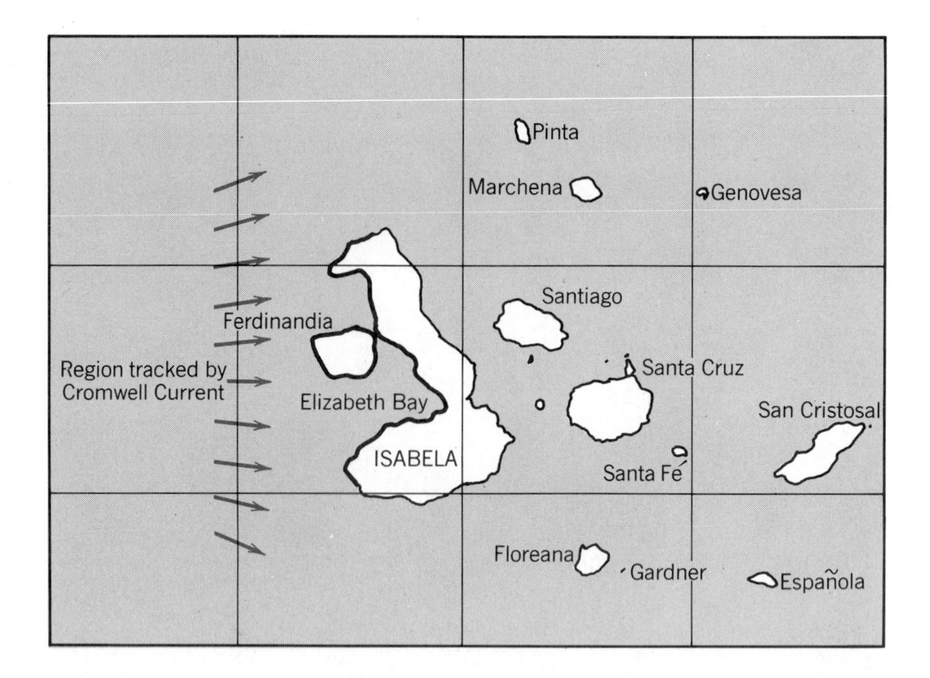

FIGURE 11.12

Breeding sites of Galapagos penguins (*Spheniscus mendiculus*).

The penguins breed only on the westernmost coasts in the archipelago where waters of the Cromwell Current (Equatorial Undercurrent) upwell from time to time. Adult penguins can be found throughout the waters of the archipelago, but they do not come ashore on other islands. (From data of Maxwell, 1974, and Boersma, 1977.)

raise more than one chick. The second egg is an insurance policy against loss by accident—incubated or hatched and ready to go if needed. The premium paid on the insurance policy is the calories invested in one egg, roughly the same number of calories needed for three days' metabolism of the mother. So two eggs can be the optimum clutch even when only one chick is to be raised. This is quite a common tactic among seabirds and others that face uncertain food supplies for the young, a tactic that ensures a chick is coming along at minimum cost, ready to exploit chance good times.

All five of the penguin habits listed above make sense as adaptations to exploit very short-lived and unpredictable gluts of resources. Molting before breeding (an unusual trait of Galapagos penguins), for instance, ensures that the mother is in her best shape to go fishing in the cold water of the upwelling. After that, the habits are simply suited to rushing one chick through as fast as possible, but with the best chance for cutting losses and getting out if this proves impossible. Yet this behavior confers fitness only if the mother lives to breed again another day. A long, safe life for the mother is vital to the habit of abandoning chicks in lean times.

The Galapagos penguin clearly has a cautious, conservative attitude toward reproduction that is an adaptation suited to exploiting an intermittent, yet ample, flux of food. Many other seabirds behave in a

FIGURE 11.13
***Sula sula:* conservative breeding on short rations.**
Red-footed boobies nest in bushes on Galapagos Isla Genovesa, but wander the ocean at other times. Food for reproduction is always scarce. Like many other seabirds their adaptation for breeding is a long juvenile period of learning (5 years), long life, one egg at a time, slow growth of young, and long adult life. Half the young die in the nests, apparently from lack of food.

similar way, or they may show the behavior in more extreme form. Some, like most albatrosses, lay only one egg at a time and seem to be under more strict limits from food even than the penguins. The red-footed booby (*Sula sula*) of the Galapagos is of this kind (Figure 11.13; Nelson, 1968). These boobies do not breed at all until they are five years old, then lay one small (compared with body weight) egg at intervals of 14 months or so. Adults are long-lived, perhaps living ten, twenty, or more years. Half of the chicks die or are abandoned, so the reproductive rate of this species is extremely low. What seems to have happened for these birds is that food is always in short supply so that the rapid feeding of young is impossible; likewise, the chance of a chick's starving must always be high. Since food is scarce, the chicks grow slowly, and this increases the cost of a chick because its metabolic needs must be met for a longer time. As chicks become more difficult to feed, and more expensive in the long run, there is a greater value in keeping the parent alive for later reproductive episodes.

Many animals with a large-young strategy can be forced into this pattern of life history. Scarcity of resource first decrees that clutch size be small and growth rates slow. After this the mathematics of demography show that survival of parents is an important component of fitness. Figure 11.14 shows a plot from Goodman's (1974) life-table calculations of the extra young needed to compensate in fitness for the death of a parent, assuming there is only one reproductive event in a lifetime (**semelparity**). The calculations show that for breeding in the first year the parent can be replaced by a single egg, but as the breed-

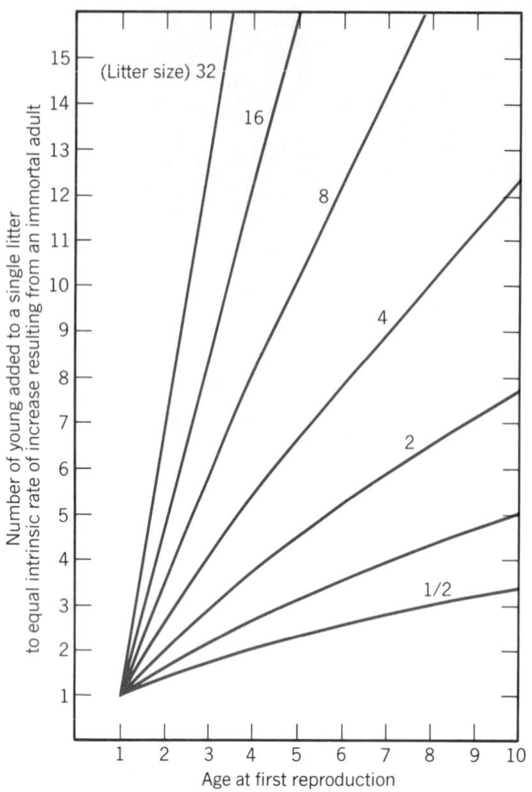

FIGURE 11.14
Increased clutch needed to compensate for adult death.
Fitness represented by survival of a parent goes up with the age at first reproduction. Data are plotted for a number of litter sizes, showing the extra eggs needed by a species that breeds only once (*semelparous*) to compensate for the death of the parent. (From Goodman, 1974.)

ing parent grows older its replacement cost goes up until a five-year-old parent with a litter size of one egg gives the same fitness value as three additional eggs, and so on. This plot reveals the hard facts behind the life history of typical seabirds. Food is hard to get, which forces clutch size to be small. Furthermore, the difficulty of finding food requires that birds be experienced before breeding, so they cannot breed in their first year. Living longer than a year before reproduction increases the value of the adult as a unit of fitness, which increases the selection pressure for adult survival still further. The result is a strategy of long life and low reproductive effort.

This logic provides a ready hypothesis to explain some common facts of natural history. Familiar seabirds of the north, like herring gulls (*Larus argentatus;* Figure 11.15), live as immature birds for several years, and these immatures are marked so that they can be easily distinguished from breeding adults. Long juvenility is understandable if immature birds are learning to forage under all circumstances until they have sufficient skills to be able to become parents with minimum risk to themselves. Their distinctive markings probably free them from wasting energy and time on sex encounters while they learn.

In a sample of 50 fulmar petrels (*Fulmarus glacialis*) ringed as chicks, the modal age at first breeding was eight years, with a range from six to 19 years (Ollason and Dunnet, 1988). That these long juvenile periods are indeed a time of learning is consistent with histories of breeding success, which for females rose until about the tenth breeding season, by which time the birds were in the range of 16 years old.

More important are the population consequences of long life and conservative breeding. As Mertz (1971) was first to point out, a species with this strategy is threatened with extinction as soon as any significant mortality of adults occurs. This is why efforts to save the California condor from extinction show so little success. Condor breeding strategy is like that of the red-footed boobies, perhaps actually an extreme version since the adults can live 50 years. They breed very sparingly and are likely to abandon their young in adversity in the expectation that they will live to breed another day. Continued increased

FIGURE 11.15
***Larus argentatus:* juvenile status proclaimed by color.**
In herring gulls, as in many long-lived seabirds, the non-breeding juveniles are clearly marked. Juvenile years of learning skills at foraging apparently are needed before care of offspring can be attempted and the dark coloring may serve to spare the juveniles unrewarding social encounters.

mortality from firearms cannot be offset by reproduction in birds with this breeding strategy.

Condor, booby, and penguin life histories make them poor colonists. They cannot exploit new environments by rapid reproduction. Therefore, it is safe to say they are not opportunist species. On the face of it, they certainly seem to fit a general description of an equilibrium species. Calories are diverted from reproduction to persistence, and there is a stable population in the absence of strong environmental change. In fact, the population will be stable for any environmental change less than that which imposes excess mortality on adults. And yet there must be a little discomfort in calling these birds equilibrium species, and even more in suggesting that they are *K* selected. Their life histories are perhaps better understood as a result of selection to cope with a particular pattern of resources than as a response to crowding or competition.

TO BREED ONCE OR TO BREED REPEATEDLY?

In both animals and plants some species breed but once in a lifetime, ("**semelparous**" species), and others breed repeatedly, ("**iteroparous**" species). When resources for reproduction are scarce, iteroparity can give an advantage in fitness (Figure 11.16). But a single breeding episode must give the advantage in different circumstances, because many species are semelparous in nature: annual weeds with a single flowering episode before death of the parent and the Pacific salmon, which return to the rivers of their birth but once, lay eggs, and die, are the most spectacular examples of semelparous species.

FIGURE 11.16
***Nyctea nyctea:* flexible clutch for fluctuating food supply.**
Snowy owls breed in arctic tundra, raising their young largely on lemmings and other rodents. Clutch varies from one to ten, depending on the supply of rodents. This owl is feeding its single chick on eider ducklings, letting an arctic naturalist suggest that the photograph was taken in a low lemming year.

When an organism matures, reproduces once, and dies, perfectly acceptable logic suggests that the death of the parent can be compensated by a single addition to the clutch: from the point of view of fitness, one female equals one egg. This was first formally shown using the mathematics of life tables by Cole (1954), and the conclusion is now referred to as "Cole's result." This, however, leads to disturbing conclusions, such as saying that a ten-kilogram female salmon represents no more chance to the fitness of its genotype than a tiny egg. The resolution of this paradox is that fitness is secured only by survival to the next reproductive episode, and whether survival to another reproductive episode is more likely for egg or fish may not be clear-cut.

Whether individuals of a species are semelparous or iteroparous must depend on the resources available for reproduction and the chance of survival to the next reproductive period. Birds like boobies and condors, which use large-young strategies and face scarce resources, are forced into iteroparity. Iteroparity probably is nearly always necessary for animals with a large-young strategy because of the high cost of the young and the high survival power of adults.

But for a small-egg strategy the issue is not clear-cut. For a salmon, for instance, the choice is not actually between one fish and one egg, but between one fish and very many eggs. The advantage then seems to swing toward the eggs. But some salmon are semelparous and others (the common Atlantic salmon, *Salmo salar*, for instance) are iteroparous. Whether one strategy or the other gives the advantage must depend on factors like the local changes of seasons, or local predators, which influence the survival of young salmon.

In plants the change from semelparity to iteroparity can be seen to represent a continuum of strategies, essentially parallel to the opportunist–equilibrium continuum itself. Extreme annual weeds are semelparous while perennials are iteroparous, and iteroparity seems to be prolonged through a variety of strategies roughly mirrored in the progress of a secondary plant succession. Prolonged iteroparity in plants may be accompanied by delays in age of first reproduction, just as it is in long-lived seabirds. Many trees, for instance, do not set seed until they are several years, or even decades, old. Plants with life spans of considerable length may be semelparous, too, like a Pacific salmon. Green algae of the reef-building genus *Halimeda* may pass through a number of asexual generations of six or more months each before a sexual episode, when the green plant turns white overnight as all its tissue is mobilized to form a cloud of zoospores that, when released, leave behind only the skeleton of white carbonate (Figure 11.17).

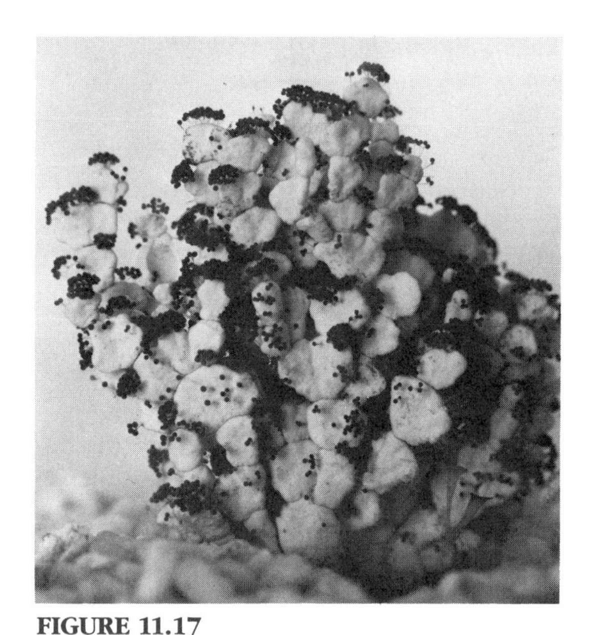

FIGURE 11.17
***Halimeda simulans:* semelparous ocean plant.** Green algae of the genus *Halimeda* are important benthic plants of tropical coasts. The photograph is of a plant in culture about to release gametes. The entire organic content of the thallus has been mobilized to produce the gametangia, seen as dark, grape-like clusters on the tips of segments. Only a dead, white, calcareous skeleton remains after gamete release.

These variations on a theme of semelparity and iteroparity provide a very large number of possible ways in which life histories may be fitted to patterns of environment and resource. They present not so much a continuum of possibilities as endless permutations. And each variation must have consequences both for the history of the population and for the structure of the community in which the individual lives. Large young, small young; semelparous, iteroparous; early reproduction, delayed reproduction; helpless young (nidicolous), precocial young (nidifugous); all are strategies that work in some circumstances, all are selected to confer fitness on individuals, and all have consequences for populations and communities.

CONTINGENCY, SELECTION, AND STRATEGY

Many patterns of life-history phenomena were fixed by selection long ago; the mammalian practice of parental care, for instance. Natural selection can only modify such preexisting designs. This is **contingency**. The habits of species populations were also fixed within narrow limits by natural selection, also usually long ago. This also is contingency. As species populations are shuffled around by history, climatic change, and fate into different communities, their relative abundance, or their order of appearance in the community, is a function of the strategy originally forged for the species, with chances for modification being minor. It is these contingent strategies, set as they were by ancient speciation events, that we may profitably rank along an opportunist–equilibrium continuum. But within each species population exists wide genetic variation on which selection can work according to local circumstance. Fitting organisms from a species population to ever-changing environments according to the Red Queen hypothesis (see Chapter 9) is one consequence of this intraspecific selection. The r and K continuum is another; a selection working purely within the contingency of constraints inherited by a species population.

Preview to Chapter 12

Latrodectus mactans: the female black widow is more powerful than the male.

NO satisfactory theory for the prevalence of sexuality exists. Scenarios for the first origin of sex in unicellular eukaryotes plausibly attribute primitive sexual systems to be associated with DNA repair in an environment flooded with ultraviolet light. The difficulty is in explaining how sex could still be maintained by natural selection 2 billion years later, despite the immense fitness costs sex imposes on females. Some cell biologists dismiss this problem with the claim that sex is immune to selection, being an unalterable holdover of the distant past. For evolutionary ecologists, however, a satisfactory theory must provide a selective advantage that more than compensates individual females for the half interest in offspring lost to the male, as well as various other costs to the female of sex. The ecological hypothesis depends on the advantage in fitness to the female of having varied offspring to contest for niche space in an ensuing generation when conditions for life necessarily will be different. This argument is powerful for highly fecund organisms whose breeding strategy allows the loss of unsuitable genotypes from selective winnowing, but has difficulty explaining sex in low-fecundity animals like mammals and birds. An alternative genetic hypothesis provides the necessary fitness advantage by the masking of deleterious mutations through genetic recombination, essentially a DNA repair argument. This genetic hypothesis, however, invokes group selection of a kind that has not been demonstrated. The ubiquity of sex sometimes opposes male and female interests, which combine with patterns of resource for breeding to produce such different results as polygamous and monogamous mating systems.

Chapter 12

The Cause and Consequence of Sex

Most living things come in two sexes: the egg-laying, young-producing sex and the other one. The reason for this ubiquity of sexual systems is the subject of vigorous, even polemical, debate among geneticists, cell biologists, and ecologists, with no obvious end to the argument in sight (Williams, 1975; Maynard Smith, 1978; Bell, 1982; Shields, 1982; Margulis and Sagan, 1984; Halvorson and Monroy, 1985; Michod and Levin, 1988; Stevens and Bellig, 1988). Part of the debate dwells on the origin of sex in Archaen seas flooded with ultraviolet radiation more than 2 billion years ago. But ecologists are more concerned with how two sexes can continue to be preserved by natural selection 2 billion years later. This problem can be succinctly stated with the question, "Who needs males?"

The female is the reproductive unit. She gains fitness through reproduction, and only through reproduction. Natural selection requires that she reproduce to the utmost, thrusting as many copies of herself into the next generation as possible. But sex requires that she give a half genetic interest in all her eggs to another, the male.

This loss of a half interest in the genes of a female's offspring can be called **the 50% cost of meiosis** (Williams, 1975) because of the deliberate halving of the mother's contribution to her progeny by splitting her chromosome pairs in a meiotic division. But this is only a portion of the total cost. A second cost bound to seem important to an ecologist is the recombination of genes that is the immediate consequence of sex. The scrambling of genes means that the offspring will not be like the mother, which is quite contrary to ecological theory explaining conformity of individuals to the species niche as resulting from ruthless selective removal of all deviants. Offspring should be like their mothers. But indulgence in sex puts the comforting carbon copy of a mother out of reach by imposing a "**recombination genetic load.**"

Nor are the costs of sex to females ended with recombination of genes, for the cost of mating may also be heavy (Daly, 1978). Sex display and the other activities of mating require energy and involve danger. These costs fall most heavily on the male, but the female has her share, including energy costs of (1) sexual mechanisms, (2) mating behavior, and (3) escape from unwanted sexual attentions, together with risks from (4) predation, (5) disease transmission (AIDS is our latest human illustration of this cost), and (6) injury inflicted by the males.

The total adverse effects of sex to a female, therefore, include the 50% cost of meiosis, recombination genetic load, and the cost of mating. For sex to be preserved by natural selection, the benefits must outweigh all three costs combined for sexuality to exist.

In seeking a causal mechanism for sex, it is first necessary to free the mind of an old heresy: that the value of sex is somehow connected with the promotion of evolutionary change. More rapid evolution probably is indeed a consequence of sex, because shuffling of genes during the exchange of chromosomes seems inherently likely to multiply the varieties on which selection can operate. But evolutionary change is a side effect of the working of natural selection, not a target. As long as a species exists, natural selection works primarily to preserve the status quo, not to change it.

AN ECOLOGICAL MODEL FOR THE SELECTIVE ADVANTAGE OF SEX

Ecological explanations for sexuality are based on the idea that the fitness of an individual female can be maximized if she has a variety of offspring among which are some that might be better suited than she herself would be to cope with the changed conditions of life certain to be encountered in the next generation. This is the underlying effect on which Leigh Van Valen's (1973) Red Queen hypothesis was based (see Chapter 9). But for this hypothesis to be acceptable, the concept of coping with changed conditions has to be so phrased that the advantage can be seen to be immediate. The female who engages in sex must be expected to benefit herself so that her personal fitness is boosted more than enough to offset the massive direct cost to female fitness implied by sexuality.

The basic ecological model of sex was supplied by Williams (1975), whose analysis follows.

Consider a very fecund organism like a codfish laying thousands of eggs or an elm tree scattering tens of thousands of seeds. Each fish or tree will saturate the habitat with so vast a number of her own young that the chances of survival for each are almost vanishingly small. Typically only one can replace the mother in time and the rest must perish. This is particularly clear if we forget the codfish and think only of the elm tree, because an elm can saturate the ground underneath itself with baby elms growing side by side. Only one, however, will make it to the canopy, and ruthless competition will do for the rest. Yet this competition will not only be with siblings (brothers and sisters), but also with the dispersing young of other elm trees and other tree species. It is in this competition with the offspring of other trees that the advantage of variety can be seen.

An asexual elm would saturate the ground beneath itself with a clone of identical copies. Williams suggests that this approach to the pending competition is like trying to win a lottery by photocopying your one lottery ticket over and over. For the uncertainties of the struggle to come it is much better to enter the game with many different tickets. With thousands of tries at its disposal, a highly fecund organism can afford to waste half its stake if this provides many chances of winning a single game.

If we enlarge the game to include all the habitats to which elm or codfish might disperse, the logic still holds. Occupying any habitat should involve competition between a host of applicants. Entering these many lists with a variety of different combinations, each subtly different, should be better than having all one's entries be identical (Figures 12.1, 12.2).

Sex, therefore, makes sense for the highly fecund. Most animals and plants are highly fecund. Thus the phenomenon of sex is satisfactorily explained for the majority of organisms by a natural-selection-based ecological hypothesis. Clearly this explanation is less convincing for low-fecundity animals like mammals and birds

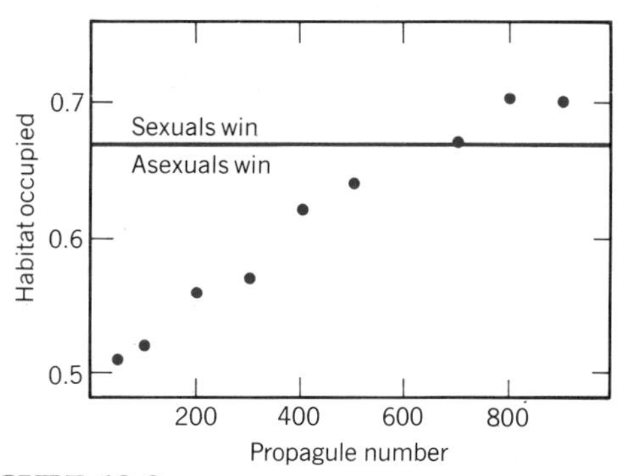

FIGURE 12.2
Computer simulation of elm–oyster model.
As the number of propagules increases, so the proportion of the habitat won by sexual progeny goes up. The horizontal represents the level above which sexual progeny have more than twice the success of asexual progeny and so compensate for the 50% cost of meiosis. (From Williams, 1975.)

not given to wasting offspring on lotteries, an objection to the hypothesis that is discussed below. But for the highly fecund the hypothesis offers a prediction that can be tested. If any animals can be found that reproduce asexually, they will be found to live in constant or predictable environments where variety in the next generation should give no advantage. Such animals do exist because of the phenomenon of parthenogenesis.

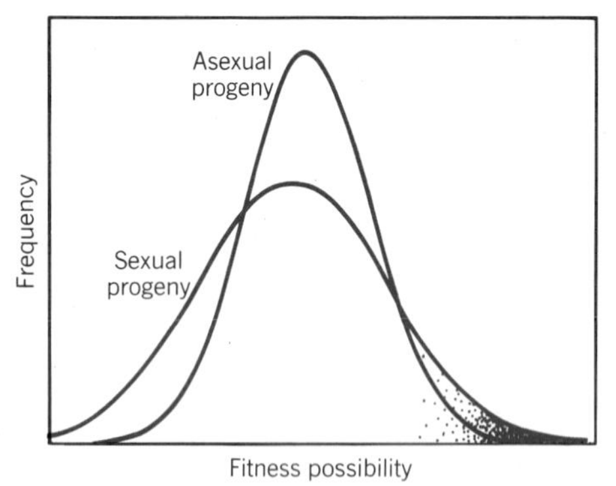

FIGURE 12.1
Model of advantage and disadvantage of sex.
Sex yields progeny with a lower average possibility of fitness but also a low frequency of progeny with a very high possibility of fitness (black stippled area). Sex will be advantageous if there are enough progeny to give a high probability that some with a very high possibility of fitness are included, otherwise asexual reproduction gives the advantage. (From Williams, 1975.)

Parthenogenesis and Environmental Predictability

Parthenogenesis (from the Greek *parthenos*, or virgin) is the development of an egg into an embryo without its being fertilized. This essentially is the avoidance of sexual recombination of genes by animals that are otherwise equipped for sex. Common examples are aphid clones and many planktonic animals of freshwater lakes, particularly rotifers and cladocerans. These animals exist for many successive generations entirely as clones of females.

They lay diploid eggs that were formed by avoiding the meiotic division. The 50% cost of meiosis is not paid, and the young are all carbon copies of the mother. There may, of course, be some mutation, but this is a very small source of variety.

But a characteristic of these parthenogenetic animals (with very few exceptions) is that they do engage in sexual recombinations at intervals after several asexual generations. Thus they alternate sexual and asexual generations (for which the technical term is a "heterogonic life cycle"). The Williams hypothesis then predicts that the asexual, parthenogenetic generations will live in predictable, unchanging environments, and that the onset of a sexual episode should presage environmental uncertainty. The way in which these predictions appear at first sight to be fulfilled can be judged by the natural history of common parthenogenetic animals. Life cycles of these begin with a colonization event.

A colonist may be a winged aphid settling on a growing shoot or an individual water flea (*Daphnia pulex*) hatching from an egg in a puddle that has filled with rainwater. The new habitat is not crowded, so food is abundant and competition is at a minimum. Good times are at hand, though they will not last indefinitely. Evidently it should be good strategy to reproduce as rapidly as possible, and the simplest way to do this is asexual reproduction, which has every offspring a daughter, herself creating more daughters. Fecundity is doubled, and a large array of descendants is available to contest the pond or shoot when the crowded days of competition do appear. If the habitat persists after it is crowded, and the shoot or pond is colonized originally by two or more females, then there will be competition among clones. The best clone will come to occupy the habitat all to itself and one founder female will have achieved locally that ultimate fitness of filling a universe with copies of herself.

Gardeners know how rapidly a population of aphids can overwhelm the succulent shoots of prized plants. This is reproduction triumphant, the product of females achieving maximum fecundity by thrusting perfect copies of a perfect mother into a known future, with no costs of sex, and with each offspring a daughter as fecund as her mother.

But a plant shoot will die and a puddle will dry up. Aphids produce another winged generation and water fleas make ephippial eggs that can lie resting in dry mud until the rains return. From these a new generation of colonists must come, but conditions in the habitats of the future may not be like those of the past habitat, or the emigrants may have to compete with others who were there before them. The emigrant generation, therefore, is predicted by the Williams hypothesis to be produced by sexual recombination. This is what happens. One of the parthenogenetic daughter generations produces male offspring as well as daughters, they mate, and their varied offspring fly or swim their way into an uncertain future.

Parthenogenesis is particularly common among zooplanktonic animals in temperate lakes, being demonstrated by cladocerans and rotifers (but not copepods). For cladocera and rotifers in lakes, winter ends the run of parthenogenetic generations and is passed in dormancy following a sexual episode. This is predicted by the hypothesis because conditions in the open water of next year's spring will certainly differ from this year's, thus releasing the Williams lottery again. According to the hypothesis, sexual episodes should be avoided completely in lakes where conditions change only minimally from year to year. This may be true in the arctic where one short summer is like another. Hibbert (1981) has suggested that clones of *Daphnia* in arctic lakes may be as old as the lakes they dominate, perhaps having persisted for several thousand years. These arctic *Daphnia* make their resting eggs parthenogenetically too. Hibbert's evidence is that each of several arctic lakes holds a distinct clone of *Daphnia*, the members of which can be recognized by electrophoresis.

Objections to the Ecological Hypothesis

A powerful objection to the ecological hypothesis is the weak explanations it offers for sex in birds, mammals, and other low-fecundity animals. How can an animal with a clutch of four, or two, or even one pay the 50% cost of meiosis, the cost of genetic recombination, and the cost of mating? Logic predicts, therefore, that birds and mammals should be without sex. But the truth is that they are sexual animals without exception.

Two kinds of answers are offered to this paradox. The first is that birds and mammals have inherited sex systems from highly fecund ancestors and are now stuck with them. It certainly seems clear that all tetrapods evolved from fish stocks in which the high fecundity of small-egg strategies was usual, and it can be argued that the elaborate system of meiosis, gamete production, and fusion to form a zygote is not easily lost. Parthenogenesis, however, is possible in many animals, including such vertebrates as salamanders, making this argument of inherited sex less than satisfactory.

More encouraging is an argument based on the kinds of habitats in which low-fecundity sexual vertebrates live (Glesener and Tilman, 1978). These are habitats with tolerably stable physical conditions but where the animals are exposed to strong biotic stresses, or where they live in "biological accommodation." When the young engage in intraspecific competition they do so in a universe largely defined by the other animals and plants of the place. The variety of biologic stresses or limits imposed in this way may be complex, in fact resulting in those subtly changing selection pressures that lead to gradual population changes posited by the Red Queen hypothesis. Thus competitive success for complex animals like mammals or birds may turn on subtle variations that give an extra premium to the recombinant advantages of sex. It must be admitted, however, that this argument is less than totally convincing to account for such large direct fitness costs to females being as ubiquitous as is bisexuality in complex organisms.

This difficulty of explaining sex in complex animals remains by far the most damaging problem to the ecological hypothesis. But other arguments have been marshaled against it as well. Comparative studies of some species of parthenogenetic rotifers, lizards, and grasshoppers have found difficulty in showing that the animals encountered more stressful environments than similar sexual species, letting in the argument that parthenogenicity is not after all associated with constant environments as Williams predicted (Bell, 1982). To an ecologist with a firm grasp on the antiquity of species and a knowledge of changing climatic systems and geography through which they have passed, however, contemporary data from a few seasons are unlikely to be persuasive.

THE GENETIC REPAIR HYPOTHESIS

The hypothesis most favored by geneticists and evolutionary theorists is that sexual recombination maintains fitness by removing deleterious genotypes (Maynard

Smith, 1978). If a population reproducing only asexually had no means of genetic repair, then the only way deleterious mutations would be removed from the population would be by selection against the phenotypes carrying them. This would be effective as long as the mutation rate was less than the rate of reproduction. A high mutation rate, however, should soon result in accumulating deleterious genes with consequent loss of fitness for all members of the clone.

In sexually breeding populations, deleterious mutations are swapped with relatives and recombined. The genetic repair hypothesis suggests that natural selection then shuffles through the entire common gene pool of the population, thus always having adequate material to preserve genotypes in which the deleterious mutation has been paired with a beneficial mutation. This hypothesis has strong appeal, though a major drawback is that it necessarily requires a form of group selection (see Chapter 15).

THE PRIMORDIAL TRAIT HYPOTHESIS

The primordial trait hypothesis (my term) is of a different order from the genetic and ecological hypotheses in that it denies that natural selection has anything to do with sexuality (Margulis and Sagan, 1984). The pairing of chromosomes in meiosis fundamental to all bisexual systems is first shown to be a trait possessed by the original eukaryotic ancestor or ancestors of all multicellular organisms. Meiosis served these organisms in the gene repair mode, just as various systems of DNA pairing served prokaryotic organisms before them. These systems of gene repair had added importance in the early earth before the atmosphere acquired oxygen and its ozone shield (see Chapter 28) because of the intense ultraviolet radiation to which the organisms were subject. Eukaryotes first acquired their paired chromosomes by a form of cannibalism, in which the undigested genetic system of the prey paired with the genetic system of the host to provide the superior genome repair and maintenance system that became meiosis. Thus the origin of sex itself is explained by this hypothesis.

To explain the persistence of meiosis, and hence of bisexuality, for two thousand million years thereafter, the hypothesis invokes two mechanisms: first that the system could not be removed once installed, and second that without a regular pairing of chromosomes the complex tissue differentiation needed to grow a multicellular organism should not be possible.

Perpetual meiosis that cannot be shed equates the doubling of chromosomes in reproduction with other systems that have apparently come down unchanged from the first eukaryotes—with Krebs cycles, Calvin–Benson cycles in photosynthesis, and even with reliance on ATP, amino acids, or even carbon chemistry. Some things are too fundamental to be trashed by natural selection. The primordial trait hypothesis treats meiosis as akin to these.

To the logical retort that meiosis seems in a different order to other surviving traits of primeval eukaryotes, and one that surely could be avoided, the hypothesis appears to answer with the suggestion that meiosis serves an essential function without which histogenesis (the growth and differentiation of complex tissues) cannot be undertaken. The pairing of chromosomes is likened to "a roll call or game plan ensuring that all genes are in formation prior to the unfolding process of development of the multicellular animal" (Margulis et al., 1985). The short paper from which that quote is taken uses the word "believe," as in "we believe," at least four times in 12 pages of text. Though perhaps still a less passionate statement of faith than the Nicene Creed, I take this to show that in understanding the origin and maintenance of sexuality science may still have some way to go.

AN ECOLOGIST'S BIAS

In a subject where weighty monographs and fat edited volumes are beginning to be fired in salvos, no real summary is possible. Ecologists will accept hypotheses of sex derived from primordial traits immune to natural selection only with strong reluctance. Instead, selection for traits that could manipulate varieties of offspring in ways that increase fitness is surely inevitable. A not improbable final resolution of the debate is the identification of a complex array of selection pressures working both to maintain genetic repair systems and to provide limited variation among progeny.

Whatever the ultimate cause, the separation of organisms by sex is a fundamental fact of existence. Selection then works on each sex separately. The result is added complexity to the diversity of life on earth and the basis of most social systems among animals and plants.

CAN THE COST OF SEX BE REDUCED BY INBREEDING?

One obvious way of avoiding the cost of meiosis is for a female to fertilize herself. The system of meiosis is then intact, but the offspring have only the mother's genes. Some plants do this to some extent, but those that are self-fertile usually have systems to keep "selfing" to a minimum. Birds and mammals, organisms with invariant bisexuality, never fertilize themselves, nor do they normally mate with close relatives. The simple explanation for this is that "selfing" or mating with close relatives is likely to make a proportion of the offspring homozygous for deleterious recessive genes. With selfing, one quarter of the offspring will be homozygous for each recessive lethal allele, with commensurate loss of fitness. Mating with unrelated individuals, on the other hand, will yield heterozygotes with low probability of pairing deleterious alleles. Natural selection, therefore, should preserve mechanisms that prevent selfing or incest (Ralls et al., 1988).

Despite the hazard of pairing deleterious genes, the theoretical benefit to a female of increasing the proportion of her own genes in her offspring through reduced outcrossing remains. William Shields (1982) argues that a trade-off is inherent between costs to the genome of inbreeding and benefits to the female of a larger investment in her offspring. This might condition one of the selective forces to **philopatry**, which is the familiar habit of many animals of breeding close to their own birthplaces, the best known examples of which are migratory birds that return to the ancestral spot to breed after journeys of thousands of kilometers. Some selective advantages of this behavior are breeding in familiar terrain for which the animal's own genes and upbringing are peculiarly suited. But a consequence is a strong likelihood of breeding with cousins.

Optimal inbreeding on the Shields model would minimize segregation loads of deleterious genes while maximizing the mother's genetic contribution. The hypothetical benefits of inbreeding can also be expressed in purely genetic terms, since inbreeding should protect successful ancestral genomes against disruption. This effect ought to be more important to low fecundity animals because recombinations in highly fecund animals should likely produce a proportion of genomes like the ancestral ones which can then be preserved by natural selection. An additional advantage to outcrossing for the highly fecund would be their potential for success in the lottery described by Williams's model. Thus the inbreeding optimum should shift toward inbreeding for small populations of low-fecundity animals and toward outcrossing for large populations of highly fecund animals (Figure 12.3).

A test of the Shields hypothesis comes from the prediction that philopatry should be common in low-fecundity animals like birds and vagrancy common in high-fecundity animals like insects. On the basis of anecdotal evidence, at least, this seems to be true. A conclusion is that birds do indeed fly back to where they are likely to mate with their cousins.

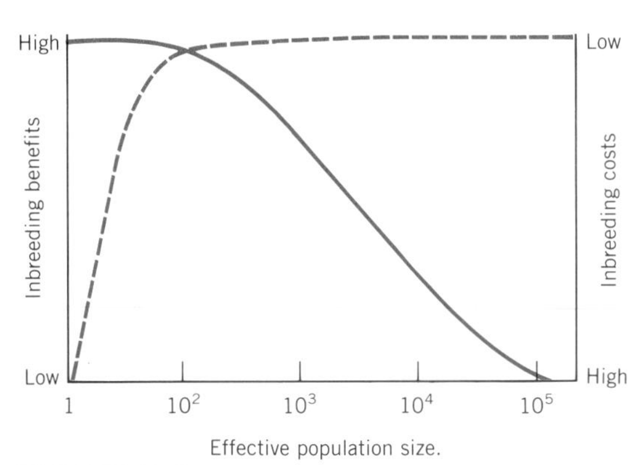

FIGURE 12.3
Optimal inbreeding.
Benefit from inbreeding in conserving the parental genome is likely to be highest in low-fecundity animals. Conversely, the highly fecund have many chances of repeating the parental genome from random outcrossing, and are thus unlikely to derive benefit from inbreeding. Both groups face the cost of inbreeding depression from the release of deleterious homozygous recessive genes. A cost-benefit analysis suggests, therefore, that inbreeding will be tolerated most in animals with large young strategies. (From Shields, 1982.)

WHY NOT BE HERMAPHRODITE?

In Greek myths Hermaphroditus, the son of the love goddess Aphrodite and Hermes, the messenger of the gods, was united in one body with the nymph Salmacis while bathing in her fountain. Thus hermaphrodites have both sexes at once. They exist abundantly in plants and quite commonly in invertebrate animals. On the face of it, the hermaphrodite is the complete solution to the mother's fitness dilemma: be both mother and father, thus claiming back the 50% price of meiosis.

But hermaphrodites should not self, since this would release segregation of deleterious recessive genes without sufficiently compensating benefits (see Shields' analysis above). Then, whether an individual should produce both eggs and sperm for outcrossing purposes depends of the relative costs of each (Charnov, 1982). When a female can achieve a certain amount of male function by giving up a smaller amount of female function, then she should do so, becoming a hermaphrodite. Similarly males should add female functions only if the cost to their male function is limited. Other cost–benefit considerations can determine the outcome as well. For instance, a powerful selection pressure for plants to add maleness to their female flowers is that a small production of pollen could be an excellent lure to insects that otherwise might not arrive to deposit outcrossed pollen on the stigma. But in most circumstances hermaphroditism does not offer a solution to the cost of meiosis problem.

NATURAL SELECTION AND SEXUAL SELF

Eggs and sperm of unequal size have powerful consequences for the lives of animals. This is because the investment of a parent in each of its offspring is so different, ranging perhaps from the micrograms of organic matter in a typical sperm cell for a male to the kilograms in a mammalian baby for a female. Much in the sex and social lives of animals can be traced to this fact. Males and females use each other selfishly for the purpose of promoting their own individual fitness, but, because of the initial disproportionate investment in young, they do so from different points of view.

The initial inequality of investment brings a selective pressure for males to seek to be polygamous. A male can produce many more sperm than a female can produce eggs, so the obvious solution to his fitness problem is to spread his sperm over the egg production of many females. The need of the female to have all her eggs fertilized can be served by any virile male. From the narrow point of view of fertilizing eggs alone, therefore, the rest of a male's sex life is unimportant to a female. Thus males can win extra fitness by mating with many females, but a female needs only to be mated by one male. This is the reason why many mating systems are **polygynous**, in which one male mates with several females. Both **monogamy** and **polyandry** (in which one female mates with several males) are much less common.[1]

But fertilizing an egg to form a zygote is not all that is needed to gain a unit of fitness—the zygote must first grow into a reproducing adult itself. The mating behavior of both males and females is influenced by this fact. Females should choose to mate with a male only when his personal qualities, and local circumstance, are such that there is a high probability that their offspring will survive. Conversely, a male should persevere with his promiscuous polygynous behavior only if this yields more surviving offspring, not just more inseminations.

Polygyny in Resource Gluts

A polygynous male with a harem can both gain and lose fitness by adding females. Inseminating one more female directly increases the number of his potential surviving offspring, but this potential can be realized only if resources are available to rear the infant. The added female may mean more competition among females for resources, which may cost the male dearly, reducing the chances of many of his offspring for survival. If the male contributes to take care of the offspring, then there will be the extra penalty from spreading his services more thinly.

The interests of the male, therefore, suggest that the size of a harem should reflect the flux of resources available for raising young. With copious resources, males can serve merely as sperm banks. The males probably are not needed for parental care, nor might female competition be serious. Then the male can afford to be promiscuous and ambitious as he seeks females, probably limited only by competition for females from other males.

If resources are abundant and a harem is relatively small, it may well be that a female suffers no loss of fitness from harem living. But it is not to the advantage of a female to join a harem where females already compete for resources. A female offered the services of a male with overcommitted resources should reject him

[1]These names for mating systems are all from Greek words: *polus* (many); *gamos* (marrying); *monos* (single); *gyne* (woman); *andros* (male).

and find another male not so committed. Relative size of a harem, therefore, should be as much a result of female choice as it is of male choice.

The trade-off between male and female interests that can result is nicely illustrated by a pioneering study on marmot social systems by Downhower and Armitage (1971). Yellow-bellied marmots (*Marmota flaviventris*) make homes in burrows, and their young are nursed in the burrows (Figure 12.4). The animals must forage within a short distance of their burrows and so are dependent on the resources of

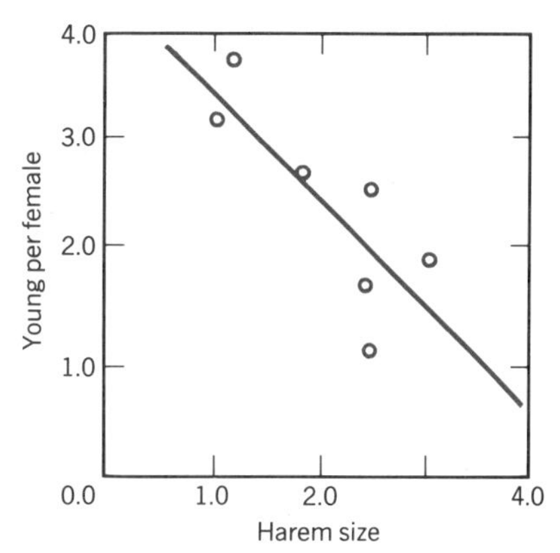

FIGURE 12.5
Effect of harem size on litter size in yellow-bellied marmots.
Females had the largest litters when they had a male to themselves, and litter size fell progressively with increased size of harem. (From Downhower and Armitage, 1971.)

FIGURE 12.4
***Marmota flaviventris:* polygyny as male and female compromise.**
Yellow-bellied marmot females appear to have maximum fitness in monogamy, males maximise fitness in polygamy, a common compromise appears to be bigamy.

the immediate neighborhood for raising young. They are polygynous. Males establish themselves in an area, seek to exclude other males, and mate all females within the area. A male can maintain a harem for from one to three breeding seasons, and he may have from one to several females in his harem.

Female marmots had larger litters when in small harems, as shown by monitoring the numbers of young per female for seven breeding seasons (Figure 12.5). Reasonably enough, a female could attempt a larger family when all the resources within foraging distance were available for herself and her offspring alone. The large families of these females did indeed result in proportionally greater fitness, as was shown by measuring the number of yearlings raised per female (Figure 12.6, straight line). These data leave little doubt that the ideal state for a female yellow-bellied marmot is monogamy.

Males are not well served by monogamy, however. The number of yearlings raised for a male shows that his highest fitness

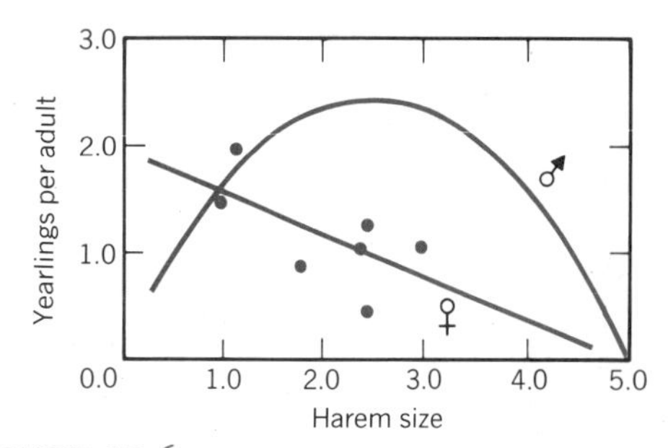

FIGURE 12.6

Effect of harem size on yearling production by yellow-bellied marmots.

Yearlings per female were highest for monogamy as was littler size (Figure 12.4). But yearlings per male given by the curve were highest for intermediate sizes of harem. (From Downhower and Armitage, 1971.)

was achieved when he could mate with three females (Figure 12.5, curve). The data also hint that overly promiscuous males mating with four or five females actually might end up with less fitness than would accrue to a monogamous male.

What is optimum for females in the marmot circumstances is not optimum for males and vice versa. Females should choose monogamy, and males should choose to live in a harem with three females. Both males and females have coercion at their disposal. Males establish themselves on favorable ground, so that a male with a nice piece of real estate freed from other males coerces with the promise of resources. The female coercion is her ability to choose both male and real estate with which to live. Since each sex can exert itself to try to achieve its personal optimal mating arrangement, it is not surprising that a trade-off should often result. Downhower and Armitage report that the most frequent arrangement was in fact bigamy.

This analysis of marmot behavior reveals why polygamy is so common. The best arrangement for females may very frequently be one male, or one patch of resources, to herself, but the best arrangement for males will usually be to mate a number of females. If resources are relatively easy to come by, it should nearly always be possible for a male to raise his fitness through polygamy, though it is unlikely that a female can. The actual polygynous arrangement that results will come about from a trade-off between selection for females best able to reduce harem size and males best able to increase it.

Monogamy

Monogamy apparently represents total victory for females in what is usually a trade-off. Since total victory in any dealing is an improbable state, it seems likely that monogamy must actually result when it is in the male's best interest also. This is likely when resources are uncertain in time or space. Not only should scarce resources increase female competition, but they might actually put a premium on the contribution of the male to parental care. With the need both to reduce female competition in his own interest and to maximize his return on parental care, it may be that a male is best served with a harem size of one. Male and female choices are then identical and monogamy results.

The most abundant and conspicuous monogamous animals are birds with helpless (altricial) young. Their monogamy probably results from the demand on resources caused by their own breeding strategy. All of them require their young to grow extremely quickly, either because the flux of resources they use is short-lived, like the insects of a northern spring, or to keep as short as possible the time when nests would be at risk from predators. Not only must female competition be reduced to the minimum, but males must often be required as foragers for the young. For a bird like a robin, therefore, males and females have an identical optimal harem size of one. Comparable patterns of available resources and needs of the young should be the cause of the few cases of monogamy in mammals also.

Polygyny and Sexual Dimorphism: The Pribilof Fur Seal

Pribilof fur seals (*Callorhinus ursinus*) breed in harems on beaches with 20 to 80 cows per bull being possible (Figure 12.7). Male harem masters weigh up to 500 kg though mature females weigh only 80 kg. The spectacular oddity of this reproductive arrangement is a result of an equally unusual distribution of the resources needed for reproduction. Fur seals are animals wonderfully insulated against the cold of the arctic waters in which they live, but their pups must be born on land. The critical resource for breeding, therefore, is a beach shrouded in fog, where the insulated parents will not overheat in the sun. But the beach must also be in easy swimming reach of rich fishing to support the mother's milk supply during the weeks of nursing. And these beaches are in very short supply.

For the dense populations of fur seals, made possible by the food supply, to exist,

FIGURE 12.7
***Callorhinus ursinus:* sexual dimorphism.**
The fur seals are polygynous (Chapter 12) with large bulls maintaining harems on the breeding beaches. Management is to kill surplus males for skins but data for females from a half century of counting give guidance to management.

females must congregate on the limited beaches to give birth and nurse. Males apparently have two functions: sperm bank and baby-sitter while the mother is fishing. Male fitness in these circumstances becomes entirely a matter of how many females he can inseminate, and natural selection will have preserved the traits that lead to numerous inseminations.

Fur seal females can be mated only immediately after they give birth, and birth itself follows shortly after arrival on the beach. Males, therefore, dare not go fishing. Accordingly, fur seal bulls must be able to starve from the day they haul ashore before breeding begins until the last mated cow has left the beaches with her pup. A primary selection pressure resulting in large size for male fur seals, therefore, is to provide a massive fat reserve. Being large also means being old, and male fur seals do not breed until they are six to nine years old, whereas females breed at age three. The growing years are accommodated by males living together on bachelor beaches during the breeding season, until they become sexually active. A fur seal bull is a highly sexed, and somewhat aged, huge package of blubber. A subsidiary effect of large size is undoubtedly that it confers an advantage in male to male competition for the privilege of occupying a section of breeding beach before the females arrive. These male contests often result in bloodshed.

The females have a clutch size of one, but infant survival is high and rapid population growth is possible. Females increase their fitness by prolonging their reproductive life, which they do by reaching breeding age when much younger than the males and when much smaller. Evidently their smaller size also is set by minimizing metabolic demands of the mother to allow for carrying and nursing a large baby. Females give birth on the breeding beaches and are ready for mating shortly after. Implantation of the fertilized egg into the uterine wall is delayed so that the next birth will be exactly a year away.

A female swimming toward the breeding beaches with her baby to be born within a few days, and surveying the line of males along the beach, must find a good safe spot for the pup she is about to deliver. This same place must be good for her own health too, safe on land from social stress and with safe access to the sea as she comes and goes to nurse the pup. And she needs to ensure as best she can that her next egg will get good sperm, well supplied with genes that lead to survival and future fitness. She probably chooses both the properties of the beach and of the bull who guards it, and cows can be seen swimming up and down offshore for some time before they land, as if surveying the beaches. We do not know how much the final choice is one of bull or beach, but the point is moot because the bulls themselves will have tried to choose the best beach. If she chooses a bull with a large harem she may be choosing sons that themselves will be successful harem masters.

This highly unusual breeding system of the fur seals is a direct consequence of an equally unusual pattern of resources needed for reproduction. Among its consequences are extreme sexual dimorphism that adds diversity to the ecosystems of the oceans with populations of both large and small seals.

MALES AND FEMALES USE EACH OTHER

Despite our difficulty in producing a unified theory for the cause of sex in existing populations, sexuality is clearly necessary for the reproduction of virtually all multicellular plants and animals. In an open breeding system, the numbers of males and females must remain constant, as R. A. Fisher (1930) showed long ago. This conclusion may not be self-evident, but it stems from the trite observation that every organism has just one father and one mother. If one sex became rare for some reason, say because females managed to produce only daughters, then the rare sex would enjoy great fitness because of having to compete with fewer of its own kind. In a universe skewed toward daughters, any genotype resulting in more males would be heavily favored, and the balance in numbers would be redressed.

The diversity of the earth is, in a sense, multiplied by two by the reality of sex. More importantly, the needs of male and female, sometimes common, sometimes contrasting, decide much of the structure of communal life. But the details of what that structure shall be is critically determined by patterns of resource; even to whether an animal shall live in monogamy or polygamy.

Preview to Chapter 13

"Keep away from me": cock yellowhammer on his territory.

SONGBIRDS in the northern spring space themselves out to breed, the prime spacing mechanism apparently being aggressive behavior to conspecifics that approach. This is one form of territorial behavior with several advantages for breeding, the most important of which is sequestering food supplies near the nest. Spacing through aggressive behavior is also a feature of polygynous mating systems in which breeding males establish themselves in conspicuous mating positions. The behavior operates a sperm bank but affords none of the selective advantages of songbird territoriality. A third spacing system is based on mutual avoidance, as when mountain lions spend winter in private hunting reserves. All three behaviors are called "territorial" but have nothing in common other than spacing. Comparison of any with human behavior is absurd. Animals that lose in ritualized, aggressive breeding encounters preserve fitness on the "run to fight another day" principle. Classic territorial behavior of birds has the side effect of setting population limits, even though the behavior boosts individual reproduction. This explains the constancy of bird numbers at spring census. Some precocial birds with minimal parental care mate in arenas called "leks." Gaudy males are characteristic of lek species, probably principally as a result of selection by female choice. Sexual dimorphism results because different selection pressures apply to males and females. In most species females are the larger animals, primarily because males need fewer energy reserves for reproduction. Mixed-species flocks give individual advantage to their members through enhanced foraging efficiency and making attack by predators more difficult. Group selection to control populations is impossible, and selection never applies at the group level except on small scales and with tight constraints that can rarely be met in nature.

Chapter 13

Social Diversity: Territory, Mating Systems, and Group Living

All living is social living. Ubiquitous sex conditions much of society, for animals obviously, for plants with no less exigency. But even without sex animals and plants must accommodate to their neighbors as they compete for resources, for such is a necessary consequence when natural selection presses reproduction until animals or plants crowd in on each other. Social consequences inevitably follow.

The social systems of animals are remarkable and various. Some space themselves, either as loners or in groups; we call them "**territorial**." Some live in huge assemblies, like shoals of the menhaden fish more than a million strong; a habit that still puzzles us. Often animals meet in ritual combat in which one protagonist meekly admits defeat. Other animals submit to low status in dominance hierarchies. Animals gravitate to social living even with other species, as birds do in their winter flocking. And sex drives peculiar breeding arrangements in seemingly endless diversity.

Among the more spectacular triumphs of modern biology have been successes in explaining social diversity from the principle of natural selection acting on individuals. Group phenomena are always individual phenomena writ large, as it were. Whenever an individual takes part in a group activity, even if it "loses" an encounter or lives low on a pecking order, we can show that fitness is better served by accepting the social rules than by flouting them.

HOWARD'S CONCEPT OF TERRITORIALITY

Arguably, a beginning to what would one day become a new discipline called sociobiology was the publication in 1920 of a book called *Territory in Bird Life* by an English ornithologist, Elliot Howard. Howard invented the concept of territoriality, having set himself to explain well-known behavior of breeding birds.

In the northern temperate spring, familiar songbirds pair off to breed, in the process becoming enchantingly apparent. The males in particular, brightly colored and visible, sing from conspicuous perches or, as the European skylarks do, fly ever higher while trilling their songs, to the delight of uncounted generations of poets. The birds even engage in combat, twittering balls of fluff, perhaps shedding a feather or two. Howard set out to explain all aspects of this behavior in terms of selective advantage to the individual participants.

If a bird was a winter resident, like the common English yellowhammer (a sparrow-sized finch, *Emberiza citrinella*), the first signs of the behavior were when males began to leave the mixed flocks in which they had wintered. Each male yellowhammer took to spending more and more time around one particular perch, from which he sang, loosely transcribed as "a-little-bit-of-bread-and-NO-cheese." Because every male kept to a favored perch to sing about cheese, they were soon spread out across the countryside.

The male yellowhammers became irascible, rushing at any other male that came near the favored spot. Often this resulted in one of the aggressive displays that others had interpreted as combat for a mate, but Howard noted that early in the season no females were present to watch the combat. These combats could not, therefore, be jousts of cavaliers before blushing brides-to-be, as analogy with male-dominated contemporary human society might suggest. The fighting was both ritualized and purely male-to-male. And the remarkable outcome of every encounter seemed to be victory for the home team. The newcomer always retreated in the end, although apparently unhurt, leaving the original bird in lonely splendor to sing about cheese.

As the early spring progressed, each male yellowhammer was joined by a female, and each pair nested close to the perch where the male had first experimented with song. Probably the singing had helped with this coupling, acting as a beacon necessary for motile animals like birds. But the singing and irascibility to other males continued after the nests were made. Moreover, females shared in the aggression against interlopers, against both intruding females and intruding males. Now it was pairs of birds, separately spread out over the countryside, each pair with a nest full of chicks, each pair showing ritualized aggression to all comers. The essentials of this behavior were repeated by virtually all the familiar monogamous songbirds of the English spring.

To find the real selective advantage of the behavior, Howard asked what the immediate consequence was, finding the answer to be that the birds were spread out. This was spacing behavior. Howard saw four possible advantages to the spacing and the way it was achieved. Each pair:

1. Had a private feeding area near the nest. This would keep journey time

down to a minimum. In that the behavior keeps other birds out of the feeding ground, the behavior also is an effective mechanism for competition.

2. Spent most of its time on familiar ground, which should be a significant help when coping with predators. Not only would the birds learn where to hide and what the escape routes were, but their constant patrolling of their own neighborhood would let them keep watch for enemies approaching their nest.

3. Mixed little with others of its kind, which might prevent their contracting diseases and infecting their young.

4. Was kept together as a breeding unit for as long as it was necessary to raise the young. The habit provided a pair bond, almost serving as a marriage contract.

Selection should act to preserve the behavior on all four counts. Modern ecologists tend to stress the importance of the advantage of a private food source, but Howard suggested that the most important was the forging of pair bonds. It should be impossible for territorial songbirds to raise their helpless young without some device to keep the sexes together, and here was one such device—shared hostility toward all comers. It is hard to imagine a simpler one.

The behavior needed a name, and Howard called it "**territorial behavior**." He did so with misgivings, including a forceful passage in his book warning of the dangers of using anthropomorphic language to describe what animals do. Howard's fear was that the unwary would liken animal "defense of territory" to human battling for land, when it was nothing of the sort. Time was to show that the opposite vice was more of a menace, as the unwary try to explain regrettable doings of human societies as due to animal natures that make humans act with 'territorial imperative.' This is even more mischievous nonsense than Howard feared.

A *territory* came to be seen as *the defended part of the home range*, or the *area within a defended perimeter*. Much mischief might have been avoided if Howard had called the behavior "**keep-away-from-me behavior**," which describes more accurately what he and his successors actually saw.

Howard's inspiration that pair bonding provided the strongest selective advantage for the behavior seems less important now that it is known that pair bonds can be fixed in other ways. Brewer's blackbirds (*Euphagus cyanocephalus*), for instance, are monogamous but nest in large colonies in the Pacific Northwest, where they show little behavior that a modern ecologist could call "territorial" (Orians, 1980). They spend their winters in Mexico, then migrate northward in the spring to breed. But when they arrive on the breeding grounds the Brewer's blackbirds are already paired off into monogamous couples. Clearly they have managed to forge excellent pair bonds without the rituals of spacing that Howard saw in territorial birds.

Yet Howard had shown that the fetching ways of many birds in the spring, from the singing in a dawn chorus to ritual displays, could be understood as adaptations for breeding success from which both partners benefited. His analysis of the general advantages that could accrue, from increased food to social relationships, was sound. Seventy years later we know that there can be more complexities in territorial behavior (Klopfer, 1969; Davies, 1978; Orians, 1980), but the thrust of Howard's pioneering work remains valid.

Territoriality in the Red-Winged Blackbird

The red-winged blackbird (*Agelaius phoenicius*) is an abundant bird of North American farmlands, though it breeds in marshlands in dense populations. Red wings are both territorial in Howard's sense and polygynous (Orians, 1980; Horn, 1978; Figure 13.1). In spring, males establish territories in marshes and along lakeshores in the classic manner: singing, displaying, and engaging interlopers in ritual combat. Females arrive in the territory, build nests, and mate, but one male may attract up to a dozen females to his territory. Some males are not so fortunate and may even be monogamous. Because the sex ratio is one to one, other males must be more unfortunate still and have no mates at all.

Orians (1969) suggested that this pattern resulted because marshlands were patchy environments such that some territories held much more food than others. A female ready to mate might find herself with the choice of going to a male with a poor territory and no mate or to a male with a good territory who already had several mates. Her behavior should be to go where she can breed with most hope of success. Her best choice might very well be to go where the food is, even though other females were there already.

Orians put this explanation forward as a hypothesis, suggesting that there should be a "polygyny threshold" over which the fitness of a female is greater if she accepts polygamy (Figure 13.2). To test this hypothesis Orians needed to show that females did achieve more fledglings when in shared territories and that these territories were of higher quality than territories with

FIGURE 13.1
***Agelaius phoenicius:* Territorial harem in rich habitat.**
Red-winged blackbirds are both territorial and polygynous. This combination of traits is possible because they nest in marshes, like this nest in the cattails, and marshes are both of limited area and sufficiently productive of insects that one male can sequester sufficient resources for several families to be raised.

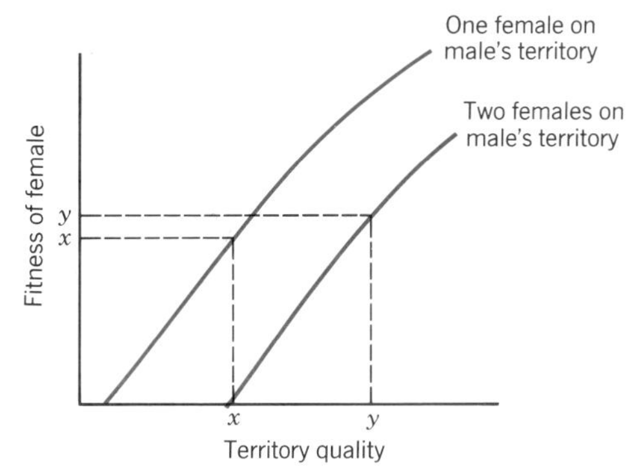

FIGURE 13.2
The polygyny threshold model.
The model assumes that fitness of a female is a function of the quality of a territory, of the parental contribution of the male, and of the number of other females already there. (From Orians, 1969.)

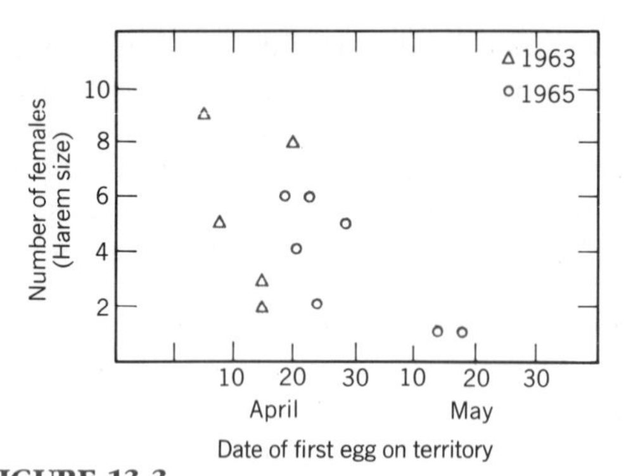

FIGURE 13.3
Correlation between harem size and date of first laying.
Data are for red-winged blackbirds near Seattle, Washington, in 1963 and 1965. (From Orians, 1980.)

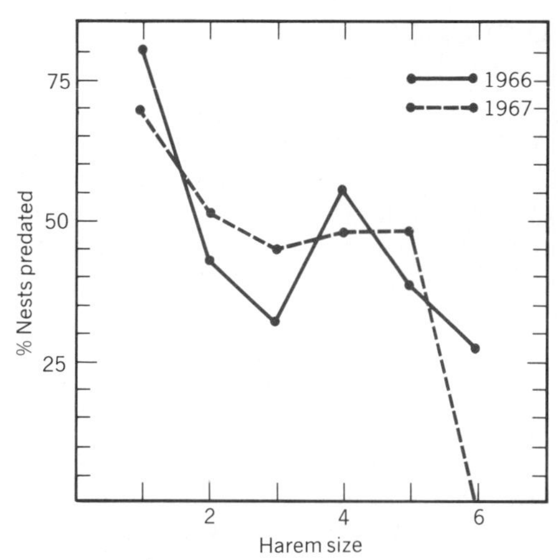

FIGURE 13.4
Effect of harem size on predation of red-winged blackbird nests. (Data of C. R. Holm, from Orians, 1980.)

only one or two nests. Confirming data have come from many years of work (Orians, 1980). The first eggs laid are always on the territories that come to house the most nests (Figure 13.3), and the risk of predation was markedly less in the densely settled territories (Figure 13.4).

Measuring the quality of a territory directly was dauntingly hard, because this would mean assessing food supply in a way that would be convincing to a redwing. But Orians was able to show that a patch of ground holding a many-nested territory also held many nests in other years. The individual males might change but the relative density of nests in recognizable acreage remained the same from year to year. Evidently it was indeed the resources of the territorial space that determined density of nests.

Thus the density of nests around the male in these polygynous birds demonstrates the correctness of Howard's first proposed advantage of territoriality. Local resources, meaning food supply near the nest, is the prime determinant of breeding success. Therefore provision of a private feeding area near the nest gives selective advantage to the behavior.

Territoriality in the White Rhinoceros

Males of the white rhinoceros (*Rhinoceros unicornis*) set about their arrangements for breeding in a manner reminiscent of males of monogamous songbirds. A sexually active male establishes a perimeter he will defend against other males. He fails to sing as charmingly as a songbird, but instead dribbles urine and establishes piles of dung around his perimeter. And he patrols. If another mature male enters his territory the two meet head to head and there is what Owen-Smith (1971) calls "a tense but silent confrontation" (Figure 13.5). The intruder backs away and leaves the territory.

All this follows the songbird pattern nicely, but the sequel is quite different. Young males, old males no longer sexually active, and females not in heat wander in and out of the rhinoceros territory at will. This freedom to wander is doubtless necessary because all available habitat is di-

FIGURE 13.5
***Rhinoceros unicornis:* A tense but silent confrontation.**
Territoriality in the white rhinoceros serves males by providing mating stations. The behavior spaces out males in a way similar to the spacing of territorial songbirds but details of the selective advantage of the behavior are quite different.

vided into comparatively small territories by the breeding males, whereas females with young need a much larger home range, 10 to 15 km^2, in which to wander.

Yet all mating takes place in a territory. Cows in near-estrus interest males for two to three weeks. When a near-estrus cow enters a territory, the resident male endeavors to prevent her leaving, nudging her away from the periphery with his horn and becoming vocal to the point of making a special squealing sound. The cow can escape if she makes a determined run for it and will be pursued only 2 to 300 meters outside the territory. But usually she can be persuaded to stay until she is mated. Once mated, however, she goes her own way, carrying on with her wanderings across the territories of many bulls.

This story of the white rhinoceros is particularly valuable because none of the advantages set out for territoriality in monogamous songbirds by Howard and others seem to apply. Cows and young get no food advantage from the behavior, risks from predation (if there are any for this animal) are not changed, and they have no pair bonds to worry about. Yet the pattern of establishing and maintaining a territory is remarkably similar.

In the rhinoceros the behavior serves male fitness by resolving competition between males for mates with the minimum of risk to males. Benefit to females would follow because choosing to mate with a male on his territory should result in her having sons likely to have the genetic requirements for winning a territory and thus to carry her genes into the next generation.

Territoriality in rhinos and songbirds has little of real substance in common. In both animals the behavior has something to do with breeding, but to equate them for that reason is like saying that communism and capitalism are the same thing because both talk of improving human welfare. The only other thing rhino and bird behavior have in common is that males space themselves out from other males as part of the breeding ritual. But selective advantage from this spacing behavior is given in utterly different ways: private energy supply in the one; public semen bank in the other. A skeptic might question the propriety of giving the same name to both behaviors.

Winter Territoriality of Mountain Lions

North America has one native big cat, variously known as a catamount, cougar, puma, or mountain lion (*Felis concolor*; Figure 13.6). In the cold winters of the State of Idaho, the lions have habits legitimately called "territorial," which have nothing to do with breeding as far as we can tell, and scarcely seem to involve aggression, ritualized or not. M. G. Hornocker (1969) revealed this winter spacing of the lions by tracking known animals through the snow. Figure 13.7 is a map of the areas roamed by some of the individuals he came to know best.

FIGURE 13.6
***Felis concolor:* private range for winter hunting.**
The mountain lions of Idaho keep to themselves in winter, possibly letting each hunt where deer are least disturbed. The separate areas can be mapped by following tracks in the snow, when the tracks also show that the spacing is achieved by avoidance of each other, rather than by confrontation.

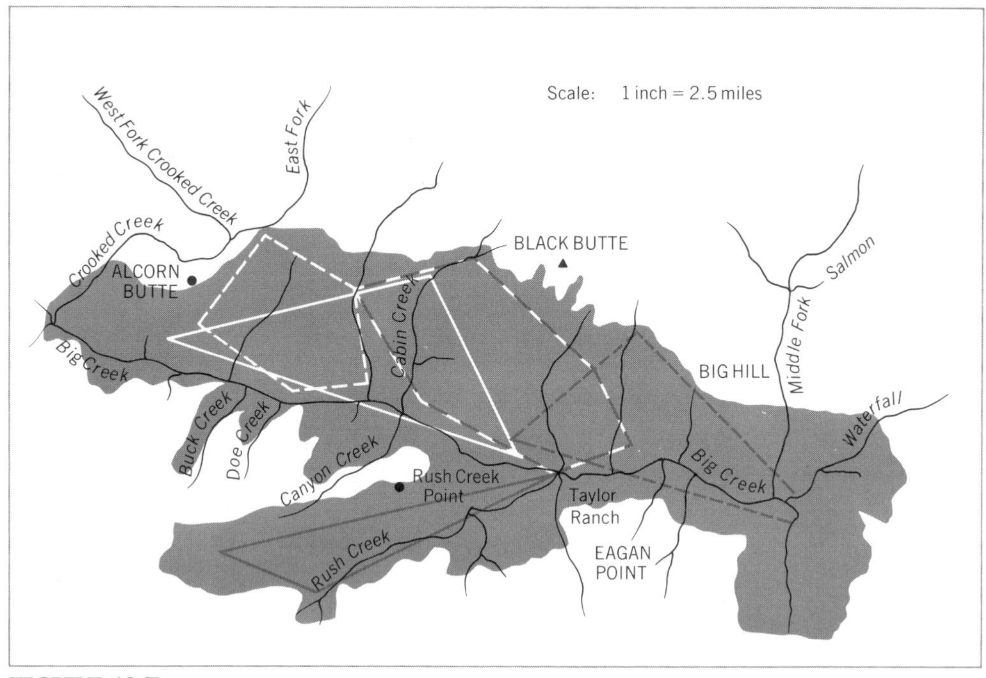

FIGURE 13.7

Minimum winter home ranges of mountain lions in Idaho.

These home ranges were mapped by following the lions' tracks in the snow. The ranges were fairly discrete, and it seemed that the animals were seldom in the overlapping portions at the same time. One home range was used by two male lions, and another by a male and three females, but the other ranges had solitary occupants. The spacing of the lions in winter can be explained as a mechanism for ensuring that each is able to hunt an area where the deer have not been disturbed by other lions. (After Hornocker, 1969.)

These lions apparently lived separate lives in winter, the range of each being as separate as should result if the home ranges were in fact defended territories. And yet there was no evidence that the lions did defend their territories. Younger lions in particular wandered widely across the territories of older, more established animals in their hunting. Once Hornocker actually caught a young animal at its kill in the territory of an old established male, and found tracks of the owner first approaching, then turning away, from where the younger and weaker beast was feeding.

The lions marked their passage with scrapes and excrement, as do domestic cats. The trails revealed that all the lions, young and old, males and females with cubs, turned away from the presence of others. The territorial spacing of individuals, apparently so similar to those of songbirds in the spring, was achieved by mutual avoidance. Established animals had home ranges that were as distinct as the territories of songbirds, but they did not defend them.

Hornocker postulated that the survival value of this system of spacing by avoidance was that it let each lion hunt deer that had not recently been molested by other lions. The lion's technique of hunting by stealth requires that the target deer be unsuspecting. Keeping away from other lions lets the attack be launched against deer that have not been disturbed recently, thus increasing chances of success. The winter spacing of mountain lions, therefore, results because the lions avoid each other as a necessary requirement of the hunt.

Territoriality of Howling Wolf Packs

Wolf packs are social units that hunt and live together for at least parts of a year. Each pack occupies a home range from 125 to 155 km^2 with an overlap of about 1 km between ranges occupied by different packs. Scent marking is frequent at the periphery of each area and in the zone of overlap. The wolves of each pack, however, spend most of their time in an inner core area where other wolves never venture except in times of food shortage. When food shortage is critical, members of a pack will attack alien wolves that intrude. This package of traits describes group territories for wolf packs, apparently marked out on the ground by scent and actively defended only in extremis (Peters and Mech, 1975).

Wolves also howl (Figure 13.8). Harrington and Mech (1983) investigated the social function of howling by simulating wolf howls from various points around the adjoining, and partly overlapping, home ranges of two wolf packs. They found that the best way to get a response from wolves was not to play back recorded wolf howls but to imitate wolf howling themselves. Because wolves in both packs carried radio transmitters, the team knew where the wolves were and thus could seek responses from wolves in different parts of their home ranges. Wolf response was

FIGURE 13.8
***Canis lupus:* howling as a territorial signal.**
Howling by members of wolf packs is acknowledged by members of other packs within hearing, but the packs do not approach. Apparently wolf-howling helps keep packs separate to avoid disturbing each other's hunting.

quite independent of where the wolves were. When the wolves answered, they either remained where they were or retreated. Wolf howling thus serves to keep hunting wolf packs separated by mutual avoidance.

But if wolf packs space themselves out by keeping to themselves, in the manner of mountain lions in winter, how important is scent marking in defining territorial boundaries? Harrington and Mech (1983) concluded that intense scent marking at the edge of a home range results because wolves scent-mark whenever they encounter the scent of a strange wolf. When packs keep to themselves, strange wolf scents are mostly encountered at the periphery of the range. Thus scent marking of the boundaries of a wolf-pack territory may be a consequence of the spacing rather than a cause.

TERRITORIALITY RECONSIDERED

Classic territoriality as described by Howard was spacing aggressively achieved, the main selective advantage of which was sequestering energy supplies for the feeding of young. Territoriality in animals like the white rhinoceros is again spacing aggressively achieved, but with the vastly different selective advantage of sexual advertisement and mate choice. Winter territoriality in mountain lions and wolf packs, however, is achieved without aggression but by the self-interest of keeping to a hunting preserve unmolested by others. Davies (1978) offers a definition that encompasses all three by saying that *territoriality* is "*any spacing of animals that is more even than random.*" At least this definition should take the sting out of some of the more absurd linkings of human behavior to imagined territoriality of humankind.

FITNESS FOR THE LOSER IN RITUALIZED COMBAT

Most territorial encounters are ritual combats with minimal actual damage inflicted, and yet one individual breaks off the engagement. One protagonist wins and the other loses. This means that the loser breaks off the combat without gaining his objective and without being hurt, raising the important point of the possible advantage of territorial encounters to the loser.

When an animal finds itself up against a determined defense, it has the options of fighting or quitting. If it presses into battle, the chance of victory complete enough to allow breeding must be small. This is most obvious if we think of an intruding male of a monogamous species of songbird. The defender is ensconced, perhaps already with nest and eggs. His stakes are high and he cannot afford to quit. The intruder cannot afford to continue the attack against a defender who cannot afford to quit. Continuing the attack might mean being hurt, and certainly means strife with no end in sight. Young cannot be raised in this vision of the future.

The intruder that quits when faced by a determined defense, however, has a definite chance of finding a territory and a mate somewhere else or some other time. For an intruder it is nearly always "fit to quit" when the going gets tough. Perhaps the only exceptions are for polygynous, resource-sequestering males with large harems at stake. Numerous surplus males may be driven to gamble and to risk destruction. In these species fighting may be less ritualized and involve some chance of wounding. As with fur seals, however, an aspiring young male will seek out a harem master well past his prime if possible. The old defender himself usually quits when it is clear he is losing.

It is vital to realize that success in a territorial encounter is not genetically preordained. All individuals are equipped with the full complement of the behavior. This behavior requires that a defender be confident, and that a challenger always retreat from a show of confidence. All learn by experience; all have the gene for "run away and live to fight another day." This normal behavior must be maintained by natural selection since both the overaggressive and the overly pusillanimous will lose fitness by not knowing the proper time to quit.

THE POPULATION CONSEQUENCES OF TERRITORIAL BEHAVIOR

Territorial behavior in songbirds and rhinoceros serves as an aid to reproduction. The result must be to boost population growth. But if space for territories is limited, the behavior must set a limit to numbers and block population growth. As G. S. Gilbert would have put it, "A most intriguing paradox."

For a time ecologists argued over this paradox, some declaiming that population limit by denial of territory required some birds to give up breeding, which was impossible to conceive of in a universe moderated by natural selection. And yet, as Peter Klopfer (1969) put it, "if the size of the territory cannot be reduced beyond a certain point, and if successful reproduction requires that the bird possess a territory, the regulatory function of territories becomes a function that is beyond dispute."

The hypothesis that bird territories impose limits on populations makes a clear prediction that can readily be tested: that in every breeding season there is a population of surplus birds that are fully capable of breeding but that do not do so because they have been denied a territory. Demonstrate the existence of these surplus virile birds, and the hypothesis may stand.

The test is to shoot all the breeding birds and see if their territories remain vacant (the null hypothesis) or if they are quickly reoccupied by members of a surplus population. I think of the classic experiment as "the evidence of the Maine gunners." It came about because foresters in the State of Maine wanted to measure the effect of birds as predators on spruce budworms, and so commissioned a "removal experiment." Budworm populations were to be compared in forests with and without birds.

The experimenters first made a census of all the breeding pairs of birds on a 40-acre section of forest. This could be done accurately because the birds were territorial, made themselves conspicuous, and remained in place. The team counted 148 pairs. Then they went out to shoot them all. But when they had shot 302 males out of the 148 that were originally on the territories they stopped the slaughter, though birds were still singing in the woodlot. The work was repeated in the following spring, when they counted 154 breeding pairs to start with and had shot 352 males by the time they quit (Stewart and Aldrich, 1951; Hensley and Cope, 1951).

The Maine gunners had demonstrated conclusively that a floating population of male birds existed that was fully capable of occupying territories but was precluded from doing so by other birds. Yet the results held an oddity in that the surplus birds were nearly all male. The original complement of females was shot along with the males, but few extra females came in to take their places (Brown, 1969). Of the ten commonest species present, not a single female was replaced, or, at any rate, not one female more than could be accounted for by the original census was collected. The best explanation for this seems to be that, by the time the territories were made vacant by death, surplus females were no longer in breeding condition although surplus males were.

Other removal experiments followed. Figure 13.9 gives the result of one such experiment for the great titmouse (*Parus major*) in England. The best great titmouse habitat was a patch of woodland that was completely parceled out into territories. When six pairs were shot on their territories, space was reparceled with four more pairs moving in from suboptimal habitat in nearby hedgerows (Krebs, 1971).

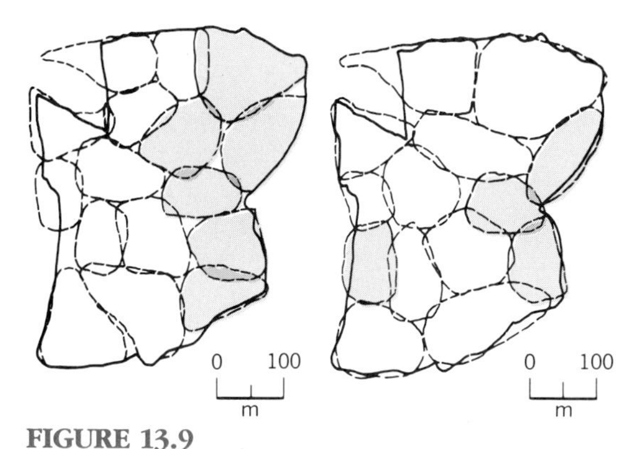

FIGURE 13.9
Removal experiment to test the prediction that surplus birds are denied territories.
The male great tits (*Parus major*) holding the six territories shown shaded in the left-hand diagram were shot. Four new males moved in (shaded, right), and the remaining residents expanded their territories. (From Krebs, 1971.)

Removal experiments have now been tried on mammals, fish, dragonflies, butterflies, and limpets as well as on birds (Davies, 1978). As a result, it now seems certain that a side effect of territorial behavior is to set a limit to population. The paradox of population control resulting from behavior selected to promote reproduction remains elegant, but is resolved by the "fit to quit" argument. Birds of the floating populations denied a chance to breed have been the losers in this year's territorial disputes. Next year perhaps they will get in first.

The remarkable constancy of bird numbers from year to year is explained by the territorial habit, requiring no subtleties of density-dependent death in winter (see Chapter 8).

LEKS

For birds with precocial young, males are not needed for parental care. They revert to their role as strutting sperm banks, like a white rhinoceros. In ten different families of birds, display by such males is communal on central breeding grounds called "leks" or "arenas" (Rubenstein and Wrangham, 1986). Males show themselves off with other males, while females come to inspect.

Characteristics of lek species are extreme physical difference between the sexes (sexual dimorphism) and unequal breeding success. Males are showy and large; females small and dowdy—like the differences between peacock and peahen or between male and female birds of paradise (Figure 13.10). Despite crowded ranks of males at the lek, just a few of them do all the mating. A lek of sage grouse (*Centrocercus urophasianus*) displays perhaps 400 males on a hectare of prairie, each bird in a tiny territory. Between them they will mate 400 females. But 10% of the males will in fact do 75% of the mating, and no more than one or two will do most of the mating even of the favored 10% (Figure 13.11; Wiley, 1973).

Females come to the lek to mate, pass into the mass of waiting males and, usually somewhere deep inside the arena, each stays with a male until mated. But how is the lucky male picked? One possibility is that a top male gets the honor by reason of his holding the best territory against other males, perhaps toward the middle of the lek. Probably this mechanism does work to some extent as males experienced in successful encounters from other years establish themselves early in favored places at the center of the lek. But more evidence is accumulating to suggest that female choice is usually decisive.

In leks of a bird of paradise (*Parotia lawesii*) in Papua-New Guinea, male mating success was clearly correlated with his propensity to display to the females (Figure 13.12; Pruett-Jones and Pruett-Jones, 1990). It is difficult to see, however, why female choice should be so consistent that

FIGURE 13.10
***Parotia lawesii:* male finery for whose benefit?**
In birds of paradise the males are spectacular, the females dowdy, as in many polygynous bird species. If female choice is the selection pressure driving male gaudiness, it might be that females copy each other in choosing a male.

FIGURE 13.11
***Centrocercus urophasianus:* sexual dimorphism in the arena.**
All mating in the sage grouse is by a few males occupying tiny territories on a mating ground called an *arena* or a *lek*.

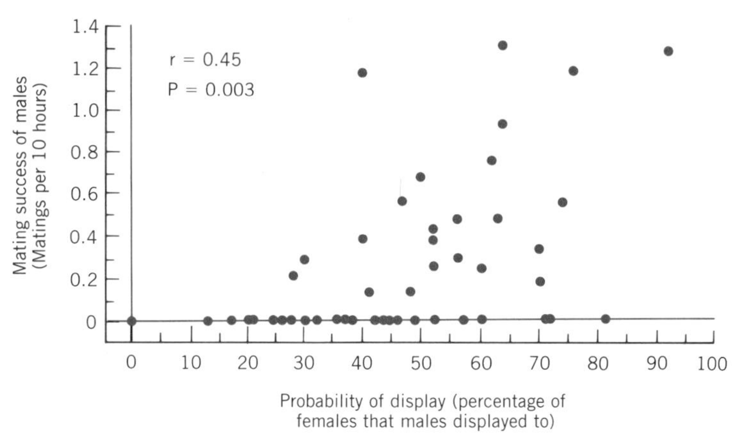

FIGURE 13.12

Male mating success a function of propensity to display.

In *Parotia lawesii* (Figure 13.9), males mated most often if they displayed most often. (From Pruett-Jones and Pruett-Jones, 1990.)

all females choose the same one or two males when presented with hundreds, as at sage grouse leks, unless of course the females tend to copy each other. But this may be precisely what they do (Pomiankowski, 1990). Copying others might be a particularly good strategy for young and inexperienced females, letting them benefit from others' decisions of what is a good male.

Female choice of bizarrely beautiful males implies that these gorgeous creatures nevertheless carry genes for high-quality dowdy mothers and competent precocial young. But so they must because natural selection winnows both their offspring. A feedback nevertheless must exist between sexually successful male extravagance and foraging and survival competence of the offspring (Maynard Smith, 1987).

ON SIZE AND SEX

Males are larger than females in a few conspicuous animals, like deer, fur seals, and some lek-mating birds. In these animals, niche and patterns of resource combine to make competition for mates by breeding males the final arbiter of male size. Excessive male size then may result from the need to store food when males cannot afford to eat, territorial display, or female choice. In most other animals, however, females are the larger sex: spiders, insects, fish, birds of prey, etc. Because females are the reproductive unit, and males basically no more than costly sperm banks, the prevailing large size of females seems intuitively reasonable.

Because the primary reproductive individual is the female, it is prudent first to look at the possible advantages of large or small size to females before taking the male point of view.

The benefits of large size to females are as follows:

- Maximize fecundity
- Early births in seasonal environments
- Maximize survival through short periods of food shortage
- Maximize fat reserves for lactation
- Maximize success in female-to-female conflict

Maximize success in female-to-male conflict

The benefits of small size to females are realized through lowered metabolic requirements that allow:

Survival through prolonged periods of food shortage

Minimize maternal metabolism to maximize resources for infants in species that care for their young

These lists could be extended, but they probably include the commonest factors leading to selection for either largeness or smallness in females.

Maximizing fecundity probably is the reason for the large females in most fish and spiders, since the number of eggs laid is a function of the mass of the female, a point that Darwin noted. Large size also allows early egg laying in seasonal environments because the large maternal body allows energy to be stored. This may be the reason for the large body size of female hawks that migrate long distances: the large female has fat reserves both for the long flight and to make eggs when she arrives on the breeding grounds. Male hawks, having no eggs to lay, can be smaller (Downhower, 1971). That large size also allows increased supplies of milk in mammals, and increased success in conflict, needs no elaboration.

Possible advantages of smallness for females seem fewer and probably all reflect the fact that a smaller body requires less food to maintain it. Small size thus should permit survival through long periods when food is in short supply. Periods of extreme food shortage, however, might promote large size, for then the time of shortage can be survived on fat reserves. The food supply will thus result in selection for smallness or largeness depending on both the absolute amount available and the intermittency of its availability. For females that care for young, particularly mammals with long lactation periods, the food supply that each female can win must support both mother and infant, producing a selection pressure to minimize the mass of the mother.

Males have fewer reasons to be large since the biomass and maintenance costs of sperm are minimal. This is probably one reason that males tend to be small. Small males also have a selective advantage because they become sexually mature early (Vallrath and Parker, 1992). Even where social and sexual selection favors large size in males, relative size of the sexes must still depend on separate selection for optimal size in females as well as males.

In polygynous primates, the relatively small size of females can involve large costs to fitness for which compensating benefits must exist. A band of one large male and several smaller females with their young may hold together for years, until, in fact, the male dies or is displaced by another (Struhsaker, 1977). The replacement male begins his tenure of the harem by killing the young babies sired by his predecessor (Figure 13.13). This serves his fitness well because the babies have none of his genes, whereas he can sire their replacements. It does not serve the fitness of their mothers, but the mothers cannot adequately defend their young because they are too small to stand up to the male.

Obviously this reveals a selection pressure for the females to be large enough to fight off the male. The fact that the females are not large, therefore, demonstrates that the opposing selection pressures for smallness are stronger. Hrdy (1981) suggests that in primates the important selection for smallness is the need to keep the mother's maintenance costs down in the inter-

FIGURE 13.13
***Presbytis entellus:* male dominance and infanticide.**
Polygyny in Hanuman langurs results in group living with one large, sexually active male to each group. When eventually he is replaced by a younger male, the newcomer kills the babies of the troop, the selective advantage of which (to the male) is that group resources go to raising his own offspring. This behavior is possible because the males are larger than females, who cannot successfully defend their babies.

ests of prolonged lactation. The female primate must forage for both herself and an infant, but the male only for himself in the equation: one female + one infant = one male.

Whether males or females are larger depends on separate selection on each sex. No general rules apply, other than those of allocation of resources in time and space. But the likelihood is that selection for small size in males requires that females usually be larger. The largeness, and hence probable social dominance, of females in most species, however, has been masked by social attitudes in male-supremist human cultures, attitudes that have actually led some writers to refer to largeness in females as "reversed sexual dimorphism."

SEXES OCCUPYING DIFFERENT NICHES

Males and females of different size or habit could, in some circumstances, reduce pressure on resources needed for breeding. This mechanism may be among many operating for the Galapagos marine iguana. Figure 13.14 shows basking marine iguanas (*Amblyrhynchus cristatus*) at Punta Espinosa on Isla Fernandina (Narborough). All the animals in the photograph are male. P. D. Boersma counted the animals in this herd, finding that there were usually more than 2000 present, though they came and went. Females sunned themselves in separate herds.

Figure 13.15 is Boersma's sketch map of all the iguanas on the point. Also on the map is the depth of water close inshore. The large herd of male iguanas lies close to deep water and the females close to shallow. The iguanas feed on benthic algae, either at the bottom of the sea or when the algae are exposed at low tide. The male iguanas are much bigger and heavier than the females, so it may be that the big, heavy males can forage in deep water more successfully than can the smaller females, perhaps because the larger males will cool more slowly in the cold water.

It is appealing thus to argue that the iguanas are equipped to feed in different places as an adaptation to increase the resources available to them, but we cannot be certain about this. For instance, why shouldn't males cheat and use their weight to elbow females aside from the shallows? One possible answer is that there are better pickings in the deep water where the females cannot go. We do not have data to test this hypothesis, however.

All we can be sure of about the Espinosa iguanas is that the males are indeed larger, that the sexes are separated, and that the males do forage in places that might be hard for females to reach. It seems prudent to suggest that this sexual dimorphism evolved from selection for re-

FIGURE 13.14
***Amblyrhynchus cristatus:* male segregation.**
The crowded Galapagos marine iguanas on this rock at Point Espinosa are all males. The rock fronts on deep water, suggesting that the heavy males can reach resources on the sea floor that are denied to lighter females. Assemblies of females are found close to shallower water nearby.

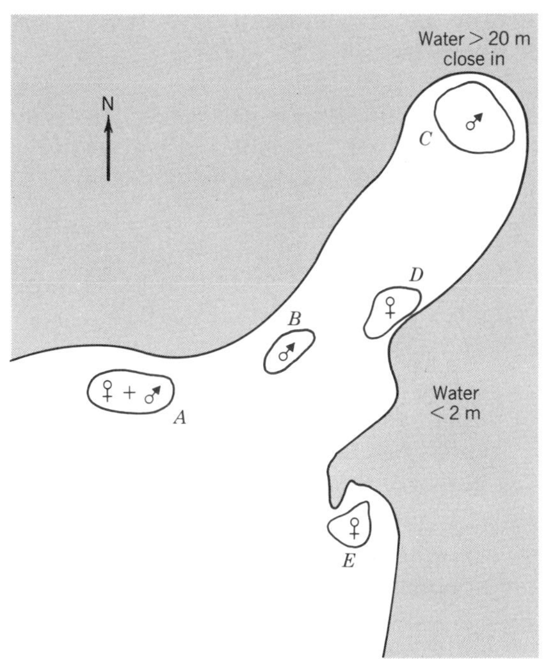

FIGURE 13.15
Subpopulations of male and female Galapagos marine iguanas.
The large assembly of males (Figure 13.13) is at the edge of deep water, whereas many of the smaller females are close to shallow water. (Original data of P. D. Boersma from Pta. Espinosa, Fernandina.)

productive prowess within each sex, the males becoming large to overawe other males and the females remaining small to optimize the energy available for their huge eggs. Occupying different parts of Punta Espinosa may be as accidental a consequence of different size as is infanticide by male primates.

SOCIAL STRESS AND REPRODUCTION

Overcrowding as a result of successful reproduction is inherently likely for all populations, raising the possibility of physiological responses of the kind that physicians call "stress." In humans, a complete syndrome of physiological changes in response to stress is recognized, called the **general adaptation syndrome** (GAS). It is identified by a high titer of adrenocorticotrophic hormone (ACTH) in the blood and enlarged adrenal medullas (Selye, 1950). GAS apparently reflects a general mobilization of the body's resources to overcome stress, and thus is adaptive. But since GAS requires changes in hormonal balance, all functions of the individual are influenced, undoubtedly including the reproductive function. Stress, therefore, may affect breeding. Moreover, it can be shown that extreme stress produces an extreme GAS reaction that, far from being adaptive, is called "stress disease."

Stress disease can be induced in laboratory rats and mice by crowding them, as in a typical laboratory experiment on population growth. When the population is dense enough, a whole generation of rats fails to reproduce, either by having no sexual activity, by resorbing embryos, or by showing abnormal behavior like eating their young or being hyperaggressive against mates and offspring. These crowded rats exhibit physical signs of the general adaptation syndrome in extreme form, the most obvious being grossly swollen adrenal glands (Christian, 1950).

This discovery suggests the hypothesis that crowded animals in the wild might sometimes suffer shock disease from the effects of crowding itself, behave abnormally, and so suffer population losses. Field biologists frequently observe population "crashes" of animals like rodents or deer when populations suffer sudden heavy losses after a period of sustained growth. Shock disease in the crowded animals offers an alternative explanation to outright starvation. Testing this hypothesis involves searching for GAS symptoms in dead or crowded animals.

But the GAS usually must be truly

adaptive, being provoked to ensure life and breeding, not to end it. One telling set of data on this point comes from comparing adrenal glands of placid laboratory rats with those of wild rats trapped in sewers. The sewer rats are found to have the larger adrenals. Apparently the stresses of sewer life provoke the truly adaptive response of the GAS, letting rats endure the vicissitudes of living in sewers. But this does not exclude the possibility that gross overcrowding in the wild can produce the diseased form of the GAS with consequent population collapse (Christian, 1950).

A direct test of the hypothesis was attempted on a herd of sika deer (*Cervus nippon*) by Christian et al (1960). The deer had been introduced onto a small island in Chesapeake Bay, Maryland, and had grown to a density 20 times what is thought to bring on starvation in deer populations in comparable habitats, suggesting a population crash might be imminent. Animals were sampled and autopsied before, during, and after the crash that followed. Adrenal glands of crowded animals before the crash were twice the weight of those in the depopulate aftermath, yet no signs of disease were detected. The most provocative observation, however, was that none of the animals that died gave evidence of starvation, all apparently having adequate fat reserves. The actual cause of death of these deer remains unknown, but outright starvation is excluded. Shock disease remains a possible cause of the deaths but is by no means demonstrated.

SOCIAL FORAGERS IN DIVERSE FLOCKS

Birds the world over can be found in mixed-species flocks. When they are feeding together, a number of possible advantages to each individual in the flock can readily be imagined. (Cody, 1974; Bertram, 1978.)

1. By feeding where others feed, patchily distributed food is found efficiently (information exchange).
2. Traveling with a flock means that the individual will not waste time hunting over the ground that other birds have already covered.
3. The feeding of other birds lets them act as "beaters" that expose prey. For instance, insect feeders might do well catching insects disturbed by birds ahead of them.
4. Being in a flock, both single species and mixed, can serve as a defense against predators.

The three benefits for foraging suggest powerful reasons for joining a flock. Some ways of feeding, of course, preclude group foraging, as apparently does the hunting of mountain lions. Otherwise joining a group may save considerable time and effort in the food quest, provided that competition for food within the group is not so intense that the losses from going alone are not balanced by the occasional extra benefit of a good food find. The decision to forage with the flock or alone, therefore, will be a function of food size, food abundance, food dispersion, and competition for food. In many circumstances, feeding with the flock should pay the individual in foraging efficiency.

FLOCK FOR SAFETY

Defense against predators is a different kind of reason for foraging in a flock. In one sense joining a flock may seem a poor thing to do because the flock is conspicuous so that a predator may find it easily.

Yet there are many ways in which this danger is offset by protection within the flock. Four such advantages are given below.

Vulnerable Individuals are Hidden in Flocks Wildebeest hide their young from marauding hyenas by passing them to the opposite side of a herd (Kruuk, 1972).

Individuals Get Warning of Predators Kenward (1978) demonstrated the effect directly by flying a trained goshawk at English wood pigeons. The goshawk's strategy was to attack on the ground as the pigeons fed, and the pigeons could escape if they got up flying speed in time. Kenward assumed, therefore, that the hawk had been spotted at the instant the pigeons took off. His data (Figure 13.16) show that hawks were spotted much more quickly by flocks than by single birds. Moreover, the kill rate of the hawk was significantly higher when it was launched against single pigeons or small flocks (Figure 13.17). This higher kill rate was probably due to the lack of early warning, but it may also have been influenced by the fact that solitary birds tended to be social outcasts, either young or weak.

Safety in Numbers Musk oxen rings face a wolf pack (Figure 13.18). Tinbergen (1951) long ago suggested that hawks at-

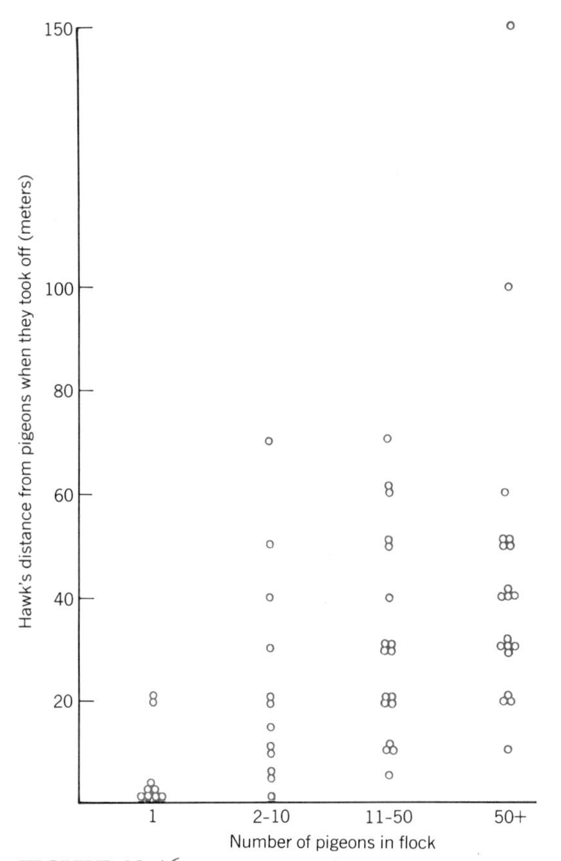

FIGURE 13.16
Hawk detection by pigeon flocks.
When a trained goshawk was flown at English wood pigeons on the ground, the pigeons took flight earlier if they were in larger flocks, suggesting that they detected the hawk more quickly when in flocks.

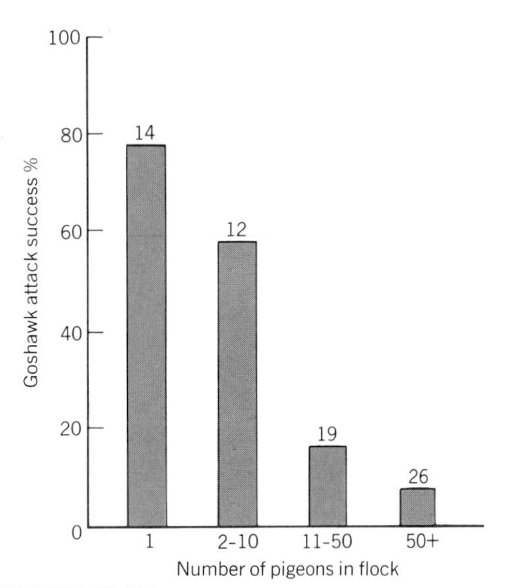

FIGURE 13.17
Success rate of goshawk attacking pigeons in flocks.
Attack by a trained goshawk rarely resulted in capture of a pigeon from a large flock, although most attacks on single pigeons were successful.

FIGURE 13.18
***Ovibos moschatus:* response to uncertainty.**
When musk oxen huddle together this may benefit each individual by giving an appearance of menace or perhaps simply because the numerical chance of being a victim decreases in a crowd. Or the behavior may have more subtle origins and the appearance of a defensive ring is deceptive.

FIGURE 13.19
Possible confusion of a predatory fish by shoaling.
The behavior of the prey (*Stolephorus purpureus*) in scattering may so confuse the predator (*Euthynnus offinus*) that the attack is frustrated. Alternatively, the advantage to each individual may be that the chance is increased that neighbors will be the victim.

tacking in flight could be frustrated if the prey was so bunched up that one bird could not be struck down with the hawk's talons without a real risk of colliding with a second. A midair collision at the speed with which a hawk attacks probably would mean debilitating, and hence fatal, damage to the hawk. The closed or massed formations of birds like starlings might protect individuals from hawks. But it may be that "confusing" the predator is the more important effect: a sustained chase and strike by a fast-moving predator at a dissolving cloud of numerous prey may reduce its success rate (Figure 13.19).

Attack Statistics When a predator strikes at a group of four prey, the chance of being the victim is one in four. When the prey is alone the chance of being the victim is 100%.

These various properties of groups for defense are also problems for an attacking predator. Bertram (1978) offers the model of probabilities given in Figure 13.20 that look at the problems of group defense from the predator's point of view.

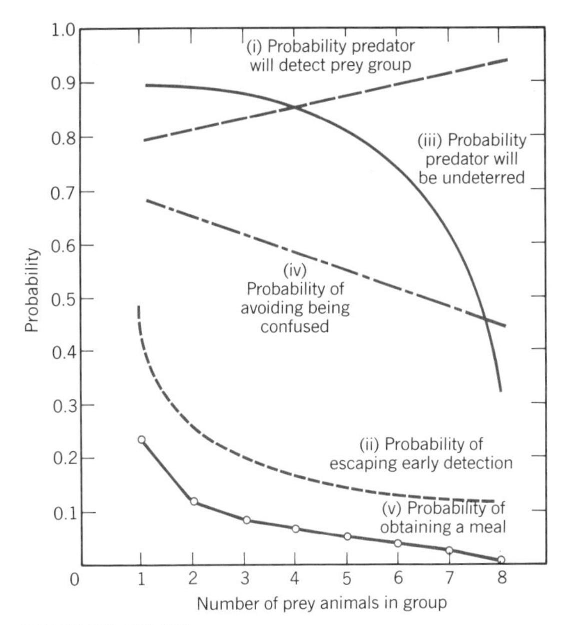

FIGURE 13.20
General hypothesis of the effect of group size on predation.
A solitary predator is imagined attacking groups of varying size. The probabilities shown in the models are all hypothetical. (From Bertram, 1978.)

GROUP SELECTION

Social groups can always be fully and sufficiently explained as properties emerging as individuals pursue individual fitness. Groups are, as it were, composed of individuals that use each other for their personal advantage. Ecologists have learned to resist the temptation of looking for group advantage instead of for individual advantage. To think otherwise is to allow that a system of group selection exists. But groups do not reproduce; no group genes or group offspring can be found to be winnowed by natural selection. Nor is group inheritance by some mathematical sharing of genes in the next generation a practicable proposition (Weins, 1966).

But 30 years ago one ecologist argued otherwise. V. C. Wynne-Edwards (1962) postulated that nearly all crowds of animals in nature had the ultimate function of population regulation for the good of the group. When starlings flew in close formation, this was a way of sensing the density of the crowd so that they might adapt the breeding effort accordingly. When menhaden swam together a million strong, this told how many menhaden there were, with consequences for reproduction. Singing at the dawn chorus was a device to tell all birds their population density, as was the singing of frogs, crickets, and cicadas. Wynne-Edwards made a special point of studying leks, failing to see the purpose of all that display if the strutting males were not really showing off to count heads among themselves as a step toward family planning. He said that all these "epideitic" displays were not just for individual benefit but for group benefit.

Wynne-Edwards's proposed group benefit was to be won through population control. Dense crowds produced hormonal responses of the kind known from the general adaptation syndrome, reducing reproductive effort. The proposed benefit of this response was that the whole group avoided starvation by keeping the population below food limits. Coming during the environmental awareness of the 1960s, these writings enjoyed some vogue.

Neither Wynne-Edwards nor anyone else was able to explain how individuals could be prevented from cheating in such a system of group birth control. Natural selection persists in encouraging individuals to breed. Any family lineage that reduces its reproductive effort below the maximum, as required by the hypothesis, will soon have no descendants. Group selection birth control is simply impossible, and the hypothesis was rapidly discredited.

But Wynne-Edwards had shown biologists real problems of group living that had to be answered. Leks were oddities; so were flightings of birds and all the other social activities of animals that make up the modern discipline of "sociobiology" (Wilson, 1975). Finding better answers to Wynne-Edward's lists of strange behavior played no small part in founding sociobiology. A second consequence was a search for circumstances in which some form of group selection could be possible. The crop is meager, but there is a crop.

When Group Selection Can Operate

Suppose that groups of animals, both predators and their prey, live in isolated patches, a condition in fact common in nature. Always there will be strong selection pressures for predators to become more efficient, as, of course, there are opposite pressures for the prey to achieve a better defense. But in some group, somewhere, an adaptation will appear that gives individual advantage by improving the foraging efficiency of predators that have it. This adaptation will be selected strongly within the confines of the group. Gilpin (1975) shows mathematically that this will result in rapid extinction of the predator population within the group as the predators reduce the prey to numbers so low that predators cannot find enough food to reproduce. A whole group of predators then goes extinct and the patch is recolonized by less efficient predators from elsewhere.

The whole proceeding relies not only on life in patches, but on slow rates of gene exchange between groups (slow colonization or immigration) and rapid effects within the groups. A nice balance is required between the rate at which a group suffused with a fatally efficient genome goes extinct and the rate at which colonists with genomes that allow persistence arrive.

Another way of explaining the Gilpin model is to think of the system set up by Gause (see Chapter 10) in which *Didinium* preys on *Paramecium*. In Gause's experiments the *Didinium* was a devastatingly efficient predator that always eliminated its prey, and hence itself, from any culture tube. Gilpin's model requires that there be other strains of *Didinium* that are less devastating, living in culture tubes (patches) elsewhere. *Didinium* individuals of this less devastating kind will survive and their groups with them. These individuals can then recolonize patches (culture tubes) where a whole group of super-efficient *Didinium* has gone extinct.

The sequence of events in the Gilpin predator–prey group selection model is:

1. Mutation or recombination produces an extra efficient predator.
2. Selection favors this efficient predator such that all predators within the group come to be of this kind.
3. All individuals of the group become extinct.
4. The vacant patch is recognized with persistent, less-efficient predators from other groups.

Even this model for group selection in which whole groups are winnowed by the consequence of their own devastating predation clearly has some narrow constraints to meet. An alternative group selection model based on direct aid to neighbors appears to be actually impossible. Consider two earthworm strains. Type A, a community benefactor, acts in ways that improve the environment for earthworms at some cost c, giving a gain in fitness g to itself. But type-B worms in the same community have an equal gain but no commensurate cost, and thus should leave more surviving offspring. Selection will remove the community benefactor's genes from the population rather rapidly. This will be the result of any community function that incurs a cost in any system with freely assortative mating.

Yet there is a single circumstance in which the A-type community benefactor can be favored by selection: when living in a very small subpopulation in a generation free from intense competition. In this very small group, all individuals, both type A and the rest, will gain from the presence of type A. The group as a whole will leave more surviving offspring than other groups of similar size with no type-A individuals. If at the end of a single generation there is an episode of dispersal, the costs of the benefaction never catch up with type A (Figure 13.21).

The key to this form of group selection is life in isolated groups without a continuous and general mixing of genes throughout the population, what D. S. Wilson (1980) calls a system of "**structured demes**." "Deme" is the geneticist's term for a panmictic population in which crossbreeding is general and random. "Structured demes" describes the fact that in many real populations potential inbreeders live much of the time in small subpopulation isolates. Wilson uses the example of a small cavity of water on a tree trunk where mosquito larvae live for a generation. All larvae in that one cavity would benefit from habitat improvement by one strain of larvae, after which adults would disperse to other cavities for the next generation. In every generation the community improvers would derive benefit from the improvement, but whether the other types derived their even larger benefits would depend on whether an improver was

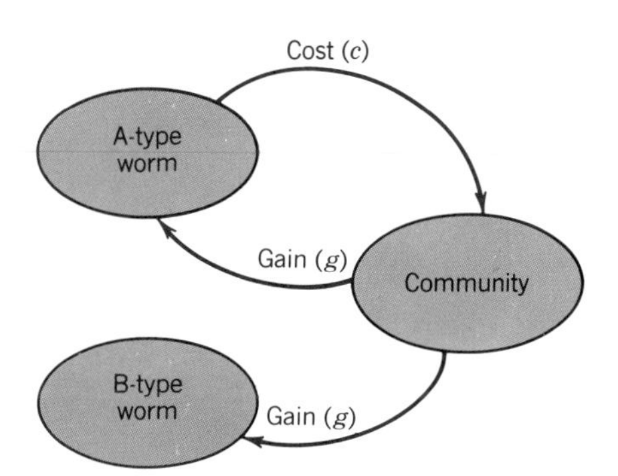

FIGURE 13.21
Model of indirect community effects for an individual adaptation.
The A-type must be selected against in a continuous crossbreeding system because a cost falls to the A-type but not to others. However, if the small group disperses shortly, the A-type has a gain in fitness. (From D. S. Wilson, 1980.)

present or not. Mathematical models leave no doubt that this mechanism would work if the limiting conditions are met.

Wilson (1980) makes convincing arguments that this mechanism of group selection by structured demes is important in the evolution of small, tightly knit communities that disperse as whole entities. That is a major catch, though; the whole community must be able to disperse as one. But some tiny communities can be carried intact by a single traveling organism. The community associated with bark-tunneling beetles is of this kind. Another is the community of mites and other organisms that travel with scarabs or other dung beetles, using the flying beetles as transport for the whole community, so-called "phoresy."

The Gilpin and Wilson models for group selection are fun. But they work by continuing to allow individuals to act in ways that maximize individual fitness. And they have no generality beyond the systems of tight constraints for which they were designed. The great diversity of social activity that makes natural history so interesting can be explained when all participating individuals act selfishly to maximize their own fitness.

Preview to Chapter 14

Aquila harpyja: top predator in Amazon canopies.

PREDATION removes individuals from a lower trophic level, the most important consequence of which is to prevent monopoly competitive success among the prey. Predation thus allows increased diversity through what might be called the "cropping principle." This effect has been demonstrated experimentally by removing top predators, resulting in drastic reductions in prey diversity as successful competitors, freed from predation, preempt resources. Herbivores acting as plant predators can similarly increase diversity in plants, as when overgrazing allows the invasion by weeds of pastures originally planted with one or a few pasture grasses. High seed predation near parent trees in tropical forests is important to the maintenance of high tree diversity, preventing the preemption of canopy space by a few successful trees. Jaguars and harpy eagles of the Amazon rain forest are probably vital to the maintenance of diversity through their manipulation of diversity among seed predators. Wolves and lions preying on ungulates, however, have their take limited by the effective defenses of prey animals, such that their predation cannot interrupt rapid population growth of the prey when food and population dynamics produce exponential increases, but relatively high predator densities accentuate population crashes that follow.

Chapter 14

Predation as Diversity Inducer

A predator is an organism that uses other live organisms as an energy source and, in doing so, removes the prey individuals from the population. This definition allows the concept of predation to be extended to include herbivory as well as carnivory. Working ecologists now often talk of "predation" when describing sheep hunting grass or squirrels searching for nuts. Also included in the definition are "**parasitoids**," which are organisms that, although parasitic, kill the host. The important parasitoids are animals like hymenopterans, whose larvae feed on the body of the host, which does not die until the parasitoid's development is complete. True parasites are not included in the definition, since the body of the host is not removed from the population, and the same argument excludes herbivores when the damage they do to the plant is minimal. These parasites and herbivores, however, do have effects on the population of their prey, and can be considered alongside true predators.

When predators kill, they remove contestants in an ecological game. This changes the rules for all the other players. If a competitor is taken out, those that are left benefit. If a seed dispersal agent is removed, a plant is not transported. If an enemy is killed, an old victim flourishes. Predators are, in a sense, arbiters of community structure and local diversity.

The first rush of ecological theory saw predators as potential controlling agents for populations. Certainly predators can utterly transform population histories (see Chapter 10). But the more interesting effects are probably on diversity and structure, as predator winnowing of populations alters patterns of competition.

PREDATORS ON UNGULATE HERDS

It is a truism of natural history that much of the food of wolves and big cats consists of the old and the sick. This was certainly so for wolf predation on the Isle Royale wolf herd when corpse yields of moose and wolves were used to calculate the low ecological efficiency of wolf predation (see Chapter 4). Isle Royale wolves ran down young moose and ate the old. But when a moose in its prime turned to stand its ground, the wolves retreated (Mech, 1966; Figure 14.1). This immunity of prime moose from attack certainly contributed to the low efficiency of energy transfer because the moose dispersed most of the energy of their trophic level as respiration during the prime years of their middle age, leaving little to pass on to the wolves in their aged carcasses.

Wolf restraint when hunting so powerful and dangerous an animal as a fit moose is understandable; a dead wolf, even a well-kicked wolf, has a reduced chance of fitness. Wolves have been found with damaged skulls, evidence of severe blows on the head, and an autopsy on a dead wolf in Minnesota concluded that it had been killed by a deer that gored it (Nelson and Mech, 1985).

A classic study on Mount McKinley in

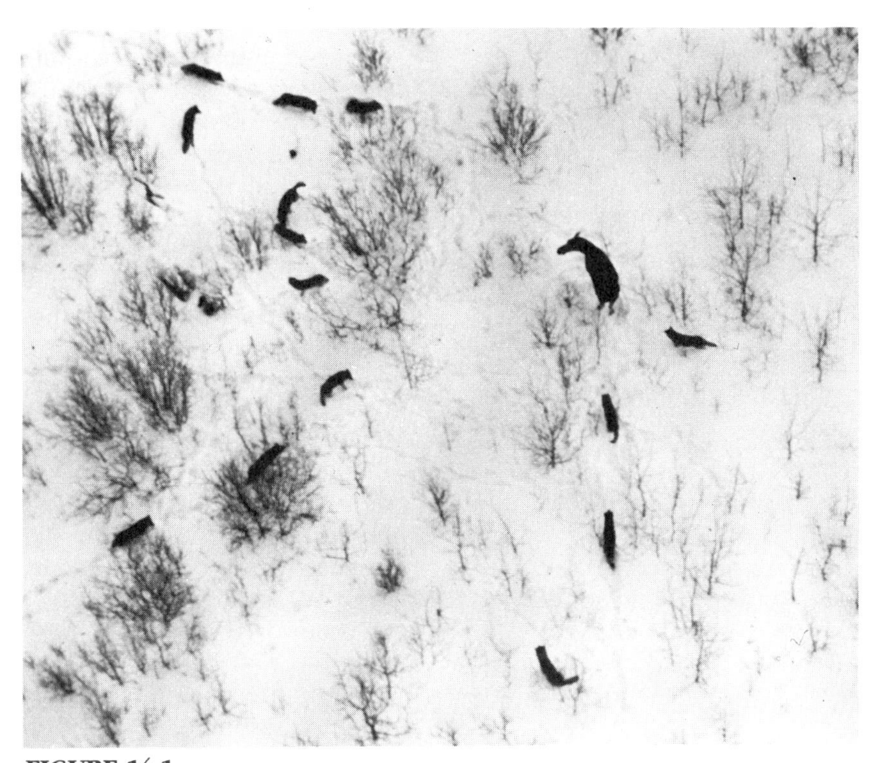

FIGURE 14.1
***Alces americana:* stands its ground under wolf attack.**
The wolves left the moose shortly after this photograph was taken. The moose was in the prime of life and too dangerous a proposition for the wolves. Immunity of prime moose to wolf attack is reflected in their life table.

Alaska has long suggested that wolves hunting Dall mountain sheep failed to kill many sheep in their prime (Figure 14.2). This result comes from the combined field observations of the pioneer Alaskan naturalist Adolph Murie (1944) and the first life table for a wild animal constructed from the evidence of sheep skeletons by Edward Deevey (1947).

Deevey built his sheep life table after the manner of the life insurance industry, but with less tractable data. Insurers have census data for different years that give numbers born and dying, together with age of death, and their object is to predict age of death for the population still living. Deevey's data set was almost the inverse, a record only of deaths. But Dall sheep bones carried a record of their owner's age at death, mostly in annual rings on the horns. Muir had gathered 608 skeletal remains, had determined age of death for each, and was able to declare that all had been eaten by wolves. Deevey's life table for Dall sheep under wolf attack then demonstrated that few sheep had been killed in the prime of life (Table 14.1.).

FIGURE 14.2
***Ovis dalli:* immune to wolf attack.**
The evidence of bones on the tundra of Mount McKinley shows that the local population of Dall mountain sheep were killed by wolves when old, but not when in their prime.

Even when avoiding prime-of-life prey, however, wolf predation on the young can have important population consequences, depending on the density of wolves in an area. In a Minnesota wilderness, 71 out of 100 white-tailed deer fawns were lost to wolves between fall and late winter one year (Nelson and Mech, 1981). Predation at that rate should have drastic consequences for the deer population if maintained over several seasons. But wolf densities were thought to be high; the phrase "saturation densities" is used in the report. Adult deer were killed in winter, particularly in deep snow, which gives relative advantage to the lighter wolves, although yearlings and adults continued to be safe in summer. The report notes, "Not one of our radiotagged yearling and adult deer was killed by wolves in summer during 60 deer years of radiotracking" (Nelson and Mech, 1981).

Wolf predation has important consequences when wolf densities are high or prey densities are low. In the Minnesota study it was thought that deer could be wiped out locally from the center of a hunting ground defined by a pack's territory while the wolf density remained high, although not regionally. The wolf packs establish hunting territories for the pack as a social unit, scent-marking the periphery (Harrington and Mech, 1979). The territories are defended by more than ritual aggression in times of food scarcity, providing a "no-wolf's land" between packs

TABLE 14.1
Life Table for Dall Mountain Sheep
A small number of skulls without horns, but judged by their osteology to belong to sheep nine years old or older, have been apportioned **pro rata** *among the older age classes. This table was constructed solely from examination of a collection of sheep skulls from Mount McKinley. It reveals that most mortality was suffered by the very young and the very old, which was as expected because wolves were the main source of mortality and they were known to hurt the weaker animals. (From Deevey, 1947.)*

x	x'	d_x	l_x	1000 q_x	e_x
Age (years)	Age as Percent Deviation from Mean length of Life	Number Dying in Age Interval out of 1000 Born	Number Surviving at Beginning of Age Interval out of 1000 Born	Mortality Rate per Thousand Alive at Beginning of Age Interval	Expectation of Life, or Mean Lifetime Remaining to Those Attaining Age Interval (years)
0-0.5	−100	54	1000	54.0	7.06
0.5-1	−93.0	145	946	153.0	—
1-2	−85.9	12	801	15.0	7.7
2-3	−71.8	13	789	16.5	6.8
3-4	−57.7	12	776	15.5	5.9
4-5	−43.5	30	764	39.3	5.0
5-6	−29.5	46	734	62.6	4.2
6-7	−15.4	48	688	69.9	3.4
7-8	− 1.1	69	640	108.0	2.6
8-9	+13.0	132	571	231.0	1.9
9-10	+27.0	187	439	426.0	1.3
10-11	+41.0	156	252	619.0	0.9
11-12	+55.0	90	96	937.0	0.6
12-13	+69.0	3	6	500.0	1.2
13-14	+84.0	3	3	10.000	0.7

where deer got through the winter almost unmolested.

The importance of wolf predation thus depends on the relative abundance of predator and prey. Low wolf populations, content to avoid the more dangerous prey, possibly have minimal population consequences, but crowds of wolves in years that are already hard for the ungulate prey can produce local devastation. The devastation should be short-lived, however, because it imposes food limits on the wolves. Wolves starve at such times, lose their pups, and even die in territorial encounters, producing a population crash (Nelson and Mech, 1981).

Deer and moose populations can prosper and become dense despite wolf predation. This prospering is inherent in the possibility of runs of good seasons and bad, but is perhaps even more inherent in the dynamics of density-dependent systems (see Chapter 10). A plunging prey population at a time when wolf numbers have gradually built up is likely to change the quality of wolf predation and can lead to locally devastating effects.

At Isle Royale, moose numbers did begin to rise after the period of stability used for the efficiency study (see Chapter 4), and signs of malnutrition appeared (Peterson et al, 1984). Wolf numbers finally started to rise even as moose numbers fell (Figure 14.3). We have no evidence that the initial

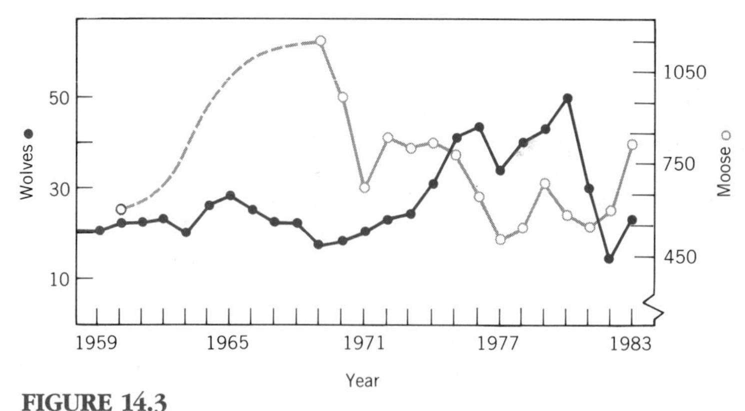

FIGURE 14.3

Effect of oscillating moose numbers on wolf population.
A rise and fall of moose numbers of Isle Royale between 1960 and 1985 caused a slow rise in wolf numbers, followed by rapid collapse of the wolf population. (From Peterson et al., 1984).

fluctuation in moose had anything to do with wolves; possibly the oscillation depended on density-dependent dynamics from intraspecific competition within the moose herd (see Chapter 10). But a consequence was some hard years for the wolves when moose were scarce but wolf numbers relatively high. As the supply of old moose ran out, the wolves began attacking more moose of breeding age, thus tending to accelerate the decline.

Lions and Wildebeest of the Serengeti

Seven hundred lions hunt the Serengeti plains of East Africa, feeding almost entirely on a herd of 200,000 wildebeest, or they did in the 1960s (Talbot and Talbot, 1963). A guild of leopards, cheetahs, hyenas, and hunting dogs take wildebeest in their different ways, though these are less important than lions (Figure 14.4). All the predators have alternative prey, ranging from the marvelously diverse guild of ungulates to primates and rodents.

If each lion kills a wildebeest every week, the cost to the wildebeest herd is between 12,000 and 18,000 animals a year, probably about 8% of wildebeest productivity (Figure 14.5). To this must be added estimates of predation from the other predators, together with deaths from disease, but the total death rate from all these sources was clearly insufficient to offset the birth rate.

The Talbots discovered, however, a powerful density-dependent death rate in the dense wildebeest herd in the way that the young got separated from their mothers. Wildebeest calves must follow their mothers within minutes of being born, and they tend, quite simply and literally, to get lost. They starve and become prey to hyenas, jackals, and vultures. Loss of calves is density dependent since the babies seemed to be safe by their mothers' sides in small herds. Only in the larger, crowded herds was this loss significant.

No run of data available to us from the game herds of Africa can be sufficient to tell us much about the long term dynamics of the populations of large animals. Fluctuations of large amplitude are to be ex-

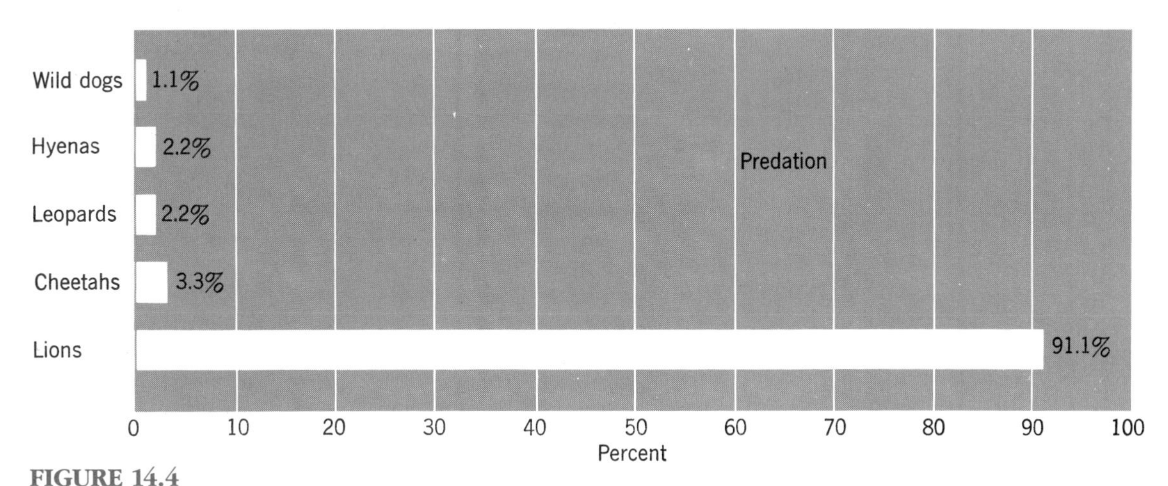

FIGURE 14.4
Predation on western white-bearded wildebeests.
Lions were the only serious predators of wildebeests in the Serengeti, but it took only a relatively small fraction of the wildebeests' dying every year to support the known lion population. More important causes of death in wildebeests were disease and young animals' getting detached from their mothers in crowded herds. (After Talbot and Talbot, 1963.)

FIGURE 14.5
***Felis leo* culling: short term population effect.**

FIGURE 14.6
***Crocuta crocuta:* hyenas in the predatory guild.**
Despite the great variety of predators in the East African large mammal fauna, numbers of both predators and prey fluctuate radically in ecological time measured in millennia.

pected, however. The climate is strongly seasonal with monsoon-based rains, and the climate changes on historical time scales, making inevitable runs of good breeding years and bad (see Chapter 11; Figure 11.5). Moreover, the large herds must be subject to intraspecific competition and density-dependent effects, either of the baby-losing kind described by the Talbots or more orthodox food shortage, letting in the variety of density-dependent dynamic fluctuations predicted in May's (1986) models (see Chapter 10).

Killing by a guild of large predators should vary widely over time as both predator and prey numbers fluctuate from their own internal dynamics (Figure 14.6). If we had a continuous but short record for populations in these herds, say spanning just the last 25,000 years, we would probably find times of great rarity for some species, but also times of great abundance. But the prey are all grazing animals. When dense predators encounter falling prey populations, as at times should be inevitable, new opportunities for establishment and spread would be opened for the food plants.

MINK ON MUSKRAT

A lifetime's work by Paul Errington (1963) outlines the relationship between mink (*Mustela vison*) and muskrats (*Ondatra zibethica*). Errington described his data collecting for muskrats as "studying the meaning of tracks and trails, of diggings and cuttings and heapings, of food debris and droppings, or miscellaneous traces, of blood, fur, wounds, and carcasses." He mapped the presence of individuals of both species by signs, trapped them, marked them, and released them. He was given sample carcasses by the local fur trappers for autopsy, particularly for estimating the number of offspring that had been born to females from the number of scars of old placentas in their ovaries. He examined mink scats for muskrat remains. And he concluded that nearly all the muskrats eaten by predators in his part of Iowa were individuals that were doomed to die from some other cause, such as disease, if the predators had not eaten them first.

The lives of Iowa muskrats were strongly influenced by the marked seasonal changes of the local climate, by the alternation of warm continental summers and bitter winters, by the chances of floods, and the chances of drought. Muskrats established in fine summer weather in their home ranges seemed almost immune to predation, probably because they had somewhere to run to when approached by a mink or because they faced the potential aggressor so confidently that they made the risks of combat unacceptable to it. But if they were flooded out, or left exposed by drought, or suffered epidemic disease, the minks killed many of them. Sometimes the onset of some calamity attracted predators to such an extent that whole populations of muskrats were almost wiped out.

For these mid-sized animals, therefore, the inference is strikingly similar to what is now being suggested for big predators on ungulates—there are intervals when the prey population supports the predator partly with its discards of sick and old, taking some loss of young withal, along with intervals of catastrophe. The dynamics of the prey population, as set by environmental stress of intraspecific competition, is probably responsible for bringing on conditions in which predators can devastate.

A PREDATOR REMOVAL EXPERIMENT

On rocky substrates of intertidal or subtidal stretches of coast, population limits to both animals and plants tend to be set by space. Seaweeds live anchored to rocks against the waves and maintain productivity comparable to that of land plants of mesic habitats by extracting nutrients from the flowing water. But these plants must compete with filter-feeding animals for space on the rocks (Figure 14.7), as various bivalves and barnacles feed by extracting plankton or debris from the flowing water. The flux of food reaching the rock surface typically is in super-abundance, and the total number of filter-feeding animals that can live is set by space, not by food.

Contest competition for space can be ruthlessly strong in this system, as Connell (1961) demonstrated for barnacles on the European side of the Atlantic. Intraspecific competition within single-species

FIGURE 14.7

Intertidal competition.

Anemones, brown and green algae, the sea grass *Zostera*, and others compete for space on the coast of Washington and a predatory starfish crosses.

stands is by the simple process of crushing or prying off neighbors. Separate bands of different species like those often seen in intertidal zones also result from interspecific competition.

In the eastern Pacific Ocean on the shores of Washington State, a particularly complex fauna occupies the rocks (Figure 14.8). A large starfish, *Pisaster*, is a top carnivore able to prey on all the filter feeders and herbivores that form the base of food chains, as well as on the only other important predator, the snail *Thais*. The bivalves, acorn barnacles, and *Mitella* are anchored filter feeders. The limpets and chitons are grazing herbivores, but they too require space for anchoring themselves when passing the hours during which they are exposed to the air at low tide.

Paine (1966) and teams of diving students removed all the *Pisaster* from a section of shore 8 meters long and 2 meters in vertical extent, and kept *Pisaster* away for six years through frequent visits. An adjacent area was kept as a control, and both were surveyed at frequent intervals. The appearance of the control plot did not change.

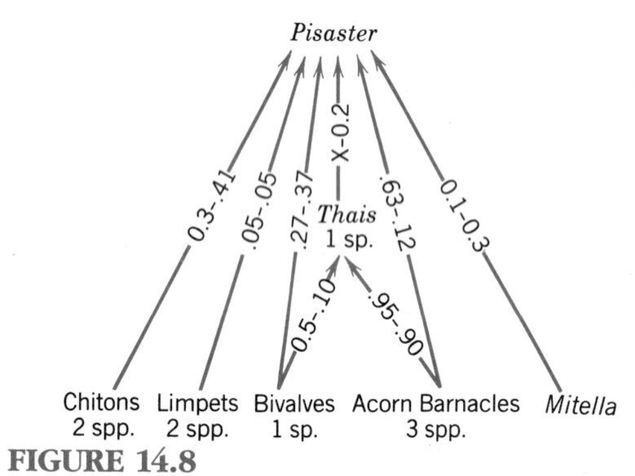

FIGURE 14.8

***Pisaster*-dominated subweb in Washington intertidal.**

When *Pisaster* is removed from a section of reef, species diversity of the prey species falls. Numbers are calories ($x = <.01$). (From Paine, 1966.)

Where the predator was gone, however, the community progressively changed. Within a few months a barnacle population of *Balanus glandula* spread to take up most of the space, after which the bivalves *Mytilus* and *Mitella* proceeded to displace the barnacles in their turn. After several years it is clear that the rocks are settling down to being mostly the home of a long-lived population of *Mytilus*.

The part of the rocks covered by animals also has grown at the expense of seaweeds that are being scraped off or denied space in which to grow. Perhaps even more revealing is that animals that can move, like the larger chitons, have left the area.

What was originally a 15-species system was reduced to an 8-species system, merely by removing one predator species. In a sense, the predator was controlling the populations it attacked. Of far more importance to the running of the system was the fact that without the predator, competitive exclusions in the prey populations drove species to local extinction, reducing diversity.

One way to look at these results is to see that changes in one population in a complex food web can cascade through the web to alter the living conditions, and hence population size, of all species in the web. An intellectual consequence has accordingly been a renewed interest in the structure and maintenance of food webs (see Chapter 22).

The Cropping Principle

The *Pisaster* removal experiment was of historical importance in alerting ecologists to what might prove to be the most important function of predation; not so much population control, as diversity control. In setting population limits, predators are but allies of competition and natural catastrophe, their own populations often radically destabilized by the dynamics of the populations of their prey (see Chapter 10). But many predators, particularly those high on food chains, can, by reducing the numbers of prey below levels set by competition, massively increase diversity in the prey trophic levels.

This diversity-inducing function of a predator can usefully be called "the cropping principle."

When predation allows increased diversity in the prey trophic level, increased diversity of predators is at once possible, because the new potential prey species afford opportunities to specialized predators. The cropping principle, therefore, offers an almost unlimited contract with increasing diversity: more kinds of prey leading to more kinds of predator, whose depredations make room for more kinds of prey, and so on.

Sheep and Rabbits as Predators and the Diversity of Pasture

Sheep, like all large grazing animals, have their preferences about what they like to eat. They do not go across a field like a mowing machine, but hunt out their preferred food plants. If a good pasture of palatable grasses is overgrazed, sheep make bare spots that are invaded by "weeds" unpalatable to sheep. Overgrazing a pasture with sheep is thus a splendid way of increasing the diversity of plants that live there, as the pasture becomes stocked with interesting plants that are totally inedible to sheep. Adding extra predators in the form of other grazing animals should increase the diversity still more.

Rabbits (*Lepus cuniculus*) are exotic introductions to Britain. They may have been there since the Norman invasion, but only became abundant in the eighteenth century (Sheail, 1971). In a classic experiment, Tansley and Adamson (1925) fenced off some plots on old pastures on chalk in southern England, and maintained the rabbit-proof fences for six years. Their results are illustrated in Figure 14.9. The rich diversity of the chalk pasture changed into a tall stand of one grass, *Zerna* (*Bromus*) *erecta*, with a few scattered other plants. This result foreshadowed Paine's *Zoroaster* removal experiment by some 40 years, which is perhaps understandable from Tansley, the inventor of the ecosystem concept.

Tansley's tiny plot experiment has since been "performed" over the whole of Britain when the disease myxamatosis almost eliminated the rabbits from the entire island for a number of years. The growing up of the tall grass stands became a common sight.

As with sheep, however, the effects of rabbits depend on their food preferences. The removal of rabbits from a small island off Wales by myxamatosis resulted in the discovery of 30 plant species not reported before in some 300 years of observations. Presumably, invasions of the island by seeds of these plants from the mainland had previously been suppressed by the rabbits (Lancey, in Harper, 1969).

A 900-Year Predator-Control "Experiment"

The effects of prolonged cropping were demonstrated elegantly when it was shown that pastures along the river Thames in England had been managed continuously in the same way for 900 years (Baker, 1937). One set of meadows had always been grazed by varying herds of cattle, horses, and geese, the only break being during three years of the British civil war in the 1600s. Adjacent meadows have been allowed to grow into hay to be cut in the summer, after which the stubble was grazed, also for 900 continuous years.

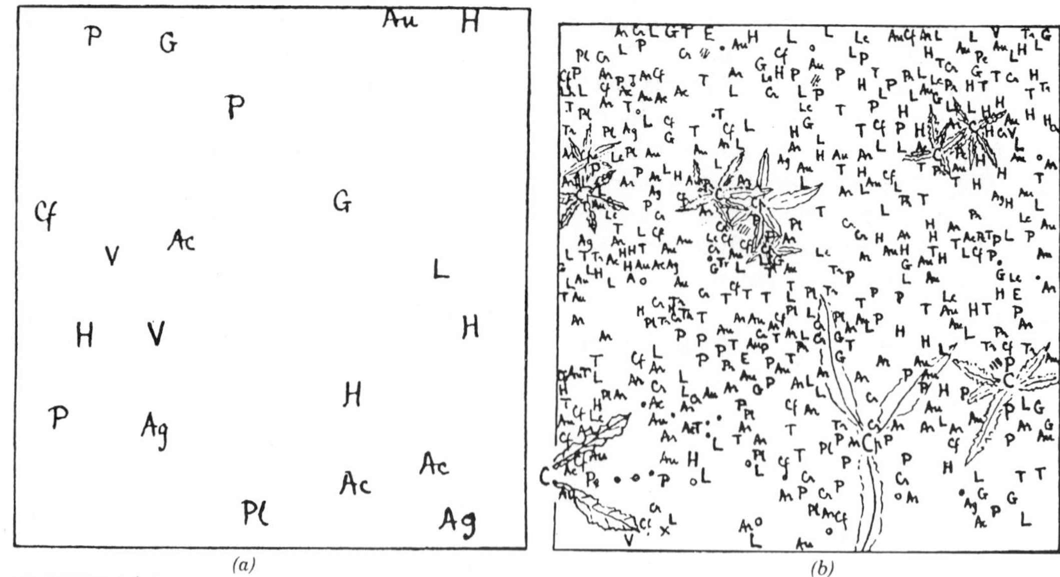

FIGURE 14.9

Effect of rabbit grazing on English pasture.

The two diagrams record all the principal plants in two plots, except that the plain background in plot (*a*) was a pure stand of the tall grass *Zerna erecta*. Rabbits had been excluded from plot (*a*) but had access to plot (*b*). (From Tansley and Adamson, 1925.) The diagrams are photocopies of the originals, hand-drafted as typical of the journals of the 1920s.

The haymakers, and the grazing livestock, loosed on these neighboring Thames meadows act as predators with different hunting techniques. Haymakers prey on tall plants, but they do so only after the plants are fully grown, often after they have set seed. A consequence is that the tall grasses establish a species equilibrium and can eliminate short plants and rosette plants by competitive exclusion. Haymaking leads to a pasture of tall species (Table 14.2).

In the other Thames meadows, the grazing cattle, horses, and geese prevent tall plants from growing high, selectively eat them, and eliminate them from the community. Under this grazing pressure there is selective advantage in being a short rosette plant and it is this life form that is common in the grazed meadows (Table 14.2).

The lists of species in the adjacent pastures are different after these 900 years of different predation. Of the 95 species of plants existing on both meadows in 1937, only 30 were common to pastures under both managements.

TABLE 14.2

Plant Diversity in Ancient English Meadows

The management of these meadows essentially has not changed for 900 years, suggesting that an interactive species equilibrium, S, is achieved. Diversity and life form depend on whether a meadow has been grazed or cut for hay. (Data of Baker, 1937.)

	S	Predominant Life Form
Total species pool	95	
Restricted to hay meadows	39	Tall
Restricted to grazed meadows	26	Short rosette
Common to both meadows	30	—

Even greater diversities should result if more kinds of grazing animals increased specialized grazing pressures. Pasture in the Middle East that is badly overgrazed by cattle, donkeys, horses, goats, sheep, and camels combined has a species richness of more than a hundred herbs, although productivity is extremely low (R. H. Whittaker, personal communication).

Seed Predation and Tree Diversity in Tropical Forests

Forest trees cannot easily be destroyed by grazing animals when they are adults, but they are vulnerable to predation as seeds or seedlings. Janzen (1970, 1971) suggested that the area under the canopy of a tropical tree is a particularly dangerous place for that tree's offspring because seed and seedling predators would congregate there, the more mobile of them, such as rodents, actually using the parent tree as a "flag" to guide them to the food supply. The result might be a death zone under the parent tree in which no regeneration could take place (Figure 14.10).

Tropical trees on Janzen's model should be able to regenerate only far from the parent, putting a premium on dispersal. Establishment of seedlings would probably be in a ring round the parent tree, far enough away to avoid most of the waiting seed predators, and yet near enough for a significant number of dispersing seeds to arrive (Figure 14.9, population recruitment curve—PRC). Meanwhile, space under the parent should be colonized by other species from seeds not eaten by the specialized seed predators waiting there. The result is well-spread-out trees and a diverse forest.

Granted the high diversity of insect seed predators in the tropics as well as the high diversity of trees, Janzen's proposed grazing effect almost certainly applies. Seed predators should be acting in the way described, suggesting that removal of insect pests would be disastrous for the diversity of the forest. Our contemporary rain forest

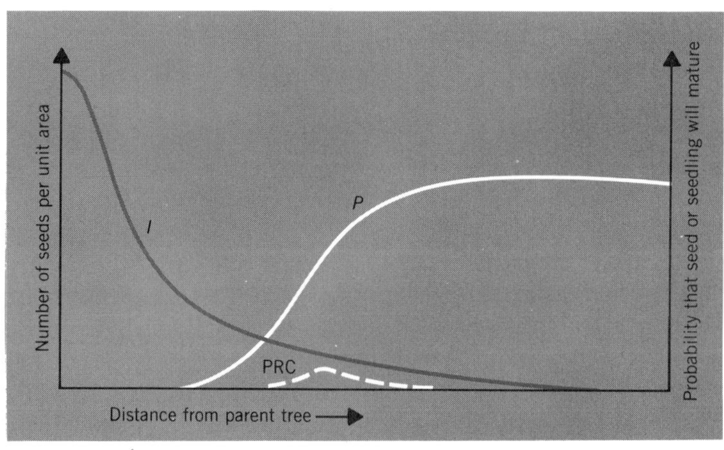

FIGURE 14.10

The Janzen hypothesis that tropical trees are dispersed as a result of predation on seeds and seedlings.

Curve *I* is seeds reaching unit area; *P* is the probability of survival, and PRC is the resulting recruitment curve.

experiment, however, is the contrary one of removing the trees.

Unwitting Predator Removal in Tropical Plantations

Despite the massive diversity of insects in the tropics, oil and rubber plantations in Malaysia had remarkably little trouble with insect pests, until, that is, the invention of chlorinated hydrocarbon pesticides. In the late 1950s, many estates growing oil palms in Malaysia suffered plagues of defoliating caterpillars, and the estate managers called in entomologists to advise them. One of the entomologists was able to work out a detailed history of the insect plagues and to explain how they came about (Wood, 1971).

The start of the problem had been a small alarm, the appearance of sufficient cockchafers on one plantation to worry the local manager. The cockchafers were not doing much damage, but the manager thought he ought to spray against them as an insurance, so he sprayed the plantation with DDT and noted the fact in his monthly report. Shortly afterward there were outbreaks of leaf-eating caterpillars on that part of the estate, so the manager sprayed again, and also sprayed the neighboring areas as well "to contain the pests."

The next thing that happened was that caterpillars became troublesome over the entire sprayed area. Other planters learned of it, became convinced that they were faced with a general outbreak of insect pests, and began spraying too. The caterpillar plague then became general and Wood was called in.

Wood (1971) found that the only change in the environment or management of the estates that was synchronous with the pest outbreaks was the spraying with insecticide itself. He postulated that the pesticide had killed natural predators of the system, thus allowing the herbivorous pest insects to increase in number. This hypothesis received immediate encouragement from the fact that the worst of the defoliators was a bagworm, a species of moth whose caterpillars lived in shared cocoons. It seemed likely that the communal cocoon would protect caterpillars from the spray. Hymenopteran parasitoids of these caterpillars (about 20 species of these were present) would be more exposed to the sprays than their prey and thus would be selectively killed. The test of this hypothesis was, of course, to stop spraying and see what happened, which Wood recommended. To a large extent his stratagem worked.

But Wood's remedy of stopping spraying was not always completely successful. Plenty of the hymenopteran parasites soon reappeared, but the bagworm populations sometimes remained at a disturbingly high level for some time. Doubtless numerous competitive and predator relationships had been undone by the spraying, and some population dynamics had to work themselves out.

SPARE THE TOP PREDATOR

The most important effect of predators appears to be to stimulate diversity. They prevent successful species in the trophic level below from squeezing out the rest. Conservationists should look upon top predators as economists look upon antimonopoly laws. They are essential to preserving diversity in the system.

Top predators might be particularly important in rain forests, which perhaps owe much of their astonishing diversity to the peculiar working of the cropping principle in a productive land of great trees. Tree diversity is critically dependent on seed dispersal and seedling growth, both of which are influenced by seed predation (Figure 14.9). Among the most important seed predators are the small and mid-sized animals of the forest, ranging from rodents to pigs and primates. And the numbers and diversity of these are set by the top predators of the rain forest: big eagles and bigger cats.

In Amazon rain forests the top predators on the ground are a guild of large cats, with jaguars and pumas the most conspicuous. In the canopy are harpy eagles and their kind, preying on a complex guild of browsing primates, sloths, and rodents. These predators turn out to be generalist hunters, not specializing in particular prey as the big cats of the Serengeti do, but searching for many kinds of food. This catholicity of taste may well be a necessary adaptation to the relatively small size and diversity of rain forest herbivores.

The most pristine community left in the Amazon is probably that at Cocha Cashu in Peru, where Emmons (Terborgh, 1988) finds that jaguars and pumas between them take 8% of the total herbivore biomass annually, about the same as the big predators of the Serengeti. But scat analysis shows that all of the rain forest herbivores are included in the diet, including relatively low-fecundity animals like agouti, paca, and coati, all of which are seed predators. The data at least invite the hypothesis that diversity and selectivity of seed predation, so vital to forest diversity, is itself maintained by the top predators.

This hypothesis can be tested by comparing tree and seed predator diversity at Cocha Cashu with Baro Colorado Island (BCI) in the Panamanian rain forest, where climate and seasonality is closely comparable. BCI is a patch of lowland rain forest that was turned into an island by the building of the Panama Canal at the beginning of this century. The BCI reserve is large, but not large enough to support the top predators, which have accordingly been absent for 80 years. Compared with Cocha Cashu, the number of small mammals is huge (population densities up to ten times as great). Large effects on future tree diversity seem inevitable.

Harpy eagles and jaguars are probably essential to the maintenance of Amazonian diversity. Exterminating these animals, a task now probably nearly complete over much of the Amazon, makes long-term preservation of diversity impossible. Forest reserves need to be large enough for viable populations of the big predators to persist.

Preview to Chapter 15

Feeding the ant guardian.

ADDING new species to a diverse community requires reallocation of resources, and this must follow whether the new arrival comes as competitor, predator, or friend. Reallocation can evolve into mutualisms as organisms gain fitness from the presence of the other. Bees gain fitness by feeding on plant nectaries, and plants gain compensating fitness when bees transfer pollen to anthers. Selection then refines the behavior of both parties to the mutualism for individual advantage, but all resulting adaptations are contingent on preexisting properties, suggesting unlimited possibilities for change and counter-change. A similar pattern of alternating advances can be seen in coevolutions between specialized predators and their prey, as each new defense makes possible selection for a more novel attack. An important resource to be reallocated by coevolution is time, as activities are timed to avoid or use others (so-called phenology). Resources can be preempted through the use of allelopathic chemicals, and organisms can use each other by interpreting or abusing their signals, including luring prey to predators with false signals. Some of the tightest coevolution is found in the chemistry of plant–insect interactions, as highly specific, toxic plant defenses promote selection for specialized feeders that can detoxify them. Optimal defense theory suggests when quantitative and physiologically costly defense by tannins gives protection against diverse herbivores, or when qualitative, alkaloid defense is the cost-effective solution. An alternative defense of plants against herbivores follows when ants use plants as both habitat and source of nectar as nutrient, the so-called ant-guard symbiosis. Animals that are dangerous to eat because they are poisonous are at a selective advantage if potential predators can learn to recognize them, leading to the phenomenon of warning coloration. When a warning signal is recognized by predators, selection should promote use of the same signal by other dangerous animals, leading to a convergence of traits called "Müllerian mimicry." Ecologists have long speculated that harmless prey might evolve false warning coloration, thus letting them act as parasites of the dangerous originals. Recent data, however, suggest that this proposed "Batesian mimicry" may not be an important natural phenomenon.

Chapter 15

Coevolution: Mutualists, Defenders, and Mimics

Individuals of new species must inevitably make their way in established communities; they come as competitors, predators, or friends. In whatever guise they come, the members of the new species must use resources that have previously been used by other species populations, even if only by the decomposers that deal with waste. Adding a new species to a community, therefore, is a process of reallocation of resources.

The opportunities for a new species to enter an already diverse community are narrowly constrained by the strategies of species already there; the newcomer must in a sense fit in. At the same time the options of the newcomer are even more narrowly constrained by the history of its parent stock. The feats of engineering or food opportunism possible for natural selection at any one time are contingent on past history.

Bats, for instance, which are guided in flight by sonar images rather than by sight, enter communities as pollinators, fruit eaters, or insect hunters of the night. Success for a bat requires that the target foodstuffs be available at night. Bat success, therefore, is tied to plants that fruit or flower by night and to insects that fly by night (Figure 15.1). Birds, on the other hand, navigate by sight, and their success is dependent on food availability by day. Thus the options for increased diversity through evolution are always contingent on the engineering of the would-be newcomer and the engineering of the others already in place.

Much of the evolution that compounds preexisting diversity is necessarily **"coevo-**

FIGURE 15.1
Winged mammal: Night hunter.
Equipped by ancestry with sound location as a prime sense, new species of bats enter communities as night foragers.

lution." The newcomer uses other organisms for resources, which at once makes for selection pressure adapting the others toward maximum fitness despite their being used. Bee–plant pollination systems illustrate this relationship.

BEES AS POLLINATORS

Many plants use bees as pollinators, thus turning bees into a resource, but the plant must feed the bee pollen or nectar. For a plant to maximize fitness, bees must be attracted for frequent visits. Yet donating nectar and pollen to bees must divert calories and nitrogenous nutrients that might otherwise have been used for making seeds and fruits, requiring that plants attract the most bees with the least possible bait of nectar and pollen.

Plants that use bees as pollinators, therefore, have a problem in optimal resource allocation between bee bait and reproduction. The optimization problem is compounded when a plant must compete for bees with other species of plants, perhaps forcing it to spend extra resources on bee bait. The bees have comparable problems in optimal resource allocation, needing to maximize their return in calories and nitrogenous nutrients with the minimum number of flower visits and minimal journey time. Bee-plant pollination systems, therefore, are likely to be highly coevolved.

A simple structural adaptation in flowers of the desert willow *Chilopsis linearis* results in the maximizing of bee visits (Witham, 1977). Fine grooves run out from the main pool of nectar in the flowers; both grooves and pool are filled with nectar in the early morning. Bees that visit flowers early can maximize their retrieval of nectar by draining pool nectar only and then leaving to collect pool nectar from another flower. Later in the day when all pool nectar has been collected, all flowers still retain nectar in grooves where it is held by capillary action. It pays the bees to make a second visit to collect this nectar, a groove at a time, although their take is much less than on the earlier foraging visit. This adaptation of the plant therefore uses the bees for multiple visits, thus maximizing the chance of cross-pollination. In this coevolved system both plants and bees can be seen to maximize their returns.

A familiar habit of bees seen in gardens is the working of tall spikes of flowers like *Delphinium* or *Aconitum* from bottom to top. For the bees, this is a good search pattern because the lowest flowers develop first and hold more nectar. Taking the flowers in order on a vertical climb minimizes the search effort and maximizes the return in nectar. Bees abandon their climb before they reach the top of the spike because the undeveloped terminal flowers are without nectar.

But flowers on the spikes are arranged spirally, which means that the vertically moving bee misses some flowers in its ascent. Pyke (1978) has shown how the spiral arrangement of flowers serves to exploit bees for efficient pollination. An arriving bee does not carry sufficient pollen to pollinate all the female flowers on a spike, and the spiral arrangement, which forces bees to miss some flowers, serves to present bees with no more flowers than it can pollinate in a single ascent.

More subtle still in the use of bees by *Delphinium* spikes is the order in which anthers and stamens develop. The flowers are "protandrous," meaning that each flower changes from male to female as it ages. Old flowers at the bottom of the spike, therefore, are female and receive pollen brought from outside, whereas the last flowers visited high on the spike are male, thus recharging the bee with pollen to take to another flower.

FIGURE 15.2
***Selasphorus platycercus* and *Campsis radicans:* Coevolved.**
Both hummingbird and trumpet creeper allocate resources, in self-interest, for the other. For this plant a bird, not a bee, is the "flying penis."

OPTIMIZATION PROBLEMS

Coevolved mutualisms between bees and plants can be understood as giving optimal returns on energetic investments; the bees act as we would expect from optimal foraging theory and the plants are adapted to use the bees efficiently as, in Krebs's (1979) words, "a flying penis" to effect cross-fertilization (Figure 15.2). However, although the bees are clearly well used by the plants, it is difficult to say that this use is optimal. Perhaps a clever engineer could think of better solutions for getting service out of bees. The groove and pool nectar system, for instance, seems rather crude. The device of spiral arrangement of flowers need not necessarily be ideal. Both these arrangements obviously result from slight modifications of preexisting flower structures; doubtless they make better use of bees than earlier modifications of those same structures, but this does not mean that the designs are the best possible.

Natural selection, like politics, is a process constrained by what is possible. The result always is relative extra fitness compared to the fitness granted by some alternative adaptation. We cannot expect any community to be made up of species whose arrangements are truly optimal. Oster and Wilson (1978) make this point with an elegant analogy to economic theory. Economists of the Friedman school have argued that business firms evolve by a process of natural selection for the most efficient, which leads to the conception that the present distribution of business economic power is both ideal and the outcome of inevitable forces. This conclusion seems quite unsatisfactory, since it is obvious to most people that the present distribution of economic power is neither efficient nor inevitable. Likewise with mutualisms. Individual species evolving into them are selected to make the best possible solution to cost–benefit problems, though these solutions are unlikely to be truly optimal.

Direct evidence that even the most complex of communities is made up of species of less than optimal efficiency is suggested by the fact that new species can be made by competitive exclusion and character displacement. Squeezing out a new niche by displacement of character to minimize competition should be impossible if the preexisting species already were the product of ideal optimization theory, making perfect use of available resources. Granted historical and structural constraints, selection should make all individuals into optimal foragers of a sort, but those constraints are real and prevent ideal solutions.

It seems, therefore, that all communities can be invaded by new species offering fresh solutions to old problems. But because the new solutions require ever more sharing of resources, adding new species to communities that are already diverse often must involve mutualisms, most commonly plant–animal interactions but also interactions between animals higher on the food chains.

Empirical study of plant–animal interactions is one of the most interesting and satisfying developments of modern ecology. Theory for these studies tends to be weak, since optimal solutions can be neither predicted nor expected. Instead, the habits of animals and plants reveal to the astute naturalist the selection processes that have been important to their selection.

Cost–benefit analysis and optimal foraging theory can serve to demonstrate the utility of particular adaptations, like the arrangements of flowers and the flying patterns of bees, and they may also suggest where more subtle adaptations are to be found. If a habit should seem less than optimal at first analysis, perhaps it has been preserved because of circumstances that appear with low frequency but which must be survived when they do occur.

PLANT-HERBIVORE INTERACTIONS

Much coevolution concerns endless variations on attack and defense. Neither perfect attacks nor perfect defenses are possible, and new variants of first one and then the other are constantly jury-rigged out of old adaptations. One long train of thrust and counterthrust is suggested by data from passion flowers and the caterpillars that eat them.

Passion flowers (*Passiflora*) are vines that live scattered through many kinds of tropical vegetation. A primary defense mechanism of passion flowers against herbivores seems to be alkaloid toxins, but the plants are nevertheless attacked by butterflies whose larvae can sequester or otherwise detoxify the alkaloids. Butterflies of the genus *Heliconius* in particular feed on *Passiflora* and may become specialized to feed on only one or a few species. A good hypothesis to explain this is that ancestral butterfly stock able to detoxify the alkaloids of a range of *Passiflora* species came to specialize on a single species for ecological reasons. Factors like time of vegetative growth or shade tolerance of the plant led to preferential egg laying thereon. The *Heliconius* population thus restricted to a single species of passion flower would be selected for rapid growth. This process in turn would select against unnecessary mechanisms for detoxifying alkaloids of other passion flowers. Thus the general defense of toxins in the *Passiflora* genus is overcome, and a specialist *Heliconius* evolves that can eat one single species of *Passiflora* with impunity.

This victory of offense sets the evolutionary stage for the construction of a new defense. *Passiflora adenopoda* has fine, hooked hairs on leaves called trichomes, which give it total protection against *Heliconius* caterpillars (Gilbert, 1971). The tiny hooks tear into the pseudopods of the caterpillars and rip holes in the body wall through which hemolymph leaks, with fatal effect (Figure 15.3).

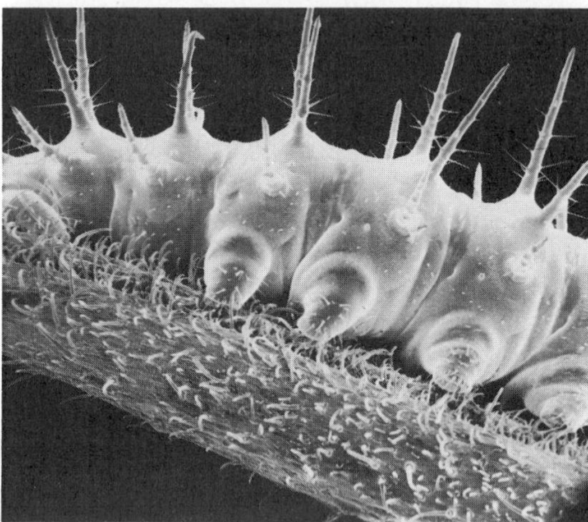

FIGURE 15.3
Passiflora adenopoda: **Caterpillar defense.** Fine spines, called trichomes, inflict fatal wounds on caterpillars, except those selected for the communal defense of silken mats laid over the spines.

Trichomes may not be as good a general defense as toxins for passion flowers because they would not work against other kinds of herbivore, but if a specialist caterpillar immune to the toxin appears, then the trichomes are apparently the most cost-effective defense against this one species. In the system studied by Gilbert, the trichomes apparently give total victory for the defending caterpillars. It may be, however, that this victory lasts only for a short span of evolutionary time before countermeasures are found by natural selection to unleash a fresh lethal attack.

Rathcke and Poole (1975) have shown that caterpillars of butterflies of the genus *Mechanitis* cope with the dangerous barbed-wire entanglements of trichomes on other plants with a communal spinning of a silken mat that lets them walk with impunity above the hooks, and thus to feed safely on the edges of leaves. The mechanism is an elegant demonstration of natural selection as the art of the possible and as an improviser. Virtually all butterfly larvae routinely spin silk as an attachment for themselves or their cocoons, and the main evolutionary step necessary for this answer to the trichomes is gregarious living that makes spinning a complete surface mat on the leaf a cost-effective undertaking.

So far this silk-matting method of attack

does not seem to have been applied by *Heliconius* against *Passiflora* trichomes. Possibly it never will be. But if it is, then doubtless a *Passiflora* will jury-rig a new defense in the fullness of evolutionary time, but what this would be we cannot predict. Naturalists of the future will have to discover the next stage in passion flower defenses and then rationalize the behavior with an after-the-fact cost–benefit analysis.

INCREASING DIVERSITY: A MAKESHIFT PROCESS

Adding diversity to preexisting richness is an endless process of makeshift adaptations being piled on top of one another. We can identify major themes in this process, themes that occur again and again, either because habitat properties make them likely or because the engineering or energetic designs of major classes of organisms make them necessary.

One such theme depends on how escape or opportunity often can be manipulated by restricting activities in time; ecologists call this "**phenology**." A second theme is that coexisting species interact through chemicals, either as toxins or as guides to behavior, letting us recognize a subdiscipline of ecology called "**allelochemics**," which loosely translated means "the chemicals one throws at another." Another grand theme is that of dispersal, as plants, being stationary, display an endless array of behaviors that coerce animals to transfer plant genetic material or disperse propagules.

Once the main phyletic lines of animals and plants were fixed long ago, whole families of interaction had become contingent on basic engineering plans. Birds should be largely workers by day, bats workers by night; spiders are constrained to hunt by entrapment on silken snares. But each set of engineering designs are permutable by phenology, allelochemics, dispersal, and even mimicry, to pile diversity on diversity.

Diversity Through Fragrance

The massed organisms of any community give it a chemical environment. Many chemicals in the environment are used as social cues recognized as signals by members of the same species, chemicals that are collectively called "**pheromones**." Males of some moth species can home in for several kilometers on pheromones emitted by an uncopulated female. Other organisms mark their trails chemically as ants do, territorial mammals mark their territories, and so on.

The chemical cues can be used by other animals, particularly by predators or parasites. Dogs hunt by scent. Perhaps more importantly, invertebrate predators and parasites do so also. A tick waits on branches at places marked by the facial scent glands of deer as a suitable point of embarkation (Rechar et al., 1978). Bark beetles (*Ips confusus*) use terpenes released in their own excrement (frass) as a homing signal to take advantage of a food supply found by other beetles. But the scent in this same frass is used by predators to find the beetles (Whittaker and Feeny, 1971). Some caterpillars expel their frass explosively, so that it falls clear of the leaf on which the caterpillar is feeding to the ground underneath, a habit best explained as a defense mechanism against parasitoid hymenoptera that track prey by the scent of frass (Rhoades, 1979). Into the chemical environment of intraspecific pheromones, therefore, come selection pressures for coevolution between species.

More Diversity Through Chemistry

The true allelochemics are compounds, the prime selective advantage for which comes from their influence on other species. Called by plant physiologists "**secondary compounds**," these are substances like the alkaloid toxins and the protein-fixing tannins for which no obvious physiological purpose can be found. A general hypothesis of ecology is that these compounds give selective advantage either in competition with other plants ("**allelopathy**") or as defense mechanisms in plant–herbivore interactions. The general hypothesis states that plant secondary compounds have been manufactured at some cost in fitness to the plant for a compensating benefit in fitness (Price et al., 1991).

Some examples of allelopathy are well known, like the impossibility of growing tomatoes near walnut trees in the American Midwest because of a chemical agent, called juglone, that is washed from walnut leaf surfaces to the soil (Bode, 1958). A number of examples are also known in pasture grasses, one mechanism of particular subtlety being that of the grass *Aristida oligantha*, whose exuded chemicals attack not rival grasses directly but nitrogen-fixing bacteria. This tactic maintains the habitat at a low state of productivity, which in turn prevents invasion by plants competitively superior to a pioneer *A. oligantha* (Blum and Rice, 1969).

More widely studied are the coevolutionary systems between herbivores and "secondary plant metabolites" (using the name for secondary compounds now commonly used in the literature of plant–herbivore interactions; Rosenthal and Janzen, 1979). That these substances can act against herbivores is demonstrated in feeding trials, both with natural foods and with artificial mixtures. Chemical and physiological data also suggest that heavy costs are accepted by plants to maintain these compounds because they tend to be autotoxic and thus have to be rendered chemically harmless within the plant, as when a glycoside is combined with glucose. The general hypothesis that these compounds are maintained at a cost for herbivore defense, therefore, is plausible.

Qualitative and Quantitative Chemical Defense

The defensive compounds work in two distinctly different ways. One class provides "**quantitative defenses**" by making food inedible. Tannins are of this class, working by being tightly bound to proteins (which is why they work for tanning leather). They make plant proteins largely indigestible to all herbivores, but their effect is dosage-dependent, since the plant must have sufficient tannin to render most of its protein inedible (Figure 15.4).

Other quantitative chemical defenses employ simpler compounds like silica, which is deposited in plant cells as **phytoliths** (literally "plant stones," likely to be bad for the munching or digesting apparatus), and the lignin of wood. Lignin appears to have the special advantage in its role as a defense compound in that it does not require chemical inactivation to prevent it from harming its producer. In addition, the cost of its manufacture can be allocated to other budget accounts, like competitive advantage through large size, as well as to the defense budget.

The second class of defense compounds is the active toxins, called "**qualitative defenses**." Compounds like the alkaloids of this class interfere with the internal metabolism of animals directly and, like other poisons, are effective in low concentrations. They can, however, be detoxified by intended victims that have the appropriate

FIGURE 15.4
Appalachian forest: Quantitative chemical defense.
Tannins make leaves scarcely edible to all herbivores because close-packed trees of the canopy have no cheaper defense in dispersal. Uneaten leaves send a large part of the flux of net productivity to the litter and decomposers.

chemical mechanisms. These facts lead to the many exciting sagas of coevolved systems of chemical attack and defense that have been discovered.

OPTIMAL DEFENSE THEORY

Optimal defense theory states that the secondary compounds are allocated as defenses in ways that maximize inclusive fitness and that such allocation always involves costs to the plant (White, 1969; Rhoades, 1979). It will be noted that the predictive power of this theory is constrained by species histories or mechanical possibilities, like all optimization theories in ecology, as discussed above. But the theory does lead to some general predictions that are testable.

First is the prediction that less-defended organisms have higher fitness in the absence of enemies. Many data from the agricultural literature support this. Different strains of crop plants have different toxic properties, and it can be shown, for instance, that nontoxic morphs of clover (*Trifolium*) grow and reproduce more vigorously than toxic morphs. Tobacco (*Nicotinia*) plants especially rich in alkaloids are stunted, suggesting that energy allocated to the manufacture of nicotine has been abstracted from energy available for growth (Whittaker and Feeny, 1971). Many pest problems of agriculture may result from our habit of selecting for rapid growth at the expense of chemical defenses.

Two other predictions are that chemical defenses should be allocated to different tissues in proportion to risk and defenses ought to be reduced when enemies are absent if possible. Many data can be understood in light of these predictions. Tannins are concentrated in leaves, which are prime targets for herbivore attack. Secondary compounds are transferred from leaves to seeds or other reproductive structures at the end of the growing season by annual plants, or at annual leaf-fall for perennials.

In a number of commercial forestry trees, secondary compound concentration depends on the history of pest outbreaks. In fir, pine, spruce, larch, and birch, resin and phenol concentrations increase after significant infestation by pests (Edmunds and Alstad, 1978). Plants stressed by drought tend to be more nutritious because they have less toxin, which would be expected if the plants under stress have fewer resources to allocate to toxin production. White (1969) suggests that some pest outbreaks in arid regions actually are brought on by the increased palatability of plants in droughts.

Chemical Defense as a Function of Plant Spacing

Optimal defense theory also predicts general circumstances in which qualitative, as opposed to quantitative, chemical defenses should be used. Scattered, opportunistic weeds receive considerable protection from herbivores by reason of their separation in space. They will, however, be subject to attack from various herbivores coming from neighboring plants of other kinds. A good defensive strategy in these circumstances is to use toxic, qualitative, chemical defenses, because these will be lethal to most attackers at minimal cost.

Although defenses against toxins can be evolved, the defense must be toxin-specific and is unlikely to be possessed by many of the varied herbivores finding scattered plants. We do indeed find that herbs and other early-succession plants tend to have alkaloids and other toxin defenses. It is on these plants that we find the intense coevolution systems between herbivore and host of the kind described above for *Heliconius* butterflies.

Large plants of climax communities growing closely together cannot defend themselves by spacing out and are easily found by herbivores. For these the theory predicts the quantitative defenses of tannins and the like that make the tissues inedible, because these defenses are almost impossible to overcome by biochemical means. Accordingly we find that forest trees have leaves defended by tannins (Figure 15.4). Eighty percent of woody dicotyledonous plants possess tannins, but only 15% of herbs do.

Large plants with quantitative defenses are said to be "apparent," and the scattered plants bearing toxins are termed "unapparent." From considerations of resource availability alone, we should expect that unapparent plants would be attacked by generalist herbivores because they are hard to find. Optimal defense theory, however, shows correctly that the unapparent plants should be attacked only by specialist herbivores able to find their scattered prey, because only specialists can coevolve the detoxification systems necessary to overcome the chemical defenses of unapparent plants.

CARRIERS OF POLLEN AND PORTERS OF SEEDS

Plants use animals in reproduction to transport both pollen and seeds. They also use wind and water for these functions, but probably most plants use animals some of the time, making mutualisms between plants and animals as prevalent as the predator–prey relationships of herbivory.

To exploit animals for pollination, a plant must provide both signal and reward. The signal must be appropriate to the sense systems of the animals used. Scents made from flavenoids and volatile terpenes are signals for insects for which olfaction is more important than sight, which is why flowers smell nice. Birds use visual cues, which is why bird-pollinated plants like orchids tend not to smell at all, although their shape and color may be remarkable.

The rewards offered by plants, often but by no means always nectar, vary considerably, suggesting that manipulation of the rewards provides a control on the animals visiting. A high-quality reward should have not only large volume and high sugar concentration, but also contain amines and other protein derivatives. Scattered plants that require long journeys by pollinators might be expected to offer large rewards (and do so), thus making visits cost-effective for animals (Baker and Baker, 1975).

But a rich reward of nectar in a scattered plant could be taken by visitors from nearby plants bringing the wrong pollen, a circumstance that leads to the somewhat surprising discovery that high-quality nectar typically is poisoned with toxins. Only specialized pollinators equipped with the appropriate detoxification mechanism, therefore, can win a large reward. Their toxin specializations then force them to serve the plant species with which they have coevolved.

Plants also manipulate pollinators by structure, as with the groove and pool nectar system, and spiral flower arrangements. Selection among pollinators is also arranged by flower geometry, making access difficult for all but the target insect.

But if plants must coevolve with their pollinators, they must also sometimes coadapt with neighboring plants with which the pollinator pool must be shared. A common result of such coadaptation is staggered times of flowering, itself one of the important mechanisms driving phenological events, particularly in tropical ecosystems. Figure 15.5 shows the times of flowering over two successive years of ten species of plants in a tropical forest, all of which are pollinated by hermit hummingbirds. It seems inescapable that the plants have been selected to avoid as far as possible flowering at the same time as others in the hummingbird-using guild. Since the incidence of rain, the most important environmental cue, is very variable in this system, it seems likely that some plants of

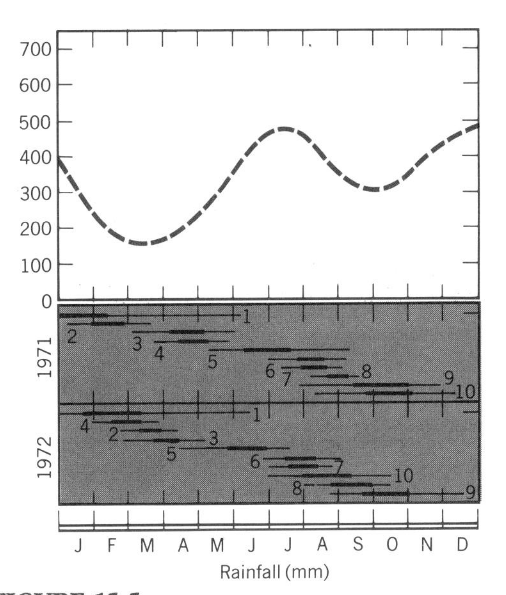

FIGURE 15.5
Phenology of ten species of *Miconia* bushes. Flowering times of the ten species shown for two successive years. All are pollinated by hermit hummingbirds. Flowering times are staggered, even though rainfall (top) is variable. (From Stiles, 1977.)

the guild pay a high cost in staggering their flowering to times that are not physiologically optimum. The value of reduced competition for hummingbirds, however, apparently makes this cost worthwhile.

A similar set of relationships exists between plants and their coevolved or coadapted seed dispersers. Signals, both olfactory and visual, must be accompanied by appropriate rewards, hence the soft parts of many fruits contain sugars and emit aromas. A common arrangement is for the fruit to be edible but the seed to be toxic. The edible fruit rewards visitors, and the toxins in the seed are an adaptation to reduce seed predation by less welcome visitors, notably insects. A tight web of coevolution between a plant and different sets of animals, some to encourage mutualism and others to discourage predation, can result. The logic that a particularly rich reward offered by a well-dispersed plant should be laced with toxins to discourage all but specialized visitors applies equally to fruit and pollen dispersal. An example is the poisonous tree *Hippomene mancinella*, whose apple-like fruits are temptingly luscious but depressingly poisonous to humans, though presumably not to the dispersing animal for which they were made.

Plants make even more devious use of animals in their reproductive endeavors when they rely on them to deactivate seed defense mechanisms emplaced against insect seed predators. This happens when seeds are provided with lignified coats of extreme toughness to resist insect boring, which cannot, as a consequence, germinate unless the seed coat is abraded, usually in the gizzard of a bird. It has recently been suggested that a tree endemic to the island of Mauritius, *Calvaris major*, is now nearly extinct because its very thick seeds relied on passage through the gizzard of a dodo before they could germinate. The tree was so closely coevolved with the dodo that extinction of the bird spelled extinction for the tree (Temple, 1977).

HOW TO USE ANTS

Ants are among the most abundant of animals, small as individuals yet living in colonies of thousands or millions whose aggregate mass may range from a few grams to 20 kilograms for a giant colony of African driver ants (Wilson, 1971). They patrol large areas, often aggressively, with diets ranging from carnivory to herbivory, and they make forays into such exotica as gardening. They tend to build nests full of organic litter or food stores. They live in social systems in which morphologically discrete castes achieve divisions of labor. The behavioral responses of individual ants, however, are few and simple: ability to follow a pheromone trail, recognition of conspecifics and food, and self-sacrificing aggression if necessary.

Ants turn out to be a resource to be exploited by other organisms, particularly as a patrol of ant sentries provides an excellent defense for a plant against herbivore attack. Many tropical trees (*Acacia* are particularly well known) both house and feed ants with this result. Special cavities in the branches are used by ants to build nests, and the tree has extrafloral nectaries (that is, nectaries on stems and outside the flowers) on which the ants feed. The ants are thus nurtured by the tree. Experimental removal of the ants with insecticides results in destructive attacks on the tree by defoliating insects, showing that the ants have the coevolved function of tree defense.

Travelers in the tropical rain forest know that it is unwise to rest a hand on any tree because of waiting ants, suggesting a generally close relationship between ants and trees. The symbiosis is not restricted to trees and tropics, however, and some temperate sunflowers also attract ants with extrafloral nectaries. The trumpet creeper (*Campsis radicans*) of the American Midwest also exhibits the "**ant-guard symbiosis**" with no less than four kinds of extrafloral feeding points for ants (Figure 15.6). Nectar for the ants in these trumpet creepers is quite different from the floral nectar used to attract hummingbirds and bees for pollination (Elias and Gelband, 1975).

If plants can be adapted to use ant sentries, then so can herbivores. Caterpillars of Lycaenid butterflies, among many others, are tended and herded by ants, which will carry the caterpillars from food plant to food plant. This transport by ants may be particularly valuable to Lycaenids, which are often food specialists on small or scattered herbs where the ants can take them. The caterpillars feed the ants directly for these services, giving them nectar-like secretions called honeydew. Lycaenid species seem to have coevolved with their own species of ants and their own specialist food plants. The symbioses may explain why the family Lycaenidae is one of the most species-rich of butterfly families (Gilbert and Singer, 1975).

More familiar to people of the temperate zone are ants that tend aphids for honeydew rewards. These systems may not be entirely at the expense of the host plants because the ants certainly serve to protect the plants against all herbivores other than the chosen one. These plants are in effect rewarding the ants for their defense services with honeydew channeled through the herbivores instead of making it themselves in extrafloral nectaries. Various systems probably exist, ranging from ones in which the herbivore load to the

FIGURE 15.6
***Formica* sp. feeding: payment to the guardians.**
The ants patrol the trumpet vine *Campsis radicans*, feeding from nectaries on the lower parts of petioles. This is an example of the ant guard symbiosis.

plant is small to others where the ant-enhanced herbivory is critical.

Army and driver ants also represent resources to be exploited by ingenious adaptations. They serve as lines of beaters from which other small animals, particularly arthropods, flee, leaving their shelters. Whole guilds of ant-shrikes exist, birds that follow army ants and feed on the insects that expose themselves. At the same time, some arthropods take advantage of the army ants more directly, marching with them, following the ant pheromone trails, and escaping attack from the ants either by mimicry, as with staphylinid beetles, or perhaps by chemical means (Figure 15.7).

These various adaptations to take advantage of the ants use their fighting prowess as a resource and their simple, stereotyped behavior so that they can be duped. Another resource offered by the ants is their untidy, food-rich nests with their own reproductive factories inside. An immense array of social and nest parasites of ants is known, as beetles, flies, and other organisms feed on ant litter, food reserves, and larvae, or actually induce the ant workers to feed them themselves, like insect versions of cuckoos (Wilson, 1971).

MIMICRY TRUE AND FALSE

A diverse community is a place of myriad signals. Pheromones are emitted, songs and chatters are sounded, and every movement or shape is a signal to watching eyes or listening sonar. Every organism will both give signals and receive them, and its survival may depend on how it does its signaling.

The evolution of mimicry is the evolution of false sets of signals, and the opportu-

FIGURE 15.7
***Calymmodesmus* sp. marching by pheromone with hosts.**
The millipede is symbiotic with the army ants (*Labidus praedator*) and follows their odor trails on the march.

nities for this in a diverse community are many. Mimicry can serve to hide the weak and eatable, to give cover to predators, and almost any purpose for which false signals can give selective advantage. The more diverse the community, the greater the opportunity for mimicry, which accordingly seems to be most prominent in ecosystems like the tropical rain forest or coral reefs.

Mimicry requires three actors: the model, the mimic, and the dupe. The mimic is a counterfeit copy that plagiarizes the model. The dupe is the enemy or victim of the mimic. Enemies and victims are perhaps more safely called "signal receivers," although the expressive term "dupe" is widely used by ecologists (Pasteur, 1982).

We know a disproportionate amount about mimics that counterfeit light signals, since we are so dependent on light ourselves. This bias can let us imagine that mimicry is more perfect on land than in the sea when we see marvelously exact copies of butterflies (Figure 15.8) and compare them with what seem to be crude copies of some marine organisms. But the butterfly copy must satisfy birds, which have eyes perhaps even better than ours, whereas the marine mimics might have to satisfy only fish eyes, which are not very good.

Much mimicry is mere camouflage in which the model is indifferent to the dupe. The many imitations of plant parts by insects that serve to hide both predator and prey are of this kind. But in some offensive mimicry the model may be agreeable to the dupes. This happens with the saber-toothed blenny, a fish that is a mimic of a cleaner fish (Figure 15.9). Cleaner fishes have a mutualist relationship with client fishes by taking ectoparasites from their bodies, and are thus allowed to approach within nibbling distance. The saber-toothed blenny is, however, a predator that bites the clients of the cleaner fish once its

FIGURE 15.8
The hypothesis of Batesian mimicry.
The model (monarch) is above, the mimic (viceroy) below. On the assumption that the viceroy is edible but the monarch is not, the dupe is any bird learning that a butterfly looking like these tastes nasty. Recent work, however, suggests that both butterflies taste nasty.

FIGURE 15.9
Treacherous mimic: disguised aggressor.
The saber-toothed blenny mimics a cleaner fish, is allowed to come close, and takes a bite of flesh instead of removing a parasite.

disguise allows it to get within nibbling distance of its dupe. A saber-toothed blenny is both a parasite of the cleaner fishes that it mimics and a predator of the client fishes.

Issuing false signals is a common tactic of predators, even down to the twitching tip of a cat's tail that may serve to distract the quarry's attention. A more dastardly distraction is offered by firefly females of the genus *Photuris*, which prey on males of another firefly, *Photinus macdermotti.* Male fireflies find females with light signals, the flashes of which are species-specific. The *Photuris* females mimic the flashes of *Photinus* females, thus luring *Photinus* males to a deadly embrace. He homes on the light expecting sex but gets eaten instead. (Lloyd, 1975).

Müllerian Mutualists with Danger Signals

The mimics with most fascination for biologists have always been the groups of species that seem to share warning coloration. The basic premise is that species with effective defenses can gain selective advantage by advertising them. Species that give warning are called "aposematic," as illustrated by the stripes on a wasp, the sound of a rattlesnake, or the brilliant orange of a monarch butterfly. The warning signal, once learned by the hunting bird or mammal, spares the conspicuous one from the damage of even an aborted attack.

If one species with a potent deterrent against attack develops an effective warning coloration, a selective pressure should then exist for another species with an equally potent defense to give the same warning signal. Letting the equally well-defended show the same markings should give selective advantage to all the animals involved. The predator needs to learn but one set of signals. All the defended prey species would reduce the number of attacks on the learning curve by their mutual use of the same warning signals.

For dangerous species to look alike, therefore, makes good evolutionary sense. This is the Müllerian hypothesis to explain the similarity of many brightly colored species. They are mimics of each other to the human eye, although the actual selection involved is that of convergence on a common signal from which all benefit. In a system of so-called **Müllerian mimicry**, the look-alikes are mutualists that give similar signals that fool nobody. The signal receivers are not dupes, but instead are animals that have no call to tell the aposematic ones apart.

Ecologists as Batesian Dupes

Intellectually more fascinating was the hypothesis of the nineteenth-century naturalist Henry Walter Bates, who suggested that a harmless animal might gain immunity from attack by mimicry of the well-defended. Look like the uneatable or dangerous and you will be safe.

In the Batesian hypothesis the mimic is eatable, the model is poisonous, and the dupe is a vertebrate capable of learning that the model is unpleasant to eat. The coevolved species of tropical America for which Bates put forward the idea are monarch butterflies of the family Danainae and butterflies that look like them (Figure 15.8). Monarchs and their kind store toxins—bitter-tasting heart poisons that are also powerful emetics (Brower and Brower, 1964). Birds quickly learn to avoid the danaid butterflies after a few unfortunate tries. The monarch's spectacular orange and black fashion statement is thus aposematic.

Bates suggested that look-alikes of the monarch, such as the unrelated viceroy butterfly (Figure 15.8), were perfectly edible "sheep in wolf's clothing." They were warningly colored although not themselves dangerous. In this **Batesian mimicry**, therefore, are dupes a'plenty, both the receivers of false signals that missed out on a meal, and the models whose own immunity should be diminished when signal receivers found some of the targets good to eat after all. A Batesian mimic should be a parasite of its model.

That look-alike species of butterflies should indeed be Batesian mimics was encouraged by the discovery that toxins stored by monarch butterflies had been sequestered by their caterpillars from their milkweed food plant, but that the look-alikes fed on plants without comparable toxins. Thus it was reasonable to expect that the mimics could collect no toxins and were all safe to eat.

Probably every beginning biology textbook now describes Batesian mimicry as an elegant example of coevolution, in which a harmless mimic copies a warningly colored model for protection against a dupe. Yet the hypothesis has long held glaring deficiencies. The mimic was engaged in a bluff with no stick to back it up, and it acted as a parasite of the model (Vane-Wright, 1991). Surely natural selection should work to let the so-called dupe see through the ruse, at the same time as the host model in the proposed "host parasite" system was selected to be distinguishable from its parasite's mimicry. The bluff should have been called. We now know, in fact, that the warning color of the viceroy was not a bluff. The viceroy is poisonous too (Ritland and Brower, 1991).

True, viceroys do not sequester toxins from their food plants. Instead they synthesize their own toxins. And having done so, their evolutionary convergence on the monarch's warning system becomes understandable as simple Müllerian mimicry. The real error that allowed Bates's hypothesis to persist for a hundred years was the belief that insects primarily acquired their toxic defenses by sequestering plant alkaloids, with its corollary that all insects that fed on nontoxic plants were themselves necessarily nontoxic also. Far more likely is the hypothesis that insects develop their own toxins just as plants do. The habit of monarchs and some other insects of sequestering plant toxins is likely to have been a secondary adaptation of insects already equipped with toxin-handling chemistry, and which therefore were preadapted to feeding on toxin-laden plants (Brown et al., 1991).

Preview to Chapter 16

Amazonon diversity: why so many kinds of plant?

THE first modern attempt to explain the number of species, the Santa Rosalia model, assumed that local resources determined how many species could be packed into a habitat. More successful have been subsequent hypotheses that rely on allowing more species to coexist because successful predators or recurrent environmental hazards prevent successful species from preempting resources through competition. This "cropping model" is most powerful in accounting for high diversity in partly barren systems like the Sonoran desert or arid land pastures, but also may account for high tree diversity in tropical rain forests. For the more numerous organisms like insects, understanding diversity is severely handicapped by the fact that most species are not yet described and so cannot reasonably be counted. A second handicap is the practical difficulty of measurement, because many species are rare. A wholly satisfactory statistic that conveniently allows quick measurement of species richness and relative abundance at the same time is not yet available. Despite these difficulties of documenting local species richness, it is clear that a cline of species richness from equator to poles is a characteristic of the earth in all geological epochs, and theorists have concentrated on explaining this cline. Correlates of species richness with productivity, seasonality, land area, and other correlates of latitude have not yielded generally satisfactory explanations of the cline. Missing so far are accepted hypotheses that demonstrate how these correlates of latitude could affect speciation rates.

Chapter 16

Toward a General Theory of Species Diversity

For many of us in the profession, the grandest task remains to explain why there are so many species. We do realize that, in one sense, the task is impossible, because of the randomness in the evolutionary system. The earth has not always held the same number of species. Mass extinctions at intervals of tens of millions of years leave the earth depauperate until the speciations of millions of years more make the earth diverse again (see Figure 9.12). At no time in geological history can we assert that the earth was "full" of all possible species.

But we also have good reasons for thinking that numbers of species are far from purely random, the most exigent evidence being the enduring geographical patterns of species diversity. Oceans have fewer species than the land; dry places less than wet; small areas less than large areas. Most striking of all is the decline of diversity from equator to poles. Tropical ecosystems are richly diverse, both the tropical rain forests of the land and the coral reefs of the sea. Ecosystems of the arctic are species-poor, and a long gradient of diversity runs between these extremes. Fossils show this cline from equator to poles to have been present in all geological epochs (Figure 16.1). Constant verities, therefore, establish the relative numbers of species that can coexist, showing that a general theory for the limits to species diversity should be possible.

We are hampered by our gross ignorance of the number of species actually inhabiting the contemporary earth. We have cataloged only about 2 million species, but suspect the actual number to be anywhere from 5 to 50 million (Wilson, 1988; see Chapter 1). Furthermore, we look out on

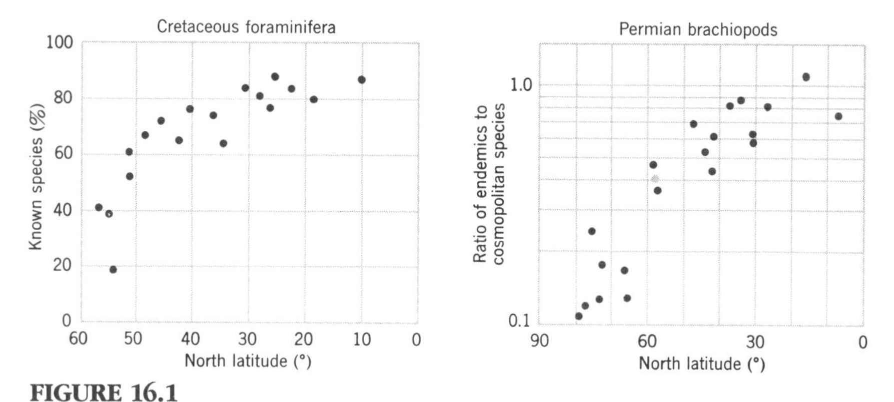

FIGURE 16.1
Diversity and latitude in the geological past.
(*a*) Planktonic foraminifera in the Northern hemisphere. (*b*) Family diversity of Permian brachiopods. (From Stehli et al., 1969).

the earth at a time of geological upheaval and differential extinction. For more than a million years the earth has been subject to oscillating ice ages, with no end in sight. And for the last 10,000 years or so a plague of intelligent hominids has progressively removed the larger animals from all the ecosystems of the earth, undoubtedly with profound consequences for the numbers of smaller species (Terborgh, 1988).

Yet the grand patterns of diversity are still plain to see. If we can explain them, then we shall be far on the way toward a general theory. Also, although our ignorance of the numbers of all species combined is abysmal, our knowledge of larger taxa like birds is respectable. We can, therefore, test hypotheses of causes of diversity against geographic patterns, using the taxa that we know best.

THE SANTA ROSALIA PAPER

For those of us now growing gray in the service of ecology, our pressing urge to know the limits to species numbers can be traced to a paper by G. E. Hutchinson (1959), in which he summarized the state of theory on the causes of diversity. Hutchinson wrote the paper in Palermo, Sicily, and, in a fit of scholarly whimsy, dedicated it to Palermo's patron saint, so that the paper ever since has been called the Santa Rosalia paper.

In Santa Rosalia, Hutchinson tried to answer the question, "Why are there so many different kinds of animals?" by assuming that food energy set the limit. This was the period when the energy flow paradigm had finally become established as an important intellectual tool in ecology (see Chapter 2). It was reasonable, therefore, to seek in energetics an answer to the number of species. Energetics should limit complexity of food chains. Competition for energy should decide how many niches were possible and hence how many animals could coexist.

The possible number of niches is multiplied in a food chain by as many times as there are links in the chain. When a herbivore species results from natural selection, an opportunity for a carnivore niche is at once created. Appearance of a carnivore then creates the opportunity for another niche at the next trophic level, and so on. But this multiplication of niche number obviously is narrowly constrained since most food chains have only three to six links.

Hutchinson recounted the energy flow arguments restricting the lengths of food chains (see Chapter 2), noting that an animal at the fiftieth link of a food chain would have a population of 10^{-49} times the population of the first link, assuming an unrealistically high ecological efficiency of 20% and a modest doubling of individual size with each link in the chain.

Actually, as Hutchinson argued, food chains may be even shorter than is decreed by thermodynamics in this way, because selection works to compress the chains. Selection at the highest trophic level should always work to favor the most effective predator, which will therefore be in constant danger of exterminating its prey. If the animal at the nth link exterminates, or seriously depletes, the $(n - 1)$th link, then it must resort to cannibalism or concentrate on the $(n - 2)$th link. In this way, selection will work to shorten food chains as well as to lengthen them.

The Santa Rosalia paper thus recognizes that increasing diversity through food chains is not limited only by the resource energy available. Energetics may set an upper limit to the number of links, but this is unlikely to be reached. The actual number of links probably results from an equilibrium between the rate at which food chains are both shortened and lengthened by selection.

NICHE-PACKING

Having shown that the possibilities of multiplying diversity through food chains were strictly limited, Santa Rosalia then explores the possibilities for dividing energy between pairs of competing species. This was an argument of "niche packing," and was the crux of the Santa Rosalia argument.

Crucial to the Lotka–Volterra–Gause model of speciation is that, when allopatric populations are brought together to live in sympatry, they can do so only if their niches are separated by some minimal distance (see Chapter 9). Hutchinson proposed that this minimum separation of niches might set a limit to the number of species. If evolution has had long enough to saturate the earth, then these finite differences would determine how many species there could be.

Studies of closely related sympatric species (see Chapter 9) had shown that animals could divide resources by different behavior or through having beaks or jaws of different sizes that let them feed on food of a specific size or shape. In Santa Rosalia, Hutchinson reasoned that if species had evolved to the possible limits of similarity, then these limits would be reflected in dimensions of the feeding ("trophic") apparatus. Table 16.1 gives his data for skull length of pairs of mammal species and bill lengths for pairs of birds. The mean ratio of lengths between sympatric pairs is 100 to 128, and the variance from this ratio is not large. Finding these rather similar ratios in the size of trophic structures between pairs of sympatric birds and mammals gave encouragement to the idea that they had evolved to be as similar as possible and that many communities might in fact be saturated with bird and mammal species.

These arguments were applied in Santa Rosalia only to animal numbers. Hutchinson was careful to keep the argument to animal species only, though acknowledging that a prime cause of animal diversity was, of course, plant diversity. If every plant species had the potential of being the base of its own food chain, then a prime determinant of the number of kinds of an-

TABLE 16.1
Trophic Measurements of Closely Related Species in Allopatry and Sympatry
Measurements are skull length of female mammals and culmen length on birds. (Modified from Hutchinson, 1959.)

Species	Measurement when Sympatric (mm)	Measurement when Allopatric (mm)	Ratio when Sympatric
Weasels			
Mustela nivalis	33.6	34.7-36.0	100:134
M. erminea	45.0	41.9	
Mice			
Apodemus sylvaticus	24.8	25.6-26.7	100:109
A. flavicollis	27.0		
Nuthatches			
Sitta tephronota	29.0	25.5-26.0	100:124
S. neumayer	23.5		
Darwin's finches			
Geospiza fortis	12.0	10.5	100:143
G. fulginosa	8.4	9.3	
Camarhynchus parvulus	7.0-7.5	7.0-8.0	100:128:162
C. psittacula	8.5-9.8	10.1-10.5	100:127
C. pallidus	11.2-12.6	10.8-11.7	
			Mean ratio 100:128

imals was the number of kinds of plants. However, the arguments for niche separation should be directly applicable to plants also, though the important niche axes are not so easily measured as is, say, beak length in birds.

With the hindsight of knowing about how coevolution works to accommodate increasing numbers of species (see Chapter 15), it is clear that the Santa Rosalia conception of numbers of species being limited principally by interspecific competition or lengths of food chains does not allow for all the possibilities. Species can also be packed together through the dynamics of attack, defense, cooperation, mutualism, timing, or dispersal. That niches may be packed together in different ways does not, however, invalidate the argument that a limit to numbers of species in any ecosystem requires a limit to the number of niches that can be packed together.

THE PROBLEM OF MEASUREMENT

Counting the number of species in a place is vexingly hard, because most species are uncommon. A trained investigator can quickly list the common species, but soon the pace of discovery falls as she patiently searches out the less common. Most species are comparatively rare, so the patient search continues to be rewarded. Perhaps additional species would continue to be found if the search went on for ever. When do we stop counting? How can we know what local diversity is?

This difficulty of counting species reveals diversity actually to be in two parts: number of species and relative abundance. A general theory of diversity,[1] therefore, must explain relative commonness and rarity as well as absolute number of coexisting species. Ecologists call the actual number of species present "**species richness**" and different relative abundance the "**equitability**."[2]

Consider two simple communities, each with only one trophic level and each with 100 individuals spread among four species A, B, C, and D.

Community 1 has	25 A	25 B	25 C	25 D
Community 2 has	5 A	5 B	1 C	89 D

Exactly the same species list can be constructed by diligent collecting in each of these two communities, yet they clearly are very different. Both have the same species richness, but community 1 is more equitable than community 2. As a result of the different equitability, however, anything less than a complete census of community 2 would describe its species richness as less than that of community 1.

Comparing diversities of different communities would be easier if we could devise a statistic that measured both species richness and equitability at the same time. No generally satisfactory statistic has been devised, however, although a number have been proposed. Of these the most widely known is the Shannon–Wiener[3] information theory index, H', where

$$H' = -\sum p_i \log p_i$$

where p_i is the proportion of the total number of individuals in the ith species. This actually serves as a statistical measure of the probability of guessing the identity of an individual taken from a sample at random, and ecologists have come to realize that the Shannon–Wiener index has serious statistical shortcomings (Hurlbert, 1971; Goodman, 1974). Table 16.2 shows a range of calculations of H'.

TABLE 16.2
Sample Calculations of the Shannon–Wiener Diversity Index, H'
Notice that to produce strikingly different values the "communities" are given extremely different relative abundances. (From Price, 1975.)

No. of Species	Species 1	Species 2	Species 3	H'
2	90	10		0.33
2	50	50		0.69
3	80	10	10	0.70
3	33.3	33.3	33.3	1.10

[1]The word "diversity" comes from the Latin words *dis* and *vertere*, which put together can be read as "to turn away," hence stressing the difference between objects. When we talk of roads "diverging" we build on the same Latin roots.

[2]Ecologists have difficulty defining equitability in a way that is applicable to all systems. Whittaker (1977) draws on Lloyd and Gelardi (1964) to offer the comprehensive but cumbersome definition of equitability as the "relative evenness of the importance values of adjacent species in the sequence from least important to most important."

[3]The Shannon-Wiener index often is called the Shannon–Weaver index because the most accessible review of its mathematics is in a chapter by Shannon in a book by Shannon and Weaver (1949). This is incorrect, Wiener having published the statistic independently.

The older Simpson's index, λ, may be the safest alternative to the Shannon–Wiener index when making a simple species list is considered an insufficient measure. The index

$$\lambda = \sum p_i^2$$

is the probability that any two individuals picked at random will be of the same species. The index is thus a measure of how individuals in a sample are concentrated into a few species, for which reason this is sometimes called a dominance index. It is, however, an inverse measure of diversity.

A semantic tradition among some ecologists is to distinguish between the terms "diversity" and "species richness," arguing that "diversity" should be reserved for formal measures, like H' or Simpson's index, which combine equitability with richness. However, this distinction may have had its day. Measures of species richness alone, S, with control for equal sampling effort, serve to compare different communities as well as the various indexes (Grigg and Margos, 1974), and without the possibilities for being misled inherent in some indexes (Hurlbert, 1971; Pielou 1975). The advice of Connell (1978) seems sound, to use S, species richness as the general measure of diversity.

DIVERSITY CONSEQUENCES OF STRESSFUL ENVIRONMENTS

Hot springs, salt flats, mountaintops, and the like are rigorous or stressful habitats with few species (Table 16.3). They do, however, support some species, inviting the question, "Why not more?" The basic answer we give depends on these habitats' being rare, scattered, and sometimes ephemeral. Migration to one of these places from another of its kind is difficult and slow. At the same time, small size of the habitats (or extreme unproductivity) makes population density low, so that extinction rates are high. If arrival, whether by immigration or speciation, is slow and local extinction is high, then the number present at any one time is bound to be low.

TABLE 16.3
Diversity in Rigorous Habitats

Low diversity, low population densities
Hot springs
Salt flats
Caves
Mountaintops
Extreme sandy deserts
Low diversity, high population densities
Brackish estuaries
Abandoned fields
Polluted lakes
High diversity, variable population density
Moderate deserts
Serpentine soils (with toxic ions)
Overgrazed pastures in arid regions

This answer to low diversity in small areas of unusualness (stress by definition) thus depends on an equilibrium between arrival and departure. This is quite different from the niche-packing arguments of the Santa Rosalia model, in which habitats were postulated to fill with species until the process of adding more was resisted. These small areas of stressful habitat are in effect outside the game of filling the earth, acquiring their species lists as a function of the species richness of the larger world around them.

The dynamics setting species number as an equilibrium between immigrations and extinction now has its own body of theory known as "island biogeography" (see Chapter 21). Low diversity in patches of rigorous environment, like a hot spring in a verdant continental expanse, results from similar "island" effects.

More familiar or widespread patches of low diversity are brackish estuaries, polluted lakes, and abandoned fields (Table 16.3). Like hot springs, these places are islands of unusualness in a sea of normality. All three are stressed by frequent changes of state, as in a tidal estuary, or by a short existence, as forest closes over an old field or pollutants are buried in lake mud. Extinction rates, therefore, should be high and the equilibrium species number low. But they differ from the hot spring group by being highly productive, which invokes quite different species limits.

Because old fields or polluted lakes are highly productive, some species populations can be so dense as to secure a competitive advantage. In a polluted lake, for instance, high nutrient loading lets a few species of algae attain very high population densities. Diversity is low in the sense that nearly all individuals belong to a few species, but it is likely that a very intensive sampling effort will discover rare individuals of many other species. Thus, when a habitat is both unusual and extremely productive, competition can keep apparent diversity below what it would in a place of more normal productivity.

ENHANCED DIVERSITY THROUGH RESTRAINT OF COMPETITION

Predators that reduce the numbers of their prey allow more species of the prey trophic level to coexist. This was introduced in Chapter 14 as the "cropping principle," and was classically demonstrated by the starfish removal experiment in the rocky offshore environment of the State of Washington. Likewise, overgrazing a pasture with a guild of grazing animals increases diversity when the heavy grazing opens up the pasture to invasion by new kinds of weeds.

Any process that would reduce the population of resident organisms should have this same effect of removing the blocks to invasion imposed by successful competitors. Weather and seasonality could work as well as predators to produce this same effect, because they can impose wide fluctuations on short-lived organisms like insects or annual weeds (see Chapter 10). In variable weather, a community must frequently be open to invasion as resources are left unused by local populations that are far below crowding densities.

Openness to invasion also follows as a necessary consequence of density-dependent population dynamics, when this leads to irregular population histories. When density-dependent interactions are strong, the effect on populations is not constancy of the kind suggested by the old logistic model but either oscillations or irregular population histories indistinguishable from chaos (see Chapter 10; May, 1986b). The population lows of these histories should remove competitive barriers to invasion, even as an efficiently cropping predator does.

Thus diversity must always be a function of any process tending to reduce populations below levels set by energetic considerations alone. As populations are reduced by predators, the weather, or their inherent dynamics, so resources are freed to support additional species populations. The equitability component of diversity should, perhaps, be most affected, as rare organisms are allowed to become common. But species richness itself should also be enhanced by systems that make communities vulnerable to invasion, whether by immigration or by speciation.

SPECIES RICHNESS OF ARID LANDS AND TOXIC SOILS

Broad regions of the earth have high diversity but to the human mind seem to be rigorous enough (Table 16.3). The Sonoran desert, toxic soils frustrating to agriculture, and ruined pastures of arid lands all contain many species despite stress by drought or chemistry.

The arid lands are "marginal" in the sense that they rank between well-watered "mesic" habitats and barren lands such as complete deserts. Typically they are supplied with rain only intermittently or are otherwise strongly seasonal. Part of the high species diversity of arid lands, therefore, can be explained by the restraint of competition argument developed above. Space is not always preempted by resident plants so that invasion by seedlings following rain is always possible. The desert blooms with a diversity of flowers seen at no other time.

But these marginal habitats have in addition a biota of long-lived equilibrium species. Desert bushes and small trees live spaced out, probably reflecting competition for water and suggesting a stable population equilibrium (Figure 16.2). This arrangement leaves an open canopy and bare ground that can be invaded by weeds able to complete their life cycles in the short weeks of a chance favorable season. Marginal desert communities, therefore, add a resident and unique flora of equilibrium plants to their large number of opportunists. They win high diversity in both ways, and their consequent spectacular species richness is a result.

High diversity of plants on bizarre soils like those on serpentine rocks can be explained by an analogous argument. Only generalist weeds have sufficiently broad tolerances to live there. These suffer high local extinction, but they are always resupplied from surrounding habitats. In short, the unusual chemistry lets opportunists live with minimal competition: chemistry is the cropper.

BETWEEN-HABITAT AND WITHIN-HABITAT DIVERSITY

An elementary observation of natural history is that searching ever-larger areas yields ever-larger lists of species. The explanation of this result is intuitively obvious: the larger search takes in new habitats with new species included. But this leads to another obvious inference: some species share a habitat whereas other species live in separate habitats. Continually increasing a species list always depends on sampling ever more habitats with ever more species in them.

Small, uniform habitats must be shared, but relatively large patches of land may be partitioned among species. MacArthur (1965) introduced convenient terminology to describe these two conditions: "**within-habitat diversity**" and "**between habitat diversity**." Earlier Whittaker (1960) had pointed out the same essential dichotomy, calling the two conditions "**alpha diversity**" and "**beta diversity**." If a large empty habitat is colonized by animals with considerable resource plasticity, they will share all parts of the habitat, resulting in an initial within-habitat diversity (α diversity). But if more colonizing animals arrive, the greater crowding might well cause the habitat to be divided into subhabitats by species keeping themselves apart. Between-habitat diversity (β diversity) has now been added to the original within-habitat diversity.

Cody (1975) extended the classification for bird species diversity to allow compar-

FIGURE 16.2
***Carnegiea gigantea:* spaced out in a desert.**
The saguaros in Arizona are more evenly spaced than random (over-dispersed), demonstrating intraspecific interaction, probably competition for water. Theirs is an equilibrium strategy for the desert, but the bare ground between can be used by opportunists when it rains.

ison among whole geographic regions. "Point diversity" describes the complete overlap of bird ranges over very small areas. "Gamma diversity" describes the species replacements that occur over very large geographic regions. Adding these two concepts results in the following scheme:

within-habitat	*Point diversity*	=	found together in very small samples
	Alpha diversity	=	found together in small homogeneous habitats
between-habitat	*Beta diversity*	=	diversity across a variety of habitats
regional	*Gamma diversity*	=	regional diversity including geographical replacement

Diversity on Three Continents Compared

Almost certainly, the most complete of our world species lists is for birds, and it is to birds that we must look for such large undertakings as comparing diversity among continents. Figure 16.3 compares double logarithmic plots of species–area curves for the bird faunas of California, Chile, and South Africa (Cody, 1975). These three regions have much in common, with Med-

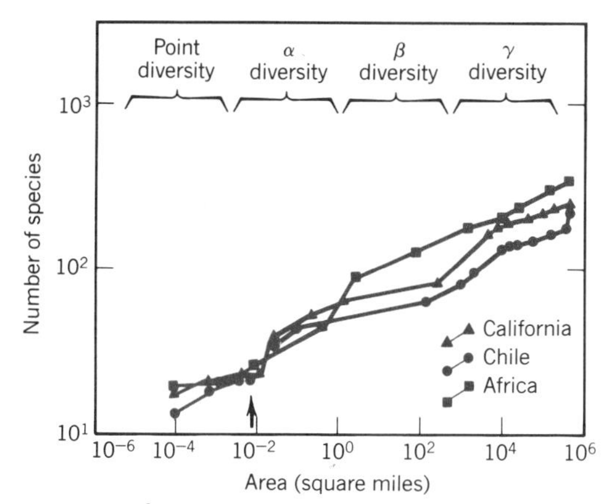

FIGURE 16.3
Species–area curves on three continents.
Data are numbers of species of birds of broad-leaved sclerophyllous vegetation (chaparral) in California, Chile, and South Africa. (From Cody, 1975.)

iterranean or coastal climates, areas of chaparral-type vegetation, and mountainous relief. They have also been isolated from each other for so long that their bird faunas are taxonomically quite different. The three regions thus provide us with a virtual experiment in comparative bird diversity, in which similar regions were filled with birds three times over.

The slopes of the species–area curves differ slightly within the three regions over different increases in area, showing that the determinants of diversity are not strictly the same on each of the three continents. Nevertheless, species richness of similar areas on each continent are in the same order of magnitude. This is true despite the fact that different species live on the three continents. Evidently bird species diversity is set in a general way by processes acting in the same way independently on all three continents.

Diversity on the three continents was most similar when measured for areas of about 10 square miles, suggesting that within-habitat (α) diversity is similar within these three southern continental regions. **Point diversity** (left of the curves) varies, however. This is confirmed by field data that show Chilean birds to have more interspecific interactions and less tolerance of overlapping territories than the others. Chilean birds are thus qualitatively different from birds in California or South Africa, a perfectly plausible result since they come from different phyletic stock. The resulting difference in point diversity is another of the results produced by natural selection that is contingent on traits present in ancestral populations.

The continental species–area curves diverge most sharply for larger areas, revealing that both between-habitat (β) and regional (γ) diversities differ somewhat from continent to continent. A parsimonious explanation is that this reflects different topographic diversity of the landscapes of Chile, California, and South Africa, but evolutionary contingency may also be involved since many of the birds are quite unrelated. Even on this regional scale, however, the similarity in species richness is more striking than modest differences fairly regarded as variations on a common theme.

That comparable areas on different continents, with unrelated birds, should have comparable species richness gives strong encouragement to the hypothesis that natural limits to species number do indeed exist, at least for birds. This conclusion is encouraged by Cody's (1975) further observation that increasing the topographic diversity or climatic region included in the sampled area increases the number of species in collections from each of the three continents. Whatever sets bird species richness seems global in its action.

DIVERSITY CLINES WITH LATITUDE

The general proposition of a latitudinal cline of diversity from the high Arctic to the equator is undoubted, and data are easily marshaled for its support. In the state of Ohio, for instance, about 2000 species of vascular plants are known, whereas in the Ohio-sized country of Ecuador about 20,000 are believed to occur. The 300 species of largish trees recently reported in a single hectare of Amazon forest (Gentry, 1988) is in the order of the combined lists of tree species from temperate North America and Europe combined.

With many animal groups, the latitudinal clines also appear self-evident. The large array of frogs and toads in the tropical rain forest can be contrasted with a fraction of that number in the U.S. Midwest, three species on the island of Britain, and none in the Arctic. All the primates are tropical, reptile diversity seems to be correlated with latitude, most bats and birds are tropical, and all naturalists suppose that the place to find lots of insect species is the tropics. Accurate documentation of the cline of insect numbers from the tropics is not possible because of the parlous condition of the insect catalog. Most of the 3 million to 40-some million species still to be described are insects, and most of those are thought to live in the tropical rain forests (Wilson, 1988).

Thus the cline of diversity with latitude is real and dramatic, even though our catalog of the living species is not adequate for proper documentation. The cline appears to be a permanent feature of life on earth as revealed by the fossil record (Figure 16.1).

The cline is one of total species richness, not of richness within related taxa. Phyletic lineages can be adapted to the environments of different latitudes so that within a lineage the latitudinal cline may not appear. Bears, seagulls, shorebirds, and seals are either confined to high latitudes or have their greatest diversity there. Distributions of these animals is contingent on their evolutionary history, which makes them variations on a theme of arctic living. Their higher diversity toward the poles in no way conflicts with the observation that species richness of all animals combined is greatest toward the equator.

Evolutionary accident has also been important for local diversity within related species. Tropical Africa has between 50 and 100 species of palm trees, but more then 1100 palms are known from South America. The New World island of Jamaica has as many species of orchid as the whole of Africa (Richards, 1973). Different genera, families, or even orders have flourished on different continents and at different latitudes. The cline of all plants, all trees, all herbivores, all carnivores, etc. combined remains.

The cline of total species richness from equator to pole remains one of the most striking facts about the distribution of life on earth. In devising hypotheses to explain this cline in richness, two other diversity clines are important, those on high mountains and in the deep sea.

Diversity Clines on High Mountains

The decline of diversity with altitude on a mountainside is readily demonstrated. Figure 16.4 gives data for one cline in plant diversity up high mountains in Arizona. The transect starts at about 700 meters in hot desert, passes through the Sonoran, and thence on through successive belts of live oaks and pines to a high alpine forest with Douglas firs (*Pseudostuga*). The data show an overall decline of diversity with altitude but with a large peak at middle elevations coinciding with the species-rich Sonoran desert.

Figure 16.5 shows a longer cline in plant diversity encountered in an ascent of the Himalayas. In this run there is no aberra-

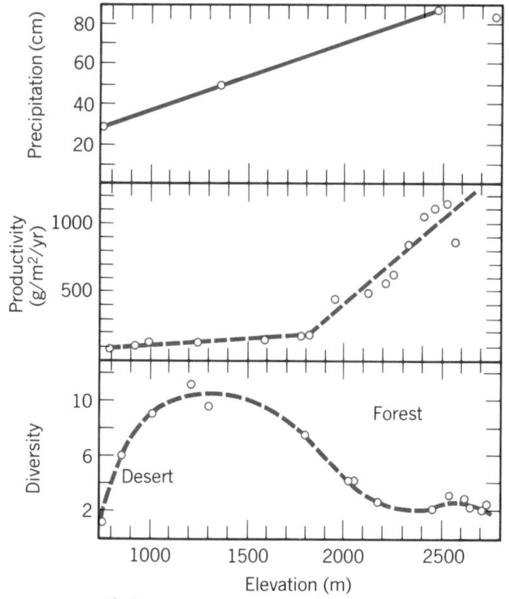

FIGURE 16.4
Decline of plant diversity with altitude on Arizona mountains.
Diversity is expressed as an index that relates species richness to relative productivity as well as relative abundance, but a plot of simple species richness would be of similar shape. (From Whittaker, 1977.)

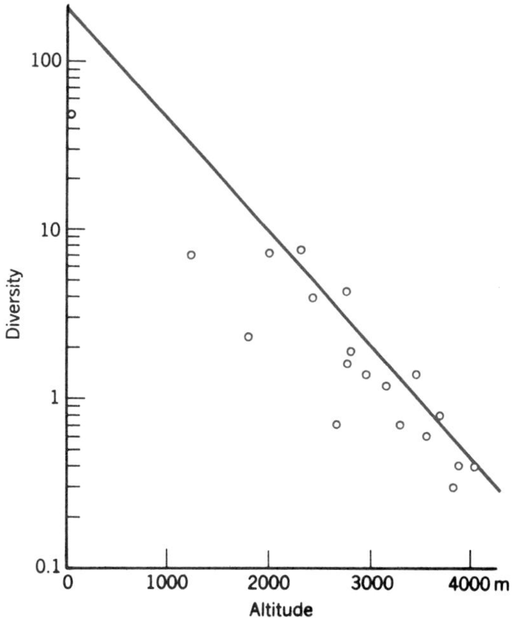

FIGURE 16.5
Decline of plant diversity with altitude in the Himalayas.
Diversity is expressed as an index allowing for relative size of trees based on measurements of breast height diameter (dbh). (From Yoda, 1967.)

tion of a seasonal, high-diversity desert like the Sonoran to interrupt the steady decline in diversity from the North Indian Plain to the tree line.

Diversity Increases with Depth at the Bottom of the Sea

A truly remarkable cline of diversity exists in animals of the bottom mud of the oceans, where data suggest a cline from low diversity in shallow water to high diversity in deep water. Figure 16.6 shows data for the benthic fauna that can be sampled with a grabbing device lowered from a ship to the bottom mud. More species can be collected in this way from the deep-sea floor than from continental slopes, shelves, or estuaries. The cline is

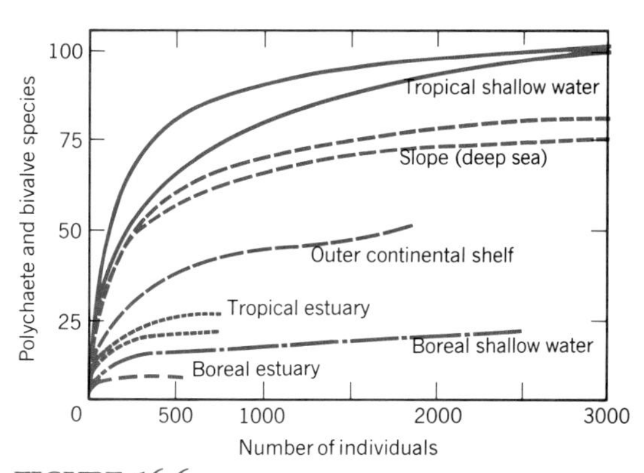

FIGURE 16.6
Increase of diversity of marine benthos with depth.
The deep sea mud of the abyss has more species of polychaete and bivalve than any other ocean bottom except tropical shallows. (From Sanders and Hessler, 1969.)

intriguing because diversity is low in the well-lighted, warm, productive bottoms of shallow water, but high in the dark, cold, unproductive bottoms of the deep sea (Sanders and Hessler, 1969).

The animals in the samples are collectively known as "**infauna**," because they live actually buried in the bottom mud. All are detritus feeders, relying entirely on debris from the lighted life zone overhead. The energy flux to the abyss from the fall of debris is extremely small and the bottom waters of the oceans are always cold; apparently a poor place in which to live. And yet the data show that the infauna of the deep sea is species-rich, certainly much richer than surface mud under productive fisheries of temperate coasts.

As Figure 16.7 shows, only in the tropics does species richness in the shallows equal species richness at depth. On all other coasts a steep cline of infaunal diversity runs from few species in the shallows to many species in the deep sea.

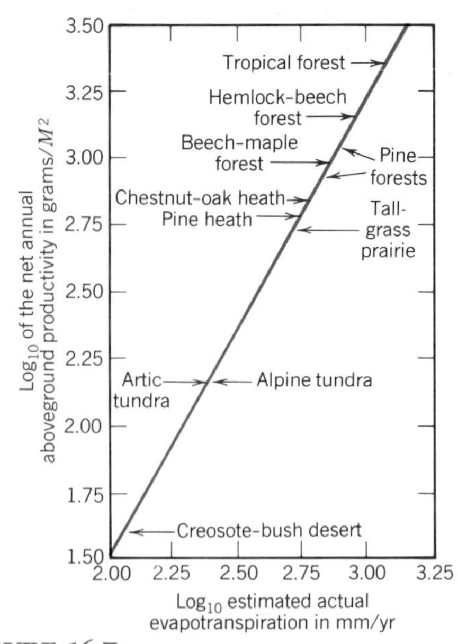

FIGURE 16.7
Indirect correlation of primary production with temperature.
Evapotranspiration depends on moisture and temperature. (Diagram of M. Rosenzweig from MacArthur and Connell, 1966.)

THE PRODUCTIVITY HYPOTHESIS

Correlation between high diversity and high production seems obvious on a global scale. At one extreme, the richly productive rain forests are highly diverse; at the other, the barren tundras of the Arctic or mountaintops have comparatively few species (Figure 16.7). Energy supplies set the size of populations: Why, therefore, should not energy also set a limit to species richness?

High productivity should mean a larger total population of individuals, and hence the chance to divide the available energy among more species populations. More energy for plants can mean dense local populations of the kind required to promote between-habitat diversity. Diversity of plants provides a diversity of niches for animals, thus increasing animal diversity in every trophic level.

At the equator, the absence of winter cold means that plants and animals need not budget energy for systems to get them through the winter, thus freeing up still more production. Thus life in low latitudes can produce year-round, while having to meet minimal costs for year-long maintenance. It might seem almost self-evident that equatorial life should be the most diverse (Connell and Orias, 1964; Figure 16.8).

The productivity hypothesis has gained appeal recently from studies demonstrating a particularly clear correlation between species richness and net primary produc-

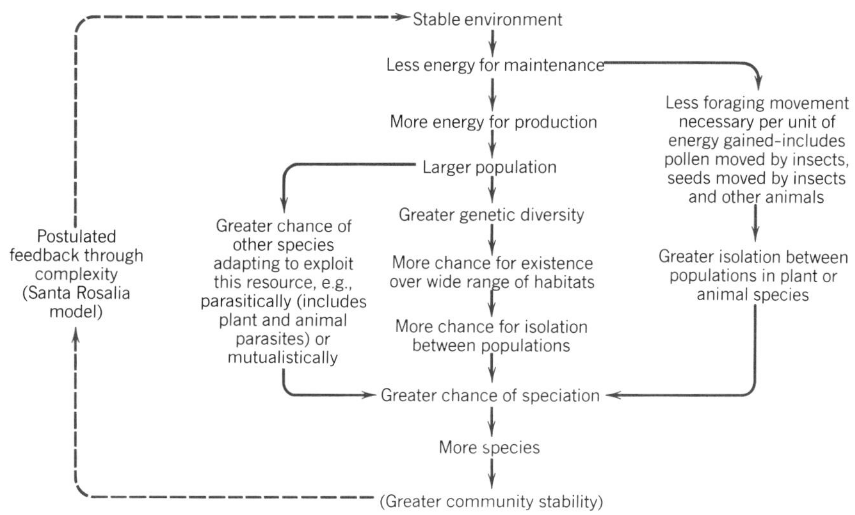

FIGURE 16.8
Postulated way in which diversity could be sponsored by productivity.
The scheme includes a feedback from species diversity to stability, which has now been shown to be untrue. (Modified from Connell and Orias, 1964, and Price, 1975.)

tivity, not just across lines of latitude but within northern temperate forests also. As Figure 16.9 shows, the correlation between the production of organic dry matter in tons and the number of tree species per unit area is fairly robust within these northern forests (Currie and Paquin, 1987; Adams, 1989).

Whether this correlation is causal is, however, far from certain. That both diversity and productivity are correlated with latitude is basic to our problem. Productivity is known to be dependent on temperature and seasonality, leaving no mystery about the clinal fall in tons of dry matter produced toward the poles. The question to answer is, Why is diversity also linked to latitude?

Figure 16.10 shows how tree diversity is a function of geography in North America, showing, as expected, that highest diversity is in the warm southeast and lowest in the cold arctic. Here is the latitudinal cline skewed a bit to the right by rain shadow effects of the western mountains. Nothing in the data set requires that productivity, rather than temperature, or seasonality, or some other correlate of latitude is primarily responsible for setting species richness of American trees.

The difficulties of the productivity hypothesis are formidable. For a start, correlations between high diversity and high productivity are in fact far from general. Species richness in seasonal deserts like the Sonoran or in arid land pastures is extremely high, particularly of plants. And particularly productive systems, such as estuaries, salt marshes, and polluted lakes, are noted for their low diversity. The record of the deep sea directly conflicts with the general hypothesis correlating diversity with productivity because the infauna are highly diverse in the extremely unproductive abyssal bottoms (Figure 16.6).

But the theoretical difficulties are probably the most serious. High productivity

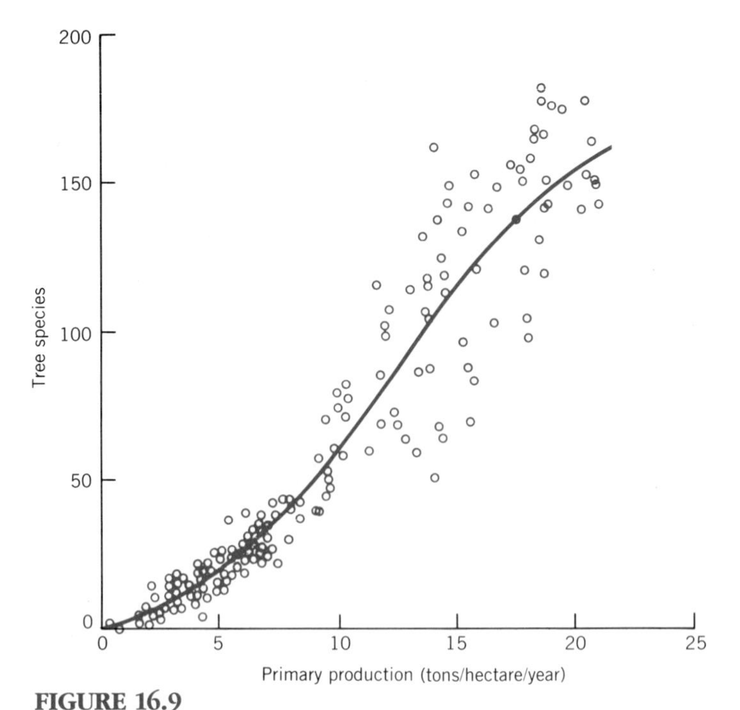

FIGURE 16.9
Tree diversity correlated with productivity.
Diverse forests also tend to be the most productive. This, however, leaves the possibility that both diversity and productivity are correlates of climate, whether reflecting latitude or not. (From Adams, 1989.)

allows dense populations, with all the perils of competitive success and competitive exclusion that this implies. The "cropping principle" explanation of the high species richness of the Sonoran desert and of overgrazed pastures is precisely that dense populations of winner-take-all competitors are prevented by intense seasonality or heavy predation. Parallel to this is the apparent cause of low diversity in fertile estuaries and salt marshes where high tonnage of dry matter production allows successful competitors to preempt the habitat, resulting in competitive exclusion of the many by the few.

Productivity in the tropical rain forest might well be the highest on earth (see Chapter 24), raising the possibility of extremely dense populations of successful tree species. To an ecologist this compounds the problem of high tropical diversity rather than offering a solution. Dense populations suggest low species richness due to competitive exclusion of the kind seen in estuaries and salt marshes. Obviously this "success of the few" is prevented in some way in the tropical rain forest. But how this is achieved is part of the puzzle of explaining the high species richness of the wet tropics.

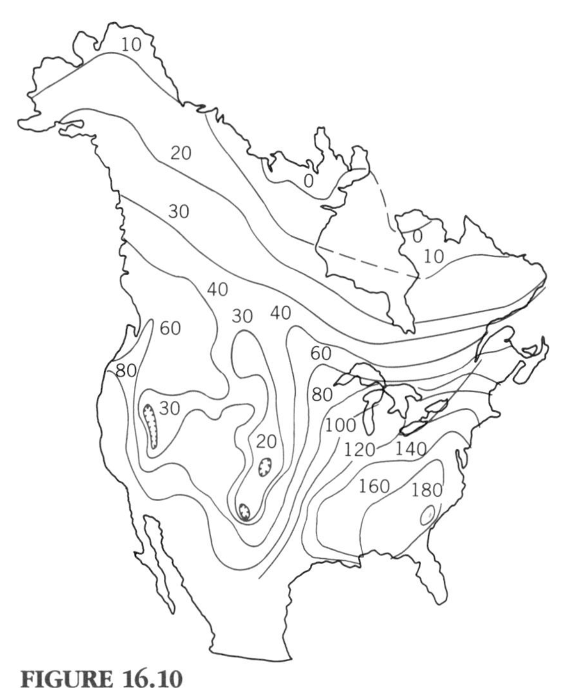

FIGURE 16.10
Tree diversity in North America.
Numbers are of woody plant species more than 3 meters high found in a standard large area in each region. Cold and desert places have fewer trees per unit area than the moist and productive southeast. (From Adams, 1989.)

THE HYPOTHESIS OF DIVERSITY CONTROLLED BY HABITAT STRUCTURE

Complex vegetation, as in a tropical rain forest, suggests spatial multiplication for animal niches. MacArthur (1965) tested the idea that extra layers of canopy in tropical forests accounts for the larger number of bird species there by comparing bird species diversity with a measure of relative layering of forests, which he called "foliage height diversity." Throughout the United States he obtained a good correlation between bird and foliage-height diversities, suggesting that birds at least are more various in structured forests (Figure 16.11).

A similar study in grassy, shrubby, and woody areas of southern Chile, Africa, and California by Cody (1975) found similar results (Figure 16.12). The study areas in the three continents were all within the broad-leaved sclerophyllous biome (chaparral) and thus should be comparable. Within-habitat diversity increased sharply along a gradient of increasing structure in each region, the most disparate data being from the South African woodlands known to have a notably impoverished bird fauna.

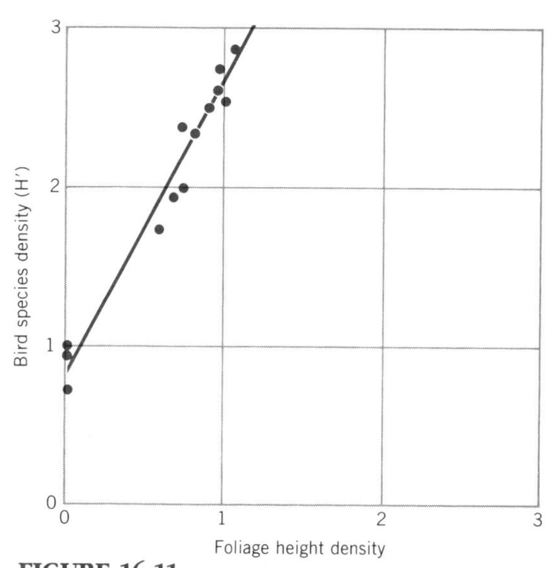

FIGURE 16.11
Bird diversity and foliage height diversity.
This is MacArthur's (1965) original figure. Bird diversity is plotted as the Shannon-Wiener index, H'. The graph shows so close a correlation between bird species and plant structure as to suggest that these habitats are saturated with birds. The "fit," however, is much less good if species richness is plotted instead of H'.

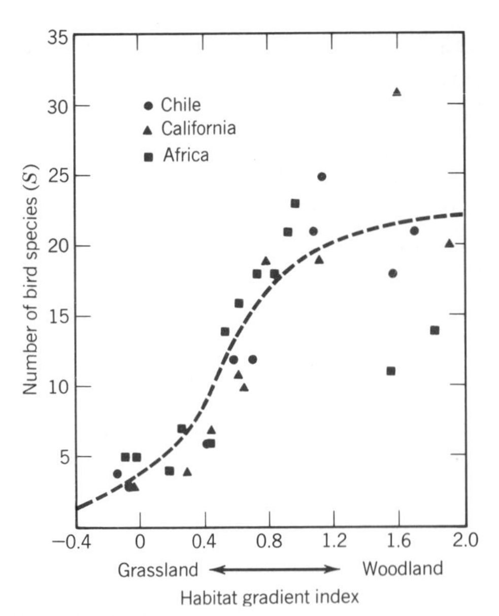

FIGURE 16.12
Correlation of bird diversity with habitat structure in "chaparral."
The habitat gradient index is a function of vegetation height and density. The curve is a fitted logistic curve. (From Cody, 1975.)

Some portion of high tropical bird diversity, therefore, can be assigned to higher structural diversity, but probably nothing like enough to account for the observed richness of tropical birds. A minimal calculation of the number of discrete layers needed to explain the numbers of birds in a rain forest is seven, whereas botanists can recognize only four or five (Lovejoy, 1975).

For animals other than birds, structure seems even less important. Terry Erwin (1982) gassed the canopies of a single species of tree in a Panamanian forest, bringing down 1200 species of beetle. It is unthinkable that diversity on this scale could be attributed to mere structural complexity.

NEW GUILDS IN THE TROPICS

The year-round living of tropical places allows specialized ways of life that are quite foreign to high latitudes. This is particularly clear for the animals we know best, like mammals and birds. For these animals, for instance, a tropical rain forest may be likened to a prairie set on stilts 20 meters high. Whole guilds of animals exist to exploit the year-long resources provided by this elevated and evergreen prairie. Primate guilds, fruit-eating bats, and guilds of fruit-eating birds like toucans or hornbills, exist, coevolved to the needs of the forest trees to have their progeny dispersed. Other primate and mammal guilds are present as arboreal herbivores. And this canopy life supports guilds of canopy-hunting predators, like monkey-eating eagles.

The press of insect life has its own repercussions up the food chains. A characteristic of tropical forests is army ants, which beat their way across the forest floors in a seemingly endless hunt. Other insects flee them, so that the passing of the army ants is marked by arthropods of every kind breaking cover in their efforts to escape. A guild of antshrikes exploits these driven arthropods, birds for whom an essential resource is fugitives from the army ants.

Thus some part of high tropical diversity of the kinds of animal most familiar to us is surely explained because of the wholly new guilds that are possible (Terborgh, 1992). It is possible that the vastly greater diversity of small invertebrates similarly reflects a hugely greater array of possible guilds in the tropics. If this possibility were to be proved correct, it would suggest that conditions of year-round productivity are of critical importance. Certainly the extra guilds of tropical birds are critically dependent on year-round production, which keeps the canopy stocked with supplies and lets the army ants thrive.

THE REJUVENATING CATASTROPHE HYPOTHESIS

When biogeography was in its infancy, Alfred Russell Wallace (1878) put forward the hypothesis that species poverty of the high latitudes followed ice age extinctions (Fischer, 1960). Glaciers scraped the more northerly land masses bare of all species; the temperate belt south of the ice suffered massive climatic change and extinctions also. Only 10,000 years have been available for recovery. No comparable disaster, as it seemed in Wallace's day, could have afflicted the tropics, and the observed latitudinal cline is the natural consequence.

Probably this history does explain some part of the species poverty of the North. European forests of the Tertiary, for instance, included plants like *Liriodendron* (tulip tree) and *Liquidamber* (sweetgum) as part of a species-rich forest now known only in the eastern United States and China. The subsequent loss of species can be explained most parsimoniously because the Fennoscandian ice sheet pressed against the east–west–trending mountain chains of southern Europe and Asia, denying habitat to many species of the old arcto-tertiary forest (Figure 16.13).

Against this conclusion, a somewhat tenuous argument can be raised to suggest that the remaining European forests are in fact not significantly less diverse than temperate forests elsewhere occupying comparable areas. This relies on the observation that the correlation between tree richness and productivity in Europe is the same as in America and eastern Asia (Adams and Woodward, 1989). Figure 16.14 shows the supporting correlations for Europe and eastern America. The argument then is made that, since Europe has the "right" number of trees after all, the coming and going of the Fennoscandian ice sheet had no effect on modern tree diversity. The analysis holds true, of course, only if tree diversity is really set by productivity, which can be doubted.

The explanatory power of the rejuvenating catastrophe hypothesis, however, is more closely limited in other ways. Particularly difficult for the hypothesis is the maintenance of latitudinal clines in all geological epochs, those without ice ages as well as those with them. Figure 16.2 shows data from the Cretaceous for foraminifera and from the Permian for brachiopods.

Finally is the nearly insuperable objection that significant climatic change in ice age times was experienced at the equator as well as in arctic latitudes. In tropical Africa, for instance, glacial climates were both colder and drier than now, with trees of the tropical rain forest restricted to the remaining moister sites (Livingstone, 1993; see Figure 11.5; Chapter 11). Probably climatic changes through glacial cycles were equally severe in all equatorial regions. Thus the latitudinal decline of diversity toward the poles may well be exaggerated during epochs with ice ages, but the fundamental cause of the cline must lie elsewhere.

THE TIME-STABILITY HYPOTHESIS

The **time-stability hypothesis** was put forward to explain the cline of infaunal diversity with depth in the deep sea (Sanders,

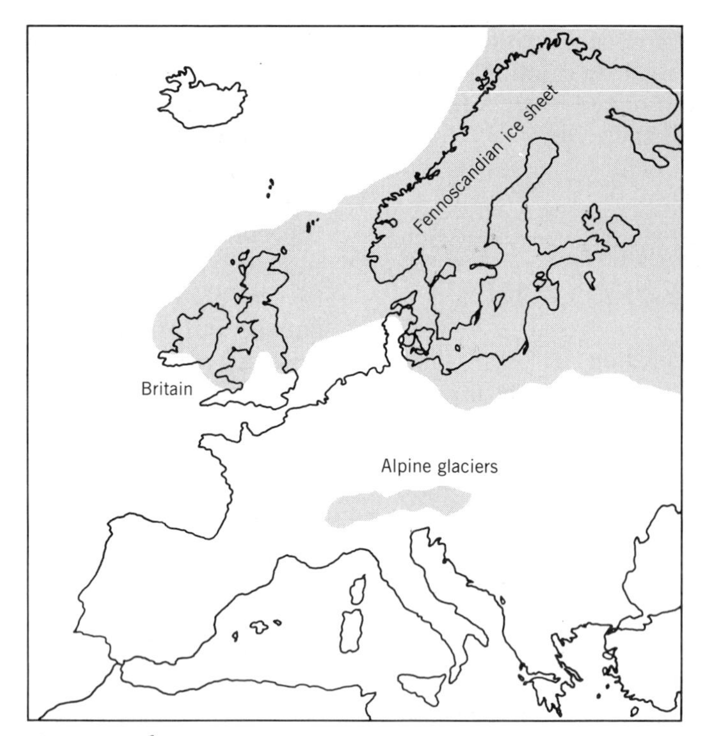

FIGURE 16.13
Reduction of European tree habitat at the last glacial maximum.
Forest habitats in Europe were almost eliminated as central Europe was squeezed between the Fennoscandian ice sheet and glaciers in the Alps. Pollen analysis shows that much of the land left free of ice between the glaciers held only tundra. It is arguable that Europe lost many of the tree species from its old Tertiary forests because of this reduction of forest habitat with each successive ice age.

1968; Figure 16.7). The dark, energy-starved floors of the abyss have highly diverse mud-living organisms, the apparently richer continental slopes have fewer species, and rich, productive estuaries have very few indeed. Obviously productivity could have little to do with this data set. Sanders asked what other property of the deep sea might be causally correlated with high species richness, and found a candidate in the profound stability of the deep-sea environment.

A deep-sea floor is without seasons. It might have the same environment for the whole of the 100 million years or so between its creation in a mid-ocean ridge until its final destruction in a subduction zone (see Chapter 18). In this stable environment populations should not fluctuate, suggesting that extinction would be low. Whatever new species are produced in the deep sea would thus collect and diversity would continue to grow.

Estuaries and continental shelves, on

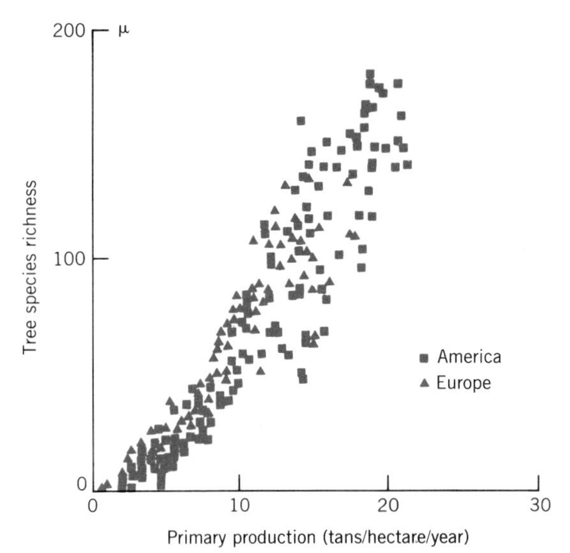

FIGURE 16.14
European and American tree richness as a function of productivity.
On the hypothesis that productivity sets tree diversity, these data suggest that low tree diversity in Europe reflects low productivity of the northern continent rather than the history of ice ages. On the alternative hypothesis that low European tree richness reflects ice age history, these data suggest that tree diversity is not in fact controlled by productivity. (From Adams and Woodward, 1989.)

the other hand, have frequent changes of environment, both seasonal and catastrophic. They have high extinction rates and consequently a low equilibrium number of species. The species number of any place, therefore, should be proportional to the time a habitat has been in existence and its physical stability (Figure 16.15).

The time-stability hypothesis was attractive to ecologists in the 1970s because it then appeared that the hypothesis might explain the cline of diversity on land as well as on the bottom of the sea. For the changing environment of estuaries and coasts read the changing seasons and weather of the Arctic. For the everlasting sameness of the deep-sea floor read the supposed equable climate of the timeless, immemorial tropics. Tropical rain forests, on this hypothesis, acquired their huge species lists because the absence of physical calamity ensured that no species ever became extinct, just as Sanders argued for the cold, dark depths of the sea bottom.

We now know that the concept of the unchanging tropics was a myth. Equatorial regions have both weather and seasons, and their communities are in continual change with the coming and going of

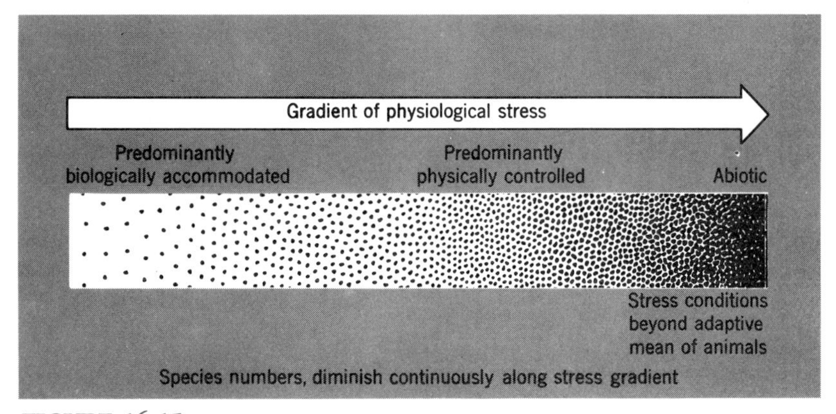

FIGURE 16.15
The time stability hypothesis.

ice ages. The time-stability hypothesis thus lost both its generality and its appeal. Finally, other explanations for high diversity in the deep-sea benthos are possible. The deep-sea floors are of vastly greater area than the continental shelves and slopes with which the hypothesis compares them, so that any hypothesis relying on larger area can explain the data as well as this one, which relies on antiquity and stability.

THE LATITUDE-AREA EFFECT

A degree of latitude close to the equator makes very little environmental difference, whereas a degree of latitude in France or Ohio makes a big difference. Only 3 degrees of latitude separates northern France from the Mediterranean Sea, embracing an almost massive climatic change, yet 3 degrees in the Amazon merely shifts you from one patch of lowland rain forest to another. Thus equatorial regions provide areas of uniform climate that are far larger than equivalent areas at high latitude.

The spherical shape of the earth ensures that the actual land area on equatorial continents is large, despite the impression given by maps drawn on Mercator's projections that most land is in the North. Therefore, habitable areas of uniform environment are largest in equatorial regions. If a large area means a large population, and if a large population means a lowered extinction rate, then the cline with latitude may be the necessary consequence of the shape of the earth (Osman and Whitlatch, 1978; Terborgh, 1992).

This was a view put forward earlier in a different context by Darlington (1959), who marshaled evidence to show that many facts of zoogeography could be explained as a result of speciation in the tropics, followed by subsequent migration to higher latitudes. In Darlington's evolutionary view the earth is a large area of tropics, where species collect, and a peripheral temperate region, to which species migrate. This does accord with the geometry of the earth rather well.

Terborgh (1973) argued that the huge expanse of equator plus tropics effectively was larger still. Equatorial regions generally have temperatures that are lower than might be expected because of heavy cloud cover. At the edges of the equatorial belt, cloudless skies make for temperatures quite as warm as those over much of the equator itself (Figure 16.16). The area occupied by equatorial species thus can be extremely large.

Large area is a property of the deep sea as well as of equatorial forest. Thus species richness correlates with area of constant environment on a global scale. If all the large areas of tropics or deep sea floors are available for the production of new species that might then survive in any part, it is at least possible to imagine numerous species collecting.

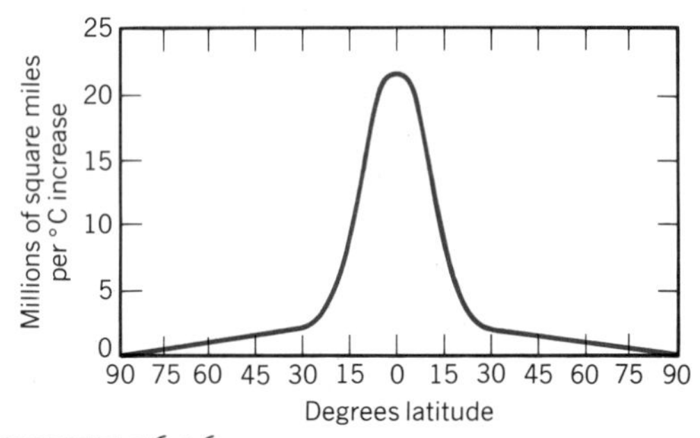

FIGURE 16.16
Area available at equal temperature at different latitudes.
(From Terborgh, 1973.)

CROPPING RAIN FORESTS: THE INTERMEDIATE DISTURBANCE HYPOTHESIS

A highly productive, year-round ecosystem like a tropical rain forest might well be expected to be species-poor. What should happen is that dense populations of forest trees would lead to competitive success by the few, competitive exclusion of most, and a species-poor system. Clearly this does not happen.

It is a paradox that the most productive of ecosystems has by far the highest tree diversity, a paradox that can be resolved only by showing how the expected competitive exclusion is prevented. One possibility is seed predation that forces tropical trees to live spaced apart from each other, an invocation of the cropping principle for forest trees (see Chapter 14; Janzen, 1970). But environmental "cropping" of the forest might be as important as actual herbivore attack. This is the basis of what has come to be known as the "**intermediate disturbance hypothesis**" (Connell, 1978).

Competition between forest trees is readily apparent in the later stages of ecological succession, as a long series of ever more shade-adapted trees replace light-adapted designs (see Chapters 1, 3, and 20). The theoretical end point of this process of competitive replacement is total success for the one or the few in the Lotka–Volterra manner (see Chapter 8). If, however, many or most patches of forest are knocked down before this competition runs its course, forest successions will forever oscillate around their early stages, and local competitive exclusion will almost never happen.

Current ecological thought sees tropical rain forests as ecosystems facing almost perpetual disturbance, in which gaps are torn in the canopy with such frequency that successions are never allowed to run their course to the logical end point of dominance by the few. This is the intermediate disturbance hypothesis (Connell, 1978). Disturbance is not so severe as to cause extinction itself, as the coming of ice-ages has been supposed to do in the North, but it is sufficient to disrupt competition between species. Numerous gaps in the tropical rain forests, caused when canopy trees are blown down in storms, are evidence that the required intermediate disturbance does exist.

Physical properties of tropical rain forest ecosystems make them vulnerable to local stresses. They have grown quickly because of the warmth, moisture, and prolonged growing season of their habitat, and tend to be top-heavy or have minimal rigidity. Deeply weathered soils lead to tight nutrient cycles in which retrieval of nutrients from surface litter is vital, which in turn leads to shallow rooting systems (see Chapter 25). Climbing plants, creepers and llianas thrive in environments without winter, so that tall trees, shallow-rooted in wet ground, carry a heavy load of creepers. These laden trees are fit to be toppled when assaulted by the frequent thunderstorms characteristic of the rain forest climate.

The intermediate disturbance hypothesis is intellectually satisfying as an explanation of why tropical rain forests are not simple systems like polluted estuaries. Allied with the hypothesis of tree spacing dependent on seed predation, it explains why no tree occupies the habitat as a supercompetitor in winner-take-all competition. To explain why rain forests have more species of tree than do temperate forests, however, it would be necessary to show that temperate forests suffer less from intermediate disturbance.

SANTA ROSALIA 30 YEARS ON

The contemporary earth is between great extinctions, although of course it is well into the great extinction that we ourselves are bringing about. If we stretch our time to include the last million years or so of which we have tolerable knowledge from the fossil record, it is fair to say that contemporary patterns of distribution and abundance describe the earth when well-filled with species.

Thirty years after publication of the Santa Rosalia paper we still have no satisfactory general theory to explain species diversity (May, 1986*a*). More mechanisms promoting diversity are known, however, the most powerful of which are the effects of cropping agents, whether environmental stress or foraging animals, in promoting high diversity by preventing competitive exclusion.

Possibly the intellectual field in which least progress has been made is in mastering the processes controlling rates of speciation. If, for instance, rates of both speciation and extinction were functions of seasonality, then the latitudinal cline of diversity persisting through all geological epochs should be explained. The earth might then never have filled with species in the way envisaged by Santa Rosalia, but rather would have an equilibrium number dependent on the size of land masses and their distance from the equator.

A satisfactory theory of diversity, however, could well remain a vain dream unless an effort is made to catalog the enormous part of contemporary diversity of which we know so little.

PART THREE

COMMUNITY

Preview to Chapter 17

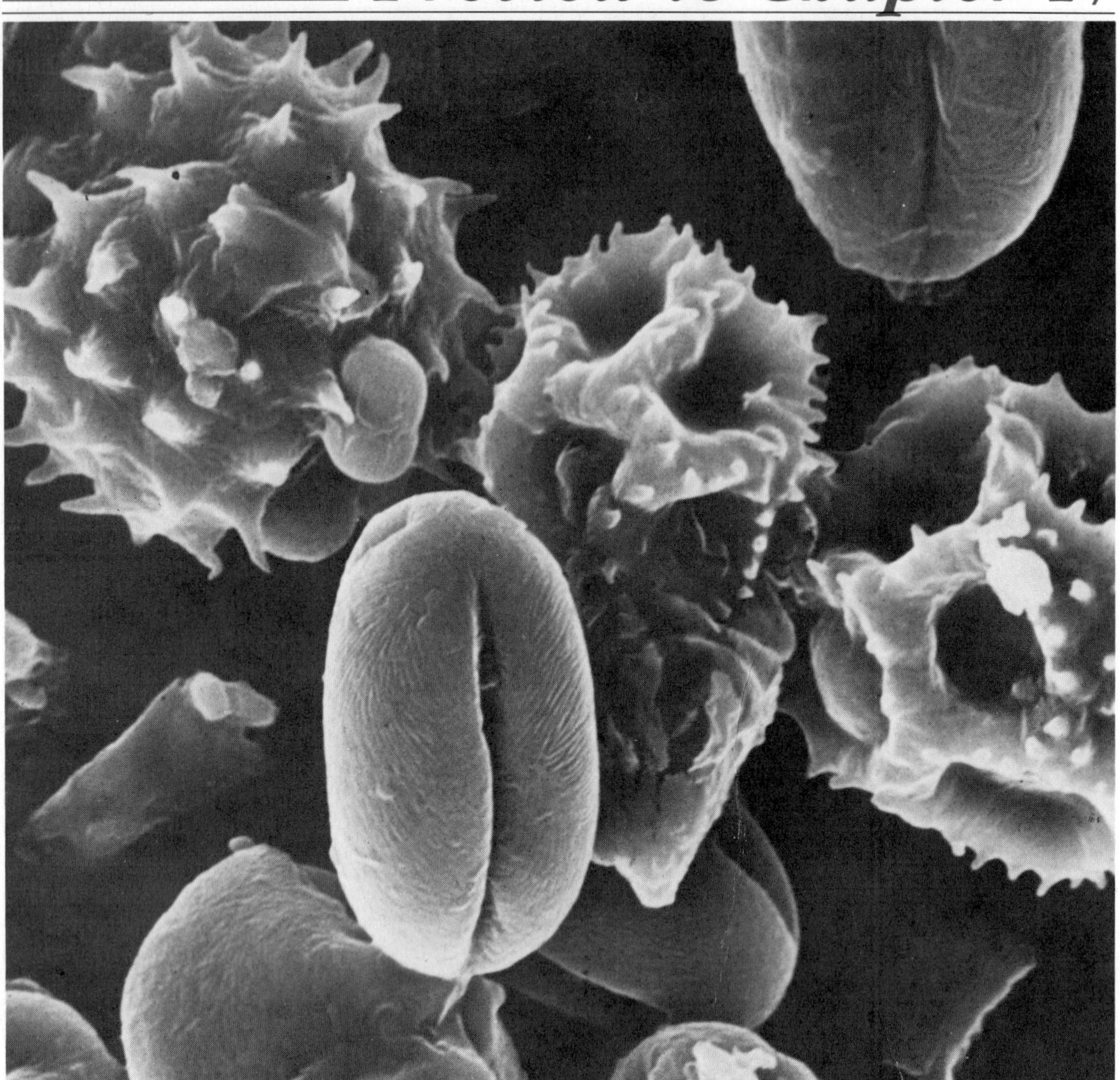

Taraxacum, Senecio, Potentilla: **pollen that testifies to ancient presence.**

FOR ecologists the past may be the key to the present. Ten thousand years ago the last ice age ended. Before that all the world was different, with each part having a different climate. Paleoecologists study the distribution and abundance of life in the long ice age in order to understand community structures on the modern earth. Their methods overlap with those of geology, coaxing the fossil record to reveal the results of long-term experiments we might like to run but cannot, because of the constraint of human lifetimes. A prime resource is the mud of ancient lakes, with thin layers laid down each year like the pages in a book. History is written in the book of mud as pollen grains, as parts of planktonic animals or diatom frustules, as chemistry, minerals, and even magnetism. Analysis of the tens of thousands of pollen grains typically preserved in a gram of lake mud describes the changing plant communities of a landscape, together with the climates in which the plants lived. Other microfossils or chemical analyses yield the history of the lake and its watershed. In a like manner, ocean sediments yield histories of climate, and they provide a clock of relative age by the history of glaciations preserved in oxygen isotope ratios. Radiocarbon dating puts ages on fossils up to 35,000 years old, and an arsenal of other dating techniques is being developed. In these ways, paleoecology reconstructs the onset of the modern earth, a time of poignant interest as the epoch of the rise and spread of the human species.

Chapter 17

The Ecologist's Time Machine

This is a methods chapter. But the methods of paleoecology afford the best, or only, means ecologists have for answering vital questions in community theory. Thus, How did the Canadian forests (or their individual trees) survive an ice age? Was the Amazon forest always there, or did its warmth-loving species have to take refuge in some forest Shangri-la when the whole earth was cooled? And how do communities react to grand environmental changes like the pending doubling of the carbon dioxide supply? Questions like these need answers from the past that can be obtained only by appeal to the fossil record.

A modern ecologist looks at communities that are massively young by the standards of their component species and in terms of the climatic changes through which the earth is passing. The Holocene (Greek for "wholly recent") period in which we live is just 10,000 years old. Before then was the 80,000 years or so of the last ice age, the platform of normality, as it were, in which all important modern species lived so many generations more than in the brief Holocene that we know. The methods described here are those used in reconstructing that ice age starting-off point for modern life.

THE MEANINGS OF PALEOECOLOGY

Paleoecology literally means "the ecology of the past," but different disciplines give their own shades of meaning to the word. Geologists use fossils as "time-stratigraphic indicators" and as tools for reconstructing past environments. Standard divisions of geologic time are still based on fossil assemblages, as the strata are assigned ages according to the fossils they contain (Figure 17.1). But more than this, geologists use their fossils to infer such things as temperature, habitat, and climate of the past epochs they name, basing these reconstructions on the **principle of uniformity**.

Subdivisions Derived from Strata				Age $\times 10^6$ yr
	Systems	**Series**	**Important Fossils**	
CENOZOIC	Quaternary	Holocene		0.001
		Pleistocene	First humans	2.5
	Tertiary	Pliocene	Elephants, horses, and large carnivores	13
		Miocene	Mammals diversify	25
		Oligocene	Grasses and grazing animals become abundant	36
		Eocene	Primitive horses	58
		Paleocene	Mammals prominent for first time	
MESOZOIC				63
	Cretaceous		Dinosaurs become extinct; Flowering plants	
				135
	Jurassic		Dinosaurs reach climax	
				180
	Triassic		Birds Primitive mammals appear; conifers and cycads become abundant Dinosaurs	
PALEZOIC				230
	Permian		Many reptiles; conifers develop	
				280
	Pennsylvanian (Upper Carboniferous)		Primitive reptiles appear; insects become abundant Coal-forming forests widespread	
				310
	Mississippian (Lower Carboniferous)		Fish diversify	
				340
	Devonian		Amphibians, first known land vertebrates Forests	
				400
	Silurian		Land plants and animals first recorded	
				430
	Ordovician		Primitive fish, first known vertebrates	
				500
	Cambrian		Marine invertebrate faunas with hard skeletons	
PRECAMBRIAN Complex assemblages of rocks, largely metamorphosed			Procaryotes long present; first metazoa with soft skeletons	

FIGURE 17.1
Geologic time scale.
The stratigraphy was erected using fossils as time-stratigraphic markers. Dating in years came later by radiometric methods. (Modified from Longwell *et al.*, 1969.)

The uniformity principle is invoked whenever the claim is made that *the present is the key to the past*, and rests on the assumption that processes in the past were the same as processes on the contemporary earth. Charles Lyell and his contemporaries in the nineteenth century developed the uniformity principle into the key tool for reconstructing earth history, turning geology into a great detective story in which the clues were traces in rocks of familiar processes working long ago. Charles Darwin had a copy of Lyell's book with him on his Beagle voyage.

Applying Lyell's uniformity principle to the study of fossils is to say that the nearest living equivalent to a fossil form lives in such and such an environment, and that therefore the same environment pertained in the time of the fossil. This is the paleoecology taught in many a geology course: the reconstruction of past environments from the evidence of fossils.

A biologist can get more from a fossil if the environment in which it lived can be deduced from other evidence. We want to know what the animal or plant from long ago was up against and how it lived. For all biologists, paleoecology is the reconstruction of life histories, life strategy, and niche of extinct species by combining physical and chemical evidence of past environments with the life-form data of the fossils.

But ecologists have a more particular and precious use for fossils. We appeal to the fossil record to test hypotheses of population or community ecology. Population changes that go, for instance, into the building of a forest must last many human lifetimes and can be studied only from the fossil record. We cannot run a thousand year experiment, but we can, as one of the founders of our subject, E. S. Deevey (1969), put it, "coax" history to conduct the experiment for us. Paleoecology then becomes the use of the fossil record to test hypotheses of population and community changes in long-lived organisms.

In one special way ecologists must part from geology even to the extent of reversing Lyell's logic of uniformitarianism. Lyell said the present was the key to the past, but the key to some modern distributions is to be found in ice age patterns of life from which life on earth has so recently emerged. To this extent, the past is the key to the present.

MICROFOSSILS IN CORES OF SEDIMENT

Reconstructing population or community histories requires quantitative samples of the past. Small fossils that can be sampled by the thousand, therefore, have particular advantages. The empty husks of pollen grains of trees, for instance, are often present in lake mud in concentrations of more than 100,000 grains per milliliter. In anoxic mud of lakes or bogs these pollen husks are beautifully preserved, alongside such other microfossils as diatoms, remains of zooplankton, phytoliths, and sometimes beetle fragments. Similar fossils are found in ocean mud, though the most useful marine microfossils are the tests of calcareous foraminifera. To obtain these quantitative proxies for past communities, ecologists have learned to sample mud under water.

The mud of a lake or ocean must be sampled as a **core**, a long, continuous section through the sediments. Moreover, the sediments must not be disturbed in the process. The sampling is done with piston samplers, using principles first introduced into oceanography by Kullenberg (1955) and then adapted for use in lakes by Livingstone (1955). A piston sampler works by using the hydrostatic head of water against a stationary piston to overcome friction on the inside of a sample tube as it is pushed into the sediment (Figure 17.2).

The difficulty in coring mud is not penetration, which is usually easy. Much lake sediment, for instance, can be penetrated by a good push. A 5-kilogram hammer sometimes helps, but nothing like a rotary drill is needed. The real problem is that mud will be prevented from entering the sample tube by friction on the inside. A good shove on an open tube merely results in bunging up the end. The captive piston solves this.

In lake work with a Livingstone sampler, the piston is held by clamping its cable to the raft (Figure 17.3). The tube is then pushed down around it. What happens can be visualized by thinking of a hand-held bicycle pump, which is used by pushing the piston and holding the pump tube stationary. In piston sampling you hold the piston and push the tube, but the relative motion is the same.

In lake work it is necessary to anchor a raft, usually a pair of rubber boats, to three anchors in a "Y" formation so that the raft cannot move, and to lower a pipe (the casing) through the water to the mud. The piston sampler is then used inside this casing, hauling up the sediment a meter at a time, then going down the old hole for the next meter, and so on. In ocean work it is not possible to work with rods and casing, and so a Kullenberg sampler typically takes 10-meter cores at one bite, the long tube being driven down by a heavy weight

FIGURE 17.2
Coring Lake Manacaparu in the Central Amazon of Brazil.

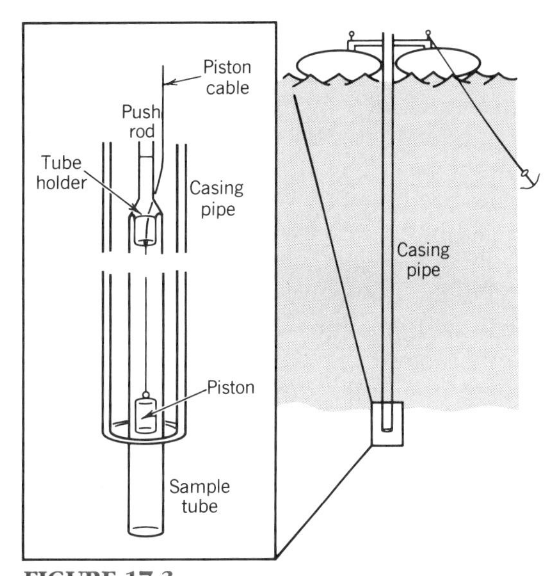

FIGURE 17.3
Piston Coring of sediments under water.
A raft is anchored and connected to the bottom of the lake by casing. Push rods are used to insert a sample tube down the casing and into the mud. A piston inside the sample tube is held stationary at the mud surface by a cable clamped to the raft, and the sample tube is forced down past the piston, if necessary using a drop hammer. Hydrostatic pressure acting against the piston overcomes friction on the inside of the sample tube, letting the tube cut into the sediment without disturbing it.

(half a ton of iron), and the whole thing being hauled back to the ship on a cable. By these means long cores of undisturbed sediments are raised from both lakes and oceans. Microfossils can then be extracted at intervals to work out histories of populations and communities.

THE POLLEN TOOL

Over any community in spring and summer there hangs a cloud of pollen undetected, save by sufferers from hay fever, yet composed of an almost unthinkably vast multitude of tiny, drifting grains, gently settling as a pollen rain. Pollen grains from different plants are easily identified to genus with a light microscope (Figures 17.4 and 7.1), although not to species. Pollen that settles in waterlogged places is preserved, for the outer coat of pollen grains is made of one of the most resistant materials produced by living things. The pollen contents quickly rot, but the outer shell may persist in the sediments of a bog or lake for thousands of years, and hundreds of thousands of years if the sediments remain wet and anoxic.

It is a simple, although time-consuming, matter to sort a few thousand pollen grains from sediments, destroying the organic matrix with sodium hydroxide and a sulfuric acid–acetic anhydride mixture and dissolving minerals with hydrofluoric acid (Faegri and Iversen, 1989; Moore et al., 1991). The pollen extract is mounted on a microscope slide, after which it is a comparatively simple task to traverse the slide at a magnification of 400 diameters and to identify and tabulate every pollen grain until enough have been counted to calculate the percentage composition of that ancient pollen cloud.

The Pollen Percentage Diagram

Figure 17.5 is a pollen percentage diagram from the sediments of a small kettle lake in Vermont (M. Davis, 1965). Depth in sediment is plotted on the ordinate and percentage of each pollen type on the abscissa. Each horizontal line of histograms (called **pollen spectra**) represents a single analysis of the past vegetation at a discrete interval of time, giving a complete list of the common taxa that make up most of the pollen rain with an importance value (percent pollen present) for each. A glance at Figure 17.5 shows that vegetation has changed radically around the lake. In the early days there was much spruce and pine pollen but little beech and hemlock. Since the lake was made by glacial ice, it is obvious that in the spruce–pine episode we are looking at a trace of the landscape of Vermont in early postglacial time, and that much was to change before final establishment of the forest that the first New Englanders were to see.

FIGURE 17.4 (opposite page)
Pollen of European tree genera drawn to a common scale.
The scale is in microns, which means that the 100 divisions shown represent one-tenth of a millimeter. The largest pollen grain in the figure, the pine, would be just visible to the naked eye as a minute speck. Notice how distinctive are the grains of different genera. Even when built on a common plan, as are the 3-pored grains of birch and hazel, there are distinctive differences (between these two the shape of the pores) with which the analyst quickly becomes familiar. But it is seldom possible to distinguish between species within a genus. Alder grains, for instance, may have 4 or 5 pores (as shown) or even 6 or 7, yet there seems to be no correlation between the number of pores and the different species of alder. Sometimes you are lucky, however, and two species of lime can be told apart by the shape of their pores, as shown. (Redrawn from Godwin, 1956.)

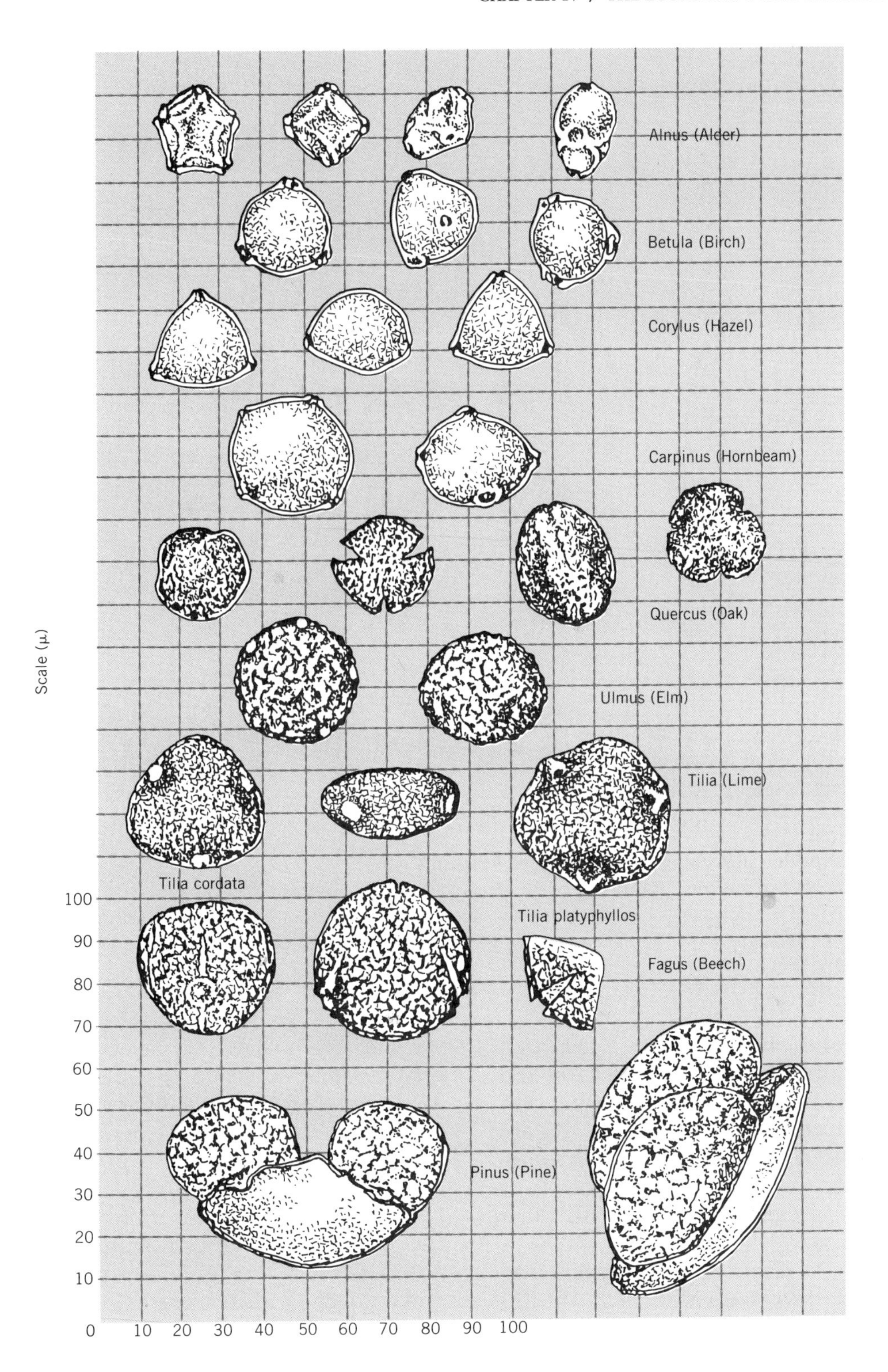

Alnus (Alder)
Betula (Birch)
Corylus (Hazel)
Carpinus (Hornbeam)
Quercus (Oak)
Ulmus (Elm)
Tilia (Lime)
Tilia cordata
Tilia platyphyllos
Fagus (Beech)
Pinus (Pine)
Scale (μ)
100
90
80
70
60
50
40
30
20
10
0
10 20 30 40 50 60 70 80 90 100

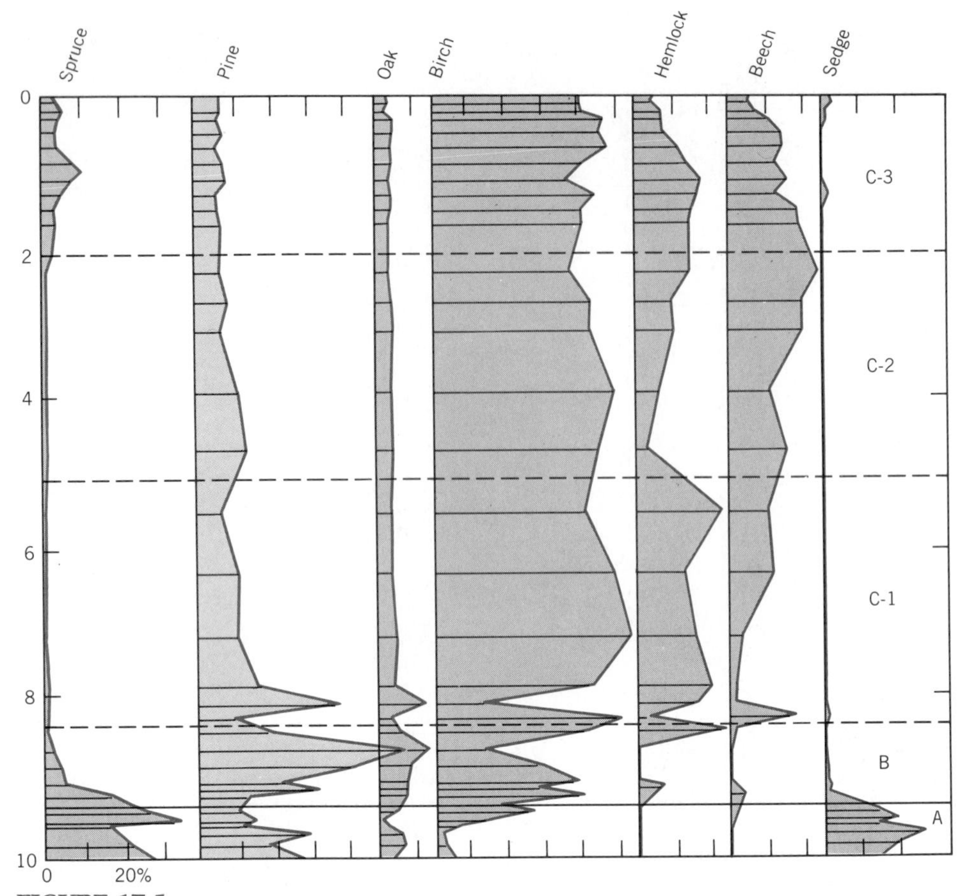

FIGURE 17.5
Partial pollen percentage diagram from Vermont.
The record extends back to the time of the glacial retreat, and records the development of the deciduous forest of modern Vermont from a time of open vegetation with spruce and pine. For interpretation of the pine percentages (dark shaded area), see Figure 17.6. (From Davis, 1965.)

To control the large amount of data in a pollen diagram and to help interpret the record, pollen analysts divide the diagram into **pollen zones** (zones A–C in Figure 17.5). This can be done by a multivariate analysis that identifies regions of comparative homogeneity, but it is usually done by eye.

Most of the thousands of pollen diagrams that have been drawn in the 60 years of palynology have been **pollen percentage diagrams** like the one in Figure 17.5, but they have one very great disadvantage—there is no effective way of deciding if a small change in the percentage of a single pollen taxon has any statistical validity. The variables are as numerous as the pollen types, and a change in real numbers of any one taxon must affect the percentages of all the rest. Conversely, a high percentage of one taxon might represent a lowering of pollen production in the forest as a whole rather than a real increase in the population of the taxon in question. We now know, for instance, that the high percentage of pine at the bottom of zone A in Figure 17.5 is mostly a reflection of a landscape with few heavy pollen producers other than pine. The percentage of pine pollen was high because other pollen to mix with it was scarce, despite the fact that pine trees were scarce too. But this was not discovered from the percentage pollen diagram. The full analysis on which Figure 17.5 is based has 24 dependent variables (the pollen taxa), and is completely intractable for a statistician. Measurement of pollen influx (Figure 17.6) was needed to show the true status of the ancient pines.

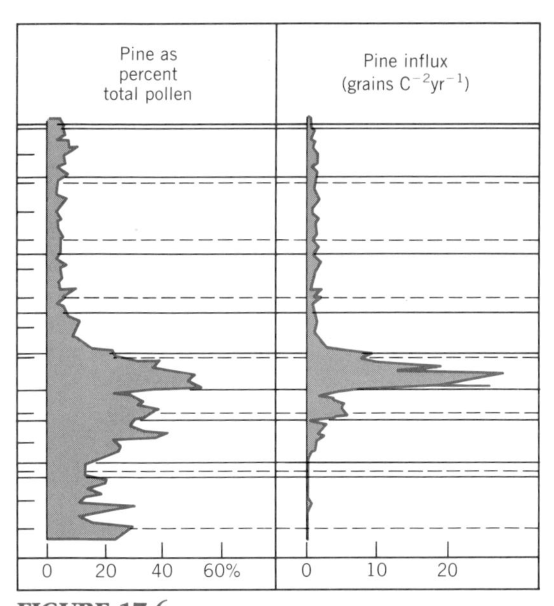

FIGURE 17.6
The history of New England pine populations as revealed by pollen influx measurements.
Calculation of pollen influx shows that there were few pine trees 14,000 years ago in New England, despite high pine percentages. A true pine population event happened later, but lasted less than 1000 years. (From Davis, 1969.)

Pollen Concentration and Pollen Influx

The truth about the Vermont pine was found by calculating the flux of pine pollen falling per square centimeter per year throughout the record. To calculate **pollen influx** it is necessary to measure the concentration of pollen in the sediment and to know the sedimentation rate. Then

$$\text{Pollen influx (grains cm}^{-2}\text{ yr}^{-1}) = \frac{\text{Pollen concentration (grains ml}^{-3})}{\text{Sedimentation rate (yr cm}^{-1})}$$

To measure pollen concentration it is necessary to start with a measured volume of fresh, wet sediment. It is possible to extract every pollen grain from this measured volume, make a small subsample, and then count every grain in the subsample, but this is very laborious. Instead, pollen analysts add to the measured volume a "spike" of a known number of grains of exotic "marker" pollen. Temperate-zone pollen analysts use the distinctive pollen of the Australian genus *Eucalyptus* for this purpose. Then a pollen preparation is made from the spiked sample in the usual way, resulting in a slide containing both the exotic marker and the fossil pollen. Both are counted, and the concentration of each of the fossil pollen types can then be calculated from the ratios of fossils to marker as follows:

$$\frac{\text{Concentration of fossil pollen}}{\text{Concentration of marker pollen}} = \frac{\text{Fossil pollen counted}}{\text{Marker pollen counted}}$$

With these concentration data alone it is possible to plot a **pollen concentration diagram**[1] in which the pollen importance values are grains per milliliter or grains per gram. This diagram escapes the statistical constraints of a pollen percentage diagram in that the numbers given to any pollen taxon are not dependent on the quantities of other pollen present. Pollen concentration diagrams, however, introduce a fresh uncertainty in that they are ratios of pollen number relative to mass of organic or inorganic matter in the sediment. Obviously, the rate at which debris other than pollen can collect can be highly variable, and it is easy for pollen concentration to vary by a factor of ten up and down a section of peat or lake sediments. Pollen concentration diagrams, therefore, tend to be no more informative than percentage diagrams, and sometimes less so. The value of a measure of concentration lies in the possibility of calculating pollen influx.

If the sedimentation rate is known from all parts of a sediment column, then the influx calculation can be made. Figure 17.6 is a pollen influx diagram for the sediments of glacial lakes in New England, showing that populations of pines had a complicated history that was not evident in Figure 17.5. It is clear that the percentage diagram gives a misleading account of the history of pine (M. Davis, 1969).

Special Problems with Pollen Data

A pollen spectrum is a very incomplete species list of a plant community because some plants produce much more pollen than others. Plants are pollinated either by wind (anemophily) or by animals, principally insects (entomophily), though hummingbirds and bats serve for orchids and other tropical flowers. Wind pollination disperses very large amounts of pollen; insect pollination very little. Furthermore, pollen dispersed by insects is not loosed into physical circulation except on an insect corpse, so that the pollen analyst rarely finds the pollen of entomophilous plants. A pollen spectrum, therefore, tends to list mostly wind-pollinated plants.

In temperate latitudes most forest trees are wind-pollinated, and the pollen spectra in these places are mostly of trees (Figure 17.5). In the Arctic and on prairies, wind-pollinated grasses and sedges are well represented in pollen spectra, though sometimes much tree pollen blows over open country from distant forests and shows up in the pollen rain.

In the wet tropics most of the trees join the herbs in using animals for pollination. One of the many reasons that pollen analysts have done less work in the tropics than elsewhere is the resulting expectation that pollen influx in the tropics should be low. A pleasant surprise of the last few years is that, on the contrary, tropical lake sediments like those of the Amazon rain forest can contain even more pollen grains than temperate sediments, making possible subtle histories of tropical forests (Liu and Colinvaux, 1985; Bush, 1991).

Pollen can usually be identified only to genus by light microscopy (Figure 17.4). Sometimes it is possible to do better with an electron microscope but the practical difficulties of surveying thousands of pollen grains with this instrument means that we do not often try. So a pollen percentage diagram like that shown in Figure 17.5 lists mostly generic names only. For some kinds of plants, most frustratingly grasses and sedges, it is possible only to identify

[1]Pollen concentration diagrams are sometimes called "absolute pollen diagrams" or the concentration is referred to as "absolute pollen frequency" (APF). Some pollen analysts believe that this term is misleading and should be abandoned (Maher, 1972; Colinvaux, 1978), but the term still appears in the literature.

pollen to family, so that the taxa in a pollen analyst's list become "Gramineae" and "Cyperaceae."

If a pollen sample is taken from the ground under a tree it should not be surprising to find the sample rich in the pollen of that particular tree. Pollen analysts say that this is a problem of **local overrepresentation**. This is always a problem with palynology of bog or peat deposits because the sediment collects pollen of bog plants as well as of the surrounding community. The problem is avoided with cores from lakes of modest size, though these present other problems: pollen is swept by waves and currents, and so may be concentrated by size in different parts of the lake, or it may be mixed down into the mud away from its original layer by burrowing animals (R. Davis, 1967). Though these various problems can lay traps for the unwary, they can be avoided (Birks and Birks, 1980).

CLASSICAL PALYNOLOGY: THE EUROPEAN SEQUENCE

Pollen analysis was invented early in this century to test hypotheses about climatic change in northern Europe. The hypotheses came from studies on the gross stratigraphy of bogs and suggested that there had been a series of widespread climatic episodes, some six to ten in all, since the end of the last ice age. The proposed sequence of climatic events is called the Blytt-Sernander sequence (Figure 17.7). The extent to which this climatic reconstruction is correct is central not only to the history of European climate and vegetation but also to the archaeology of Europe.

The Swedish bogs may be more than 5 meters thick. Where peat cutters have been working it is possible to observe a cross section of peat, where some remarkable layering becomes obvious. Most striking are layers of tree stumps, erect and in place, showing that trees once grew on the bog. These traces of ancient woodlands might, of course, mean no more than that a particular bog had been drained by some local geographical event and later flooded once more, drowning the trees. Geographical changes of this kind occur all the time, as streams cut out valleys and landslides block them again. But the pattern seemed common to many Swedish bogs, suggesting that changes in climate were responsible for drying and wetting bogs of a large area (Sernander, 1908; Flint, 1971).

Climatic Periods
The Subatlantic Period. Climate humid, and, especially at the beginning, cold.
The Subboreal Period. Climate dry and warm, much as in central Russia.
The Atlantic Period. Climate maritime and mild, probably with warm and long autumns.
The Boreal Period. Climate dry and warm.
The Subarctic Periods of Blytt. The climatic conditions more or less undetermined.
The Arctic Period. In Scandinavia a climate like that of South Greenland.

FIGURE 17.7
The climatic sequence constructed by Blytt and Sernander from bog profiles in southern Scandinavia.
Tree stumps define the Boreal and Subboreal periods, and leaves of arctic plants like *Dryas* define the Arctic period. The Atlantic and Subatlantic periods have peat of relatively moist and warm times. (From Sernander, 1908.)

Under the bottom layer of stumps was marshy peat, but included among the leaves and seeds of sedges were leaves of the little arctic plant *Dryas*, a small, white, herbaceous rose with a yellow center, a plant exclusively of the arctic tundra. This bottom layer thus represented vegetation of what could be called an arctic period. Higher up was the first line of tree stumps, commonly birch, but sometimes other trees too. It seemed certain that a bog must have been partially drained to allow trees to root, suggesting climatic drought, the Boreal period. But there is peat between the *Dryas*-bearing bottom mud and the first line of stumps, a gradation probably, but one that could be used as a stratigraphic unit—the subarctic period. Above this was more wet peat overlying the tree stumps, doubtless due to flooding and a wetter climate—the Atlantic period. Then there was another line of stumps, commonly pine tree stumps, thus another dry time, called the subboreal period. Next was the modern peat that continues to the treeless top of the bogs, which represents the modern climatic epoch, a wet time, the sub-atlantic period. In this way (Figure 17.7) Blytt and Sernander used plant fossils to infer climate and then to separate postglacial time into a series of discrete climatic periods.

From the first moment that this scheme was put forward, there was room for doubt. What did we really have other than a record of water levels in bogs? When the bogs dried a little, trees grew on the tops of them; when they flooded again, the trees were drowned. A climatic explanation certainly seemed necessary to synchronize the water levels in bogs all over Scandinavia, but there was nothing in this to show that climate was divided into discrete epochs. All we really had was evidence that the bogs had dried out twice, slowly each time, and enough each time for trees to flourish for a generation or two. There was quite a leap from this to the impression of climatic succession given by Figure 17.7. Pollen analysis was invented to discover what was happening outside the bog itself (von Post, 1916).

Sweden is a fine country for pollen analysis because it is vegetated with only a few species of trees, all wind-pollinated. Von Post found that zones in his percentage pollen diagrams matched the periods of the Blytt–Sernander sequence. Since then hundreds of percentage pollen diagrams have shown that this sequence can be identified all over northern Europe, not just in Sweden. Furthermore, extra pollen zones allow the sequence to be refined. In all, nine (ten granting modern times a thin zone at the top) pollen zones have been identified since the retreat of the continental glacier. They are known by Roman numerals, counting from the bottom up (Figure 17.8).

European pollen diagrams like that in Figure 17.7 extended the old bog record of Blytt and Sernander with the three zones at the bottom, recording what is now known as the Allerød oscillation. This was a brief warm period, shown as a small blip of tree pollen, showing that trees briefly invaded the tundras alongside the retreating ice before the cold returned for the last time. Paleoclimatologists are now, in the 1990s, checking to see how widespread this Allerød warming was.

The completed European pollen sequence is as follows:

IX sub-Atlantic
VIII subboreal
VII Atlantic (later)
VI Atlantic (early)
V boreal
IV preboreal
III younger dryas
II allerød
I older dryas

(III–I: Late glacial, the original Arctic period)

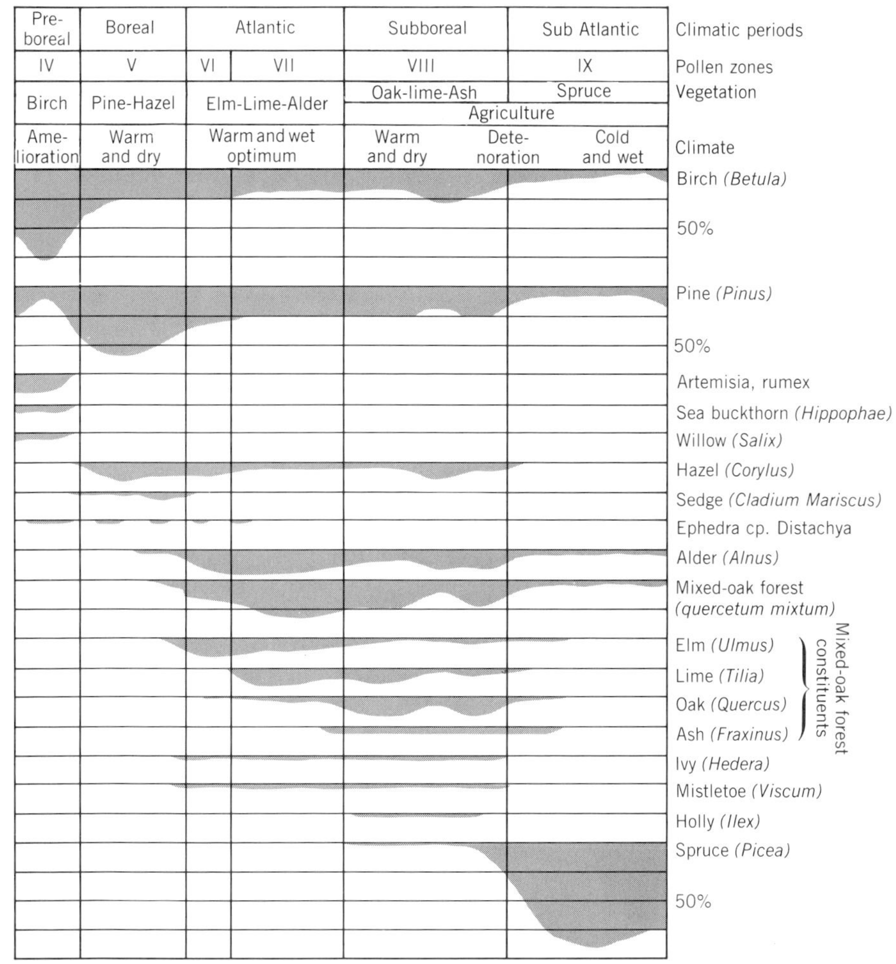

FIGURE 17.8
Blytt–Sernander climatic periods and pollen zones.
The pollen are shown as percentage total arboreal (tree) pollen, and each division represents 25%. The few herbs in the total count (nonarboreal pollen, or NAP) are shown in darker shading. (After Hafsten, redrawn from Faegri and Iversen, 1964.)

The hypothesis of a sequence of large-scale climatic changes since the last ice age, therefore, could not be falsified by pollen analysis and became established as a working hypothesis (Flint, 1971). This early success of palynology established the usefulness of the pollen tool in paleoecology according to the geological definition. Traces of fossil communities had been used to infer past environments. The march of logic had gone:

1. From pollen spectra describe a vegetation type.
2. From the vegetation describe a climate.

At the same time palynology came to be useful as a tool for dating. Over the first 40 years of palynology there was no radiocarbon dating; indeed, there was no method at all of putting an accurate age on anything older than the invention of writing. But a pollen zone could be a time-stratigraphic marker, just as geologists used older fossils to organize their stratigraphy. Up to about 1950, therefore, European pollen analysts often were called in to "date" an archaeological dig or a late glacial event by collecting pollen from surrounding mud and matching it to one of the nine pollen zones. This did not give ages in years, but assigned a dig to its relative position in the archaeological sequence.

In these early circumstances the use of pollen to test ecological hypotheses themselves was obviously limited. Also, without independent dating, calculations of pollen influx were impossible. More seriously, the habit of using assumptions about pollen to construct climates became so established that reversing the logic was not easy.

DATING

Dating in paleoecology can have the same dichotomy of opposed objectives as the rest of the subject. In dating, this conflict takes the form of, "Shall the fossils be used for dating or is there an independent way of dating the fossils?" In classical European palynology, pollen zones were used as a rough system of dating the events of the Holocene, as we have seen. Current practice in North America is to use pollen for dating in one very special instance—the dating of first European settlement. First settlement appears as an abrupt rise of ragweed pollen (*Ambrosia*) from a trace to 10% or more of total pollen. Since the date of first settlement can be found from historical records, the *Ambrosia* rise can date a layer near the top of lake sediments to the year. This can be very valuable to a paleoecologist needing time scales over the last two centuries. But for most purposes in paleoecology and paleoclimatology we want a date that is independent of the fossils to be used. Radiocarbon is the mainstay within its range, but other isotopic and comparative techniques are available as well.

Radiocarbon Dating

The isotope carbon 14 is synthesized in the upper atmosphere out of nitrogen 14. The energy source is cosmic radiation, which induces a flux of neutrons in the upper air. A neutron striking a nitrogen atom is the primary synthesizer of ^{14}C:

$$^{14}N + n = {}^{14}C + H^-$$

The radiocarbon so formed decays spontaneously back to nitrogen, emitting a beta particle in the process:

$$^{14}C = {}^{14}N + \beta^-$$

Beta decay of ^{14}C proceeds at a constant rate, giving a **half-life** of 5730 years. If cosmic bombardment of the earth was really constant, then the atmosphere would have a constant composition of ^{14}C depending on the rate of synthesis due to cosmic bombardment and the rate of beta decay. It is now known that the intensity of cosmic bombardment has fluctuated somewhat over the last 10,000 years, with peak intensities 5000 or 6000 years ago, which cause radiocarbon ages to be about a thousand years too young in this period. For many dating purposes we can assume constant bombardment and the consequent constant concentration of ^{14}C in the atmosphere. This ^{14}C is oxidized and remains in the air as carbon dioxide 14, in which form it is incorporated by plants in photosynthesis. All living things, therefore, contain ^{14}C in their tissues, the concentration being a function of the concentration of $^{14}CO_2$ in the contemporary air.

Radiocarbon dating is merely an estimation of how long organic matter has been dead by a proportional measure of how much ^{14}C remains undecayed. We know the concentration of ^{14}C in the living tissue, the half-life of ^{14}C, and the residual ^{14}C in the corpse, so it is a simple matter to calculate the time of death. But there is a big catch. Until recently it was not possible to measure residual ^{14}C directly because the concentration is so low. What can be measured with comparative ease is the beta decay rate. Instead of measuring all the ^{14}C atoms in a sample, therefore, we have been reduced to measuring that tiny fraction of atoms that are actually emitting their beta particles in our laboratory.

The standard method of radiocarbon dating is to burn the sample to convert all the carbon into gas, which is then collected. Some laboratories insert the resulting $^{12}CO_2$–$^{14}CO_2$ mixture directly into their counting chambers; others convert the carbon to methane. The actual counting operation involves placing the flask of gas within a ring of Geiger counters to record the rate of beta emission. This rate is, of course, a function of the ^{14}C remaining in the carbon mixture and allows the age calculation to be made.

But measuring only the tiny fraction of included ^{14}C that actually decays during the time of measurement is crude. Much more precise measurement is possible if the whole mass of ^{14}C present can be measured and compared with the ^{12}C. What is needed is a mass spectrometer so sensitive that it can identify ^{14}C. This is at last possible using an accelerator as a mass spectrometer, and the first specially built facilities are now operating. They can date tiny samples in the milligram range.

Other Methods of Dating

In principle, any system of radioactive decay can be used as a clock, always provided that the initial conditions at the time the clock was started are known. For ^{14}C the assumed initial condition was an atmosphere in equilibrium for ^{14}C with constant cosmic bombardment. From this assumption it follows that ancient plants held ^{14}C in the same proportion as ^{14}C in living plants. With other dating systems the biggest difficulty usually is in knowing the initial conditions when the clock was set.

Several methods have been devised using the decay products of uranium, known as the **uranium daughter series**. The world oceans are so large that the concentration of uranium in seawater is either constant or else changes negligibly within a few hundred thousand years. Uranium 238 decays to a series of daughter nuclides, of which thorium (^{230}Th) has a very long half-life (8×10^4 years). But thorium is precipitated in sediments so that the sea is swept clear of thorium. Any crystalline process that traps uranium from seawater, therefore, carries within it a private clock. The initial conditions are an input of ^{238}U and no thorium. Provided the crystal is properly isolated from its environment it begins to collect ^{230}Th from the decay of its own ^{238}U. To date such a crystal all we need do is measure the uranium/thorium ratio. This method has been applied to calcite in oceanic sediments, most notably in the carbonates of coral reefs. It gives ages in the range 5000 to 350,000 years, and is most useful in the 100,000 to 300,000 bracket, which cannot be reached by radiocarbon.

Uranium daughters can also be used for dating bones. Living bone is essentially without uranium, but dead bones trap uranium from groundwater. Bones serve as a uranium trap, however, only as long as they contain organic components as well as the bone mineral apatite. The organic matter in buried bones quickly rots, thus leaving only a short window of time in which uranium is collected from solution. This sets the clock, and the initial conditions are dead bones with uranium but no daughter products. As with oceanic calcite crystals, the $^{238}U/^{230}Th$ ratio can be measured to yield a date. A particular hazard with bone dating, however, is that ^{230}Th can be deposited into the sample from outside. Uranium provides an alternative clock, though one with a shorter range, which can be used as a check against this. In addition to ^{238}U there is the isotope ^{235}U in the bone samples, and this decays to protactinium (^{231}Pa). If the $^{235}U/^{231}Pa$ and $^{238}U/^{230}Th$ clocks give the same age, there has been no contamination. This method has been used recently to demonstrate that human remains in California, which were once thought to be so ancient as to be puzzling, are in fact no more than 11,000 years old, like other traces of early man in the Americas (Bischoff and Rosenbauer, 1981).

Young sediments formed over the last few decades can be dated by isotopic clocks also. The radionuclide cesium 137 was produced by bomb testing in the atmosphere and reached its highest concentration in lake sediments around 1963 (Pennington et al., 1976). With a half-life of only 8 hours, its usefulness should soon end. The lead 210 method remains, however. Lead 210 is in the uranium daughter series, the immediate product of the gas radon 222, and has a half-life of only 22 years. Since ^{210}Pb reaches lake sediments both from atmospheric fallout and from uranium series decay within the sediments, care has to be used in determining the initial conditions (Brugam, 1978). Dates back to 150 years are possible, and the method has value in studies of human impact on lake systems.

Other dating methods continue to be invented. For example, amino acids in bone collagen change spontaneously from *l* isomer to *d* isomer depending on temperature.

OXYGEN ISOTOPES AND ICE AGE CHRONOLOGY

The isotopes of oxygen have been used to record temperatures of the past directly or to supply chronologies of ice ages in marine sediments. In both roles, these isotopes provide some of the most elegant instruments in the paleoecological tool kit.

There are two isotopes of oxygen, ^{16}O and ^{18}O. The heavy isotope, ^{18}O, is stable, just like the more common ^{16}O, and does not decay. Its usefulness comes from the fact that it enters into chemical reactions as a function of temperature. Thus the ratio of the two isotopes can, in principle, be used as a thermometer that records the temperature at which a chemical synthesis or change of state took place.

As with using isotopes for dating, it is essential to choose materials that have remained unaltered since the original synthesis, which has directed much ^{18}O work to marine carbonates, particularly the mineral calcite. Over temperatures of interest to biologists the changes in concentration of ^{18}O in carbonates are extremely tiny. A change of 0.5°C, for instance, results in a change in the abundance of ^{18}O of only one part in 25 million. Extremely sensitive mass spectrometers are required to detect such changes accurately, although several laboratories now achieve these results almost as a matter of routine.

The first demonstration of the possibilities of ^{18}O for ecology was made by Urey et al. (1951), who examined the shell of a belemnite from the Jurassic. Belemnites were free-swimming mollusks related to squids, but they had rather massive external shells of calcite. Sections of these shells reveal a pattern of concentric rings, almost certainly growth rings. Urey and his colleagues made a slice through a belemnite shell and assayed the $^{16}O/^{18}O$ ratio for each of the broader rings (Figure 17.9). They put their conclusions as follows:

This Jurassic belemnite records three summers and four winters after its youth, which was recorded by too small amounts of carbonate for investigation by our present methods; warmer water in its youth than its old age, death in the spring, and an age of about four years.

This is a postmortem on an animal that had been dead for more than a hundred million years.

This demonstration had, of course, also revealed the temperature of that part of the Jurassic ocean: a mean temperature of 17.6°C with a seasonal variation of about 6°C. Obviously there are powerful conclusions that can be drawn about any marine environment that left suitable calcite fossils as we describe temperature and seasonality in advance of collecting community and population data.

^{18}O *and the Ice Age Oceans*

The ^{18}O method was next applied to foraminifera tests. These calcite skeletons of planktonic and benthic animals are found in large numbers in piston cores from ocean sediments. The planktonic species that live near the sea surface can be separated from the benthic forms. Measures of the $\delta\ ^{18}O$ of planktonic foraminifera tests from deep sea cores thus should record the temperature of the ocean surface throughout ice-age times. However, this temperature record is at least in part obscured by a stronger signal, that of the extent to which the two isotopes of oxygen are stored in glacial ice.

Snow has less of the heavier oxygen isotope than does warmer water, and the great ice sheets of glacial times, made as they were from snow, were reservoirs of the lighter isotope ^{16}O. When ice sheets are present on the land, therefore, the surface oceans that evaporated to make snow have relatively more of the heavy isotope ^{18}O.

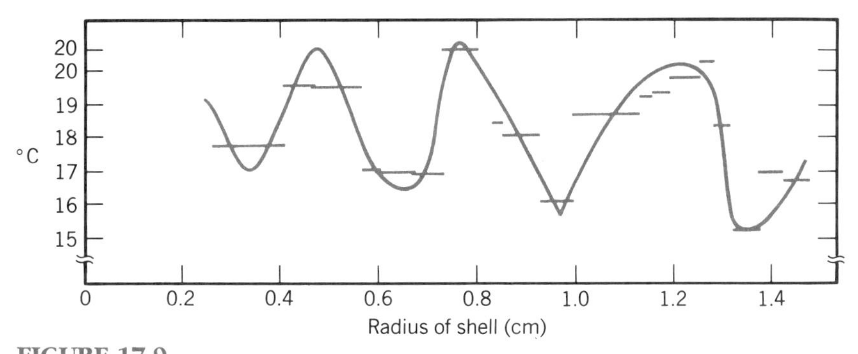

FIGURE 17.9

Seasonal temperatures of the environment of a Jurassic belemnite. Temperatures computed from $\Delta^{18}O$ in successive growth rings in the shell. Horizontal lines show thickness of sample used. (From Urey et al., 1951.)

Thus high $\delta\ ^{18}O$ of planktonic foraminifera from ice-age times must be due both to cooler water and to the fact that ^{16}O had been differentially extracted from the oceans and stored on the land in glacial ice.

Figure 17.10 shows the oxygen isotope history of ocean surface water for the last half million years. The isotope mixture is relatively light in modern mud at the left of the diagram, was heavy in the immediate past about 14,000 years ago, and was not relatively light again until somewhere between 100,000 and 150,000 years ago. The diagram is divided along the top into isotope stages in what is now a universally accepted time-stratigraphic scale. Isotope stage 1 is the Holocene, and stage 2 records the maximum extent of ice sheets in the last ice age. Isotope substage 5e records the brief time of the last interglacial.

The isotope stages were first defined by Emiliani (1966) upon the supposition that the strong signal in the isotope ratio of planktonic formanifera was from the temperature of the water: colder oceans in glacial times and warmer oceans in interglacials. It was soon concluded, however, that the overriding signal reflected the storage of light oxygen in glacial ice (Shackleton, 1968). As a record of the relative volume of glacial ice on the earth, Figure 17.9 provides a history of ice ages. This history has been continually refined over the last 25 years with more cores, more detail, and more dates until it provides the standard reference for the chronology of ice ages.

Because water temperature was difficult to infer from isotope ratios that were heavily influenced by the ice-storage effect, ocean temperatures have been interpreted by using foraminifera as index fossils for temperature, as described below. Recently, however, Emiliani (1992) has reasserted the importance of the ocean temperature effect in determining the $\delta\ ^{18}O$ content of sea water. This does not alter the value of the record as a chronology of ice ages, but it is important for ecologists reconstructing the temperature of the lowland tropics in ice ages. Emiliani's interpretation requires cooler sea surface

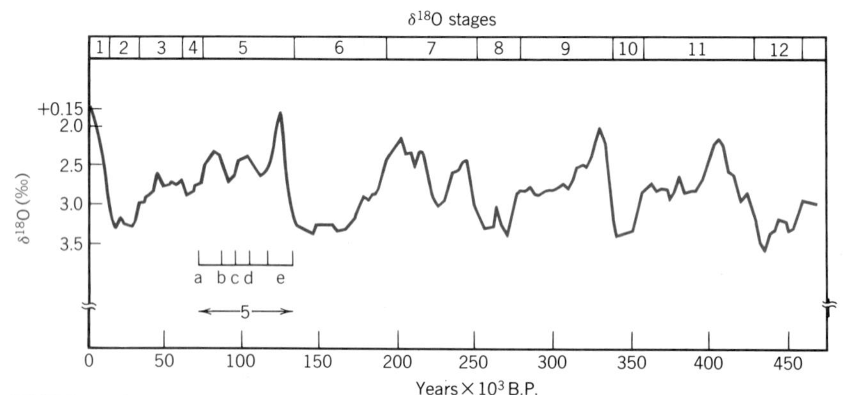

FIGURE 17.10

$\delta^{18}O$ record from deep-sea cores.

Low $\delta^{18}O$ represents interglacials like the present and high $\delta^{18}O$ glacial times when ice sheets differentially stored ^{16}O. The last interglacial is divided into stages 5a–5e. The chronology is derived by extrapolation from various radiometric dates. (Data of Hays et al., 1976.)

temperatures at the equator in ice-age times, thus making temperature reconstructions from deep sea cores compatible with estimates of pollen analysts working in tropical rain forests.

The Possibilities of a C4 Thermometer

It may be that a land paleothermometer can be found that has an accuracy comparable to that of oxygen isotopes in marine calcites. The possibility relies on the fact that different photosynthetic pathways tend to be used by plants at different temperatures (see Chapter 3). The C4 pathway is predominant in hot, dry places and the C3 pathway in cooler, moister places. Livingstone and Clayton (1980) suggest that we use plant microfossils from lakes to infer the photosynthetic pathways of local plants, and hence the local temperature.

It seems particularly reasonable to apply the method to grasses, since whole genera or subfamilies within the grass family Gramineae use either one pathway or the other, but not both. Livingstone and Clayton show that in East Africa the importance of these grass genera in the local vegetation depends on altitude and temperature. The C4 grasses are in the hot lowlands and the C3 grasses in the cool highlands, thus suggesting that relative presence of the two groups of classes change with temperature. However, the mixture in ice-age grasslands might also have been influenced by the low atmospheric oxygen of ice-age times, requiring that temperature and carbon dioxide concentration effects be separated.

Separating C3 and C4 grass remains in sediment cores cannot be done by pollen analysis, since all grass pollen looks alike. But the East African sediments have many charred grass fragments that were blown into the lakes as dust from old fires and these fragments of tissue often can be identified to genus. The paleoecologist must hunt for microscopic, charred fragments of grass cuticle to use this thermometer, and also must learn to recognize grass genera from cuticular fragments.

TRANSFER FUNCTIONS ON FORAMINIFERA

Maps can be made of the distribution of planktonic foraminifera in the oceans. The result is something like a vegetation map in that broad groupings of foraminifera can be found in different oceans in the way that broad formations occupy areas on land. The most plausible hypothesis to explain this pattern is that it reflects water temperature. This should mean that a species list of fossil foraminifera from a deep sea core holds a record of the temperature of the sea surface.

Imbrie and Kipp (1971) showed how this temperature record might be read using the technique of transfer functions. The method depends on finding best-fit correlations between the distribution of modern foraminifera and isotherms of the sea surface. Species lists of foraminifera from an ocean can be quite large, with perhaps four to 600 species being present. Whole species lists are examined, species by species, in the hunt for those whose distributions do coincide with measures of temperature. Many species do not show clear correlations and are accordingly rejected. In the end a grouping of species is left, the maps of which are good fits to maps of water temperature. From this list are calculated the transfer functions that derive water temperature from community composition.

These transfer functions are then applied to species lists of fossil foraminifera from the cores, again rejecting the species that are known to fail to meet the requirements of the model. The result is a measure of the surface water temperature at the core site at the time the foraminifera lived. The method has been applied to core material with radiocarbon ages of 18,000 years from all the world oceans to yield a map of surface temperature at the last glacial maximum (see Chapter 7; Figure 7.9). This is the map on which models of global climate of the last ice age are now based (CLIMAP, 1976).

It is important to note that this transfer function method of Imbrie and Kipp makes no claims about communities of animals. There is no suggestion that discrete communities of foraminifera exist in nature, nor that communities have boundaries set by temperature, nor even that temperature is an overriding limiting factor for individual species. All that is claimed is that many species are affected by temperature, a proposition that no ecologist is likely to resist. The investigators put together artificial groupings by their computer selection program, seeking to amplify the temperature signal hidden in a species list. Then they sort out analogs of these artificial groupings in the fossil assemblages.

Transfer Functions on Pollen Data

In principle, the transfer function method can be applied to generate climate maps from pollen data as well, but special difficulties threaten the precision that can be obtained. The most serious is that the species list of a pollen assemblage is much smaller than the species list of foraminifera: all the foraminiferans in the parent community may be represented in the fossil collection, but not many members of a plant community turn up in a pollen spectrum. A second difficulty is that perennial one of palynology, the lack of precision in taxonomy that gives us the names of genera only: obviously a genus can have species with widely different tolerances to temperature. And a third difficulty is that most existing pollen data are expressed in percentages rather than as influx.

Nevertheless there are some striking demonstrations of the possibilities inherent in the method (Webb and Bryson, 1971). The method begins by scanning surface samples and the fossil pollen spectra from the region to be worked in a subjective quest for pollen taxa that look as if they might have fairly reliable correlations with climatic variables. All taxa that are scarce in pollen diagrams, or that appear at the surface as a result of human activity, or that are common in fossil assemblages for unknown reasons, are excluded from further consideration. In addition, a further list of taxa is excluded on the grounds of botanical intuition: the belief that the parent plants are not closely limited by climatic variables. What is left is a set of artificial groups of species. Transfer functions are then calculated between these lists and climatic variables such as temperature and moisture, using multivariate methods as Imbrie and Kipp did for the foraminifera. Figure 17.11 shows the kinds of results that can be obtained.

THE USES OF TREE RINGS

The most familiar use of tree rings is as a clock to date a living tree: in temperate latitudes the rings are annual and it is a simple matter to count back to year one (Figure 17.12). But where long-dead trees are available alongside living trees, it is possible to match rings of the living and dead where their lives overlapped and to count back far into the past. This is the technique known as **dendrochronology**.

A tree-ring calendar in Arizona now spans the last 9000 years, taking advan-

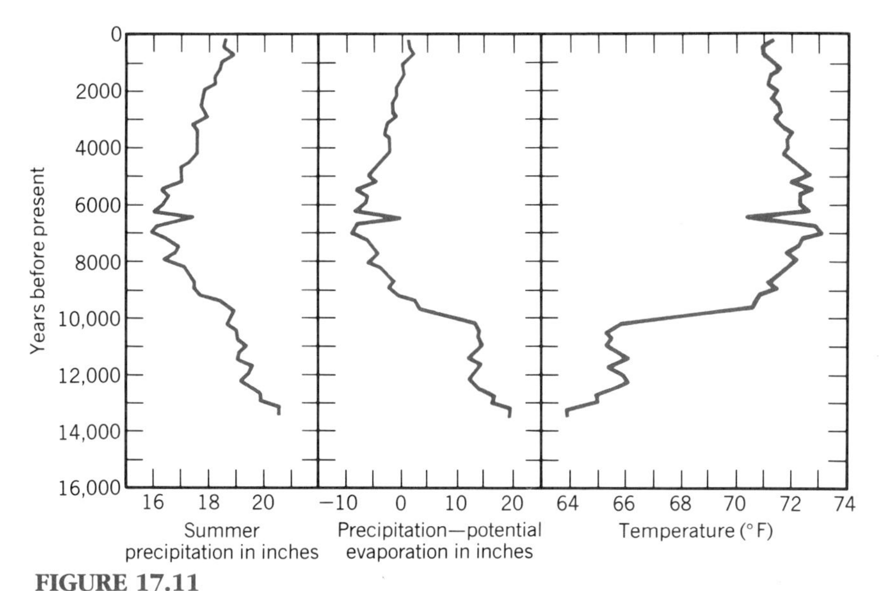

FIGURE 17.11
Climate from pollen data using transfer functions.
Transfer functions correlating percentage total pollen of selected taxa are correlated with modern climatic variables and then applied to pollen percentages from cores from Kirchner Marsh in Minnesota. (From Webb and Bryson, 1971.)

tage of the very long lives of the bristlecone pine (*Pinus aristata*) and a desert environment that preserves dead wood for long periods (La Marche, 1974). A similar chronology of bog oaks from Northern Ireland and Germany has now been completed to yield a calendar for Europe spanning the last 7000 years (Pilcher et al., 1984).

A tree-ring chronology can be used to date wooden artifacts or ruins if rings in the specimens can be matched to a block of rings in the chronology, but a more general use in the service of dating has been to calibrate the radiocarbon time scale. Wood from a single year or small run of years can be taken and its ^{14}C content determined in the usual way. By this means it has been possible to discover periods when atmospheric concentration of ^{14}C was not at the assumed constant level because of temporary changes in the cosmic ray flux. At one point in postglacial time a carbon date can be as much as 1000 years in error for this reason. However, because of calibration of ^{14}C against tree rings, this can be allowed for in assigning the age of a specimen over these spans of time (Figure 17.13).

Perhaps more exciting for paleoecology, however, is the opportunity these long series of tree rings give for their own environmental records. A tree ring is a sample of living material from a known past year. The $^{16}O/^{18}O$ ratio in the ancient wood has been investigated for its possible use as a direct thermometer recording past air temperatures (Libby et al., 1976; Gray and Thompson, 1977). Figure 17.14 shows how well temperatures reconstructed in this way from oak trees in Germany compare with the historical record of temperature from England over the last 600 years.

FIGURE 17.12
Tree rings: annual samples of old atmosphere and climate.
Trees of high latitude forests deposited rings of characteristic wood in low and high growth times of the year, making possible tree ring chronologies and identifying organic samples of ancient ecosystems for analysis. Unfortunately trees of the wet forests of low latitudes grow continuously, leaving no rings that can be used for histories of the tropical rain forest or of equatorial air.

So far, however, the most thorough knowledge of past climates from tree rings has come from applying transfer functions that relate simple tree-ring widths to climatic variables (Fritts, 1991). It has long been known that a tree will put on a wider ring in a "good" year than in a "bad" year, showing that there is a correlation between ring width and temperature or precipitation. The individual record, of course, is confounded by both the health of a tree and accident. The contribution of Fritts (1991) and the Tree-Ring Laboratory at Tucson was to quantify the relationship between ring widths of many trees at each of many sites and a series of environmental measurements. Multivariate analysis of these data sets yields correlation coefficients between ring widths and specific environmental variables that can then be used as transfer functions to apply to sets of tree rings from the past. Figure 17.15 shows a specimen of these results, where climate is determined in terms of departures from mean width by rings of different periods.

THE USES OF BEETLES

Beetle fragments are common constituents of sediments, though it takes large samples of mud to get a good collection. A 4-kg bag of mud from a Pleistocene de-

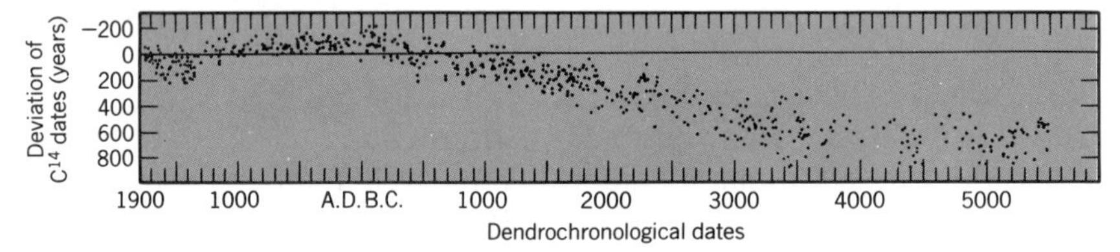

FIGURE 17.13
Calibration of radiocarbon with tree rings.
Data points are the departures of radiocarbon ages from dendrochronological ages of wood samples. (From Ralph and Klein, 1979.)

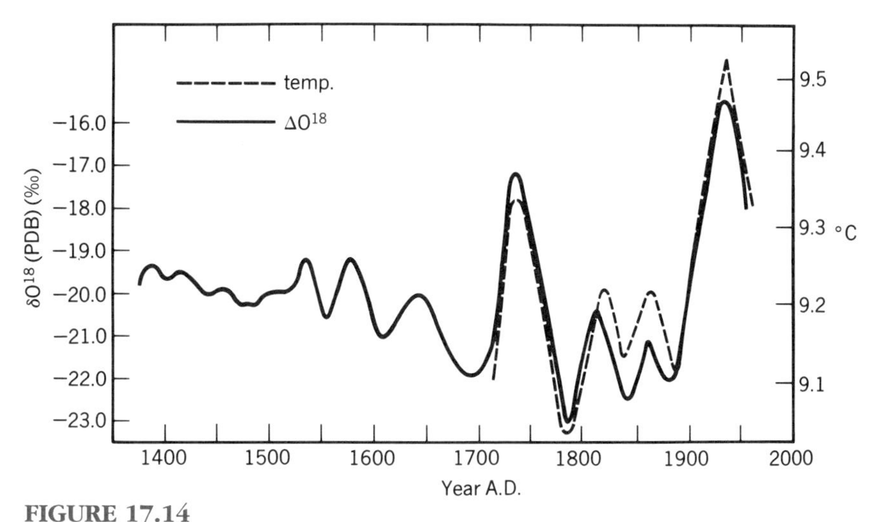

FIGURE 17.14

Correlation between $\Delta^{18}O$ in tree rings and historical records of temperature. The ^{18}O was measured on oak trees from Germany. The temperature records are from England. (From Libby *et at.*, 1976.)

posit, for instance, may yield fragments of 2000 individual beetles. Beetle analysis cannot be applied to typical lake cores raised with a Livingstone sampler because these yield only grams, rather than kilograms, per stratum. But where there are quarries, road cuts, or bluffs through old deposits, it is possible to collect large samples of old beetle communities without much difficulty. Typically the mud is sieved to remove large particles and then dispersed, and the beetle remains float

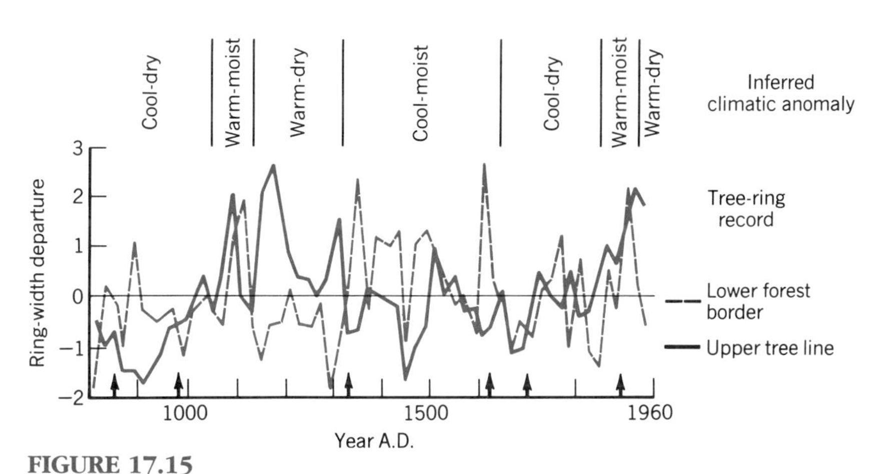

FIGURE 17.15

Climate inferred from widths of tree rings.

Departure from mean width of rings is correlated by transfer functions to the various climatic parameters and these transfer functions are then applied to past tree rings to yield a climatic history. (From La Marche, 1974.)

out. Beetle analysis was invented by G. R. Coope (1959), and has since been developed by him and others into a major tool of paleoecology (Coope, 1977, 1979; Morgan, 1973).

Perhaps the most striking novelty that Coope introduced was the idea that a beetle fragment could be identified to species. The remains consist of elytra (wing cases), the thorax, which is often in one piece, heads, fragments of all of these, and many disarticulated legs. Nothing can be done with the legs, but bits of elytron, head, or thorax usually contain enough ornamentation to be identified to species. The thing that makes this possible is, in fact, this ornamentation (Figure 17.16), which is often highly characteristic of a species. Beetle parts are made of chitin, which preserves very well, so that all this fine structure is maintained.

Beetle faunas are very species-rich, so that the first essential for beetle paleoecology is a thorough taxonomic knowledge of local taxa. Even the depauperate fauna of Britain, where Coope started his studies, has more than 3000 species of named beetles. Elsewhere the lists of possible species are many times larger. To know a fauna so well that the species can be identified from fragments obviously implies much learning, and fossil beetle specialists will only work where, in Coope's words, they "know the vocabulary." Because of this need for special knowledge, beetle studies so far are restricted to Britain and related areas of northern Europe, to Russia, and to a few regions of North America. For communities in these northern latitudes, striking conclusions about both community ecology and climatic history have been made (Figure 17.17).

FIGURE 17.16
Beetle fragments washed from mud.
Fragments like these can be identified to species by a worker who "knows the vocabulary." When fragments can be concentrated from about a kilogram of mud or peat, the species composition of ancient beetle communities can be reconstructed.

PALEOLIMNOLOGY

Lake mud holds traces of the lake's own inhabitants, as might be expected. Most

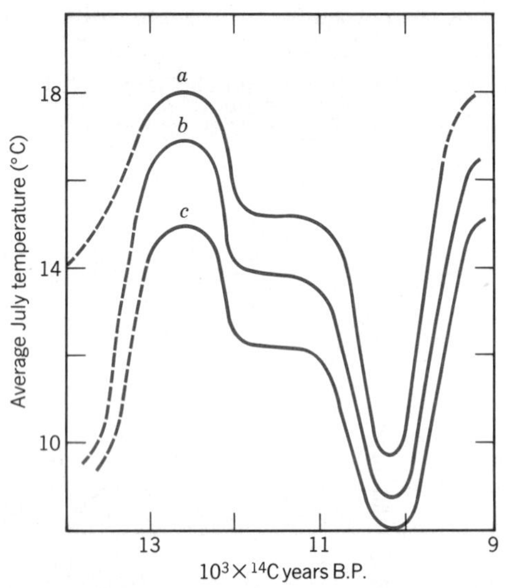

FIGURE 17.17
Climatic reconstruction of ice-age England from fossil beetles.
Primary data are species lists of co-occurring beetles identified to species from subfossil fragments. July temperatures are inferred from present-day distributions. (*a*) southern, (*b*) central, and (*c*) northern England. (From Coope, 1977.)

lake biota are small, and the fossils are even smaller. One milliliter of lake mud may hold many thousand claws from cladocera, together with diatom remains and algal cells even more abundant than pollen grains. All these remains are of select subgroups of the lake's inhabitants, because not all are preserved in the lake mud.

Copepods and rotifers generally leave no traces. Insect larvae, particularly midge larvae (Chironomidae), leave pieces of integument and their jaws. These have been of only moderate use because the taxonomy of the modern forms is difficult and the taxonomy of the remains is even more difficult. There is nothing in lake mud that can be identified with the clarity of beetle fragments, but cladocerans of the family Chydoridae do leave traces that can be identified to genus and sometimes to species (Birks and Birks, 1980).

A chydorid analysis proceeds in much the same way as pollen analysis, namely, extraction from unit volume of mud, counting measured subsamples, and calculating fossil influx if possible, otherwise expressing results in a percentage diagram. Every recognizable fragment is scored as an individual.

Probably the most complete record of a portion of a lake's biota is given by diatoms. These are the best fossils of all for taxonomists, since the complete silica skeletons collect in the mud. They can be extracted and cleaned with nitric acid or strong oxidizing agents, like 30% hydrogen peroxide, yielding a preparation of fossils that looks no different from a preparation of a modern plankton sample. Diatoms of the open-water plankton tell of an ancient open lake; diatoms that can stand exposure on the soil surface tell of a drying episode, and so on. Diatoms are even sensitive to pH, thus yielding histories of water acidity, allowing tests of the hypothesis that the acidity of modern lakes is due to industrial acid rain (Figure 17.18).

More can be learned from the mud itself, rather than just its fossils. The organic content is easily measured by weighing an oven-dry sample and burning it before weighing it again. The second weight is

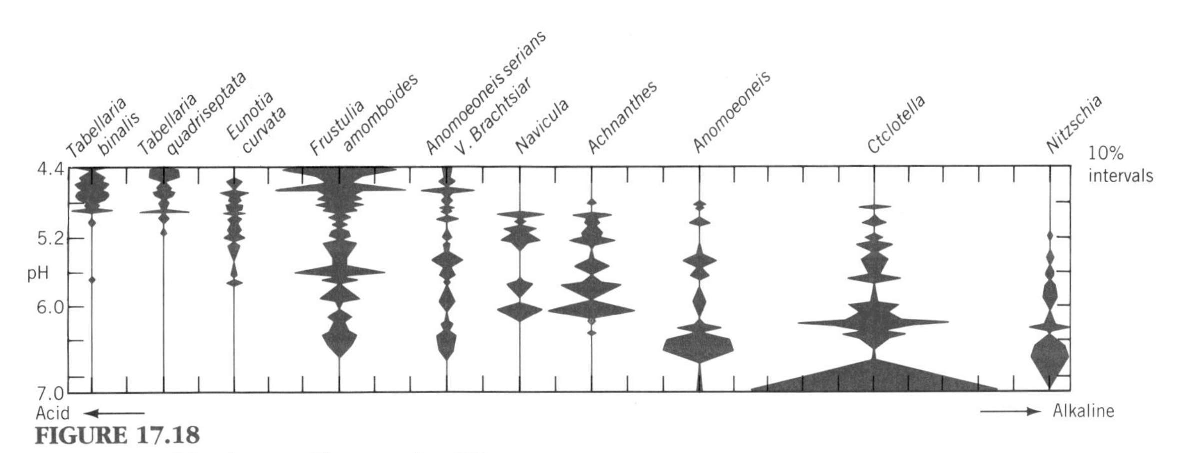

FIGURE 17.18

Sensitivity of freshwater diatoms to acidity.

Data are percent of each taxon present in samples from 44 Norwegian lakes. Vertical axis is pH of lake water. Diatom taxa are arranged according to the mean pH in which they were found. Based on this data set, the percentage composition of diatom frustules in lake sediments can be used to measure the pH history of the water in the past. Diatom studies have been used to demonstrate the acidification of lakes by acid rain. (From Davis, 1987.)

less by reason of the organic matter that was lost as CO_2, and the "loss on ignition" is a crude measure of the total organic content.

The chemistry and mineralogy of sediments can be used to reconstruct much of the physical state of the ecosystem. The concentration of solutes dissolved in the water of a lake is a function of the burial of nutrients in the mud as well as of the rate of input from the watershed (see Chapter 23). It follows that sediment chemistry allows inferences to be made about past nutrient states. The mineral species in mud tell of the rate of weathering and the source of the mineral fraction.

The record in mud, therefore, can be used to reconstruct the history of a lake ecosystem and its external environment on a scale that is not possible for any other ecosystem type. Pollen and macrofossils give the surrounding vegetation. Minerals and chemistry tell of watershed inputs. Sometimes ash layers tell of volcanic eruptions. Oxidized layers tell of low water or drought. All these can be dated. Then organic measures by loss on ignition and pigments may tell of productive state, and fossils of diatoms, chydorids, and other biota tell of the response of the lake's inhabitants.

One of these studies is described in

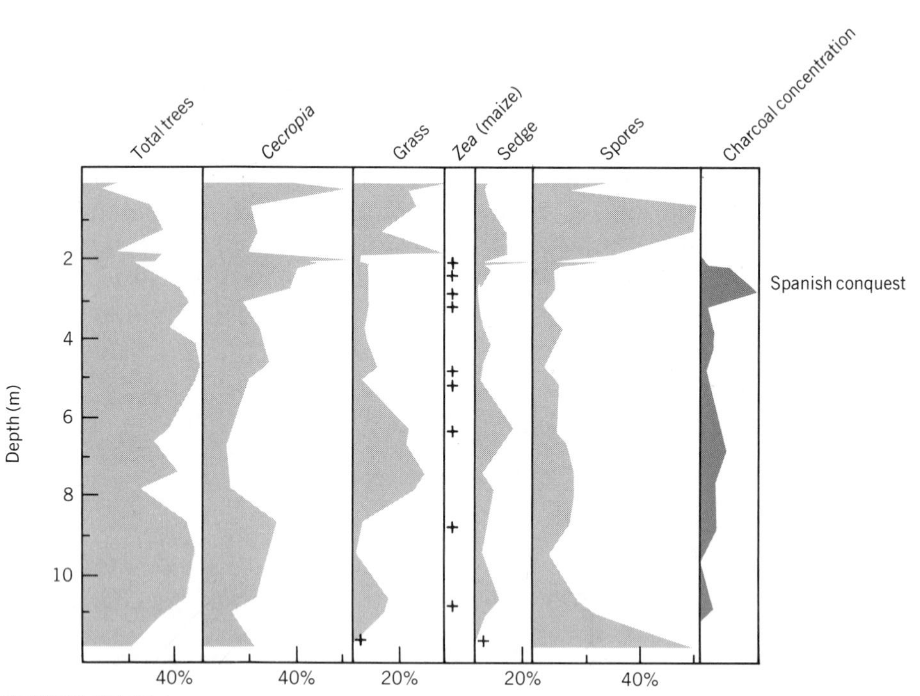

FIGURE 17.19
Four-thousand-year history of the Darien.
Data are pollen percentages of the most abundant taxa from recently discovered Lake Wodehouse, in the Darien rain forest of Panama, together with concentration of charcoal fragments from landscape fires (dark shaded area). Pollen from Indian corn, coupled with the evidence for burning, shows that the region was cultivated for all the 4000 years spanned by the deposits, until the time of the Spanish conquest 350 years ago. *Cecropia* is a pioneer tree of disturbed sites throughout tropical America.

Chapter 26 (Figure 26.11), that of Lago de Monterossi, where pollen describes vegetation and various sedimentary studies give the nutrient states of the ancient lake, the result being a demonstration of how a lake ecosystem is controlled by its watershed (Hutchinson, 1970). This lake story crossed the history of ancient Rome. The use of paleolimnology to look for the traces of the human presence in even earlier times, before the the onset of written history, is increasing.

If people burn a forest, grow crops, or change drainage, lake mud will record those facts. In current attempts from my laboratory to reconstruct the environmental history of tropical America we had the good fortune to find pollen and phytoliths of corn (*Zea mais*) from a 6000-year-old sample of lake mud from the Amazon, making this the oldest trace of human cultivation in the Amazon basin (Bush et al., 1989). Now we have a history from the Darien rain forest of Panama showing that people were farming for 4000 years before the Spanish conquest. But they vanished with the arrival of the conquistadors, after which the forest closed over their lands (Figure 17.19).

Preview to Chapter 18

Treefern (*Dicksonia*) rainforest of coastal Brazil.

A vegetation map, describing the principal plant formations of the earth, is also a map of climate. The principal plant shapes of a formation, adapted as they are to allow cooling, heating, or water conservation in the various climates, do much to define conditions for life, so that plants, animals, and climate are linked into grand ecosystem types called biomes. Maps of biomes made by ecologists essentially plot regions of common adaptation of plants and animals regardless of the relatedness of individual species. A quite different geography is used by evolutionary biogeographers, who map together into regions plants and animals showing obvious signs of relatedness, regardless of adaptation. Important to this evolutionary biogeography is the long time scale on which land masses have been separated or merged by continental drift. The ecological patterns of the contemporary earth have developed only in the 10,000 years since the last ice age, so that the pattern of modern life is still profoundly influenced by the patterns of the ice age earth. Ice sheets to the 40th parallel and below compressed climate zones toward the equator, displacing biomes. Lowered sea level connected islands or drained shallow seas between continents. The tropics were colder, or drier, or both. This geography of the ice age earth lasted for most of the last million years and is to be regarded as normality for most species still living. Modern biogeography is the result of species' having altered range or abundance to accommodate the unusually warm time in which we live.

Chapter 18

The Geography of Communities

In 1807 the naturalist and explorer Friedrich von Humboldt classified plants of different parts of the world by their characteristic shapes: spruce tree shapes, palm tree shapes, cactus shapes. He was remarking the now familiar phenomenon that vegetation in different parts of the earth actually looked different. Soon the naturalist travelers of the nineteenth century had built on this idea of plant forms characteristic of whole continental regions to map the world into plant "**formations**": tropical rain forest, deciduous forest, boreal[1] forest, tundra, and the rest.

The vegetation map of the earth included in most atlases is a map of these formations (Figure 18.1). The maps do some violence to reality, for they show a world more neatly parceled out between formations of plants than it really is. Mapmakers draw boundaries at convenient places, taking advantage of an isthmus, mountain range, or tree line, where such exist, drawing lines where best they may across regions of transition. Nevertheless, the maps do reveal that immense continental regions support vegetation of characteristic form. This is the first datum of community ecology.

The characteristic plant shapes of the formations can be explained as adaptations that make for efficient balancing of heat budgets in different temperature and moisture regimes. Rain forest trees keep cool as year-round evaporators, plants of the arctic tundra keep warm by remaining low in warm air trapped close to the ground, needle leaves of the boreal forest are kept at the temperature of surrounding air by convection, and so on (see Chapter 5). This is the modern understanding, but a correlation of plant form with climate was intuitively obvious to the early ecologists.

It can be argued that the real beginning of modern ecology came in 1855 when

[1]In Greek myth *Boreas* was the god of the north wind.

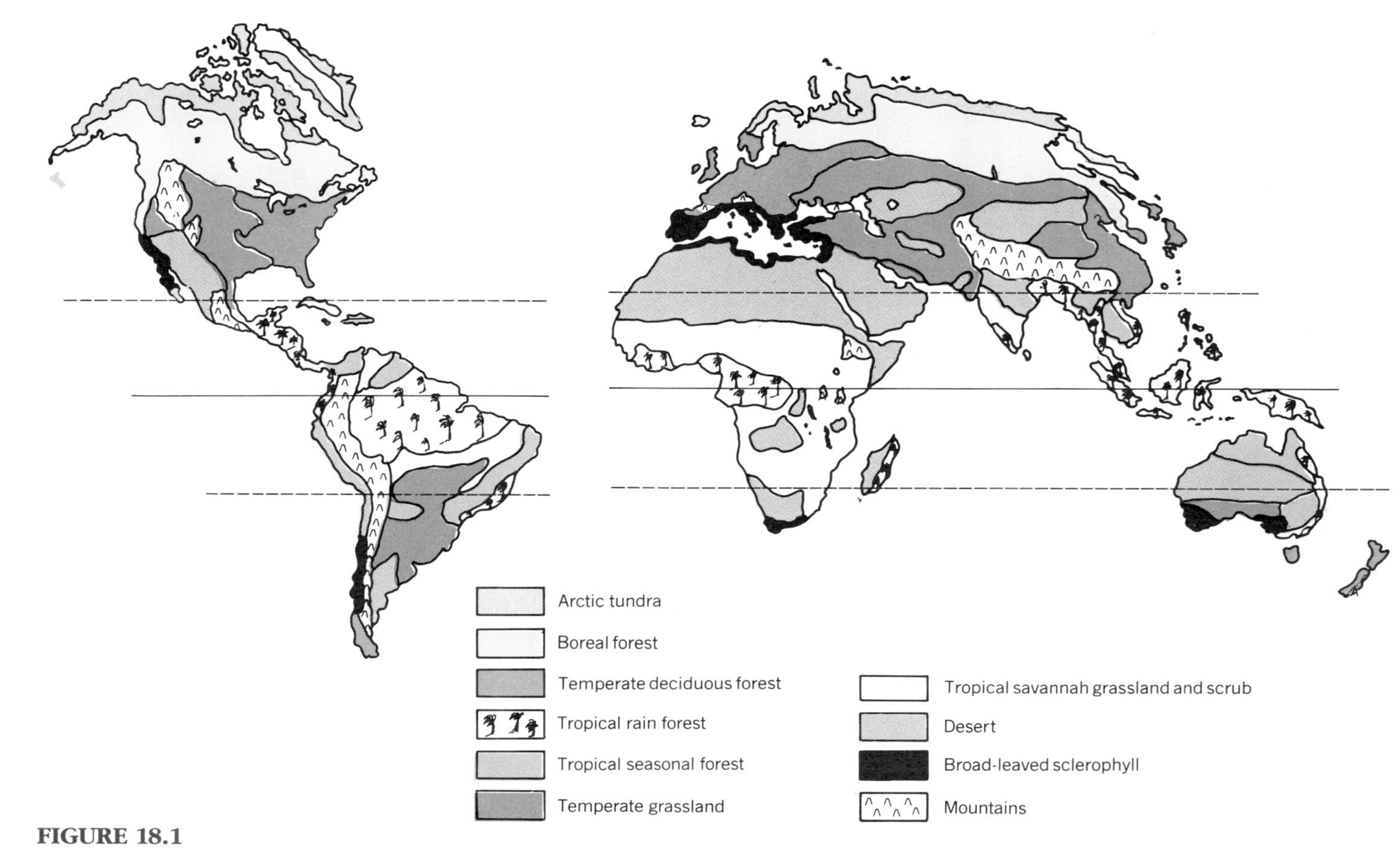

FIGURE 18.1
Major formations and biomes of the world.
(From Cox and Moore, 1980.)

TABLE 18.1
Correspondence Between Classification of Life Forms and Climate

de Candolle Plant Life Forms	Köppen Climatic Type
Megatherms	A
Xerophiles	B
Mesotherms	C
Microtherms	D
Hekistotherms	E

Alphonse de Candolle[2] put forward the hypothesis that life forms of plants, as described by a map of plant formations, were set by moisture and temperature. Five principal community life forms were, for him, enough to describe the vegetation of the whole earth, from the "megatherms" ("much heat") of the tropical rain forest to the "hekistotherms" ("least heat") of the arctic tundras (Table 18.1). Here was the forerunner of the modern concept of biomes, a large communal entity where lifestyles are set to respect their common climate.

THE CLIMATE HYPOTHESIS

A test of the de Candolle hypothesis came when Vladimir Köppen (1884) based climatic maps of the earth on vegetation maps. Köppen set out to classify climatic types in a world over much of which climate data were almost lacking. He used maps of the great plant formations as his base maps, then assigned to each plant formation the climate that seemed appropriate to it. Köppen elaborated on the temperature and precipitation suggested by the early ecologists for each formation, and called the climates A, B, C, D, and E (Tables 18.1 and 18.2). He then took the boundaries shown on maps of plant formations and used those same boundaries to delimit climates, the result being the familiar map of global climates (see Chapter 7; Figure 7.7).

Thus Köppen and the other founders of climatology mapped the plants and called them climate. Maps of vegetation and climate in standard atlases, at least until satellite maps of weather were available, were the same map. But these climatic maps proved to be generally accurate, describing within tolerable limits the actual climate of different parts of the earth. The de Candolle hypothesis that the world was set out

TABLE 18.2
The Köppen Climatic System

Zone	Symbol		Explanation
Tropical rain climate	1.	Af	Tropical rain forest climate
	2.	Aw	Savanna climate
Dry climate	3.	BS	Steppe climate
	4.	BW	Desert climate
Warm temperate rain climate	5.	Cw	Warm, dry-winter climate
	6.	Cs	Warm, dry-summer climate
	7.	Cf	Moist temperate climate
Boreal or snow-forest climate	8.	Dw	Cold, dry-winter climate
	9.	Df	Cold, moist-winter climate
Snow climate	10.	ET	Tundra climate
	11.	EF	Perpetual frost climate

Symbols: B = inclined to be dry; S = steppe climate; W = desert climate; E = warm only in summer; T = tundra climate; F = frost climate; A, C, and D = location and timing of the dry season; s = main dry season in summer; w = main dry season in winter; f = perpetually moist (rain in every month).

[2]Alphonse de Candolle, when in charge of the Paris museum collections, made the last attempt by one person to describe all plant species known to science, resulting in his "Prodromus" of 1824–1873. The formation data crossed his desk in the process.

in a patchwork of different plant formations whose domains depended on temperature and precipitation, therefore, had been tested and found to be correct.

This exercise necessarily raises questions about the reality of the boundaries between formations or climatic regions shown on the maps. Climate is fluid, so that there should be no discrete edges between its patches. Likewise, formations should blend as the driving force of climate blends with the next climate over great distances. Most boundaries are indeed really diffuse, appearing distinct merely as a mapmaker's convenience. Figure 18.2 gives an ecologist's impression of how clines of plant shape merge across formation boundaries. But some, like the arctic tree line in North America, do their merging over remarkably steep gradients from one formation to the next. These are weather boundaries.

Boreal Forest and the Arctic Frontal Zone

The arctic frontal zone over eastern Canada marks the transition between the cold arctic air mass and warmer air coming in from the Pacific. The front moves, both with the random perturbations typical of all weather and more predictably with the seasons, going south in winter and north in summer. Bryson (1966) mapped the mean positions, in summer and winter, from a ten-year accumulation of measurements of air temperature and of wind speed and direction. He found that the frontal zone in summer coincided with the arctic tree line and that in winter it coincided with the southern edge of the boreal forest (Figures 18.3 and 18.4).

Bryson's data provide an elegant demonstration that gradients between significantly different climates can sometimes be sharp enough to produce visible boundaries between prevailing life forms of plants. Face the arctic air in summer, and selective advantage goes to life forms pressed to the ground. Facing the arctic air only in winter allows large trees to grow, but imposes the curious mathematics of energy budgets for a needle-leaf design (see Chapter 5). The time in the annual cycle when the arctic air is encountered is critical, and plants of the wrong design cannot sustain viable populations.

Mean positions of climatic fronts must move over the millennia, in what climatologists call **"secular time scales."**[3] In a colder earth, the heat engine of climate (see Chapter 7) should push the front farther south in winter; conversely, a warmer earth should force the summer front, and the treeline, to the north. Paleoecological data now show that in fact frontal positions have wandered by hundreds of miles with the changing climates of the last few thousand years. Pollen diagrams show how the cold of the little ice age squeezed the boreal forest to the south. Both edges of the boreal zone in which the softwood forests of Canada grow to supply pulp for paper will move north by comparable distances with the coming greenhouse warming.

BIOMES AS USEFUL ABSTRACTIONS

Because the range of plant shapes yielding high fitness depends on seasonal temper-

[3]Use of the term "secular" in science is not to be confused with its use in religious affairs to refer to lay activity or civil administrations outside the church. This ecclesiastical usage is a special corruption of the origins of the word as used by ancient Romans to describe rare events, like the "secular games" that occurred once in a century. Climatologists and paleoecologists find the word useful to describe long-lasting or indefinite processes. Ice ages come and go in "secular time."

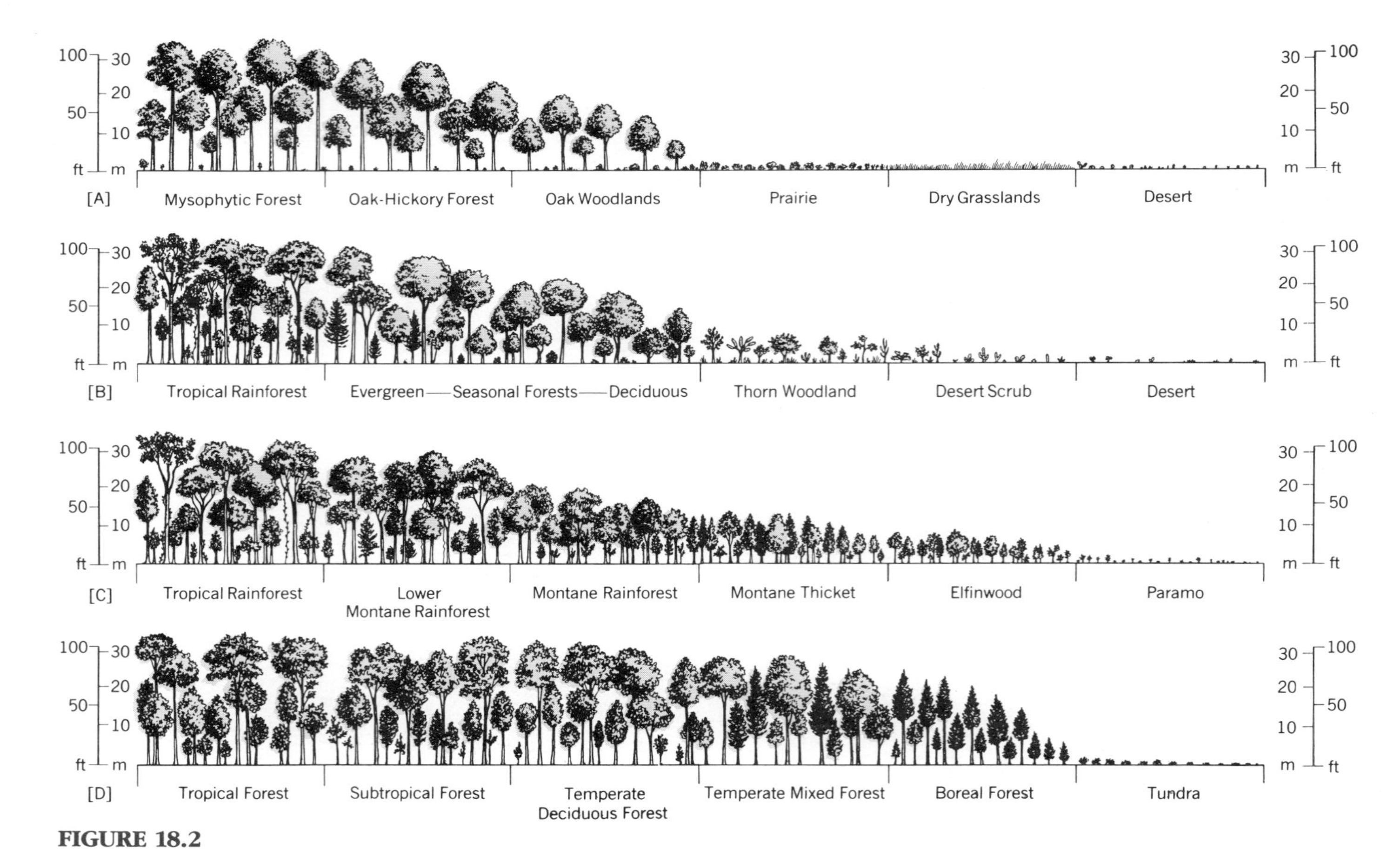

FIGURE 18.2
Profiles of merging formations.
(*a*) From moist Appalachian mountains westward to the desert. (*b*) Rain forest to desert in South America. (*c*) Up a tropical mountain in South America, from forest to treeless upland. (*d*) South to north from tropical forest to tundra. (From Whittaker, 1975.)

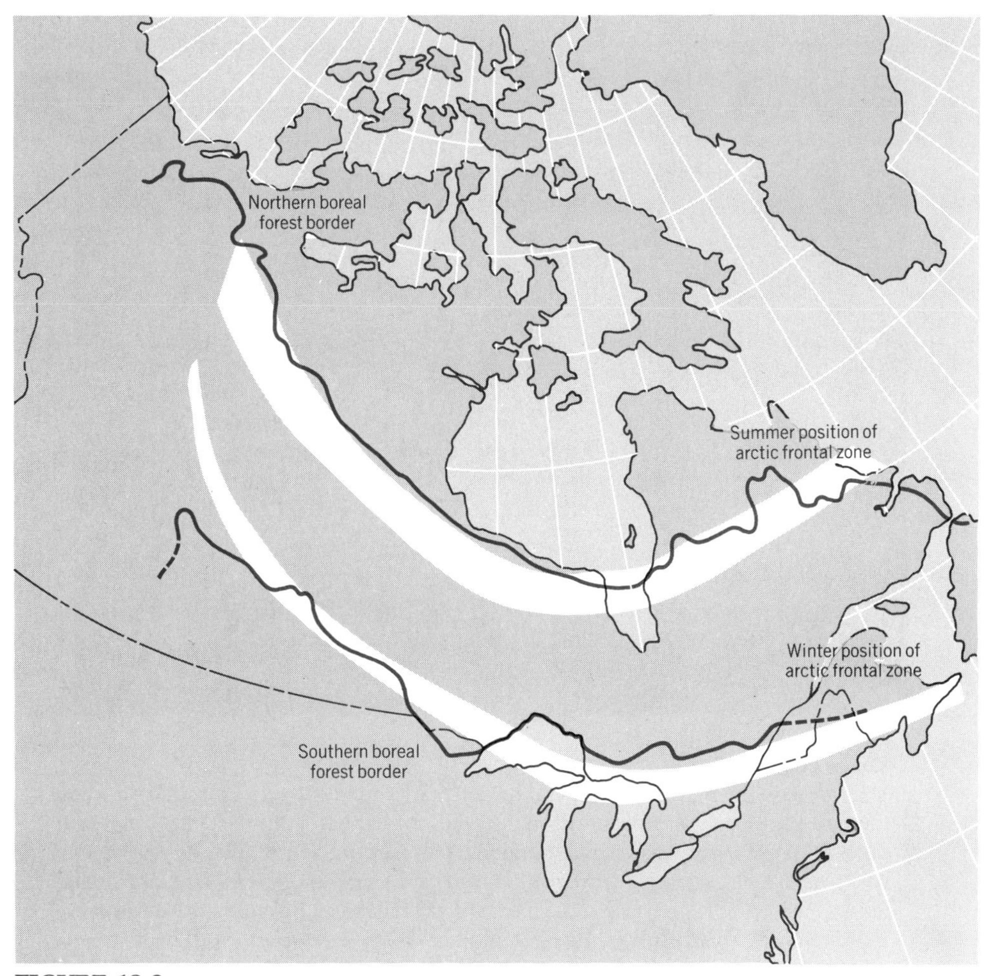

FIGURE 18.3
Forest boundaries and air mass fronts to eastern Canada.
The tree line and the southern edge of the boreal forest are perhaps the most distinct of continental vegetation boundaries. The plot shows that they closely follow the mean positions of the front of the arctic air mass at different seasons of the year. This study of Bryson provides direct evidence that extents of the Canadian boreal forest and tundra are directly influenced by the sway of an air mass. (Redrawn from Bryson, 1966.)

atures and precipitation, life on earth is coerced into great contrasting domains of plant design by climate. These domains are the great plant formations. The regional spread of plant design with each formation, in turn, sets limits to lives in all other trophic levels. Thus the world is set out by climate into a patchwork of living systems, each with its own rules of existence. The pioneer ecologist V. E. Shelford gave a name to these great living systems, calling them **biomes**.

A biome includes all animals and plants adapted to common climate, which thus

FIGURE 18.4
The arctic tree line.
The forest on the right, the tundra on the left, and a good enough separation between them to please any mapmaker. In such places formations almost seem to face each other like rival armies at a disputed frontier. This part of the tree line is in Alaska. The tall trees are black spruce (*Picea mariana*), but there are several kinds of broad-leaved bushes in the forest also. The lines of small bushes in the tundra will be growing along drainage channels, and the open tundra, which looks grassy, will be humpy, a mixture of grass and sedge tussocks with prostrate woody plants like dwarf birch and blueberry.

share habitats and must accommodate to each other. The animals of a biome are further constrained by the physical structures of plant design: arboreal animals need trees, and so on. Thus a biome is a community in which the living things offer solutions to a common problem; they are environmentally linked.

Biomes do not have threads of relationship among their characteristics. Thus the tropical rain forest biome is easily recognized in both Africa and South America, although the species are quite different on the two continents.

When ecologists talk of biomes, they refer to abstractions, because infinite varieties of mixtures of animals, climate, and plants are possible (Figure 18.2). No section of this endless progression of change can be defined in a way that is clear-cut. Various ecologists produce various lists of biomes. Accepting that there should be eight principal biome types on land, described as follows, is conservative.

Tundra

A tundra is a biome type without trees or other tall perennial plants and where this absence of trees is associated with low temperature or short growing season (Figure 18.5). Lack of tall plants may be explained as a consequence of the requirements to balance a heat budget and to conserve moisture (see Chapter 5). The principal region of tundra is the circumpolar lands north of the Arctic Circle, but there are smaller areas in the southern hemisphere, and a variant of tundra (so-called arctic-alpine) occurs on high mountains at all latitudes. Circumpolar tundras have climates dominated by the polar zones of high pressure. Tundra soils are deeply frozen for all or much of the year, and are poorly drained and shallow. Climates and productivity are strongly seasonal. Animals hibernate, migrate in the colder season, or live as the lemmings do under the snow. Characteristic mammals are lemmings, hares, musk oxen, caribou, foxes, and wolves. Except for ptarmigan and a few predators and seabirds, tundra birds are migratory, using the short productive period for reproduction and surviving the rest of the year elsewhere; examples are jaegers, geese, shore birds, and insectivorous songbirds. Tundra communities are variable, probably reflecting different summer temperatures, length of growing season, and precipitation. Alpine tundras of high mountains do not look very similar to arctic tundras to the naturalist familiar with both. What they have in common is absence of trees and the sharing of many species, or at least genera, of plants and animals. Important causes of differences are the different lengths of growing seasons and the different irradiance, for a low-latitude mountaintop may be almost without seasons and have many hours of sunlight each day and yet be cold enough, or physiologically dry enough, to be without trees. These sites are colonized by relatives of plants of cold arctic regions, or they are invaded through evolutionary time by cold-adapted endemics, the relatives of plants in the forests below.

Coniferous Forest (Boreal Forest, Taiga)

Characteristic coniferous forest is the broad northern belt that fronts the tundra, usually called the boreal forest, after the Greek name for the north wind, or taiga, which is the Russian name (Figure 18.6). The climatic limits are discussed in Chapter 5. These forests exist essentially where winters are very cold, like those of the tundra, but where summers are longer, perhaps with a short period of warm continental weather. These climates lie along the line where cold, high-pressure air of the Arctic meets air traveling northward from about the fortieth parallel. Most of the forest trees are evergreen and needle-leaved, this being a design that apparently is suited to achieving useful working temperatures in temporarily productive periods at a minimum cost of maintenance (see Chapter 5). Broad-leaved species within the forest are deciduous (maples), and some have special adaptations to restrict heat loss (for example, trembling and vertical leaves, as in *Populus tremuloides*; see Chapter 5). The evergreen gymnosperms that have prevailing adaptations of this biome type are flammable, with the result that the boreal forest is subject to periodic fires. A burn–regeneration cycle in the forest is an important secondary characteristic. Apart from seed-eating rodents like mice and squirrels, the principal herbivores of older parts of the forest are insects, particularly caterpillars of Lepidoptera (butterflies and moths) and saw flies (herbivorous Hymenoptera). These herbi-

FIGURE 18.5
Tundra in Alaska.
The upper photograph is a landscape on the Alaskan coastal plain within 50 miles of the Arctic Ocean: sedges, grasses, prostrate willows, diminutive heaths, saxifrages and composites like *Petasites*. The lower photograph is Alaskan tundra further south, within forty miles of the tree-line on Seward Peninsula at the latitude of Bering Strait: tall tussocks of cotton grass (a sedge, *Eriophorum vaginatum*) define the landscape, with prostrate woody plants, particularly dwarf birches (*Betula nana*) between and below the tussocks.

FIGURE 18.6
Boreal forest: land of the north wind.
Needle leaves and evergreen habit adapt conifers to high latitudes with cold winters and short summers. The result is a biome where the properties of conifers combine with climate to define the conditions for life. Most abundant trees in this example are white spruce (*Picea glauca*) and balsam fir (*Abies balsamea*).

vore populations are subject to irruptions when a section of forest has not been burned for a long time. The preponderance of insect herbivores in this seasonal environment causes abundant insects to be predictable in early spring, a property of the boreal forest that is used by many migratory songbirds as a resource for the rearing of young. The large mammalian herbivores include several species of deer and bears, all of which are adapted to make use of regenerating forest in burned areas for browse or as a source fruits (bears feeding on raspberries are a feature of Autumn in eastern Canada). Carnivores are essentially similar to, or the same as, the cold-adapted carnivores of the tundra: wolves, foxes, and medium-sized cats, like lynx and mountain lions. Patches of coniferous forest very like the boreal forest are found in the southern hemisphere, but little southern land lies in the appropriate weather system, so the extent of this southern forest is limited. More extensive are the moist coniferous forests of the western coast of Canada and adjacent parts of the United States, a forest that is often described as a biome in its own right (Figure 18.7). These forests consist almost exclusively of very large coniferous evergreen trees like Douglas fir (*Pseudotsuga*). These trees live in a climate that is cool and moist year-round from the influence of warm coastal seas against colder land.

FIGURE 18.7
***Pseudotsuga* fallen: old growth forest in the Pacific North-West.**
A rainforest of giant conifers like Douglas fir (*Pseudotsuga*) develops where climate is moist and cool all year. Even the fallen Douglas Fir log is covered with greenery in the year-round moisture. Exceptional trees have exceeded 100 meters in height, taller than anything in the tropical rain forest.

No completely satisfactory explanation seems available of why a coniferous evergreen strategy prevails under these conditions over a broad-leaved evergreen strategy, though the niceties of balancing heat budgets at a suitable operating temperature should be responsible (Chapter 5). Coniferous woodlands are known in other climates, like that of the southeastern United States, but, except when on high mountains, these are usually transient, successional communities and are not considered part of the boreal forest biome type.

Temperate Forest

This is the biome type to which the temperate deciduous forests belong, but there are variants even though the total range is not great (see Figure 18.1). Temperate forests cover northern Europe, eastern China, the eastern and midwestern United States, a small area in South America, and part of New Zealand; in total not a large part of the globe. These are places of seasonal climates with cold winters but longer summers than in the land of the boreal forest. All patches are either adjacent to coasts, escape rain shadows, or are otherwise well watered in the growing season. In some places nearly all the trees are deciduous (Figure 18.8), but there may be mixtures of deciduous trees with evergreen conifers or evergreen broad-leaved trees like holly (*Ilex*). In the wetter or warmer parts of land covered with temperate forest, evergreen broad-leaved trees are a more important part of the forest. The predominant tree strategy, however, is to carry broad leaves and to drop them in winter, a strategy that can be understood in energetic and heat budget terms (see Chapter 5). It should be clear that many individual solutions to manipulating heat budgets in seasonal environments exist as well as several strategies for overwintering. Evergreen and deciduous variants recombine differently over the range of the temperate forest. Possibly this biome type is less homogeneous than others, like boreal forest or tundra, because the climate is less homogeneous. Local weather depends on rains and air mass movements of midlatitudes where seasons are unpredictable. Temperate forest mammals are comparable to those of the boreal forest, but they are more abundant and include additional forms like badgers, wild pigs, and more species of squirrels and small predators. Bird species richness is high com-

FIGURE 18.8
Temperate deciduous forest in Europe.
Most of the trees in this example are beech (*Fagus silvatica*).

FIGURE 18.9
***Rana temporaria* in temperate Germany.**
Amphibians are abundant but not diverse in the temperate decidious forest: three or four species of frog, a toad or two, and a few salamanders.

pared to that of the boreal forest, including both migrants that rear young on spring gluts of insects and many resident birds, some of these seed or fruit eaters (frugivores). The most important herbivores are insects. Amphibians (salamanders and frogs) are present, whereas they are nearly, though not quite, absent from boreal forests (Figure 18.9). This distribution of amphibians may be the clearest evidence for a quantitative change in climate from boreal to temperate forest.

Tropical Rain Forest

Tropical places with copious rain in most months are productive year-round and are always green (Figure 18.10). The typical trees are broad-leaved, dicotyledonous plants of wide taxonomic diversity, nearly all of which are pollinated by animals (insects, birds, or bats). The trees, therefore, have conspicuous flowers, unlike the wind-pollinated flowers of trees in deciduous and boreal forests. The pea family, Leguminosae, is widely represented so that a significant percentage of the trees may be legumes. Most of these trees carry leaves year-round, but they do replace leaves and some may shed leaves synchronously, thus being almost leafless for a short time. Buttress and stilt roots are common, adding to structural diversity in the forest. The forest also holds numerous tall, monocotyledonous plants, particularly palms, as well as tree ferns and other large cryptogams. These plants are usually of subordinate status in the forest, though in parts of the Amazonian rain forest the taller palm trees may reach to the canopy.

FIGURE 18.10

Tropical rain forest of Amazonian Ecuador.

The productive layer is an evergreen canopy raised 30 meters in the air. 300 tree species have been counted in 10,000 square meters. A few huge trees have the habit of emerging above the rest, giving the canopy its uneven appearance when seen from the air.

Plants of the understory typically have very large leaves, suggesting a monolayer strategy (Chapter 3) to cope with the dim light there (Figure 18.11). Vines of many kinds (llianas) are present as well as many epiphytes, particularly orchids and bromeliads. Like the main forest trees, these have conspicuous flowers and use animals for pollination. Many rain forest plants also use animals to disperse seeds, so that fruits are large, succulent, or showy. This forest can be found in many subtle variants, depending particularly on the seasonality or evenness of the rains, but the essential properties for animal life are warmness and wetness. Most food production occurs many meters above the ground; indeed, one way of looking at a tropical rain forest is as a prairie standing on 30-meter stilts. Heavy browsing animals of the ground are thus not numerous and are replaced by browsers that can climb, principally primates but also including sloths in South and Central America (Figure 18.12 and 18.13). A diverse array of fruit-eating and pollinating animals provides specialists on particular plant species. Climbing or flying amphibians, reptiles, and insects are present in great variety. Important predators are large carnivorous birds, like harpy and hawk eagles, that hunt the browsing animals of the canopy (Struhsaker and Leaky, 1990). A particular property of the biome type is the role played by animals acting as detritivores, particularly ants and termites. As a consequence little litter collects on the forest floor, and soils can be almost bereft of organic matter.

FIGURE 18.11
Inside tropical rain forest where the canopy is less complete.
Palms, ferns, and other plants with large leaf, shade-adapted strategies are prominent where the canopy trees are less dense, as on the poorer soils.

FIGURE 18.12
***Bradypus:* Cow of the rain forest canopy.**
Sloths may be the ultimate browsers of the rain forest vegetation. Like the primates, they are members of food chains that end with eagles.

FIGURE 18.13
Browsers of the rain forest canopy.
A tropical rain forest may be likened to a prairie perched on stilts more than 20 m high. Guilds of primates browse the canopy, with mixed diets of leaves, fruits and insects: *Alouatta seniculus* (howler monkey) largely vegetarian and large (left), *Ateles geofroni* (spider monkey) and—*Saimiri sciurus*—(squirrel monkey) are more omnivorous. This guild of canopy herbivores and omnivores is hunted by guilds of eagles.

Tropical Savanna

Where it is warm year-round but with a very long dry season (several months), trees are stunted and grow spaced out, which allows grasses to grow between them (Figure 18.14). Broad belts of the resulting savannas flank the rain forests to the north and south where the equator bisects the continents of Africa and South America. This corresponds to the latitudes where air descends and begins its travels back to the ITCZ (see Chapter 7). Usually a gradual transition spans tree-studded savanna, dry woodland with grass, and tropical rain forest so that the boundaries of these biome types may be particularly difficult to map in detail. Sometimes, though, a sharp discontinuity forms because the savanna burns and the wetter forests do not. Fire, an important fact of the savanna environment, can set down a scorch line at the edge of the savanna so that, as in northern Nigeria, forest patches stand like cliffs out of the savanna. Tropical savanna grasses may be 1 to 3 meters tall, but they represent productivity within the reach of heavy grazing animals. Accordingly, savannas are inhabited by herds of grazing mammals and their large mammalian carnivores—lions and other big cats, hunting dogs, jackals, and hyenas. The large mammals in turn provide a living for large scavengers, so that savannas are characterized by vultures. The C4 photosynthetic pathway is widespread in the herbs of this biome, where all the grasses are C3. Trees have small leaves, are multilayered (see Chapter 3), and have flat tops—shapes that can be understood as representing a compromise between reducing the maintenance costs of unnecessary height with the requirement of being tall enough to reduce mammalian browsing from grounded animals. The result is the familiar appearance of tropical savannas (Figure 18.12), representing a total system of life starkly different from tropical rain forest and resulting from the simple fact that rain comes at wide intervals, though it may be ample when it does come.

FIGURE 18.14
Tropical savanna in East Africa.

Temperate Grassland

The prairies and steppes of the world lie in temperate, seasonal, and dry climates (Figure 18.15). A little wetter climate allows the deciduous forest to close over it; a little less rain in the dry season and bare ground between the grasses spreads until a desert results. This can be sensed by looking at the extent of prairies in North America in Figure 18.1 where boreal forest, deciduous forest, and desert are the boundaries. Boundaries from prairie to desert are extremely gradual, but boundaries with forests can be distinct because of fire. Like the tropical savannas, the temperate grasslands burn, and fire is an expected property of the environment. Unlike tropical savanna, however, the temperate grasslands are without trees. Perhaps we do not have a sufficient explanation for this. Lack of trees can be equated with drought, as was discussed when considering what sets a treeline on high mountains, the mechanism for which is now tolerably understood (see Chapter 5). Scattered trees in tropical savanna presumably reflect the absence of winter, allowing an evergreen and broad-leaved strategy not applicable to prairies with winters. Whatever the detailed mechanisms that ban trees from prairies, the reality is that subtleties of climate produce a land free of trees and thus dominated by the inclined leaves of grasses (see Chapter 3). Many of these grasses are C4 photosynthesizers. They are alternately green and brown with the seasons and have

FIGURE 18.15
Temperate grassland.

among them numerous species of low, flowering plants. A vital difference between these temperate grasslands and the treeless expanse of the tundra is that the ground is not frozen. Deep organic soils collect, and a thick bed of organic matter is a prime parameter of this biome type. The fauna of a temperate grassland has much in common with that of tropical savanna: grazing ungulates, large carnivores, and scavengers. This can be seen most clearly in South Africa, where the temperate grassland, known locally as the "veldt," shares many animals with the tropical savannas of East Africa to the north.

Desert

The ultimate desert with no water has no plants, merely bare rock or a sandy sea of shifting dunes. But by "desert biome" ecologists mean land too dry to support prairie or savanna, but with enough moisture to allow specially adapted plants to live (Figure 18.16). These deserts are not only dry but also hot. Typical perennial plants have swollen stems rather than leaves, with heavy cuticles, sunken stomates, C4 or even CAM metabolism (see Chapter 3), and spiny defenses against large browsing animals, and they live spaced out (see Chapter 3). All these adaptations are for coping with heat stress, water stress, competition for water, or the necessity to avoid damage from herbivore attack when the energy costs of repairs are so hard to meet (see Chapter 5). Where the mixture of heat and water stress is less severe, perennial bushes of the Chenopodiaceae or Compositae form more or less regular arrays with bare ground in between (the saltbush and sagebrush familiar to fans of Western movies). Numerous annuals grow briefly following rain, and some perennial bushes may be opportunistically deciduous like the *Bursera graveolens* trees of the Galapagos (see Figure 11.2; Chapter 11). A particular property of desert life is an alternation of hot days with cold nights, resulting from the absence of clouds to impede irradiance by day or to stop the warm

FIGURE 18.16
Vegetated desert near 30 North.
Descending dry air, hot sun and infrequent rain determine life forms that shed heat while conserving water.

ground from radiating heat to the cold black body of space by night. Low productivity of plants, temperature extremes, and shortage of water narrowly restrict the lifestyles of animals. The low-energy systems of reptiles are well adapted to this regime (see Chapter 6), and hot deserts accordingly have a wide variety of reptiles. Amphibians are almost absent because of lack of water. Mammals are restricted to taxa that produce concentrated urine or take refuge from the sun in burrows, or those with special tolerance to desiccation and with heavy fur insulation like that of camels (see Chapter 5). Insect life histories tend to be strongly seasonal, with diapausing eggs or adults able to survive the dry times. Plague grasshoppers or locusts are almost characteristic of deserts, like the Australian grasshoppers discussed in Chapter 10.

Chaparral (Broad-Leaved Sclerophyll)

The very distinctive life form of the vegetation of the chaparral biome type is shown in Figure 18.17. The strategy used by the plants that make up nearly all the vegetation is that of woody and dense bushes with permanent, thick, pubescent (hairy), leathery leaves. Neither trees nor

FIGURE 18.17
Broad-leaved sclerophyll vegetation in California.
Hot dry summers and cool moist winters lead to shrublands like the Chaparral of California and the Maquis of the southern France.

grassland replace these bushes; they can, and frequently do, burn, but then they are replaced by their own kind. Many different families of flowering plants contribute species with this strategy and life form to "sclerophyll" vegetation in different parts of the world wherever the peculiar climatic pattern for which it is adapted occurs (Figure 18.1). The climate must be an alternation of hot, dry summers with cool, moist winters. These climates occur where cool ocean currents turn away from continents as they complete their gyres, causing cold upwellings along the coast (see Chapter 7). When this happens at middle latitudes with high irradiance, the land is hotter than the ocean in summer, preventing both clouds and rain; the result is known as a Mediterranean or Southern Californian summer. But lowered irradiance in winter makes the difference between land and sea temperatures less so that both clouds and rain are possible. The sclerophyllous bush strategy is apparently to grow during the winter rain and to sit tight in summer. Possibly trees are prevented from taking over by the hot summers, and grasses are outcompeted by the bushes that can overtop them in the cool winter. It certainly seems that a hypothesis of both physical adaptation and competition is needed to account for the success of this peculiar life form. Seed-eating rodents and birds are common, and a number of ungulates such as goats and deer can live in chaparral. These herbivore populations naturally support wolves or big cats, but Mediterranean regions are so favored by people that the larger predators were usually removed long ago.

It is possible to apply the biome concept to aquatic systems, though the concept loses much of its original meaning. The open waters of both lakes and oceans, for instance, support only planktonic plants and the food chains that come from them. Properties for life are set by the microscopic sizes of these plants, so that some ecologists talk about the "planktonic biome." Different patches of coastline can be called biomes; rocky shores with anchored seaweeds (rocky-shore biome), muddy estuaries without large seaweeds, coral (algal) reefs, and so on. In each of these places physical constraints of fluid flow and substrate determine the life forms of plants, and hence the strategies of consumers. The differences between them, however, are of a different order than those set by climate on the terrestrial earth. These biota are best discussed in the context of the physical systems of which they are a part.

THE GEOGRAPHY OF EVOLUTIONARY REGIONS

An alternative to mapping the world into biomes of common adaptation is mapping by relatedness of animals and plants. The result is grouping by historical affinities among populations, so that similar biomes on different continents are cast into different biogeographic "**regions**" (Figure 18.18).

Mapping by biogeographic regions records the fact that animals or plants of any

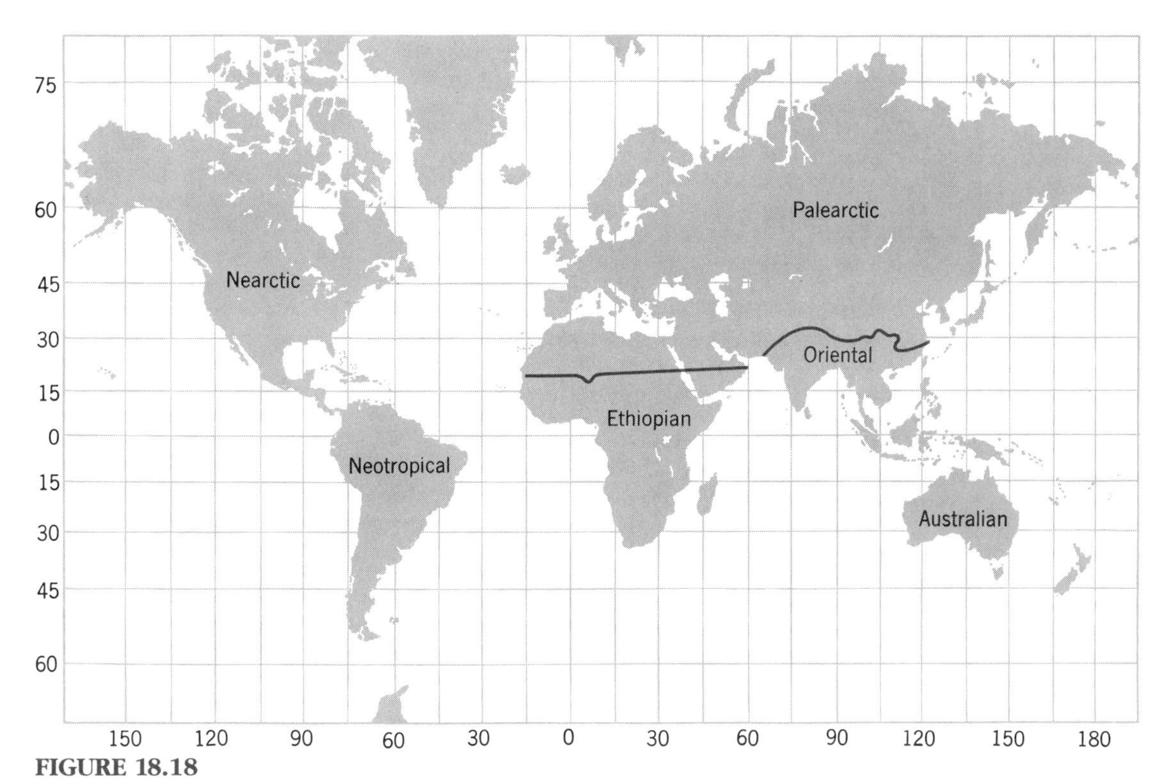

FIGURE 18.18
The zoogeographic regions of Sclater and Wallace.
These regions were established because the birds and mammals of each continent or subcontinent were closely related to each other, even though they might have adapted to different ways of life. Use of this system by Wallace was one of the clues on which the theory of evolution was based.

one place are more closely related in a taxonomic sense than they are to the animals or plants of more distant places, regardless of their adaptations. For instance, the rain forests of Africa, South America, and India all look alike, representing the same biome. Yet the species of the Amazon forest are more related to the species of Patagonia or the Argentine pampas than they are to the species of the African or Indian forests.

Mapping regions of relatedness was es-

Objections to the system of Circumpolar Zones.—Mr. Allen's system of "realsm" founded on climatic zones (given at p. 61), having recently appeared in an ornithological work of considerable detail and research, calls for a few remarks. The author continually refers to the *"law of the distribution of life in circumpolar zones,"* as if it were one generally accepted and that admits of no dispute. But this supposed "law" only applies to the smallest details of distribution—to the range and increasing or decreasing numbers of *species* as we pass from north to south, or the reverse; while it has little bearing on the great features of zoological geography—the limitation of groups of *genera* and *families* to certain areas. It is analogous to the *"law of adaptation"* in the organisation of animals, by which members of various groups are suited for an aerial, an aquatic, a desert, or an arboreal life; are herbivorous, carnivorous, or insectivorous; are fitted to live underground, or in fresh waters, or on polar ice. It was once thought that these adaptive peculiarities were suitable foundations for a classification,—that whales were fishes, and bats birds; and even to this day there are naturalists who cannot recognise the essential diversity of structure in such groups as swifts and swallows, sun-birds and humming-birds, under the superficial disguise caused by adaptation to a similar mode of life. The application of Mr. Allen's principle leads to equally erroneous results, as may be well seen by considering his separation of "the southern third of Australia" to unite it with New Zealand as one of his secondary zoological divisions. If there is one country in the world whose fauna is strictly homogeneous, that country is Australia; while New Guinea on the one hand, and New Zealand on the other, are as sharply differentiated from Australia as any adjacent parts of the same primary zoological division can possibly be. Yet the *"law of circumpolar distribution"* leads to the division of

F 2

FIGURE 18.19

Photographic facsimile of a page from Wallace's *Geographical Distribution of Animals* (1876), in which he attacks the Allen ecological classification for obscuring evolutionary truths.

sential to the discovery of the process of evolution by natural selection. If the rain forest birds of the Amazon were more closely related to the prairie birds of the Argentine pampas than they were to rain forest birds of Africa, a reason had to be found. A parsimonious explanation was that birds of Amazon and pampas were true relatives, derived from common stock. Geographical facts like these triggered the minds of Darwin, Wallace, and others of their generation to the realization that animals had evolved locally to fit local conditions of life, into what we should call different niches. And biogeographic maps then became a powerful tool in explaining the new discovery of evolution to others.

In the early days of both ecology and evolutionary biology the different world views represented by the two systems of mapping came into conflict. On one side were ecologists studying the effect of environment on form and lifestyle. Allen (1871) stated this view of geography in the days when it was fashionable to write opinions as "laws," as "the law of the distribution of life in circumpolar zones," the "**zones**" being roughly what we now call biomes. But evolutionary biogeographers like Alfred Russell Wallace (1876), the codiscoverer of evolution by natural selection, objected that ecological maps obscured the all-important hereditary origins of species (Figure 18.19).

It is curious that ecology, the most evolutionary of subjects, should have had to begin with a conflict with evolutionists. The development of ecological thought was in fact held back for nearly half a century by this conflict, until after the process of evolution by natural selection became the accepted working tool of biology. After that it was possible to study the developing relationship between organism and environment as an evolutionary process itself.

CONTINENTAL DRIFT

Ecologists' understanding of geography has been changed by a new paradigm that has risen in geology, that of plate tectonics and sea-floor spreading. The continental masses of the lands, together with the continental shelves, are now to be thought of as plates of rock 700 kilometers thick. These continental plates are nudged along by motions of the sea floor so that the continents drift on a scale of evolutionary time.

The parting of drifting continents leaves a tear in the earth's crust, along the length of which hot rocks pour to fill the gap and form the lines of the mid-ocean ridges (Figure 18.20). Where the sliding plates collide, the crust buckles into a mountain range as one plate slithers under the other. The line of a plunging plate (subduction) yields a deep ocean trench. As the descending plate is driven down, rocks melt and a line of volcanoes marks its passing.

Plate tectonics explains many things at once: the abyssal trenches; the island arcs of volcanoes; the fitting shapes of distant continents; the lines of mountain building (orogeny); the high heat flux along the mid-ocean ridges; the intense magnetization of these ridges resulting from the rapid cooling of their lava in seawater; and the long "transform faults" like the San Andreas (Figure 18.21).

The drifting, colliding continents set and reset the physical stages on which communities are built, altering the relative importance of arctic and equatorial climates in different stages of earth history. On the assumption that each continental plate remains intact, the relative position of continents can be plotted from data on the direction of the earth's magnetic field recorded in ancient rocks, and dated by ra-

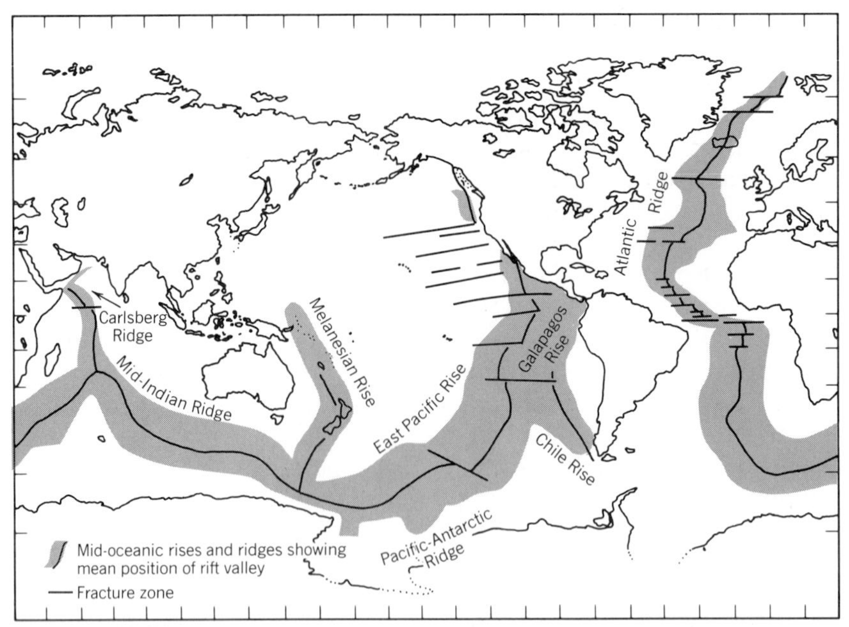

FIGURE 18.20
Mid-oceanic rises and ridges.
The rises and ridges are where new crustal rock is formed. Continents are moved away from the ridges.

diometric means, to yield maps of ancient continents (Figure 18.22).

Some of these maps may seem to be of times impossibly remote, and yet their traces are with us still. The early Cenozoic map of Figure 18.22, for instance, tells of a time perhaps 50 million years ago, and yet some of our forest trees, or species extremely close to them, were already present in the contemporary forests. The communities in which they lived might have been more than subtly different from our familiar forests, but some trees were the same or closely similar. The gene pools of large populations of their modern descendants may well include instructions for adaptations that have preserved these species through all the climate and habitat changes of those unimaginable numbers of generations.

Alfred Wegener (1966) put forward the original theory of continental drift early this century, long before remanent magnetism or plate tectonics were imagined, because he saw that the shapes of Africa and South America were so complementary that they could be fitted together to make a super-continent (Figure 18.22). He found some evidence that the two were once joined by matching rocks from either coast, but even more in the evidence of "disjunct distributions" of species. The continents had floated apart within evolutionary time scales, carrying ancestors to distant continents. For instance, the old super-continent called Gondwana[4] had split in the middle before the first placental

[4]Often called Gondwanaland, as it appears in the slogan favored by geologists "Reunite Gondwanaland." The suffix "land" is in fact a redundancy, since "Gondwana" by itself means "land of the Gonds."

FIGURE 18.21
San Andreas: Mark of shifting crust.
Transform faults, like the San Andreas fault of California, are part of the evidence that continental parts of the earth have shifted position within evolutionary time scales.

mammals evolved. The bit of Gondwana that became Australia was left with only the more ancient mammalian marsupial forms. Placental mammals did not reach Australia until people took some there in boats.

Wegener might just possibly have lived to see his brilliant insight confirmed by modern geology had he not lost his life during field work on the Greenland icecap, where he was last seen driving a dog team into a blizzard to fetch supplies for his beleaguered winter camp (Georgi, 1934).

GEOGRAPHY OF THE ICE AGE EARTH: "IN THE BEGINNING"

The most immediate origins of modern distributions must be found in the geography

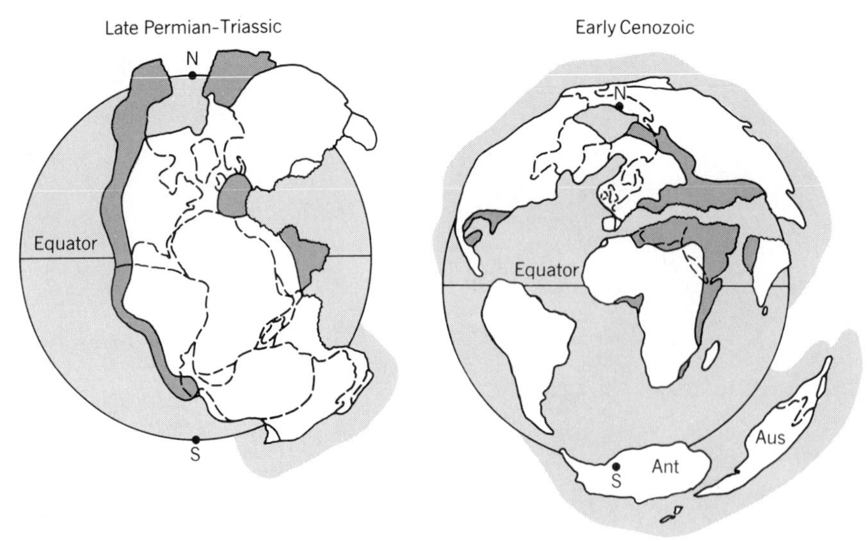

FIGURE 18.22
Continental drift: data for biogeography.
Relative positions of the continents at two stages in the earth's history.

of the ice age earth, that almost recent time ending some 150 human lifetimes ago. Livable land was different then, vegetation was different, and climate was different.

Thick ice sheets covered the northern hemisphere down to roughly the fortieth parallel (Figure 18.23). These ice sheets represented a very large volume of water that was held in cold storage on the land instead of being allowed to run back to the sea. The first consequence of this is that world sea level was about 100 meters lower during the ice age than it is now. All shallow seas were drained, connecting continents and islands in novel ways. Alaska and Siberia were fused by the draining of the Bering and Chukchi seas flanking the Bering Strait. The island of Britain was part of continental Europe. Islands, like the Galapagos, were more than twice their present size.

Many of the islands of the Indonesian archipelago were fused, some being joined to Australia and some to Asia. Between the two landmasses was a last deep channel of ocean, which has long served as a barrier between Australia and the old-world tropics. The faunas of islands on opposite sides of this channel are so different that the division was recognized as a biogeographic barrier, "**Wallace's line**," long before the history of low sea level in ice ages was discovered.

All the climates of the earth were changed. In part the earth was cooled directly because it received less effective sunlight. This modest loss of solar heating appears to have been due to compounded effects of changes in the earth's orbit and tilt, which fluctuate rhythmically with periods of their own (the so-called Milankovitch hypothesis; Pielou, 1991). More important for contemporary life than this modest reduction in solar heat were displacements of climates due to the presence of the ice sheets themselves.

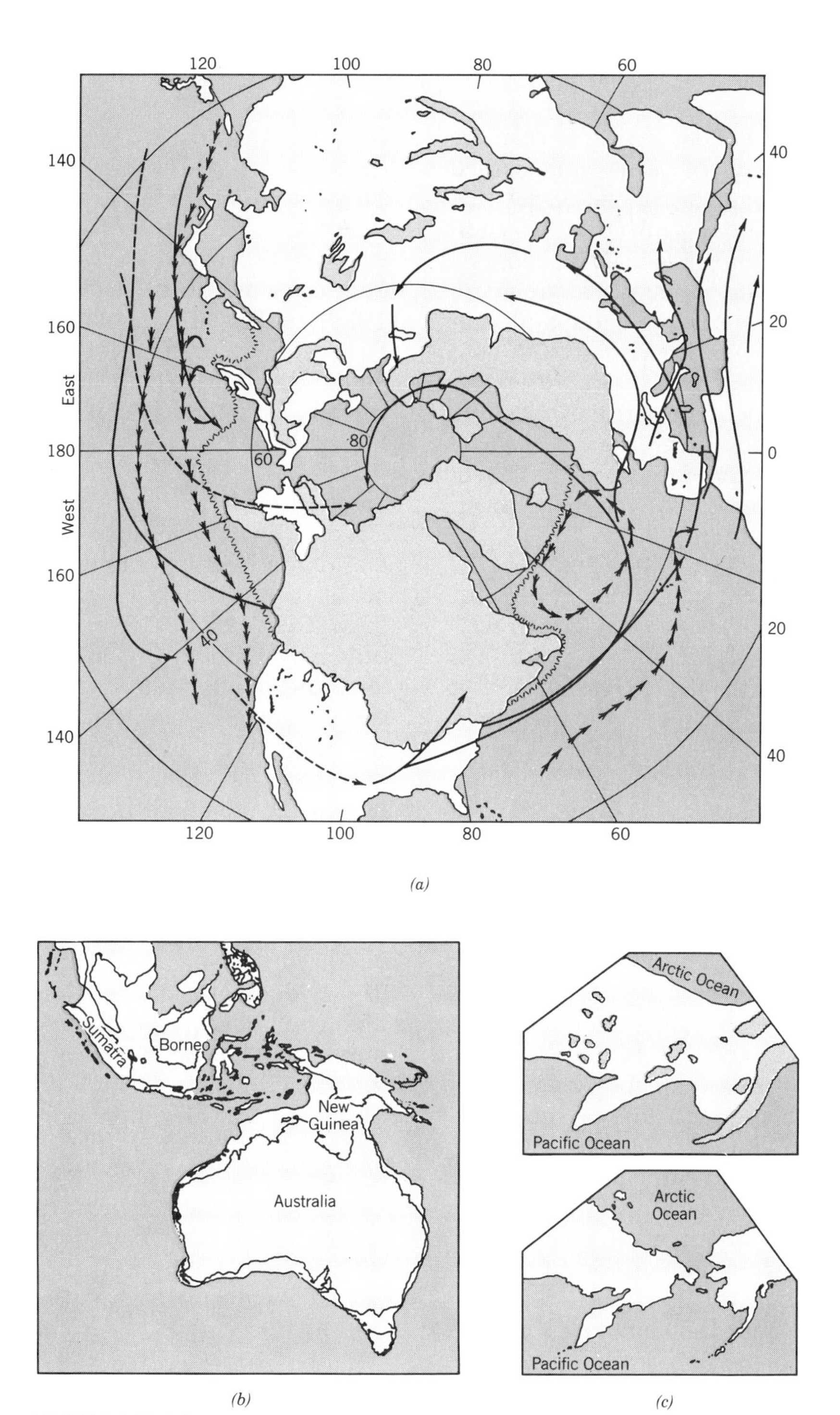

FIGURE 18.23

The ice-age earth.

(*a*) Glaciation in the northern hemisphere at the last (Wisconsin-Würm) glacial maximum. (From Flint, 1971.) (*b*) Australasia in the last glacial maximum showing coastlines when eustatic lowering of sea level of about 100 meters was in effect. (From Quinn, 1971.) (*c*) The Bering Strait region (Beringia) in glacial times and today. (From Pielou, 1979.)

The circulation of the atmosphere in the northern hemisphere was compressed southward as the polar zone of cold, high-pressure air extended south to the fortieth parallel along with the ice. This produced a domino effect on air masses the world over as local patterns of wind and rain changed. Air did not descend at 30°N as it does now (see Chapter 7), but rather descended variously to the south of that line, moving deserts southward as it went. The intertropical convergence zone (ITCZ) of heavy rains was probably well to the south of its present position.

An accurate model of the climate of the ice age earth is critically important to students of climatic change as well as to ecologists. Accordingly, a significant effort has been made over the last two decades, first to reconstruct then to model the ice age climate. Because climate change never ceases, an arbitrary date of 18,000 years before present (18,000 yr B.P., or "18 K" in the vernacular of paleoclimatology) was taken as the target date for a global model. Pollen and other proxy data for climate are being collected from the land, but data from the oceans have been gathered more rapidly. The transfer function technique applied to foraminiferal fossils in deep sea cores (see Chapter 17) taken from the world oceans yields maps of the temperature of the sea surface (see Chapter 7; Figure 7.9).

Figure 18.24 is an early and provisional map of the biomes at 18 K, based on both the sea temperature data and paleoecological data from the land. Notice that no subdivisions are given for forests, a consequence of the limited pollen data available for all but northern latitudes when the map was made in 1976. But even this early map shows some of the almost astounding differences of the spread of biomes in the ice age earth, like the great forests that then spread over much of Australia.

Ice Age Africa

Africa south of the Sahara was a drier continent at 18 K, as shown particularly by evidence that lakes with small catchments had less water in them (Street and Grove, 1979). Pollen diagrams of East African lakes show a progression from dry woodlands or savannas to the modern forests as the ice age ended (Livingstone, 1993; see Figure 11.5).

Ice age aridity in Africa is easily accounted for by climate models. The modern climate of East Africa is dry by tropical standards, because rains are strongly seasonal, being brought by monsoon winds from the Indian Ocean. Monsoons are driven by temperature differences between land and sea that allow winds to blow from wet ocean to hot land at critical times of the year. When large climate models are developed for a world with ice sheets and sea surface temperatures (see Figure 7.9), they show a reduction in monsoon rains of 10 to 20%, sufficient to account for the falling lake levels and changed vegetation of Africa (Kutzbach and Guetter, 1986).

But Africa was also cooler in an ice age. The evidence is only just becoming available, but it seems clear enough (Livingstone, 1993). The evidence is from pollen and macrofossils of high-altitude plants found in low-elevation deposits of glacial age, both from West Africa and Madagascar (Maley, 1991). Finally, parts of what is now the southern Sahara desert received sufficient rain to be vegetated, and even to hold lakes, through at least part of the glacial interval (Bonnefille et al., 1990).

Thus for most of the last 80,000 years, African climates have been drier on the plains, wetter in the desert, cooler all over, and with areas for wet forest biomes much more restricted than they have been in historic times.

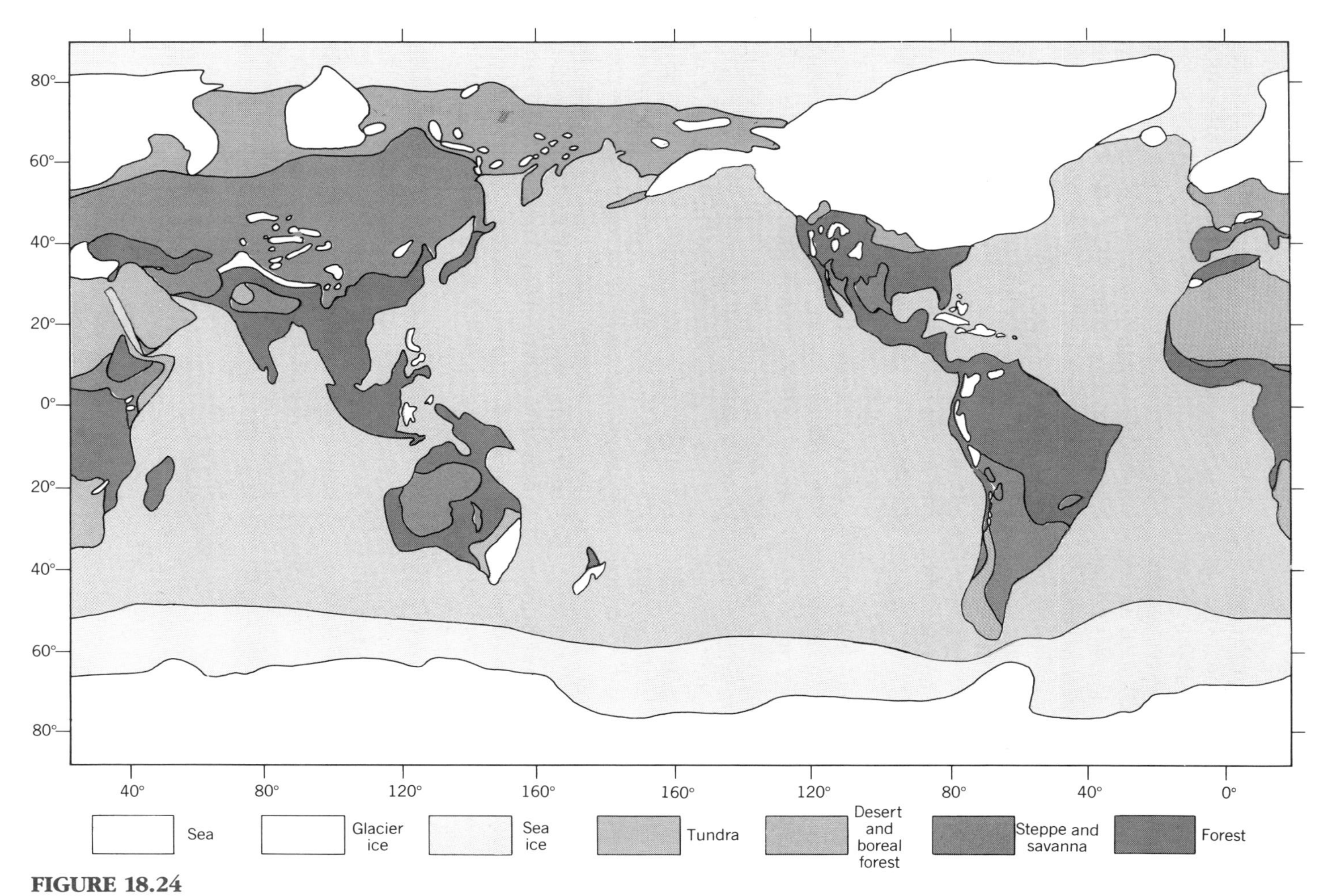

FIGURE 18.24

Vegetation and climate of the ice-age earth.

Map shows the results of climatic modeling based on reconstructions of ocean temperatures from fossil data from deep sea cores, supplemented by other geological and paleoecological data. Broad vegetation types are chosen for similar albedos rather than for the community type a botanist would choosen—for instance boreal forest is hidden in the desert category because the albedo of snow-covered conifers is similar to that of sand at 30–39% (based on CLIMAP, 1976 and modified from various sources).

The Ice Age Amazon

The condition of the ice age Amazon is still one of the great unknowns of geographical exploration. The first radiocarbon-dated fossil evidence for vegetation types of the Amazon in an ice age was published as recently as 1985. Even now, I know of only five such sites for the entire Amazon basin, two of them as yet unpublished. This is for an area about the same size as the continental United States, in which hundreds of dated pollen sections were needed just to reconstruct the history of the eastern forests alone (Webb, 1988; see Chapter 19).

For want of data, theory has filled the void, and a scenario of the ice age Amazon has been written to account for some curiosities of modern biogeographic distributions. The resulting "**Pleistocene refuge theory**" conceives of the Amazon in an ice age as being arid like ice age Africa, with the rain forests restricted to local patches that received the most rain, the so-called refuges.

The basis of refuge theory is the extraordinary observation that birds and butterflies in selected groups are represented by different species in different parts of the Amazon basin (Figure 18.25). If the Amazon basin is thought to be occupied by continuous forest, this is a remarkable finding. Species do not have disjunct distributions within single habitats, least of all volant organisms that might be expected to travel. The data for helioconid butterflies, some toucans, and a few other bird genera, however, appear to show just that (Whitmore and Prance, 1987).

A petroleum geologist and brilliant naturalist, Jurgen Haffer (1969), first pointed out the oddity of these disjunct distributions, and put forward the refuge hypothesis to explain them. He reasoned that this pattern of distribution should be fully explained if the Amazon had been drier in the past so that only the areas of endemicity had sufficient rain to support tropical forest with its attendant birds and butterflies. In that imagined past, the rain forest was fragmented, persisting only in wetter, more elevated regions, as if in islands in an arid sea. The moist "islands" would have been refuges against the aridity, hence the name of the hypothesis.

Haffer's postulated arid time had to be the ice age. Even in the late 1960s when Haffer wrote, the first evidence for aridity in Africa and other parts of the tropics was coming in. His own wide geological experience of the Amazon basin revealed plenty of evidence for different environments over the course of the Amazon's long history. Thus Haffer's reconstruction of the past was wholly plausible, even if it lacked direct supporting data.

Hard evidence for the imagined ice age aridity has never been forthcoming (Salo, 1987; Irion, 1984), Instead, refuge theory has remained a biogeographical construct, guiding nearly all biogeographical research in the Amazon for two decades, with only what Terborgh (1992) calls "faint voices" doubting its reality. Mine is one of the faint voices of opposition singled out by Terborgh.

Our first paleoecological evidence for the vegetation and climate of the ice age Amazon come from the western side of the basin, where the forest laps against the foot of the Andes mountains of Ecuador. This is the wettest, most species-rich portion of the forest, treated as the great "Napo forest refugium" in refugial theory. Two sections with radiocarbon dates for the interval 26,000 B.P. to 33,000 B.P. come from these Napo forests, both with pollen, wood, and phytolith records of the ice age vegetation. The ancient Amazon forest so revealed was quite unlike any modern Amazon forest, being a mixture of

FIGURE 18.25
Endemism in neotropical butterflies.
Maps like this have been the basis for the hypothesis that tropical rain forests were fragmented in the last ice age, the so-called "refugial model". Black areas are centers of most overlap of endemic subspecies, with lesser overlap in crosshatched areas. Squares are sampled quadrats without concentrations of endemics. (Modified from Brown, 1982.)

rain forest trees still growing in the region with Andean trees that had descended at least 1500 meters from their present levels in the mountains (Figure 18.26; Liu and Colinvaux, 1985; Bush et al., 1990).

We interpret the pollen record to require cooling over the Amazon of between 6 and 9°C at various times in an ice age. Cooling of this magnitude would require the forest species most in need of warmth to have been confined to the lowest elevations, down in the great bottom lands of the Am-

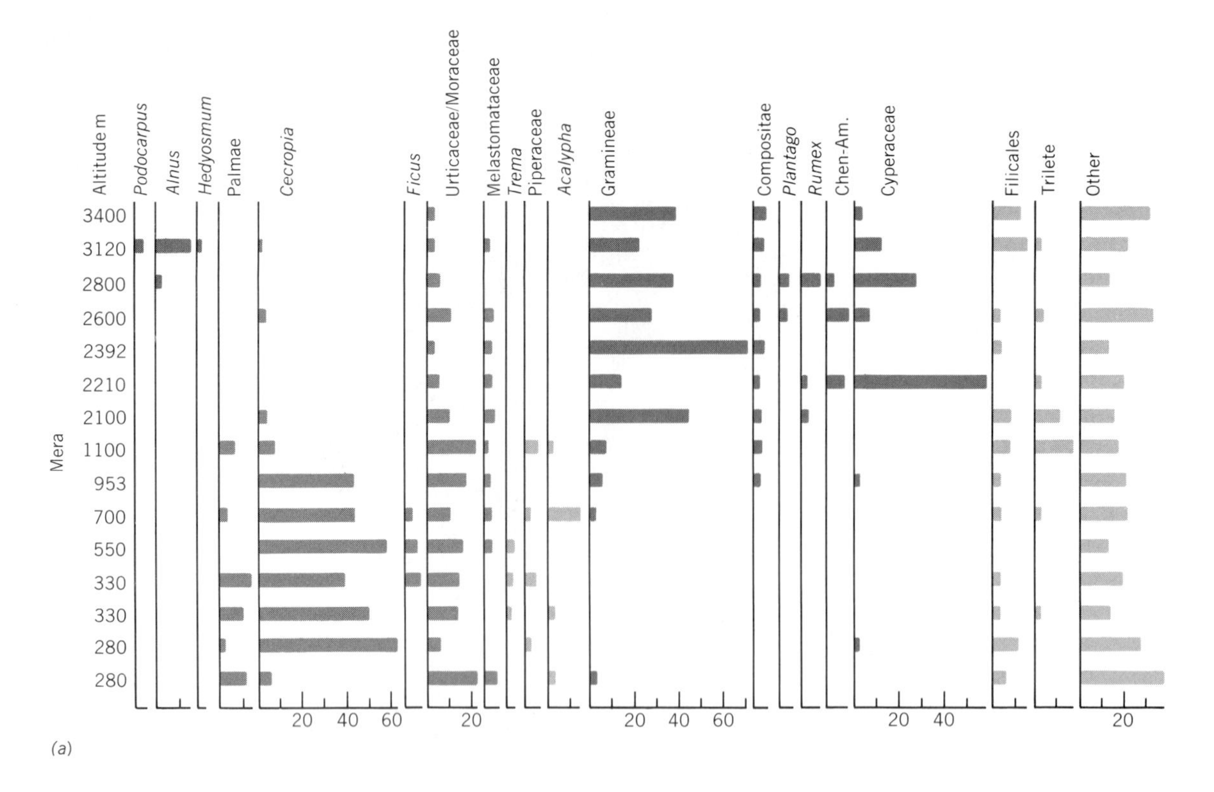

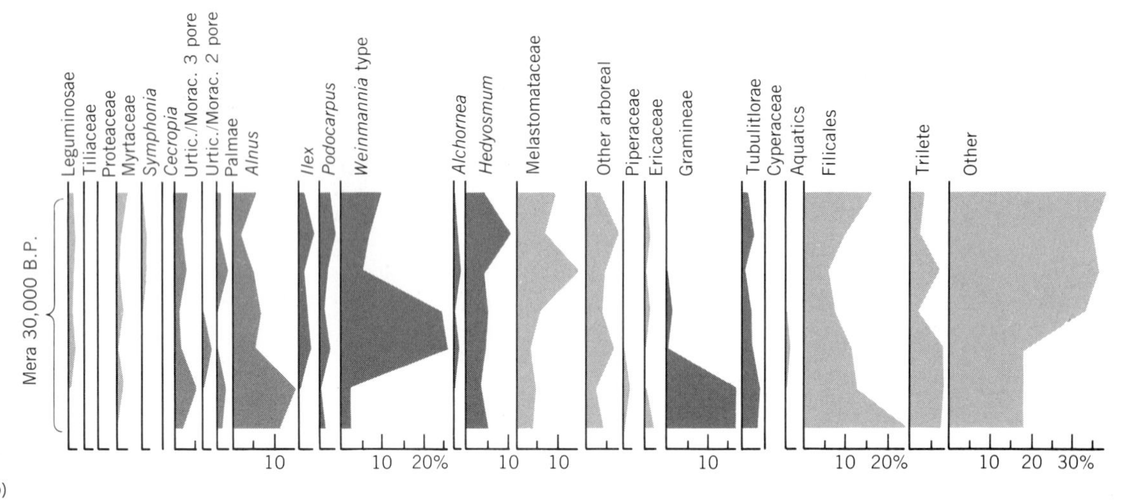

FIGURE 18.26

Pollen evidence for the environment of the ice-age Amazon.

Data are from Mera in the Amazonian rain forest of Ecuador, at 1100 meters elevation at the foot of the Andes. The top figure (*a*) gives surface (modern) pollen spectra of a transect from the Amazon lowlands at 280 meters elevation to above the Andean treeline at 3400 meters. The bottom figure (*b*) shows pollen from sediments at Mera radiocarbon dated to between 26,000 and 31,000 B.P., when continental glaciers were present in the northern hemisphere. High Andean taxa like alder (*Alnus*) evidently grew down near Mera in the ice age, as did plants of the modern elfin forest like *Weinmannia* (dark color). On the other hand, pollen characteristic of the rain forest (light color) were still present. (From Bush et al., 1990.)

azon between 200 and 300 meters above sea level. From these first paleoecological data for the ice age Amazon, therefore, we conclude that the Amazon basin held forest and thus could not have been arid (Colinvaux, 1987, 1993).

We shall not understand the biogeography of the ice age Amazon until we have at least a few dozen long pollen diagrams scattered across its immense expanse. To a paleoecologist, the attempt to depict particular past climates from the fleeting distributions of postglacial time has a resolution comparable to that of the telescopes that once saw canals on Mars.

Cooling at the Ice Age Equator

A striking result of the 18 K map of sea-surface temperatures given in the CLIMAP map of Figure 7.9 is that the tropical oceans are shown with only modest cooling, perhaps as little as 2°C, whereas sea temperatures at high latitudes cooled by as much as 6°C. Paleoecologists of the land are now getting quite different results, claiming that equatorial continents cooled by much more than the amount suggested by the oceanographic data.

Pollen diagrams from mountain forests in New Guinea suggest the descent of mountain species of more than 1000 meters at 18 K, suggesting a much colder Southeast Asia (Stuijts et al., 1988). This closely parallels the inference for South America.

Evidence that equatorial America cooled has long been available from the longest pollen record in the world, that from Bogota, Colombia (Hooghiemstra, 1989). The Bogota site is close to the modern tree line at 2600 meters. Pollen data accordingly let the passage of the tree line up and down the near-equatorial Andes be recorded throughout the glacial period. Tree lines were lowered by 1500 meters in ice-age time, suggesting that temperatures dropped by about 6°C. This is fully consistent with evidence that mountain glaciers on the highest peaks of the Andes also descended about 1500 meters (Clapperton, 1987; Hastenrath and Kutzbach, 1985).

Thus from the mountains of Africa, Southeast Asia, and South America, pollen and glaciological data allow the inference of an equatorial cooling in an ice age of 6°C or more, whereas reconstructions based on foraminifera in deep-sea cores show minimal cooling in the neighboring oceans. This is currently the cause of controversy.

When the discrepancy between land and sea data first appeared, it seemed that it could be explained by the hypothesis that only high altitudes in the tropics had cooled (Rind and Peteet, 1985). The tropical lowlands, from which (obviously) none of the montane records came, would in fact be found to have remained warm to correspond to a warm sea surface. This view no longer appears tenable, following our demonstration of cooling at low elevations in both the Amazon and in Panama (Bush et al., 1990, Bush and Colinvaux, 1990) and similar results from Africa (Maley, 1991; Livingstone, 1993).

The paleoclimatic reconstructions from land and sea use quite different organisms: from the land, forest trees; from the oceans, planktonic foraminifera (see Chapter 17). Livingstone (1993) suggests that the oceans are so rarely as warm as they have been in the last 10,000 years of Holocene time that true warm water foraminifera have never evolved. Thus the modern seas of the equator are inhabited by the same foraminifera as in ice age times, which are now living at the extreme warm end of their temperature tolerances. Paleooceanographers necessarily find the same foraminifera in equatorial seas of the ice age as live there now, but this does not mean that the temperature of the sea has not changed.

BIOMES THROUGH TIME

Biomes surely change with time, such that the vegetation maps of the past must differ from modern maps, even though most available species are the same. But the changes must be fairly closely constrained by energetics that give each biome its characteristic form. Cold has always been in the North, warmth always at the equator. Accordingly, tundras were in the North and tropical rain forests at the equator, regardless of what detailed maps may eventually show ice age climates to have been.

What must change with secular climate change is the species compositions of communities within the biomes. Forests or tundras might look the same, but diversity alters, particularly as common species become rare or rare species become common. The contemporary earth has biomes probably representative of all the last million or more years, though its communities are all fleeting products of the recent past.

Commonness, rarity, and distributions of species still reflect the disruption of normality caused by the end of the last ice age 10,000 years ago. Their earlier distributions had lasted for 80,000 or 90,000 years since the previous interglacial, and probably was representative of 80% of the last million years (Berger et al., 1989). Life in an ice age has been the norm for the ancestors of all modern populations.

Preview to Chapter 19

Beech-Maple: classic association of the American mid-west.

THE study of communities began with attempts to identify distinctive plant associations that could be given rigorous descriptions and then classified into a fundamental ordering of nature. Associations could be identified intuitively, sought out, and described, leading to practical field techniques for describing vegetation. Alternatively, species lists could be made by sampling successive quadrats until the rate of addition of fresh species was deemed negligible to identify the smallest units of coexisting species, called "sociations." Neither method produced convincing classifications. Instead, the discovery that the imagined social communities of plants were arbitrary led to the ecosystem concept, which provided a more useful unit to study. The expectation that complex plant communities must nevertheless be ancient and permanent assemblages continued even after ecosystems became the preferred units for study and had to be explicitly falsified. The apparent discreteness of communities on mountainsides was shown not to survive direct analysis of plant importance along gradients. More importantly, the hypothesis that plants live in lasting associations has been falsified by reconstructing vegetation histories through pollen analysis. Assembly of hundreds of pollen histories for eastern North America shows that all contemporary plant associations have been assembled in the last 10,000 years. The first pollen records of ice age times from Panama and the Amazon rain forests suggest that even the most species rich forests in the world are young as communities, though the species that make them are ancient.

Chapter 19

Phytosociology

The first sustained studies of communities were attempts to identify and describe units of vegetation smaller than the plant "formations." Within the eastern deciduous forest of North America, for instance, were oak–hickory woods, beech–maple woods, or woods where deciduous trees were mixed with the conifer hemlock. The plant communities of these woods were distinctively different yet related as part of the temperate deciduous forest, a formation, or as we should now say, a biome.

This observation of community living raised immediate questions about how the individuals of communities are held together, essentially of how the communities are "organized" and of how communities may be classified. Plants seemed to live in societies of sorts, inviting a sociology of plants. The first sustained efforts at community analysis accordingly came to be called "phytosociology."

Botanists of the phytosociology schools worked out the first formal methods of describing plant communities. An outcome of their work was the invention of the ecosystem concept. They were also responsible for much of the sense of "holism" that hovers around ecological studies, of the sense that the whole is more than the sum of the parts, that whole communities have independent lives of their own. All this came from attempts to classify plant communities and to study their development through succession (see Chapter 20).

THE PLANT ASSOCIATION OF ZURICH-MONTPELLIER

When confronted with land plants in the mass, ecologists tend to talk of "plant associations." Beech–maple forest of the American Midwest, for instance, is referred to as "the beech–maple association." The phrase connotes much more than a "stand" of beech or maple trees in that it conveys the idea of a whole community of plants of which the beeches and maples are the most prominent. It goes even further, because inherent in the statement "beech–maple association" is the idea that any forest to which you can apply that name will be pretty much the same. See one beech–maple forest and you have seen the lot, right down to knowing all the dozens of other plants expected to be found in a beech–maple woodlot.

At the dawning of modern ecology, this abstract idea of plants living in characteristic **associations** was powerful, setting ecologists to wondering if all life was lived in societies of obligate associates. Particularly important was work done by European botanists working in the south of France, and centered on institutes in Zurich and Montpellier. They had rich vegetation to work with, though it was harried and dissected by the agriculture of 5000 years.

The Zurich–Montpellier (Z–M) method of discovering the social order of plants is to decide in a purely subjective way what is a representative piece of vegetation or stand, and then to describe it according to a formal set of ground rules. The process is repeated in other stands, and the separate descriptions are compared for similarities. Out of this exercise emerges a list of characteristic species believed to be held in common by all stands of an association (Braun-Blanquet, 1932; Whittaker, 1962).

First, choose a piece of vegetation believed to belong to a distinct type. The piece should be large enough to be representative of all facets of the community.

Second, identify the "ecological dominants" subjectively. These are the most common or important plants: the oaks in an oak forest, the beeches and maples of a beech–maple woodlot, the *Festuca ovina* of a sheep's fescue pasture in Wales.

Third, make a list of all the species of plant present in the chosen stand. This is not a census or sampling exercise. The botanist searches with diligence and a naturalist's skill to make sure that no single species is missed.

Fourth, assign a set of **importance values** to each species on the list. These record whether a species is common or rare, small or large, dispersed or clumped. It is evident that these are all values that could, in principle, be measured in an objective and quantitative way. Some of the measures would not be easy, however, such as the number of grass plants in a meadow sample. Sometimes it is more convenient to measure the percent cover as the importance value in place of numerical abundance. But the Z–M method sidesteps this issue by assigning a **cover-abundance index**, with 5 being the most important (covering more than three fourths of the sampled area), 4 the next most important (covering one half to three fourths of the sampled area), and so on down to 1. Dispersion is described with another 1-to-5 scale, the **sociability index**, as are **condition** and **vitality** (Figure 19.1). It should be noted that all these indexes are assessed subjectively and depend on the good judgment of the botanist doing the work.

Fifth, write the complete list of species found with the cover-abundance and sociability indexes after each name in that order, thus: *Zea mays* 3.1, which describes the role of a corn plant in a typical field of corn (the corn community). The cover-abundance index of 3 means that corn plants cover one fourth to one half of the field (the rest being bare ground between the rows), and the sociability index of 1 means that the corn plants are evenly dispersed. The complete list written up in this way is called a "**relev**."

Sixth, repeat the whole process at a number of other sites chosen for their apparent conformity to the type of community being described and tabulate the results.

Seventh, compare all the relevs in quest of common denominators other than the ecological dominants. The dominants, of course, will be in every list because the botanist chose to sample only communities where the dominants grew. Now the botanist needs to find other, more humble plants that are faithful to the postulated community. Such faithful plants should be in every relev. They are called the "character species."

The exercise, summarized in Figure 19.2, is now complete. The investigator considers that an association of plants has been demonstrated. It is characterized by

Cover-abundance ratings according to following scale:

- 5 covering more than ¾ of the sampled area
- 4 covering ½ to ¾ of the sampled area
- 3 covering ¼ to ½ of the sampled area
- 2 with any number of individuals covering 1/20 to ¼ of the sampled area, or very numerous individuals but covering less than 1/20 of the area
- 1 numerous, but covering less than 1/20 of the sample area, or fairly sparse but with greater cover value
- \+ sparse and covering only a little of the sampled area
- r rare and covering only a very little of the sampled area (usually only 1 example)

(n.b. "+" is always spoken "cross")

Sociability is estimated for each species in terms of another scale.

- 5 in large solid stands; very dense populations
- 4 in small colonies or larger mats; rather dense populations
- 3 in small patches or polsters; distinct groups
- 2 in small groups or clusters or tufts
- 1 growing singly

Further symbols relating to the **condition** and **vitality** of the individual species, as those below, may be recorded beside the cover-abundance and sociability estimates.

- oo - very poor and especially not fruiting (e.g., +^{oo} or 2^{oo})
- o - poor vitality (e.g., 1°)
- g - germinating plant
- Y - young plant
- st - sterile
- bu - budding
- bl - blooming
- fr - fruiting
- no notation—normal growth
- · - luxurious growth (e.g. 4·)
- e - being driven out (by other plants)
- d - dying
- def - defoliated
- dd - above-ground organs dead or dried out
- s - present only as seed
- # - specimen collected

FIGURE 19.1
The descriptive indexes used by the Zurich-Montpellier school of phytosociology as interpreted by Benninghoff (1966).
This is an arbitrary set of indexes, but one that has had such wide use in the literature that it should be used in preference to ad hoc systems. For the purpose of pure description is it the most expedient method, but it is well to remember that the labor involved in using it may be very great and that such description might only be worthwhile when the object is to record for posterity the present condition of changing vegetation.

the presence of both the dominant species and the character species. The association is then given a name based on the names of the dominants, like the Beech–Maple Association (various latinized systems of nomenclature are still in use also, particularly in Europe; Mueller-Dombois and Ellenberg, 1974). Easy access to computers now makes possible the ordering of relevs with multivariate statistics, but the sampling process itself remains subjective.

QUADRAT CENSUS: THE UPPSALA SOCIATION

A grand criticism of the Z–M philosophy and technique is that the whole exercise is

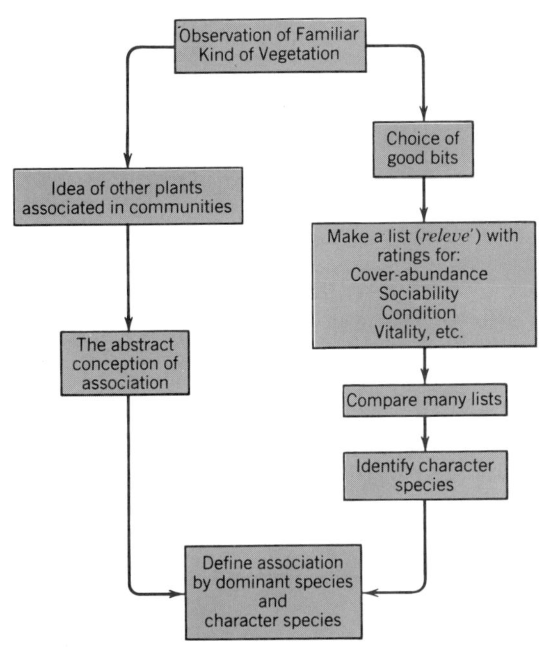

FIGURE 19.2
Procedure of the Zurich–Montpellier school of phytosociology.

subjective. The proponents believed in typical organized societies of plants, they chose stands that fitted their preconceptions, they described these stands, and, lo and behold, the expected society was revealed.

Botanists of a more quantitative turn of mind found alternatives. Typical were those at the Swedish University of Uppsala, who worked in the much more homogeneous vegetation of Scandinavia. They used a system of random subsamples when describing vegetation. These subsamples are called "**quadrat samples**" or simply "**quadrats.**" The essence of quadrat sampling is marking out small plots at random, then listing, measuring, or counting all the plants found in each plot.

In the Uppsala method the species lists from series of quadrats are added and the results are plotted as a **species–area curve** (Figure 19.3).[1] As the area sampled increases with the number of quadrats added, so the expectation of finding new species falls and the species–area curve tends to level off. This curve will never become truly horizontal, however, unless the whole earth is sampled at once, because increasing area always must bring in other habitats with different species. Yet reasonable rules can be written for deciding when a curve inflects, showing when all but chance plants of a locality are included. The area under the inflection point was called the "**minimum area.**"

By this means a species list of co-occurring plants is derived and the smallest patch of ground on which this community can exist is defined. When the process has been repeated in a number of localities, the community lists can be compared to detect the regularity with which plants appear in the lists, expressed as "**percent constance.**" The community type, eventually to be called a "**sociation,**" is then defined in terms of the ecological dominants and "constant" plants (those more than 80% constant) (Figure 19.4; Du Rietz, 1930; Whittaker, 1962).

It is evident that a piece of vegetation defined by minimum area and constant species in this way is likely to be much smaller than the piece of vegetation chosen for description by a follower of the Z–M school. Yet the goals of both methods are the same: to categorize fundamental community types. Both schools were in quest of fundamental associations of plants, and both talked of ordering and classifying associations. But since the actual pieces of vegetation were quite different, an Inter-

[1]Species–area curves are used more widely now in the different context of diversity studies in which the goals are to understand how the number of co-occurring species in unit area comes about (see Chapter 16).

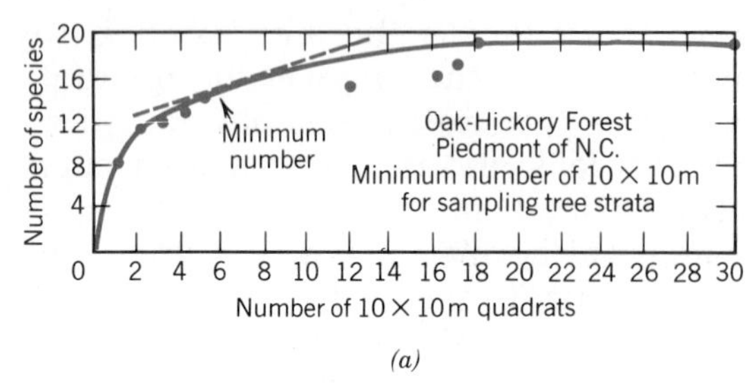

(a)

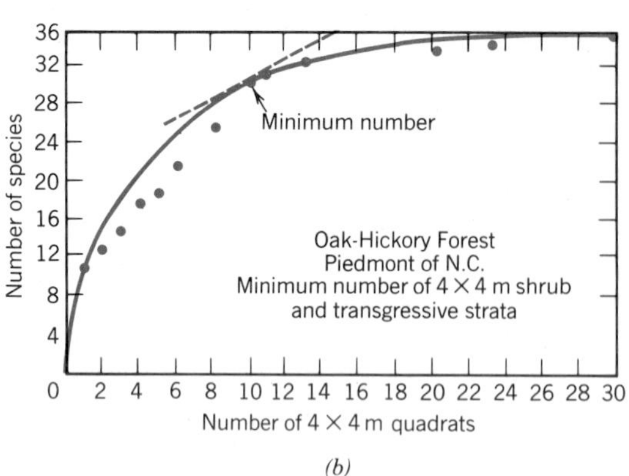

(b)

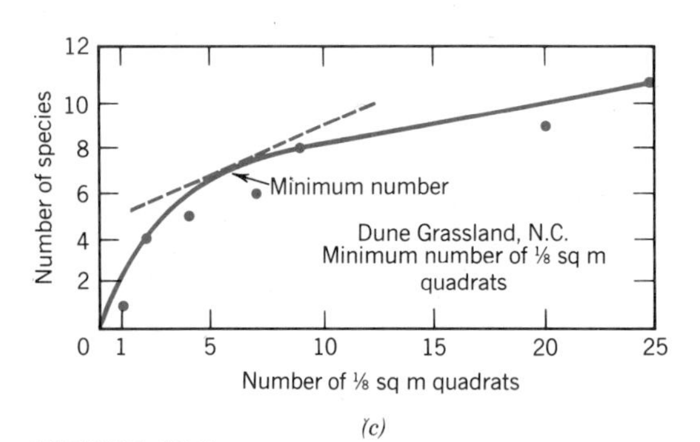

(c)

FIGURE 19.3
Species–area curves for North Carolina vegetation.
Sizes of quadrats have been chosen in each example to be convenient. This "break" in the curve was chosen such that a further 10% increase in area would yield additional species equal only to 10% of those already present. This is a convention. Minimum areas are (*a*) (forest trees) 100 m^2, (*b*) (shrubs) 16 m^2, and (*c*) (grassland) 0.75 m^2.

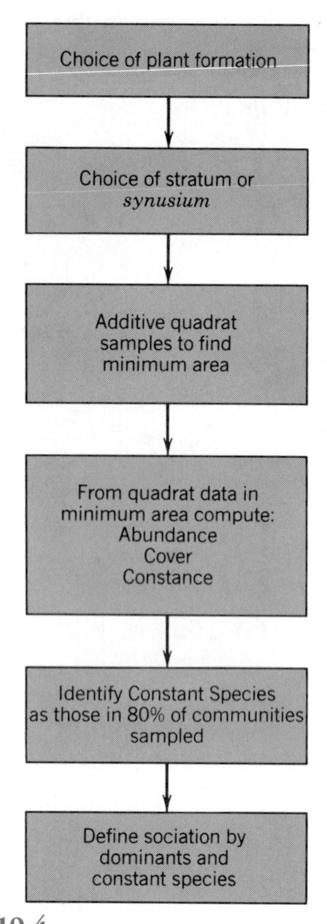

FIGURE 19.4
Procedure of the Uppsala school of phytosociology.

national Botanical Congress ruled that communities defined by minimum area should be called **sociations** to distinguish them from associations found by subjective means. It may be that little attention was paid to this ruling by botanists of the time because "association" continues to be used as a loose term to describe familiar bits of plant land, however this is described.

THE INTELLECTUAL POVERTY OF COMMUNITY CLASSIFICATION

Many variants on the two opposed systems of Z–M and Uppsala were tried. Most were compromise methods that involved choosing stands in the Z–M manner, then using quadrats or other plot-sampling techniques within the stands. But all had in common a primary purpose that was taxonomic. The associations were to be classified into higher orders or subdivided into parts. Figure 19.5 shows some of the classifications proposed. At the left of the figure is a list of the subdivisions used in classical Linnaean taxonomy of species populations. The various schools of phytosociology produced their several names to be comparable. Uppsala saw their sociation as comparable to a species. Later both Uppsala and Zurich–Montpellier thought of associations as if they were biological families. Naturally enough, considering the variable properties of vegetation, the units sampled by botanical intuition or estimates of minimum area were of flexible size and so could be arranged to appear in different levels of the hierarchy.

A vital difference between the community classifications and Linnaean taxonomy appears when we consider the criteria used for separating units at different levels in the hierarchy. Linnaean taxonomy groups related species together. The only act of differentiation is a species separation; the affinity between members of a phylum or class reflects a distant common ancestor, many acts of speciation ago. But the acts of separation of the higher orders proposed for communities are quite unrelated. The lowest unit is in fact a layer, called a "synusia" or "socion." In plain language this means the bit of vegetation that happens to be made up of trees, shrubs, or herbs. At the other extreme, the largest unit is the "formation," the vegetable part of a biome set by climate. All the divisions in between depend on the botanical methods used to describe them, though they may also reflect large changes in habitat, like soil type. This is shown at the right of Figure 19.5.

This analysis suggests that vegetation can be divided up by layers, or by life forms adapted to different climates, or by collections of species adapted to particular soils and patterns of drainage, et cetera. But these separations are all of different kinds and use different criteria.

Two widely used botanical terms speak to this dilemma of classification—"floristic" and "physiognomic." "Floristic" refers to the species composition of vegetation and "physiognomic" to the shapes or structures of plants. Both layering and the life forms of formations are recognized by physiognomic characters, but all the inter-

Linnaen Taxonomy	Phytosociological Classifications						Cause of Separation	
	Uppsala 1921	Uppsala 1928–30	Uppsala 1930–35	Zurich–Montpellier	Rübel	Gams		
Variety	—	Socion	Socion	—	Synusia	Synusia	Layering	Physiognomic
Species	Association	Sociation	Consocion	Facies	Sociation	Phytocoenosis	Botanical methods or physical habitat	Floristic
Genus	—	Consociation	Associon	(sub-association)	Consociation			
Family	Association complex	Association	Federion	Association	Association			
Order	—	Federation	Sub-formien	Alliance	Alliance			
Class Phylum	Formation	Sub-formation	Formien	Order	Formation		Climate	Physiognomic
		Formation	Pariformien	Class				

FIGURE 19.5
Classifications of plant communities proposed by phytosociologists.
For discussion see text. (Based on a compilation by Shimwell, 1971.)

vening units (associations and the like) are set by differences in floristics.

The phytosociological classifications are rather what we would get if we classified all flying animals together into the "flight formation" on the basis of physiognomic characteristics, and then separated them into mixes of insects, birds, and bats by specific differences—a sort of animal floristics. The result would be a most un-Linnaean and un-Darwinian classification. But something like this is what classifying plant communities as units of formations implied.

Braun-Blanquet (1932), a wise botanist of the Z–M school, reviewed the efforts of plant sociology to find and classify natural communities in a book that summarized the work of his school. He noted that community boundaries were always habitat boundaries: a change of soil, exposure, or moisture. The "*plant community*" was actually *the collection of plants that shared a habitat*, and physical, not biological, factors were decisive in ordering communities.

The observation that habitats were crucial determinants of plant communities was, of course, made independently by every botanist who went out to sample. All were equally aware that the fortunes of many a plant also depended on the animals that lived with it, to eat it, to pollinate it, or to carry its seeds. Even as attempts to classify vegetation met with indifferent success, these thoughts of habitats and animals led botanists to a different idea, that of the ecosystem.

THE CONCEPT OF THE ECOSYSTEM

The ecosystem concept has deep roots in ecology. Möbius (1877) often is given credit for the first formal statement of the idea, for which he used the term "biocoenose" (or biocoenosis). Möbius reflected on an oyster bank: the things the oysters ate, oyster predators, and the narrow set of physical requirements that must be met for an oyster bank to persist. Doubtless many naturalists before him thought in like ways about a community and its needs, but Möbius coined a formal term to define the unit. Even now the term "biocoenosis" is used in preference to "ecosystem" in central European and Russian writings.

In that same year Forbes (1877) proposed the same idea with the name "microcosm."[2] Like Möbius, Forbes worked with an aquatic system, thinking of a lake as a self-contained unit with its own plants, animals, and nutrient supplies all dependent on each other. A tradition of thinking of environment, nutrients, and food spread as naturalists everywhere arrived at positions close to those of Möbius or Forbes. But these ideas came to be stated as the central concept of ecology most clearly when botanists reached a general consensus on the nature of the association.

The one factor that determined the species list of every association was the physical habitat. The unit for study, therefore, must be the whole tangled mixture of plants and animals and their physical surroundings. An idea was born of these studies, and scholars groped for new words to describe it: the "landscape unit" or "raume," the "biogeocoenosis," the "biochore," and the "biotic district" all had

[2]Forbes's essay was published in the *Bulletin of the Peoria Scientific Association*, and cannot have been widely read. It was rediscovered by later ecologists as the ecosystem concept was developing independently.

their devotees. Terms like "landscape" or "districts" show the influence of the terrestrial botanists on whose work this central theme of community ecology was based. And one of them, Tansley (1935), proposed the term "ecosystem" for the new idea (Figure 19.6).

Tansley's **ecosystem** would be considered to include all the animals, plants, and physical interactions of a defined space. Natural limits would be identified for the ecosystem under study, like the changes of soil or drainage defining plant associations, or the edges of streams and lakes. An ecosystem could be of any size depending on the communities to be studied—as small as the cup of a pitcher plant or as large as a biome or the whole biosphere.

Modern ecologists think easily of ecosystems in terms of energy flow, carbon flow, or nutrient cycles. But these concepts came later; the idea of energy flow, for instance, was not formally stated until seven years after Tansley clarified the phytosociological debate with his ecosystem (see Chapter 2).

THE ECOSYSTEM

I have already given my reasons for rejecting the terms "complex organism" and "biotic community." Clements' earlier term "biome" for the whole complex of organisms inhabiting a given region is unobjectionable, and for some purposes convenient. But the more fundamental conception is, as it seems to me, the whole *system* (in the sense of physics), including not only the organism-complex, but also the whole complex of physical factors forming what we call the environment of the biome—the habitat factors in the widest sense. Though the organisms may claim our primary interest, when we are trying to think fundamentally we cannot separate them from their special environment, with which they form one physical system.

It is the systems so formed which, from the point of view of the ecologist, are the basic units of nature on the face of the earth. Our natural human prejudices force us to consider the organisms (in the sense of the biologist) as the most important parts of these systems, but certainly the inorganic "factors" are also parts—there could be no systems without them, and there is constant interchange of the most various kinds within each system, not only between the organisms but between the organic and the inorganic. These *ecosystems*, as we may call them, are of the most various kinds and sizes. They form one category of the multitudinous physical systems of the universe, which range from the universe as a whole down to the atom. The whole method of science, as H. Levy ('32) has most convincingly pointed

[3] If this statement is applied to the individual organism, it of course involves the repudiation of belief in any form of vitalism. But I do not understand Professor Phillips to endow the "complex organism" with a "vital principle."

FIGURE 19.6
Photographic facsimile of the first appearance of the word ecosystem. (From Tansley, 1935.)

THE FALLACY OF DISCRETE COMMUNITIES

If the ecosystem concept was a massive achievement of phytosociologists, the concept of discrete communities with independent existence of their own was a compensating disservice. Despite the extreme difficulty of reconciling organized communities with the process of natural selection working to individual advantage, this notion of discrete plant associations as obligate and lasting arrangements between species persisted well into the modern period.

As a hypothesis, however, the existence of discrete communities had the virtue that it was testable against the fossil record. The general hypothesis predicted that important plant communities should have a prolonged existence in time. This was elaborated for the species-rich deciduous forests of eastern North America in the 1950s and finally falsified with a massive array of pollen data in the 1980s.

The deciduous forests of eastern North America are equaled in diversity only by those of eastern China, and they are composed of species many of which are known as fossils from the Tertiary. The northern habitats of the modern forest were close to the glacial front, posing the biogeographic and ecological question, "What happened to these diverse forests in an ice age?"

Three hypotheses account for all possibilities:

1. The forest associations remained intact and in place over much of the modern range.
2. Forest associations persisted in refugia in the southern part of their range, or to which the whole community had migrated.
3. Modern forest associations did not exist in the ice age, but the component species were dispersed individually to different habitats throughout North America.

The first of these hypotheses, derived from the community concepts of phytosociology, was the prevailing view 30 years ago (Braun, 1955). Those who accepted early pollen evidence of Deevey (1949) as requiring tundra landscapes immediately south of the ice sheets fell back on the second hypothesis of life in refugia, often requiring vegetation to have migrated southward (Martin, 1958). And yet we now know that the third hypothesis is correct.

Demonstrating the complete absence of forest refugia in glacial times in North America has been the greatest triumph of pollen analysis. Davis (1986) and Webb (1986) have used a large number of pollen sections from throughout eastern North America to show that the forests of glacial times were quite different from those of the later Holocene. Their maps of successive Holocene distributions show the modern forest communities being built species by species as the various trees converged on their present distributions from different directions (Figure 19.7).

By the evidence of pollen analysis, therefore, we can say quite definitely that the eastern deciduous forest of North America is a purely ephemeral accommodation between plants. Indeed, the complexity was still being put together when the post Columbus invasion by Europeans ended it all with axes.

This is but the most elegant demonstration of the fallacy of thinking that communities are organized entities, preserved together and existing through long spans of time. Associations always represent mere ephemeral accommodations between species, a reality that was long obscured by the habit of looking at plant communities as if they were social systems.

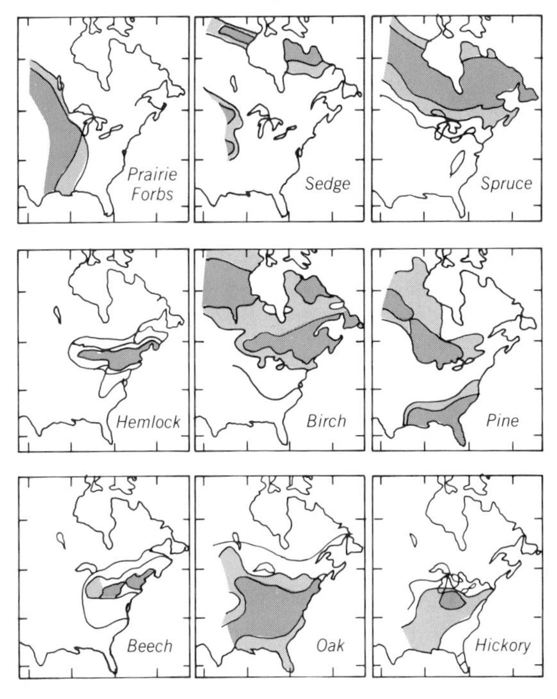

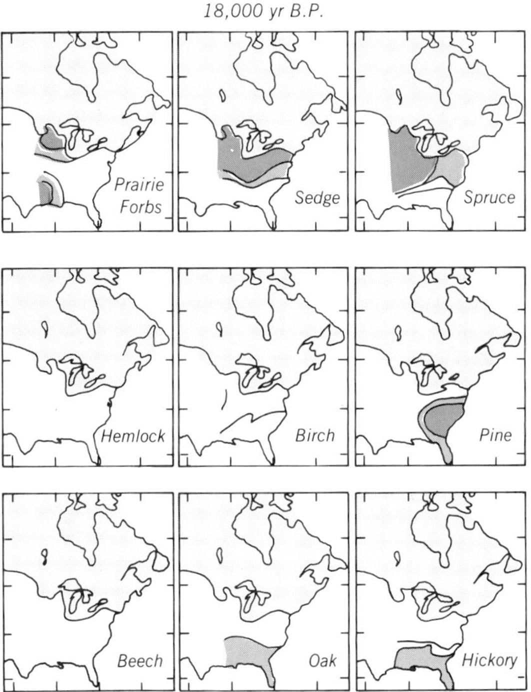

THE FALLACY OF DISTINCT COMMUNITY BOUNDARIES

Mapping associations or sociations required identifying boundaries. Failure to find these boundaries led the wise Braun-Blanquet to admit that the only boundaries were those of habitats. But those stubborn about the discreteness of communities were able to point to the remarkable phenomenon of bands of vegetation on mountainsides.

To the eye of a spectator from across the valley, the sides of mountains truly seem to be banded with recognizable communities. Often a high treeline is visible, with a change in color below it where vegetation changes at the cloud base. Below this, other shades of color can often be seen. These apparent boundaries could be used in defense of the discrete community hypothesis: they could be seen with the naked eye! And yet the proposed disjunctions always turn out to be illusory on closer examination, unless a real physical boundary is found.

The range of possibilities for distributions of species on environmental gradients like a mountainside is given in Figure 19.8. Two forms of species distribution required by the discrete community hypothesis are shown at the top of the figure, with the corresponding distributions from the independent plant hypothesis at the bottom.

FIGURE 19.7
Eastern North America in glacial and historic times.
Maps are of pollen percentages in sediments for each species, density of shading being proportional to pollen percentage. Successive maps show the forest communities of eastern North America being built species by species throughout postglacial time. (From Webb, 1988.)

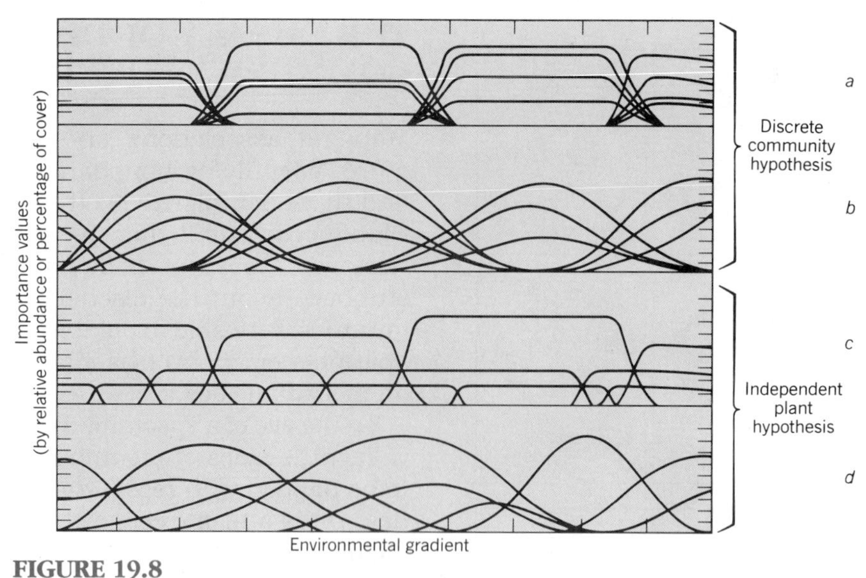

FIGURE 19.8
Theoretical distributions of plants along environmental gradients.
Either (*a*) or (*b*) is required by the ***discrete community hypothesis*** of classical phytosociology. (*c*) and (*d*) are both consistent with the ***independent plant hypothesis***. (Based on a figure of Whittaker, 1975.)

Whittaker (1956) measured the distribution of individual tree species up and down mountainsides by setting out transects and measuring the percentage composition of trees in the forest at intervals. A revealing set of his data from the Great Smoky Mountains of eastern Tennessee is given in Figure 19.9, where the distribution is clearly that of the independent plant hypothesis in the fourth alternative of Figure 19.8.

The Great Smokies are mountains that look to be nicely banded in the fall, having colors as different as the dark green of hemlock and the autumn tints of turning maple leaves, yet the survey data show no disjunctions in the ranges of individual species let alone of whole communities.

Each species of tree in Whittaker's census had an optimum elevation at which it lived in peak numbers, but the distributions overlap. There are actually no precise bands or edges in this system, though the human eye can pick out any of the peaks *a–d* in Figure 19.9 when viewing from across a valley and see it as a band. The eye picks out prominent colors or textures and fits these together as bands in the way that we talk of bands of color in a light spectrum or rainbow (Figure 19.10). Thus the apparent phenomenon of discrete communities on mountainsides that can be seen as bands of color is an optical illusion.

CHANGING FORESTS OF THE NEW WORLD TROPICS

The idea of permanent associations has lingered longest in thoughts about tropical rain forests. These are the most complex and most coevolved of communities, and

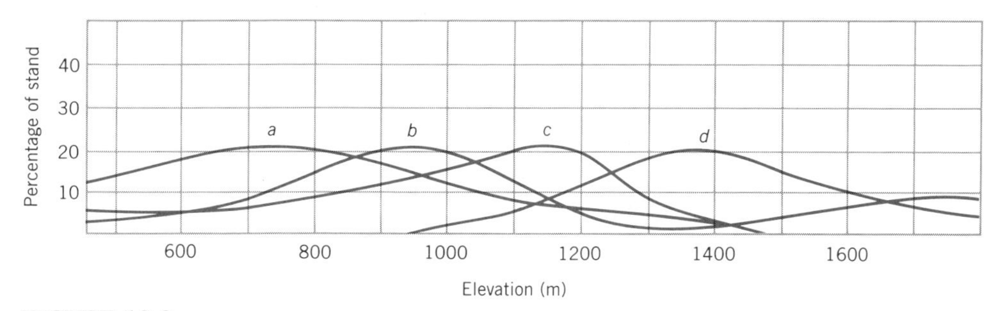

FIGURE 19.9
Distribution of tree species on a slope of the Great Smoky Mountains in Tennessee.
The percentage compositions of the forest were calculated from sample counts of trees more than 1 centimeter in diameter 4 feet from the ground. The forest was sampled at 100-meter elevation intervals. Four tree species out of the sample are shown here. There are no distinct vegetation belts, for the distribution of each species grades into that of its neighbors. A distant observer might well resolve points *a*, *b*, *c*, *d* as the middle of separate vegetation belts: *a*, hemlock; *b*, *Halesia monticola*; *c*, lime; *d*, mountain maple. (Modified from Whittaker, 1956.)

FIGURE 19.10
Trees on the Great Smoky Mountains.
Tree populations merge together in a smooth cline up the slopes as in Figure 19.9, yet the woods appear to show bands of color in the fall as the eye picks out levels where plants of a particular hue are concentrated.

their siting far from the haunts of northern glaciers or European and North American professors has given them an aura of immemorial antiquity. This aura, however, has come from lack of data, not from evidence. The first pollen diagrams of forests of the ice age Amazon and Panama show otherwise.

Figure 19.11 is the first transglacial pollen diagram from tropical forests in the New World. The site is a drained lake basin at 500 meters elevation, at El Valle on the Pacific coast of Panama. The record spans from near the beginning of the last ice age until the early Holocene at 8000 yr B.P. (before present), but other samples from

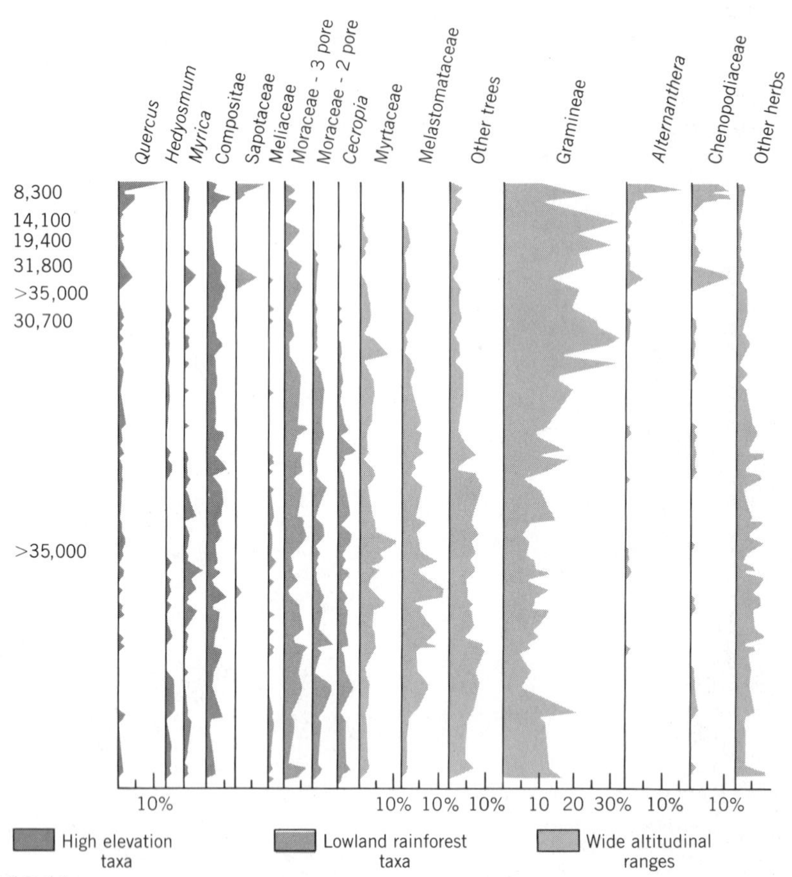

FIGURE 19.11

Transglacial pollen history for lowland (500 meters) forest in Panama.

Data are percent total pollen in a 55-meter core of sediment from an extinct lake on the floor of which the city of El Valle is built. Lake sediment stopped collecting about 8000 years ago. The historical vegetation of the region was lowland tropical forest, which is represented by taxa colored green. This forest has evidently been present in the Panama lowlands throughout a complete glacial cycle. But the diagram also records the descent of oaks (*Quercus*) and other high altitude taxa, which can be found in modern Panama only at elevations above 1200 meters (gray shaded area). The pollen diagram shows that rain forest species of ice age times (color shaded area) lived in the Panama lowlands as they do today, but in different communities that included plants now confined to cooler uplands. (From Bush and Colinvaux, 1990.)

the region show that the top spectrum is already approaching those of modern forests disturbed by human activity. The presettlement vegetation was tropical seasonal forest, with many species shared with the rain forests of nearby Darien.

In Figure 19.11 tropical forest taxa are printed in color. Other taxa, such as oaks, are plants of high elevations that descended on the site from the interior highlands in response to ice age cooling. Simple inspection of the diagram reveals a history of mingling and remingling of various lowland and highland taxa with the secular climatic changes of a glacial episode.

A similar history is shown by the two Amazonian rain forest pollen diagrams dated to about 30,000 yr B.P. (see Chapter 18, Figure 18.24). The ancient Amazon forests at the foot of the Andes are shown to have been built from what are now lowland species mixed with what are now highland species.

Those old rain forests of Amazonian Ecuador have no modern analogs, suggesting that tropical rain forests are made and remade as climates subtly change, just as easily as temperate forests. The modern successors to those ancient Ecuadorian forests are the most species-rich in the world (Gentry, 1988), possibly as a result of the number of potential invading species in the varied habitats of the Andean foothills. Amazonian forest associations are temporary affairs, like those in all other habitats.

Preview to Chapter 20

Mt. St. Helens aftermath: laboratory for primary succession.

SPECIES-RICH communities are put together in increments by multiple invasions, resulting in ecological successions of communities. In many plant communities, particularly in abandoned agricultural fields, the invasions appear to be sequential and predictable. The predictability could result from progressive changes in habitat or from a series of competitive replacements, or both. In early stages of some primary successions, as on sand dunes or glacial till, habitat changes resulting from plant growth itself are important, but most habitat changes are due to geographic processes like drainage or weathering. Plants merely respond to these externally induced habitat changes. Nearly all successional change in wetlands is of this kind. Secondary successions in old fields are probably little influenced by habitat changes and can be explained by a hypothesis of contest competition in which opportunist or ruderal weeds are replaced by species diverting ever larger proportions of resources to competition. Successions do not proceed in ways that make communities more energetically efficient, but actually tend toward inefficiency and lowered production as communities become more complex. Fanciful scenarios of ecological successions being processes of ordered and directed community building are important to political philosophy but have no scientific substance.

Chapter 20

Ecological Succession

The idea of ecological succession has been one of the most powerful intellectual stimulants of ecology.

When a farmer in Europe or the United States abandons a field, the weeds close over it. The weed community is invaded by other weeds, then by brambles and small trees of open fields. The trees grow up to form woodland, which is then invaded by still more trees. An **ecological succession** of plant communities arriving in series has preceded the reestablishment of the ancient forests.

This much can be seen by naturalists all over the north temperate belt. Before the Second World War, old field successions were common enough on bankrupt farms (Table 20.1). More recently the prime examples are on farms near cities that were bought for development. All show that land fit for forest usually goes through successive occupations by communities of ever longer-lasting weeds before the forest comes back.

In successions of land plants, habitat quality inevitably changes even as plants replace each other. A shaded habitat is different from one exposed to the sun; soil covered with fallen leaves is different from bare ground. But these habitat changes made it possible to suggest that successions were driven by habitat changes, that the arrival of each successive community was contingent on the prior preparation of the habitat by others. A philosophy of contingency is introduced. Communities are then thought to develop by invasions and habitat changes that are contingent upon prior invasions.

The philosophical possibilities of successions are the more piquant because of the concept of "**climax**." Succession on an old field in Germany or Massachusetts might be expected to end in forests like those cleared by the first agriculturalists. If this were true (and how to deny it, without living many generations?), the succession ends with a restoration, it is a process of renewal, it has a climax.

Renewal of the climax in old field suc-

TABLE 20.1
Old Field Succession on the Piedmont of North Carolina
(From data of Billings, 1938.)

Years Since Abandonment	Most Abundant (Dominant) Species	Other Common Plants
0(fall)	Crabgrass	
1	Horseweed (*Erigeron*)	Ragweed
2	Aster	Ragweed
3	Broomsedge	
5-15	Short-leaf pine	Loblolly pine
50-150	Oaks and other hardwoods	Hickory

cession makes this a "secondary succession." The idea of succession to climax, however, lets in a still more interesting idea: that successions may be primary. Primary successions should proceed by conquest, taking over ground that had never before been used by plants, and placing on them the local climax vegetation type. A sand dune, for instance, might be colonized by specialized weeds whose shade and debris can then make possible invasion by other plants, including trees. A primary dry-site succession leading to a climax of plants quite unable to live in the original, unaltered sand dune thus seems possible (a succession called "**xerarch**").

And what about a pond's being filled with the rotting parts of the water plants at its edge? Might not what once was a belt of reeds be raised out of the water so that plants of the land could invade? This invokes the idea of a primary succession of plant communities running from wet pioneers to a dry climax, what could be called a "**hydrarch succession.**"

Thinking on these lines made the phenomena of succession, real or imagined, the most important intellectual property of ecology during its formative years. Plant associations (see Chapter 19) were of two general kinds: associations of the climax and associations that had more fleeting existence as stages (**seral stages**) in a succession. The more complex communities of the local climax were considered always to have been put together in sequential stages as if they had been the deliberate steps of a master plan.

Perhaps sadly, modern ecology has had to debunk the intellectual hyperbole of early succession theory. Most replacements of plants by others are inevitable consequences of different strategies of dispersal and reproduction (see Chapter 11). A large element of randomness means that many alternative series of replacements are possible—as is direct regeneration of communities as slow growing as forest without successional precursors. Events in old fields of western agriculture are often dependent on the ready availability of the seeds of agricultural weeds. Some proposed primary successions, like those imagined to put forests on to old ponds, simply do not happen.

But ecologists have learned much wisdom in sorting the true from the false in succession stories, and from mastering the properties that let some plants displace each other from a habitat in sequence.

OLD FIELD SUCCESSION DOCUMENTED

The details of what actually happens in an old field succession were accurately documented in the late 1930s, and the old data still serve well enough. The Piedmont of North Carolina was nicely suited to the task because farms there had been progressively abandoned for 150 years, ever since food from the western frontier states began to put eastern farmers out of business. Billings (1938) had at his disposal both newly abandoned farms and old farm sites that could be dated from court records.

The succession in outline is given in Table 20.1. As the herbal communities of the old fields succeed each other and give way to pines, the organic layer of the soil deepens (Figure 20.1), and the water retaining capacity of the soil increases (Figure 20.2). In successions where legumes are prominent among the colonizing species, it is likely that combined nitrogen also would accumulate. Thus the habitat is changing in ways likely to make the fields better places for plant growth.

Data like these let botanists argue that habitat change was a critical forcing function for successions, that in fact the progression of plant colonizations could actually be dependent on prior improvements of the habitat. In an old field this was, perhaps, a less-than-compelling argument, because farmland is in a satisfactory state for growing things from the start.

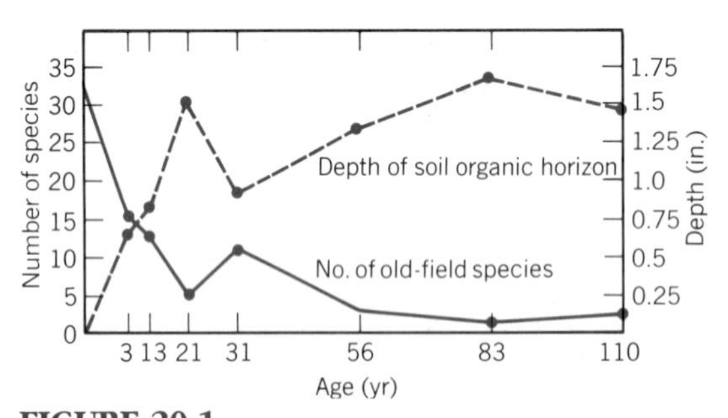

FIGURE 20.1

Changing depth of soil and number of herb species in old fields in the piedmont of North Carolina.

The organic horizon grows progressively deeper as the number of colonizing herb species falls. It happens that pines, arriving between years 5 and 15, grow after the organic layer is about as thick as it will become. (From Billings, 1938.)

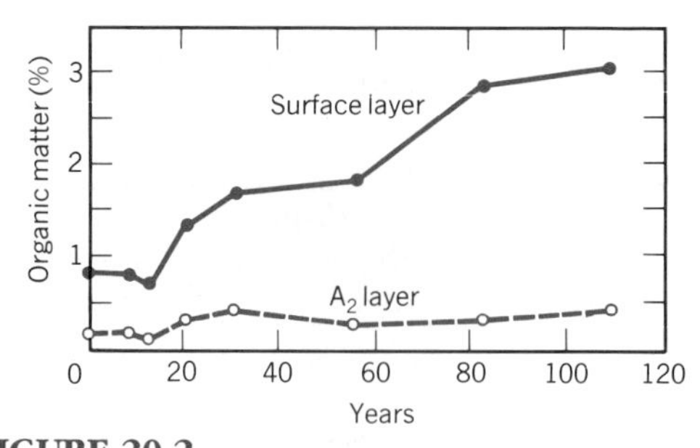

FIGURE 20.2

Evidence of changing soil moisture holding capacity in old fields in the piedmont of North Carolina.

The capacity of soil to retain water is a function of the organic content. The thickness of the elluviated "A_2" horizon is a record of relative leaching (Chapter 20). These two curves suggest that the soil capacity for moisture increased continuously for a hundred years of succession as the rate of leaching decreased. (From Billings, 1938.)

But in some primary recessions on barren ground, the habitat improvement argument should have more force.

PRIMARY SUCCESSION ON GLACIAL TILL

Valley glaciers at Glacier Bay, Alaska, have been retreating for about 200 years, and their progress has been monitored by a series of expeditions ever since 1890 (Crocker and Major, 1955). They leave behind them moraines of glacial till—bedrock ground fine and mixed with stones. The local till is alkaline (pH 8.0–8.4) because of fragments torn up from carbonate bedrock, and is without organic matter or combined nitrogen. Stages in succession up to 200 years old are present in the valley and can be dated with some precision.

The Glacier Bay succession goes from a pioneer community of arctic herbs and dwarf willows, through willow scrub, then to an almost pure stand of alder bushes 10 meters high at the 50-year mark. The alders are slowly invaded by sitka spruce (*Picea sitchensis*) and replaced, until a forest of conifers is present at the 120-year mark. The spruce forest, however, is progressively invaded by two species of hemlock (*Tsuga mertensiana* and *Tsuga heterophylla*) for the next 80 years, resulting in the climax spruce–hemlock forest of the region (Figure 20.3).

Profound changes in the physical habitat are brought about during this 200-year succession, starting with bare, basic, nutrient-deficient till and ending with an acidic podzol (spodosol, see Chapter 25) carpeted with needles, with ample nitrogen and other nutrients, organic matter, and a complex structure. These soil properties seem to be added bit by bit from the activities of the successive plant occupants.

Figure 20.4 shows changes in pH of both the litter (dashed line) and the top of the mineral soil (solid line). The rapid change in pH occurs when the alders are in possession, not when pioneer plants leave bare ground for the rain to wash. Removal of carbonates is not achieved by percolating rainwater alone. The inference is inescapable: that it is the acid residues of alder leaves that are responsible for dissolving carbonates. Lowering of pH, therefore, is a consequence of plant growth.

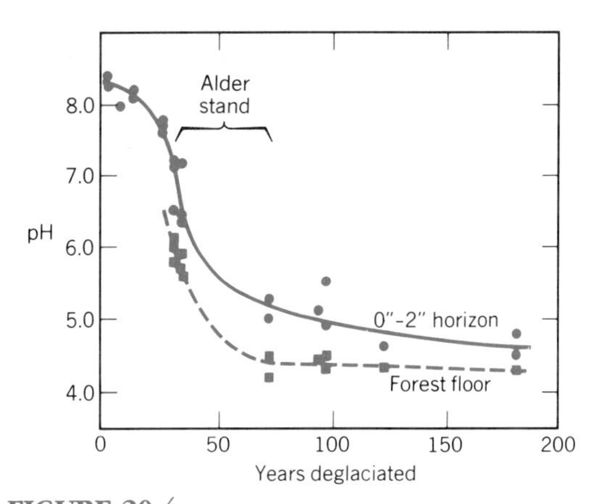

FIGURE 20.4
Changing pH in Glacier Bay succession.
The years of rapid change are those when alder occupies the habitat. (Based on Crocker and Major, 1955.)

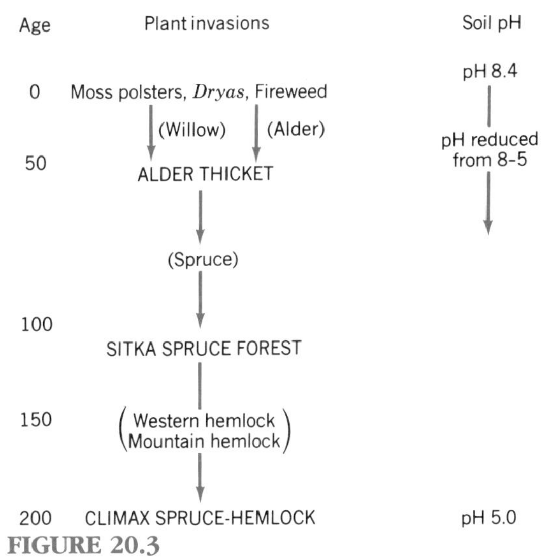

FIGURE 20.3
Primary successions at Glacier Bay on well-drained sites.
(Age in years, data of Crocker and Major, 1955.)

Figure 20.5 shows that organic matter accumulates continuously and is less dependent on the kinds of plants growing than is the change in pH. Possibly a steady-state reservoir of soil organic matter is achieved in about 100 years, with fresh inputs being balanced by losses due to respiration.

Figure 20.6 shows that nitrogen rises slowly at first in the pioneer stage, but some nitrogen fixation occurs even then. Probably *Dryas* populations (a herbal rose of the tundra) are important in this since they, like legumes, have nitrogen-fixing bacteria (see Chapter 26). The big increase in nitrogen occurs in the alder stage, which consists of a pure stand of the tall

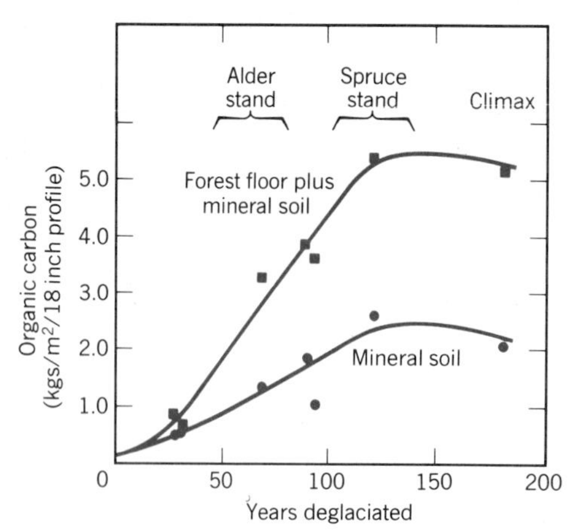

FIGURE 20.5
Accumulation of organic carbon in Glacier Bay succession.
Carbon continues to accumulate in the soil throughout the pioneer and alder stages before reaching what looks like a steady state under evergreen forest. (From Crocker and Major, 1955.)

bushes. Alders have copious nodules with nitrogen-fixing bacteria of their own. Nitrogen falls when the alders are replaced

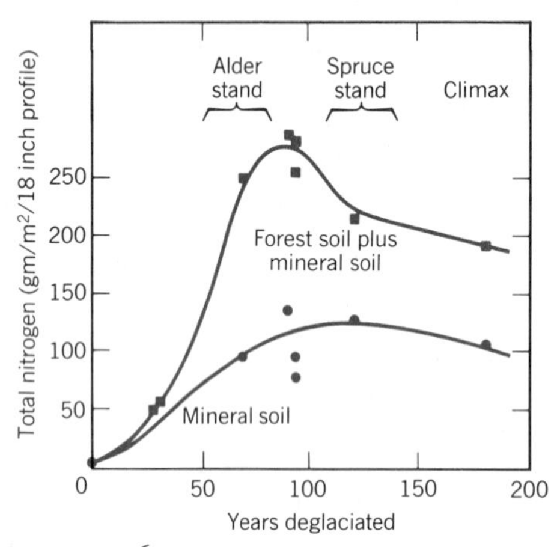

FIGURE 20.6
History of combined nitrogen in Glacier Bay succession.
Nitrogen is highest under the stands of alder before falling under evergreen forest. High inputs by nitrogen-fixing bacteria in alder root nodules explain this. (From Crocker and Major, 1955.)

by spruce, doubtless because few plants in a spruce forest, and certainly not the trees, are able to fix nitrogen.

The record from Glacier Bay shows that a spruce–hemlock forest cannot grow on the raw habitat left by the glacier, but that spruce trees and hemlocks can claim habitats that have first been lived on by pioneer plants and alder bushes. The pH must first be lowered, and in all probability nitrogen and other nutrients must be added before the coniferous trees can invade the site. Conifer niches, therefore, are dependent on the local presence of other plants adapted to high pH and low nutrients.

All the plants in the Glacier Bay succession can easily be seen to be acting in ways that secure individual fitness. Opportunist plants, called "ruderals" in botanical usage, are adapted to invade bare till and do so. Like fugitive *Eurytemora* copepods in Elton's sewage farms (see Chapter 11), they occupy that unfavored spot until it is taken away from them. Alders have another strategy for living in unfavored, poorly drained land, one that uses copious nitrogen-fixing bacteria. Using these habitats, of which the Arctic offers many, serves individual alders well. The conifers have been programmed to allocate fewer resources to cope with unsavory habitat, avoiding for instance the costs of supporting all those nitrogen-fixing bacteria paid by the alders. Their strategies too yield fitness in a landscape well-provided with old and vegetated habitats.

But it is undeniable that primary succession on glacial till at Glacier Bay is driven by habitat modification, at least in its early stages. Each stage is contingent on the stage before, and the course of succession is roughly predictable. Improving the habitat by the action of plants can be called **facilitation** (Connell and Slatyer, 1977).

THE HYPOTHESIS OF XERARCH SUCCESSION

Vegetation does grow over sand or rock. Both of these inhospitable surfaces can be the home of specially adapted plants—grasses with creeping stems (stolons) and deep roots in sand and lichens and mosses on rocks. It is thus simple to put forward a hypothesis that sand and rock are vegetated through a process of succession, starting with these specially adapted plants as pioneers. The postulated succession is called "xerarch," the literal Greek meaning of which is "dry commanding."

The xerarch succession hypothesis was applied to communities on the sand dunes of Lake Michigan long ago, with effects that reverberated through ecology for decades (Cowles, 1899). Lake Michigan has been retreating northward ever since the end of the last glaciation 14,000 years ago, leaving behind beach sands thrown up into a long array of dunes near the University of Chicago, and thus convenient for a pioneer ecologist, Henry Samuel Cowles, when he was appointed to one of its first professorships.

Cowles found local beech–maple forest growing on ancient sands, the trenching of which left no doubt that they were old beaches. On other ancient beaches were stands of black oaks, a rather simple community. But near the modern active beaches the only trees were cottonwoods (*Populus*) and pines. Cowles argued that beech–maple forest could be present on beach sands only if the others had preceded it in succession.

Cowles postulated that Michigan dune succession progressed as follows. First the blowing sand was fixed by dune grasses, the pioneer seral stage. Then cottonwood seedlings took root (they do, arriving in clouds of tiny seeds carried by the wind on fluffy "cotton" parachutes). As the cottonwoods grew, their fallen, leathery leaves covered the ground and they slowly rotted. Pine trees were the next seral stage, only to be replaced by oaks, as pines tend to be in so many seres (Chapter 3; early succession pines are light-adapted monolayers). Finally, the oak forest must have developed into the climax beech–maple forest in parallel with the final sorting of trees in secondary successions of beech–maple–covered Illinois.

Cowles's hypothesis of dune succession dominated ecology teaching in America for 60 years; I have even seen the movie (a commercial product, made for class use). Then Olson (1958), possessed of techniques unavailable to Cowles, took another look at the classic site. Olson was able to age ancient dunes by radiocarbon assay and so arrange the communities that covered them in an accurate time sequence, with results that made Cowles's basic hypothesis untenable. Figure 20.7 shows the chronology revealed by this exercise.

When very fresh, the dunes hold only the specialized grasses, but shortly thereafter, perhaps less than a century, a stand of poplars (the cottonwood, *Populus deltoides*) is in place. Nearby, or mingled with the poplars, are stands of pine. But dunes of all ages from 300 to 12,000 years support woodlands of black oak (*Quercus velutina*). Occasional small stands of poplar or pine can be found on older dunes, but in circumstances that suggest recent disturbance.

Black oaks, once established, remain indefinitely as almost single-species stands. Even in 12,000 years these stands have not been invaded by beeches, maples, or any of the 18 species of tree known to be present in beech–maple forests. Olson's explanation for this is that old sand dunes are too acid for all but the acid-tolerant

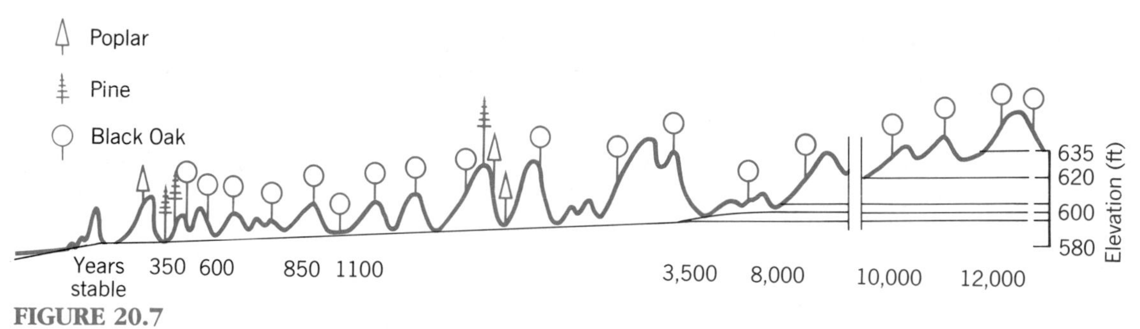

FIGURE 20.7
Time series across Lake Michigan sand dunes.
The lake is at the left. Dunes are dated by radiocarbon. The time series shows that succession stops with black oak woodland and does not continue to beech-maple forest as once postulated. (Transect near Gary, Indiana, from Olson, 1958.)

oaks. The young dunes—where grasses, poplars, and pines grow—are alkaline, with a pH above 8. Carbonate minerals are quickly leached from the well-drained surface of these dunes during the first centuries, of exposure and the pH has fallen to about 4 when the oaks are in place.

If the word "climax" is appropriate at all to what happens on the dunes, clearly the climax is a pure stand of black oaks. Very little else will grow there, and not because seral communities have done something to the habitat either. All that happens is that rain washes carbonates from the sand leaving it fit only for the suitably adapted oaks.

Thus the Cowles hypothesis that beech–maple forest can result from xerarch succession on sand dunes seems securely falsified. How then to account for the undoubted fact that patches of beech–maple forest can be found growing on what is undeniably old beach sand? Olson's hypothesis to explain this rests on the observation that these beeches and maples all grow in wet depressions. In wet bottom lands, carbonates are not leached from the sands. Moist sand in depressions could be invaded by plants of the beech–maple forest, perhaps following a succession of pioneers quite different from the dune dwellers proposed in Cowles model. Because these habitats were never dry, successions could hardly be called xerarch, though they were primary.

When the water table is very high over old Lake Michigan beaches the result may be a swamp. Even on the dry dunes, many sequences of colonization are possible, depending on chance and local circumstance (Figure 20.8). Only in the pioneer stages is there a significant modification of habitat by plants, as pioneers fix the shifting sands. Otherwise the successive invasions record no more than the play of alternating fortunes of invaders with various strategies for coping with a dry, sandy place.

Applying the xerarch succession hypothesis to rocks produces less of interest. On smooth, hard rock in exposed places, only lichens have the adaptations necessary for growth (Figure 20.9). In moister places like hollows, thick mosses grow. If cracks or depressions trap wind-blown organic debris, sufficient soil can collect for a community of drought-tolerant grasses and heath-like plants. In the deepest soil patches, trees with rock-hugging strategies can grow (Figure 20.10). It is a moot point if the mosses and small plants grow-

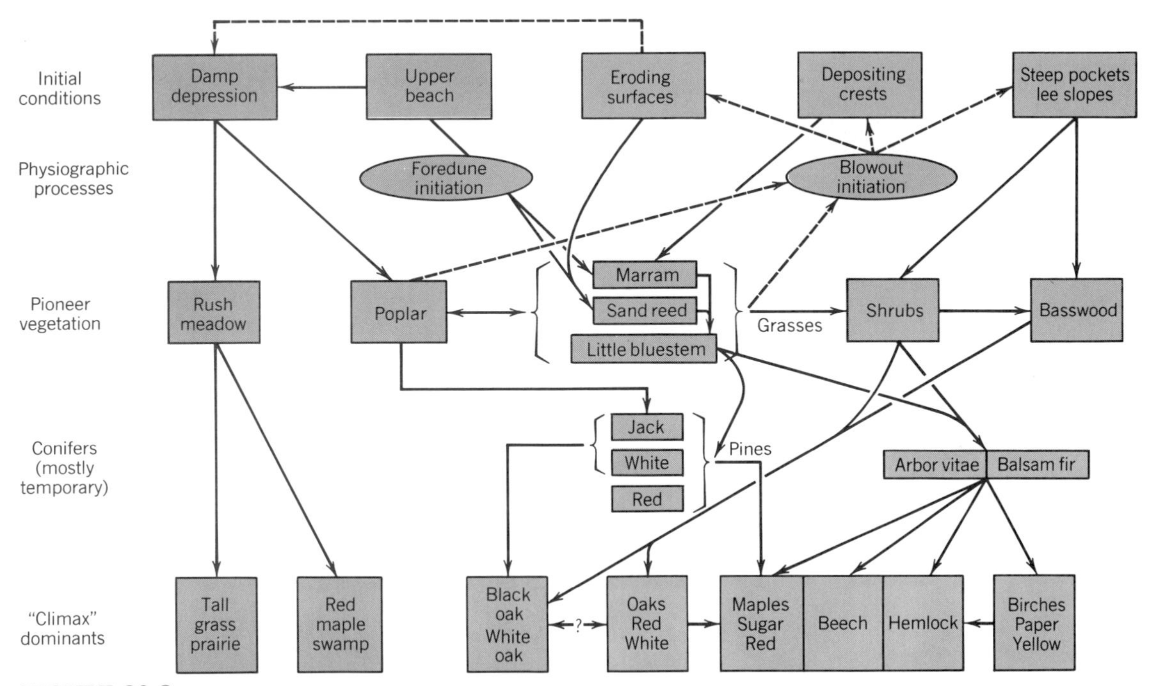

FIGURE 20.8

Alternative successions on Lake Michigan sand dunes.

What communities develop on any part of a dune is a function of aspect and physical history more than of preceding plant populations. (From Olson, 1958.)

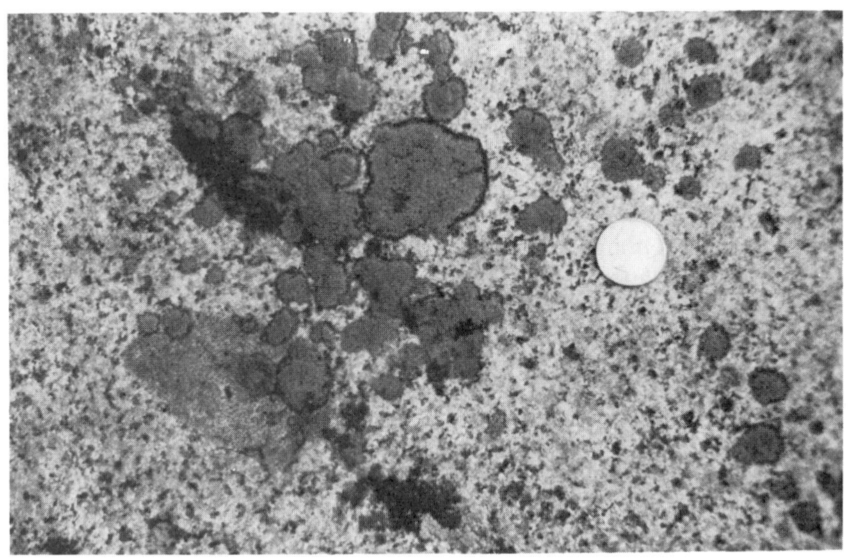

FIGURE 20.9

Lichens on rocks on Mount Rainier.

These lichens have been growing since the boulders were dropped by a retreating mountain glacier about 200 years ago. The lichens continue to grow in solitary state without taking part in successions.

FIGURE 20.10
Tree colonizes rock.
Woodland can establish itself on rocky ground when sufficient soil collects in cracks. The trees undoubtedly are preceded by smaller colonists that grow more quickly.

ing in cracks have much to do with the establishment of trees, except in supplying some of the debris that blows into the cracks. Lichen communities persist for centuries or millennia without ever being replaced, thus letting the pioneer be the climax.

THE HYPOTHESIS OF HYDRARCH SUCCESSION

The growth of land plants in wet places depends on control of the water table. Water tables are lowered by drainage, typically because streams cut their channels deeper. Raising of water tables follows the damming of outlet streams by landslips, volcanism, glaciation, or earth movements. Beside any body of water, therefore, a succession of plant communities should follow the water table. As land dries, so communities of dry land should replace wetland communities. As land floods, so marsh communities should succeed. We expect hydroseres to be driven by physical control of the water table.

It is possible to argue, however, that ponds or lakes may be filled with organic debris, following which successive seral stages of vegetation develop on the filled basin. This is the hypothesis of the hydrarch succession. First organic mud collects in the open water until the lake is shallow enough for rooted aquatic plants. When shallower still, cattails or reeds grow and a marsh results. In theory it might be possible for willow and alder bushes at the edge of the marsh to lower the water table and for forest to close over the site.

The succession imagined in the above scenario would be very much hydrarch in its literal meaning of "water commanding." Such postulated hydrarch successions are widely illustrated in biology texts by drawings like that in Figure 20.11, though no documented examples of this happening to real lakes are easily discoverable in the literature. The hypothesis of the hydrarch succession has never had its stock example, as xerarch succession has had its Lake Michigan sand dunes.

The hypothesis of hydrarch succession is a good hypothesis, in the classical sense that it is potentially falsifiable. The postulated basin fill should be perfect for preserving plant remains, particularly pollen. Thus digging a hole under the presumed climax vegetation should reveal traces of all the previous seral communities as they occupied the site in sequence. The hypothesis does not survive such tests.

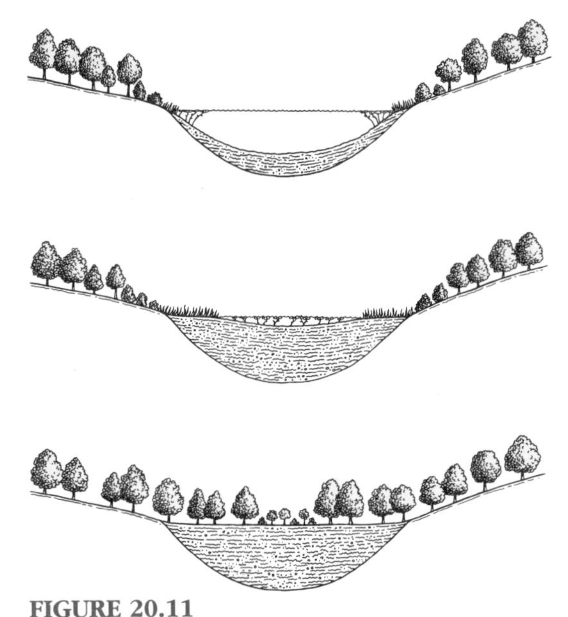

FIGURE 20.11
Conventional hydrarch succession diagram: probably wrong.
Floating mats and peaty sediments are supposed to displace water. The concentric rings of vegetation constrict until forest grows at the climax where once was water. The hypothesis pictured here, and in many diagrams like this, probably is wrong. The most likely outcome of filling a bog lake with peat is a permanent peat bog. Compare the real data of Figure 20.13.

Lakes and ponds in muskeg are often cited for evidence of hydrarch successions in progress because they are ringed with floating mats that seem, particularly from low-flying aircraft, to be constricting the open water. As with the Michigan dunes, I have seen the movie. A formal test of the hypothesis of successional infilling of ponds in peatlands reveals a history quite different from that predicted (Heinselman, 1963, 1975).

Figure 20.12 is Heinselman's reconstruction of a section of peatland in Minnesota at three stages in its history: early in postglacial time, at an intervening period, and at the present day. The reconstructions are based on more than 50 borings through the blanket peat bog that covers the landscape and through the bottom of Myrtle Lake itself. Myrtle Lake never fills in; it merely rises as the marsh landscape rises on its own accumulating peat.

That lakes persist, rising above their own mud like Myrtle Lake, is the common experience of paleoecologists who drill their sediments (see Chapter 17). A recent example worked in my laboratory is from tropical Lake Wodehouse in the remote rain forest of the Darien. In the dry season the lake is reduced to a pond some 300 by 70 meters by about 1 meter deep, but it swells over its surrounding marsh in every annual wet season to cover two or three kilometers. Our boring shows its mud to be 11 meters thick. Radiocarbon dating shows it to be 4000 years old, and diatom fossils in the mud show it to have changed its depth several times. Pollen analysis shows that once it was so shallow as to have been covered with cattails, but these were drowned by rising water. Now is one of the deep times. Tropical Lake Wodehouse is raised intact on its own debris just like boreal Myrtle lake (see also Chapter 17, Figure 17.19).

A more general paleoecological test of the hypothesis of hydrarch succession in temperate latitudes uses pollen analysis to reconstruct the histories of numerous hydroseres. Walker (1970) identified 12 plant communities of the kind that would be expected to take part in any postulated hydrarch succession in Britain. The communities ranged from pure aquatic plants to reedswamps, wet bushlands, and *Sphagnum* bogs, and could be ranked in order of expected appearance on a continuum from wet to dry.

It should not be expected that all these communities would appear in succession at any one site. For the hypothesis of a hydrarch succession to stand, however, some subset of these communities should

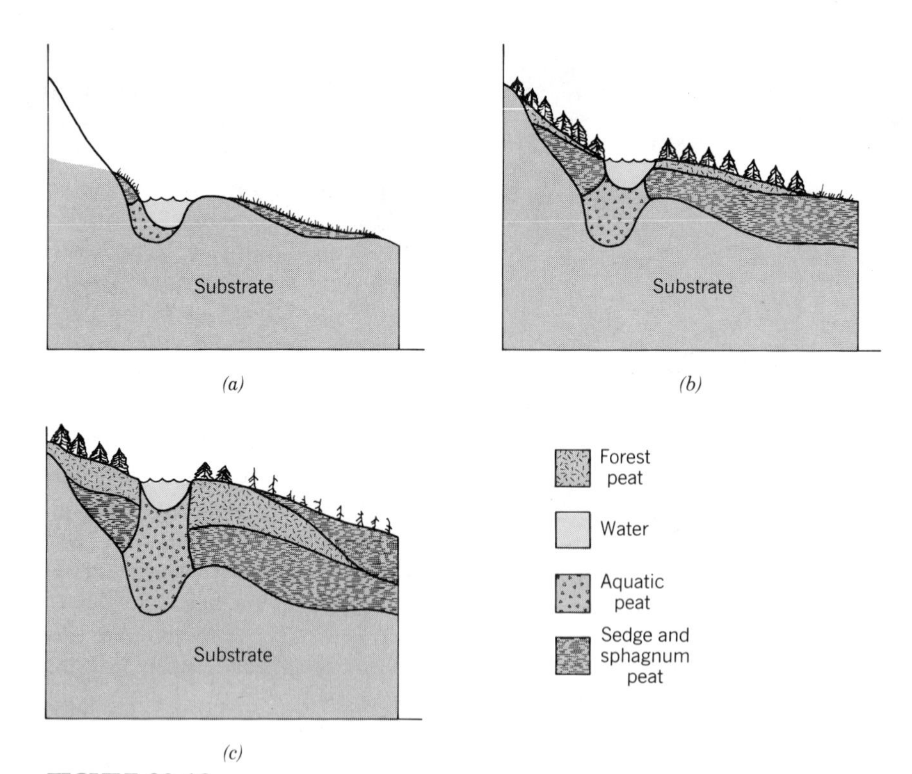

FIGURE 20.12
The history of a bog lake.
The three cross sections show the land around Myrtle Lake, Minnesota, in the early postglacial, several thousand years later, and at the present day. Growth of peat did not eliminate the lake. There was no succession to forest, merely a short-lived invasion of trees. This true history of a bog lake was quite unlike the postulated history of Figure 20.11. (From Heinselman, 1963.)

appear in the right order. To test if this was so, Walker searched the literature for stratigraphic histories where the pollen record was sufficiently unequivocal for it to be possible to describe the order in which plant communities actually had appeared. He found a total of 159 measured directions of transition between pairs of communities.

Walker's results for sets of small and large lakes are given in Figure 20.13. The data are plotted as frequencies (simple numbers of occurrence) of transition between pairs on the left-hand side of each diagram with quasi-vector diagrams showing directions of change to the right. Some community replacements are more likely than others, and there is a general tendency for "drier" communities to replace "wetter" communities. But there is a large amount of apparent randomness about the sequence of events, with many replacements actually going in the "wrong" direction (that is, from dry to wet).

Where the pollen histories were long, the concluding community of the hydrosere was always a peat bog. Even if the pollen showed that a seral community of

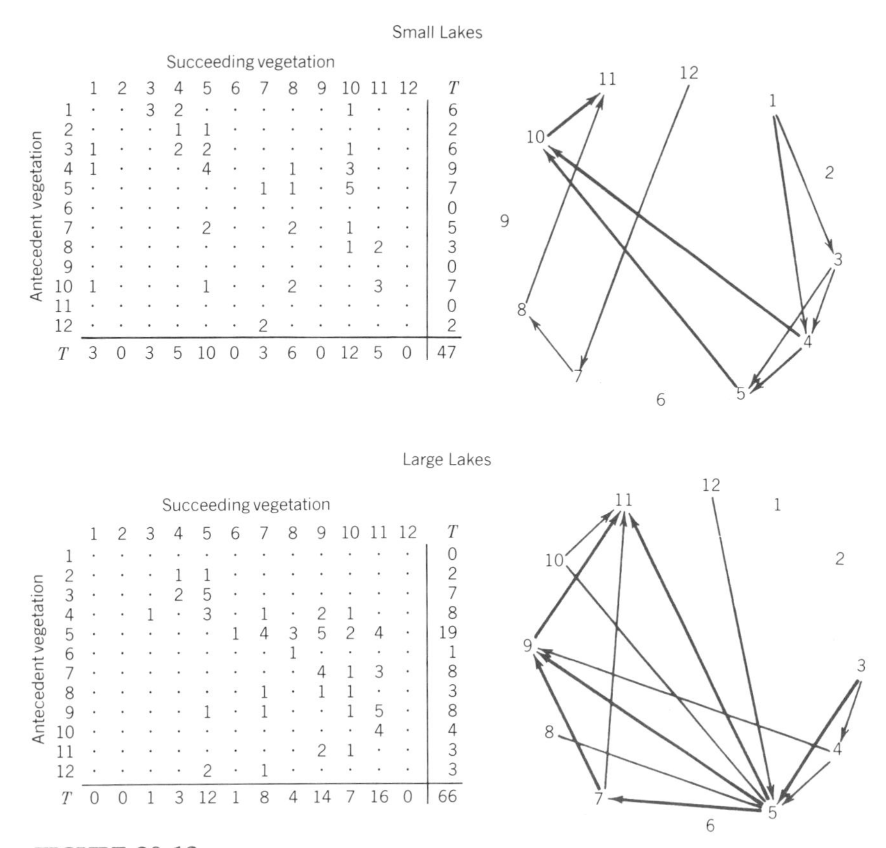

Small Lakes

Antecedent \ Succeeding	1	2	3	4	5	6	7	8	9	10	11	12	T
1	·	·	3	2	·	·	·	·	·	1	·	·	6
2	·	·	·	1	1	·	·	·	·	·	·	·	2
3	1	·	·	2	2	·	·	·	·	1	·	·	6
4	1	·	·	·	4	·	·	1	·	3	·	·	9
5	·	·	·	·	·	·	1	1	·	5	·	·	7
6	·	·	·	·	·	·	·	·	·	·	·	·	0
7	·	·	·	·	2	·	·	2	·	1	·	·	5
8	·	·	·	·	·	·	·	·	·	1	2	·	3
9	·	·	·	·	·	·	·	·	·	·	·	·	0
10	1	·	·	·	1	·	·	2	·	·	3	·	7
11	·	·	·	·	·	·	·	·	·	·	·	·	0
12	·	·	·	·	·	·	2	·	·	·	·	·	2
T	3	0	3	5	10	0	3	6	0	12	5	0	47

Large Lakes

Antecedent \ Succeeding	1	2	3	4	5	6	7	8	9	10	11	12	T
1	·	·	·	·	·	·	·	·	·	·	·	·	0
2	·	·	·	1	1	·	·	·	·	·	·	·	2
3	·	·	·	2	5	·	·	·	·	·	·	·	7
4	·	·	1	·	3	·	1	·	2	1	·	·	8
5	·	·	·	·	·	1	4	3	5	2	4	·	19
6	·	·	·	·	·	·	·	1	·	·	·	·	1
7	·	·	·	·	·	·	·	·	4	1	3	·	8
8	·	·	·	·	·	·	1	·	1	1	·	·	3
9	·	·	·	·	1	·	1	·	·	1	5	·	8
10	·	·	·	·	·	·	·	·	·	·	4	·	4
11	·	·	·	·	·	·	·	·	2	1	·	·	3
12	·	·	·	·	2	·	1	·	·	·	·	·	3
T	0	0	1	3	12	1	8	4	14	7	16	0	66

FIGURE 20.13
Paleoecological test of the hydrarch succession hypothesis.
Twelve potential serral communities are ranked from wet (1) to dry (12). Pollen and macrofossil evidence shows the order in which pairs of communities succeed one another. Almost any order was possible, though there was a slight bias toward the direction of increasing dryness. Diagrams at the right show the most common replacements of one community by another. (From Walker, 1970.)

bushland or swamp woodland occupied the site at some time, this was always replaced in succession by a peat bog that drowned out the trees. That this should happen is understandable from the known properties of *Sphagnum* peat communities, which cease growing as all available nutrients are locked up in dead peat. Drying that permits invasion by trees also results in decomposition of peat. Nutrients are released, the bog grows again, and the trees are drowned.

Walker (1970) concluded that the climax of hydrarch succession in Britain was a peat bog. Wetlands never would be drained by hydrarch successions. The actual sequence of seral stages in a hydrosere was a function of water table changes due to stream down-cutting, damming, or climatic change. Figure 20.14 is a better description of the history of a pond than is conventional Figure 20.11. Hydrarch successions cannot replace swamp vegetation with communities typical of better-drained

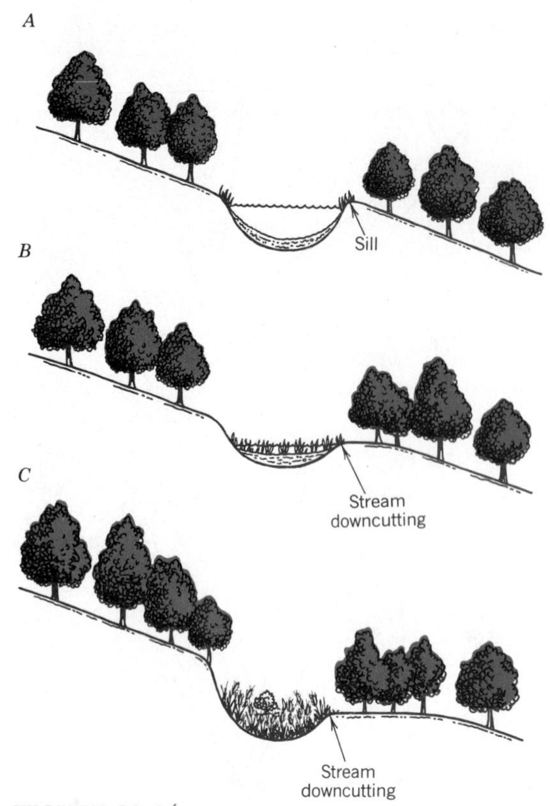

FIGURE 20.14
History of a typical pond.
Small ponds and lakes are drained by stream downcutting even as sediments collect. Seldom, if at all, are they "reclaimed" by surrounding vegetation. When stream downcutting lowers the water table sufficiently a permanent marsh is established (compare Figure 20.11).

sites. Only physical drainage can lead to this. The classic hypothesis of hydrarch succession must, therefore, be rejected.

HABITAT-DRIVEN OR COMPETITIVE REPLACEMENT?

Secondary successions in old fields must be competitive replacements. They look like contest competitions for space in the old field. Plant interactions certainly are present in primary successions: spruce invading alder groves at Glacier Bay, *Populus* taking over from dune grasses at Lake Michigan. But in primary successions the important community changes are driven by changes in habitat, most of them resulting from outside forces.

Lowering a water table due to stream down-cutting is an outside force, and the hydrosere that results is an "allogenic succession." The emplacement of black oak on acid sands, washed free of carbonates at Lake Michigan by rain, is likewise allogenic. Most of the community replacements in primary successions are allogenic in this way; habitats change and the species populations with them.

Some fraction of habitat changes is influenced by plant cover (**facilitation**). Fixing sand dunes and raising soil nitrogen are prominent examples, but the contribution of organic matter to soil is always massively important to habitat development on more barren sites (see Chapter 23). To this extent the succession is "**autogenic**," and vegetation has a physical impact on the landscape.

Most wetland succession is **allogenic**, and some part of dry-site succession is autogenic. All primary successions are best studied as landscape processes in which vegetation develops its important physical contributions to ecosystems. The plant replacements are, in the main, habitat-driven.

Secondary successions, however, appear to be entirely autogenic. Moreover, the changes in habitat appear minor—some increase in soil nitrogen or structure, as in North Carolina fields, and the inevitable shading when larger plants invade. The main part of the action in secondary successions clearly involves plant interactions, inviting a hypothesis of competition.

THE COMPETITION HYPOTHESIS

A competition hypothesis for succession must explain not just species replacements but also the predictable order of those replacements. If maple trees were to prove the final victors, why should they wait for the brambles and pines to do their bits first? Ecologists for long could not answer questions like this without referring to the habitat. They said maples had to wait for other plants to improve the habitat for them. But a hypothesis of pure competition must explain the apparent wait of the maples in a constant habitat.

Competition suddenly became a satisfactory explanation for secondary successions when the concept of a cline of possible strategies between colonist and competitive species was developed in the 1960s (see Chapter 11). The seral replacements were indeed the result of competition, but competitors arrived in the order of their powers of dispersal. Drury and Nisbet (1973) described secondary succession as the replacement of *r*-selected species by *K*-selected species. Colinvaux (1973), suspicious of the language of *r* and *K* selection, said that "opportunist species" were replaced by "equilibrium species."

The essence of this hypothesis is that both opportunist and equilibrium species are able to coexist within an area that might be called "reasonable dispersal distance." The familiar seres we see are compounded from the life histories of species already existing in the neighborhood.

The hypothesis predicts that the plants of the pioneer seral community will be extreme opportunists, adapted for dispersal and rapid growth. They should be highly fecund, short-lived, and without elaborate storage organs and the more costly predator defenses. If these plants exist in a neighborhood, it can be predicted that they will occupy vacant habitat first. This prediction is upheld by the plants of pioneer communities, which are the typical annual weeds of an old field.

Opportunist weeds can be present in a neighborhood even though surrounding land is occupied by equilibrium plants of a climax forest, occupying such sites as bare ground of a flooded stream bank, spoil from a rodent burrow, or ground torn up by a fallen tree. An old field, or a massive treefall in a tropical rain forest, is like a giant rodent burrow where a dense population of opportunists can quickly flourish.

But the pioneers will certainly be displaced as slowly dispersing, better competitors invade the site from surrounding complex communities. This is what we observe, most notably in seres that lead to a forest climax. The progressive invasions of perennial herbs, shrubs, and trees are by species whose life history phenomena are of progressively longer life, with all the diversion of resources from reproduction to storage and defense that these life histories demand.

Of particular appeal is the way this hypothesis predicts the orderly and repeatable properties of old field successions. The sere will proceed at a rate that is a function of the number of species strategies available locally, that is to say of the local species pool (Figure 20.15). As a result, a good naturalist can always tell to within a year or two when a given old field was first abandoned by farmers from looking at the seral stage then occupying the field. But the naturalist must have local knowledge, because the rates of replacement depend on locally available plants.

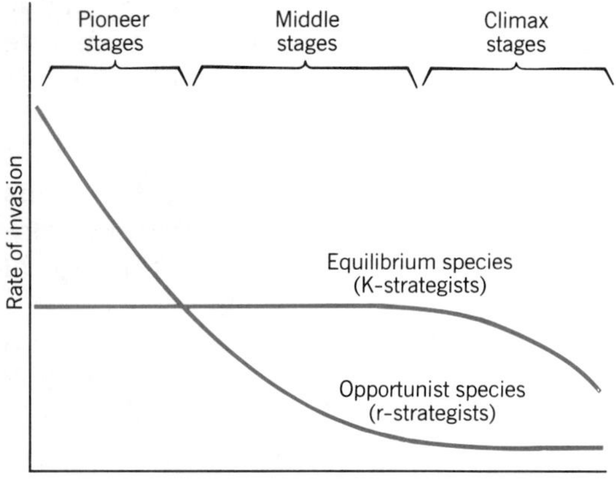

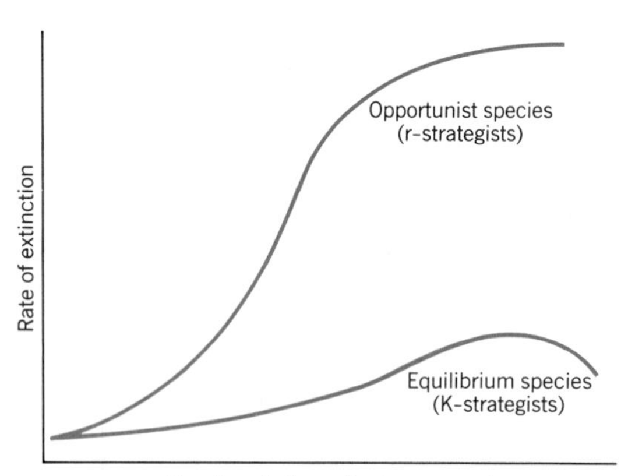

FIGURE 20.15

Model of succession as a function of rate of invasion.

Opportunist species invade early and rapidly (*top*) but their extinction by competition is also early and rapid (*bottom*). The final number of species at climax will depend on the rate of continued invasion (which may proceed for thousands of years) and the rate of local extinction, which eventually should be very slow.

Continued Replacements in Closed-Canopy Forest

When first promulgated in 1973, the hypothesis of opportunist species' being replaced by equilibrium species had no mechanistic explanation for the continued replacements of trees in temperate forests after the canopy had closed. This lack had meanwhile been made good by Horn's (1971) model of the adaptive geometry of trees (see Chapter 3). The first trees that replace the shrubs and herbs of an old field are of multilayer design. These trees cannot grow in the dense shade of their parents because their leaf geometry does not then yield a sufficient excess of net photosynthesis over respiration. Shade-adapted trees with fewer effective layers of leaves, therefore, invade. Thenceforth, trees of a successional sequence are equipped with an ever-diminishing number of effective layers until the climax trees are close to being monolayers (Table 20.2). Even in established forests, therefore, a variety of strategies allows continued replacement by competition.

A THREE-STRATEGY MODEL FOR PLANTS IN SUCCESSION

Grime (1979) has suggested that a series of three base strategies should be recognized for plants instead of just variants of opportunist and equilibrium strategies. These three strategies would be that of colonizer, competitor, and stress competitor, called by Grime **ruderal**, **competitive**, and **stress-tolerant**, or R, C, and S, strategies.

Ruderals are plants suited to frequent disturbance but not to competition. This concept is close to that of the "fugitive species" (see Chapter 11), since powers of escape to uncontested habitat are essential to the strategy. Competitors are adapted to low disturbance and low stress. They will always replace ruderals in succession by competition. To this extent, the Grime three-strategy hypothesis does not differ materially from the opportunist–equilibrium model.

TABLE 20.2
Tree Layering and Shade Tolerance
The tree species are arranged in the conventional order in which they appear in secondary succession. The order also represents a ranking by shade tolerance known to foresters. Horn's calculations of relative number of layers show that there is a relative shift from multilayer to monolayer designs as succession proceeds as predicted. The ratio of total leaf area to ground covered also is reduced as succession proceeds. (From Horn, 1971.)

Species	Number Measured	Percentage Light/Branch	Percentage Light/Tree	Number of Layers ± SE	Leaf Area/Ground Area
		Early succession			
Gray birch	10	44	3.6	4.3 ± 0.4	2.4
Bigtooth aspen	6	45	6.9	3.8 ± 0.5	2.1
White pine	13	25	0.8	3.8 ± 0.4	2.9
Sassafras	3	14	0.8	2.7 ± 0.7	2.4
		Mid-succession (on moist soil)			
Ash	10	26	3.0	2.7 ± 0.2	2.0
Blackgum	7	15	1.4	2.6 ± 0.5	2.2
Red maple	21	20	1.8	2.7 ± 0.2	2.2
Tuliptree	6	17	2.3	2.2 ± 0.2	1.8
		Mid-succession (late on dry soil)			
Red oak	19	23	2.6	2.7 ± 0.2	2.1
Shagbark hickory	12	18	1.4	2.7 ± 0.2	2.2
Flowering dogwood	13	5	2.1	1.4 ± 0.1	1.3
		Late succession			
Sugar maple	8	9	1.2	1.9 ± 0.1	1.7
American beech	16	6	1.5	1.5 ± 0.1	1.4
Eastern hemlock	13	8	2.1	1.6 ± 0.1	1.4

The third strategy of "stress tolerator" is invoked among crowded, successful competitors. In these dense populations, competition itself is a powerful stress. The stress tolerators are suited to these conditions of low disturbance with high stress. Resources allocated to competition are constrained by resources allocated to surviving stress. Horn's monolayer designs would be stress tolerators in this model.

Thus Grime's model has three patterns of resource allocation:

1. Most resources to fecundity and dispersal = ruderals (R species).
2. Most resources to competition = competitors (C species).
3. Resources divided between stress resistance and competition = stress tolerators (S species).

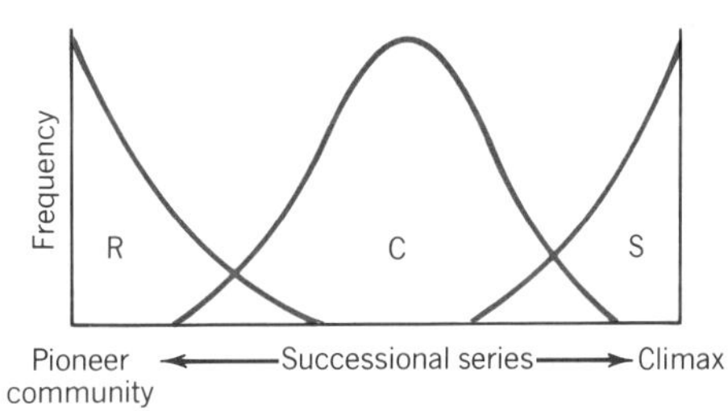

FIGURE 20.16
Three-strategy model of succession.
Ruderals (R) are pioneer herbs that are replaced by competitors (C). At the climax stress-tolerators (S) invade. For details see text. (From Grime, 1979.)

An R-C-S continuum fits the facts of succession more closely than the simpler opportunist–equilibrium continuum first proposed (Figure 20.16). The particular value of the model is that it predicts continual replacements even after dense populations of large plants are attained.

REGENERATION AND GROWTH STRATEGIES IN SUCCESSION

Life histories of plants depend on dispersal and germination of seeds. Wind, bird, or mammal dispersal of seeds are obvious alternative strategies that require seeds of particular size or number and result in different dispersal distances. At least as important, however, are alternative strategies for germination once the seed comes to rest on soil. Germination can be immediate or it can be delayed, perhaps for years, until some necessary environmental cue is received. A result of delayed germination is that "seed banks" collect so that a soil is charged with viable seeds waiting for an environmental cue.

The most familiar manifestation of seed banks is in agriculture, where plowing always results in the growth of weeds. The weed seeds may have germinated when they were brought to the surface and exposed to light, or germination may have been triggered even for deeply buried seeds by changes of temperature caused by breaking of the soil surface (Thompson and Grime, 1978). This "seed-bank strategy" of the pioneer plants presumably is an adaptation that lets the pioneers colonize gaps that form by accident in forests or other vegetation or that are left when a large tree dies. The seed bank strategy is also used by species that colonize old burns in the boreal forest, when the heat of fire is the germination trigger. Jack pine (*Pinus banksiana*) follows this strategy.

An alternative to a seed bank used by some species, particularly in tropical rain forests, is a "seedling bank." Under closed forest, seeds germinate on the forest floor and the resulting seedlings persist for a long time, perhaps for years. In the dim light, photosynthesis may be barely sufficient to balance respiration, but is enough to keep the seedling alive. It is expected that seedlings using this strategy will have well-developed chemical defenses against herbivores. The gain from this strategy is a head start in the race for the canopy when a gap appears with the removal of a shading tree.

Many plants regenerate asexually as vegetative clones. This strategy seems to work particularly well in dense carpets of vegetation like arctic tundras, where sites for seed germination are extremely rare. Herbs of forest floors also sometimes use this strategy, as do trees like aspens.

Rates and shapes of growth can be varied, allowing other strategies. Within forests, the members of a seedling bank can grow as rapidly as possible in the dim light, becoming tall and thin, or they can grow broad and slowly, perhaps increasing energy reserves for a sudden spurt later (Bazzaz and Pickett, 1980).

COMPUTER SIMULATION OF FOREST SUCCESSION

The competition hypothesis makes possible computer simulations of successional change within forests, using only data for the relative success of forest trees themselves. Although these models are strictly empirical, they are consistent with the opportunist–equilibrium and R-C-S models in that they describe properties of plants with these strategies. One such simulation, called JABOWA, uses data for invasion, growth, and death of trees of different species (Botkin, et al., 1972).

Empirical equations relating growth to size, temperature, precipitation, and site defined a "grow subroutine." Known expectations of life defined a "kill subroutine." Fresh invasions were assumed to be random, but proportional to the presence of saplings in existing stands, yielding a "birth subroutine." All interactions of one plant on another were assumed to reflect competition for light, modified by the known adaptation of each species to relative shade. These interactions were described by equations in the grow subroutine. Simulations then proceed in the order described in Figure 20.17.

Runs with JABOWA, started with early succession data, have correctly described the course of forest succession at a test site in New Hampshire (Hubbard Brook; see Chapter 23). Projections give a most realistic impression of what climax ought to be like. Figure 20.18 shows what should happen from an early forest seral stage throughout 2000 years of history in a constant climate, for six species whose ranges overlap at an elevation of 610 meters. After the initial successional events characterized by the decline of yellow birch, the community settles down to a dynamic system in which all species persist though their populations oscillate, apparently at random. New Hampshire forest stands at this elevation that have been declared to be climax by field naturalists are closely comparable to the outcome of this simulation.

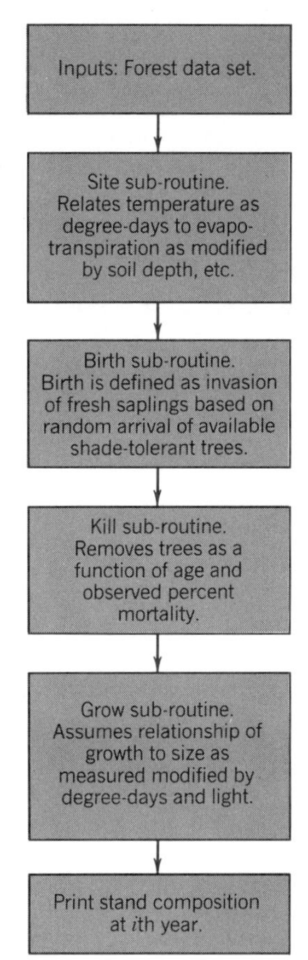

FIGURE 20.17
Partial flowchart of JABOWA program.
Much modified and simplified from Botkin et al., 1972.)

Succession as a Markovian Process

Simpler assumptions than those of a simulation model let succession be modeled as a Markovian process. This method uses a statistical procedure in which chains of random events are constrained by current states but no other. Markov chains of stochastic events eventually settle down into final stationary distributions quite comparable to the relationships of tree populations at climax (Kemeny and Snell, 1960).

Horn (1974, 1976) has applied the mathematics of Markov chains to forest data to predict future states. The initial data set is an estimate of the probability that any one tree in a forest will be replaced

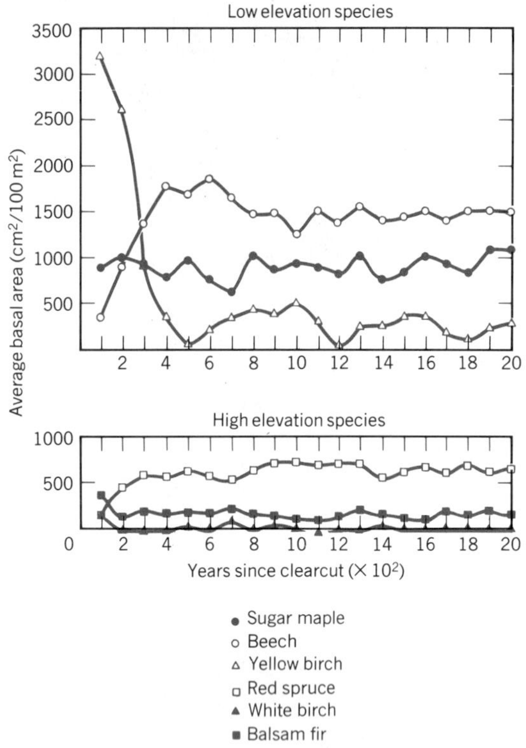

FIGURE 20.18
JABOWA simulation of 2000 years of forest growth.
The run begins with a clear-cut and describes the expected proportions of six trees in future forests at an elevation of 610 meters. (From Botkin et al., 1972.)

by one of its own kind or one of another kind. These data could be derived most easily by mapping every tree in a forest and then remapping the same forest at some future year, say 50 years later. A more practicable way to get usable data is from estimates of replacement probabilities made from measures of saplings actually growing under trees in the forest, supplemented by measures of relative shade tolerance.

When a matrix of probabilities of individual replacement for all species of a forest is known, fairly straightforward calculations allow the extrapolation of the future composition of a forest. The only data going into these calculations, therefore, are present composition of a forest and measured probabilities of individual replacements. The method has been applied to early succession data in New Jersey forests and predicts the composition of future climax communities to be closely similar to that existing in a virgin New Jersey stand now (Horn, 1975).

THE ENERGY FLOW HYPOTHESIS

By many measures, succession leads to complexity and order. First the bare ground, then a simple community of a few pioneer species, then sequential invasions leading to a more complex community. When the paradigm of energy flow became the currency of ecology, it was natural to explore the possibility that ecological efficiency should also increase through the course of succession.

Lindeman (1942) explored this possibility in his original energy flow paper. He postulated that the ecological (Lindeman) efficiency (see Chapters 2 and 3) should increase with each seral stage in all successions. In practice this meant that the gross primary productivity (assimilation plus respiration) should increase with each successional change, starting with low productivity for a field of pioneer weeds and ending with maximum productivity in the climax. Lindeman thus offered his own original hypothesis to explain the striking phenomena of succession: the communities were continually rerigged to converge on ultimate thermodynamic efficiency (Figure 20.19*a*).

Lindeman's idea had intellectual elegance. The course of succession was one of increasing complexity, hence of increasing order, from the apparent anarchy of a

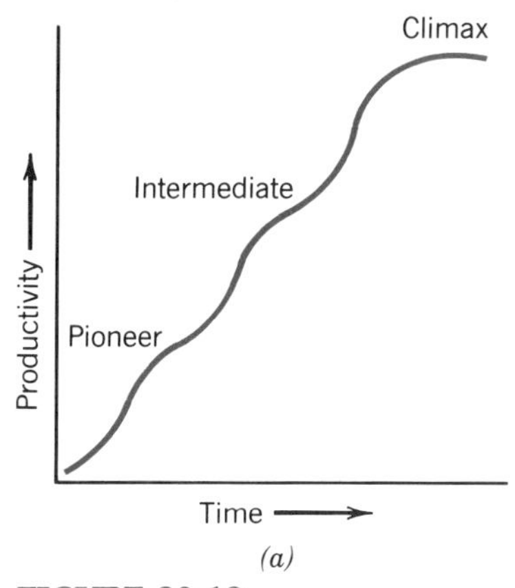

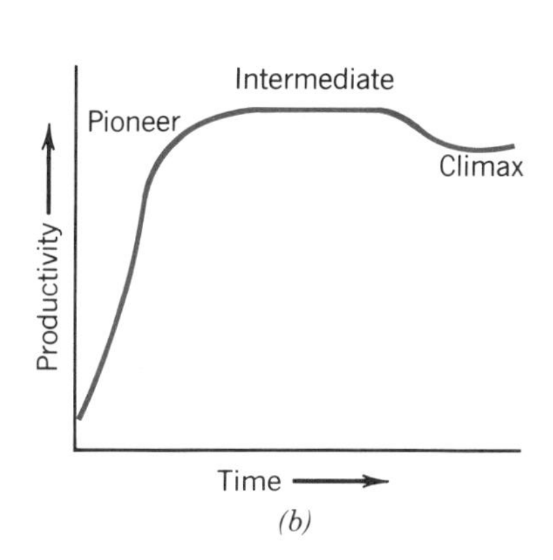

FIGURE 20.19
Productivity during succession.
(*a*) Is a statement of the Lindeman (1942) hypothesis that productivity increases throughout succession. (*b*) Is what is thought actually to happen in many successions, 40 years of measurement later.

field of weeds to the complex order of a climax forest dominated by a few masterful species of tree. Increasing order throughout the succession would result from the increasingly available free energy resulting from the increased efficiency of gross production.

However, measures of production made over the last 40 years show that Lindeman's prediction of increasing energetic efficiency through succession are quite unfounded. So much for an elegant idea! The fields of weeds are usually highly productive (see Chapter 24), contrary to the hypothesis, being perhaps the most productive communities of all. Moreover, some intriguing data suggest that productivity actually may be lowest in the climax in some successions, the exact opposite of what Lindeman predicted.

Figure 20.20 shows the results of productivity measures from Southeast Asia by the plantation-stand method (see Chapter 24). Production of organic matter increases over the first few years in the plantation. This is certainly because the young saplings require several years to close the canopy. In older stands, however, production falls even as biomass continues to increase. A concomitant fall in the leaf-area index suggests that the mature plantation

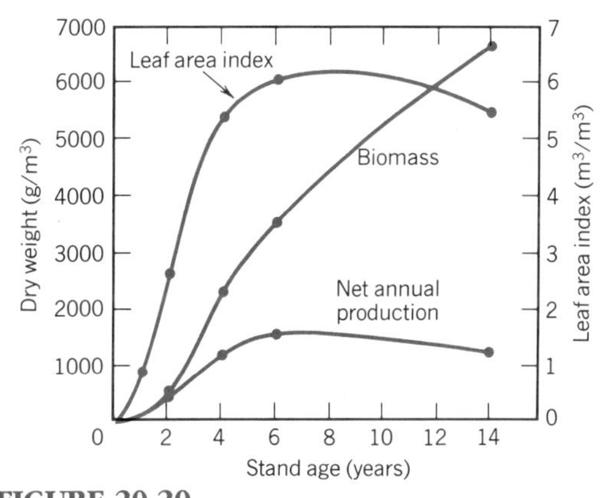

FIGURE 20.20
Productivity of aging forest stands.
Measurements are by the plantation series method applied to tropical forest plantations in southeastern Asia. Productivity falls in aging stands. (From Kira and Shidei, 1967.)

produces less because the packed trees actually offer less leaf surface to the sun.

What we now know about tree strategies in climax forest suggests that gross production should be low in mature natural forests also, although measures do not seem to be available to demonstrate this. Canopy trees of the climax are monolayers adapted to peak efficiency in shade and cannot be expected to collect carbon dioxide as rapidly as the multilayer early-succession trees that they replace (see Chapter 3). Moreover, it is likely that climax forest trees often are under nutrient or water stresses not experienced earlier in succession (Grime's stress tolerators). The likely actual change of productivity with secondary succession is given in Figure 20.20*b*.

Nutrient stress also lowers productivity in aquatic microcosms as they age (Figure 20.21). The microcosm data describe the history of production measured by the pH method (see Chapter 23) in beakers of water containing planktonic and benthic algae and supplied with artificial light. The fall in productivity in the older beakers is striking, but an inspection of the microcosms explains this easily enough. The sides of the glass beakers become covered with films and festoons of moribund benthic algae, like the walls of all neglected aquaria. In this moribund biomass is locked the better part of the nutrient supply originally in the water. Productivity falls because of nutrient deficiency (Cooke, 1967).

Information Theory and a Strategy for Ecosystem Development

Despite the failure of Lindeman's energy-flow model of succession, the appeal of energetics models was not easily abandoned in the profession. Consider Cooke's aquatic microcosms described in the last

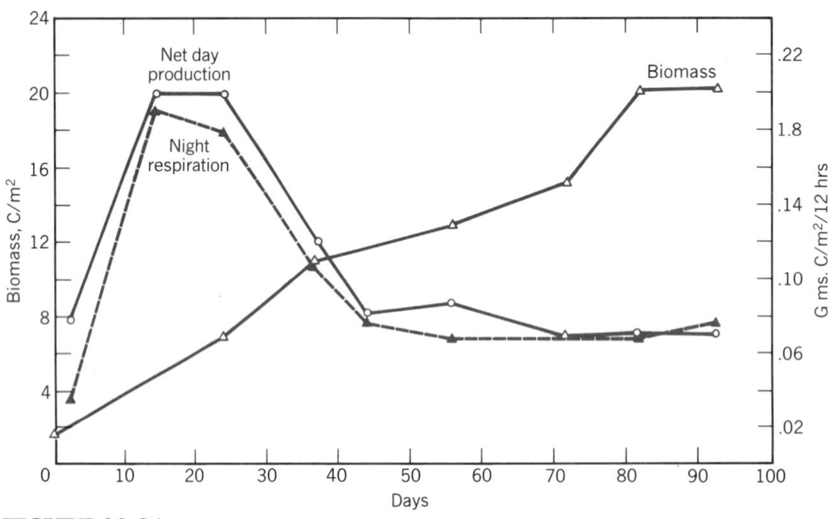

FIGURE 20.21
Productivity and biomass in developing microcosms.
The freshwater microcosms in 300-ml beakers were closed but provided with an energy source. Productivity and respiration both fell at the climax, even as biomass rose. Low climax productivity is caused by nutrient shortage as these are locked up in dead organic matter. (From Cooke, 1967.)

section. Productivity fell at the climax because most nutrients were locked in a mass of moribund biomass, but the biomass could, in theory, be burned; it represented stored energy. So, although productivity and energetic efficiency both fell during succession, the stock of biomass energy was built to a maximum and was maintained at that maximum.

Likewise, aging forests accumulate energy as biomass, even though their rates of gross production fall. A well-established forest, called "mature" by ecologists, has a tremendous, energy-rich biomass stored in the form of tree trunks alone, even without including accumulated leaf litter and other dead or moribund parts. The forest pays for this mass and maintains it, with a heavy cost in respiration. Net production is accordingly proportionately even lower, relative to a field of weeds, than is gross production. A forest, therefore, maintains a large standing stock of energy (tree trunks and dead biomass) at the cost of high respiration.

The eminent ecologist E. P. Odum, in a seminal paper, used these facts of forest growth to define properties that he suggested should characterize all mature ecosystems (Odum, 1969; Figure 20.22). The

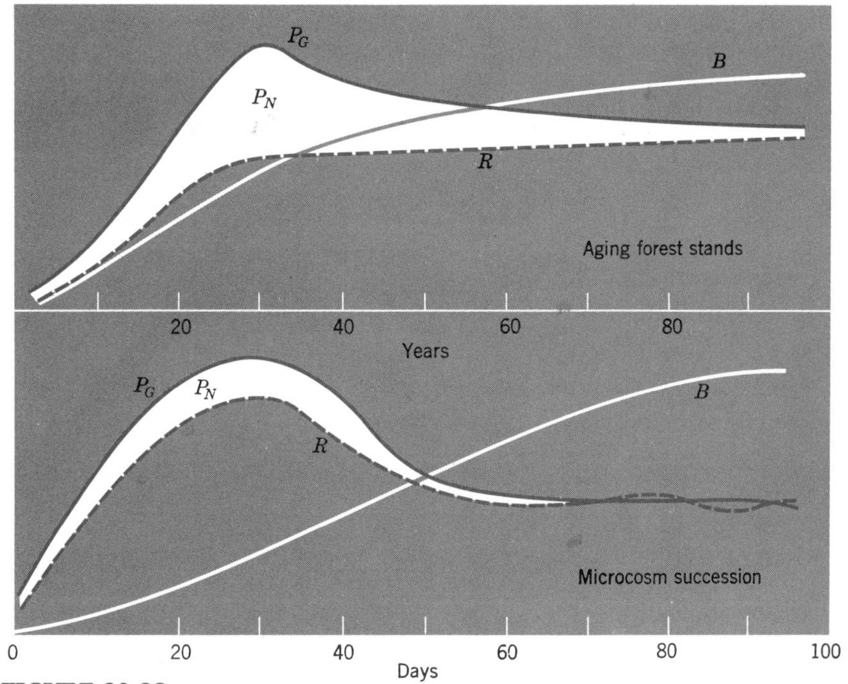

FIGURE 20.22
The increase in biomass during different successions.
The upper diagram is compiled from data from forest stands of different ages. The lower diagram results from measurements of replicated laboratory microcosms, essentially of algae in water. P_G is gross production, P_N is net production, R is total community respiration, and B is total biomass. The data suggest that as systems age they accumulate biomass (including detritus) until respiration balances production, when they may be called mature. Notice that gross productivity falls at maturity in both systems. (Redrawn from Odum, 1969.)

I. CONCEPT AND CAUSES OF SUCCESSION.

The formation an organism.—The developmental study of vegetation necessarily rests upon the assumption that the unit or climax formation is an organic entity (Research Methods, 199). As an organism the formation arises, grows, matures, and dies. Its response to the habitat is shown in processes or functions and in structures which are the record as well as the result of these functions. Furthermore, each climax formation is able to reproduce itself, repeating with essential fidelity the stages of its development. The life-history of a formation is a complex but definite process, comparable in its chief features with the life-history of an individual plant.

Universal occurrence of succession.—Succession is the universal process of formation development. It has occurred again and again in the history of every climax formation, and must recur whenever proper conditions arise. No climax area lacks frequent evidence of succession, and the greater number present it in bewildering abundance. The evidence is most obvious in active physiographic areas, dunes, strands, lakes, flood-plains, bad lands, etc., and in areas disturbed by man. But the most stable association is never in complete equilibrium, nor is it free from disturbed areas in which secondary succession is evident. An outcrop of rock, a projecting boulder, a change in soil or in exposure, an increase or decrease in the water-content or the light intensity, a rabbit-burrow, an ant-heap, the furrow of a plow, or the tracks worn by wheels, all these and many others initiate successions, often short and minute, but always significant. Even where the final community seems most homogeneous and its factors uniform, quantitative study by quadrat and instrument reveals a swing of population and a variation in the controlling factors. Invisible as these are to the ordinary observer, they are often very considerable, and in all cases are essentially materials for the study of succession. In consequence, a floristic or physiognomic study of an association, especially in a restricted area, can furnish no trustworthy conclusions as to the prevalence of succession. The latter can be determined only by investigation which is intensive in method and extensive in scope.

Viewpoints of succession.—A complete understanding of succession is possible only from the consideration of various viewpoints. Its most striking feature lies in the movement of populations, the waves of invasion, which rise and fall through the habitat from initiation to climax. These are marked by a corresponding progression of vegetation forms or phyads, from lichens and mosses to the final trees. On the physical side, the fundamental view is that which deals with the forces which initiate succession and the reactions which maintain it. This leads to the consideration of the responsive processes or functions which characterize the development, and the resulting structures, communities, zones, alternes, and layers. Finally, all of these viewpoints are summed up in that which regards succession as the growth or development

3

FIGURE 20.23

Photographic facsimile of the first page of F. E. Clements' book on "Succession" (1916).

His claim that the study of succession "necessarily rests . . . etc." though now seen to be false, was for long taken seriously.

P/R ratio (gross production/community respiration) should approach unity; that is, the hoard of biomass should grow until nearly all the product of photosynthesis is used for maintenance and almost none for growth. At this stage of maturity, a community has the most physical structure (built from all that biomass). And the community is complex in that it supports a large number of individuals, perhaps related in elaborate food webs.

Odum's (1969) penetrating description of ecosystems at maturity does nothing, however, to supply a causal mechanism for succession, being content to note that successional change stops when the load of accumulated biomass is so great that nearly all community respiration is consumed in its maintenance. Others were not so cautious, finding in the energy reserves of accumulated biomass an information source that fed instructions back to the developing succession (Margalef, 1963). Without more understanding of how the energy reserves of moribund tissue can influence the fates of colonizing species populations, most of us in the profession continue to prefer models that explain succession as the consequence of individuals of species with different properties acting in ways that maximize individual fitness.

A POINTED STAKE FOR THE SUPERNATURAL ORGANISM

A hundred years have passed since gifted ecologists first marveled over succession. One of the most inspired of that first generation, F. E. Clements, offered the brilliant, though wrong, hypothesis that plant successions were organizational properties of plant communities. The plant societies themselves were represented in their most perfect states by the climax formations, and all invasions by plants were but stages in the building of the perfect climax entity (Figure 20.23).

Clements wrote his doctorate on the "Phytogeography of Nebraska," exploring the whole wilderness state with horse and mule train in the 1890s, soon after the end of the Indian wars. He saw the trails left by the vanished buffalo being closed over by a succession of herbs that took about 50 years to replace a climax community of prairie grasses. Later he worked in the forested east and saw old field successions, where replacement of the old forest climax on cleared fields was even slower, evidently requiring a century or more to complete. He was influenced by Cowles's (1899) original work on the sand dunes of Lake Michigan, which postulated (incorrectly) that succession on the dunes could proceed almost everywhere to the regional climax of the beech–maple forest if given enough time.

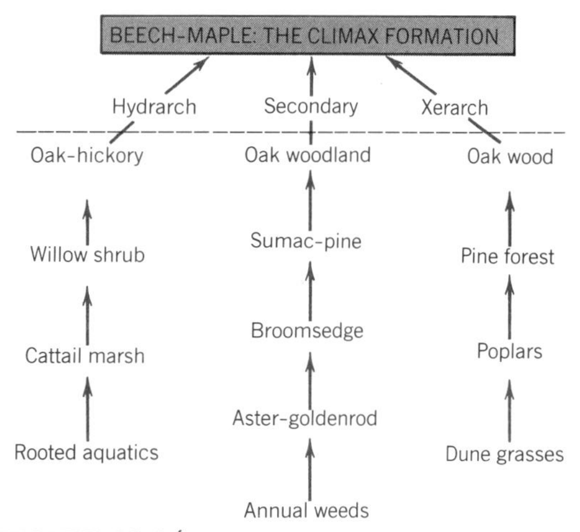

FIGURE 20.24
The kind of community taxonomy proposed by Clements.
All associations are supposed to be related to the climax formations by their roles as seral stages in successions.

Where one or both of the primary causes changes abruptly, sharply delimited areas of vegetation ensue. Since such a condition is of common occurrence, the distinctness of associations is in many regions obvious, and has led first to the recognition of communities and later to their common acceptance as vegetational units. Where the variation of the causes is gradual, the apparent distinctness of associations is lost. The continuation in time of these primary causes unchanged produces associational stability, and the alteration of either or both leads to succession. If the nature and sequence of these changes are identical for all the associations of one general type (although they need not be synchronous), similar successions ensue, producing successional series. Climax vegetation represents a stage at which effective changes have ceased, although their resumption at any future time may again initiate a new series of successions.

In conclusion, it may be said that every species of plant is a law unto itself, the distribution of which in space depends upon its individual peculiarities of migration and environmental requirements. Its disseminules migrate everywhere, and grow wherever they find favorable conditions. The species disappears from areas where the environment is no longer endurable. It grows in company with any other species of similar environmental requirements, irrespective of their normal associational affiliations. The behavior of the plant offers in itself no reason at all for the segregation of definite communities. Plant associations, the most conspicuous illustration of the space relation of plants, depend solely on the coincidence of environmental selection and migration over an area of recognizable extent and usually for a time of considerable duration. A rigid definition of the scope or extent of the association is impossible, and a logical classification of associations into larger groups, or into successional series, has not yet been achieved.

The writer expresses his thanks to Dr. W. S. Cooper, Dr. Frank C. Gates, Major Barrington Moore, Mr. Norman Taylor, and Dr. A. G. Vestal for kindly criticism and suggestion during the preparation of this paper.

FIGURE 20.25

Photographic facsimile of Gleason's reply to Clements.

The second paragraph is a succinct statement of the outlines of what would be established fifty years later. (From Gleason, 1926.)

Those were also the days when vegetation maps of the earth had triumphantly been shown to be climate maps also. Every great expanse of land had a characteristic formation of plants whose domains were limited only by climate. These were the formations that could be built on any conceivable substrate through succession, given time. Clements envisaged a classification of communities in which all the plant associations of a country were but seral stages in one of the several seres that ended in the climax formation (Figure 20.24). Fanciful, certainly, but it seemed to lock the associations together in a way that was forever to elude the phytosociologists of Montpellier and Uppsala (see Chapter 19).

At least as important as the Clementsian conception was the passion with which Clements proclaimed it. His passage, "As an organism the formation arises, grows, matures, and dies" (Figure 20.24) would not be unworthy of the language of the King James bible. He won over to his point of view not only ecologists, but public figures also. General Smuts, ex-commander of guerrilla cavalry and philosopher–prime minister of South Africa, acknowledged the ecological writings of Clements as the basis of his theories of "holism."

Clements's vision lies at the root of many of the political and social movements that take their names from ecology in the present day. Whenever activists accuse their political or exploiter adversaries of "ecocide" they invoke Clements's teachings. They borrow from him the idea that the ecosystem of the climax is an organism that can be wounded and slaughtered. They are wrong, as Clements was wrong.

Ridding ecology of the fantasies of succession theory has been a slow process, and was not completed without techniques not available to the early protagonists. Radiocarbon dating was crucial to debunking the Michigan sand dunes myth; modern advances in paleoecology were necessary for falsifying the hypothesis of hydrarch successions leading to the climax formation. Above all, the concept of clines of species strategies was necessary to explain the predictability of old field successions.

Nevertheless, some ecologists always resisted. The outspoken H. Gleason of the New York Botanic Gardens put forward the modern view in 1926, without benefit of modern data, in a powerful piece of writing in rebuttal to Clements (Figure 20.25). His "every species of plant is a law unto itself" is as elegant a way of putting the consequences of natural selection as could be asked.

No organizing principles exist in nature to order the lives of plants in communities. All plants grow and reproduce in ways that promote individual fitness. Succession is an inevitable consequence when plants with different strategies live in the same country. This is the mundane truth of the matter. And yet people tend to prefer brilliant myth to mundane fact, as witness the outnumbering of astronomers by astrologers. The Clementsian myth of supernatural organization in plant societies might require the hammering in of a sharpened stake before cock crow to end it.

Preview to Chapter 21

Surtsey: clean slate for colonists to build communities.

AN exciting model to explain species richness on islands was proposed 30 years ago and came to be known as "island biogeography." If a potential pool of colonists far exceeded the number of species that could coexist on an island, and if extinctions were random or proportional to species richness, then the model predicted that each island would have a characteristic equilibrium number of species, depending on the size and remoteness of the island. Experiments with mangrove islets colonized by arthropods and beakers of water as habitats for plankton yielded the expected equilibria. Comparing island bird diversity with island size and remoteness gave initially encouraging results. The model, however, seems inapplicable to land plants. Pollen data from the Galapagos and species time data from Krakatoa suggest that in plants, the required frequent extinction and replacement on islands does not occur. Krakatoa species-time data for butterflies and birds also give no evidence of species turnover, suggesting instead that species richness of butterflies and birds is a function of plant diversity. If significant rates of extinction are not important in establishing species richness in most animal and plant communities, others of the many mechanisms relating richness to area may be critical in defining communities on oceanic islands.

Chapter 21

Communities on Islands

Building species-rich communities in the presence of powerful competitors results in the relatively rapid and roughly predictable series of replacements that we call ecological succession. The end result is minimal further change other than that following physical disruption, like the knocking down of trees by storms. The state of little change is dubbed the climax.

But what if powerful competitors do not exist or do not have access to the colony? Under these conditions, contest competition should always be indecisive as fresh immigrants with subtly different contest strategies constantly arrive to dispute the habitat. A climax of no more change might then be impossible.

Habitats for which powerful stress competitors might be lacking are plentiful. All islands qualify, whether true oceanic islands or isolates like mountaintops or puddles of water. So, perhaps, do open-water systems where tiny organisms of the plankton scattered through a fluid medium may make habitat preemption impossible.

A theory of species richness in these conditions is known as "**island biogeography**," because it was first worked out to account for the characteristic diversity of island faunas (MacArthur and Wilson, 1967).

THE CONCEPT OF SPECIES EQUILIBRIUM

Consider a totally new oceanic island created by a volcanic explosion. It is without life and can receive terrestrial animals only by long-distance transport across the sea. Colonists begin to arrive, and observations of actual young islands like Surtsey, the young island that appeared off Iceland a few years ago, show that these first colonists are prompt (Fridriksson, 1975). Some colonists establish themselves and live on the island. This makes no difference to the rate of immigration at first because the island is underpopulated for some time, so that there is no competitive bar to the establishment of newcomers. Arrival of colonists continues at a constant rate, though the number of new species among them falls progressively.

But a small island, particularly a young one, may be an uncertain place in which to live, suggesting that some of the first-comers who became established die out. Smallness of populations also makes extinction likely. Since the island is still largely empty habitat, this chance of extinction will not depend on the presence of other animals; it will depend only on physical adversity. The rate of extinction, therefore, can be treated as constant in this early phase of colonization, just like the rate of immigration.

Figure 21.1*a* illustrates this early phase of community building on an island. Rates of immigration and extinction are both linear because neither is dependent on population density. As the number of species living on the island rises, the rate of species arrival falls as a simple function of the number already there. The rate of extinction rises as a simple function of the number of species at risk. The model suggests that an equilibrium number, S_1, will be established when the rate of immigration is balanced by the rate of extinction. This equilibrium does not depend on any interaction between populations on the island for its effect. The assumption is that there is no competition and no predation. The result is a "**noninteractive species equilibrium.**"

The noninteractive species equilibrium, S_1, should persist only until populations on the island become large enough for competition and predation to become important facts of life. After that, successful immigration requires both arrival and establishment, which depends on resistance by species already present. Both immigration and extinction then become density-dependent, and both rate curves become exponential (Figure 21.1*b*). Where these curves intersect there is a new equilibrium number of species, S_2. This number is very much dependent on interactions between species on the crowded island and is an "**interactive species equilibrium.**"

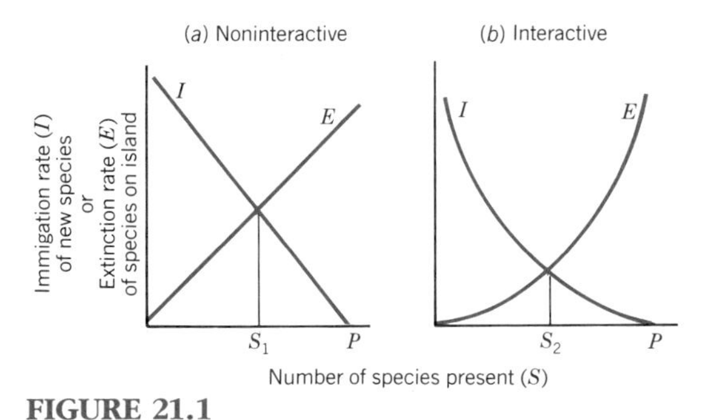

FIGURE 21.1

General hypothesis for the regulation of species number on islands.

When an island is lightly populated, so that most populations are not under density-dependent control, the rates of immigration and extinction are linear (noninteractive model, *a*. In a well-populated island (interactive model, *b*), the rates are exponential. The equilibrium numbers, S_1, S_2 are unlikely to be equal. P is the total species pool from which the colonists must come. (Modified from Wilson, 1969.)

SPECIES–AREA PREDICTIONS OF THE MODEL

The model makes predictions both about accumulations of species through time and the resultant number for given areas. Species–time data should offer the more crucial tests, because the model predicts perpetual immigration and extinction. With an adequate time machine, it should be possible to plot species arriving and species disappearing.

A few species–time data are at last beginning to be offered, (particularly from Krakatoa), but the predictive properties of the model that have been most subject to testing up to now are species–area predictions arising from the fact that numbers must be a function of both island remoteness and island size.

Figure 21.2 shows the predictions of the model for islands near and far and islands large and small. Distant islands will receive immigrants at slower rates than islands near the mainland, simply as a function of the distance to be covered by potential colonists. The extinction rate, however, is not affected by distance. The hypothesis predicts, therefore, that distant islands will hold fewer species at equilibrium than will near islands.

The size of an island, as opposed to remoteness, affects both immigration rate and extinction rate. Immigration goes up with size because the target is larger, and extinction goes down with size, because large islands provide more opportunity to escape competition or predators. Large populations on large islands also face less chance of random extinction. The equilibrium number, therefore, will be larger on large islands.

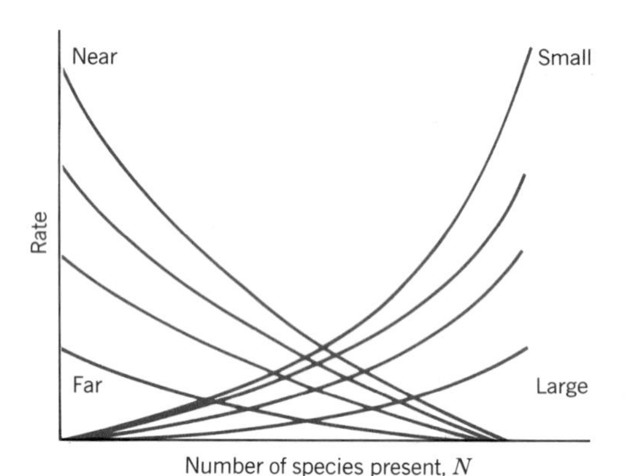

FIGURE 21.2
The effects of island size and distance on equilibrium number of species.
An increase in distance (near to far) lowers the immigration curve,. whereas an increase in island areas (small to large) lowers the extinction curve. (From MacArthur and Wilson, 1963.)

Massacre on Mangrove Islets

The red mangrove (*Rhizophora mangle*; Figure 21.3) grows in shallow water in suitable sites at the edge of tropical seas. Clumps of the trees effectively form small islands cut off from mainland swamp forest by reaches of water varying from a few meters to up to half a kilometer across (Figure 21.4). Islets can be chosen that are discretely separate and from about 10 to 20 meters across. About 1000 species of terrestrial arthropods have been collected from these islets in Florida, showing that the species pool from which the islets are colonized is $P = 1000$. Yet the actual num-

FIGURE 21.3
Rhizophora mangle: **Island in the sea for arthropods.**

FIGURE 21.4
Mangrove islets.
Although nearly 1000 species are known to be able to live on islets, the equilibrium numbers seem to be between 10 and 40 species per islet.

ber of species found on any one islet is usually less than 40.

Wilson and Simberloff proposed that this discrepancy between 1000 possible species ($P = 1000$) and less than 40 actual ($S < 40$) was due to the establishment of equilibria between immigration and extinction at a small fraction of the total pool. General support for this hypothesis was available at the start of the investigation because the species lists were different on different islets. Wilson and Simberloff chose six islets 11 to 18 meters in diameter and 2 to 500 meters from shore and killed all the arthropods on them. They did this by enclosing the entire islet in a plastic tent and injecting a massive dose of insecticide. They then had six islands that were completely "defaunated" and could watch the process of recolonization directly, making complete census of the islets every few weeks (Simberloff and Wilson, 1970).

Results for four of the islets are given in Figure 21.5. Before the animals were killed, the number of species present appeared to be a function of distance as predicted, with the least in the distant islet E1

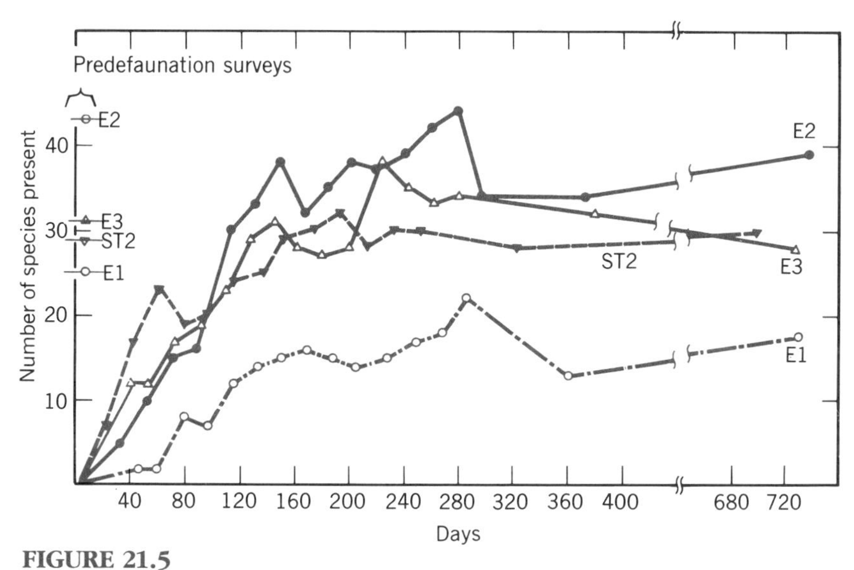

FIGURE 21.5
Recolonization of defaunated mangrove islets.
The species numbers before defaunation for each islet are shown at the left. E1 is the nearest island, E2 the farthest, and the other two are intermediate. (From Simberloff and Wilson, 1970.)

and the most in the near E2. After two years of recolonization, these approximate numbers had been reestablished on each islet. As predicted by the hypothesis, however, the actual species lists were not the same; it was only the total number that was restored. Moreover, the periodic surveys clearly demonstrated a frequent replacement of species as predicted by the hypothesis.

Of particular interest in these studies is the question of whether the demonstrated species equilibrium was interactive or noninteractive: Was it competition and predation that set the species numbers so low, or was this merely the result of random extinctions in small populations? Two lines of argument suggest that both processes are involved at the same time.

Simberloff and Wilson calculated turnover rates and were able to show that these were in the order expected from the noninteractive model. For the noninteractive model (Figure 21.1*a*)

$$\frac{dS_1}{dt} = \lambda_A(P - S_1) - \mu_A S_1 \quad (21.1)$$

where

λ_A = average immigration rate per species

μ_A = average extinction rate per species.

For the special case of equilibrium, and with a sufficient lapse of time, it can be shown that noninteractive equilibrium predicts that turnover rates must be within a narrow range (Wilson, 1969). The data from the islets do yield turnover rates within this range, showing that the results could be due to completely noninteractive processes.

However, an assessment of these same data show that separate equilibria are maintained by arthropods in each of a few major categories of habit: detritus feeders, herbivores, wood borers, etc. (Heatwole and Levins, 1973). Species equilibria were maintained in each of these broad-niche classes independently of the rest. Clearly there is a trophic structure on the island into which immigrants must fit.

One datum that is particularly suggestive of interaction is that the colonization curves "overshoot" the eventual equilibrium number before settling down to the equilibrium that existed before defaunation (Figure 21.5).

EVIDENCE OF BIRDS FROM ARCHIPELAGOS

Figure 21.6 shows how the number of species of birds on islands generally is correlated with area. Many such studies and data sets are now available, and they generally confirm that species number does

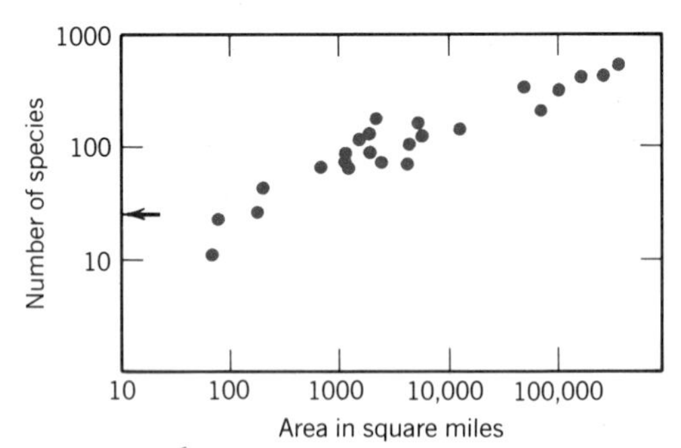

FIGURE 21.6

Bird numbers on Pacific islands.

The islands range from Christmas Island, the smallest, up to New Guinea. The islands are set in a ring around the western Pacific Ocean, all of them being close to continental land. There is thus no effect of distance on species number. The correlation with area, however, is very strong. Krakatoa birds of the post-eruptive equilibrium would place at the left end of the correlation just off the graph. (Arrow: 30 species, 8 square miles.) (From MacArthur and Wilson, 1967.)

correlate with area. The strongest correlations, however, are with islands of similar kind, particularly of similar relief. Comparing low islands with high islands, for instance, does not result in a good species-area correlation (Simberloff, 1974). This is to be expected since high islands should have more varied habitats, and hence a greater variety of resources, than low islands. But for similar islands, and over a considerable range of size and remoteness, a correlation with island area and number of bird species can often be shown.

Figure 21.7 shows the species–area correlation for small islands off New Guinea. The straight line is a calculated regression. Dispersal in the data turn out to be particularly revealing. Diamond (1973) was able to show that these islands had been disturbed in the not-too-distant past. In particular, many of the offshore islands had been connected to the mainland at the time of the last ice age as a result of the eustatic lowering of sea level (see Chapter 17). Since these islands were then easily penetrated by immigrants, they had collected more species than they would have in their present island state and are now probably progressively losing species, called "relaxation" by island biogeographers.

Figure 21.8 shows comparable data from islands in the Caribbean, where the land bridge effect is particularly noticeable. Some of the larger of these old land bridge islands hold species that have never been known on islands elsewhere, but these truly continental animals are not now found on the smaller of the old land bridge islands. This strongly suggests that this part of the continental fauna already

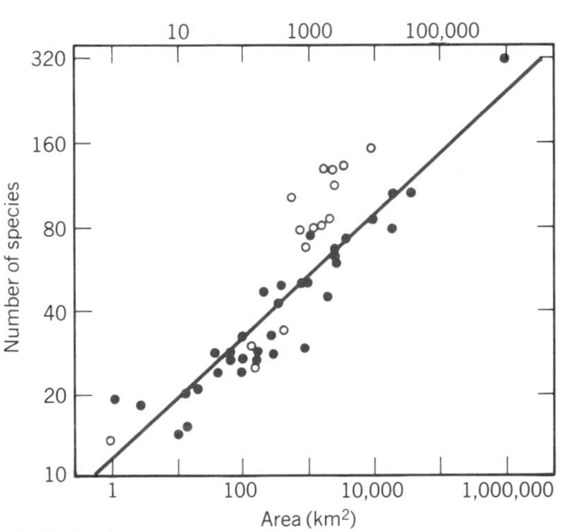

FIGURE 21.7
Resident bird numbers of islands offshore from New Guinea.
Species numbers and areas are plotted on logarithmic scales. Solid circles are islands thought to be at equilibrium. Open circles are of islands not yet recovered from disturbances, either having too many species from a recent land connection or too few following volcanism, etc. (From Diamond, 1973.)

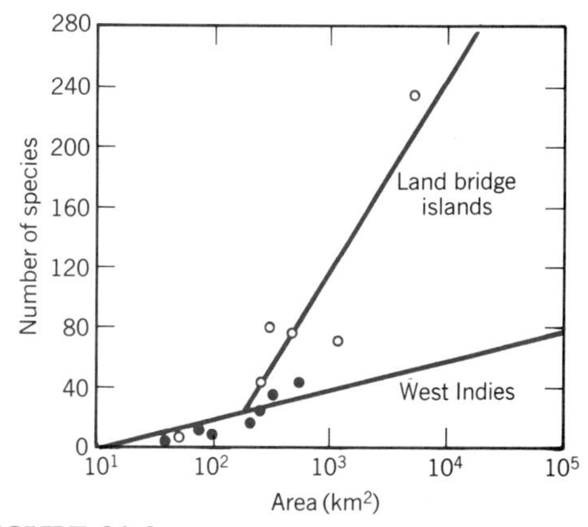

FIGURE 21.8
Species–area plot for land birds of Caribbean islands.
Open circles are islands that were connected to the mainland when sea level fell during the last ice age; solid circles are islands that have never been part of the mainland. The lines are computed regressions. The larger land bridge islands apparently are still oversaturated with birds from the days of easy immigration, though smaller land bridge islands have almost relaxed to equilibria appropriate for oceanic islands. (From Terborgh, 1974.)

has gone extinct on the smaller islands and that the process of extinction still proceeds toward the equilibrium number (Terborgh, 1974).

These and similar data show convincingly the determining role of area. Data showing the predicted effect of distance are also available. It has long been an observation of biogeographers that very remote oceanic islands have depauperate faunas. Equilibrium island biogeography predicts that this is because the immigration rate is slowed by distance, resulting in a lowered equilibrium number. However, the poverty of oceanic islands could simply be because dispersal across oceanic distances takes so long that a species equilibrium has never been reached. The theory offers a general test of the predicted effect in the predicted slope of the regression line.

Figure 21.9 shows the species–area relationship for separate archipelagos at varying distances from the mainland. Islands of the most remote archipelago are shown as squares, and it is easily seen that the slope of the species–area relationship is steeper for these islands. The area effect is in fact exaggerated because small remote islands are tiny targets for dispersal at very long range and so have "too few" species.

For birds we also have actual time-series data from some islands where observations of different dates document the changing composition, but constant number, of species present. The best known of these data are from the Channel Islands off the coast of California (Diamond, 1969). The base data are from a census of 1917 and another of 1968. The numbers of species on the different islands had not changed in this interval but the species lists had. We do not know how many species both arrived and disappeared between the two census dates, but the recorded changes suggest that the turnover rate might have been between 0.3 and 1.2% a year. Unfortunately, serious questions remain about the precision that is possible using these data (Simberloff, 1974).

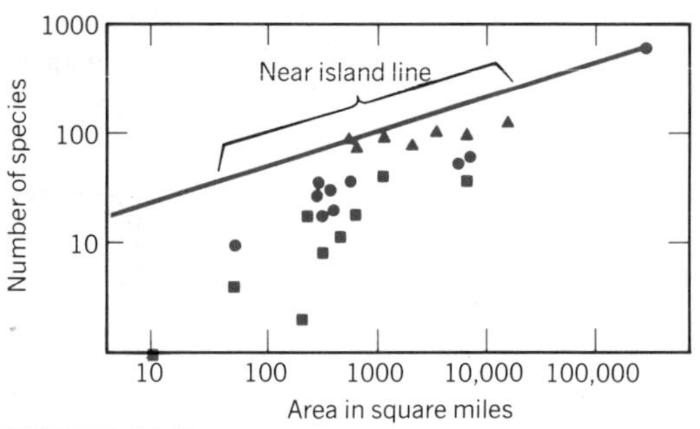

FIGURE 21.9
Effect of distance on species–area curves of archipelagos.
Squares denote islands more than 2000 miles from the mainland; triangles are islands less than 500 miles from mainlands; and circles are islands at middle distances. The line represents the slope for near-shore islands. The much steeper slope of the distant islands is predicted because rates of immigration are expected to be slowest for these islands. Some part, but not all, of the low species numbers on the remote islands of this particular set, however, may be accounted for by the fact that the remote islands are low-lying and so have little habitat diversity. (From MacArthur and Wilson, 1967.)

MAMMALS ON MOUNTAINTOPS

An early application of the theory was to the mammals of mountaintops, considered as islands in seas of valleys. Since the time of Darwin and Wallace, naturalists have speculated that many mountaintops were populated in the last ice age by cold-adapted animals that then could cross the intervening lowlands more easily than now. This makes their position very similar to the birds of old land bridge islands that penetrated when the ice age lowered sea level: extinction would continue even though replacement by immigration had stopped. Brown (1971) put forward the hypothesis that species numbers of mammals on mountaintops should have been falling since the last glacial retreat 10,000 years ago.

Brown's "islands" were small mountain ranges between the Sierra Nevada and the Rocky Mountains, the two chains that formed the "mainlands" of his study. He defined 17 of these mountain islands by arbitrary criteria of height and remoteness from each other, and then showed that the number of species of mammals on the islands was significantly correlated with area (Figure 21.10).

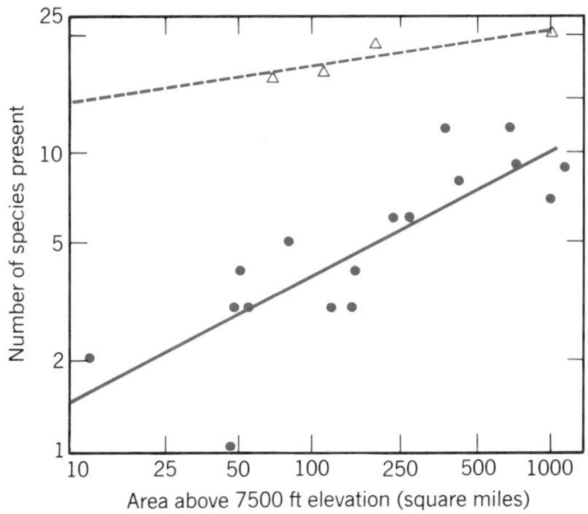

FIGURE 21.10

Species–area relationships for mammals on mountaintops.

The solid dots, and the solid regression line, are for 17 isolated ranges in the Great Basin. Triangles, and the dashed regression line, are comparable areas attached to the present-day Sierra Nevada. The steeper regression for the isolated mountain ranges suggest that they still retain extra species of mammals from the time when the ice age climate connected them with alpine-type environments in what are now dividing valleys with warmer climates. (From Brown, 1971.)

The slope of the logarithmic plot, however, was very low. Slope is a measure of the difficulty of immigration, a low slope indicating easy access. For the island biogeography model to be valid, the low slope requires ease of access to the mountaintops resulting in a fast immigration rate and a high number of species at equilibrium. But Brown examined various factors that could make immigration easy between the various islands and could find no correlations between ease of access and number of species. He accordingly put forward the hypothesis that the mountaintops were oversaturated as a result of invasions of ice-age time and that the number of mammal species has not yet relaxed to equilibrium, even after 10,000 years.

The oversaturation hypothesis is plausible, but it does not exclude the possibility that the number of mammals on mountaintops is regulated by area in ways quite different from that of the species equilibrium. Perhaps both habitat and food web structures on mountaintops accommodate large numbers of species, with species turnover of minimal importance, as apparently applies to plants on the Galapagos and Krakatoa.

PLANKTONIC PLANTS AND LAKES AS ISLANDS

Microscopic algae should meet the boundary conditions of the island biogeography model. Because all bodies of fresh water are ephemeral, phytoplankton must disperse readily. No phytoplankter can preempt space to block immigration, as can a land plant. Colonization of virgin waters is known to be exceedingly rapid.

Maguire (1963) set out beakers of water along a transect running from a freshwater pond in Texas. These beakers were islands

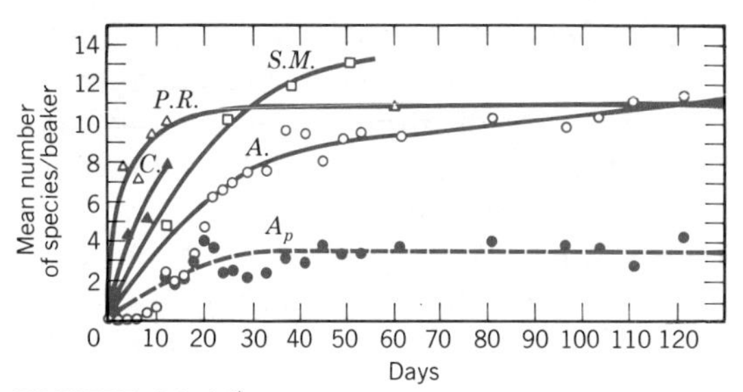

FIGURE 21.11
Immigration to beakers of water by plankton.
Sites are in Colorado (C), Puerto Rico (P.R.), and Texas (S.M., San Marcos; A., Austin). A_p is the curve for protozoa only at Austin. (From Maguire, 1971.)

for water life in a sea of land. The number of species of microorganisms (algae and larger protozoa) rose progressively over a two-month span as the rate of colonization fell, very much as predicted by the theory (Figure 21.11).

Lakes, provided they have no significant inlets through which floods of immigrants may be introduced, also serve as islands surrounded by land. Figure 21.12 shows species–area curves for closed lakes of Ecuador. Area is shown to be a good predictor of species number of algae, though such parameters as water chemistry and age of the lake basins were not (Colinvaux and Steinitz Kannan, 1981). Notice that the regression line for the Galapagos lakes taken separately is steeper than the regression for mainland lakes. This conforms to the requirement for distant islands resulting from the greater inaccessibility to the species pool.

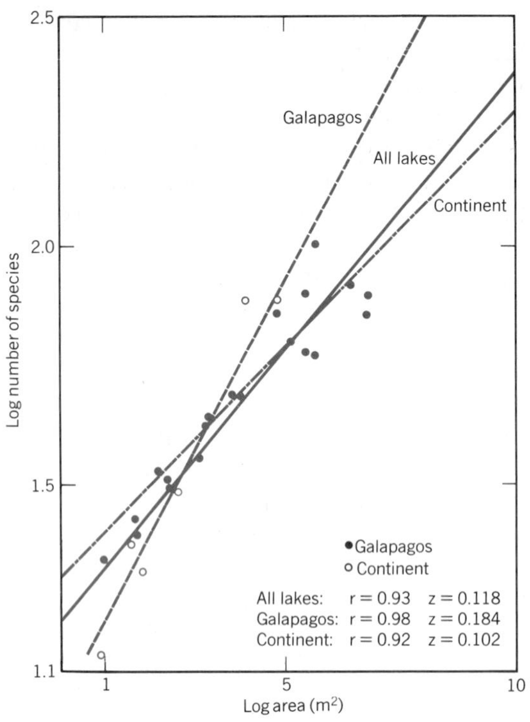

FIGURE 21.12
Species–area curves for algae of open water in lakes of Ecuador.
The lakes are all closed basins, or have minor outlets only. (From Colinvaux and Steinitz-Kannan, 1980, with additional lakes.)

Field Crops as Islands for Agricultural Pests

Agricultural fields of monoculture crops are islands from the point of view of crop pests, yielding island systems for which many data are available. Figures 21.13 and 21.14 are species–area curves on log-

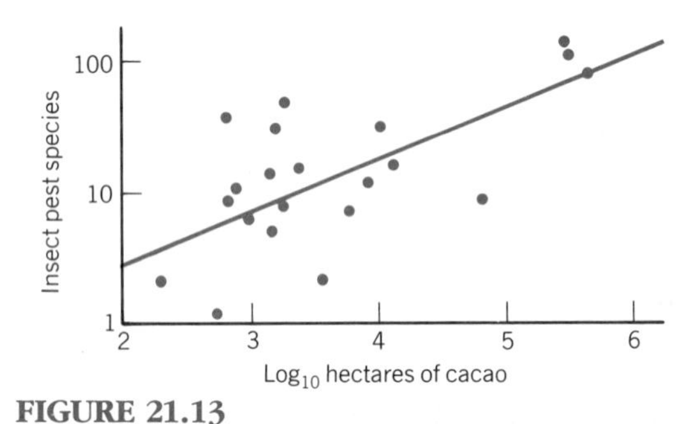

FIGURE 21.13
Insect pests of cacao as a function of hectares planted.
Data are from all parts of the world where cacao is grown. Data are numbers of species of insect pests recorded for each cacao-growing country and the acreage of the crop planted in that country. (From Strong, 1974.)

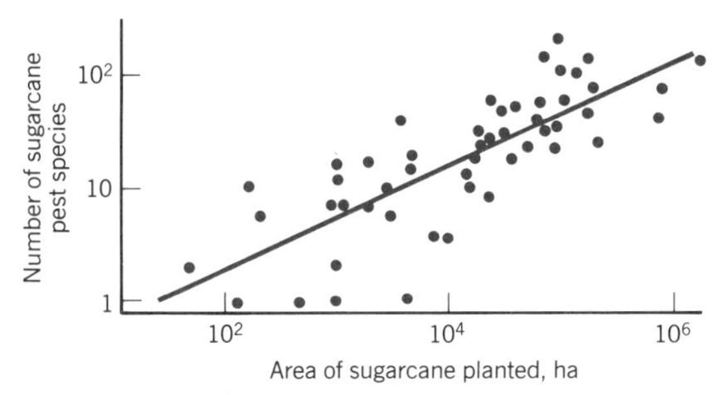

FIGURE 21.14
Insect pests of sugarcane as a function of hectares planted.
Data were gathered in the same way as for cacao in Figure 21.13. (From Strong et al., 1977.)

arithmic plots for numbers of insect pests on plantations of cacao and sugarcane. These data are perhaps particularly striking because of our intuitive belief that a field of monoculture must be vulnerable to pest attack, and yet countries with small acreage have fewer pest species than those with large acreage. The pests are not omnipotent, and local extinction is shown to be the rule. Many more pests are available than the ones actually doing damage, and more come to a farmer's notice as the island of crops increases in size.

Plants as Islands

The agricultural crop data from relatively large areas of uniform planting suggest that similar effects should be present for dispersed individual plants. Each plant then is an island, and a very small one. If the number of insects attacking at any one time (the "herbivore load") goes down sharply as planted area becomes small (Figures 21.12 and 21.13), then an isolated plant should carry few pest species, with a fair chance that it holds none at all if it falls into a zero class in the random game of immigration and extinction.

Within a population of plants, dispersal and clumping serve as distance measures that affect the rate of immigration of the animals using the plant for food or as a habitat. Clumping also serves as area in the island biogeographical sense. A very large number of animal species can be related to a single plant host as a result (Janzen, 1968).

LAND PLANTS MARCH TO A DIFFERENT DRUMMER

If data are pooled from all the islands of the Pacific Ocean, a very general correlation between number of plant genera and area can be shown, though with considerable scatter in the data (Figure 21.15). Correlation of species richness of plants with area is one of the best known phenomena of plant ecology (see Chapter 18). Many processes work together to produce this effect, notably continual increase in the possibility for between-habitat diversity as area increases.

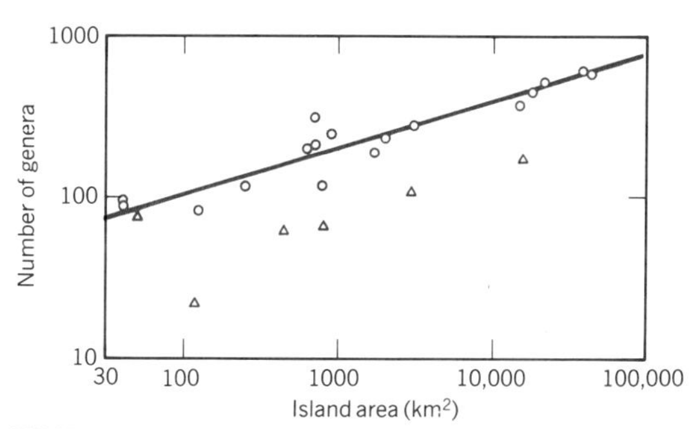

FIGURE 21.15
Species–area plot for plant genera on Pacific islands.
The line is a calculated regression for the less remote islands (circles). Triangles show the more remote islands. (Data of Van Balgooy, 1971, from Cox and Moore, 1980.)

The scatter in Figure 21.15 may result from the different physiography of islands, or it may simply reflect the fact that species lists are either not complete or not based on comparable collections. In the Galapagos Islands, for instance, there is disagreement as to whether species lists from different islands reflect actual species lists or merely the intensity of collecting effort in the different regions (Johnson and Raven, 1973; Connor and Simberlof, 1978).

But with or without scatter, correlations of land plant species richness with area does not require that the number be set as an immigration–extinction equilibrium. A process of continued species turnover for plants is highly unlikely because of their habit of preempting space. Successful competitors and stress tolerators tend to maintain themselves indefinitely, preventing further immigrations unless they are disrupted by physical force or annihilating herbivory like seed predation.

In a sense, island colonization by plants was the subject of a hundred years of succession studies (see Chapter 20). Old fields and glacial debris are islands, though they are colonized from closer ranges than islands out at sea. Species number climbs rapidly, then either arrives at an asymptote or actually falls as the competitors and stress tolerators preempt the habitat. After that species richness remains roughly constant, not as a result of turnover but because the same species remain. Succession studies, therefore, confront the concept of species number set by an immigration–extinction equilibrium with the concept of a climax number of unchanging species.

Correlations of species richness with increasing area are consistent with both the equilibrium and climax number hypotheses. For land plants, therefore, the only satisfactory test for the workings of the island biogeography model must be species–time tests.

The Plants of Krakatoa

The volcanic island of Krakatoa exploded in 1883 with such force as to sterilize what remained of the island. Research parties have surveyed the vegetation of the island at intervals ever since, providing species–time data.

The main island remnant, Rakata, has never had human inhabitants. Being only 40 or so miles from both Sumatra and Java, plants have been brought to Rakata by wind, sea, birds, and bats. After a century of colonization, about 97 species of tree are listed for the island, compared with some 211 species on an ancient island of the same size that was attached to the mainland during ice age times of low sea level (Whittaker et al., 1989; Bush and Whittaker, 1991).

Figure 21.16 gives rates of immigration and extinction calculated from the data of research expeditions. Because expedition surveys can never be complete, some minimalist assumptions were made in compiling these curves. For instance, if a plant was not recorded by one expedition but was noted on visits both before and after,

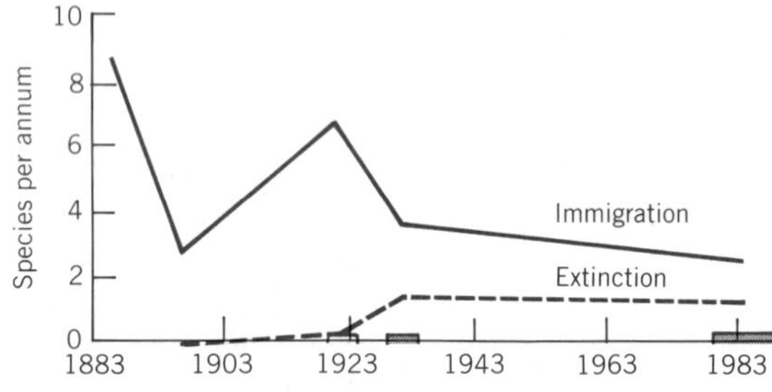

FIGURE 21.16
Immigration and extinction curves for higher plants on the island of Rakata (Krakatoa) between 1883 and 1989.
(From Bush and Whittaker, 1991.)

it was assumed to have been present all the time. Compare this figure with the MacArthur and Wilson postulate for immigrations and extinction on an island in Figure 21.1. Clearly, plant colonization does not follow the island biogeography model (Bush and Whittaker, 1991).

Extinction is low, possibly even falling, even as total species richness continues to rise (Figure 21.17). The relatively high initial rate of extinction can plausibly be explained as pioneer plants, many of them ferns, were replaced by large ground-holding plants of the forest. Possibly these vanished plants actually still exist on the island in numbers too low to be found by expedition survey methods, the old relative abundance–equitability problem of sampling diversity (see Chapter 16).

Meanwhile, the forest grows steadily richer. Because the ocean gap is only 40 miles, animal-dispersed fruits are brought in by bats and birds, though slowly. Apparently what we now see on Rakata is an ecological succession somewhere in Grime's (1979) competitor stage before the final accommodation to the stress competitors (see pages 432–433).

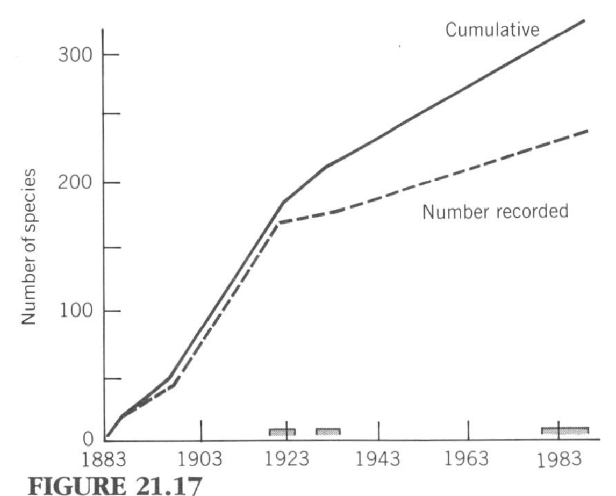

FIGURE 21.17

Cumulative species totals of plants, together with number recorded at each collection, on Rakata (Krakatoa) between 1883 and 1989.
(From Bush and Whittaker, 1991.)

A Long Species–Time Series from the Galapagos Islands

The modern Galapagos Islands support moist forest in the highlands, although the lower parts of the islands are desert. In the last ice age, however, radiocarbon-dated evidence from lake sediments demonstrates that even the highlands were dry (Colinvaux, 1972). Moisture returned to the highlands 10,000 years ago, allowing immigrations, either from the mainland 2000 kilometers away or from relict populations in moister valleys.

Pollen data from El Junco lake give a rough account of the progress of plant immigration to the upper reaches of Isla San Cristobal (Figure 21.18). From the onset of first ponding in El Junco, and hence presumably the start of the modern moister climate, between 500 and 1000 years passed before vegetation similar to that of the present day occupied the moist high ground. This longish interval perhaps reflects the progress of primary succession hampered by the necessity of many plants having to disperse over wide sea gaps to reach the site.

More significant is that very little change is recorded for the next 9000 years (Figure 21.18). Pollen data give a very poor representation of the total species list so that many species changes could take place without mark on the pollen record. Nevertheless, the data do show a remarkable constancy in the Galapagos vegetation that is not at all suggestive of constant turnover of species. As with the shorter time series from Krakatoa, the Galapagos history suggests slow accumulation of species, with only minimal extinction along the way.

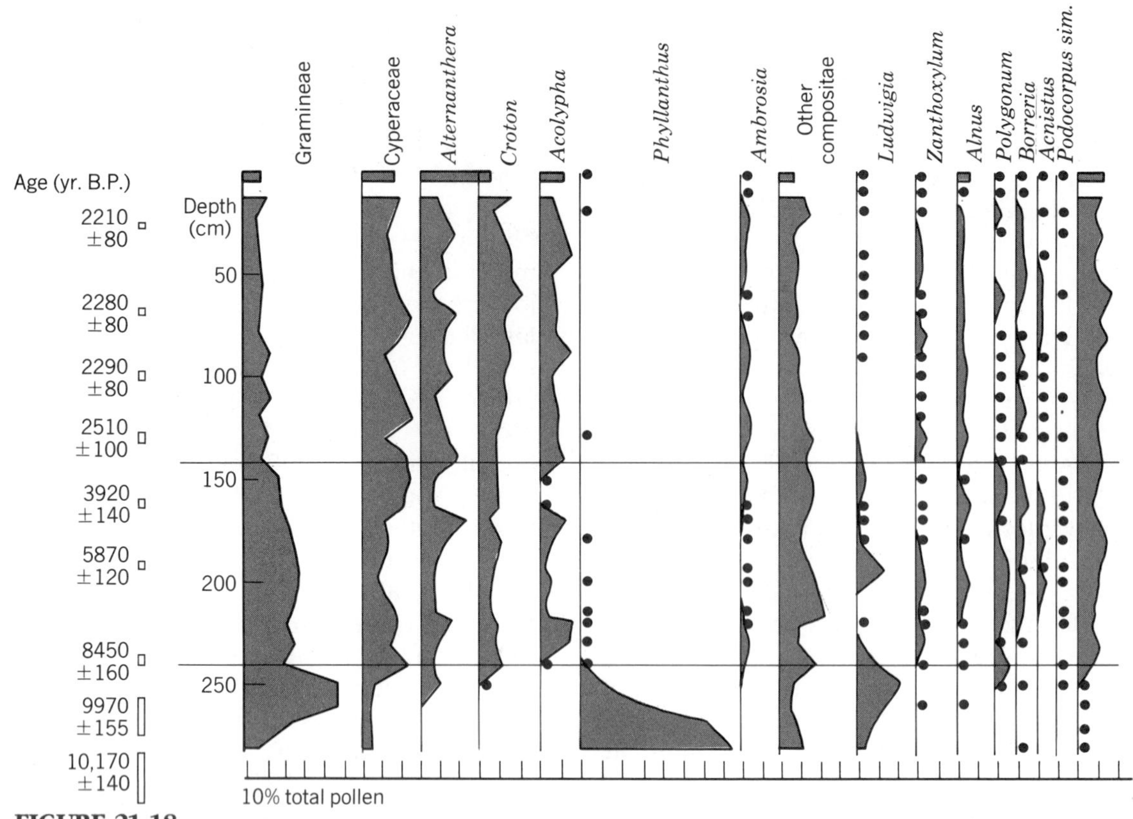

FIGURE 21.18
Resistance of Galapagos mesic vegetation to invasion.
Pollen percentage diagram from sediments of Lake El Junco on Isla San Cristobal (Chatham). The sediments below this 3 meter sequence showed that the island was dry and without mesic forest before 10,000 years B.P. The pollen of the ten thousand years of Holocene time show that there were no major changes in vegetation following the initial colonization phase. (From Colinvaux and Schofield, 1976.)

IMMIGRATION AND EXTINCTION OF KRAKATOA ANIMALS

Bush and Whittaker (1991) have compiled species-time data for both butterflies and birds on Krakatoa. Neither history is as predicted by the species-equilibrium model.

Figure 21.19 gives data for butterflies. Early colonization was rapid. A burst of apparent extinction followed 50 years later, so that the extinction curve actually intersected the recruitment curve in the MacArthur and Wilson manner. Thereafter, however, extinction fell even as total species richness continued to rise. Modern butterfly net recruitment to Rakata in fact closely follows the pattern for plants: little extinction with continued immigration.

This result is even more striking when the apparent extinction event of the 1920s is examined more closely. Some of the but-

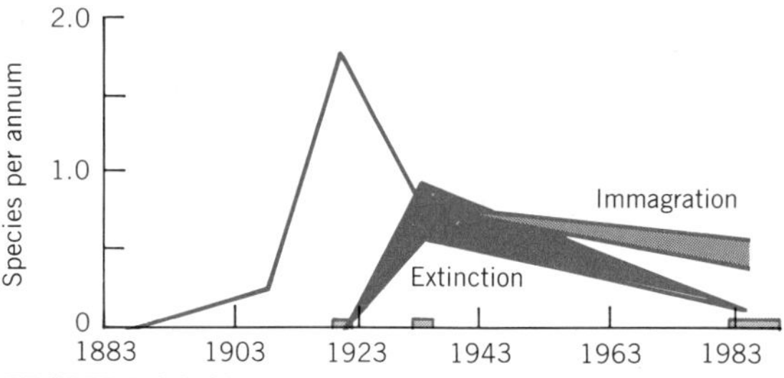

FIGURE 21.19
Immigration and extinction curves for butterflies on Rakata (Krakatoa) between 1883 and 1989.
(From Bush and Whittaker, 1991.)

terflies not recorded (deemed extinct) were migrant species whose food plants had not reached the island, meaning that they could not be resident populations. Whether they were found or not depended on breeding successes on the mainland that determined the numbers migrating, and had no bearing on the number of species inhabiting the island. With these butterflies excluded, the extinction curve should have been lower and would never have intersected the immigration curve.

Moreover, the butterflies that did truly go extinct in the 1920s were dependent on pioneer food plants of early successions, which were then being displaced by contest competition as the island canopy closed. Extinction of those butterflies had nothing to do with chance or species interactions, but rather followed the progress of plant successions.

A similar result was obtained for Rakata birds (Figure 21.20). High initial immigration quickly ceased, probably after most birds capable of living in a newly seeded, barren island had established themselves. Extinction remained low, but in the ensuing decades seven of the original 13 species of land birds went extinct. As with the butterfly extinction at about the same time, this extinction appears to have nothing to do with interactions between birds and everything to do with loss of the original open habitat as the forest canopy closed over the island.

Since the early 1940s, species have turned over, with extinction balanced by immigration in a result superficially like that predicted by the island biogeography model. But this period actually saw a 20% fall in total species richness, contrary to the model. Moreover, knowledge of the habitat requirements of the species that went extinct show that these, like the earlier extinctions, were of birds that had lost their habitats through the progress of plant succession. The latest result is that bird numbers are rising again as more forest birds begin to occupy the diversifying forest.

Species richness of both butterflies and birds of Rakata is parsimoniously explained by the progress of plant succession. Species richness of animals, therefore, is dependent on the supply of habitats or resource, and is influenced only marginally, if at all, by the number of other butterfly and bird species on the island.

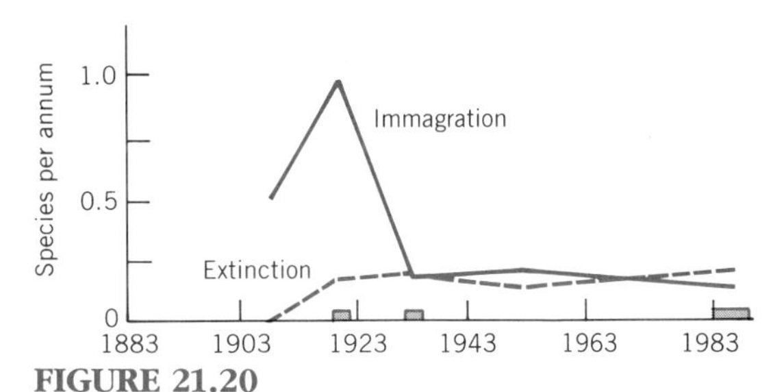

FIGURE 21.20
Immigration and extinction curves for resident land birds on Rakata (Krakatoa) between 1883 and 1989.
(From Bush and Whittaker, 1991.)

IMMIGRATION-EXTINCTION EQUILIBRIA LIKELY TO BE RARE

The immigration–extinction equilibrium was offered by MacArthur and Wilson as an alternative to the climax community model. The new model had definite and critically important boundary conditions. The first was that random extinction should be significant, and the second was that extinction should be a function of species richness as soon as richness rose above some critical (presumed low) number. A third was that the pool of potential colonists should be much larger than the number of species that could coexist on the target islands.

If these conditions could be met by real animals or plants, then the predicted equilibrium number of species should indeed result. For arthropods colonizing extraordinarily simple mangrove islets, these conditions appear to be met. The defaunation experiment has some of the properties of a computer run: input a large array of arthropods known to exist in almost infinite assortments on mangroves, together with tiny islets, each providing but a single, unchanging food plant. The predicted equilibrium is achieved, and is demonstrated by both species–time and species–area data. A system of phytoplankters colonizing beakers of water also provides the requisite boundary conditions, and yields the predicted immigration–extinction equilibrium.

For most biota, most of the time, these stringent boundary conditions are unlikely to be met. For land plants they virtually never apply. Although ruderals and opportunists are removed from community prominence in early stages of succession by contest competition, their fate is more likely to be rarity than extinction. Landholding is the name of the game for plants, but breaks are always present in the landholdings of successful competitors, so that the ruderals can coexist with stress competitors in the same landscape.

For animals in even moderately diverse vegetation, the conditions for existence are bounded first by plants as food or shelter and second by predators restricting the size of populations in ways that allow increased diversity (see Chapters 14 and 16). Thus, if plant diversity is not related to the proposed immigration–extinction equilibrium, the chance seems poor that animal diversity should be so related.

The most provocative data consistent with the equilibrium model are species–area plots—of island birds, mammals on mountaintops, phytoplankton in lakes, pests on crops, and many others reviewed in the literature but not here (Simberloff 1974). Areas influence diversity in many ways. It is possible that processes other than immigration–extinction equilibria contribute to these correlations.

Preview to Chapter 22

Of the Tribe Attini: leaf-cutters take compost to their garden.

THE food web has been reinstated as a powerful conceptual tool in ecology. Following the growing realization of the importance of top predators for the structure of communities has come the concept of a cascade of effects running down food webs and impacting on almost all populations in a community. Development of this form of "cascade" concept has been particularly vigorous in aquatic biology. Yet it is the advent of modern computing that has given the food web its greatest advantage as an intellectual tool. By simulating food webs it is possible to experiment and explore generational change in ways that were essentially impracticable before. Population resilience can be explored in the context of food chains in which real populations live, instead of in isolation. The vexing question of to what extent community persistence is dependent on food web structure can at least be addressed. A model with remarkably simple assumptions successfully generates critical properties of real food webs. This is a cascade model in that it requires predatory interactions to be in a unidirectional series, not to be confused with cascade models of the effects of predation on lower trophic levels. Alternative dynamic models construct food webs as sets of Lotka–Volterra equations, with major categories assigned and parameters within categories generated at random. With these it is possible to explore the vital processes of community persistence like invasion and extinction. A first important finding of model exercises is that such important parameters as the strengths of species interactions are yet to be measured for most natural systems.

Chapter 22

Food Webs and Community Persistence

Community ecologists again concentrate much of their effort on the analysis of food webs. The food web, originally called the food cycle, was one of the earliest ideas of animal ecology, and in the writings of Charles Elton was used as a unifying principle for the discipline in its formative years. One direct result was Lindeman's "trophic–dynamic" aspects of ecology, on which the energy flow paradigm was based (see Chapters 1 and 2).

Nevertheless, most community studies in the early days of ecology were geographically based and strongly influenced by the patterns of plant communities on the ground. When pioneer ecologists questioned the organization of species into communities, they were more likely to analyze the distribution of plants (the animals being somehow appendages of the plant communities) than to look for their primary inspiration in patterns of eating and being eaten. A "biome" set by climate, or a more local "association" of intensely coevolved species, were the intellectual abstractions resulting from these patterns of shared habitats (see Chapters 18 and 19).

A food web approach to community analysis is different. Food webs are diagrams showing which species in a community interact. Originally, the necessary interaction took only the simple and stark form of fatal meetings when one organism eats another. In food webs of the modern analyst, however, the actual web of interactions is more complex than this, as competitors exclude or mutualists assist. Even a potential predator can give a backhanded assist, if it so preys upon a potential competitor that room is made for a newcomer to invade the community. Competition and mutualism are seen to work together with predation in sharing out the communal food supplies. Thus, comparing communities as alternative food webs means comparing the results of all the dynamic interactions between species.

This new view of food webs tackles on a community scale the fundamental issues of diversity, species separation, competitive exclusion, and the like. Invasion and extinction are seen as food web properties, depending far more on interactions with other organisms of the community than with physical necessities of life.

As data on real food webs have grown, so webs in different ecosystems can be compared in an attempt to find common constraints on resilience of populations, length of food chains, or even persistence of whole communities. Easy access to computing makes possible, through modeling, repeated trials of communal histories to isolate regularities from noise. The result is at least a new perceptive look at old and vexing questions of ecology (Pimm, 1991).

CASCADE THEORY IN AQUATIC BIOLOGY

In the language of fisheries, "piscivores" are predatory fish, like pike, that hunt other fish. "Planktivores" are fish that hunt grazing animals of the zooplankton (mostly invertebrates), and the grazing zooplankton are herbivores of the tiny phytoplankton. This classification recognizes four trophic levels: top predators, secondary predators, herbivores, and primary producers. Both animals and plants live floating in a lighted void, with nowhere to hide (see Chapter 26). As we now know, predators in the open waters of lakes can have devastating effects on the populations of their prey in some circumstances.

Aquatic biologists have long known that the presence or absence of predatory fish can have profound consequences on food webs. This knowledge was drawn on in a now-classic paper by Brooks and Dodson in 1965, who showed that whether or not the larger planktonic crustacea like water fleas (*Daphnia*) could be found in a New England lake depended on whether or not the lake held planktivorous fish. Fish, hunting the zooplankton by sight, could take out all prey whose size brought them within range of their visual acuity, leaving a clear field in which smaller crustaceans, or those partially transparent, could flourish (see Chapter 26).

That piscivorous fish could treat the planktivores even as planktivores could treat plankton has been shown by various histories of introductions. Influential was the study by Zaret and Paine (1973) of the introduction of the exotic piscivore *Cichla* into Lake Gatun, the large lake built to supply water to locks of the Panama Canal by the damming of a watershed. The arrival of *Cichla* was followed by the collapse of native fish populations, these being descendants of the fish fauna of the ponds and streams of the old watershed, many of them certainly planktivores (Table 22.1).

Studies like these suggested a cascade of effects running down the trophic levels. In a lake system with four trophic levels, piscivores should keep down planktivore populations. Because planktivores would then be comparatively rare, populations of their prey in the herbivore trophic level could be expected to be large. Finally, dense populations of herbivorous zooplankton would graze down the phytoplankton so that the standing crop of chlorophyll in the lake should be low (Figure 22.1).

Productivity at each trophic level depends both on biomass and resources available. Reduction of biomass by predation releases resources, so that productiv-

TABLE 22.1
Effect of Introduction of Predatory *Cichla ocellaris* on Fish Catches in Lake Gatun
(From Zaret and Paine, 1973.)

Family	Species	Without *Cichla* Number	With *Cichla* Number
Atherinidae	*Melaniris chagresi*	200	0
Characinidae	*Astyanax ruberrimus*	160	0
	Compsura gorgonae	120	0
	Hoplias microlepis	0	1
	Hyphessobrycon panamensis	2	0
	Pseudocheirodon affinis	7	0
	Roeboides guatemalensis	195	21
Cichlidae	*Aequidens coeruleopunctatus*	10	0
	Cichla ocellaris	0	14
	Cichlasoma maculicauda	7	36
	Neetroplus panamensis	4	0
Eleotridae	*Eleotris pisonis*	4	99
	Gobiomorus dormitor	42	10
Poeciliidae	*Gambusia nicaraguagensis*	22	0
	Poecilia mexicana	17	2
Other (25)		0	1

ity per unit biomass tends to be inversely related to predation. Productivity in each trophic level is thus expected to be maximized at intermediate levels of predation, as shown in Figure 22.2.

This was the cascade hypothesis of aquatic biology, an alternation of predation pressures between pairs of trophic levels cascading down food chains. A key aspect of the hypothesis is that properties of the highest trophic level are felt all through the food webs, right down to the

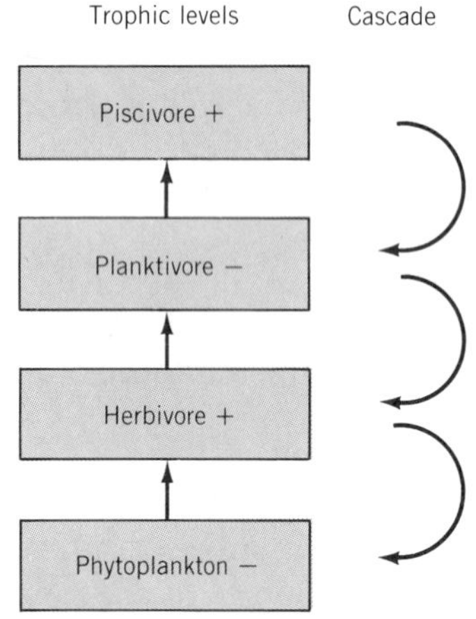

FIGURE 22.1
Postulated cascade in aquatic food web.

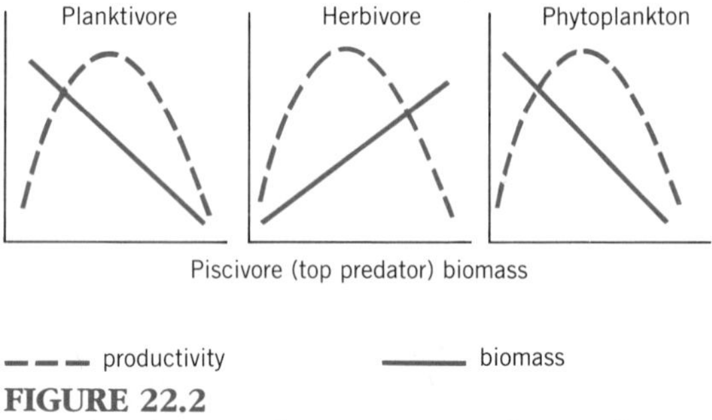

FIGURE 22.2
Postulated cascadel consequences for productivity and biomass in aquatic food webs.
In each trophic level productivity is maximized at intermediate biomass, when resources are optimal in relation to predation rate. (Based on Carpenter et al., 1985.)

primary producers. This is a top-down hypothesis. Fish predation and herbivory are seen as regulators of food webs, and hence of community structure.

The cascade hypothesis has been tested in lakes by various exclusion experiments. Plastic enclosures suspended in a lake can be used to keep fish out or to keep them in. Fish can be removed from small ponds by rotonone poisoning, which does not harm the invertebrate herbivores. Reservoirs with and without fish can be compared. An extensive literature of such studies now shows that the cascade hypothesis stands these tests well for the fish trophic levels (Carpenter et al., 1985). Prey fish biomass declines as a function of predator biomass, as required by the hypothesis. The abundance of larger crustacea of the herbivore trophic level increases as planktivore density decreases in all the studies.

The cascade also has effects inside the lower trophic levels, particularly in changing ratios of organisms of different sizes. Removal of large zooplankton by planktivorous fish results in increasing populations of small crustaceans or the even smaller rotifers. The ratios of these size classes of herbivores are further influenced when invertebrate planktivores like the large carnivorous copepod *Heterocope* are themselves preyed on by planktivorous fish (Figure 22.3). The changing sizes of

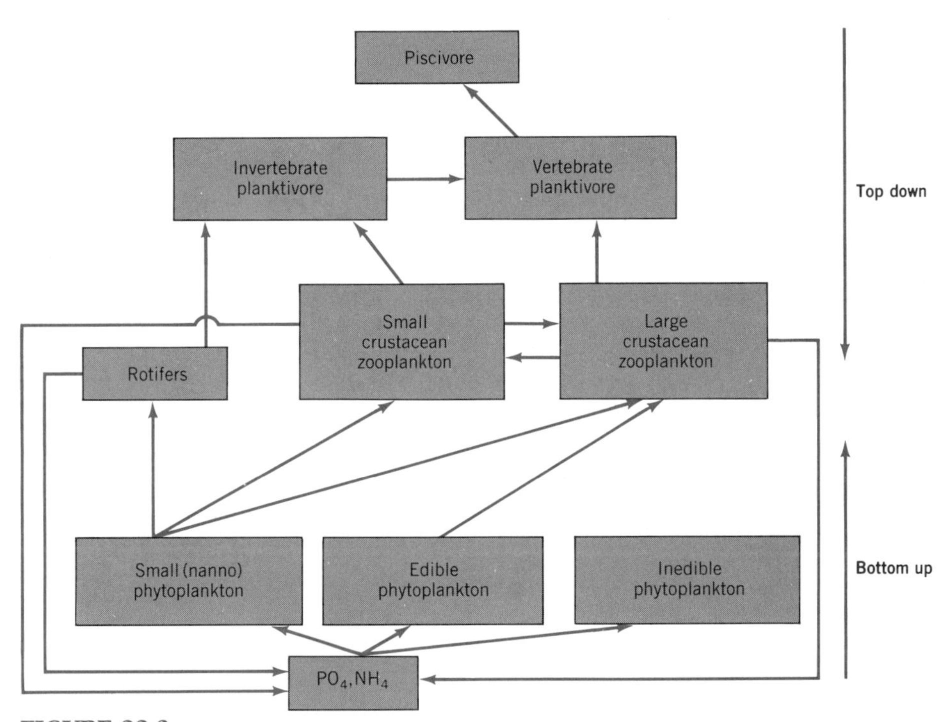

FIGURE 22.3
Outline of the food web in a lake.
A top-down cascade is expected in the upper trophic levels. In the lower trophic levels the release of nutrients by grazing down of sections of the phytoplankton can result in rapid production. These bottom-up effects can overwhelm the top-down effects of the predatory cascade. (Modified from a figure of Carpenter et al., 1985.)

herbivores then alter the rate of predation between the smallest algae (nannoplankton) and the larger edible algae.

Primary production in lakes, however, is nutrient-driven (see Chapter 24), with supplies of phosphate and ammonium ions being decisive. This might suggest that production at the base trophic level, the plants, can be only minimally influenced by the top-down trophic cascade, particularly because plant death by grazing should release nutrients for further plant growth. Yet a compelling study of 66 lakes known to have constrained phosphate supplies found that only 48% of variance in productivity could be accounted for as a function of nutrients alone (Schindler, 1978). It is likely that the rest of the variance is determined by trophic cascades resulting from the different food webs present.

In more productive, eutrophic lakes (see Chapter 26) the importance of nutrients is likely to be particularly important. This is reflected in the "bottom-up: top-down" (BU:TD) model in which the community structure of the food web is driven both by cascading predation from the top and by nutrient status from the bottom (Figure 22.3).

An opportunity to test the BU:TD model was provided by a winter kill of fish in eutrophic Lake St. George in Ontario, Canada, where the consequences of the fish kill were followed for seven years by McQueen et al. (1989). Following destruction of both piscivore and planktivore fish populations by the winter kill, zooplankton biomass immediately increased. Two years later planktivore, but not piscivore, populations had grown back to extreme densities, with consequent reduction in zooplankton. Another two years were required before the piscivore populations rose to lower the planktivore populations again.

These changes in the populations of the larger animals of Lake St. George were all of the kind predicted by top-down cascade theory, the details of the events being dependent on lags introduced by the different growth or turnover times of the organisms involved. But the data for producer biomass (measured as chlorophyll *a*) showed no year-to-year correlation with zooplankter biomass, implying that primary production was largely independent of predation in this system.

COHEN'S CASCADEL MODEL FOR ALL FOOD WEBS

Joel Cohen has used data from food webs in a search for fundamental organizing principles behind the structures of communities. He introduced the possibility with a survey of 29 food webs in 1978, after ten years of work. Partly as a response to Cohen's study, many more descriptions of food webs have been published since then so that Cohen's latest review is based on 113 food webs from many parts of the world, and from systems ranging from the open sea to continental plains (Cohen et al., 1990).

Cohen renders the published data for food webs into matrix form, as in Table 22.2 for a willow forest in Manitoba. Prey are in the vertical column and predators are along the horizontal. Some organisms appear as both prey and predators, as indeed must most organisms in the intermediate trophic levels. The natural history data of the published webs are reduced to simple alternatives, numeral 1 for "the predator in this column does eat the prey in this row," 0 for "the predator in this column does not eat the prey in this row," and −1 to express uncertainty: "the predator

TABLE 22.2
Food Web Matrix for a Willow Salix Forest in Manitoba, Canada
(From Cohen et al., 1990.)

		Predators							
		2	3	4	7	8	9	10	12
Prey	1	1	0	0	0	0	−1	0	0
	5	1	0	0	0	1	−1	0	0
	6	1	0	0	0	0	1	0	0
	7	0	1	1	0	0	0	−1	0
	8	0	−1	0	1	0	0	1	0
	9	0	0	1	1	0	0	1	0
	10	0	0	0	0	0	0	0	1
	11	0	0	0	0	0	0	1	0

1 *Salix discolor*
2 *Galerucella decora*
3 redwinged blackbird, bronze grackle, song sparrow
4 Maryland yellowthroat, yellow warbler, song sparrow
5 *Salix petiolaris*
6 *Salix longifolia*
7 spiders
8 insects, *Pontania petiolaridis*, Collembola
9 insects, *Disyonicha quinquevitata*, Collembola
10 *Rana pipiens*
11 snails
12 garter snake

in this column probably eats the prey in this row but the observations were unclear." This matrix expression of natural history data can be compared with the original food web, first published in 1930 and given as Figure 22.4.

In matrix form, data for utterly different food webs, on completely different scales, can be set in common language and can be read by machine. The data for Cohen's 113 food webs illustrate some generalizations about food webs that have come to be accepted. Cohen calls them the five "laws," but they can be collapsed into three as follows:

1. Cycles are rare: that is, the path of eating and being eaten passes straight down the web so that A eats B eats C etc. and on down the line.
2. Food chains are short. The median number of links in the 113 food webs was 4 and the longest 10 (in only one of the webs).
3. Scale invariance. The proportion of species in each trophic level is independent of the number of species in the web. This is to say that highly diverse communities have the same proportions of their total number of species in the predator and herbivore trophic levels. The total number of links is also proportional to the number of species, showing more independent and fatal meetings in a highly diverse web than in a simple one.

Cohen proceeds to make models that can be used to test hypothetical species relationships in food webs, discarding models that do not generate the requirements of the above "laws." One particular set of postulates of community organization (and as yet only one of those tried) is successful in doing this. Cohen calls this his Cascade model, and is careful to point out that this differs in important ways from the cascade model of aquatic biologists.

Cohen's cascade model has no "bottom-up" feedbacks from nutrient limitations of the primary producers. It is a top-down model only, and it respects several crucial constraints. The model requires as inputs the number of species in the food web and the expected number of interactions. And it requires that the species be ranked so that a higher ranked predator can be a predator on any species below in rank order but on none above. This is the cascade in the model, not so much a cascade of effects, as in the aquatic biologist's model,

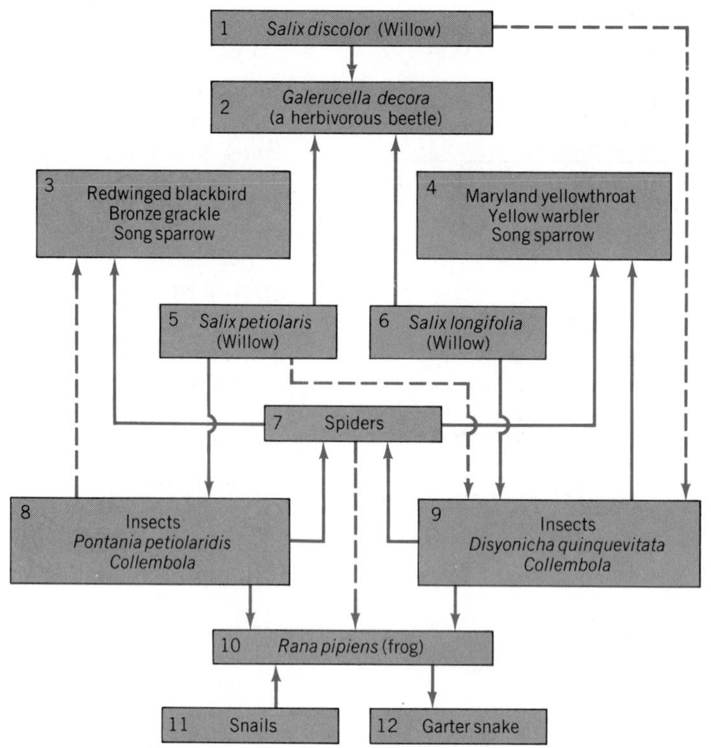

FIGURE 22.4
Food web flow diagram on which Table 22.2 is based.
(Modified from Cohen 1978.)

but rather a cascade of relationships. Beyond this, the model simply lets the organisms feed at random on the species ranked below them and within assigned groupings, or "connectance."

The data base (number of species, number of interactions, etc.) from a hundred or so of the matrix food webs is then applied to this model, with the result that many of the organizational properties of the original food webs appear to be duplicated. In particular, the number of species at the base of each food web, the number of species that are prey to some species and predators to others, the predator-to-prey ratio in the food web, and the average food chain length are scored correctly. This is a remarkable feat for a model based on very few assumptions.

Fresh from this triumphant simulation of natural systems, however, Cohen himself points out some fundamental intellectual problems. Why should there be a cascade rule? And what determines the number of species in the community, together with the numbers of their interactions? The cascade model merely shows what happens after the numbers and the scale of their interactions are set. As Pimm (1991) tells us, the success of the cascade model makes an understanding of the processes of invasion and extinction that set diversities within food webs a more pressing research need than ever.

RESILIENCE AND PERSISTENCE IN FOOD WEBS

Resilience of a species population is the rate of its rebound following heavy loss. Insect species are usually more resilient than vertebrates because of their high fecundity. Wide fluctuations in many insect population histories record their resilience.

Persistence is a community characteristic, being essentially how long a community lasts intact. A community fails to persist both when it loses species and when it gains species.

Resilience and persistence between them constrain most of the properties implied by the idea of stability of a community of animals and plants. Intuitively, both properties can be suspected of being dependent on the kind of food web on which the community is structured. The resilience of a population, for instance, must be determined in part on its position in a food chain as well as on its own fecundity and intrinsic rate of increase.

A population of effective predators should reduce resilience by removing a portion of the reproductive effort. Likewise the locking up of nutrients in large standing crop biomass should also lower resilience of plant populations by reducing rates of growth and reproduction.

Persistence should be a function of how easy a community is to invade, which is itself a function of potential competitors or predators already present in the community. Like resilience, persistence should also be dependent on resources already locked up in standing crop biomass.

Those who model food webs thus can hope to address from a fresh perspective the most fundamental ecological questions of diversity and stability (Lawton, 1989; Pimm, 1991; Pimm et al., 1991). Cohen's cascade model is one approach. An alternative favored by Pimm (1982, 1990) and others is to construct artificial food webs with systems of Lotka–Volterra equations (see Chapters 8 to 10). The resulting mathematical descriptions of interactions are certainly grossly oversimplified, essentially treating interactions in the complex daily business of living as little more than a series of meetings, like the bouncing together of molecules in an ideal gas. Simplification, however, is the purpose of models. If they can duplicate complex nature, then the chance must be good that their simplifying assumptions suggest general processes underlying seemingly complex reality.

In model runs, ranges of parameters can be set using ecological common sense, but details within those ranges are chosen at random within an assumed uniform distribution. This inevitably produces many model webs that crash in ways not permitted in the real world, as for instance when numbers of some populations become negative. These "unfeasible" models can be screened out by a sort of computational natural selection with computer programs that choose only feasible (no negative populations) and stable systems. This process is so laborious that it is the main impediment to simulating larger food webs, such that model communities with 12 species are called "large" in the trade (Pimm, 1990).

The models can be used to examine processes like the invasion of complex communities that are almost intractable to other approaches. Figure 22.5 shows results of repeated attempts to invade model communities with two levels of connectance (strongly interacting species: open squares; weakly interacting species: closed squares). The community of weakly interacting species was more easily invaded and ended up accumulating more species.

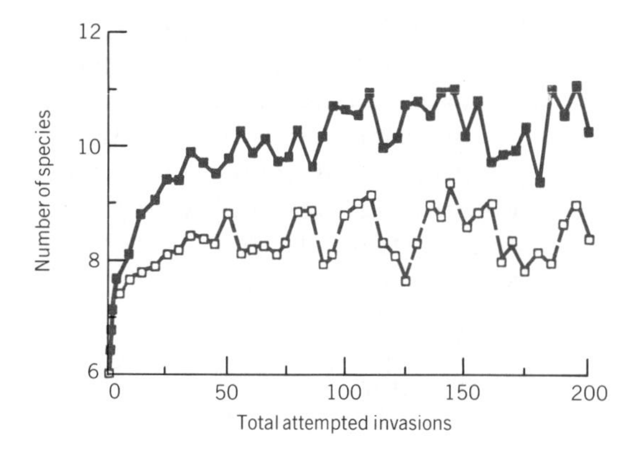

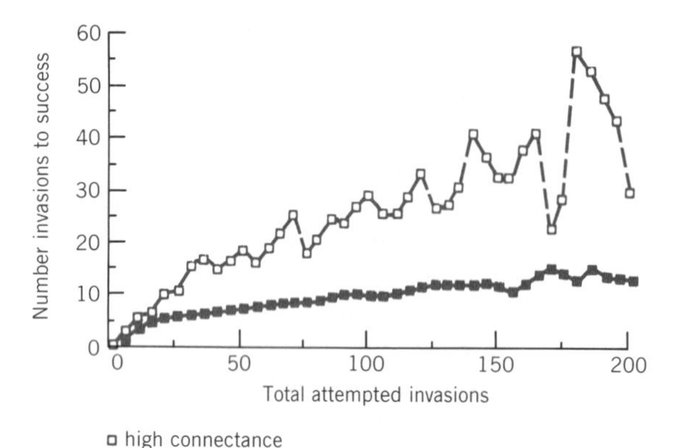

FIGURE 22.5
Simulations of invasions by a dynamic model of a food web.
Each simulation was run with high connectance (75%) and low connectance (25%) where connectance is the ratio of actual interaction to possible interactions. (For explanation, see text; from Pimm 1991.)

If invasion of real communities proceeds as in these models, then we are told something about community persistence. Invasion is evidently more difficult when species interact strongly (high connectance). Does this imply that real communities are more persistent if species interact strongly? We do not know, but the observation can be kept in mind.

One of the best "experiments" on invasions is the deliberate introduction of exotic birds to Hawaii over many decades, by what a modern naturalist would think of as a mischievous society set up for that deliberate purpose (it even collected money from school children for its "charitable" activities). Birds were released repeatedly until they were established, with the result that only exotic birds now exist in the agricultural regions of the lower slopes, the native birds having been entirely displaced. In the uplands, however, the invasions were much less successful and the native communities remain. (Moulton and Pimm, 1983). Does this mean that connectance remained strong in the less disturbed uplands, thus letting communities persist in the face of invasion? Without real data, we have no right to say so. Recent data from a different system actually suggest that the degrees of connectance accepted in food web models are in fact unrealistically high.

Paine (1992) has continued his studies on the grazing system off the Washington coast, in which he and his students made their seminal demonstration of the consequence for competition and diversity of consumers in communities under predation by a rapacious starfish (see Chapter 14). These were the studies that perhaps first set ecologists to thinking about cascades of effects tumbling down from the activities of a predator. Now Paine has investigated the seriousness of grazing by chitons and urchins on the species of brown algae at the base of his food web. Species interactions are low, lower than would result from the procedures used to assign parameters in dynamic models of the Lotka–Volterra type (Lawton, 1992).

Examining resilience and persistence in model food webs has the advantage over natural history in that time scales can be longer and experiments can be repeated until the results are statistically respectable. Modeling is vital for repeatability and for compressing the times of experiments. These are early days in modeling. Already the effort shows some of the real data that must be collected if an understanding of the limits to community persistence is to be gained—like the weakness of interactions at the bases of food chains uncovered by Paine (1992).

THE PROBLEM OF TIME SCALES

Population ecologists have a time scale of their own. Community ecologists modeling food webs comment on how short this population ecology time scale is compared to community process (Pimm, 1990). Population events quickly achieved in isolation are damped, prolonged, or altered when taking place in a matrix of interacting species within a food web. A clear example is how resilience depends on the populations of predators or competitors in the food web as well as on the innate fecundity of the species. A community ecologist must think in terms of many generations to achieve effects that a population model can produce in few.

Community time scales may well be the most important for practical ecologists to master for it is on community time scales that unfolding human history is measured. Do we design reserves to save species? If so we must understand persistence and resilience in community time. Invasions and extinctions on the contemporary earth proceed in the relict communities that expanding human populations allow to persist. They must be understood, and manipulated, in community time. An alliance between food web modelers and field naturalists may be the last best hope of conservation biology.

When addressing the grand questions of the distribution and abundance of life on the pristine earth before the modern human crisis, however, quite different time scales prevail. The ten millennia since the last ice age are an immensity of time to food web modelers, but they are a mere 40 lifetimes for a tree. And in those 40 lifetimes all the tree species of the earth have probably lived in many different communities. On this time scale of evolutionary existence communities probably have no persistence at all.

PART FOUR

SYSTEM

Preview to Chapter 23

Nutrients in an Alaskan forest: leaves to litter to roots to trunks to leaves.

LIFE concentrates selected elements of the periodic table, which elements then become critical resources to be sought out and conserved by individual plants. Even without life to select out nutrients, an empty habitat will maintain a minimal supply, dependent on inputs and outputs. Plant cover increases the nutrient stock immensely, particularly by increasing storage capacity in soil and reducing outputs. Watershed studies at Hubbard Brook show that a cover of temperate forest prevents nearly all particulate losses from erosion. With the forest intact, nutrient concentrations in soil water are constant, and nutrients accumulate in the ecosystem. Nearly half of incident solar energy goes to evaporate water from the leaves. These results were confirmed by a massive forest destruction experiment involving removal of all vegetation from an entire watershed. The ecosystem properties revealed by these studies result entirely from adaptations of individual organisms, but when individual organisms act in parallel ways the cumulative effect can transform the conditions for life on the whole temperate landscape. Similar adaptations in plants of the tropical rain forest have different results because decomposers in rain forest ecosystems remove detrital organic matter as fast as it is produced.

Chapter 23

Ecosystems Defined by Cycles

The cosmic abundance of the elements is quite different from the abundance of elements in living things. Burning any sample of life—a tree, a mouse, or a box full of bacteria—yields ashes that are remarkably similar to each other chemically, yet all with elemental compositions quite unlike the mineral crust of the earth or a solar flare (Table 23.1). Living things, therefore, select from the elements about them. They have little use for really abundant elements like silicon (30% of the earth's crust and used effectively only by diatoms), aluminum, magnesium, and iron, but concentrate the relatively rare elements phosphorus, potassium, and calcium.

Life does its collecting of most elements in solution. Carbon, of course, enters ecosystems as gas and thence is transformed along food chains as solid compounds of reduced carbon. Nitrogen can be imported as a gas by nitrogen-fixing bacteria and later vented back to the atmosphere. Likewise, oxygen is imported as a gas used for respiration, then returned to the atmosphere by the photolytic destruction of water in photosynthesis. These processes can be seen to regulate the atmosphere bathing terrestrial ecosystems (see Chapter 28). Most of the elements used by ecosystems, however, are acquired in solution, requiring each organism in an ecosystem to process water and manipulate solution equilibria.

The effects of these individual efforts combine, with profound consequences for the properties of a system or landscape. But beware the language trap that so easily follows—the trap of implying that the ecosystem "does" things, as if it were a living entity. Systems are not living things, even when they are ecosystems. Many ecosystem properties result, however, because the living things in them have common interests and common responses.

All individuals of all species seek out,

TABLE 23.1
Relative Abundance of Principal Elements
*Data as percentage dry weight. Human and alfalfa plant (*Medicago sativa*) data from Rankama and Sahama (1950); remainder from Zajic (1969).*

Element	Adult Human	Alfalfa	Solar Atmosphere	Total Earth	Crust of Earth	Seawater
Hydrogen	6.60	5.54	53.0	Trace	0.14	10.8
Helium	—	—	42.0	Trace	0.00000003	0.0000000005
Boron	—	—	Trace	Trace	0.0003	0.0005
Carbon	48.43	45.37	0.012	Trace	0.03	0.003
Nitrogen	12.85	3.30	0.031	Trace	0.005	0.00005
Oxygen	23.70	41.04	4.7	28.0	46.6	87.5
Sodium	0.65	0.16	0.0024	0.14	2.8	1.05
Magnesium	—	—	0.043	17.0	2.1	0.13
Aluminum	—	—	0.0031	0.4	8.1	0.000001
Silicon	—	—	0.029	13.0	27.7	0.0003
Phosphorus	1.58	0.28	Trace	0.03	0.12	0.000007
Sulfur	1.60	0.44	0.014	2.7	0.05	0.09
Chlorine	0.45	0.28	Trace	Trace	0.02	1.90
Potassium	0.55	0.91	0.00033	0.07	2.6	0.04
Calcium	3.45	2.31	0.0036	0.61	3.6	0.04
Vanadium	—	—	0.000031	Trace	0.02	0.0000002
Manganese	0.10	0.33	0.00086	0.09	0.10	0.0000002
Iron	—	—	0.167	35.0	5.0	0.000001
Cobalt	—	—	0.00034	0.20	0.002	0.00000005
Nickel	—	—	0.0029	2.7	0.01	0.0000002
Copper	—	—	0.000058	Trace	0.01	0.0000003
Zinc	—	—	0.00021	Trace	0.01	0.000001
Molybdenum	—	—	Trace	Trace	0.0015	0.000001
Iodine	—	—	Trace	Trace	0.00003	0.000006

and conserve, the same relatively scarce elements of the periodic table. An ecosystem-wide effect is inevitable. Thus it is honest and true to say that nutrients are conserved, and hence recycled, in ecosystems. But the language ecologists use was invented for the discourse of everyday life, for commerce, for training children, for poetry. Inevitably ecologists make an ecosystem the subject of their thoughts, saying, for example, "The ecosystem conserves" . . . or "the ecosystem cycles." This usage is natural; indeed, probably inescapable. But it is intellectually perilous. An ecosystem "does" nothing at all, though an ecosystem does have properties.

Essential nutrients are processed and cycled in large masses by ecosystems, particularly the N, P, and K (nitrogen, phosphorus, and potassium) well-known to farmers as fertilizer. They are moved in solution—nitrogen in the ammonium ion (NH_4^+), phosphorus as phosphate (PO_4^-), and potassium as the elemental cation (K^+). Work must be done to concentrate and move these radicals. Furthermore, the fact that all are soluble in water means that all must be kept against parts of the hydrologic cycle that tend to carry them away.

The biota has a relatively tiny flux of energy at its command to do this work—

some small portion of the 1 to 2% of solar energy trapped by the photosynthesis of an ecosystem—but against this is set the huge energy flux that goes to drive the hydrologic cycle. A very large portion of the total solar energy incident on any ecosystem (except a desert) goes to heat or transport water. The biota has at its collective disposal only the less than 2% of solar energy transduced by photosynthesis. The energy odds are thus heavily weighted against the life forms of an ecosystem. Nevertheless, the collective activities of individual organisms in some ecosystems do modify even hydrologic cycles, giving to a vegetated landscape properties quite different from those of bare ground.

CYCLES AND STATES WITHOUT VEGETATION

The hydrologic cycle for a world without vegetation is shown in Figure 23.1. The essence of this system is that water moves between large reservoirs formed by the oceans, surface water (including glaciers), and underground water. The actual proportion of water in transit at any one time is very small (Table 23.2), since all rivers and the atmosphere together hold only 0.0011% of the water of the biosphere. The rest is held in the various reservoirs, particularly oceans, glaciers, and deep in the ground. Only soil water is transported

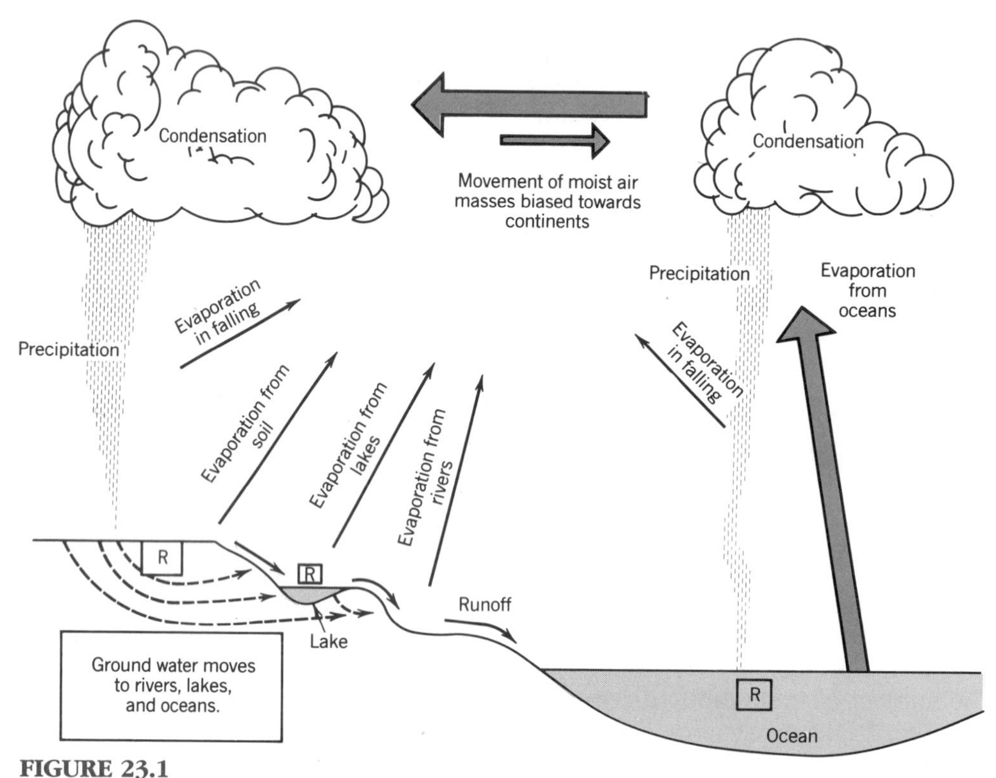

FIGURE 23.1
The hydrologic cycle in the absence of vegetation.
Three principal reservoirs of liquid water are marked "R." These reservoirs are the oceans, lakes, and the pore spaces of soils and surface rocks. The diagram omits the second largest of the earth's water reservoirs, glacier ice (see Table 23.2).

TABLE 23.2
Water Budget for the Biosphere
From Strahler (1969), compiled from data of R. L. Nace.

Reservoir	Volume Water ($\times 10^6$ km^3)	% Total
World oceans	1322.0	97.21
Glacier ice	29.2	2.15
Groundwater	8.4	0.62
Soil water	0.067	0.005
Freshwater lakes	0.125	0.009
Inland seas and salt lakes	0.104	0.008
Rivers and streams	0.001	0.0001
Atmosphere (clouds and vapor)	0.013	0.001

much by life, and this minor reservoir holds only 0.005% of the total reserve.

Even in a plantless world some water would be retained in the surface soil as the ground was wetted, as pore spaces in the soil were filled with water held by surface tension in capillaries, and as some water was absorbed on minerals, notably clays (see Chapter 25). The amount of water actually held in the soil of a plantless land would depend on the frequency and amount of rain and on the drainage.

The mass of water usually present must be a function of the rate of water input and the rate of water loss. A **steady state** would tend to be established with water constantly entering and leaving the soil reservoir. In seasons of high rainfall the steady-state volume of water in the soil would be higher than in seasons of low rainfall. With intermittent rain the reservoir would fluctuate.

Water entering and leaving a patch of mineral soil contains dissolved nutrients, acquired either from the air or by solution of minerals in passage through the ground. The reservoir of water in the soil is, therefore, also a nutrient reservoir (Figure 23.2). Where rainfall is constant enough for a steady state to be established, nutrient reserves should also be at steady state. The actual nutrient reservoir will depend on the nutrient-retaining power of the soil, and this will not be the same as the water-retaining powers. Many nutrients, notably cations, are selectively adsorbed by clays, organic matter, and minerals in the soil.

When plants are added to the habitat, three essential changes follow. First, water is pumped through transpiration streams instead of being allowed passively to evaporate or to flow away to groundwater or to streams. Second, the nutrient concentration of soil water is actively changed by plants acting as ionic pumps. Third, the nutrient reservoir is changed by the addition of both living and dead biomass. These three processes occur in different arrangements in habitats of different biomes. Together they account for the apparently ordered storage and cycling of nutrients in many ecosystems.

NUTRIENT CYCLING IN A TEMPERATE ECOSYSTEM

In a typical vegetated habitat in the temperate parts of the earth, the actual nutrient reservoir is augmented in various ways. Instead of just the dissolved nutrients and those attached to the minerals of surface rocks of the unvegetated habitat, temperate ecosystems hold the following:

1. Nutrients in living plants
2. Nutrients in living animals
3. Nutrients in detritus
4. Nutrients in soil humus

In addition to these supplies, extra nutrients resulting from covering the land with

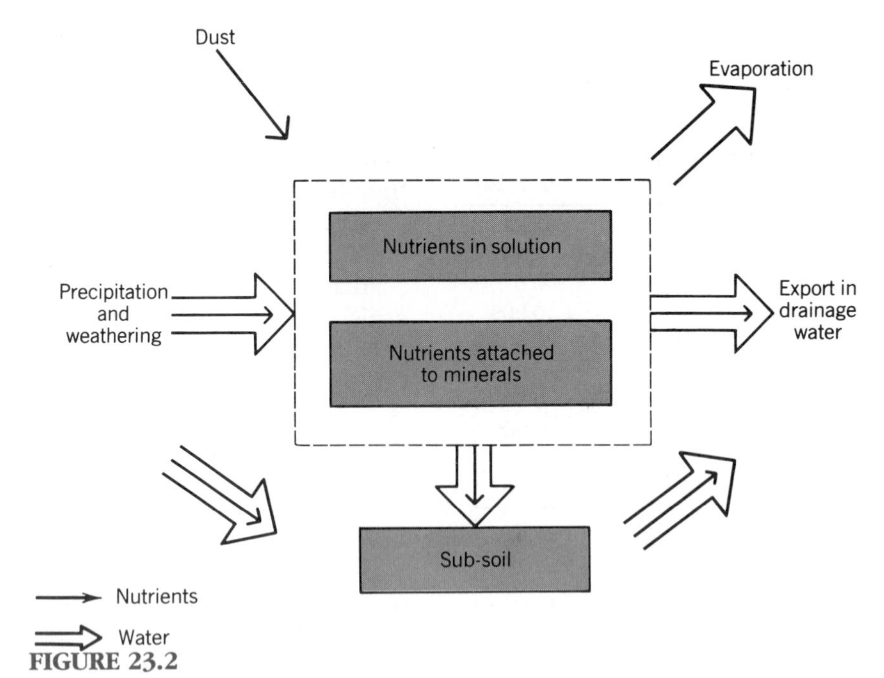

FIGURE 23.2

Water and nutrients to a lifeless habitat.
If precipitation is roughly constant, the nutrient reservoir will tend to a steady state, even without plants. The potential habitat will hold nutrients in simple solution in its nutrient reservoir and absorbed or adsorbed to minerals.

plants are held in the soil. Finely divided mineral matter is mixed into aggregates with soil humus, and the finely divided mineral matter may itself be increased in the soil-forming process. The vegetated habitat thus also holds

5. Nutrients adsorbed to soil humus
6. Nutrients adsorbed to additional clay minerals

Each of these reserves represents extra capacity. Even if the input and output of water and nutrients to and from the habitat were roughly the same as in a plantless plot, the steady-state supply in a vegetated habitat would be larger. But in fact the plants increase the inputs also. They extract nutrients from water deep in the subsoil and provide a substrate for the bacterial fixing of nitrogen. The flux of nutrients through such a temperate system is illustrated in Figure 23.3.

The dead organic matter important to soil structure is the mass of net primary production not eaten by herbivores but left to decay (see Chapter 24). Much of this material has been made as indestructible as possible by the plants, very largely as a defense against herbivores (see Chapter 15). Cellulose, lignin, and even the chitin of animals is jettisoned and buried, and is very slow to decay. It is this produce of the living part of the system that has so dramatic an effect on habitats. Habitat enrichment is thus a side effect of properties of individual animals and plants preserved by natural selection because they pro-

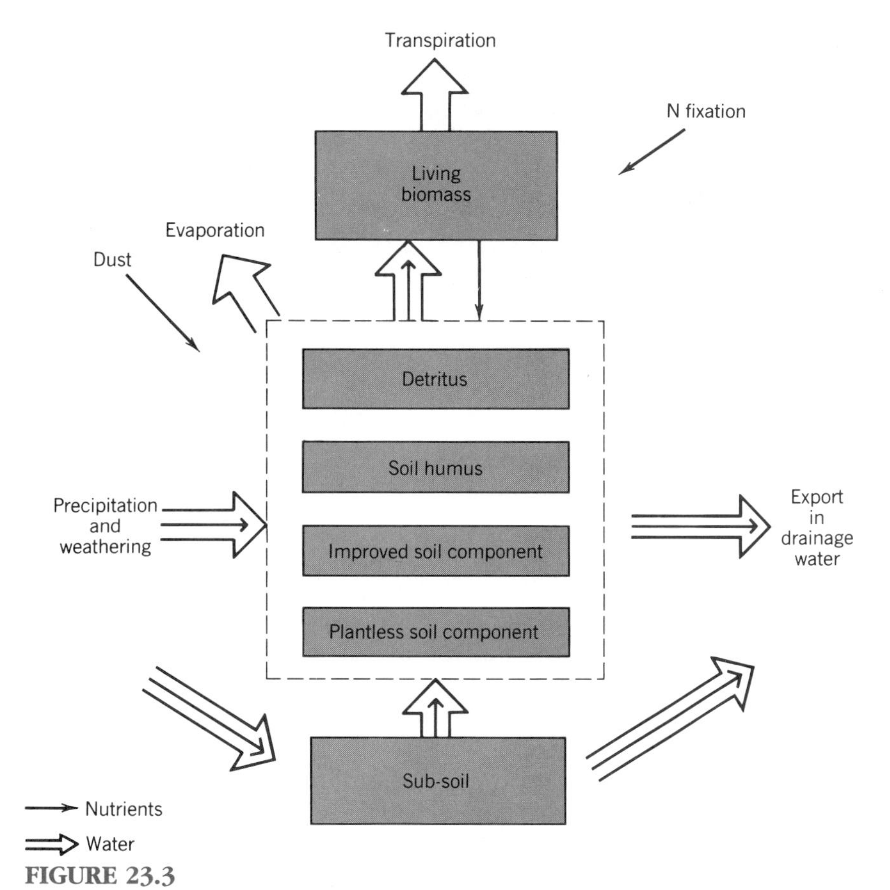

FIGURE 23.3
Nutrient flux and storage in a temperate ecosystem.
The system differs from the unvegetated habitat of Figure 23.2 by having vastly increased storage capacity for nutrients and by diverting the flow of water through the transpiration stream.

moted survival and reproduction of individuals.

The indestructible remnants of the biota in the litter do not themselves represent much in the way of nutrient reserve: cellulose and other polysaccharides, for instance, are mere arrangements of carbon, hydrogen, and oxygen and are not usually nutrient-rich. Their importance lies partly in providing soil structures in which efficient nutrient accretion and cycling is possible (Table 23.3).

A WATERSHED TO MEASURE NUTRIENT BUDGETS

A well-chosen watershed can be used to measure nutrient flows in and out of a large piece of real estate. Moreover, the measures also allow calculation of the energetics of water use in terrestrial ecosystems.

The watershed technique was first developed for forests of eastern North Amer-

TABLE 23.3
Nutrient Budget of a Belgian Oak Wood Compared with Soil Reserves
These data suggest a frugal use of the potential reservoir of nutrients held in the soil. This is in keeping with the observation that soils of temperate latitudes tend to have large reservoirs of nutrients in organic matter and clays. The high calcium content of the soil reflects calcareous parent material. (Data from Duvigneaud and Denaeyer-De Somet, 1970.)

	K	Ca	Mg	N	P	S
		In plants (kg/ha)				
Retained (increment)	16	74	5.6	30	2.2	4.4
Returned (losses)	53	127	13	62	4.7	8.6
Uptake	69	201	18.6	92	6.9	13
		Soil (top 40 cm)				
Exchangeable (kg/ha)	157	13,600	151	—	—	—
Total (metric tons/ha)	2.68	133	6.46	4.48	0.92	—

ica in New Hampshire, in the drainage basin of the Hubbard Brook (Figure 23.4). Each of the smaller tributary streams of Hubbard Brook drains a watershed that is precisely defined by heights of land. Each of the watersheds is underlain by impermeable bedrock, an important feature because it means that virtually no water or nutrients depart from the ecosystems by deep penetration and groundwater flow. Monitoring the flow of water in the drainage stream then accounts for all the liquid water leaving the ecosystem. This is done by damming each stream and making the water pass over a notched weir. Regular sampling of this water yields a measure of all the nutrients leaving each ecosystem in solution, as well as the volume of water in which they are transported.

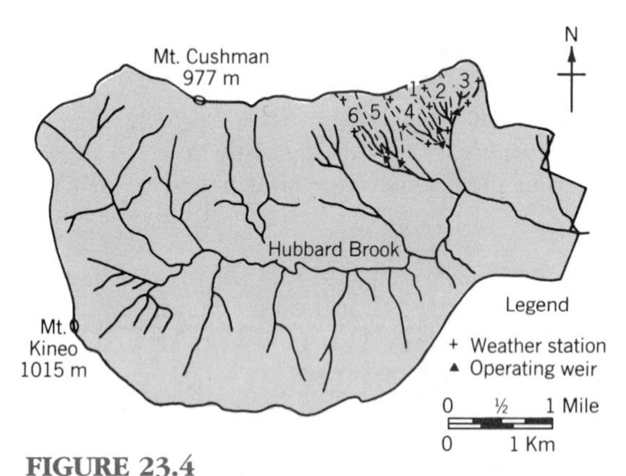

FIGURE 23.4
Watershed ecosystems at Hubbard Brook.
The six watersheds used for nutrient studies are shown by dashed lines. Each little watershed has its own stream and is divided from the next by steep terrain. A weir on each stream (indicated by a triangle) provides a site to sample outgoing water. Weather stations (crosses) are sites for sampling incoming water. (From Bormann and Likens, 1971.)

Total water and nutrient inputs to each watershed are easily measured with an array of rain gauges and dust traps, set out at the rate of one for each 1.29 hectares (Likens et al., 1977). Stocks of nutrients and organic matter can, of course, be measured directly, as can rates of turnover in plants and soil. Thus the use of a watershed underlain with impermeable bedrock lets the working of large natural ecosystems be studied directly. When Bormann and Likens (1967) organized the now classic study of the Hubbard Brook forests, they were at last giving substance to the scale of enquiry in Tansley's (1935) admonition that we should study whole patches of the earth as what he called "ecosystems."

The valley bottoms of the watersheds have forests that can be assigned to beech–maple associations (see Chapter 19), being dominated by sugar maple (*Acer saccharum*) and American beech (*Fagus grandifolia*). There is also yellow birch (*Betula allegheniensis*) and, especially toward the tops of the ridges, red spruce (*Picea rubens*) and balsam fir (*Abies balsamea*). The soils are alfisols (gray-brown podzolic), with well-leached "A_2 horizons" (see Chapter 25). The forest is second growth but the land has not been farmed, so the soils are virgin with well-developed structures. A cover of snow and litter prevents the soils from freezing despite the cold New England winters. Data from this forest, therefore, should be roughly comparable to that of European forests and the temperate deciduous biome generally.

The sampling procedure at Hubbard Brook is summarized in Figure 23.5. The important revelations of these measurements are described in the following sections.

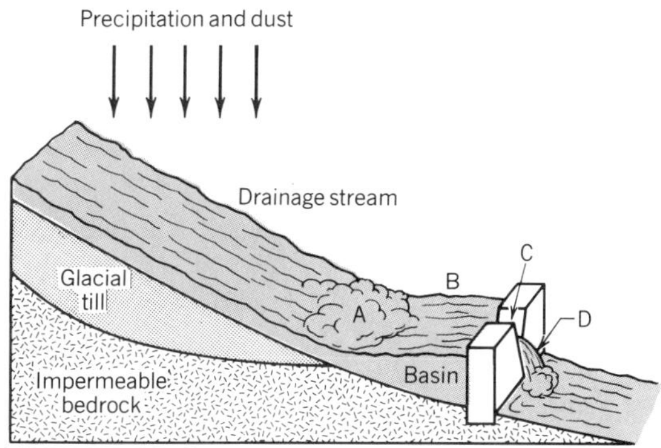

FIGURE 23.5
Nutrient measurements at Hubbard Brook. Nutrients in precipitation and dust are measured in grids of collectors. Samples for nutrients in solution leaving the ecosystem are taken at A, before the water enters the settling basin in water backed up behind the notched weir. Nutrients leaving the ecosystem as particulate matter are collected as they settle in the basin at B, and by millipore filtration and screening of samples going over the weir at D. Rate of water loss is measured at C. (Modified from Bormann et al., 1974.)

Energetics of Water Pumping

Evapotranspiration is measured easily as the difference between precipitation and stream water leaving each watershed over its dam. Over five years, precipitation varied between 95 and 142 cm, but evapotranspiration varied only between 46 and 52 cm (Table 23.4). Evapotranspiration, therefore, was constant from year to year, and quite independent of precipitation.

The water pumped each year in the transpiration streams is the cooling water whose evaporation serves to regulate leaf temperatures to maintain effective operating temperatures despite the heat loads that are inseparable from their function as solar traps (see Chapter 5). Similar evapotranspiration every year seems reasonable, therefore, granted that incident solar heat was the same. Possibly the small range in evapotranspiration between years reflects more or less sunlight received because cloudiness varies rather than different rainfall.

The energy required to evaporate the measured mass of transpired water each year is easily calculated as representing 42% of the solar radiation received by the forest during the growing season (Bormann, 1976). This is one of the more spectacular ecological discoveries of a lifetime. Ecologists are used to the somewhat depressing reality that vegetation subverts less than 2% of solar energy through photosynthesis. But forests can make use of another 42% of the total, more or less, for powering their transpiration streams. An energy budget this large makes the mod-

TABLE 23.4
Precipitation and Evapotranspiration at Hubbard Brook
Although annual precipitation varies widely, the evapotranspiration is nearly constant. (From Likens et al., 1970.)

Water Year (June 1–May 31)	Precipitation (cm)	Stream Outflow (cm)	Evapotranspiration (cm)
1963-64	117.1	67.7	49.4
1964-65	94.9	48.8	46.1
1965-66	124.5	72.7	51.8
1966-67	132.5	80.6	51.9
1967-68	141.8	89.4	52.4
Average	122.2	71.8	50.4

erating influence of a forest on landscapes entirely reasonable.

Nutrient Concentrations Remain Constant Under the Living Forest

The ionic content of the water coming over the Hubbard Brook weirs was roughly the same from year to year, despite the fact that the mass of water coming down varied widely. A slight negative correlation between sodium and water volume and a slight positive correlation between potassium and water volume are trivial compared with the changes in the mass of water coming down. It follows that some process within the ecosystem maintains the concentration of solutes in the outgoing water. Notice that only the concentration of solutes remains constant; the actual outgoing mass lost varies directly with the water flowing through, and hence the precipitation. Constant concentration must mean that a chemical buffering system is working on water percolating through the soil of the intact forest. This could be through passive soil chemistry or it could be a function of life processes. Experimental removal of a forest in fact shows that life processes are decisive for this buffering.

Particulate Losses Are Minimal from the Living Forest

Data comparing nutrient loss in solution with losses in particulate matter are given in Table 23.5. The mass of nutrients lost as particles is a small fraction of that lost in solution. Erosion losses in the intact

TABLE 23.5
Dissolved and Particulate Nutrient Loss at Hubbard Brook
When the forest vegetation is intact, the loss of nutrients in solids carried by drainage streams is small compared with the loss in solution. Data are the results of analyses at the weir of watershed number 6 between 1965 and 1967. (From Bormann et al., *1969.)*

	Percentage of Total Losses		
	Particulate		
Element	Organic	Inorganic	Dissolved
Calcium	0.7	1.8	97.5
Magnesium	2.0	3.7	94.3
Nitrogen	5.9	[a]	94.1
Potassium	0.5	17.5	82.0
Sodium	0.0	2.8	97.2
Silicon	[b]	18.8	81.2
Sulfur	0.1	0.1	99.8

[a]Not measured, but very small.
[b]Not measured.

forest ecosystem, therefore, are close to negligible.

Nutrients Are Captured from Weathered Rock in the Living Forest

The loss of nutrients in solution generally exceeds the input by precipitation and dust by a wide margin (Table 23.6), yet the ecosystem cannot be suffering a net loss of nutrients or it would be impossible for the concentration in outgoing water to remain constant year after year as it does. It follows, therefore, that the nutrients lost are balanced by a nutrient input from something other than dust or rain.

Nutrients to balance the observed losses come only from weathering of rocks. This is the best demonstration we have of the importance of deep weathering to the maintenance of nutrient reservoirs in temperate ecosystems. The energy source for nutrient retrieval in this mining operation is included in that 42% of solar energy used to power the transpiration streams.

Combined Nitrogen Is Hoarded in Temperate Forests

Less combined nitrogen goes over the dam than comes in as dust or rain (Table 23.6). This is a remarkable statistic because we know very well that the input of combined nitrogen from the atmosphere is always relatively low. Farmers must use nitrogen fertilizer, and many wild plants rely on nitrogen-fixing bacteria—evidence enough that combined nitrogen falling from the sky is of little importance to an ecosystem. And yet even this small input from dust and rain is more than the losses in solution at Hubbard Brook.

Combined nitrogen, of course, may be lost in more important ways than in solution, since nitrogen gas may be vented directly to the air. However, the breakdown of nitrate to nitrogen gas requires reducing conditions (see Chapter 28) and is not likely to be an important process at Hubbard Brook. It seems likely that most of the combined nitrogen is being stored in biomass, something made particularly likely by the fact that biomass is known to be increasing at Hubbard Brook (Whittaker et al., 1974).

TABLE 23.6
Precipitated Nutrients and Stream Outflow at Hubbard Brook
The data are for watershed 6 in 1967-68. With the exception of nitrogen, the nutrient input does not equal the nutrient output, thus demonstrating that weathering is the most important source of nutrients to the ecosystem. The data do not of themselves show that nitrogen is gained by the reservoirs of the system since the balance of combined nitrogen may be broken down within the system and released to the atmosphere. Other data, however, do show nitrogen accumulation (Table 23.7). (From Likens et al., *1970.)*

Nutrient	Precipitation Input	Stream Outflow	Net Loss or Gain
Water	171.8	89.4	—
Ca	3.0	12.2	−9.2
Mg	0.8	3.4	−2.6
K	0.8	2.4	−1.6
Na	1.8	8.8	−7.0
$N(NH_4)$	2.6	0.2	+2.4
$N(NO_3)$	5.2	2.8	+2.4
$S(SO_4)$	16.0	19.3	−3.3
Cl	5.2	5.3	−0.1
$C(HCO_3)$	[a]	0.5	−0.5
$Si(S_1O_2)$	[a]	17.0	−7.0

[a]Not measured, but very low.

EXPERIMENTAL REMOVAL OF A HUBBARD BROOK FOREST

The basic Hubbard Brook data suggest a number of properties derived from the living parts of the ecosystem. Notable among these are the regulation of nutrients in outgoing water, storage of combined nitrogen, and control of runoff up to the volume of transpired water. The plants seem clearly implicated, certainly in running the transpiration streams and probably in storing nitrogen and setting nutrient concentrations in water percolating past the roots. The importance of plants to these ecosystem properties, therefore, can be stated in the form of hypotheses that could be tested by the relatively simple experiment of killing all the plants on one of the watersheds. With six watersheds to choose from it was possible to keep some as controls while manipulating others.

On watershed 2, the forest was clear-cut and the land was sprayed with herbicide for three years to prevent all regeneration (Likens et al., 1970; Bormann et al., 1974). On watershed 4, some strips of forest were cut and others were left intact (Burton and Likens, 1973). Studies of regeneration in other New Hampshire clear-cut forests were made for comparison (Marks and Bormann, 1972). The standard measurements of the study continued to be made at each weir of each watershed. Needless to say, there were dramatic differences between the measurements for forested and treeless watersheds. When the trees went, evapotranspiration decreased, more water came down as runoff, the water was warmed as it rolled unshaded under the sun, and there was erosion and an increased leak of nutrients from the ecosystem. But the patterns of these events held a number of revealing surprises. The principal revelations of these experiments in ecosystem destruction are as follows:

Runoff Increased by About 30% This was expected because the transpiration pump that loosed water into the air had been removed. Much of the old water that was originally transpired now ran into the streams.

Water Temperature Increased by Several Degrees Celsius This was particularly apparent in the strip-cutting experiment where water warmed and cooled as its parent stream crossed strips. A secondary heating effect is that snowbanks melt earlier in cleared sections.

In the First Two Years Following Clear-cutting the Big Nutrient Loss Was in Solution This was an exciting discovery. We expect erosion losses, but what we find first is an escalation of solution losses. Over the first two years the ratio of net loss in solution to net loss as solids increased nearly four times. Killing the plants thus has the effect of increasing the concentrations of nutrients in the flushing water, and the effect is almost immediate. This is a clear demonstration that the concentration of nutrients in water leaving an intact forest ecosystem is in fact controlled by the plants themselves. Leakage remains low only as long as the forest lives.

Erosion of Solids Began to Increase Significantly Only After a Two-Year Lag This result is illustrated in Figure 23.6. The conclusion is that the once-forested landscape retains its relative immunity to erosion long after the trees are killed, presumably until roots decay, and until bacterial decomposition destroys the organic fabric of soil, litter, and stream bed.

Nitrate Was the Nutrient to Show the Largest Losses The concentration of nitrates coming over the weir of the clear-cut watershed increased over 40 times whereas losses of other nutrients were mostly increased by only 2 or 3 times (Table 23.7). This is a dramatic test of the hypothesis that combined nitrogen is being

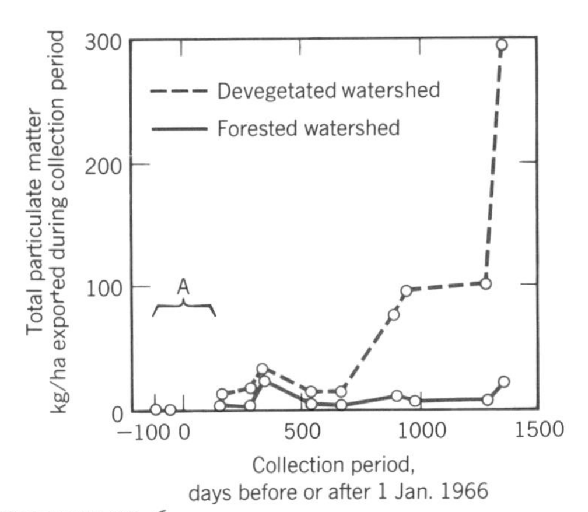

FIGURE 23.6

History of erosion in a devegetated watershed. Particulate losses from the cleared watershed at Hubbard Brook were comparable to those from an intact watershed for the first two years, after which they rose exponentially. One watershed was devegetated in time interval A. (From Bormann et al., 1974.)

TABLE 23.7

Nutrient Losses in the Clear-Cut Hubbard Brook Watershed

The largest increase in losses was in the anion nitrate. Lesser increases in each cation are probably a consequence of the nitrate flux. (From Likens et al., *1970.)*

	Metric Tons/km²/yr			
	1966–67		1967–68	
	W2	W6	W2	W6
Ca^{++}	−7.5	−0.8	−9.0	−0.9
K^{+}	−2.3	−0.1	−3.6	−0.2
Al^{++}	−1.7	−0.1	−2.4	−0.3
Mg^{++}	−1.6	−0.3	−1.8	−0.3
Na^{+}	−1.7	−0.6	−1.7	−0.7
NH_4^+	+0.1	+0.2	+0.2	+0.3
NO_3^-	−43.0	+1.5	−62.8	+1.1
SO_4^{--}	−0.5	−0.8	0	−1.0
HCO_3^-	−0.1	−0.2	0	−0.3
Cl	−0.1	+0.2	−0.4	0
SiO_2(aq.)	−6.6	−3.6	−6.9	−3.6
Total	−65.0	−4.6	−88.4	−5.9

retained and hoarded by the intact forest. Nitrifying bacteria are known to be liberating nitrates constantly, even in intact forest, but the resulting dissolved nitrates are scavenged as rapidly by plant roots. With the death of the forest, the scavenging system breaks down. Furthermore, nitrification itself proceeds many times as fast as in the living forest, both because plant inhibitions are removed and because soil temperature rises. Probably the reservoir of combined nitrogen that has been built up by nitrogen-fixing bacteria over the lifetime of the forest is leaked from the site in the first two or three years after the destruction of the forest.

The Loss of Other Nutrients Was a Consequence of Nitrate Losses Nitrification yields free hydrogen ions as well as nitrates, thus acidifying soil and stream water. Hydrogen ions then replace cations (Mg^{++}, Ca^{++}, Na^{+}, K^{+}) on the cation exchange surfaces of clay minerals and humics in the soil. The displaced cations then enter the soil water and are lost to the system. This lowered pH effect of nitrification is comparable to the consequences of acid rain, in which cations are displaced from clay minerals and soil humics in a like manner and with the same consequent losses of nutrients from the soil.

Revegetation Is Rapid Where It Is Not Prevented by Herbicides The early succession shrub or tree, wild cherry (*Prunus pennsylvanica*), grows rapidly on a clear-cut site. In three years it forms a closed canopy and is quickly accumulating nutrients. This rapid growth presumably benefits not only from the absence of shade but also from the rich supply of nutrients in the soil water.

Algae Bloom in the Drainage Streams This can be attributed both to the in-

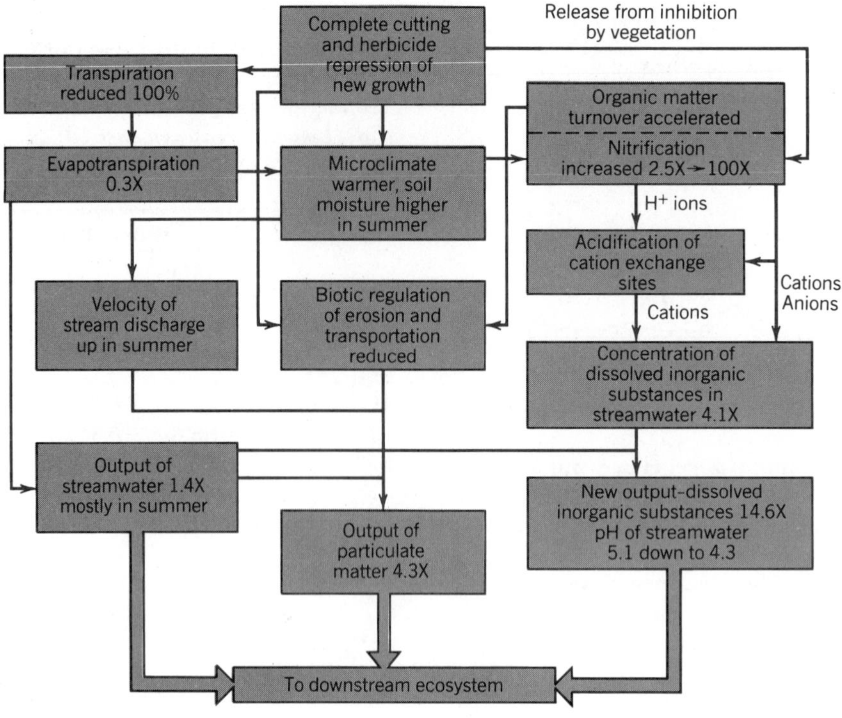

FIGURE 23.7
Interactions triggered by deforestation of Hubbard Brook watershed.
Ecologic, hydrologic, physical, chemical, and erosion consequences of removing the vegetation based on data of the first two years. (From Bormann et al., 1974.)

creased flux of dissolved nutrients and to the increased light flux reaching streams now flowing in the open.

The mutual dependence of these various effects is illustrated in Figure 23.7.

ALTERNATIVE TO WATERSHEDS: NUTRIENT AND BIOMASS BUDGETS

It is possible to calculate the nutrient budget of a piece of vegetation by measuring net production, the nutrient content of biomass, and the various nutrient losses from the plants. The losses will reach the soil in dead plant parts, in animal droppings, and in rainwater that washes through the canopy and down the stems. We can say that:

$$N_u = N_r + (N_l + N_w + N_{sf}) \quad (23.1)$$

where

N_u = nutrient uptake
N_r = nutrient retained in biomass (nutrient *increment*)
N_l = nutrient returned in litter (includes leaves and droppings)
N_w = nutrient washed off canopy in rain (throughfall)
N_{sf} = nutrient washed back to soil in *stem flow*.

Nutrient retained can be measured by monitoring net production by harvest (see Chapter 24) and making chemical analyses of representative stems, twigs, and leaves. Nutrient returned is measured in a similar way by monitoring the rate of fall of leaves, branches, and caterpillar droppings with litter traps. "**Throughfall**" and "**stem flow nutrients**" are calculated by measuring water falling from the canopy and down stems, then subtracting the chemical data for the local rainwater. The term ($N_l + N_w + N_{sf}$) is then known and is referred to as "nutrient restitution" or "nutrient losses" by different authors. Then all has been measured to solve:

$$\begin{matrix}\text{Nutrient}\\ \text{uptake}\end{matrix} = \begin{matrix}\text{nutrient}\\ \text{retained}\end{matrix} + \begin{matrix}\text{nutrient}\\ \text{losses}\end{matrix} \qquad (23.2)$$

The work involved in a study of this kind is prodigious; this is team science. But a number of nutrient budgets are now available. Data for Belgian oak forests are given in Table 23.3 (Duvigneaud and Denaeyer-De Smet, 1970).

One striking observation in the Belgian data is the relatively small portion of the total nutrient reservoir that is required by plants in any one year. This large and redundant reserve of nutrients under temperate forests partly shows why western-style agriculture works. Soils in temperate latitudes under wild vegetation accumulate nutrients so that soil reservoirs are maintained with stores far beyond the immediate needs of the plants. Farmers are able to draw on these reservoirs, which remain even when the primeval forest has been removed.

NUTRIENT CYCLING IN THE WET TROPICS

The many follies of clearing rain forest are now public knowledge, among them that agriculture on the cleared land gives poor returns for labor. This is widely true, though not universally. A huge system like the Amazon, for instance, supports an astonishing variety of soil types, depending on water tables, parent rock, periodicity of flooding, seasonality, and rainfall. Some of the blacker soils on flood plains are splendidly fertile if handled with care.

But it is also true that large stands of tropical rain forest grow on red, lateritic oxysols (see Chapter 25) that are calamitous for agriculture. The effects of removing forests from these soils are such as to give rise to the hyperbole of "red desert" (Figure 23.8). Actually, the Amazon system has its own sets of ruderals and weeds that can cope even with the mess that agriculture leaves behind on the most unsuitable soils, though the pioneers of a succession that might take centuries to work its way up to forest are poor consolation for the lost forest.

When the more calamitous tropical soils are examined to see what factors contribute to their lack of fertility, the following facts emerge readily:

1. The fine particles of the soils include few of the large-lattice clay minerals that adsorb cations to their surfaces. Clays like montmorillonite that perform this service in temperate soils are largely lacking.
2. Deep weathering has produced a soil matrix of iron and aluminum minerals from which silicates have largely been removed (see Chapter 25).
3. Surface litter or detritus is scanty and thin, with very little organic matter incorporated into the mineral soil. Soil structure is minimal by temperate standards.
4. The effects of solution from percolating water can be detected many me-

FIGURE 23.8
The clearing of Amazonia.

ters deep, far deeper than in temperate soils. Thus it is likely that soils have been leached of nutrients to great depths.

Probable reservoirs and fluxes of nutrients in an ancient red tropical soil are given in Figure 23.9. These properties of the system together result in the red mud of the red desert scenario when cleared of vegetation. Both nutrients and suitable sites for germination are scarce in the cleared land, the nutrients originally in the living biomass having largely been removed in the drainage water. Thus a remarkable conclusion: tropical rain forest can live on soils that, by any scale of agriculture, are infertile.

ON THE MAINTENANCE OF TROPICAL FOREST

The ecosystem property that makes growth on bare ground difficult on weathered red soils in the tropics apparently depends on temperature. High temperatures, and lack of a cold season, mean that decomposition is far more rapid. Furthermore, a characteristic of tropical forests is the rich fauna of detritivores, whose existence is probably itself temperature-dependent. Termites, together with detritivorous ants and other insects, dispose of litter, even tree trunks, with extreme rapidity. In addition, the true decomposers—fungi and bacteria—remove humic sub-

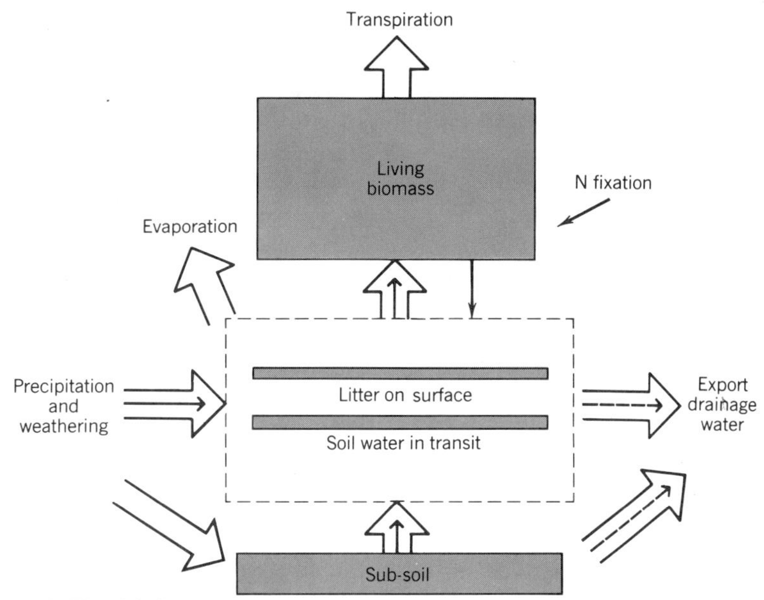

FIGURE 23.9
Nutrient flux and storage in an ecosystem of the wet tropics.

stances completely from the soil surface. Thus tropical soils have no reserves of structure or substrate provided by the organic matter of temperate soils.

But tropical forest trees, once established, are well supplied with adaptations that let them cope with short commons for nutrients. These adaptations work both to store nutrients and to scavenge them with high efficiency.

The first essential adaptation of an individual rain forest tree is a high capacity to take up nutrients. Rain forest trees must be able to sweep the water passing their roots clean of dissolved ions. They do not, as it were, bring fresh tree technology to this task, but they certainly possess the full array of uptake systems owned by trees everywhere. In particular, it has been found that they have prominent symbiotic relationships with fungi called "vesicular arbuscular mycorrhizae" (VAM).

The VAM symbiotic fungal hyphae penetrate root hairs and derive energy by ingesting cell contents. The roots, however, are able to absorb nutrient solutions from the fungus. This is possible because the fungi are extremely efficient nutrient scavengers (Wicklow and Carroll, 1981). Trees pay with lost cell sap for the services of an excellent ion pump that lets root hairs sweep adjacent water virtually clean of nutrients.

VAM are not a peculiarity of rain forest trees since they are found on some roots in virtually all terrestrial ecosystems, but they seem to be ubiquitous in the wet tropics. The symbiosis may be so developed in the Amazon that the fungal symbiont may be in direct contact with the decomposing litter, so that nutrients can be transferred from detritus to tree roots directly (Went and Stark, 1968). It looks to be a beautiful adaptation that prevents any danger of loss of nutrients by not permitting them to reach the soil water at all.

All this is no more than a thorough application of nutrient-absorbing systems theoretically available to all trees. But on an old tropical oxysol every individual tree is on short rations for nutrients, whereas the trees of a temperate forest sometimes have nutrients in abundance.

On top of these refined adaptations is one physical circumstance that aids the tropical trees—they pump more water. The adaptations of tropical trees to nutrient shortage, therefore, are to pump much water and to sweep the water they do not pump clean of useful ions, making the best possible use of symbiotic fungi in the process. The best hard data we have for the effectiveness of these ion pumps is that water leaving a tropical rain forest in streams may have electrical conductivity close to that of distilled water (Walter, 1973).

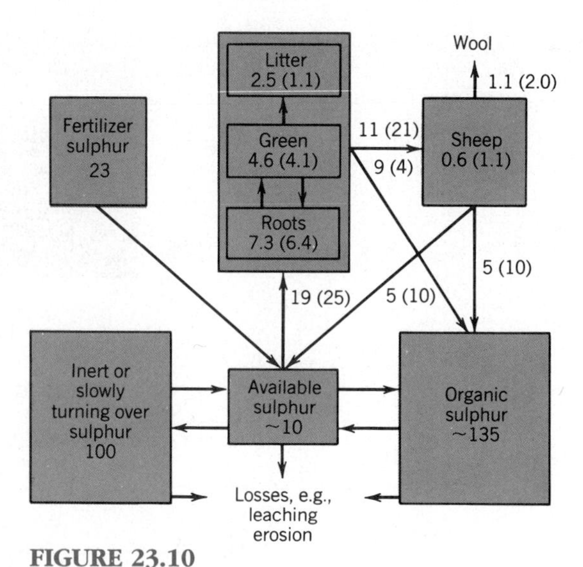

FIGURE 23.10
Sulfur cycle in Australian grasslands.
Data estimated by use of ^{35}S tracer technique. (From Till, 1979.)

NUTRIENT CYCLES IN OTHER BIOMES

In grasslands the pattern of nutrient cycling has much in common with temperate forests in that there tends to be a large soil reservoir and a lingering supply in litter and moribund parts. Figure 23.10 shows a sulfur cycle worked out for Australian grasslands by using ^{35}S as an isotopic tracer. The soil reservoir is huge by the standards of plant needs. Farmers of this ecosystem supply sulfur as fertilizer though, because sheep impose an output of sulfur on the system when it is carried off in wool. Figure 23.11 shows the phosphorous cycle on Indian grasslands, again revealing a large soil reserve. This pattern is to be expected in all natural grasslands that build organic soils (see Chapter 25).

In tundra biomes, temperature tends to impose a different constraint on nutrient cycles. Low temperature and frozen ground make the nutrients of soil and litter slow to be mobilized. In a tundra, therefore, the rate at which nutrients can be mobilized from the frozen ground in the spring is crucial to the plant supplies.

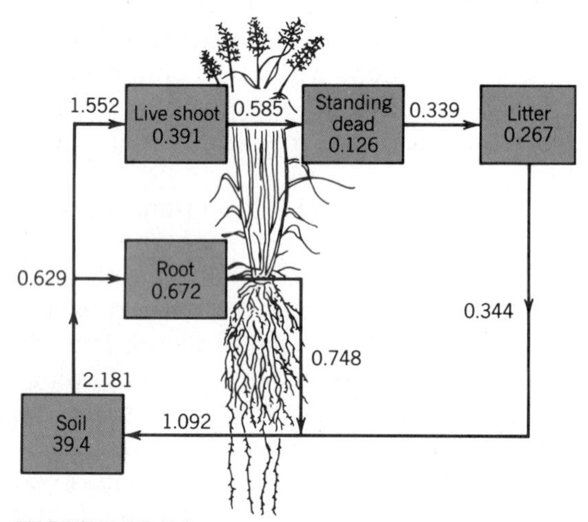

FIGURE 23.11
Phosphorus cycle in Indian grasslands.
Numbers in boxes are mean standing crops of P in g m^{-2}. Numbers on arrows are net flux rates of P in g m^{-2} d^{-1}. (From Yadara and Singh, 1977.)

TABLE 23.8
Arctic Nutrient Reservoirs
The data are for total nutrients in live plants, dead plants, and total soil and plants combined to a depth of 20 cm. It is evident that in tundra and subarctic environments the soil reserve is larger than plant demand. (International Biological Program data compiled by Dowding et al., *1981, from various sources.)*

	N (g/m^{-2})			P (g/m^{-2})			K (g/m^{-2})			Ca (g/m^{-2})		
	Live	Dead	Total	Live	Dead	Total	Live	Dead	Total	Live	Dead	Total
Alaskan arctic coast	9.1	12.8	960	0.8	0.8	63.2	1.3	0.5	15.8	3.0	5.4	—
Siberian hummock tundra	41.1	7.9	—	5.1	0.9	9.2	5.5	1.1	14.2	34.1	1.3	—
Antarctic tundra	14.0	10.3	265	1.9	1.0	8.2	10.2	3.0	17.8	1.8	3.0	14.8
Norwegian willow thicket	19.4	10.1	591	3.4	1.4	373	7.0	1.7	232	16.8	9.3	986

In extreme arctic sites the ability to mobilize nutrients from within the living tissue of the plant may be important, and plants are adapted to retain nutrients in stem or root tissue that persists for years (Dowding et al., 1981). The total nutrient reserve in a tundra soil may appear to be redundantly large, almost on the scale of a temperate forest (Table 23.8). Yet the mobilization of these reserves may be so difficult that tundra plants are nutrient-limited.

PLANTS AND THE MAINTENANCE OF LANDSCAPE

That plants stabilize landscapes has long been known. Plato wrote about the consequence of clearing forests in ancient Greece, complaining that the rich brown soil had wasted away leaving a country of "bare bones." The Hubbard Brook study shows just how necessary vegetation can be to the maintenance of habitats: particulate erosion close to zero, solution of the soil water essentially constant, nutrients massively hoarded. The destruction of watershed 2 by ecologists produced something of the same effects as the spread of Greek civilization.

These astonishing landscape effects are consequences of a system of every plant for itself. Very largely they result from the massed ranks of plant stems and roots acting as baffles to erosion. But the increments of organic matter vital to temperate soils come from the load of detritus produced when every plant defends its tissues against herbivore attack with toxic chemicals. In climates with winters, this flux of trash is disposed of slowly, with prodigal effects on the soil. In dry prairies a similar trash flux lasts even longer, setting another large nutrient hoard. But in some well-drained sites in the wet tropics, decomposer food chains recycle trash as fast as it is produced, so that the humic content of soils is minimal. An agriculture that seeks to replace old forest with crops of early succession or pioneer plants is doomed to failure on such soils.

In wet climates, the most productive plant designs rely on evaporative cooling (see Chapter 5). Their massed ranks may evaporate so much water that nearly half of incident solar energy is used to power the pumps. In these ways natural selection for individual fitness leads to systems that change the states of whole landscapes.

Preview to Chapter 24

Phoenicopteris minor: **high productivity in a Transvaal pond.**

PRODUCTION of whole ecosystems can be measured in tons of biomass, expressing the results as carbon or as calories with simple conversions. It is always possible to measure growth by increments using multiple harvests, but for forests this is usually prohibitively costly in labor. Harvesting sample trees at intervals in plantations of same-age trees has met with some success, but for wild forests empirical relationships between growth and production for each species of tree must be established before samples can be used. Gas exchange measures in enclosed samples are valuable for compact vegetation like tundra but invoke serious problems of sampling in vegetation with larger plants. Gas exchange is particularly useful for aquatic systems because algae of open water are tiny and because gasses can be measured in solution. Because all these measures are so time-consuming, our knowledge of the productivity of different ecosystems is quite unequal. Available data do, however, demonstrate that productivity is always set by simple parameters of habitat like temperature and precipitation on land or disolved nutrients in the oceans. The oceans are deserts because of nutrient shortage, as land deserts are desertic because of water shortage.

Chapter 24
Productivity

Measuring productivity, in the sense of calories fixed in unit time, has become one of the practical necessities of ecology. Planners ask for the data, wanting to know the theoretical yield of world fisheries, Siberian farmland, or fuel energy in gasoline substitutes that might be possible by planting the entire Amazon basin in sugarcane (planners do conceive such horrors). Problems of scale and complexity make the task of supplying the data difficult and invite ingenuity.

We cannot label energy to trace its passage through an ecosystem, nor do we have voltmeters to record its potential. All we can do is measure biomass and respiration. Biomass and respired carbon can be expressed fairly readily as the calories they represent, but rates of energy transfer must be computed from time-series measurements of changing biomass.

Ideal comparisons between ecosystems should compare the total productivity, called "**gross production**," which is *the energy represented by the biomass produced together with the energy that went into the work of producing it*:

$$\begin{matrix}\text{Energy} \\ \text{(biomass)}\end{matrix} + \begin{matrix}\text{Energy of} \\ \text{respiration}\end{matrix} = \begin{matrix}\text{Gross} \\ \text{production}\end{matrix} \qquad (24.1)$$

On a field scale, however, accurate measures of respiration are impracticable in most circumstances, particularly in view of the near impossibility of separating carbon input and output by photosynthesizing green plants (see Chapter 2). Accordingly, comparative measures of ecosystem productivity refer to "**net production**," which is simply the *rate of biomass production*.

HARVEST METHODS FOR MEASURING PRODUCTIVITY OF VEGETATION

The simplest way to measure net production is to collect what grows, dry it, and weigh it—a simple harvest. Harvest in unit time thus gives the rate of net production, or the "net productivity." Yet applying this technique to natural vegetation can become complex and involves multiple harvests. This is because wild vegetation seldom grows synchronously. The standing crop in any field or forest nearly always includes biomass from previous years or seasons so that at least two measures are needed: standing crop at the start of a growing period and standing crop at the end. These data can come from measures of quadrat samples taken at intervals throughout the growing season.

Herbaceous vegetation can be sampled by clipping all above-ground vegetation from the sample quadrats, being careful at the same time to collect plant parts that have died and become litter in the interim. It is then necessary to take core samples of soil to be extracted for roots in order to arrive at total net production. Table 24.1 gives some typical results.

Meticulous sequential harvest is the most accurate method of measuring net productivity of forests also, but the task is likely to be ruinously laborious. Stem thickness, shoot elongation, fruit biomass, flower biomass, and leaf mass must all be measured at repeated intervals, to say nothing of root thickness, root elongation, and root-hair biomass. Furthermore, the measurement of dead parts (litter) produced and of herbivore damage can be difficult. Litter traps must be used to collect representative samples of litter. Protection from herbivores, which may be mostly beetles or caterpillars, can be very tricky indeed. Nets over branches, perhaps coupled with chemical fumigation, offer one solution. The details of the methods have to be worked out for any particular patch of forest to be examined. But better sampling systems can be devised, particularly "mean tree" and "dimension" analysis.

The Mean-Tree Approach

Forests are made up of distinct and easily distinguished plants, unlike the vegetation

TABLE 24.1
Net Primary Productivity by Harvest Techniques
(Data compiled by Whittaker and Marks, 1975, from various authors.)

<table>
<tr><th rowspan="2">Communities and Species</th><th rowspan="2">Total ($g/m^2/yr$)</th><th rowspan="2">Stem and Branch Wood (%)</th><th rowspan="2">Leaves and Twigs (%)</th><th rowspan="2">Fruit and Flower (%)</th><th colspan="2">Root System</th></tr>
<tr><th>Rhizomes (%)</th><th>Roots (%)</th></tr>
<tr><td>Wheat</td><td>294</td><td>—</td><td>53.0</td><td>29.4</td><td colspan="2">17.6</td></tr>
<tr><td>Barley</td><td>242</td><td>—</td><td>46.6</td><td>35.5</td><td colspan="2">17.9</td></tr>
<tr><td>Zea mays (maize, high-yield)</td><td>1935</td><td>16.8</td><td>17.1</td><td>61.4</td><td colspan="2">4.6</td></tr>
<tr><td>Helianthus annuus (sunflower)</td><td>3213</td><td>37.5</td><td>17.6</td><td>36.0</td><td colspan="2">8.9</td></tr>
<tr><td>Arctic tundra, all species</td><td></td><td></td><td></td><td></td><td colspan="2"></td></tr>
<tr><td>Production</td><td>100</td><td>2</td><td>28</td><td>—</td><td colspan="2">70.</td></tr>
<tr><td>Populus (7-year-old poplars)</td><td></td><td></td><td></td><td></td><td colspan="2"></td></tr>
<tr><td>Production</td><td>226</td><td>48.6</td><td>36.7</td><td>—</td><td colspan="2">14.7</td></tr>
<tr><td>Blanket bog</td><td></td><td></td><td></td><td></td><td colspan="2"></td></tr>
<tr><td>Calluna vulgaris</td><td>351</td><td>10.8</td><td>37.0</td><td>—</td><td colspan="2">52.2</td></tr>
<tr><td>Eriophorum vaginatum</td><td>221</td><td>—</td><td>78.2</td><td>1.4</td><td>8.6</td><td>11.8</td></tr>
<tr><td>Empetrum and others</td><td>26</td><td>12</td><td>38</td><td>—</td><td>—</td><td>(50)</td></tr>
<tr><td>Sphagnum, other bryophytes</td><td>47</td><td>—</td><td>100</td><td>—</td><td>—</td><td>—</td></tr>
<tr><td>Lichens</td><td>3</td><td>—</td><td>100</td><td>—</td><td>—</td><td>—</td></tr>
<tr><td>Total bog</td><td>648</td><td>6.3</td><td>56.0</td><td>0.5</td><td colspan="2">37.2</td></tr>
</table>

of herbaceous or shrubby layers. It ought, therefore, to be possible to identify typical trees, measure the productivity of these individuals, and then extrapolate to the whole forest. This turns out to be practicable for many artificial plantations but not for wild forests.

Productivity can be measured if a series of plantations of different ages is available; say, plantations started 30, 40, and 50 years ago. In each stand the trees ought to be similar so that measures on a few could be extrapolated to the whole forest. This was the method used for the measurement of Asian forest plantation that suggested reduced production as forests age (Kira and Shidei, 1967; see Chapter 20). But even in a plantation, differences between individual trees can be such that Baskerville (1965) estimated that the error in production measurements was between 25 and 45% for plantations of balsam fir (*Abies balsamea*). If this is so for a plantation, application of the method to a wild forest is obviously hopeless.

Dimension Analysis

Despite the fact that trees vary greatly in the relative lengths and widths of their parts, there should still be some relationship between tree size and productivity. Measurement of tree sizes throughout the forest then should also be measures of production, and the problem becomes one of a suitable measure of size. This problem is solved by the technique known as **dimension analysis**. It depends on the principle of allometric growth.

As structures are made larger, the relative proportions of their parts must change. This is a well-known principle of engineering, one that requires the redesign of all the parts of, say, an airliner when its capacity is increased from 10 passengers to 300. A mature forest tree accordingly is built differently from a young sapling, just as an old mare is built differently from a foal (Figure 24.1). In biology, the process of changing shape with size is called **allometric growth** (Thompson, 1942).

The problem of estimating productivity from measures of a minimal number of trees becomes one of selecting measurements from which allometric growth can be taken into account. Dimension analysis achieves this by constructing regression equations that relate easily measured parameters of individual trees to tree size (leading to estimates of standing crop) and to individual production (leading to estimates of ecosystem productivity). Cutting down sample trees, measuring them, and estimating dry matter then yields numerical data to feed into the equations. Dimension analysis, therefore, is based on an initial harvest done in a way that will allow realistic extrapolation (Whittaker and Marks, 1975). Sufficient data for the analysis, it turns out, is provided by just the following measures of an adequate sample of trees:

Diameter at breast height (DBH)

Height

Bark thickness (by boring)

Mean current wood growth (by boring to measure and count tree rings)

Age (by counting tree rings in borings)

These data were sufficient to allow calculation of standing crop and productivity from the regression equations based on a much larger set of measurements on the specimen trees that were harvested for the eastern deciduous forest of North America (Whittaker and Woodwell, 1968; Harris et al., 1973).

The fit of the regressions between stem thickness and productivity is illustrated by data in these studies given in Figure

FIGURE 24.1
Examples of allometric growth.
Small objects have different proportions to large objects of the same kind. Regression equations can relate linear dimensions to mass and productivity of growing trees to allow computation of standing crop and net primary productivity of forests.

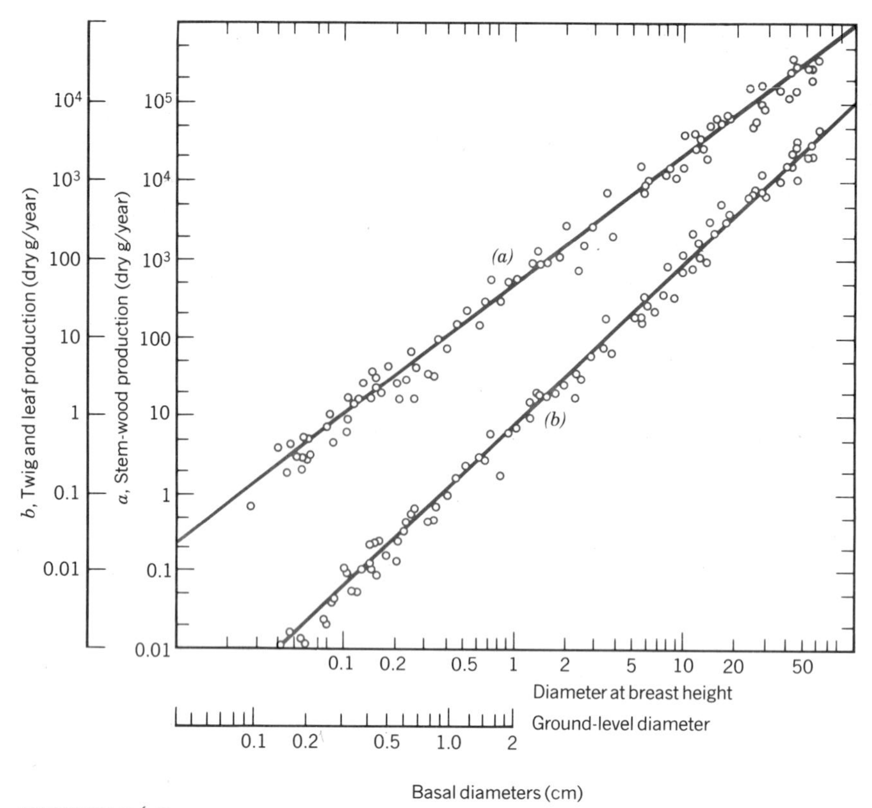

FIGURE 24.2
Regressions comparing stem thickness with productivity.
Data are from many species, ranging from small shrubs to large trees. As might be expected, individual productivity is a function of individual size, regardless of taxonomic affinity. The regressions show that stem thickness is a good predictor of both size and productivity. (From Whittaker and Woodwell, 1968.)

24.2. These data show nicely how dimension analysis does allow biomass and production to be measured in forests. The work, however, is extremely time-consuming. In effect, one measures a patch of forest by harvesting samples of everything small enough to harvest, but harvesting only a smaller proportion of the larger trees and extrapolating to the remaining trees using regression equations. As with other harvest work, litter and destruction by herbivores also must be measured, so that the total labor involved can be nearly overwhelming. It will be noted that counting tree rings is an integral part of the procedure, so the method cannot easily be applied to forests in nonseasonal climates (tropical rain forest) where trees do not make rings.

Converting Dry Weights to Calories

Table 24.2 describes the caloric content of the principal compounds making up organic matter. One way of converting dry weight estimates to measures of energy is to perform chemical analyses of sample parts of the plants and to calculate caloric contents using the data in this table. It turns out that collective plant parts of

TABLE 24.2
Caloric Content of Chemical Compounds
Estimates of productivity in grams of dry weight can be converted to calories by chemical analysis when the caloric equivalents of the components are known. (Data compiled by Lieth, 1975, from various authors.)

Compound or Matter Class	kcal/g
Starch	4.18
Cellulose	4.2
Saccharose	3.95
Glucose	3.7
Raw fiber	4.2
N-free extract	4.1
Glycine	3.1
Leucine	6.5
Raw protein	5.5
Oxalic acid	0.67
Ethanol	7.1
Tripalmitin	9.3
Palmitinic acid	9.4
Isoprene	11.2
Lignin	6.3
Fat	9.3

such widely different taxa as beech trees, spruce trees, grass plants, and legumes have very similar total caloric contents, giving ecologists confidence in using the more simply derived estimates of dry matter when comparing the vegetation of one place with another. It is, in fact, usual in field or ecosystem studies to express productivity as grams of dry matter per unit time (or sometimes as grams of carbon per unit time where carbon is assumed to be 50% dry weight) instead of the more precise estimate of calories per unit time.

CARBON DIOXIDE EXCHANGE IN TERRESTRIAL ECOSYSTEMS

If the CO_2 absorbed by the plants of an ecosystem in unit time by day could be measured, this would be a direct measure of **net primary productivity**. A similar measure for the whole community (animals included) would be smaller by the flux of animal respiration, and would yield **net ecosystem productivity**.

Figure 24.3 shows the design of a small chamber system to measure CO_2 exchange of a forest. A glance at this design should leave no doubt about the complexity of the operation, and some concerns will immediately be apparent. The photosynthesizing and respiring tissues, for instance, must be totally enclosed, which is likely to affect their function. Then it is found that fluctuation of CO_2 concentration within all but the smallest enclosures is large, making the estimates of average CO_2 concentration difficult. Very small chambers, each enclosing a single twig, have been found to be necessary (Woodwell and Botkin, 1970). When the measurements are made, it is still necessary to extrapolate from the small samples in the chambers to the whole community. This method has accordingly not been widely used for forests, but its application to Alaskan tundras, with whole pieces of vegetation in the chambers, has proved useful.

Very large chambers that enclose complete sections of forest are, at first sight, an attractive possibility. H. T. Odum (1970) built what amounted to a gigantic greenhouse over a patch of Puerto Rican forest in the 1960s, pumped air through it, and measured CO_2 in and out. The difficulties were so formidable that he does not seem to have had any successors using the technique.

On rare occasions natural phenomena may enclose an ecosystem. When an atmospheric inversion descends low over a forest, cold air is injected under warm air and trapped, almost as if it were under the glass of a greenhouse roof. This condition occurs frequently in the forest near the Brookhaven laboratory and has been used for measures of forest respiration. A series of gas sampling stations were built on

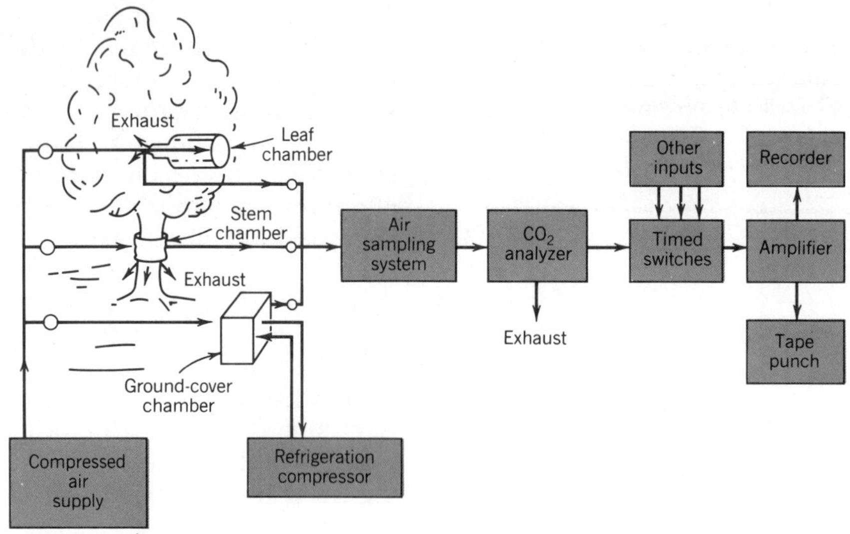

FIGURE 24.3
Small chamber measurement of CO_2 exchange in a forest.
In this analysis air must be driven through the chambers at a known rate, the temperature must be measured and controlled, and the gasses must be mixed completely and subsampled accurately. (From Woodwell and Botkin, 1970.)

towers rising through the trees so that CO_2 concentrations could be measured throughout an inversion event lasting, perhaps, for days. Figure 24.4 shows the direct results of one such episode. The change in CO_2 concentration in the forest between day and night shows up particularly clearly. In Figure 24.5 these data are reduced to estimates of total respiration regressed as a function of temperature. Ecosystem respiration was computed from the CO_2 flux at 2104 grams dry matter per square meter per year, a figure in keeping with measurements made by harvesting and dimension analysis.

Gas Measurement in Aquatic Systems

Two circumstances make the measurement of productivity easier in open water than on land: plants of the open water are small, and water is a simpler medium than air for making chemical determinations.

The light and dark bottle method was first invented for use in the sea by Gaarder and Gran (1927). They put seawater containing planktonic algae into bottles, one of clear glass and the other covered with black paint, then they lowered the bottles over the side of their ship and left them in the sea for a measured time. When the bottles were recovered, the oxygen in each was measured by titration with potassium permanganate (the Winkler method). The light bottle had *gained* oxygen, due to photosynthesis, and the dark bottle had *lost* oxygen, due to respiration. The oxygen gained in the light bottle added to the oxygen lost in the dark bottle provided an estimate of the gross primary production of the time interval.

Although the light and dark bottle method is most accurate, it has been re-

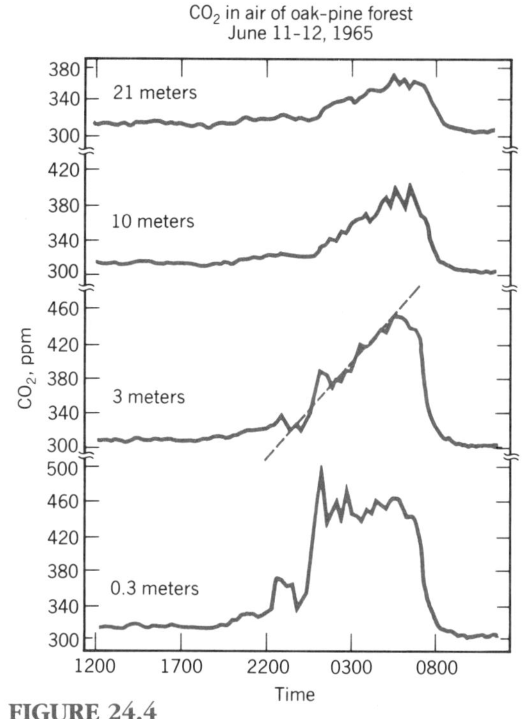

FIGURE 24.4
Carbon dioxide concentrations in a forest under an inversion.
The data are for an eastern American deciduous forest at Brookhaven. Air is trapped over the forest by inversions lasting several days. (From Woodwell and Botkin, 1970.)

placed for routine work by a method based on the uptake of labeled carbon in photosynthesis. A sample of lake or ocean water is placed in a clear bottle with ^{14}C-labeled bicarbonate and the bottle is dangled where the sample was taken. Algae incorporate ^{14}C as a function of the rate of photosynthesis. The algae can be filtered out, fixed, and the ^{14}C incorporated into the algae measured with a radiation counter (Strickland and Parsons, 1972). Estimates of productivity by this method are always lower than those by the light and dark bottle method, probably because the algae respire some of the labeled CO_2 back to the water. The method apparently gives an estimate that is closer to net production than to gross production. Because it is convenient, the method has become standard for both lakes and oceans.

A quite different approach is to work from estimates of algal standing crop provided by measures of chlorophyll. Algae are filtered from a known volume of water, their chlorophyll is extracted in acetone, and its concentration is measured in a spectrophotometer. Standing crops can be measured rapidly on large numbers of samples in this way. Moreover, samples may be stored for later measurement in the laboratory, so that many samples can be taken rapidly in the field and treated at leisure, which is a great convenience. It is then necessary to estimate productivity from knowledge of chlorophyll concentration and light intensity. This requires empirical calibration of the standing crop data with similar concentrations of live algae, whose productivity in the field is measured with the light and dark bottle technique. This has been found to be straightforward, so that chlorophyll assay is now one of the preferred methods of measuring aquatic primary productivity.

Because removal or addition of carbon dioxide to water alters its acidity, pH can also be used as a measure of productivity, provided it has been calibrated by one of the other techniques.

A peculiar virtue of all these methods (except uptake of ^{14}C) is that they can be applied to whole bodies of water, not just to subsamples. The productivity of a whole pond or small lake ecosystem, for instance, can be measured by estimating dissolved oxygen or pH by day and by night. Typically the measurements would be made every two hours or so around the clock, leading to results like those shown in Figure 24.6.

Another variant applied to moving water is to make the measurements upstream

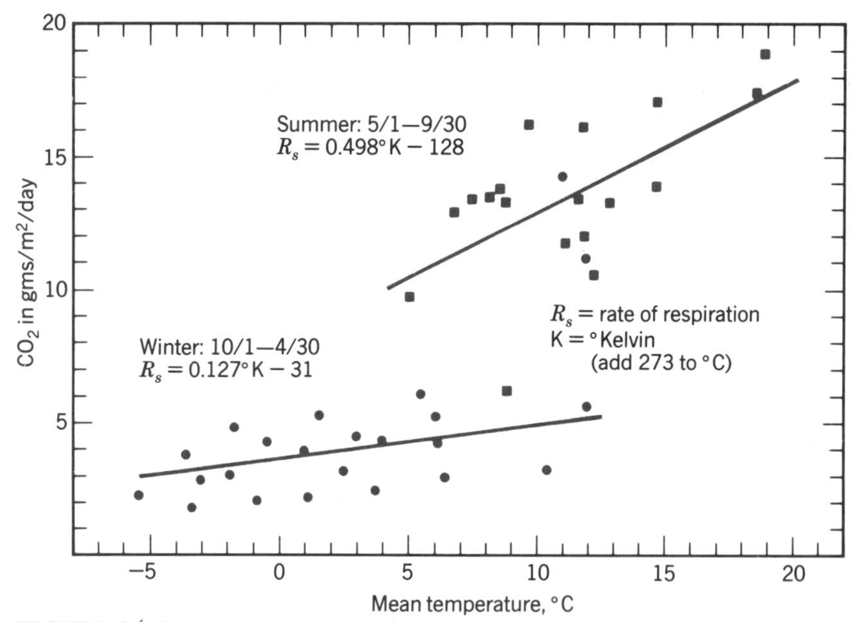

FIGURE 24.5
Respiration under an inversion as a function of temperature.
The base data are those given in Figure 24.4, temperature being the average of measures at 10 m and 3 m. R_s = rate of respiration in grams of CO_2 per square meter per day. (From Woodwell and Dykeman, 1966.)

and downstream of the community to be studied. This was done, for instance, in the classic measure of the productivity of a coral reef at Enewetak by Odum and Odum (1955), where water flows constantly in one direction over the submerged reef between two islets of the atoll (Figure 24.7).

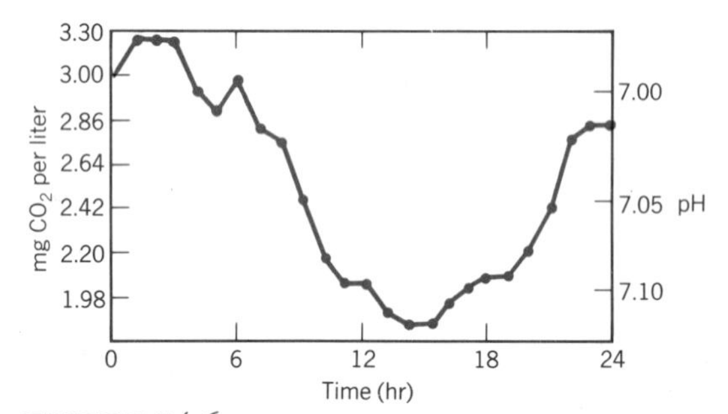

FIGURE 24.6
Diurnal changes in gas concentrations in an aquatic system.
The data are for New Hope Creek in North Carolina, where both CO_2 concentrations and pH were measured. (From Hall and Moll, 1975.)

COMPARATIVE ECOSYSTEM PRODUCTIVITY

Net primary productivity has now been measured in fairly reliable ways for samples of all types of ecosystem. Table 24.3 gives estimates for the principal ecosystem types. These estimates are also applied to the areas of each ecosystem type to compute the tonnage of net production and the net energy, which each contributes to the energetics of the whole earth. This is not quite the same as an energy budget for the biosphere, since this would require gross

FIGURE 24.7
Reef productivity laboratory.
The windward reef at Enewetak Atoll has ocean water flowing rapidly over the barely submerged reef between islets of the atoll and into the lagoon. Measuring oxygen concentrations on the ocean and lagoon sides of the reef allowed calculation of productivity of the whole reef ecosystem.

primary productivity, for which we have many fewer data. But the results are certainly closely proportional to the contribution of the biota to a true energy budget.

The truly productive places are forests, marshes, estuaries, and reefs. The unproductive places are tundras, deserts, and oceans. Other systems, including agriculture, fall in between (Table 24.4). Some of these findings must always have been expected: high productivity in wet, warm forests, for instance, and low productivity in deserts. Postulates to explain productivity are thus readily available, and are offered at the right-hand side of Table 24.4. Not all the conclusions are self-evident, however. People are still surprised to learn that the oceans are biological deserts and that the average productivity of agriculture is less than that of the typical forest.

Different ecosystem energetics are to be understood from habitat or system constraints, not from considerations of plant design. Forest ecosystems have high productivity because they are wet and warm, not because they have trees; prairies have low productivity because they are dry, not because they have grasses; lands beyond the Arctic Circle have still lower productivity because they are cold, not because they are covered with tundra.

Productivity as a systems property is nowhere better illustrated than in agriculture. We ought to expect that many forms of agriculture have lower productivity (as an ecosystems scientist calculates pro-

TABLE 24.3

Annual Primary Production of Principal Regions of the Earth

Dry biomass has been converted to carbon on the assumption that biomass is 45% C. (From Whittaker and Likens, 1973.)

	Means		Biosphere Totals			
Ecosystem Type	Mean Net Primary Productivity (g C/m²/yr)	Mean Plant Biomass (kg C/m²)	Area (10^6 km² = 10^{12} m²)	Net Energy Fixed (10^{15} kcal/yr)	Total Plant Mass (10^9 metric tons C)	Total Net Primary Production (10^9 metric tons C/yr)
Tropical rain forest	900	20	17.0	139	340	15.3
Tropical seasonal forest	675	16	7.5	47	120	5.1
Temperate evergreen forest	585	16	5.0	31	80	2.9
Temperate deciduous forest	540	13.5	7.0	39	95	3.8
Boreal forest	360	9.0	12.0	46	108	4.3
Woodland and shrubland	270	2.7	8.0	23	22	2.2
Savanna	315	1.8	15.0	42	27	4.7
Temperate grassland	225	0.7	9.0	18	6.3	2.0
Tundra and alpine	65	0.3	8.0	5	2.4	0.5
Desert scrub	32	0.3	18.0	6	5.4	0.6
Rock, ice, and sand	1.5	0.01	24.0	0.3	0.2	0.04
Agricultural land	290	0.5	14.0	37	7.0	4.1
Swamp and marsh	1125	6.8	2.0	20	13.6	2.2
Lake and stream	225	0.01	2.5	6	0.02	0.6
Total land	324	5.55	149	459	827	48.3
Open ocean	57	0.0014	332.0	204	0.46	18.9
Upwelling zones	225	0.01	0.4	1	0.004	0.1
Continental shelf	162	0.005	26.6	43	0.13	4.3
Algal bed and reef	900	0.9	0.6	5	0.54	0.5
Estuaries	810	0.45	1.4	11	0.63	1.1
Total oceans	69	0.0049	361	264	1.76	24.9
Total for biosphere	144	1.63	510	723	829	73.2

ductivity) than the wild ecosystems replaced by agriculture (Table 24.3). This is because the agricultural crop has a lesser occupancy of the land than had the original vegetation. These constraints were discussed in Chapter 3 when examining the original calculations for the low assimilation efficiency of corn. Before sowing and after harvest, a cornfield is merely bare habitat, producing nothing at all, so that the average productivity over the growing season is low. Something like this pattern occurs in wild vegetation in strongly seasonal environments, though all wild communities manage more complete cover for longer times. In a temperate forest, for instance, spring flowers cover the ground before trees carry leaves, and well before nearby fields are covered with crops. This represents production that is denied an agricultural ecosystem.

The best agriculture does much better than the averages of Table 24.3, being comparable to the highest productivity

TABLE 24.4
Relative Productivity of Ecosystems
(For details see Table 24.3)

Productivity	Ecosystem Type	Postulated Causes
High	Forests	Wet, warm, and
(>350 g	Marshes and	anchored
C/m^2/yr)	estuaries	↑
	Reefs	
Middle	Grasslands	
(70–350 g	Upwellings	
C/m^2/yr)	Lakes	
	Agriculture	↓
Low	Tundra	Cold
(<70 g	Deserts	Dry
C/m^2/yr)	Oceans	Lack nutrients

given in the table for any ecosystem. But we have better data for agriculture than for forests. A prudent ecologist would expect the best production from the best forest to exceed the best of agriculture simply because the ground is covered more completely for longer. Possibly tropical systems of farming that keep the ground covered with many kinds of crops at once would produce as well as wild vegetation on the same site, otherwise agriculture starts at a disadvantage.

There are, of course, many habitats so changed by people that the local productivity of agriculture is actually higher than that of the displaced plant communities. The most obvious examples are irrigated deserts (Figure 24.8). The increased productivity, however, is a result of the change in the nonliving parts of the ecosystem. Productivity might be improved still further by letting wild vegetation cover the site while maintaining the irrigation.

FIGURE 24.8
Just add water: Truck farm in Washington state.
The Columbia River flows through a rain shadow desert east of the Rocky Mountains. Irrigation massively increases primary productivity.

Farmers do not do this because they are not interested primarily in "productivity" but in agricultural yield of edible portions of crops.

THE MOISTURE–TEMPERATURE HYPOTHESIS

Just as the shapes of plants in formations are correlated with climate (see Chapter 17), so is it reasonable to postulate that the productivity of formations is set by moisture and temperature. Data in support of this contention are easy to come by, such as productivity measures for grasslands showing an increase of 1 gram per square meter per year for each millimeter of additional precipitation (Walter, 1962).

A more general test of the precipitation and temperature hypothesis is the correlation that is found between productivity and evapotranspiration (Currie and Paquin, 1987; Figure 24.9). Evapotranspiration is the excess of precipitation over runoff, and is, therefore, the mass of water that is returned directly to the air as vapor. In well-vegetated places we know that a large part of evapotranspiration is in fact channeled through the transpiration streams of plants, only the smaller portion being returned to the air as direct evaporation from passive surfaces. Figure 24.8 shows how close the correlation is between evapotranspiration and productivity. Various ecosystems included in the figure show their rankings in the productivity stakes as in Table 24.3. Since rate of transpiration is a function of temperature as well as of precipitation, the correlation is a test of the combined temperature–precipitation hypothesis.

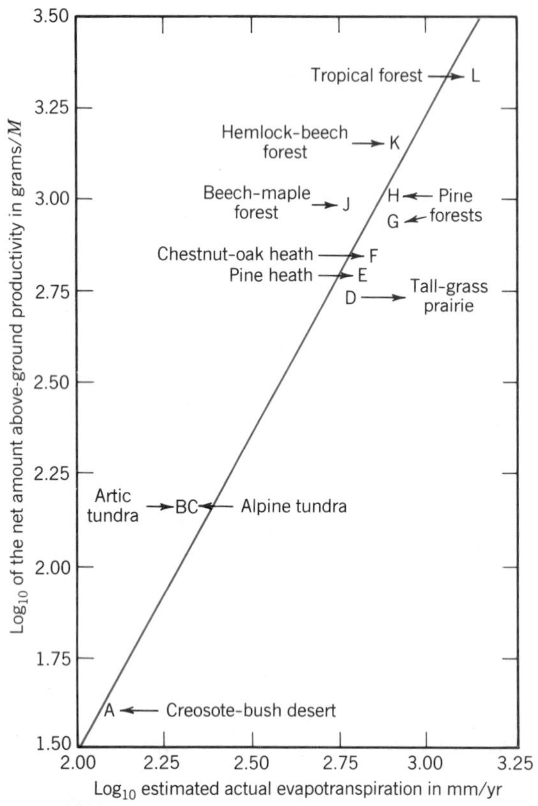

FIGURE 24.9
Correlation test of the hypothesis that precipitation and temperature determine productivity. Evapotranspiration (precipitation less runoff) depends on precipitation and temperature. The range of productivities is shown in Table 24.3. (From MacArthur and Connell, 1966.)

Model Predictions of Productivity from Climate

Lieth (1975) modeled the productivity of the whole earth from climate data. The first step in the Lieth model was to plot production data from a range of ecosystems against mean annual precipitation (Figure 24.10) and mean annual temperature (Figure 24.11; Whittaker and Likens, 1973). These data themselves show the strong underlying correlations of productivity with wet and warmth.

For each of a series of reference points on the globe, mean annual precipitation

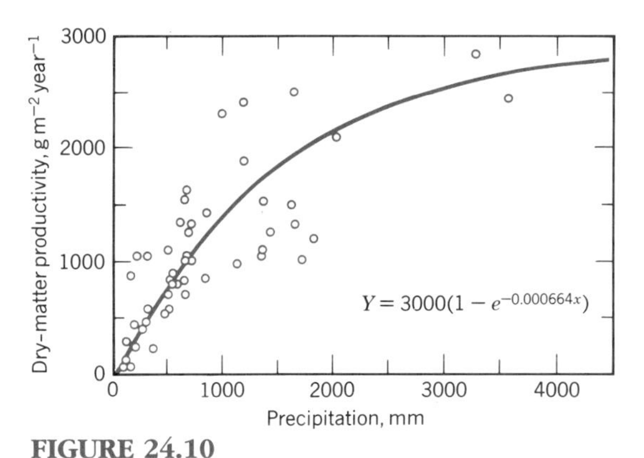

FIGURE 24.10
Productivity as a function of precipitation.
The function Y for the fitted curve was used to generate production from mean annual precipitation in the Lieth model. (From Whittaker and Likens, 1973.)

and temperature were determined independently. From the curve equations, the model then generated separately the productivity expected if temperature or precipitation at each place were the deciding factor, providing two independent estimates of productivity for each point. The model then chose whichever estimate was the lowest and accepted this figure as the

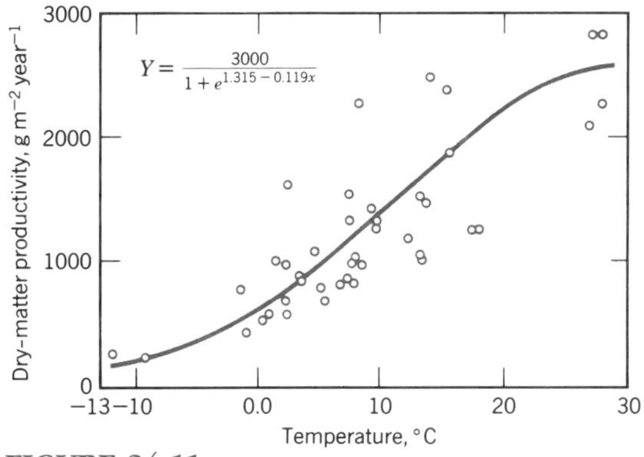

FIGURE 24.11
Productivity as a function of temperature.
The function Y for the fitted curve was used to generate production from mean annual temperature in the Lieth model. (From Whittaker and Likens, 1973, using data of Lieth.)

predicted productivity of that place. Figure 24.12 is the final productivity map of the globe that results. The productivity data of this map are strikingly similar to the direct measurements shown in Table 24.3.

It should be noted that detail is lost in going for the large scale. The global model derives productivity from temperature and precipitation alone, despite the fact that we know that many other factors regulate productivity at any one place. Nutrient supply, soil type, seasonality, plantage, plant health, grazing pressure—all these and more can influence productivity. Yet on the scale of regional ecosystems, essentially a biome scale, the influence of precipitation and temperature seems to be paramount.

PRIMARY PRODUCTION IN THE SEA

Perhaps the most striking aspect of the world productivity figures is the poor performance of the open sea. The average for the oceans of 57 grams of carbon per square meter per year given in Table 24.3 is about one sixth of the average for agriculture and not much better than one twentieth of the average for a rain forest. Table 24.5 gives a number of regional measures of ocean productivity showing variation in the open sea, but very low productivities are typical. The blue waters of the warm tropics are particularly unproductive, being truly comparable to semideserts like the Sonoran of Arizona.

Neither the supply of water nor temperature can explain low ocean productivity. Some of the most unproductive of all are tropical oceans—blue, warm, and wet, just offshore from a tropical rain forest and yet a desert (see data for the equatorial Pacific, Table 24.5).

At first sight a hypothesis of ocean pro-

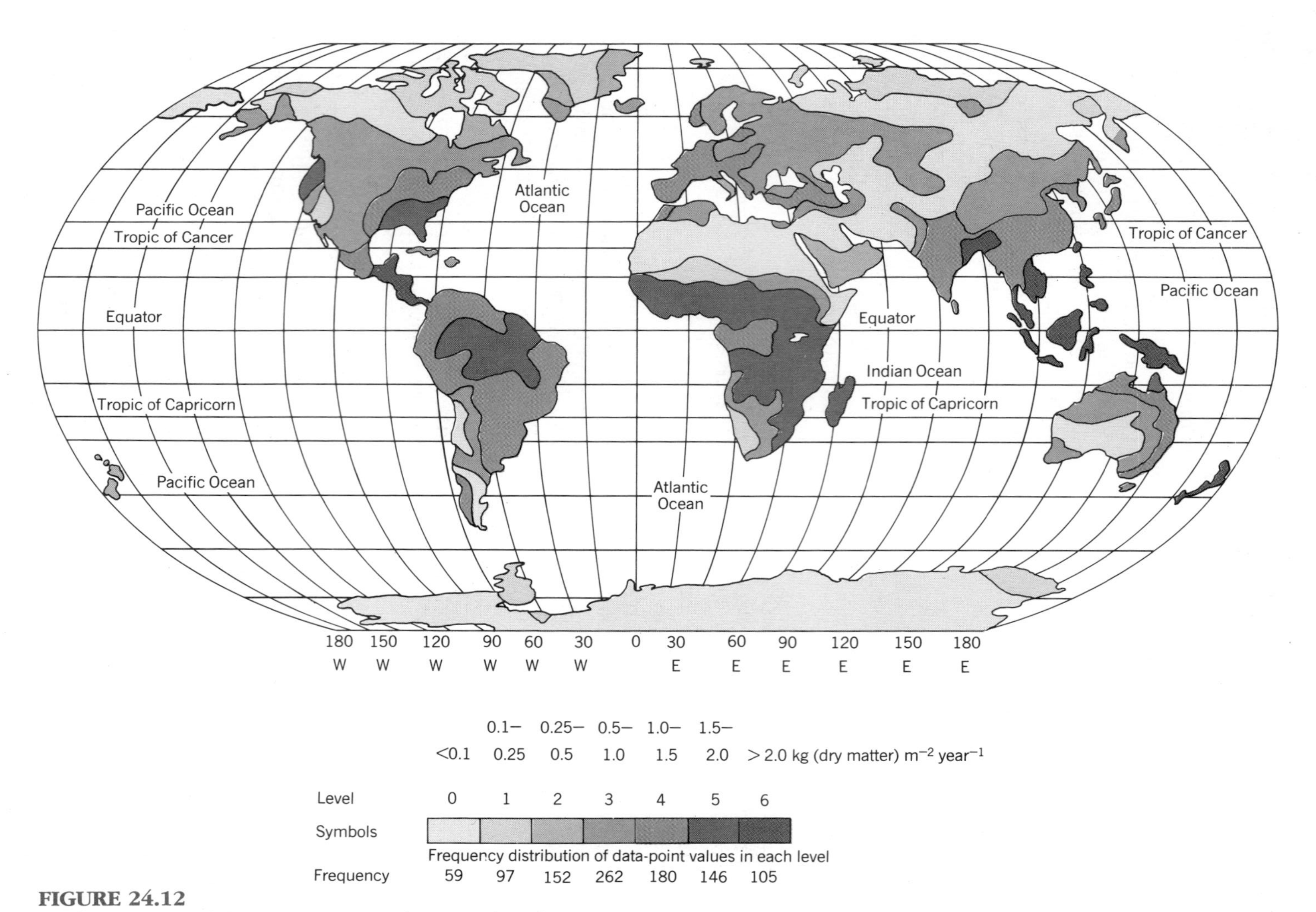

FIGURE 24.12

Lieth "Miami model" of primary productivity (1971).

The model uses data of temperature and precipitation to compute net production from the relationships in Figures 24.9 and 24.10, then accepts the lower of each pair of estimates for mapping. (Model by H. Lieth, E. Box, and T. Wolaver, from Whittaker and Likens, 1973.)

TABLE 24.5

Rates of Primary Production from Various Parts of the World Oceans

*Data for small planktonic plants are all derived from the ^{14}C uptake method (compiled by Strickland, 1960) from various sources. Data for large benthic plants include various harvest and gas exchange measurements. (*Halimeda *data from Hillis, 1980; remainder compiled by Bunt, 1975, from various authors.)*

Location	Grams Carbon (m^2/day)
Small Planktonic	
English Channel	0.50
North Sea	0.1-1.5
Danish coastal (August)	0.70
Danish coastal (March)	0.30
Danish coastal (December)	0.01
Western Barents Sea (arctic water)	1.30
Western Barents Sea (Atlantic water)	0.275
Mediterranean	0.03-0.04
Eastern Atlantic (15 miles offshore)	1.0
Eastern Atlantic (200 miles offshore)	0.15
Sargasso Sea	0.04-0.05
Pacific off Ecuador (fishery)	0.5-1.0
Equatorial Pacific (fall)	0.01
Equatorial Pacific (spring)	0.10-0.25
Sea of Japan	2.0
Off southwestern Africa (inshore)	0.5-4.0
Arctic Ocean (ice island)	0.024
Large Benthic	
Kelp beds	1.65-7.90
Intertidal (brown algae)	20.00
Tropical seagrasses	5.8
Codium (green, noncalcareous)	12.90
Halimeda (green, calcareous)	2.5
Lithothamnion (red, calcareous)	0.66
Blue-green filamentous	0.65-2.15

duction limited by light has some attractions, since light is reflected from the sea surface and some is absorbed by clear water. But the data show that light is far from limiting. Only about 5% is backscattered or reflected (Ryther, 1959) and the short wavelengths used in photosynthesis are removed in water only slowly.

The carbon supply is most unlikely to limit productivity in the oceans, because marine plants are able to use carbon from the bicarbonate–carbonate solution system as well as the limited supply from the low tension of dissolved CO_2 itself. Thus, with light, carbon, and water all apparently in excess supply, suspicion is naturally directed to supplies of dissolved nutrients.

The Classical Low-Nutrient Hypothesis for Ocean Productivity

All the nutrients needed for plant growth are present in seawater, but are usually in low concentrations (see Chapter 27 for a review of the structure and chemistry of ocean systems). Plants must live where there is light, effectively in the top 100 meters, and it is only the nutrients of these surface waters that are available to support current production. This supply can be depleted quickly. Furthermore, the open ocean does not mix very freely, being layered by temperature and salinity, so that mixing between surface and depths is impeded. This means that nutrients removed from the surface zone of light and life (euphotic zone) cannot easily be replenished from the vast hoard of the depths. Finally, the dead bodies of plants, or of the animals that have eaten them, fall through the sea, representing a continual transport of nutrients out of the euphotic zone.

The low concentration of nutrients in surface waters is compounded by the fact that oceanic plants are microscopic (see Chapter 27). This means that plants cannot build reservoirs of nutrients in their bodies. Big plants should be able to store nutrients, thus building reserves in the surface waters. This is effectively what is done on land by rain forest trees that live in nutrient-poor soils: they collect nutrients with great assiduity, then store them in their large reserves of biomass (see Chapter 17). But this is not possible in the open sea because the plants are so small.

The classical working hypothesis for the general low productivity of the ocean, therefore, is that low nutrient concentrations in the euphotic zone are the primary cause. Concentrations remain low because transport of fresh nutrients from deep water is usually impeded and because the small size of oceanic plants prevents nutrients from being stored. This hypothesis is potentially falsifiable since it yields definite predictions:

1. Productivity in the open sea will be highest where there is strong circulation of deep water coming to the surface.
2. Nutrient concentrations should increase with depth, particularly as depths greater than the euphotic zone are reached.
3. Productivity will be high whenever local circumstances permit the growth of larger plants in the sea.

Ample data are available to test all three of these predictions.

The first prediction implies that a circulation map of the world oceans is also a crude map of productivity, and the data confirm this. Figure 24.13 and Table 24.6 show data for the world oceans compiled by Russian oceanographic ships over many years (Koblentz-Mishke, 1970). This productivity map may be compared with the map of ocean currents given in Figure 24.14 (see Chapter 7). Where currents diverge, so that water rises from below, are regions of high productivity. Shallow shelf and coastal regions, where currents and storms can stir water to the very bottom, are the most productive waters of all, regardless of latitude or temperature. Deep oceans without systems of upwelling are shown as poorly productive.

Figure 24.15 shows the distribution of phosphate, nitrate, and silicate ions with depth in the three major oceans of the world (Richards, 1968). The clines of nitrate and phosphate in the surface waters (prediction 2) are particularly striking. Silicate is important to diatoms, whose skeletons are made of silicates. The importance of diatoms to the total phytoplankton flora can vary with many factors, which probably explains why there is more variation in the silicate curve from ocean to ocean than there is in the other two. But the distribution of all three essential ions with depth throughout the oceans upholds the second prediction clearly enough.

The third prediction, concerning the effect of plant size, also is abundantly upheld by the data (Table 24.5). Large plants of the sea are confined to shallow water where they can anchor, but they are always much more productive than the small plants of open water. In colder waters large plants take the form of seaweeds like the kelps, the intertidal browns and reds, or greens like sea lettuce (Figure 24.16). In warm waters the large plant biomass is associated with calcareous reefs (see Chapter 27). All these communities of large, anchored marine plants achieve productivity comparable to that of the best sites on land.

It is, therefore, a long-established general hypothesis that productivity of marine ecosystems is set by the solution or delivery of dissolved nutrients. Coastal waters and shallow banks have relatively high productivity because they are without stagnant depths to which nutrients can be lost. Regions of upwelling are fertile, because they are provided with endlessly renewed nutrient supplies. Open oceans are deserts because nutrients lost from the

(continued)

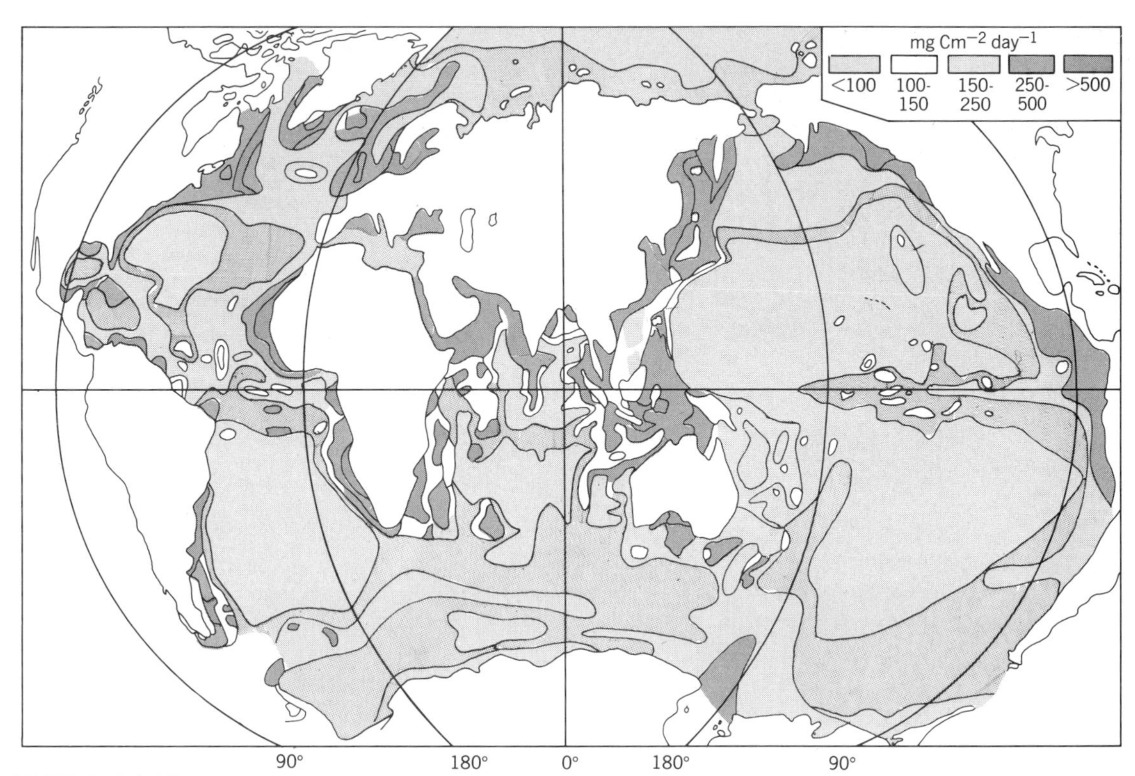

FIGURE 24.13
Productivity of the oceans.
The map is based on data gathered by Russian research ships using standardized methods. The five productivity regions mapped are described in Table 24.14. (Based on Koblentz-Mishke et al., 1970.)

TABLE 24.6
Productivity of Regions of the World Oceans
These data are based on many ^{14}C productivity measurements by Russian research ships. They have the particular merit that methods were standardized, allowing regional comparisons to be made with confidence. The data are mapped in Figure 24.14. (From Koblentz-Mishke et al., *1970.)*

Region of Ocean	mg C/m²/day Mean	mg C/m²/day Range	% Ocean	Yearly Tonnage (10 tons C/yr)
1. Blue waters of subtropical gyres	70	<100	40.4	3.79
2. Transitions	140	100-150	22.6	4.22
3. Equatorial divergence and subpolar	200	150-250	23.6	6.31
4. Inshore waters	340	250-500	10.6	4.80
5. Shallow shelves	1000	>500	2.9	3.90
Total world oceans				23.0

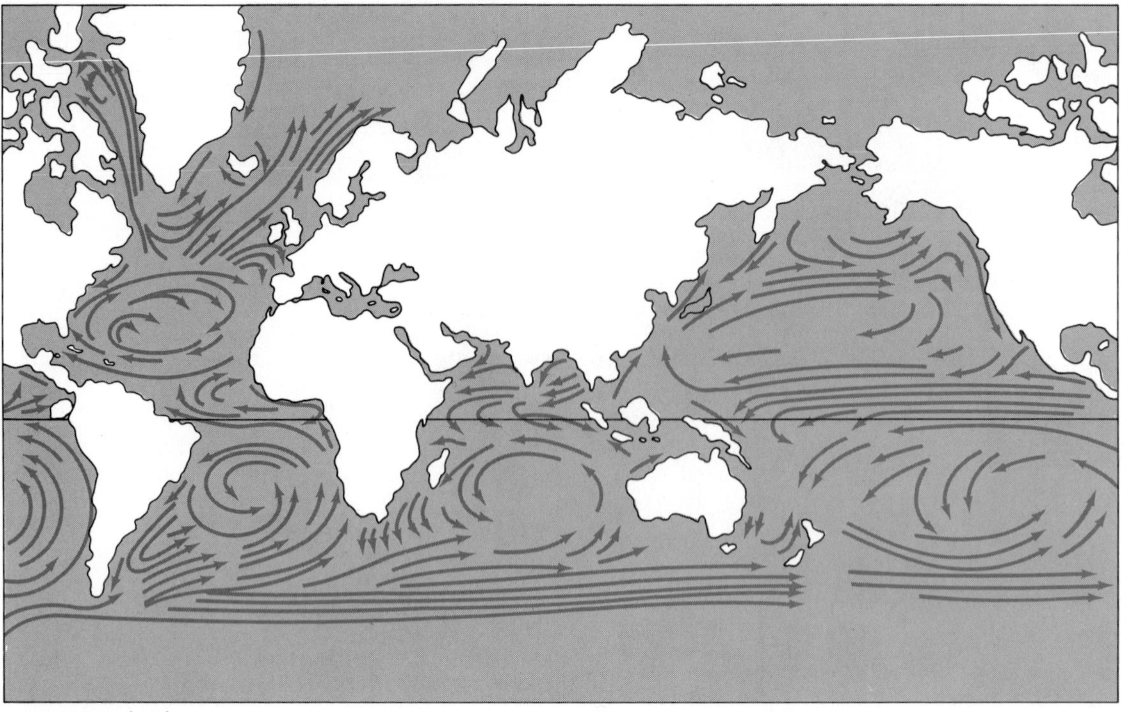

FIGURE 24.14

Winds and Coriolis effects move surface water in gyres. Water is driven round ocean basins, clockwise in the northern hemisphere and anticlockwise in the southern hemisphere.

Where currents move away from land, deep, nutrient-rich, cold water upwells, resulting in areas of high primary productivity. (From MacArthur and Connell, 1966.)

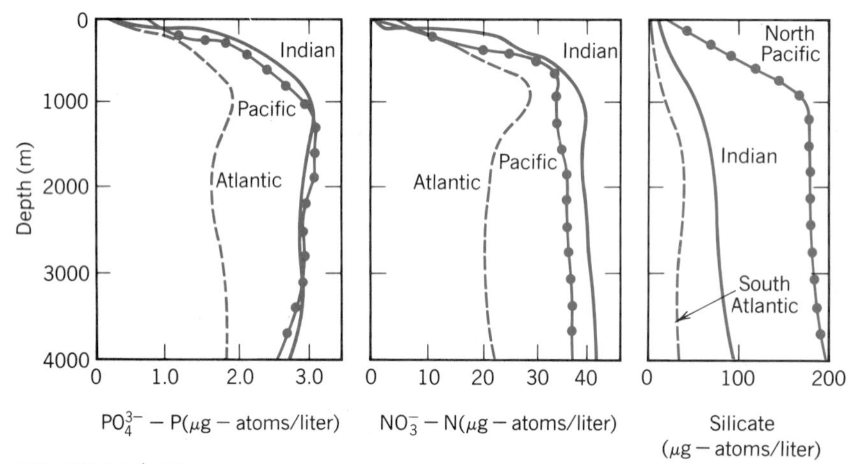

FIGURE 24.15

Nutrient concentration with depth in three oceans.

The data show that concentrations of essential nutrients are low in surface waters where light permits photosynthesis. This is predictable if ocean plants are nutrient-limited. (From Richards, 1968.)

FIGURE 24.16

***Macrocystis pyrifera:* anchored biomass in the sea.**

It is claimed that large specimens of the kelp genus *Macrocystis* can be longer than any tree on land is tall. They do grow to great lengths by being anchored in water shallow enough to be lighted to the bottom. Their anchored bulk lets them store nutrients extracted from water that flows past.

surface waters in falling organic matter are replaced only slowly.

Among the appealing features of the classical hypothesis is the way it explains the odd fact that cold oceans tend to be the most fertile. Cold water is not fertile as a result of being cold but because it is likely to be nutrient-rich, either because nutrients like phosphate are more soluble in cold water or because upwelled water from the depths is necessarily cold as well as nutrient-rich.

Iron and the Multiple-Factors Hypothesis

Phosphate and combined nitrogen are as important to photosynthetic life in the sea as they are on land. In addition, silica is an essential nutrient in the oceans because diatoms are important oceanic producers. All three nutrients can be depleted in euphotic zones (Figure 24.15).

It has been known for nearly 30 years, however, that large areas of the open oceans, particularly in the southern hemisphere, have less standing crop chlorophyll than seems warranted by the concentration of these three nutrients—particularly phosphate (Chisholm and Morel, 1991). Put simply, some other factor must limit photosynthetic life in these oceans below the limits set by these three nutrients. A recent debate among oceanographers that the critical limiting factor might be dissolved iron has spilled over into political affairs with the suggestion that fertilizing the Southern Ocean with iron would so increase productivity that the extra carbon that would be fixed could solve our greenhouse problems by taking excess CO_2 from the air (see Chapter 29).

Dissolved iron is undoubtedly a critically limiting nutrient in the sea. Fifty years ago, Hart (1934) included iron limitation among possible hypotheses to explain low productivity in the Southern Ocean, and iron has long been implicated in the termination of the spring bloom of production in places like the North Sea.

Important to the contemporary interest in iron is the suggestion that iron in the surface ocean is replenished in dust from the land rather than from ocean mixing (Martin et al., 1991). Other nutrients, like phosphate or silicate, are brought up from below, and nitrate may result from nitrogen fixation in surface waters, but iron seems to be injected from outside the system. Thus iron could be scarce at all sites far from land, even at an open ocean upwelling. This would result in precisely the observed grand pattern of open oceans having less chlorophyll than seems justified by concentrations of phophate, nitrate, and silicate (so-called high-nutrient, low-chlorophyll waters, or HNLC). Productive upwelling near shore, on this model, would have ample iron because it is within easy reach of dust supplies.

Martin and his team have tried fertilizing containers of HNLC water with iron and obtained growth rates of phytoplankton that are double those pertaining in unfertilized bottles. The enhanced growth is such that it is perfectly possible that populations of at least some part of the phytoplankton community, if supplied with iron, would grow until nitrates were depleted from the surface water.

A quite different hypothesis for the HNLC phenomenon in open oceans is that the standing crop of chlorophyll-bearing organisms is kept low by efficient grazing (Frost, 1991). This is an oceanographer's version of the Hairston et al. (1960) green earth argument, with efficient predation moved one step down the trophic pyramid (see Chapters 10, 22). Recall that in the original HSS model the earth was said to be green because herbivore numbers were kept low by efficient predation. The ocean version is that single-celled phytoplanktonic plants and photosynthetic procaryotes are likewise kept low by the efficient grazing of zooplankton.

The photosynthetic cells of the open sea are tiny, floating, and poised over a void with nowhere to hide, making plausible the hypothesis that they can suffer devastating grazing pressures from the many animals equiped to filter them out (see Chapter 27). In stable and constant conditions far removed from the turbulence of coastlines, it may be that grazing zooplankton can maintain populations so high that primary producer populations are forced down below the apparent carrying capacity of the physical habitat. Inshore, on the other hand, environmental vicissitude might prevent stable populations of grazers, letting the producers escape grazing pressures. The grazing hypothesis, therefore, is as consistent with the fact that HNLC areas are found offshore but not inshore as is the iron hypothesis.

A consequence of all enrichment experiments, iron fertilizing included, is that community compositions change. Adding iron to bottles of HNLC water favors organisms differentially, apparently larger phytoplankton at the expense of smaller producers (Cullen, 1991). How far this would go in an open system not constrained by the walls of a bottle is not known. Neither is the consequence for the system of grazing known, nor the consequence for the rate of export of nutrients in sinking organic matter.

A recent elegant symposium on this subject comprising 32 papers was printed in 470 pages of a journal, with no consensus on the general cause of the HNLC phenomenon (Chishold and Morel, 1991). The phenomenon certainly is associated with nutrient limits, though probably not just iron. Just as certainly the structure of ocean food webs is implicated, as well as the prime circumstance that primary producers of the sea are tiny. The overriding certainty is that low productivity in HNLC parts of the oceans, as elsewhere in the seas, is an ecosystem property, primarily due to physical constraints of the system.

ANIMAL PRODUCTIVITY

Secondary ecosystem productivity is much less well known in all ecosystems. Table 24.7 gives the estimates of **net secondary**

TABLE 24.7
Annual Secondary Production of Principal Regions of the Earth
These estimates are very approximate, being based on estimates of animal consumption of plants to compute herbivore consumption, and of assimilation efficiencies to compute gross secondary production. Some of the estimates are little more than intelligent guesses, probably being too high. (From Whittaker and Likens, 1973.)

Ecosystem Type	Total Net Primary Production (10^9 metric tons C/yr)	Animal Consumption (%)	Herbivore Consumption (10^6 metric tons C/yr)	Net Secondary Production (10^6 metric tons C/yr)
Tropical rain forest	15.3	7	1100	110
Tropical seasonal forest	5.1	6	300	30
Temperate evergreen forest	2.9	4	120	12
Temperate deciduous forest	3.8	5	190	19
Boreal forest	4.3	4	170	17
Woodland and shrubland	2.2	5	110	11
Savanna	4.7	15	700	105
Temperate grassland	2.0	10	200	30
Tundra and alpine	0.5	3	15	1.5
Desert scrub	0.6	3	18	2.7
Rock, ice, and sand	0.04	2	0.1	0.01
Agricultural land	4.1	1	40	4
Swamp and marsh	2.2	8	175	18
Lake and stream	0.6	20	120	12
Total land	48.3	7	3258	372
Open ocean	18.9	40	7600	1140
Upwelling zones	0.1	35	35	5
Continental shelf	4.3	30	1300	195
Algal bed and reef	0.5	15	75	11
Estuaries	1.1	15	165	25
Total oceans	24.9	37	9175	1376
Total for biosphere	73.2	17	12433	1748

production for the biosphere compiled by Whittaker and Likens (1973). These estimates are not based on measurement but are calculated from data for net primary productivity according to assumptions about the eating and digestive abilities of animals of the various places.

The procedure was to begin with plant production and to take a percentage of this as eaten by animals in each ecosystem, based on what seemed reasonable to those familiar with the appropriate plants and animals. The potential for error can be illustrated by reflecting on how little is known about the herbivore take in tropical rain forests, and yet the compilers of the table had to be bold and chose a figure (7%) as the portion of rain forest that gets eaten. The percentage of plant productivity taken by animals appears in the "herbivore consumption" column, which then has to be corrected to allow for animal defecation. The compilers took the best data for animals typical of each ecosystem to arrive at

TABLE 24.8

Ryther's Estimates for Fish Production of the World Oceans

These estimates are calculated from measures of net primary productivity. There is large uncertainty about the numbers of links in the food chains and even more about the conversion efficiencies used. The fish production estimates are probably too large. (From Ryther, 1969.)

Province	Percentage of Ocean	Area (km^2)	Mean Productivity (g C/m^2/yr)	Total Productivity (10^9 tons C/yr)	Primary Production [tons (organic carbon)]	Trophic Levels	Efficiency (%)	Fish Production [tons (fresh wt.)]
Open ocean	90	326×10^6	50	16.3	16.3×10^9	5	10	16×10^5
Coastal zone[a]	9.9	36×10^6	100	3.6	3.6×10^9	3	15	12×10^7
Upwelling areas	0.1	3.6×10^5	300	0.1	0.1×10^9	1½	20	12×10^7
Total				20.0				24×10^7

[a]Includes offshore areas of high productivity.

an assimilation efficiency for each ecosystem. From this is calculated the net secondary production in the last column. The whole exercise is highly subjective.

Very noteworthy about these estimates are the high consumption efficiencies assigned to the herbivores of marine systems, which are thought to be five times as effective at cropping ocean plants as are terrestrial herbivores at cropping land plants. Because of this, the estimate shows the oceans as having nearly as much total secondary production as the land, even though the total primary productivity of the oceans is only a third of that of the land. Something like this may be the truth of the matter, but there can be room for doubt. If it turns out that animals cannot in fact consume 40% of the primary productivity of the oceans, or 15% of reefs and anchored algae, as Table 24.7 assumes, then we shall find that secondary production does not compare nearly so favorably with that on land as some present estimates suggest.

The issue of ocean secondary productivity is important for practical economic reasons because it determines our estimates for potential yields of fish. Few commercial fish are herbivores, most feeding higher on food chains. The most-quoted estimates for potential yields of world fisheries have been those of Ryther (1969) (Table 24.8).

Ryther began with primary productivity estimates that were partly based on preliminary results of the Russian measurements and that were close to them (Table 24.6). He then estimated how many trophic levels there must be in each ocean province between plant and harvestable fish. Food chains were assumed to be long (five links) in the open unproductive oceans but only one to two links in the more productive areas.

Ryther then assigned conversion efficiencies for the transfer of energy between trophic levels. These efficiencies are not precisely defined. The efficiency used at the bottom of the food chain is between net production of plants and gross production of animals, and so is not an ecological efficiency. The efficiencies used for higher trophic levels apparently are ecological efficiencies, like that between moose and wolves calculated at 1.4% (see Chapter 5). The efficiencies used by Ryther are essentially the high efficiencies erroneously calculated by Lindeman. If game fish in the sea are more like wolves, then Ryther's estimates for ocean yields are much too high.

PRODUCTIVITY OF DECOMPOSERS

Most of the energy of net primary production goes to rot. Herbivores do not eat most plants; bacteria and fungi do. This is common knowledge to naturalists and it is given a certain formality in the estimates of animal consumption used for the secondary productivity estimates (Table 24.7). When, for instance, an ecologist assumes that 7% of the net production of a rain forest goes to the consumers, there must be the balancing assumption that 93% goes to rot. Even the optimistic assumptions of the take of plants by herbivores in the sea still leaves more than half for the decomposers. On land, there is no doubt that energy degradation by decomposers is far more important than energy degradation by all the animals combined.

But perhaps a slight eating of words is necessary here, because many of the organisms that dispose of the detritus are in fact animals. Termites in the tropics and earthworms in temperate regions are obvious examples, but there are many others. Fiddler crabs (Figure 24.17) are one of the more engaging examples, and many other benthic animals of lakes and sea find

FIGURE 24.17
***Uca pugmax:* Social equipment on a detritus feeder.**
Fiddler crabs feed on the detritus settling with mud between tide lines. The large claw of males is used in mating rituals.

TABLE 24.9
World Production of Detritus
Detritus production is calculated from estimates of net primary production by assuming how much is eaten by herbivores, etc. *(From Reiners, 1973.)*

Ecosystem Type	World Net Primary Production (dry wt.) (10^9 tons/yr)	Percentage of Production to Detritus	World Detritus (dry wt.) (10^9 tons/yr)	Carbon (10^9 tons/yr)
Swamp and marsh	4.0	90	3.6	1.8
Tropical forest	40.0	95	38.0	19.0
Temperate forest	23.4	95	22.2	11.1
Boreal forest	9.6	97	9.3	4.6
Woodland and shrubland	4.2	80	3.4	1.7
Savanna	10.5	60	6.3	3.2
Temperate grassland	4.5	50	2.2	1.1
Tundra and alpine	1.1	95	1.0	0.5
Desert scrub	1.3	95	1.2	0.6
Extreme desert	0.07	97	0.1	0.03
Agricultural land	9.1	50	4.6	2.3
Totals	107.8		91.9	45.9

much of their energy as detritus. It is because the energy flux bypassing the herbivores is so large that there are so many scavenging animals. These are properly called "**detritivores**," and they are immensely important in the economy of most ecosystems. Together with the true decomposers (the bacteria and fungi), they undertake most of the nonplant biological activity of all ecosystems.

Since plants are producing detritus continuously, a flux of this production passes through all ecosystems. Reiners (1973) has estimated the total flux for terrestrial ecosystems using a number of alternative assumptions, his extreme results differing by factors of less than two. Table 24.9 gives a representative sample of his results. He begins with standard estimates of net primary productivity, as given in Table 24.3, provides his own estimates of what animals might eat, calls the remainder detritus, and calculates accordingly. Forests are the great producers of detritus, as one might expect. The existence of numerous termites in tropical rain forests is understandable enough.

Preview to Chapter 25

Gray-brown podsolic: **fertile Alfisol of the Paris basin.**

SOILS are formed from the surface down, a process that results in recognizable layers called "horizons." Percolating water washes the surface A horizons, leaching out mineral components that may then be redeposited in the B horizons below. B horizons may also be sites of synthesis for clays and other complex minerals. The lower C horizons are oxidized and weathered parent material. The kind of soil profile is set by climate so that similar profiles may be found over areas as large as a biome. Gray or brown podzolic soils of northern regions have lost iron and aluminum, probably because they are made soluble in organic complexes. Red tropical soils with minimal organic matter retain iron and aluminum, giving them their characteristic red color. The soil profile of a climatic region is expressed regardless of the parent rock, but many local properties of soils do depend on parent material. Local soil series are defined by texture and similar physical properties, though they are usually mapped from aerial photographs showing the spread of soils by the plants they support. Soil classifications tend to be like ecological classifications of vegetation in that large groups reflect major climate types whereas smaller units of classification tend to be arbitrary.

Chapter 25

The Soil

Underlying terrestrial ecosystems is the soil, a thin layer of the earth's crust that has been remade by life and weather. Soil provides habitats with a nutrient delivery system, a recycling system, and a waste-disposal system. For plants, soils are sites of germination, support, and decay. For animals, soils are a refuge, a sewer, or even a whole environment. For decomposers, soils are a resource.

The study of soils is a discipline in its own right. It is now called "**pedology**, from the Latin word meaning "foot" ("pedestrian" and "pedal" come from the same root).

Soil is, of course, the topmost layer of the earth's crust in which plants grow. But this is an uninteresting definition. Soil is that outermost layer that has been worked over from above by processes powered by solar energy, including the work of living things. Sun, rain, and frost rework the mineral matrix from the top down in the combined process known as weathering. Plant roots and mycorrhizae change the chemistry of soil water. Living parts of ecosystems add masses of reduced-carbon compounds and complexes of combined nitrogen, both manufactured using carbon and nitrogen extracted from the air. These atmospheric products are inserted into the top meter or so of ground by plant roots and burrowing invertebrates. The result of these combined processes is the thin life-supporting layer that we call "soil," a layer often utterly different from the parent rock beneath.

Soils of different places can be amazingly different. Some soils drain, some are well supplied with nutrients, some are so acid that farmers put lime on them. Ecologists, like farmers, are aware of these sorts of differences. But ecologists with a world view find even grander differences expressed on a biome scale, as, for instance, the fact that soils in different parts of the earth have distinctive colors.

Typical soil in the temperate belt, where western agriculture had its origins, is grayish-brown. Where the wild vegetation was deciduous forest, say in northern France, the furrow turned by a plow can be richly brown when wet, turning grayish when dried. But in large parts of the tropics the soil is red, whether wet or dry, not just a reddish brown but a definite red.

Gray-brown soils tend to be associated with temperate forests and red soils with tropical rain forests or savannas. It is, on the face of it, odd that soils should be color coded by latitude. Why should the dirt underfoot be red in much of Nigeria or Viet Nam but grayish brown in New England and Europe?

An extreme version of north temperate gray-brown color is the soil found under much of the boreal forest, where the top of the ground is actually nearly white, though admittedly covered from view most of the time by a thin veneer of rotting needles and humus. Cut down the trees and plow shallowly, however, and the cleared land turns ashy white. Russian peasants call such land "ash soil" (*podzol*). Allied soldiers held prisoner in podzolic regions of Eastern Germany during the Second World War found that one of the problems of escape by tunneling was disposing of colored subsoil on top of the whitish surface of the camp compound, because colored subsoil from the tunnels was a warning signal to the guards that the prisoners were digging.

THE SOIL PROFILE

The side of a trench in any well-vegetated place reveals a succession of layers in the soil. At the surface is a litter of dead or rotting plant parts, with one or more distinctly different layers underneath separating the surface litter from the subsoil a few feet down. Sometimes these layers are sharply distinct, while other times they merge gradually into one another, but they have always been formed within the subsoil by weathering processes working down from the surface. The layers are not strata in the geological sense because they are not separate deposits. It is convenient to have a special word to describe them, and they have come to be called "horizons" by soil scientists.

Soil horizons form as rotting plant parts mix with the upper layers of the mineral soil and as drainage water percolates down through the litter to work a slow washing and chemistry on the lower horizons. The thickness of earth affected by these processes constitutes the soil. The subsoil underneath is the earth from which the soil was made and is therefore called the **parent material**. This layered appearance is called the **soil profile**, which is *the set of soil horizons between the undifferentiated parent material and the surface litter* (Figures 25.1, 25.2).

Soil scientists recognize three kinds of soil horizon above the parent material: the A, B, and C horizons. It is usually easy to separate a soil into these three horizons just by looking at it, even though there may be many subdivisions and finer layers present. The principle of this simple classification is that:

A horizons have *lost* material from leaching, though they have gained organic matter as deposits.

B horizons have *gained* material from leaching and by synthesis in situ, particularly of clay minerals.

C horizons are parent material that has been weathered, usually being oxidized or, in dry climates, having deposits of evaporites.

At the top of the soil profile (Figures 25.1, 25.2), underneath the litter of leaves, the mineral soil is colored and structured by the organic particles mixed into it by soil animals such as earthworms or by roots, and by various organic materials produced by decomposition, collectively called "humic acids." The percolating water, which is a solution of various sub-

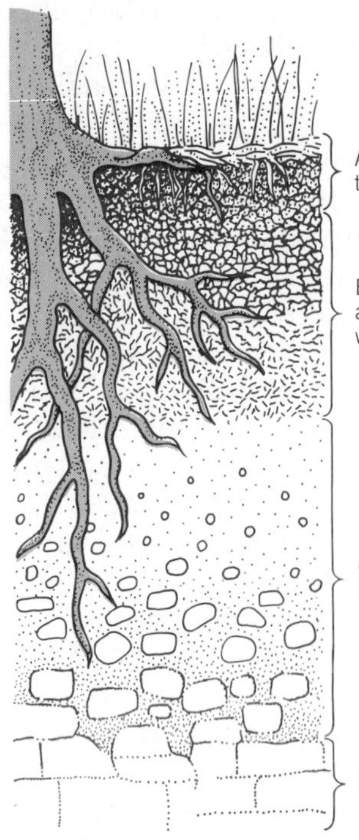

FIGURE 25.1
Idealized soil profile.
Separations between horizons are functional: material is removed from A horizons, material is added to B horizons, oxidative weathering reaches C horizons, and the R horizon is the unaltered parent material. (See Figure 25.2 for more detail.)

stances washed out of the litter, dissolves anything soluble in the surface of the mineral soil and carries it down to deeper horizons. Percolating water also removes finer particles from the top horizons and carries them physically downward. The horizons at the top of a soil profile that are being continually denuded in this way are called the **A horizons** of the soil and may be defined as *the set of layers at the top of the soil profile that tends to lose matter to the soil water or to the layers below*. A horizons are sometimes called elluvial horizons.

Underneath the A horizons is a second set of horizons that has caught matter washed down from the top, such as fine particles trapped in the spaces of the deeper soil as in a filter, or solutes redeposited over the immense surfaces of that filter bed. The result is horizons in which there has been redeposition. These **B horizons** may be defined as *the middle group of horizons in which there is redeposition or entrapment of matter brought down from above*. B horizons may also be called illuvial horizons.

A point to notice about the chemical events leading to the enrichment of the B horizons is that they involve synthesis as well as transport and redeposition. The clay particles of the B horizons, for instance, are not just broken-down fragments of rock minerals but are plate-like

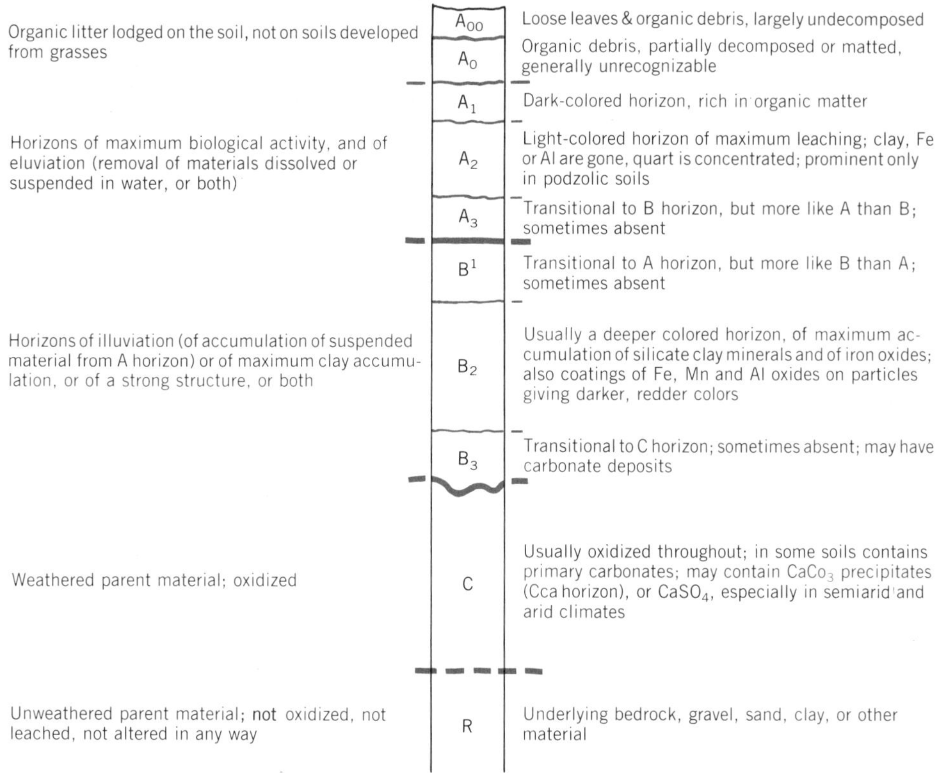

FIGURE 25.2
Hypothetical soil profile as shown in U.S. Soil Survey manuals.

aluminosilicates formed in situ. The structures of common clay minerals are given in Figure 25.3.

C horizons are weathered parent material, typically oxidized by exposure to free oxygen in the atmosphere or from oxygen in solution (see Chapter 26). In dry climates, C horizons may have concretions or accumulations of evaporites like calcium carbonate or gypsum, but none of the extensive redeposition of iron, aluminum, silicate, or organic compounds characteristic of B horizons, nor an accumulation of clay through synthesis.

Under these three groups of horizons is the unaltered parent material. In the older soil literature these were called D horizons, but they are now more usually called R horizons (R for "regolith") following the usage of the U.S. Department of Agriculture Soil Survey manuals. The parent material of an R horizon can be any surficial geological formation: bedrock, gravel, sand, clay, loess, glacial till, etc.

It will be evident that the thickness and complexity of a soil profile is a function of time and weather, as well as of the plants that grow there, themselves functions of time and weather. In a desert, for instance, the soil profile may be just a thin surface layer slightly added to by plants and slightly washed by infrequent rain, with the parent material underneath. This soil might be called an A–C soil, to record the

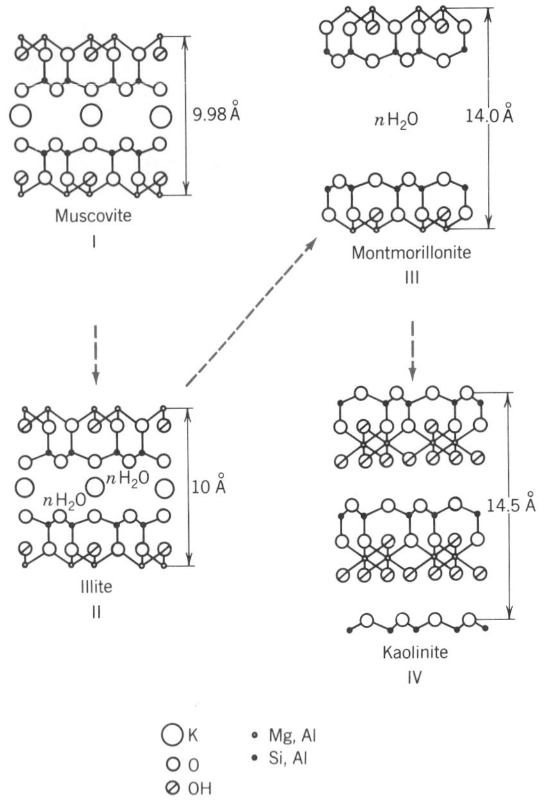

FIGURE 25.3
Structures of clay minerals.
The minerals are silicate lattices with various hydrations of magnesium, aluminum, and potassium. They may form in large plates of which only small sections are shown in edge view. Molecular water is held between the plates of the lattice of illite and montmorillonite, which makes those clays of prime importance in soils. The four lattices are shown in the sequence in which they can be weathered from the primary mineral muscovite (a mica). Lattice widths are given in angstroms. (From Thomas, 1974; Huckel, 1951.)

fact that there was no discernible B horizon of redeposition or clay synthesis.

In well-watered, well-vegetated places we should expect a thick B horizon where mineral synthesis is active. An array of clay minerals in this B horizon would serve to collect nutrients and regulate their supplies (see Chapter 23; Figure 25.4). If the

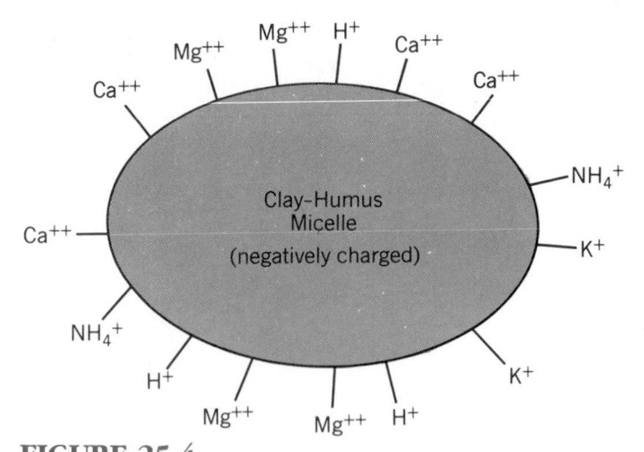

FIGURE 25.4
Clay–humus micelle.
Complexes of fragments of clay lattice and humic substances with net negative charges form in soil. At low pH the charges will be neutralized by protons (H^+), but these can be displaced by cations. Divalent ions like Ca^{++} and Mg^{++} bind strongly to the micelles, giving soils the power to retain nutrients.

soil is in a latitude where earthworms can live, then the array of minerals synthesized in the B horizon will have been moved physically up through the soil, left on the surface in worm casts, and churned back into the A horizons.

The soil profile, therefore, is an instant indicator of important ecosystem processes. These processes proceed with remarkable independence of the parent rock. It is true that quite different mineral substrates, say, alluvial clay as opposed to a granite regolith, do affect the soil profile to some extent, but the influence of mineral rock is not nearly as important as someone approaching the study of soils for the first time would expect. The soil profile peculiar to a particular climate or biome often may be expressed regardless of the type of crustal rocks.

In every soil we find traces of two quite different origins. There will always be the imprint of geology, the traces of fabric and mineral array that are derived from a particular rock type, but these traces of the mineral origins will always be overlaid by a soil profile left in the ground by forces working from above. This profile is the true reflection of the soil, the product of organic and mineral synthesis that has very little to do with crustal rocks.

PRIMARY CLASSIFICATION: THE GREAT SOIL GROUPS

Just as the first botanists who had the means to travel the globe found that the forms of plants changed from place to place, so the first people who traveled with spade in hand found that the forms of soils varied from place to place. The pedologist's "formation" is called the "**great soil group.**" It happened that the first inspired digging of holes was in Russia, where a Tsarist government thought it would be worthwhile to classify soils in order to provide an objective basis on which the land could be taxed.[1] As a result, soil names in wide use around the world are Russian names. These names are used on FAO/UNESCO soil maps and in most national soil surveys, and were also the standard in the United States until the 1960s. The U.S. Soil Conservation staff developed an alternative system in 1960, in which the great soil group was replaced with a somewhat different unit, the **soil order**, and the Russian names were replaced with new names invented from Greek and Latin roots (Table 25.1).

On a world scale, the great soil groups are the most easily mapped of any class in a soil classification. They are to pedology what the biome is to ecology, and we can talk of the characteristic profile of a soil in one of the great groups as we can talk of the characteristic profile of vegetation in a biome.

Color and color-banding of the soil profile distinguish the great soil groups in the way that plant shapes distinguish the biomes. Podzols have their bleached ashy layer under the litter surface, and the dark red and black bands of iron and humus lower down. Under the temperate deciduous forest, soils are tinted in rich browns when moist. Many deep soils under tropical rain forest are red, some of them vividly so. Soils under prairie and steppe may be the color of chocolate, sometimes flecked or banded with white carbonates. Only color plates, or the real thing, can properly show this rich variety of color.

The following sections describe the more important of the great soil groups, characteristic as they are of the principal biome types. Traditional names, some of them with Russian roots, are given first for each and the U.S. Soil Order names of their nearest equivalents (Table 25.1) follow. (For descriptions of the biomes see Chapter 18.)

Podzols: Spodosols; Boreal Forest Biome

In a well-chosen site under boreal forest the soil profile is richly banded. At the top is an edgeways view of the fragrant carpet of brown needles typical of the boreal forest. These needles retain their shape. They are the A_{01} horizon (organic horizon of litter that retains its form). Immediately under the needles is a thin layer, perhaps 1 cm thick, of black, slimy, formless humus, the A_{01} horizon. Then comes a horizon of mineral soil stained dark with humus, the

[1]The Russian scientist credited with the pioneer classification of soils was V. V. Dokuchaiev (Muir, 1961). As de Candolle had identified what we now know as biomes from the changing shapes of plants across continental distances (see Chapter 18), so Dokuchaiev's work made possible the recognition of those same biomes from their characteristic soil profiles.

TABLE 25.1

Soil Orders Recognized by U.S. Soil Survey Manual

These orders do not correspond directly with great soil groups, though there is a rough correspondence. Other orders not in the list are recognized.

Name	Derivation	Classic Great Soil Group
Entisol	Nonsense symbol	Azonal [simple profile or purely local as waterlogged (gley) soils]
Inceptisol	L. *inceptum*, beginning	Brown forest soils, some poorly drained (gley) soils
Aridisol	L. *aridus*, dry	Desert soils
Mollisol	L. *mollis*, soft	Chernozems and prairie soils
Spodosol	Gk. *spodos*, wood ash	Podzols
Alfisol	Nonsense symbol	Gray-brown podzolic soils
Ultisol	L. *ultimus*, last	Red-yellow podzolic soils
Oxisol	F. *oxide*, oxide	Latosols and laterites
Histisol	Gk. *histos*, tissue	Bog soils

A_1 horizon, and under this a striking pale gray or even white horizon that has evidently been bleached, the A_2 horizon. It is this bleached horizon that gives a podzol its name because plowed podzols look as if they have had ashes scattered over them.

The B horizons of podzols are nearly as spectacular as the A horizons, sometimes including an "iron pan" layer stained red with ferric oxide and another layer stained bluish-black with organic matter. Under this chromatic array will be the C horizon, perhaps with a quite different color given by minerals of the parent rock. People who have never seen a podzol are urged to dig a hole the first time they get into the boreal forest; it is esthetically satisfying.

No other soils are as prettily banded as podzols, which fact has resulted in hypotheses to explain the banding being generally included in the podzol description. The standard hypotheses rely on observations of the soil fauna: there are no earthworms to be found in podzols; earthworms do much digging; if the podzol had been dug over the pretty horizons would all have been mixed up to form the hue of mud; therefore the presence of the pretty bands reflects the absence of earthworms. A bit circular but probably true. The ecological question of why there are no earthworms in podzols is, perhaps, not so easily answered.

Podzols represent the result of soil-forming processes in temperate or cool regions in their most extreme expression. The principal observations are:

(a) There is an accumulation of organic matter at the surface.

(b) Percolating soil water has to pass through a layer of decomposing organic matter, where it acquires many humic solutes and a low pH (typically pH 4).

(c) An upper horizon of mineral soil is partly bleached. The pale color is found to be due to the fact that brightly colored iron and aluminum oxides have been removed, leaving the horizon enriched with silicates such as quartz.

(d) Iron and aluminum oxides and hydrates (the so-called sesquioxides) together with organic colloids are collected in the B horizons. This enrichment of iron and aluminum has apparently resulted from a process of transport and redeposition.

Tundra Soils: Included with Entisols; No Direct Equivalent in U.S. System

The profile of a tundra soil is given in Figure 25.5. The ground under the soil is permanently frozen (permafrost), and a thick blanket of dead organic matter under tussock life forms of tundra sedges and grasses insulates the frozen mineral soil from the summer sun. Soil under tundra vegetation may not thaw down to more than a few centimeters within the mineral parent material, preventing any deep development of soil profiles. Possibly one way of looking at a tundra soil is to think of it as an incipient podzol prevented by ice from developing.

On arctic sites so well-drained that permafrost is absent (gravel beach ridges), podzolic soils such as the brown arctic soil (Tedrow and Hill, 1955) develop, but these are local peculiarities.

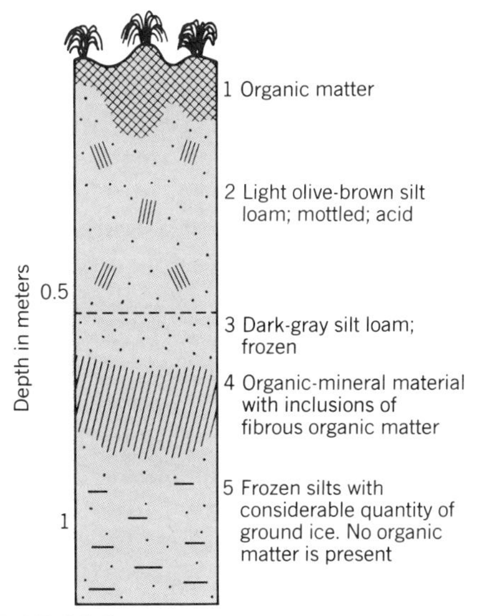

FIGURE 25.5
Profile of tundra soil.
Diagram of typical tundra soils of northern Alaska. Surface material is buried by movements of the freeze–thaw cycle more than by leaching in solution. (From Tedrow, 1965.)

Gray-Brown Podzolic Soils: Alfisols; Temperate Deciduous Forest Biome

These are the soils that underlie the brown fields of classical agriculture in Western Europe and New England, land that was originally forested with hardwoods. At the top is mulchy leaf litter, rotting underneath into humus that is intimately mixed with the surface mineral soil. The mixture of organic matter with mineral soil spans perhaps 10 or 20 cm of dark, fertile-looking earth, quite like the surface of a plowed field or a garden. Earthworms, which live on the nearly neutral leaf litter, have done the digging and mixing. The surface soil shows little visible sign of podzolic bleaching as a result, but a reddening of the B horizon farther down tells of materials brought down from above.

The absence of a distinctly bleached A_2 horizon is certainly related to the higher pH of soil under broad leaves rather than coniferous needles, as well as to the presence of earthworms. Under deciduous forest the horizons grade into each other with subtler changes of hues: brown litter, a dark mixture of mineral soil and humus, the deep brown B horizon of mineral accumulation and clay synthesis, and thence, gradually, the parent material. In its finest development, as in parts of Europe where it was first described, this soil is known as a "brown earth."

Tropical Lateritic Soils: Oxisols; Tropical Forest Biome

Laterites are the thick, red, and heavily weathered and altered surface strata of the tropical ground. Theirs is the red color of much of the wet tropics. They are defined by mineralogy and gross structure, particularly in that they have been enriched with oxides and hydrates of iron and aluminum that were left behind while other elements of the original rock were leached away. It is this accumulation of iron and aluminum, at the expense of lost silica, that gives laterites their red color. A laterite formation can be tens of meters thick, elegant testimony to the power of tropical weathering. One kind can be dug wet, shaped, and allowed to dry into excellent durable bricks in the sun; hence the name "laterite" (*later* means "brick" in Latin).

Laterites, because of their age, thickness, and the variety of processes that go into their formation, can be said to be part of the geology of an area. But plants grow in them and they weather from the top, so they are also soils, hence lateritic soils. The U.S. Soil Survey term "oxisol" clearly separates the soil interest from the geological. The profile, for as deep as an ecologist is likely to dig, consists of a few centimeters of leaf litter (A_0), an A horizon identified mostly by a slight browning of humus, and the red array of weathered residue consisting of the synthesized iron and alumina complex. It is difficult to decide where the A and B horizons divide.

Important aspects of an oxisol for an ecologist to notice are that it is extremely poorly supplied with plant nutrients, it has very little organic matter either on or in it, and it is red. The red color itself reflects the fact that iron and aluminum have been preserved at the expense of white and gray silicates, apparently the opposite of the process that leaves podzolic soils both gray and silicate-rich.

Oxisols can fairly be called the characteristic soils of tropical forests, particularly tropical rain forest. Soils on old laterites, however, are almost ubiquitous in the tropics, being found under almost every vegetation type from tropical savanna to rain forest. A wet episode sometime in the history of a tropical locality is probably sufficient to effect the necessary weathering, removing silicates and leaving iron–aluminum complexes behind. A seasonal tropical savanna, therefore, can have a red soil not very different from the red soil under a tropical rain forest.

It should be noted, however, that other soil types can be found locally under tropical forest. In the Amazon basin, rain forest can be found on outcroppings of ancient quartz sands tens of meters thick. Tropical weathering on these sands necessarily results in a silica-rich profile, to which the name tropical podzolic soil is given.

Chernozems and Prairie Soils: Mollisols, and Desert Soils: Aridisols; Grassland Biome

Under grasslands, such as the Russian steppe or the American prairie, is a more striking soil. Here rainfall is so limited that water seldom or never percolates all the way through the soil to drain away. Instead, evaporation at the surface draws water back after it has penetrated to the B or C horizons. A profile forms within the depth reached by the percolating water, but with some peculiar characteristics. Dead grass parts decompose slowly so that a thick, largely organic layer builds up at the top of the soil, forming the famous black earth of the wheat lands. This black peaty layer grades gently into mineral soil below and is mixed with it by burrowing animals, although this mixing is not done as thoroughly as in the wetter soils under deciduous forest. The result is a deep black soil, turning gray toward the bottom although still dark, and then terminating abruptly when the line of lowest water penetration is reached. A mineral band, white with carbonates, often marks the level where the drying water has left its load of dissolved matter. Below this, and sharply distinct, is the parent material.

A prairie soil is the result of this process in the wetter grassland climates, where higher production contributes most organic matter. The variant called a "**chernozem**" (Russian for black earth) is found in slightly drier climates where redeposition of calcium carbonate in abandoned animal burrows is prominent, sometimes as a distinct band called a $CaCO_3$ horizon. Versions of this soil in the American Southwest are called "**caliche**," after the local name for the thick carbonate deposits in them.

Black soils of tropical regions with strongly seasonal climates are also mollisols. The requirements are base-rich parent rocks and a hot climate with insufficient rain to wash the soil through. Like chernozems and prairie soils, these **tropical black soils** develop white concretions of carbonates, and are accordingly called "margalitic" soils (from the Latin word *marga* meaning "lime"). The famously fertile soils of central India are margalitic soils.

Where rainfall is very slight (hot deserts), the influence of percolating water is so attenuated that a B horizon of redeposition can scarcely be recognized and a simple desert soil (aridisol) having only A and C horizons results.

All these soils of dry places have in common that they are not washed completely through by percolating water. Pedologists call them collectively "**pedocals**" to distinguish them from the "**pedalfers**" of wetter places where water washes right through the soil to the water table. This separation is used as the basis of some soil classifications and is a consideration in most.

CORRESPONDENCE OF SOIL WITH VEGETATION AND CLIMATE

The world soil map that appears in every school or family atlas is a rough plot of the great soil groups (Figure 25.6). These maps have a very strong similarity to maps of climate or vegetation in the same atlas (Figures 7.9 and 18.1) This is reasonable enough, because both soil and vegetation are influenced by climate. Mapmaking itself, however, can reinforce this fundamental relationship. Once cartographers mapped vegetation and called the maps "climate" (Chapters 7, 18). In a like manner, cartographers needing a soil map tend to map the plants and call the results "soil." A certain correspondence between the resulting maps is certainly to be expected.

Cartographers making soil maps have little option but to map plants and call the result a soil map. Once the soil profile of a region is determined from sample diggings, the plants must be used as indexes of the territory covered by that kind of profile. On all scales, from satellite images to air photographs, soil surveyors are forced to infer soil type from patterns of vegetation.

Early maps of world soil were not so much sets of data as statements of a hypothesis: the hypothesis that vegetation and climate determine the soil of any particular place. This hypothesis is illustrated by a much-copied chart (Figure 25.7) by the geographers Blumenstock and Thornthwaite (1941). It shows the earth nicely parceled out into regimens or systems. If the plant formations are imagined as set out between mappable frontiers in the nation-state model, the world is set into neat system-cells. It was an idea much

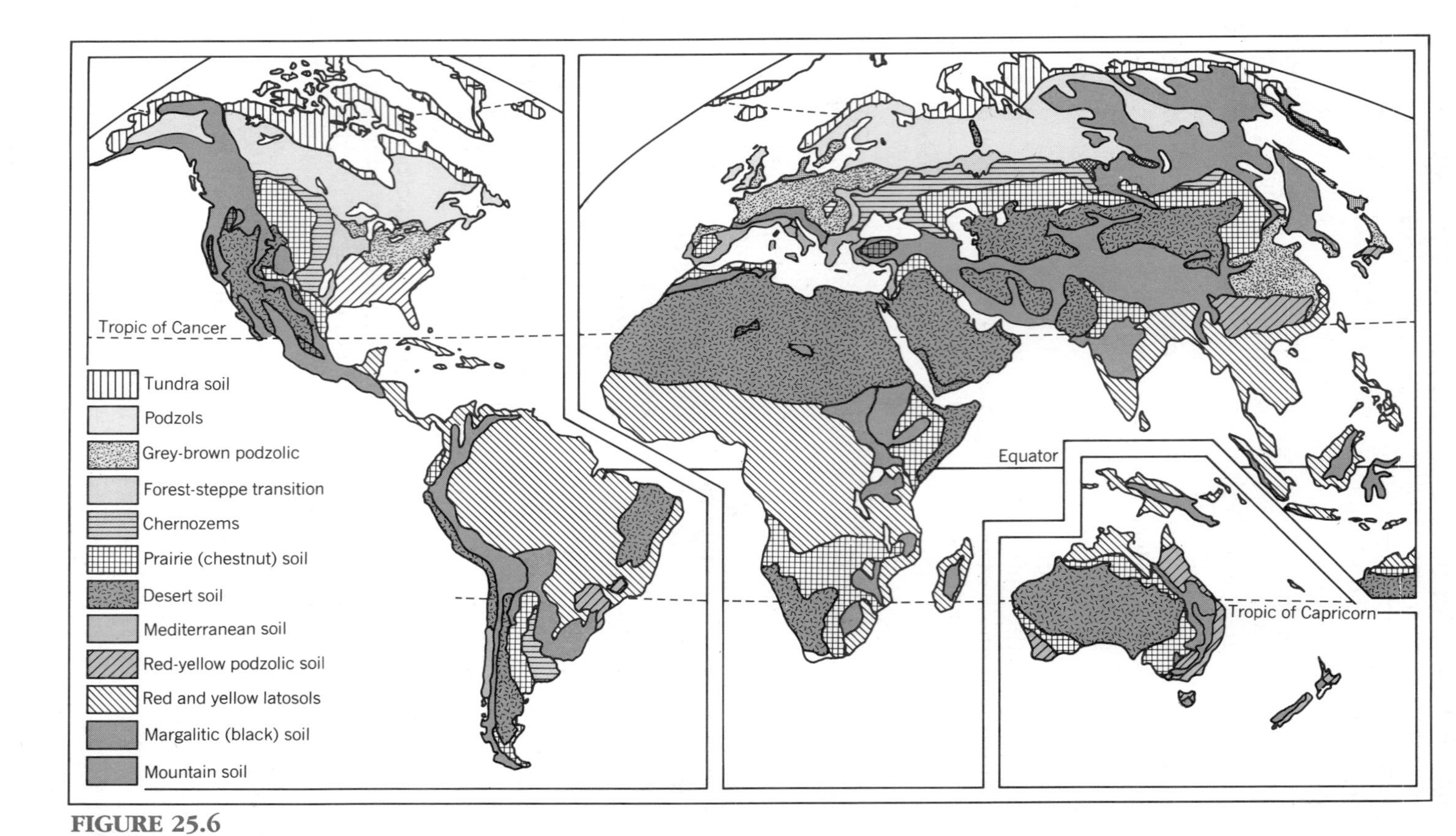

FIGURE 25.6
World soil map. (Based on Bridges, 1978.)

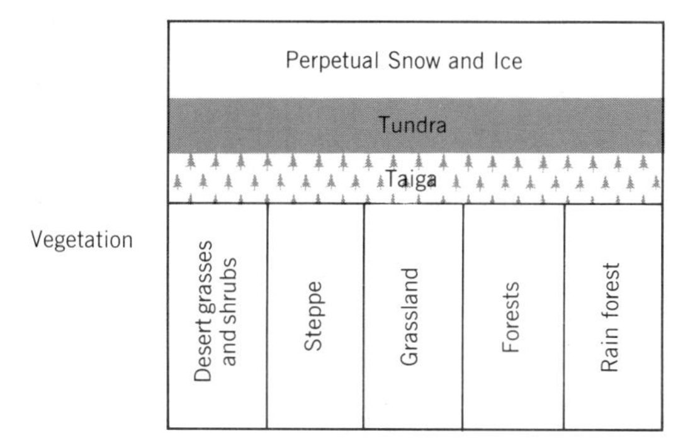

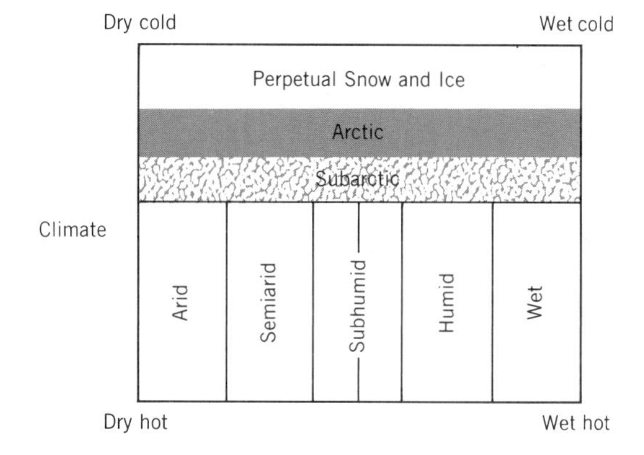

Perpetual Snow and Ice
Tundra soils
Podzols
Soils
Desert soils
Chestnut and brown soils
Chernozems and prairie soils
Podzols
Gray-brown podzolic soils
Red and yellow podzolic soils
Latertic soils

FIGURE 25.7
Conventional vegetation, climate, and soil diagrams.
These diagrams summarize the hypothesis that mapping formations yields maps of climate, and that maps of formations and climates combined yield maps of the great soil groups. As a grand generalization the hypothesis provides a useful way of thinking about the earth. But it is well that we draw the earth as square, thus accentuating the fact that we are being idealistic. Few of the boundaries shown in these diagrams can actually be found on the ground. (*Taiga*, which appears in the top diagram, describes the belt of more or less open forest that commonly lies between the tundra and the closed boreal forest; many would include it in the term "boreal forest" when speaking in very general terms.)

in the minds of ecologists when the concept of the ecosystem was invented. It is, however, a poet's view of the earth, holding a core of fundamental truth but overlooking many a contradictory detail.

Mapping world vegetation as a guide to mapping soils probably is less satisfactory than as a guide to climate. A true map of even the most distinctive of the great soil groups would not match boundaries with plants very closely. Thus a true map of all the podzols of the world might look quite different from any known map of formations or biomes.

But on the local scales in which most soil surveyors do their mapping, plant cover can be an accurate index of soil type. National or local soil surveys are less interested in great soil groups than they are in fertility resulting from fine structure, texture, acidity, mineral composition, and drainage. They must classify those soil properties that derive from the mineral composition of the parent rock or local drainage under a common climatic regimen.

SOIL SERIES, CATENA, AND PEDON

Practical soil surveyors map soils on the scale of farm fields or parts of counties. Usually only one of the great soil groups of the zonal classification will be present because all the soils of an area small enough to be mapped on this scale are likely to have had similar histories under similar climates. Yet the units called "**soil series**" that appear in a soil survey are real units. They can be mapped with reasonable precision and are usefully different. They are the expression in soil of patterns in the underlying rock.

A **soil series** is a *group of soils developed from the same kind of parent material, by the same genetic combination of processes, and whose horizons are quite similar in their arrangement and general characteristics* (Buckman and Brady, 1969).

Practical soil surveyors are soil taxonomists. When they describe their soil series they look at soil and handle it as a biologist examines a specimen. An auger brings up samples from the different horizons, a rubbing between the fingers gives the "texture" (proportions of sand, silt, and clay), a glance tells the color; the like has been seen before, it is given a series name. This works because, naturally enough, outcrops of the same geological formation in the same neighborhood yield the same physical and chemical structures in the soil.

An alternative unit of soil taxonomy is the "**catena**." This unit was developed from the observation that local topography can alter many fundamental qualities of the soil, even though the soil profile is built from the same parent material. In particular, poorly drained bottom land will develop a soil series that is distinctly different from that developed upslope on the same outcrop.

The concept of the catena of related soil series was first developed as a practical mapping unit in East Africa, where undulating topography repeatedly rang the changes of a sequence of a few soil series, but it has been found to have very general utility. Where catenas are recognized, the set of associated soil series of each will blend together, as the drainage blends down a slope from good to bad.

It is obvious that fine details of soils will vary within any soil or catena; these units, therefore, will always be somewhat arbitrary. We rely on the training of the soil surveyors to yield useful assignments of a soil to one series or another. This leaves to an individual worker the task of deciding what to describe.

A soil surveyor is forced to share with plant community analysts the difficulty of deciding which is the best piece of landscape worthy of detailed description. "Where do you dig your hole?" and "How large should the hole be?" are questions akin to "Where do we place our vegetation plot and how large an area should it cover?" (see Chapter 19). The compilers of the U.S. Soil Survey Manual make a useful contribution toward the resolution of this difficulty with their concept of the "pedon."

A **pedon** is a sample piece of the local soil. The soil surveyor decides what a typical specimen of the soil being mapped is, then describes it. The description must include the soil profile plus a number of physical and chemical measurements on all the horizons, so that the actual sample of a type soil has to be three-dimensional: a volume of ground rather than an area. The surface area of this volume must be large enough to include the minor irregularities that occur in all soils. The procedure is entirely empirical and based on the judgment of the soil surveyor. Practical experience has led to the following definition of the resulting unit: a **pedon** is a *small volume of soil large enough for the study of horizons and their interrelationships within the profile and having a roughly circular lateral cross section of between 1 and 10 meters.*

We get from the pedon to the soil series by means of the "polypedon," defined as a group of pedons contiguous within the soil continuum and having a range of characteristics within the limits of a single soil series.

Ecologists will sympathize with the difficulty of defining these various units when they reflect on their own difficulties in defining units in the endless continua of plant communities (the units association, sociation, and stand come to mind; see Chapter 19).

PROBLEMS OF SOIL TAXONOMY

Whenever a scholar classifies, whatever the subject of interest, the aim is to group the similar into categories, and then to group the categories into higher categories so that a hierarchical order results. In biology, of course, this exercise led to major discoveries, the processes of evolution and natural selection among them. But efforts at the taxonomy of soils have not resulted in any comparable discoveries.

The most ambitious attempt to classify soils is that of the Comprehensive System of the U.S. Soil Survey. This system was devised with the deliberate intention of finding natural order in the soils of the world. In a passage written in an apparent spirit of advocacy in a standard soil text (Buckman and Brady, 1969), the U.S. Comprehensive System for the classification of soils is compared directly with the Linnaean method of classifying living things, as shown in Table 25.2. The comparison in fact shows up the critical differences that make the comparison invalid.

Table 25.2 illustrates the formal classification of the Miami silt loam, a local soil of the American Midwest. The soil profile is podzolic, making it an *alfisol.* It has the deep profile that we expect in the local humid climate, making it an *udalf.* Next there is the prefix *hapl,* which separates a *hapludalf* from an *udalf,* and this prefix simply means that the diagnostic horizon is minimal. So we have the sort of profile that the podzolic process yields in a wet place: podzolic, but without the beautiful array of horizons expected in the cooler clime of a spruce forest.

Up to this point, the classification speaks only to the effects of climate. To go further in the classification, we must abandon classifying the effects of climate to a large extent and classify by texture, as

TABLE 25.2
Linnaean and Soil Taxonomic Systems Compared
The soil taxonomy is that of the comprehensive system of the U.S. Soil Survey. This system is NOT comparable to the Linnaean taxonomy of species because different kinds of criteria are used at different stages in the hierarchy.

Plant Classification	Soil Classification	
Division: Pterophyta	Order: Alfisol	Climate
Class: Angiospermae	Suborder: Udalf	
Subclass: Dicotyledoneae	Great Group: Hapludalf	
Order: Rosales	Subgroup: Typic Hapludalf	
Family: Leguminosae	Family: Fine loamy mixed mesic	Parent material
Genus: *Trifolium*	Series: Miami	
Species: *repens*	Type: Miami silt loam	

in the lower half of Table 25.2. Only by using properties principally derived from the parent rock can we separate the ultimate soil series, the pedologist's apparent equivalent to a biologist's species.

The similarity of the classification of the Miami soil series in Table 25.2 to the taxonomy of sweet clover is superficial. All the taxa of higher order than species in the Linnaean system have the logical support that they reflect the workings of the same processes that separate the species themselves. Genera, families, and orders are somewhat arbitrary groups, but all can be defended as reflecting ancient splits in species populations comparable to the splits that make modern species. Species, families, and orders all originate in acts of speciation. The pedologist's families, groups, and orders have no like relationship to their soil series.

A more valid comparison is between soil taxonomy and phytosociology. Figure 25.8 shows how both soil and vegetation taxonomists can separate biome-sized groupings by climate, whereas both must use arbitrary criteria to separate lower-order groupings. Soil classifications are bound to remain as artificial as the attempt to order plants into natural societies (see Chapter 19).

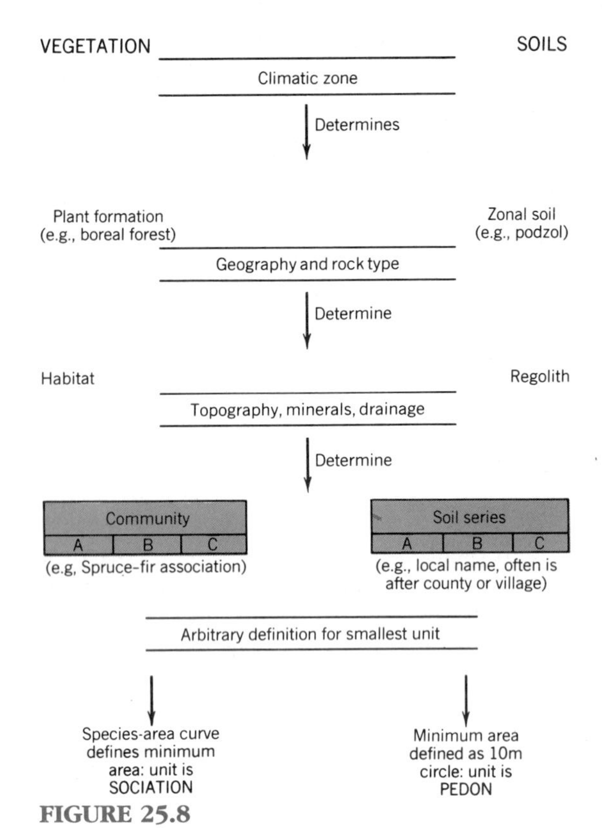

FIGURE 25.8
Soil and vegetation taxonomy compared.
Making a hierarchical classification of soils involves decisions comparable to making a classification of plant communities.

THE PROBLEM OF THE RED AND GRAY

North-temperate regions are gray or brown underfoot because their soils are podzolic. Gray silica minerals have accumulated and been washed clean of iron and aluminum compounds. We may say, therefore, that north temperate lands typical of the European west are tinted toward the gray or gray-brown because washed silicates collect at the surface of the ground. Extra pigment is supplied to this silicate paint by organic matter and the clays synthesized in the B horizons.

The wet tropics are red because silicate minerals, far from being conserved in the soil, are washed away. Oxisols are made of aluminum and iron oxides or complexes of the two. These minerals are usually referred to as "sesquioxides" in conversations between pedologists, being typically of the forms Al_2O_3 and Fe_2O_3. Their color is red-brown or actually red. Where the base pigment of the north is grayish silicate, the base pigment of the tropics is red sesquioxide. And since humus is removed from tropical soils by voracious decomposers, there is little organic pigment to blend into the base color, so tropical lands are colored red.

A simple answer to why temperate and tropical lands are of different colors, therefore, is that the local soils collect different minerals. But then we must ask why these different minerals collect. The question actually resolves into, "Why are sesquioxides removed from the surface in cold regions of podzolic soils whereas silicates are removed from warm regions of oxisols?"

Attempts have been made to suggest that soil pH decides whether silica or sesquioxides are removed. The appeal of pH is that silicates are soluble at high pH (pH 8+) but insoluble at low pH. Podzolic soils tend to have lower pH than oxisols (pH 3–5 versus pH 5–6), the difference being largely due to the accumulation of organic acids in podzols, together with carbonic acid of soil respiration trapped in the surface layers. But in fact the pH of both soil types remains in the range in which silica is largely insoluble, making the pH hypothesis less than convincing.

A more likely explanation is that sesquioxides are removed from the surface horizons of podzolic soils in organic complexes. The absence of accumulations of humic substances in the tropical oxisols prevents this removal, allowing sesquioxides to accumulate.[2]

If the organic content of the soil is indeed the arbiter of sesquioxide transport and regional soil color, as this explanation suggests, then ecologists have the satisfaction of knowing that the very color of the ground underfoot depends on living processes in ecosystems. The temperate belts are gray-brown because accumulating organic matter removes sesquioxides while silica collects. The tropics are red because decomposers expunge humic substances from the soil, thus preventing the removal of the red pigments.

[2]This explanation was suggested by F. Ugolini.

Preview to Chapter 26

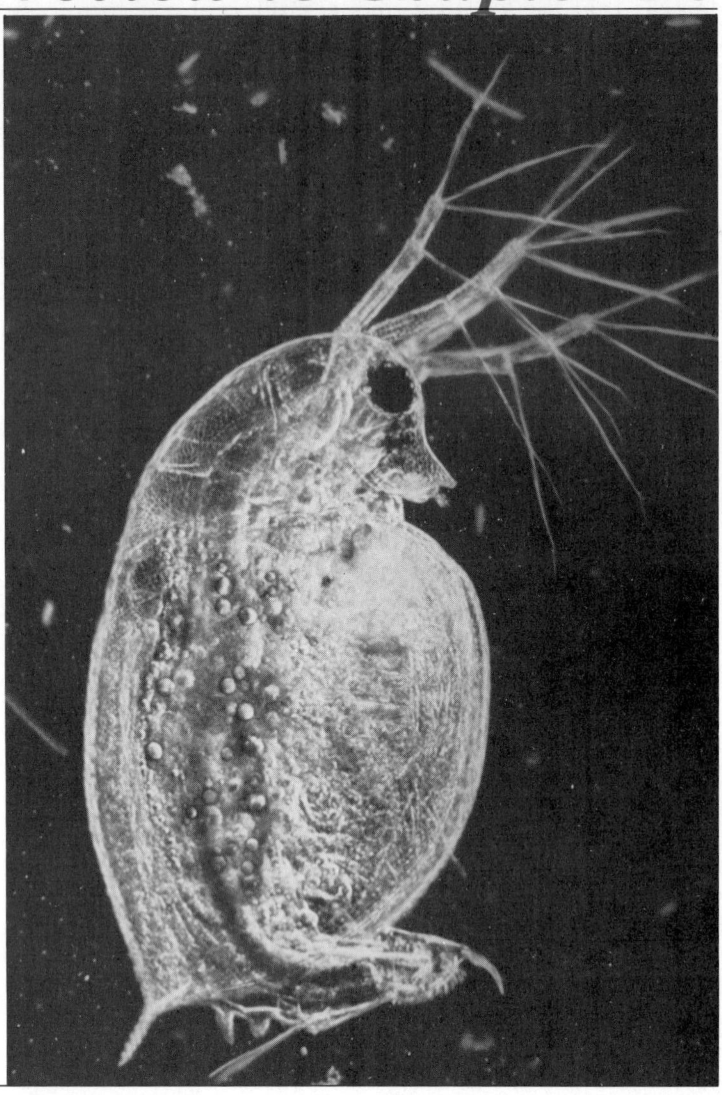

Daphnia spp: **filtering plants at low Reynolds number.**

IN summer, or in warm climates, lakes become thermally stratified as a warm upper lake (epilimnion) floats over a cold lower lake (hypolimnion). The warm upper water then serves to separate the bottom water from contact with the air, denying fresh oxygen to the depths, except in lakes so unproductive that plants grow in the deep water. Biological oxygen demand (BOD) in a stratified lake can be so large that the bottom water, together with the mud in contact with it, becomes anoxic. Nutrients precipitated to this reduced mud in falling debris can be redissolved so that the hypolimnion can become charged with nutrients. In seasonal climates at high latitudes a regular annual progression follows in all lakes: ice-covered in winter, mixed in spring, stratified in summer with collection of nutrients in the hypolimnion, mixed in autumn, with consequent surfacing of nutrients accumulated in the hypolimnion and a late bloom of productivity. Overall lake productivity depends on nutrients delivered from the watershed, the ultimate arbiter of lake fertility. But depth, shape, and temperature also determine seasonal nutrient supplies to the surface, thus constraining lakes to be fertile (eutrophic) or infertile (oligotrophic). The small size of planktonic plants requires that large nutrient reserves cannot be maintained in the biota of the open water, so that productivity is set by ambient nutrient concentrations. The paradox of the plankton is that many species can coexist when mixed in an apparently uniform fluid. Life in a transparent medium gives special importance to predators that hunt by sight, one response to which is the development of transparent storage organs by zooplankton.

Chapter 26

Lakes: Type Specimens of Aquatic Ecosystems

The study of lakes, "limnology" (from the Greek *limne*, "lake"), has had a special importance to ecologists from early in the development of our subject. Lakes are convenient ecosystems in that they are bounded by distinct edges. Inputs and outputs are clearly defined (Figure 26.1). The primary producers are tiny phytoplankton sifting through a void of water. This is one system where animals are not hidden by the forest, but can be caught at will with nets of appropriate sizes. Lifetimes are short, making possible experiments spanning several generations. Nutrients are held in solution, where they are measured easily. Even past events can be studied from fossil traces in the mud.

Lakes also exhibit some features characteristic of all aquatic ecosystems, including the great oceans, as life is constrained by the physical properties of water, held in holes in the ground and lighted from above. These constraints, and their consequences, are the subject of this chapter and the next.

A lake is a hole filled with water, lined with mud, and heated and lighted from above. Water is a transparent fluid that is also a nearly universal solvent. Among the remarkable properties of water is that its maximum density occurs in the liquid phase so that solid water and warm water both float over cold water (Table 26.1). A lake holds this remarkable fluid solvent between chemically active mud and air supplied with free oxygen. The chemical

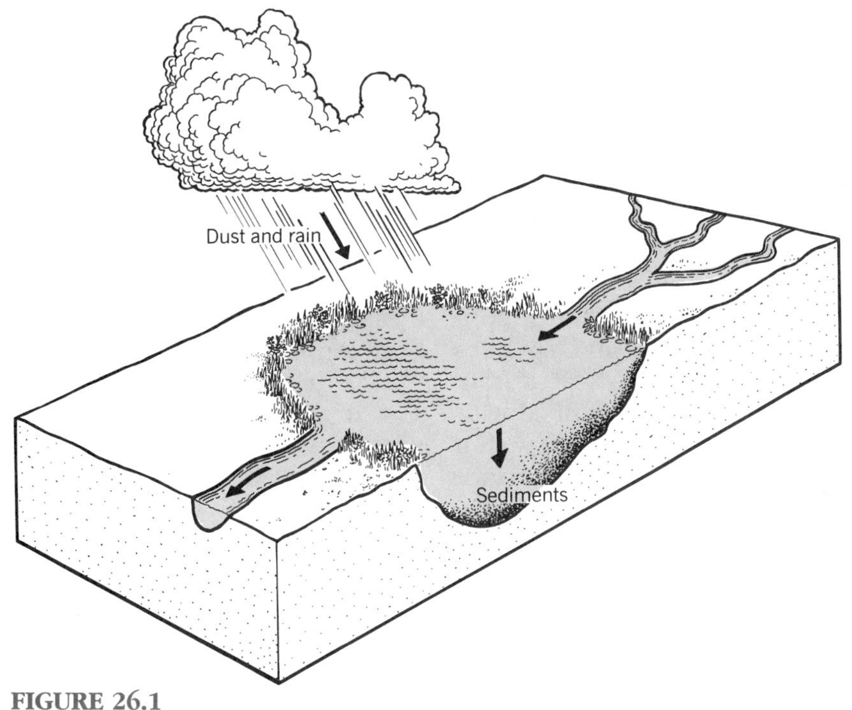

FIGURE 26.1
A lake as a flow-through system.
Inputs of dissolved substances from the atmosphere or the watershed will tend to be balanced by outputs in drainage streams and to bottom mud.

TABLE 26.1
Density of Water as a Function of Temperature
Water is densest at 4°C. The steepest temperature gradient is between cold water and ice, but the gradient is also relatively high at the warmer temperatures like those encountered in tropical lakes. (Data from Hutchinson, 1957.)

	°C	Density
Ice	0	0.9168
Water	0	0.9999
Water	4	1.0000
Water	7	0.9999
Water	10	0.9997
Water	15	0.9991
Water	20	0.9982
Water	25	0.9971
Water	30	0.9957

and physical properties of this mud–water–air sandwich have profound consequences for life in a lake.

THERMAL PROPERTIES OF LAKES

In warm weather the surface of a lake is heated by the sun. The warmed surface becomes less dense, and so remains at the surface, floating on the colder water beneath. The surface continues to gain heat from the sun, while the bottom water remains cold. If this surface heating continues for some critical period of days, without storm winds to stir the lake, a marked temperature difference can develop between top and bottom water. In lakes of the north temperate belt the temperature difference between surface and deep water can be as much as 10 or 15°C.

Warm water floating over cold water cannot easily be forced under by wind, having something of the stability of a rubber duck in a bathtub. The rubber duck persists in floating despite the application of force to sink it, and so does warmed surface water. A lake heated from above thus becomes stably stratified with warm water floating over cold, and with a layer of transition between the surface warm and the bottom cold.

A lake so stratified is effectively divided into two separate compartments, an upper lake, the "**epilimnion**," and a lower lake, the "hypolimnion."[1] The gradient between the two is the "thermocline." The middle layer across which the thermocline runs can be treated as a separate compartment in its own right, the "metalimnion" ("middle lake"), although many limnologists use the word "thermocline" to refer both to the water mass (metalimnion) and to the temperature gradient that crosses that mass (Figure 26.2).

The upper (epilimnion) and lower (hypolimnion) lakes of this stratified system really are very different places. The hypolimnion is not only cold, but is also entirely cut off from the air by an immovable lid of warm water with which it does not mix. Accordingly, the hypolimnion cannot obtain oxygen from the air, and its life must rely on oxygen reserves already held in solution, or oxygen from deep-water photosynthesis. This isolation from the air lasts until and unless the stratification is upset. However, the hypolimnion is in continuous contact with the bottom mud, potentially the largest nutrient reservoir.

By contrast, the warm and floating epilimnion is in permanent contact with the air and thus can be continuously resupplied with oxygen. Furthermore, waters of the epilimnion are closest to the light and thus are likely to receive more oxygen from photosynthesis. Yet the only contact between epilimnion and lake mud is at the lake margins, providing a lesser contact with potential nutrient reserves than has the hypolimnion. As long as they last, therefore, the epilimnion and hypolimnion of a thermally stratified lake are very different places in which to live.

In freezing weather lakes are stratified differently. Because water is at its densest at 4°C (Table 26.1) and freezes at 0°C, deep water, at 4°C, may be the warmest, with a temperature gradient from there to the surface. The surface is then covered with water in the solid phase (ice), effectively separating the entire lake from its oxygen supplies.

THE SEASONAL CYCLE IN TEMPERATE LAKES

In seasonal climates with cold winters and hot summers lakes have corresponding seasonal histories of stratification. In winter they are stratified with ice floating over warmer water. This episode ends when the ice melts and strong winds completely mix the waters of the lake—the "**spring overturn.**"

In early summer the lake goes into its second period of stratification with a warm epilimnion floating over a cold hypolimnion and a thermocline between. Surface cooling and strong winds at the end of summer mix the water once more—the

[1]The terms "epilimnion" and "hypolimnion" are literal Greek translations of the phrases "upper lake" and "lower lake." For once Greek terms are usefully evocative, emphasizing as they do the idea of separate lakes stacked on top of each other.

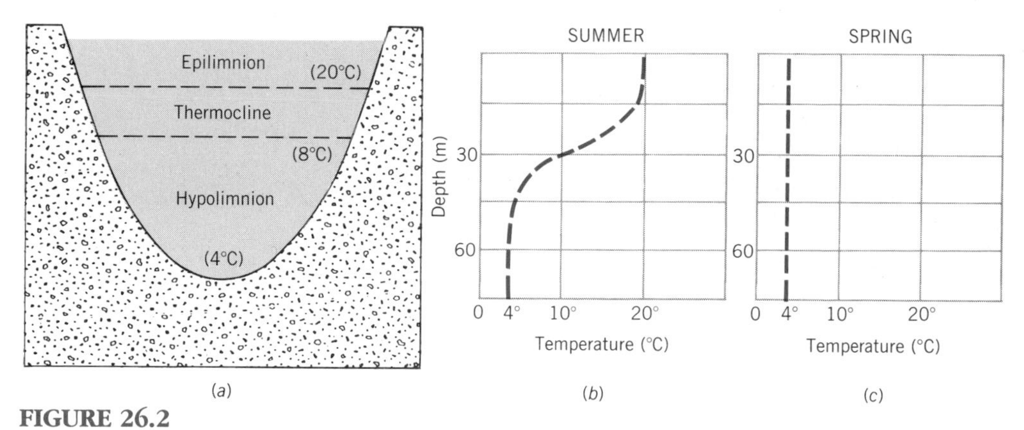

FIGURE 26.2

Temperature stratification of a lake of temperate latitudes in summer.

(*a*) Diagrammatic cross section of a stratified lake. A warm epilimnion floats over the cold water of the hypolimnion below, and the two are separated by a layer, the thermocline, of rapid temperature change. (*b*) Temperature profile of the very deep Seneca Lake in New York as it appears in August, drawn to the same scale as (*a*). The practical effect of such stratification is to isolate completely the bottom water from the air until cooling and the winds of the fall effect a mixing of the lake once more. In the following spring there will be a time when the lake water has a temperature of 4°C at all depths, as in (*c*). (Date for Seneca Lake from Birge and Juday in Ruttner, 1963.)

"**fall overturn.**" The autumn and early winter time of mixing continues until the first layer of ice forms, when the annual cycle is complete.

Very large temperate lakes like Lake Erie can be mixed by storm force winds even in summer, so these lakes sometimes alternate periods of stratification with periods of "mixis." But most temperate lakes remain stratified throughout the summer.

BIOLOGICAL OXYGEN DEMAND (BOD)

When lakes are stratified in the temperate summer, the differential oxygen supplies to the epilimnion and hypolimnion become critical, particularly in fertile lakes where nearly all photosynthesis takes place in the epilimnion. The epilimnion receives oxygen throughout the summer but the hypolimnion does not.

The demand for oxygen continues in the hypolimnion, however, even though supplies are cut off, because animals living there must respire. Even more important is the respiration of decomposers, because all dead matter from the more productive surface regions of the epilimnion falls down into the hypolimnion where it is food for decomposers. The **biological oxygen demand (BOD)** from all this respiration in the hypolimnion must be met by oxygen already held in the water before the lake stratifies. If the lake is so fertile that the amount of dead matter to be decomposed is large, the BOD can exceed the oxygen reserve so that the hypolimnion becomes entirely anoxic in late summer. Then not only does a warm upper lake float over a cold lower lake, but a lake with oxygen floats over a lake without.

Whether these trends matter to the inhabitants of a lake depends on the fertility of the lake water (Figures 26.3 and 26.4). If a lake is very infertile (Figure 26.4), some photosynthesis is carried on in the depths

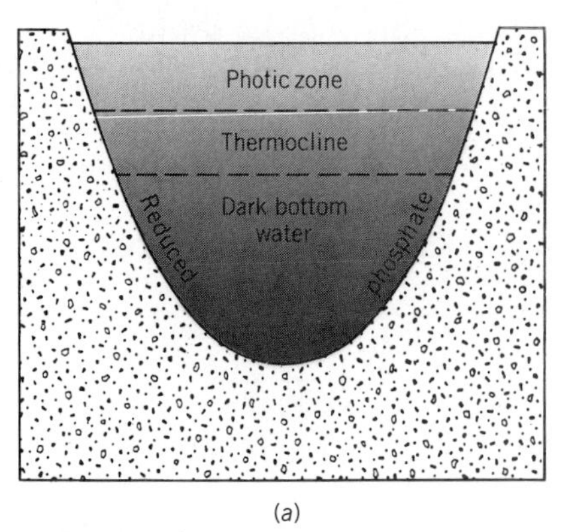

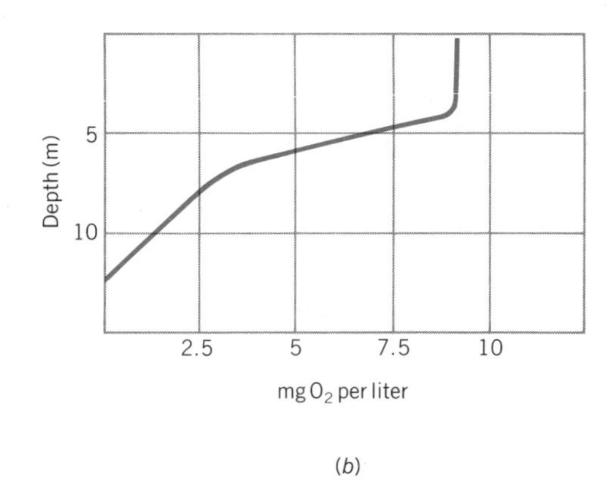

(a) (b)

FIGURE 26.3

Light and oxygen profiles of a stratified fertile (eutrophic) lake in summer.
(*a*) Section through the lake showing that light has been almost completely absorbed by the plankton of the top few meters so that too little light penetrates to the thermocline and beyond to support photosynthesis. But there is a rain of corpses into the deep water, whose decomposition requires oxygen. Since the deep water is cut off from the air until fall overturn, there develops an oxygen deficit in the deep water, and the bottom mud is reduced. An oxygen profile typical for such a lake is given in (*b*).

and little dead matter falls to the bottom so that the BOD is small. The hypolimnion never becomes anoxic. Fish like trout thrive in these lakes, spending much time in the cold water of the hypolimnion and making only brief foraging excursions to the warm epilimnion for food. If a lake is very fertile, however (Figure 26.3), the hypolimnion becomes anoxic in summer. Fish of the deep and cold bottom waters,

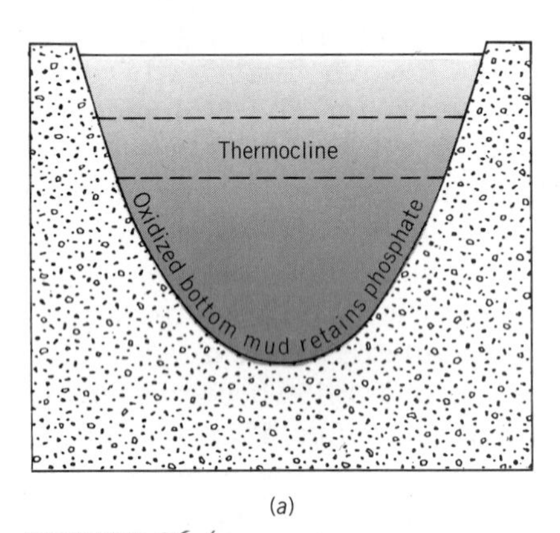

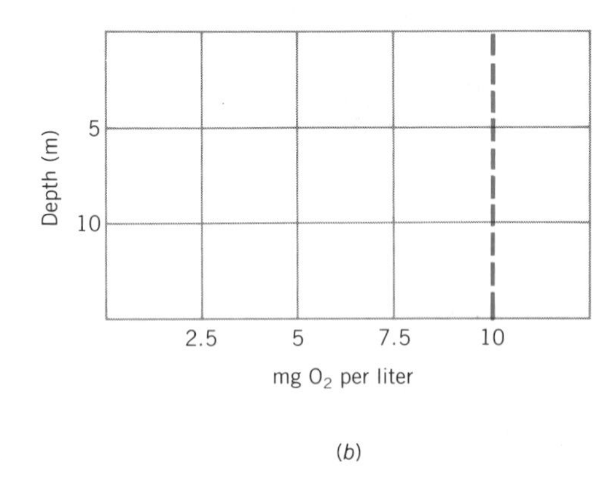

(a) (b)

FIGURE 26.4

Light and oxygen profiles of a stratified infertile (oligotrophic) lake in summer.
An infertile lake may stratify in summer, with consequent isolation of the bottom water from the air, but there is so little plankton floating in the top water that light penetrates deep into the lake, permitting photosynthesis and oxygen generation even in the hypolimnion. Also there is less oxygen demand in the deep water, since there is less detritus coming from above to be decomposed. (*a*) Cross section of such a lake; (*b*) hypothetical oxygen profile.

like trout, would suffocate and only surface-living fish could survive.

EUTROPHY AND OLIGOTROPHY

Limnologists call on their Greek to define properties of fertile and infertile lakes, calling them "**eutrophic**" and "**oligotrophic**," respectively.[2] Fertility of water is, of course, set by the concentration of dissolved nutrients. The concentration of nutrients in different parts of a lake, however, can change with the seasons, depending on such things as intensity or duration of stratification and the free oxygen available near the bottom mud. These properties, in turn, might depend on such physical parameters as temperature or lake depth. Whether a lake functions as a fertile eutrophic system or an infertile oligotrophic system thus depends on a variety of factors that actually control the concentration of nutrients in the open water where plants live.

Nutrient Cycling

The fate of nutrients entering a lake is always fairly rapid descent to the mud attached to some particle or other, if the nutrients are not actually washed away in the outlet stream. What happens after the nutrients reach the mud depends on whether the mud in contact with the water is oxidized or reduced—in chemical language, on the "**redox potential**" of the mud.

The solubility of important nutrients like iron and phosphorous is critically dependent on whether they are oxidized or not. This reality can be noted by reflecting that ferrous iron salts are easily soluble, but ferric iron compounds precipitate from solutions. The most familiar ferric iron compound is rust, a substance notoriously hard to wash away. Iron deposited at the bottom of an oligotrophic lake with free oxygen in the hypolimnion lies in the surface mud in its ferric state, fixed and insoluble. But iron falling through the anoxic hypolimnion of a eutrophic lake lies on the reduced mud as ferrous iron, and is easily redissolved into the bottom waters.

When a lake is not fertile, the passage of nutrients through the system is simple and direct. Nutrients enter from the watershed, are taken up by plants, whether plankton or rooted aquatics, then passed through food chains or back to the water through herbivory or death. But throughout this short cycling period a rapid attrition of nutrients takes place as corpses and debris fall into the dark depths of the lake. And in an oligotrophic lake with oxidized bottom mud, on the bottom of the lake is where the nutrients stay. The system has a fast throughput: from the watershed, a few cycles between short-lived planktonic plants and animals in the open water, then down to the mud for keeps (Figure 26.4).

But when nutrient inputs are large enough to make the epilimnion of a stratified lake very fertile, a quite different nutrient history results. Nutrients enter from the watershed, are cycled, and fall down to the depths as in oligotrophic lakes. But the falling particles may be so numerous that limnologists talk of a "**brown snow**" of corpses and debris. BOD is high, the hypolimnion is reduced, and the surface mud is reduced also. Accordingly, nutrients are rapidly redissolved so that the hypolim-

[2]"Eutrophic" and "oligotrophic" are Greek words for good nursing and few nursing (*trophos* being the word for "suckle," hence feed). The terms were introduced into limnology by the pioneer German limnologist August Thienemann (Hutchinson, 1957), who set up a classification of lakes based on their fertility.

nion becomes charged with nutrients (Figure 26.3).

Throughout the summer period of stratification, dissolved nutrients collect in the hypolimnion of eutrophic lakes, to be mixed back into the rest of the lake at fall overturn. Thus fertile (eutrophic) lakes have a nutrient retrieval and feedback mechanism denied to oligotrophic lakes. The states of oligotrophy and eutrophy, therefore, mean more than just "barren" and "fertile." Oligotrophy and eutrophy have echoes of self-fulfilling prophecies. Oligotrophic lakes rapidly bury nutrients, preserving their infertility. Eutrophic lakes scavenge their nutrients from the mud, maintaining nutrient reserves and their consequent fertility.

Nutrient cycling in a eutrophic lake is illustrated for the phosphorous cycle in Figure 26.5.

Lake Depth and Temperature

Temperature always influences fertility. If a lake is cold, therefore, a high nutrient loading can still fail to make it eutrophic, and a relatively high influx of nutrients from the watershed merely results in rapid deposition of those same nutrients to oxic mud under an oxygen-rich hypolimnion.

Lake shape is also critical, because deep narrow basins have large volumes of water in the hypolimnion, compared with broad, shallow basins. A large volume of water implies a large oxygen reserve. Thus a very deep lake can retain oxygen in the hypolimnion all summer, even though the surface waters are fertile, and deep lakes can retain the essential properties of oligotrophic lakes despite significant nutrient loadings from the watershed. Conversely, broad shallow lakes, in which the hypolim-

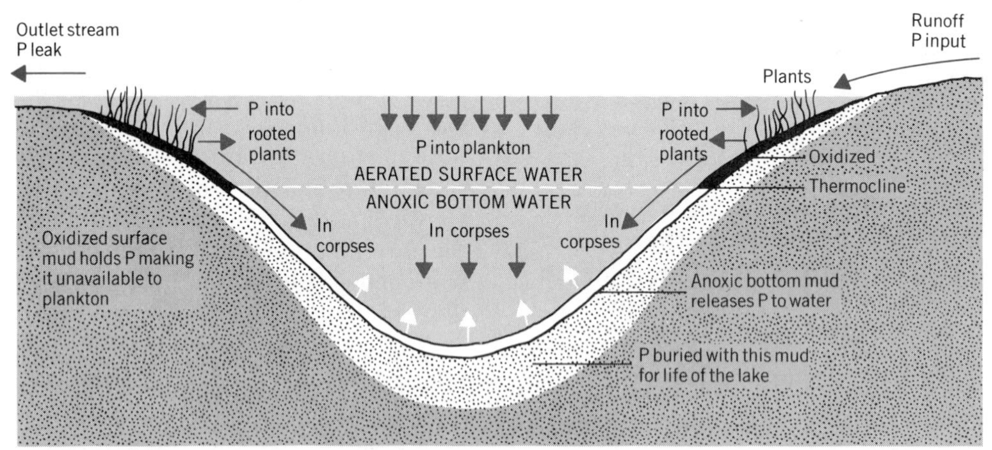

FIGURE 26.5
The phosphorus cycle in a lake.
Phosphorus enters all lakes continuously in runoff water, in inlet streams, and from the air. Phosphorus is also continuously lost to the lake in the outlet streams and by incorporation in lake mud. When a lake has anoxic bottom water, as when a lake stratifies in summer, the top few millimeters of mud are chemically reduced, a condition that allows the mud to release phosphorus back to the water. The bottom water thus becomes phosphorus-rich. Stirring of the lake by winter storms brings the phosphorus-rich water to the surface, completing an annual cycle and fertilizing the lake for a spring plant bloom.

nion is of small volume compared to the volume of the epilimnion, are easily made eutrophic by increases in nutrient loadings.

Oligotrophy and Eutrophy Summarized

Oligotrophic Lakes Concentration of dissolved nutrients is low; water appears transparent or blue; algal blooms at the surface are not visible; productivity is low; deep water retains oxygen at all seasons; sediment at mud/water interface is oxidized at all seasons; accumulation of nutrients in the hypolimnion during summer is low; injection of nutrients to surface water at fall overturn is low; there is some plant production throughout the water column and on the mud surface (Figure 26.4).

Eutrophic Lakes Concentration of dissolved nutrients is high; water is turbid, opaque and green or brown; there are dense algal blooms in the surface water with the possibility of floating mats of algae; dense shade cast by plankton of the surface water prevents photosynthesis in the lower parts of the water column; productivity is high; community respiration and BOD are high; deep waters may become anoxic in summer; sediments at mud/water interface may be reduced in summer and settling nutrients may be redissolved; hypolimnion is enriched with nutrients in summer; there is a strong fertilizing effect at fall overturn when nutrient-rich water is brought to the surface (Figure 26.3).

The terms "oligotrophic" and "eutrophic" are in part relative, but only in part. The rate of nutrient inputs obviously can be anything from low to high, suggesting a gradual cline of fertility states. But the mechanism of nutrient retrieval in an anoxic hypolimnion introduces a threshold into this cline. As soon as BOD is high enough to remove free oxygen from the bottom water, the whole nutrient regimen of the lake is changed from the oligotrophic regimen of almost total nutrient burial to the eutrophic regimen of nutrient recovery. Once increasing nutrient inputs have pushed a lake across this threshold to eutrophy, the internal dynamics of the lake system should, therefore, tend to maintain eutrophy.

Review of a Year's Events in a Temperate Lake

The annual cycle of events in lakes in temperate seasonal climates may be summarized as follows.

Winter Water is evenly mixed and tends to the same temperature and chemical composition at all depths. Deep water is never colder than 4°C, the temperature at which water has the greatest density. If the lake freezes, there will be a temperature gradient downward from 0°C at the bottom of the ice to the maximum temperature of 4°C. The nutrient concentration of the water will be at a maximum. Oxygen concentration will remain high but can be depleted under the ice in shallow lakes. Sediment at the mud/water interface usually will be oxidized.

Late Spring/Early Summer The lake stratifies and a thermocline separates a warm epilimnion from a cold hypolimnion. There is a bloom of phytoplankton in the surface water. Nutrient concentrations begin to fall in the epilimnion and oxygen concentrations begin to fall in the hypolimnion. Sediment at the mud/water interface is still oxidized.

Late Summer Maximum thickness of epilimnion and maximum temperature difference between epilimnion and hypolimnion. Epilimnion may be depleted of nutrients and hypolimnion may be depleted of oxygen. In very fertile lakes the hypolimnion may actually be anoxic and the sediments of the mud/water interface may be chemically reduced. Waters of the hypolimnion may be nutrient-rich, both from solution of particles settling down from the surface water and from solution of nutrients from reduced bottom mud.

Fall Overturn Evaporative, convective, and radiant cooling of the surface water reduces the stability of the stratified system. Strong autumn winds then mix the lake waters, imposing uniform temperatures and chemical composition throughout the water column. Bottom mud and water are resupplied with oxygen. Surface water is resupplied with nutrients, which may result in a fall bloom of phytoplankton at the surface. The lake continues to cool toward its winter condition.

LANGMUIR CIRCULATION AND THE DESCENT OF THE THERMOCLINE

Within the epilimnion there is no change of temperature with depth (Figure 26.2), showing that mixing currents must penetrate from the surface to the thermocline. Apparently these currents determine the depth of the thermocline, and they force it deeper as a season progresses. An important generator of these currents is "**Langmuir circulation**," which was first discovered in the sea.

Debris, particularly the torn-up remains of seaweeds from rocky shores, floats in rows on the sea surface, the rows being parallel to the direction of the wind. Langmuir (1938) reasoned that water currents must be moving at right angles to the wind and nudging the debris into rows, and he imagined a pattern of circulation that would provide for this. The wind would tend to push surface water along in the direction in which it was going, but this moving water must be replaced. Part of the replacement water would come in from behind, but some would be expected to come in from *below*. It must thus happen that a horizontal current will be accompanied by a vertical, ascending current. And the water moved upward in the vertical current must be replaced by water that sinks, meaning that a system of up and down circular cells must be associated with the horizontal movement of water. When such a system develops in the three dimensions of a body of water, a series of helical circulation cells results (Figure 26.6). It is this pattern of circulation cells that we call Langmuir circulation.

The system of helical cells that Langmuir postulated would explain the lines of seaweed, because these would collect

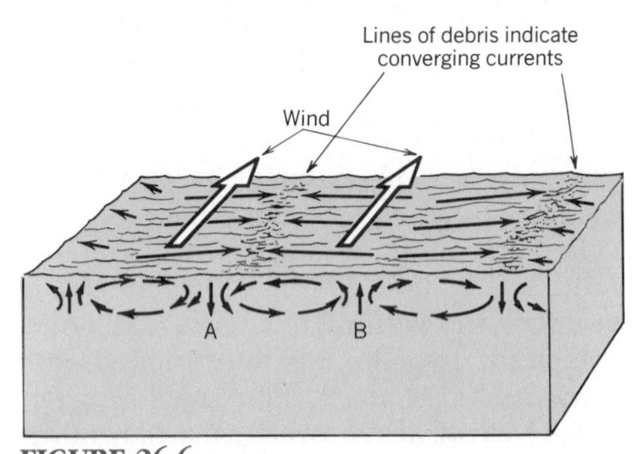

FIGURE 26.6
Langmuir circulation cells.
Turbulence in the surface water, which is driven by the wind, apparently becomes organized by drag and wave effects into helical cells; where outward-flowing currents converge, debris collects. The descending and ascending currents are probably important in mixing the epilimnion of a lake and in fixing the depth of the thermocline.

where the horizontal water vectors met. This hypothesis was testable because it predicted the current systems generated in all open water. Langmuir tested the hypothesis in a large lake. He examined the current systems directly, using dies or neutral density floats to plot the paths of currents and their velocities. Not only was the postulated pattern of currents shown to exist, but the velocities of the ascending and descending currents were found to be considerable. Downward currents where the cells merge (Figure 26.6A) were as high as 4 cm per second, and the corresponding upward currents where the cells parted (Figure 26.6B) were as high as 1.5 cm per second.

The downward currents were found by Langmuir to extend to depths of up to 7 meters. Here then is a mechanism that will account for the mixing of the waters of the epilimnion. When the wind blows strongly, Langmuir circulation cells are set up in the surface waters of lakes and they mix the warmed surface water with a thin layer lying immediately underneath. Because the underlying water is denser, the mixing process is resisted. But as the season progresses, a repeated pressure is exerted on the metalimnion by the descending currents of the Langmuir cells and the thermocline is driven down (Figure 26.7).

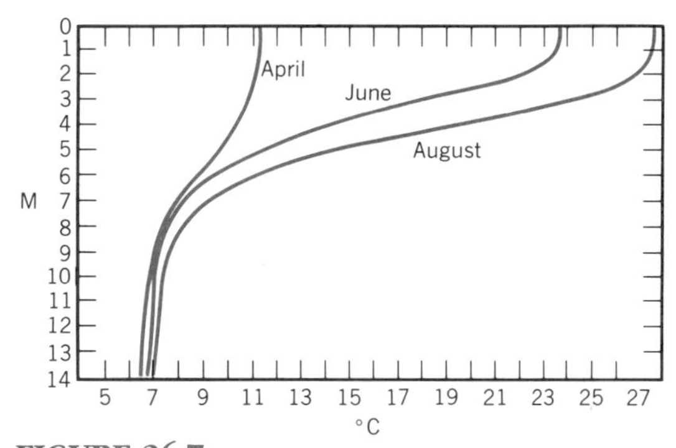

FIGURE 26.7

Development of the thermocline in a temperate lake.

Data are from Linsley Pond, a kettle lake in Connecticut. As the summer progresses the temperature stratification intensifies. (From Hutchinson, 1957.)

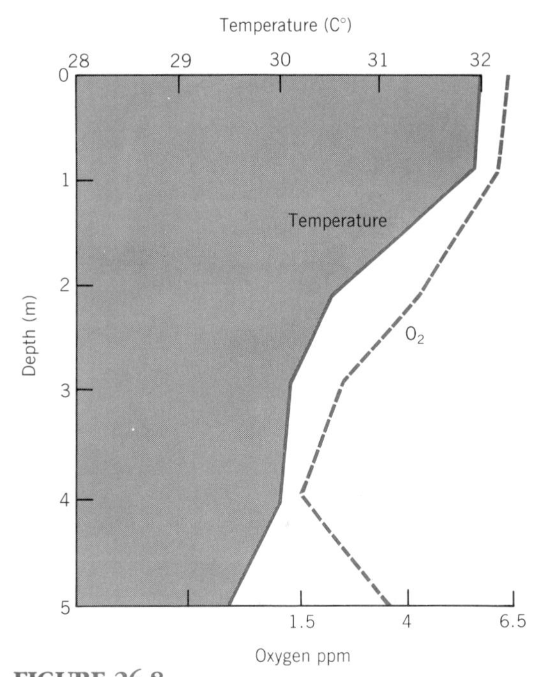

FIGURE 26.8

Temperature and oxygen profiles from a eutrophic lake of the Amazon lowlands.

Temperature change with depth is minimal, though sufficient to make the lake stably stratified. High BOD causes oxygen tensions to be low even near the surface, in the epilimnion (Data from Limoncocha by Mike Miller).

STRATIFICATION IN TROPICAL LAKES

In tropical lakes, the range of temperature from top to bottom of the columns of water is not great but the position of a thermocline can still be detected. The reason for this lies in the fact that the density of water changes more rapidly at higher temperatures (Table 26.1). The loss in density of water between 25 and 30°C is more than twice the loss between 10 and 15°C, so that a vertical difference of 1 or 2°C is sufficient to stratify the lake. In places like the Amazon forest, where strong winds rarely reach a lake, thermal stratification may last for long periods (Figure 26.8).

The hypolimnion in a lowland tropical lake is nearly always without oxygen, and oxygen deficits can be pronounced even in the epilimnion (Figure 26.8). These generally low oxygen tensions are due partly to high productivity consequent on high temperature, but also to biological decomposition of the large inputs of organic matter from productive tropical ecosystems on the banks. The typical jungle lake, therefore, is turbid, murky, and opaque with biological activity (Figure 26.9). Even these lakes, however, may be overturned by exceptional storms, or because they have been flooded by cold rain or groundwater. Accordingly they properly are called "oligomictic" lakes (few mixing). They may be stratified for weeks or months before there is a short-lived episode of mixing.

A different pattern is found at high elevations in the tropics. In the equatorial Andes mountains, for instance, lakes exist in cool climates without seasons, but with 12-hour days that alternate with 12-hour nights. Surface heating by day is partly

FIGURE 26.9
Lago Agrio: Eutrophic lake in Amazon lowlands
Opaque, anoxic below half a meter, with food chains based on inputs from the rain forest on the banks. All Amazon lakes seem to have a dugout canoe stashed away, even remote lakes unknown to geography and with no names outside those of indigeous languages.

counterbalanced by surface cooling by night, so that temperature gradients are both slight and in the low temperature range where density changes are least (Table 26.1; Figure 26.10). Such lakes turn over frequently, with short-lived times of stratification perhaps measured only in hours. Called "**polymictic**" for their habit of frequent mixing, these lakes are usually well-supplied with oxygen throughout the water column. Cold and well-oxygenated, they are easy targets for those who seek to introduce trout, a practice roughly equivalent to deliberately introducing rats to oceanic islands.

THE WATERSHED DETERMINANTS OF EUTROPHY OR OLIGOTROPHY

The fertility of a lake is a function of the whole watershed or regional ecosystem of which the lake is a part. This is because the rate of input of nutrients to the lake is of overriding importance in determining its trophic state. The usual concentration of nutrients in the water is, of course, also a function of the rate at which they are exported via the outlet or buried in the mud,

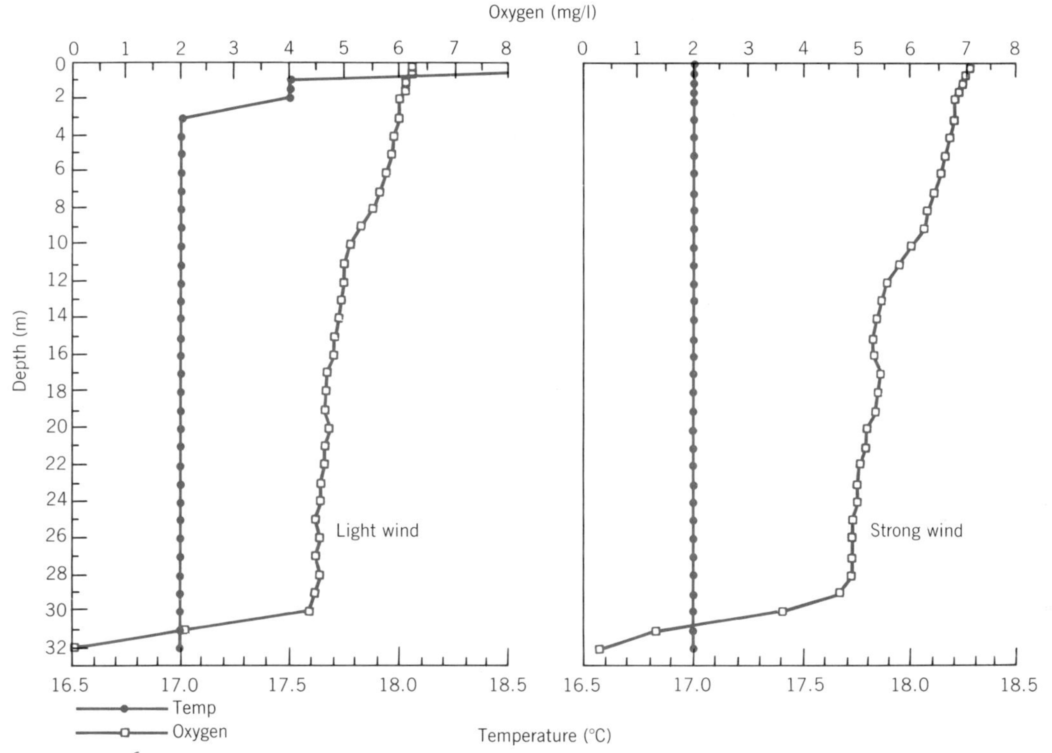

FIGURE 26.10
Temperature and oxygen profiles from a polymictic lake in the Andes.
Lake San Pablo in the Inter-Andean Plateau of Ecuador stratifies intermittently. Cold nights and strong winds easily overturn the lake, recharging the hypolimnion with oxygen. Only close to the bottom mud do oxygen tensions fall. (From Scheutzow, 1992.)

but these processes are not nearly as flexible as the inputs. If the inputs from the watershed change little, the fertility of a lake should be constant.

The overwhelming importance of nutrient inputs to the trophy of a lake can sometimes be shown by the data of paleolimnology. A striking example is the history of a lake near the city of Rome as deduced from sediment cores (Figure 26.11). Lago di Monterosi is a maar lake formed by a volcanic explosion about 26,000 years ago (Hutchinson, 1970). For the first 24,000 years of its history, right up to the building of imperial Rome, chemistry and fossils of Monterosi mud suggest very few changes in lake trophy. The lake was moderately oligotrophic and remained so. This is really remarkable because those 24,000 years saw enormous changes in the surrounding land. For the first 10,000 of these years an ice age climate prevailed in Italy and the glaciers were not far away. Then came the climatic catastrophe of the lateglacial, a 4000-year period of irregular and drastic climatic change that ended the glacial period. Then the first 8000 years of postglacial time brought the familiar warmth of the Holocene to Italy, but still the fertility of Lago di Monterosi remained constant.

Pollen analysis (see Chapter 17; Figure

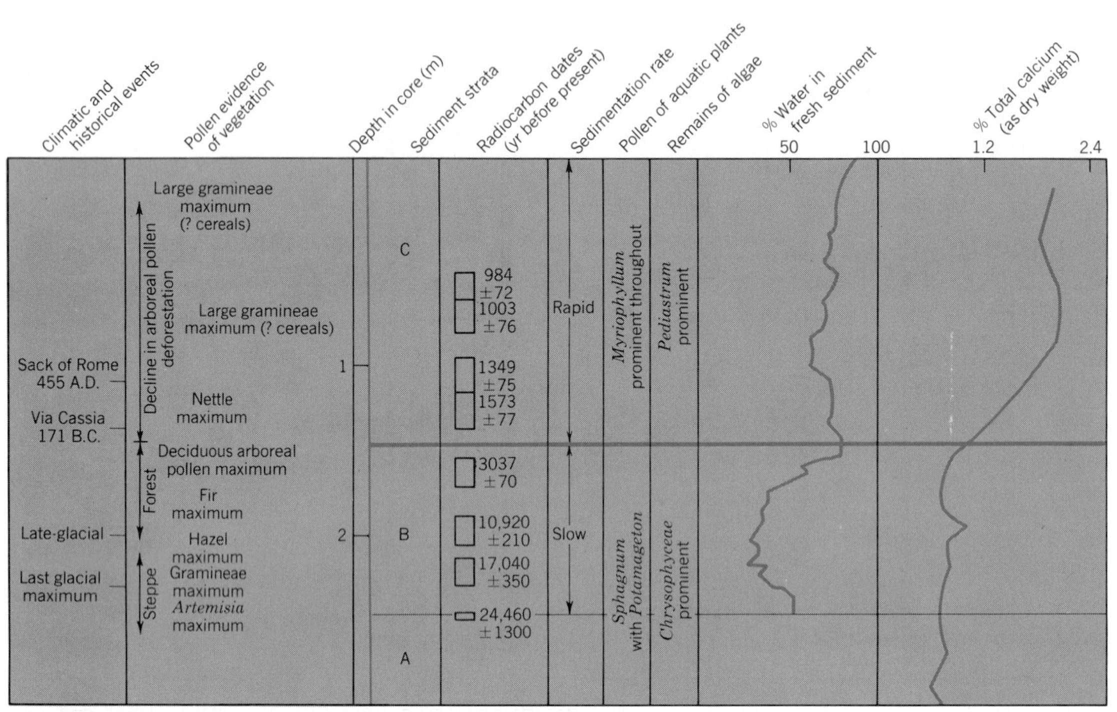

FIGURE 26.11

Reconstruction of history of Lago di Monterosi from sediments.

Three strata, A, B, and C, are recognized in the sediment cores. All the data show that the B–C boundary (heavy type) marks a profound change in the state of the lake ecosystem. Sedimentation rate, aquatic plants, water content, and sediment chemistry changed in a way that reflects the establishment of a fresh steady state. Radiocarbon dating shows that this change was synchronous with the building of the via Cassia by the Roman Republic. (Redrawn diagrammatically from figures in Hutchinson, 1970.)

26.11) of the Lago di Monterosi sediments showed that local vegetation changed from sagebrush steppe to grassland and then to the Italian forests of prehistoric times during those first 24,000 years of its history. But the life of the lake went on, supported by the old nutrient steady state, without much change.

It took the Roman Republic to alter the trophic status of Lago di Monterosi. The Romans built a road, the Via Cassia, and this piece of engineering changed the direction in which groundwater flowed. Afterward water reaching the lake percolated through volcanic debris and limestone that had been bypassed for the first 24,000 years of the lake's history. The lake water hardened (more calcium carbonate), and the lake became more eutrophic, complete with summer episodes when the hypolimnion became anoxic. This condition has persisted ever since the Via Cassia was built in 200 B.C. (Figure 26.11). But notice that the Romans only manipulated drainage in the watershed, though inadvertently. This is not a history of pollution. The Romans had merely altered nutrient inputs, something more important to lake trophy than all the temperature and vegetation changes of a glacial cycle.

This propensity of lake fertility to be critically dependent on inputs is the reason that lakes are so easy to pollute. Pollution with agricultural runoff, detergents, or sewage is actually a massive change in nutrient inputs. The response in lake fertility is immediate, being visible in algal blooms followed by increasingly anoxic bottom waters as the dead algae and other remains fall to the bottom waters of the hypolimnion, there to rot. Since, however, the new fertility is dependent on enhanced nutrient inputs, the phenomena of pollution ought to be easily reversible. Returning any polluted lake to its pristine condition ought to be possible simply by ending the inputs of fertilizers and sewage. This recipe does work, the most celebrated example being the restoration of Lake Washington by diverting sewage (Edmondson, 1991). But recovery can be less than complete if the bottom mud is so chemically reduced under an anoxic hypolimnion that it recycles nutrients from the sink of bottom mud and back to the open water.

LAKE AGING: EUTROPHY AND THE OBLITERATION OF THE HYPOLIMNION

Ancient lakes fill with mud. The process may be slow, but it is very steady—typical kettles[3] from the last glaciation, for instance, are now about half full. This is the true process of lake aging—filling with mud. But this infilling sometimes alters the trophic status of the lake, because it shrinks the hypolimnion while having little effect on the epilimnion. Figure 26.12 shows how sediment can gradually obliterate the hypolimnion of even a deep lake. Suppose that the lake is oligotrophic with a small BOD in the deep water and an original hypolimnion that retains oxygen throughout the summer; as accumulating sediment replaces the old hypolimnion the BOD will not change, but the oxygen reserve by which this BOD can be met will lessen progressively. When the hypolimnion is no more than a thin layer between mud and thermocline (Figure 26.12), the oxygen reserve held by this

[3]In limnology, "kettle" means a lake formed by the melting of a block of glacial ice. Blocks of ice, hundreds of meters or more across, are buried in glacial deposits. After the glacier has melted, the buried ice itself slowly melts, leaving a deep hole in the ground that fills with water. Such lakes are common in onceglaciated terrain.

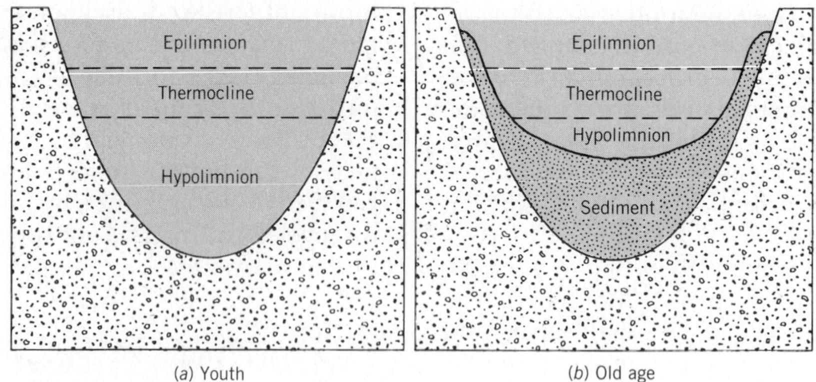

FIGURE 26.12
The effect of aging on an infertile lake.
The young lake at the left (*a*) is infertile (oligotrophic), retaining oxygen in its deep water all summer. The drawing at right (*b*) is the same lake when infilling of the basin is far advanced. The thermocline is in the same place, and the epilimnion is scarcely altered. But the hypolimnion is a vestige of its former size. The oxygen reserve of the reduced hypolimnion may be too small to support as much decomposition as before, and the bottom becomes anoxic. As defined by the oxygen content of the bottom water, the oligotrophic lake has become eutrophic by aging.

small volume of water may be so small that the BOD of even an oligotrophic lake cannot be met. The hypolimnion then becomes anoxic, and an oligotrophic lake has taken on some of the properties of eutrophy.

An anoxic hypolimnion, of course, also means a reduced redox potential in the mud with consequent release of nutrients to solution. This means that shrinking of the hypolimnion by mud can in the end actually fertilize the lake. The lake becomes more eutrophic in its old age.

When this process was first pointed out by Deevey (1955) it was an interesting theoretical curiosity, nothing more. But it had very powerful repercussions in limnology and beyond when others extrapolated from this observation to a quite unjustified inference—that pollution was artificial aging. According to this argument, since lakes were more fertile in their old age, making lakes more fertile must make them old. Students of philosophy will recognize the classic intellectual fallacy of this argument. But "artificial aging" became an environmentalist slogan. Eventually it led to the absurdity of the news magazine headline, "Who Killed Lake Erie?"

What had happened to Lake Erie was that the bottom water of the shallow west end of the lake became anoxic during short periods in summer. This eutrophication was a direct result of pollution, particularly of agricultural runoff and phosphatic detergents, which increased the productivity of the west basin. Far from dying, the lake was highly fertile and full of life. The life was undesirable life, which was why the public became interested in stopping the pollution. High BOD caused fish kills by removing oxygen, and dead fish are common in highly eutrophic lakes. But talk of the lake itself dying is rubbish.

LAKE CHEMISTRY AND THE LIMITS TO PRODUCTION

Primary production in the open waters of a lake is directly dependent on the concentration of essential nutrients in the ambient water. In this, production in lakes differs from production in terrestrial ecosystems. In a tropical rain forest, for instance, nutrients can be almost vanishingly scarce in a sample of soil water though the photosynthetic rate of the forest is extremely high. This paradox follows because the trees of the rain forest hold, and maintain, a reservoir of nutrients in their own huge biomass. But the biomass of the open waters of lakes is almost trivially small compared to a forest, being merely the tiny cells of the phytoplankton.

Thus the productivity of aquatic systems is set by the steady-state nutrient supply, being a function of inputs from the watershed, outputs to drainage or bottom mud, inputs of recycled nutrients from the hypolimnion at overturn, and the solubility of nutrients in water.

The most abundant cations of lake water are sodium, potassium, magnesium, and calcium; the most abundant anions are carbonate, sulfate, and chloride. These are present in remarkably consistent proportions in surface fresh water of different origin and in different parts of the world (Table 26.2). Typically their dissolved masses are much greater than those of all other ions. These most abundant ions are the least likely to be limiting though their supply, particularly that of sulfur, can influence productivity.

But other essential nutrients are scarce in lake waters. Table 26.3 shows the chemistry of a suite of equatorial lakes in South America, ranging from the Galapagos Islands to mainland Ecuador, across the high Andes, and down into the Amazon basin. It is evident from the table that, whether the lake is cold in the mountains, warm in the jungle, or in between, such essential nutrients as phosphorus or iron remain in short supply.

The crucial importance to lake productivity of small amounts of phosphorus and nitrogen is also suggested by data correlating measures of lake productivity with concentrations of these elements (Table 26.4). In addition, other elements found in living things, like molybdenum, iron, co-

TABLE 26.2
Mean Equivalent Proportions of Important Ions in Natural Waters
(From Hutchinson, 1957.)

	Mean Igneous Source Material	Mean Sedimentary Source Material	Water from Igneous Rock	Wisconsin Soft Waters	Uppland	Central Europe	Mean River	N. German Soft Waters
Na^+	20.1	4.8	30.6	10.9	13.6	4.5	15.7	43.0
K^+	9.8	8.0	6.9	4.8	2.2	1.9	3.4	6.7
Mg^{++}	33.1	34.0	14.2	37.7	16.9	25.4	17.4	14.3
Ca^{++}	37.1	53.2	48.3	46.9	67.3	68.2	63.5	36.0
CO_3^{--}	—	93.8	73.3	69.6	74.3	85.4	73.9	42.4
SO_4^{--}	—	6.2	14.1	20.5	16.2	10.8	16.0	14.1
Cl^-	—	0(?)	12.6	9.9	9.5	3.9	10.1	43.5

TABLE 26.3
Chemistry of Lake Waters
Data from the suite of lakes from the Andes, the Amazon, and the Galapagos Islands of Ecuador show the general similarity of fresh waters. Data are from lake surfaces; deep water below the thermocline often has very different chemistry. Notice the extremely low concentrations of phosphate. All data unless otherwise indicated are in parts per million; —, no measurement made; T_r, trace. (From Steinitz-Kannan et al., *1983; Colinvaux, 1968; and Steinitz unpublished.)*

	pH	$CaCO_3$ (mg/e)	Ca	Mg	PO_4	SiO_2	NO_3	SO_4	Cl	F	Na	K	Zn	Cu
						Galapagos								
El Junco	6	2	0.0	1	0.04	1	18	5	7	—	67	6	—	—
Tortoise	6	80	7.3	Tr	0.02	8	17	—	70	—	—	—	—	—
						Andes								
Cuicocha	8	226	5.0	25	0.01	100	7	26	73	—	—	—	—	—
San Pablo	8	94	1.6	13	0.16	150	35	10	5	—	30	6	0.32	0.011
Yaguarcocha	8	170	2.7	25	0.13	46	21	18	20	—	103	55	0.04	0.002
Conru	7	32	0.6	4	0.05	180	20	9	5	—	15	13	0.03	0.005
Yambo	9	636	2.0	150	0.27	110	131	160	112	1.4	263	56	0.05	0.005
Caricocha	6	16	0.3	2	0.02	15	28	7	5	—	—	—	—	—
San Marcos	7	43	0.0	1	—	—	1	1	<5	0.1	3	1	0.05	—
						Amazon								
Agrio	6	—	7.0	1	0.09	—	—	<10	(3)	0.1	3	3	—	—
Sta. Cecilia	5	—	9.0	1	—	—	—	—	(3)	0.1	<5	3	—	—

balt, gallium, and silicon (for diatoms), are far from prominent in analyses of lake waters (Table 26.5). These data allow a general hypothesis of lake productivity being set by concentrations of nutrients likely to be in critically short supply.

A general test of the hypothesis comes from the phenomenon of seasonal changes of productivity in temperate lakes. In these lakes algae bloom in spring and in autumn, with a time of lowered productivity in high summer (Figure 26.13). Seasonal changes in nutrients apparently explain these histories as follows. Productivity is

TABLE 26.4
Relationship between Nitrogen, Phosphorus, and Productivity of Lakes
(Modified from Likens, 1975.)

Trophic Type	Mean Primary Productivity (mg C/m^{-2} day)	Phytoplankton Biomass (mg C/m^{-3})	Chlorophyll *a* (mg/m^{-3})	Total Organic Carbon (mg/l)	Total P (μg/l)	Total N (μg/l)
Very oligotrophic	<50	<50	0.01-0.5		<1-5	<1-250
Oligotrophic	50-300	20-100	0.3-3	<1-3		
Moderately oligotrophic					5-10	250-600
Mesotrophic	250-1000	100-300	2-15	<1-5		
Moderately eutrophic					10-30	500-1100
Eutrophic	>1000	>300	10-500	5-30		
Hypereutrophic					30->5000	500->15,000

TABLE 26.5
Nutrients Required by Algae
It is considered that most phytoplanktonic algae will grow in water provided with these 17 nutrients plus the three vitamins, though many are needed only in very low concentrations. (From Hutchinson, 1967.)

Carbon	C	Manganese	Mn
Nitrogen	N	Zinc	Zn
Phosphorus	P	Copper	Cu
Sulfur	S	Boron	B
Potassium	K	Molybdenum	Mo
Magnesium	Mg	Cobalt	Co
Silicon	Si	Vanadium	V
Sodium	Na	Vitamins: Thiamin	
Calcium	Ca	Vitamins: Cyanocobalamin (B_{12})	
Iron	Fe	Vitamins: Biotin	

low in winter due to short days and low temperatures. Algal growth accelerates in the spring with the warming that precedes summer, causing the first bloom. But the lake then stratifies and the process of exporting nutrients to the hypolimnion via precipitating organic matter proceeds to lower nutrient concentrations in the epilimnion. Productivity then falls during the summer, despite the fact that this is the warmest time of the year with long days. The tiny, short-lived algae of the spring bloom are not replaced and the standing crop is lowered, ending the bloom. But production goes up again with the coming of autumn, despite lowered temperatures. This appears to be a consequence of the fall overturn, which brings to the surface nutrient-charged water formerly held in the hypolimnion.

That nutrients are limiting is also suggested by a phenomenon known to limnologists as "luxury consumption" (Lund, 1965). Algae will take up excess nutrients,

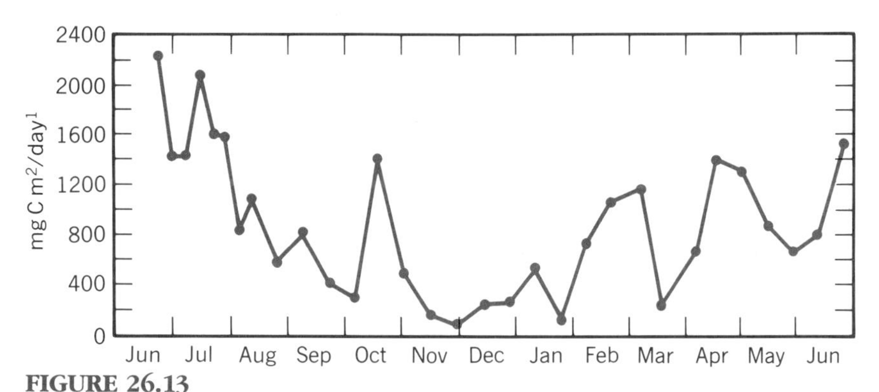

FIGURE 26.13
Seasonal productivity in a seasonal eutrophic lake.
Data are for the very eutrophic Wintergreen Lake in Michigan. The October peak of productivity coincides with the fall overturn when nutrients regenerated in the hypolimnion are brought to the surface. (From Wetzel, 1975.)

particularly phosphorus, as fast as they are added to their water and apparently in excess of their immediate needs. They are then able to draw on their body stores for growth later when low concentrations in the water might be expected to be limiting. This behavior can be understood as an adaptation to restricted or uncertain supplies of essential nutrients. It does not, of course, result in a massive hoard of nutrients, because the total biomass available is so small. But the principle is the same as that on which the tropical rain forest operates.

THE PARADOX OF THE PLANKTON

Planktonic life in the epilimnion provides one major puzzle, the answer to which is still not clear. It is hard to understand how there can be so many species of phytoplankter coexisting in the open water. Planktonic plants all acquire energy in the same way, by photosynthesis; they all require a similar set of dissolved nutrients, such as the limiting phosphorus; and they live stirred up together in a transparent medium that they can hardly divide up into private spaces. Why, therefore, do most of them not suffer competitive exclusion? On the face of it one would expect phytoplankters to be engaged in as fierce a competition as the paramecia in Gause's centrifuge tubes when he performed the experiments that led to the exclusion principle (see Chapter 8). And yet we find scores to hundreds of species of phytoplankton coexisting in the open water of a lake. Hutchinson (1961) called attention to this remarkable feat of coexistence, calling it the "paradox of the plankton."

Several alternative hypotheses have been put forward to explain this paradox, though none has been tested to complete satisfaction.

Hypothesis 1: Competitive Equilibrium Is Never Attained

Hutchinson (1961) suggested that in temperate lakes there was never time for a competition to eliminate species before the next seasonal change in the lake. Rather than competitive exclusion there was seasonal succession, as different species became abundant in turn though none became extinct. What we know of seasonal changes in lakes (Figure 26.13) is compatible with this view. Indeed, many phytoplankton species are provided with resistant cysts or spores that can lie dormant on the mud surface to pass an unfavorable season. These resting times might also serve to get species through the worst competitive episodes. One appealing aspect of this hypothesis is that it makes at least one broad prediction that is potentially falsifiable: that lakes in less seasonal places like the lowland tropics should have fewer species than temperate lakes. This is because phytoplankters at the equator should get closer to competitive equilibria, with consequent increased chance of extinction. Really good comparable data of plankton diversity in tropical and temperate lakes to test this prediction are not available, but there are indications that the species list may be somewhat smaller in equatorial lakes.

Hypothesis 2: Niches of Phytoplankton Are So Similar That Competition Works Very Slowly

Exclusion happens only if individuals are sufficiently different for selection of one over the other to be possible. All the members of a single population, for instance, coexist. Riley (1963) suggested that phytoplankton niches have so converged in their long histories (perhaps 3×10^9 years) that competitive exclusion works only slowly. Apparently no suitable test for this hypothesis has yet been devised.

Hypothesis 3: Some Species Coexist in Cooperation

If neighbors are symbiotic or commensal, there obviously would be no exclusion. The few data available suggest, in fact, that there is little such cooperation in the algae (Hutchinson, 1967).

Hypothesis 4: Adequate Niche Differentiation Is Possible, Despite Appearances

Although stirred together, the algae might have different nutrient requirements. One model assumes different nutrient limits, either as a single nutrient or nutrient sets, for each alga, and allows a large enough number of permutations of the known limiting nutrients to account for the observed diversity (Peterson, 1975).

Hypothesis 5: Competitive Exclusion Does Take Place in the Open Water, but Lost Populations Are Continually Replaced by Immigration from Refugees at the Peripheral Mud

This hypothesis arose when evidence was found that equilibrium conditions probably did occur in tropical lakes as predicted by hypothesis 1, but that phytoplankton species were still so numerous that the paradox of the plankton remained (Colinvaux and Steinitz Kannan, 1980). It was suggested that competitive exclusion would be much less likely at the lake periphery where physical separation of niches on mud and among rooted plants

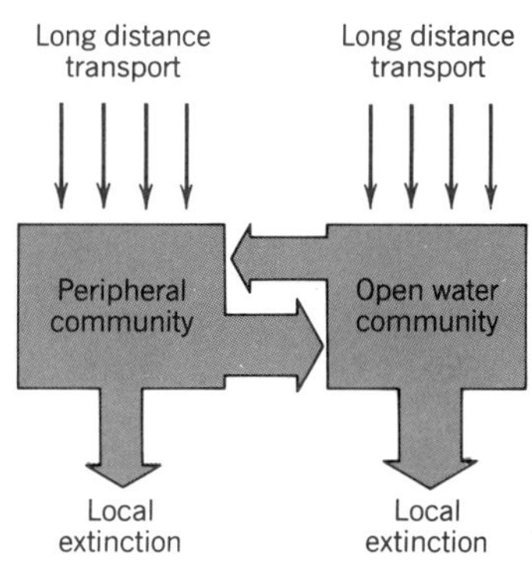

FIGURE 26.14

Mechanism for maintaining high species richness in an equilibrium phytoplankton community.

The model postulates extinction by competitive exclusion in the open water balanced by immigration from populations in mud refuges at the periphery and by long-distance transport. (From Colinvaux and Steinitz-Kannan, 1980.)

was more possible. The periphery also would continually collect species by long-distance immigration in the wind. From this refuge of the peripheral mud, populations would colonize the open water as fast as other populations went extinct there by competitive exclusion (Figure 26.14).

Since all these proposed mechanisms are plausible, it is likely that all operate to some extent. The result is one of the remarkable properties of lake systems: open water crowded with great numbers of species. A similar pattern is found in the sea.

CYCLOMORPHOSIS: WHEN SHAPE CHANGES IN THE EPILIMNION

An oddity of lake life is that the shapes of some planktonic organisms change with the seasons. A species may have one shape in the spring, but members of the same population a few generations later in high

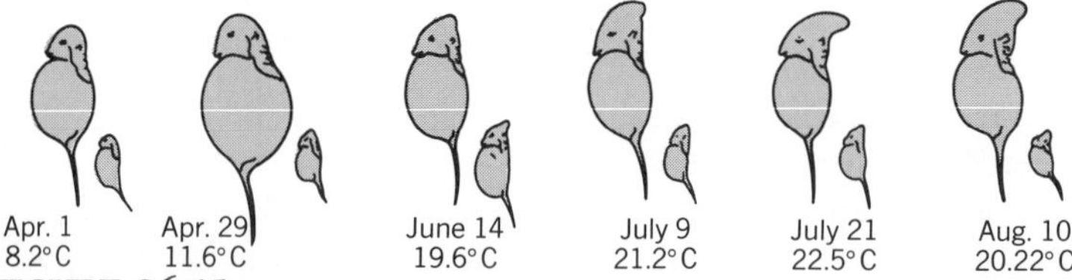

FIGURE 26.15
Cyclomorphosis in *Daphnia retrocurva*.
The drawings are of young and adults at each date, drawn to scale. The changing shape of the helmet is probably a stratagem to increase size by storing reserve calories in a transparent structure, thus minimizing the target offered to predators that hunt by sight. (From Brooks, 1946.)

summer will carry structures that make them look quite different. This phenomenon is called "**cyclomorphosis.**" It is most noticeable in zooplankton, both in crustaceans like *Daphnia* (Figure 26.15), *Bosmina*, *Ceriodaphnia*, and *Chydorus* and in rotifers (phylum Aschelminthes). That species in different phyla show essentially the same phenomenon argues for a fundamental adaptation to lake life. Even many plants, notably the dinoflagellate *Ceratium* (Figure 26.16) and some diatoms, show something similar.

The prevailing explanation of cyclomorphosis in *Ceratium* is that successive generations of the alga are shaped so that their sinking speeds are adapted to the changing viscosity of the water (Hutchinson, 1967). As the epilimnion warms in summer its viscosity falls. The extra spines (Figure 26.16) act as brakes. Certainly the possession of extra spines can be correlated with water temperature and can be induced by growing cultures in water of appropriate temperature. Other possibilities must remain; perhaps that extra spines interfere with filter-feeding animals when these are at their most abundant, or that there is some other advantage as yet unknown.

Cyclomorphosis seems much less likely to control sinking speeds in animals, however, because their relatively large size makes any change in speeding negligible. The most convincing explanation for zooplanktonic cyclomorphosis was offered by Brooks (1965), who suggested that the new shapes reduced fish predation in summer. *Daphnia* and other planktonic animals are the quarry of fish that hunt by sight. The *Daphnia* are poised in a glassy void, suggesting that they should be easy targets for a predator that hunts by sight. The Brooks explanation for the strange shapes of high summer is that they alter the size of the visible target.

Daphnia and the rest are largely transparent, yet they must be partly visible. The larger they are the more visible a target they will make for a fish, and summer is a time of rapid growth, or else of the rapid

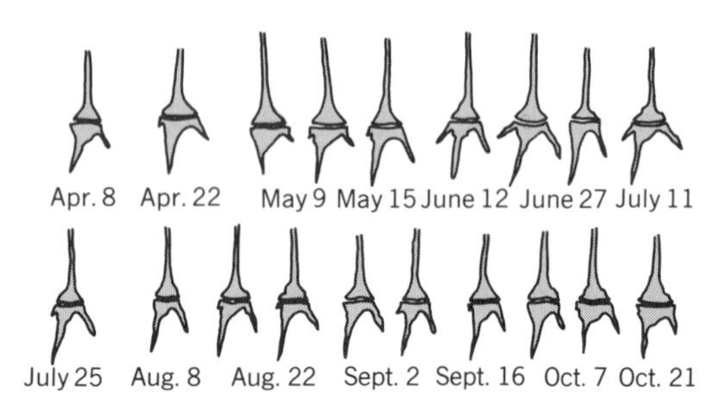

FIGURE 26.16
Cyclomorphosis in a dinoflagellate.
Ceratium hirundinella generations living in warmer water (June to July) tend to be made up of individuals with more spines. (From Hutchinson, 1967.)

storage of food reserves to be used in reproduction. Brooks put forward the hypothesis that "helmets" and other protuberances of cyclomorphic crustaceans were transparent storage structures that let the animal grow without increasing the visual target for a hunting fish.

Brooks's predation hypothesis conforms to modern knowledge of the potency of fish predation on zooplankters. Small fish can prey so heavily on large zooplankters that these can be completely removed from a lake, or nearly so, leaving the water to smaller species that escape fish. This first became clear when Brooks and Dodson (1965) showed that lakes in Connecticut with populations of a small fish of marine origin, the alewife (*Pomolobus pseudoharengus*), were without larger species of cladocera, whereas similar lakes with no alewives had the large species in abundance.

Since then it has been shown that planktivorous fish react to the presence of plankton depending on the apparent size of the target. The apparent size of a plankter is a function of its actual size and its distance away, leading to the concept of "reactive distance." Fish can see, and strike at, larger plankters from farther away (O'Brien, 1979). The advantage to a cyclomorphic plankter of faking a small size is readily apparent. Moreover, there are sometimes equally strong advantages of actually being large. The increased bulk of a helmeted form of *Daphnia* reduces its vulnerability to attack by the predacious copepod *Heterocope* in arctic lakes (O'Brien et al., 1979).

It is now, therefore, a reasonable conclusion that cyclomorphosis in zooplankters adapts the animals to minimize predation. The cyclomorphic forms allow bulk to be increased without also increasing the visual target. The new shape and size is the best mix to foil both invertebrate predators that hunt the small and vertebrate predators that hunt the big. The strength of this hypothesis must increase the plausibility that predation is involved in selection for cyclomorphosis in algae also. Perhaps extra spines on a *Ceratium* increase the chances of rejection when it thumps against the mouthparts of a filter-feeding herbivore.

Preview to Chapter 27

Sphyraena barracuda: **Sea predators floating in the lighted void.**

OCEANS have some of the properties of lakes in that water is stratified by density so that a relatively thin surface layer of less dense water floats stably over a vast denser mass below. Only the top 100 meters is lighted effectively, yet the dark depths retain oxygen because of low biological oxygen demand. Large plants are absent from the open sea—possibly because loss by drifting would reduce fitness of large plants. All photosynthesis in the open water is by tiny organisms, imposing low Reynolds numbers and laminar flow. Filter feeding, in which laminar flows of water are manipulated to bring tiny food particles against feeding apparatus, is a necessary response practiced both by planktonic animals and by large animals of the benthos like clams. In shallow nearshore waters, large attached photosynthetic structures are possible. Complex communities of large attached algae are the primary producers in colder latitudes, but are replaced by carbonate-depositing systems in tropical waters. Coral reefs derive their energy supplies from in situ *photosynthesis so that, like seaweed communities of the North, they are the basis of structured ecosystems, in many ways comparable to those on forested land. Ocean life is lived in a salt solution that appears to have changed little since before the start of the fossil record of marine organisms. A complex chemistry maintains the sodium chloride of the ocean, cycling sodium with the land, conserving chloride, precipitating carbonates, and sweeping out many elements on mud surfaces. Nutrient concentrations are kept low, resulting in the low productivity of the open oceans.*

Chapter 27

An Ecologist's View of the Oceans

In some ways the oceans are giant lakes—depressions in the crust of the earth filled with a transparent fluid. Because light and heat come from above, both lakes and oceans tend to have warmed, low-density water that floats over the cold and darker depths. In both, nutrients are delivered in solution, but at low concentration.

Yet oceans differ from lakes in being salt and very large. Where the epilimnion of a lake is typically two or three meters thick, the oceanic equivalent, called the "mixed layer," is more than 100 meters thick. In tropical oceans, a principal cause of low density in the surface water is thermal expansion as in lakes, with surface heating in places driving the thermocline down as much as 500 meters. But in cold oceans of high latitudes, surface waters can float for reasons not found in normal lakes—because of reduced salinity. In these waters, rain and snow dilute the sea surface, resulting in a mixed layer of fresher water floating over saltier bottom water. A "**halocline**" (from the Greek *hals*, "salt") at 100 meters or so separates the mixed layer from the depths in these latitudes (Figure 27.1).

In practice, the lowered density of the ocean surface is nearly always a function of both thermal expansion and differential saltiness, with temperature more important in the tropics and salinity more important in the North. Accordingly, oceanographers tend to talk directly of density gradients, referring to a "**pycnocline**" at the base of the mixed layer, whether the density gradient is really a thermocline, a halocline, or both.

Light seldom if ever penetrates through to the bottom of the mixed layer. If water is clear, effective light penetration is in the 100–150-meter range, with 10% of the more penetrating blue light left at 90 meter depth, but only 1% penetrating to 150 meters (see Chapter 3; Figure 3.15). Such

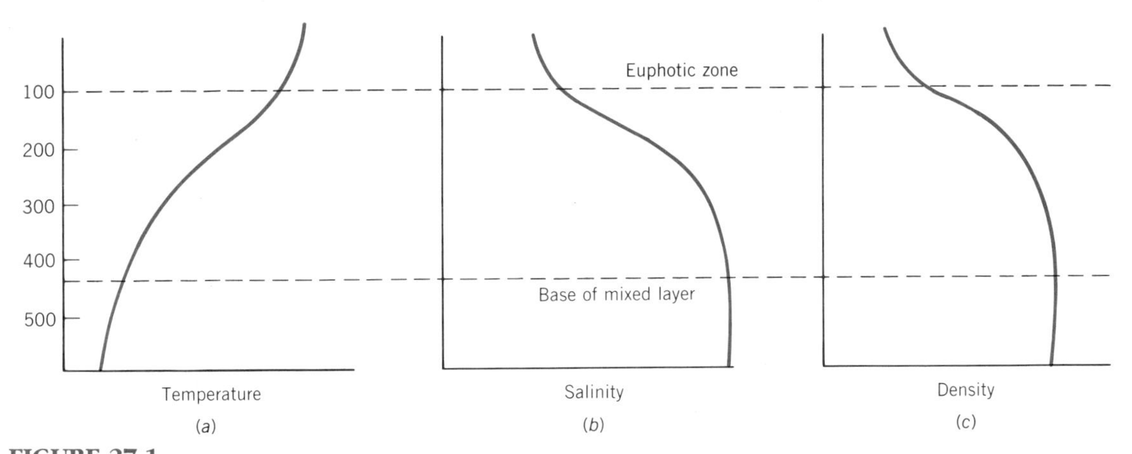

FIGURE 27.1
Density strata and light in the ocean.
The mixed layer may be less dense because warmer (*a*) or less salty (*b*). The combined temperature salinity gradient is called a "pycnocline" (*c*). The euphotic zone is the upper part of the mixed layer where light is sufficient for photosynthesis to exceed respiration.

deep penetration is, of course, realized only in the clearest of ocean water. The desert waters of blue, tropical oceans allow such penetration because the density of organic particles in them is so exceedingly low that little light is intercepted (see Chapter 24). These are the waters where the thermocline can be several hundred meters deep, however, so that the 1% light level is perhaps only a quarter of the way down in the mixed layer.

The 1% light level is typically at about 15 meters in the more productive, colder waters where dense arrays of organic particles, both living and dead, intercept the light. Mixed layers in these colder waters are thinner too, perhaps only 100 meters above the pycnocline. Nevertheless, the 15-meter penetration means that less than one sixth of the mixed layer is lighted well enough for significant photosynthesis.

Thus ocean ecosystems are based on the productivity of the top quarter or less of the mixed layer. Below are the dark depths, typically descending several kilometers. Unlike the bottom waters of a lake, however, the dark depths of the sea have virtually inexhaustible oxygen supplies. This is a simple consequence of the oceanic scale. As in a lake, oxygen in the deep water must come from contact with the air or from photosynthesis at the surface, so that water below the mixed layer is cut off from its supply. But the volume of bottom water is so enormous compared to the biological oxygen demand (BOD) that the oxygen supply lasts until the next time the water surfaces, often centuries or millennia later.

Oxygen does change with depth in the oceans (Figure 27.2). A high at the surface, under the waves and in the euphotic[1] zone records maximal oxygen input. Tensions are comparatively low in the region of the pycnocline, though still adequate for respiring animals. This is because consumer populations are concentrated close under

[1]The euphotic zone is the upper part of the lighted zone where photosynthesis exceeds community respiration.

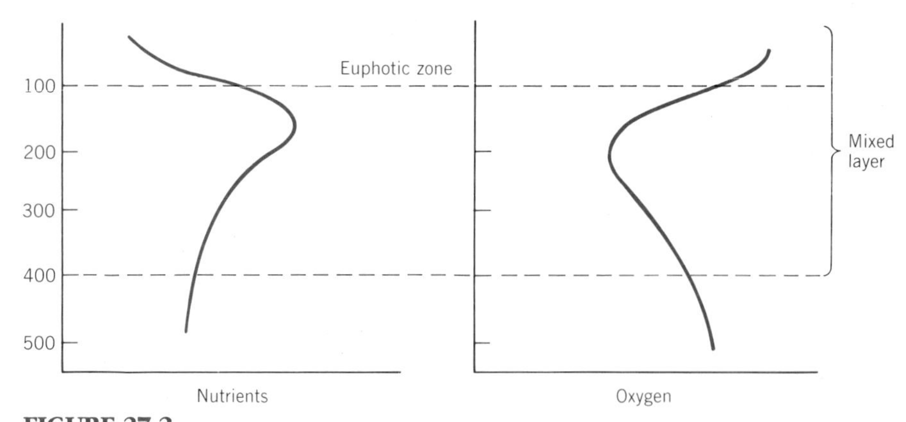

FIGURE 27.2
Oxygen and nutrient profiles in the ocean.
Concentration of ambient nutrients and oxygen through the mixed layer are inversely correlated, reflecting uptake and production by photosynthetic organisms.

the productive euphotic zone, there to digest between feeding forays overhead, to play the part of ambush predator in the dark, or to feed on the rain of corpses. From there to the distant bottom of the sea, oxygen tends to be more plentiful as respiring life is more scarce. Only in shallow embayments, like estuaries or fjords, is the bottom water of such limited volume that oxygen tensions fall as they would in a real lake.

One peculiar circumstance of the ocean is thus that oxygen for respiring life is available at all depths, but that reduced carbon fuel is synthesized only in the top few tens of meters (forgetting trivially important but massively interesting oddities like chemosynthesis in thermal vents in the deep sea). Animal life is prominent at all depths, even the deepest bottoms, because carbon is exported by gravity as detritus from the life zone overhead, and free oxygen is always available. Energy flux through the consumer trophic levels of the depths, however, must be set at some tiny fraction of the desert production overhead on which it depends (see Chapters 4 and 24). This does not prevent the consumers of the depths existing in astonishing diversity (see Chapter 16).

ON THE SMALLNESS OF THE PHYTOPLANKTON

Perhaps the greatest peculiarity of the oceans is shared by lakes. It is that plants of the open water are small, actually microscopic, a circumstance with profound consequences for aquatic ecosystems. Tiny plants, together with their small herbivores, drift with water masses, and are collectively called "**plankton.**"[2]

Whatever selection pressures serve to drive down the sizes of ocean plants must be very strong, being sufficient to overwhelm the many advantages of large size to individuals. In all other ecosystems,

[2]From the Greek *planktos*, "wandering." Planktonic plants are phytoplankton and planktonic animals are zooplankton. Individual animals and plants of the plankton are called "plankters."

large size confers fitness in obvious ways so that they are ubiquitous on land. Among the many benefits to individual plants conferred by large size are:

1. The ability to store large reserves of nutrients (rain forest trees overcome the nutrient poverty of some tropical soils by hoarding nutrients in this way).
2. Reproductive strategies based on stored energy and mass release of propagules at favorable times.
3. Herbivore defense.
4. Shade out competitors.
5. Quick growth from stored reserves after periods of dormancy.

These are formidable advantages. They seem sufficient to ensure that large plants exist in all terrestrial habitats and in all lighted benthic (bottom) habitats of lakes or oceans where plants can anchor. Only in open water are the large plants absent. There, the tiny drifting algae and photosynthetic relatives of bacteria are not only denied all those advantages of bigness but they also pay an extra high cost of smallness resulting from their large surface areas and thin cell walls. They lose a significant part of net photosynthetic production by excretion through the cell walls.

It has been argued in the older oceanographic literature that small size, by providing a large surface area in contact with the water for each cell, promotes nutrient uptake and prevents sinking (Hardy, 1956). Both of these functions can be met as well by adaptations to large shapes, by use of floats (as in the seaweed *Sargassum*), and by adopting convoluted edges. A large, spongy mass of a plant floating at the ocean surface, should be admirable at taking up, and storing, nutrients, rather like the root hair and mycorrhizal mat under a rain forest. But such ocean plants do not exist. The nutrient uptake hypothesis, therefore, is less than convincing.

For want of an acceptable general explanation for the smallness of open water plants, I suggest the anti-drift hypothesis. This states that the selective advantage of small size is that smallness minimizes loss by drifting and promotes rapid recolonization of water newly brought to the surface trophogenic zone. Large floating plants would be swept clear of the oceanic environment to which they were adapted.

The anti-drift hypothesis allows individual fitness in tiny drifting organisms to be maximized in the same way that an *r*-strategist weed maximizes fitness (see Chapter 11). Dispersal and recolonization by tiny propagules ensures descendants, and more fitness accrues to the parent from the gambler's strategy of casting huge numbers of copies of its genome into the environment than would be possible by seeking to persist as a large individual. Large size for an oceanic plant is probably a recipe for hereditary oblivion, whatever the short-term gains might be.

The Mechanics of Sinking

Most phytoplanktonic algae are nonmotile, or nearly so, and they are denser than water—typically between 1.01 and 1.03 times the actual density of the water solution in which they live (Hutchinson, 1967). A few algae have neutral or positive buoyancy, possessing oil or air flotation devices or gelatinous sheaths, but these are in the minority. Most are dense and must sink. Since this is the norm, and since there appears to have been ample time for natural selection to have favored floating forms if this habit led to greater fitness, it must be assumed that it is advantageous to sink.

The most obvious advantage to sinking is that this is a cost-effective way of moving through the medium in quest of nutrients. A perfectly stationary alga permanently in one micropacket of water would soon sweep this clean of all dissolved nutrients and be unable to obtain fresh supplies other than by the fatally slow rate of ionic diffusion.

The obvious disadvantage to sinking is that the algae might be carried below the lighted zone to a dark death below. But this fate is mitigated by turbulence within the water column that tends to return algae to the surface. There should inevitably be a continued reduction of a static algal population by sinking, of course, however turbulent the water, but real algae reproduce. The active algal population of the lighted water gains from reproduction even as it loses from the net excess of sinking loss over turbulent return. Each individual alga apparently stands to gain more fitness by slowly sinking, and thus winning nutrients needed for photosynthesis and reproduction, than it stands to lose by falling out of the productive region.

Other selection pressures must also work on the sink-or-float dilemma. Sinking, or just being heavy, might affect the rate of predation, for instance. But a strong case can be made that sinking generally is an adaptive response to the needs of nutrient uptake.

Objects of the size of typical phytoplanktonic algae (less than 0.5 mm) sink according to a classical formulation of hydrodynamic theory known as Stokes law, which says that spherical objects fall through fluids as a function of the square of their radii:

$$V = \frac{2gr^2 (\rho_1 - \rho_2)}{9\mu}$$

where

V = velocity (cm per sec)

g = acceleration due to gravity (cm per sec^2)

r = radius of sphere

ρ_1 and ρ_2 = density of the falling object and of the medium (g per cm^3), respectively

μ = viscosity of the medium (poises, or dyne-sec per cm^2)

Calculations based on this relationship show that sinking velocities for typical phytoplankters are likely to be very slow, on the order of hours per millimeter (Table 27.1).

Hydromechanics also tells us that sinking speeds can be modified as a function of shape, in particular, long, rod-like objects fall more slowly than spheres. This may be the reason for the prevalence of filamentous forms among algae. Protuberances and fan-like structures also lower sinking speeds, and these may be particularly important in the smaller zooplankton that otherwise would have faster settling velocities because of their larger size. It is possible, therefore, to put together a convincing general theory that explains the shapes of many planktonic organisms as being partly devices to retard or promote sinking velocities (Hutchinson, 1967).

Some phytoplankton, however, are self-motile, and most of the animals of the plankton move. Other animals, the filter feeders, move algae in streams of water that they generate. Because these motions are more rapid than sinking, these activities come up against the hydromechanics of laminar flow, imposing on life the peculiar constraints of operating at low Reynolds number.

TABLE 27.1
Sinking Speeds of Spheres in the Size Range of Phytoplankton
The velocities (mm/sec) are calculated on the assumption that Stokes' law applies. The dashed lines denote the domain where size and velocity are such that Reynolds numbers are less than 0.6, where Stokes' law no longer holds.

	Density of Sphere Less Density of Water						
Diameter	**0.0001**	**0.001**	**0.002**	**0.005**	**0.01**	**0.02**	**0.05**
0.01 mm	0.00000545	0.0000545	0.000109	0.000273	0.000545	0.00109	0.00273
0.10 mm	0.000545	0.00545	0.0109	0.0273	0.0545	0.109	0.273
0.50 mm	0.0136	0.136	0.273	0.681	1.36		
1.00 mm	0.0545	0.545			$Re > 0.6$		

LIFE AT LOW REYNOLDS NUMBER

The water a human knows is turbulent. Push it with your hand and the water parts, filling in behind the moving hand with spinning eddies. Swimmers thrust water aside, pressing back the eddies and hurling themselves forward as the water pours in behind. A fish does it better, causing less commotion but still riding on the turbulence its motion creates. But to the small and the slow, water has very different properties. For them flow is laminar; the water glides by as an intact layer, smoothly, without the rapid circular displacements of turbulent flow.

One of the strangest properties of laminar flow is the reversibility of position: the moving particle or water mass can be moved back the way it came and the whole system is replaced in its original position with each parcel of water back where it was. If you were to stir a mixture under conditions of laminar flow, you could unstir it by reversing the motion exactly. The smaller animals and plants of the plankton actually live under these bizarre circumstances.

Whether flow is turbulent or laminar is a function of size, velocity, density, and viscosity. In hydromechanics the formal relationship is called Reynolds number (*Re*), and is given as:

$$Re = \frac{lV\rho}{\mu}$$

where

l = length of moving object (cm)
V = velocity (cm per sec)
ρ = density of liquid (g cm^3)
μ = viscosity of liquid (poises, or dyne-sec per cm^3)

At high Reynolds numbers flow is turbulent, and at very low numbers it is laminar. The dividing line is a Reynolds number of about 1.0. From the Reynolds number equation it is obvious that all large animals will have large Reynolds numbers, because length is in the numerator. Humans swimming in water have Reynolds numbers of more than a million, and small fish have numbers in the thousands (Table 27.2). Small cladocerans like *Bosmina* or *Daphnia* live close to the dividing line, but the motile algae live well below.

TABLE 27.2
Reynolds Numbers of Animals in Water
Below Reynolds numbers of 1.0 flow tends to be laminar. (Data of R. Zaret, 1980.)

Animal	Length	Velocity	Reynolds Number
Human	1.5 m	1 m/s	1.5×10^6
Fish	5 cm	1 cm/s	5×10^3
Cladoceran	—	Sinking	0.1
Cladoceran	—	Swimming	3
Copepod	—	Swimming	500

One of the obvious manifestations of life at low Reynolds numbers is the way that flagellates swim: the undulating flagellum goes first and the *Chlamydomonas* or *Euglena* is dragged along behind. Larger swimmers have their propelling structures behind them, driving the turbulence they make backward as does the tail of a fish.

Zooplankters as a class span the Reynolds number divide at 1.0. The herbivores among them must catch phytoplankters. One-on-one encounters by simple reaching for the algal prey would be difficult in conditions of laminar flow, since the target is pushed away in an intact packet of water. The response of these zooplankters is to generate with cilia flows of water that bring the prey to the predator. This is called "**filter feeding**," and is a principal method of herbivory in the world oceans.

In filter feeding by zooplankton, the phytoplankton prey come smoothly as part of the laminar flow of water, thump against sensitive parts of the zooplankters' feeding apparatus, and are seized. In high-speed photographs by J. R. Strickler (for example, in Alcaraz et al., (1980) and Gilbert (1980), this process has been witnessed directly. A rotifer, for instance, drawing a long filamentous diatom colony into reach of its mandibles can be seen to crunch the filament in sections like brittle spaghetti, the parts being swept down its open mouth.

This procedure actually gives filter feeders some selectivity over what they eat because the photographs show some algal cells being rejected whereas others are seized. Presumably chemical cues are used to choose food, although this is not certain. These observations do suggest that filter-feeding animals do not act in quite the indiscriminate way suggested in the theoretical predatory model of the type 1 functional response (see Chapter 14).

Hunting individual prey, as opposed to filter feeding, in conditions of low Reynolds number brings special difficulties for the carnivores of the zooplankton that must attempt it. Detection of prey is a major problem, though it has been suggested that laminar passage of water transmits pressure from nearby objects that can be detected (Zaret, 1980). Seizing prey of any bulk also is difficult as it tends to be pushed away by the appendage thrust out to grasp it. High-speed photography shows that attacks can fail if the aggressor is unable to retain a grasp on the prey.

A special pattern of activity in some zooplankters comes about because they can work on both sides of the Reynolds number divide. In hop-and-sink behavior the animal launches itself with a stroke of appendages in a turbulent flow jump, then sinks quietly in what may be near laminar flow conditions (Figure 27.3). This pattern possibly serves both the search for prey in the sinking episodes and as an avoidance reaction to larger predators.

An elegant elaboration on hunting technique at the Reynolds number divide is shown by the predacious insect larvae of *Chaoborus* in lakes, which hang suspended in the water as if in ambush and are able to cross the Reynolds number divide for a rush at prey with the speed of an

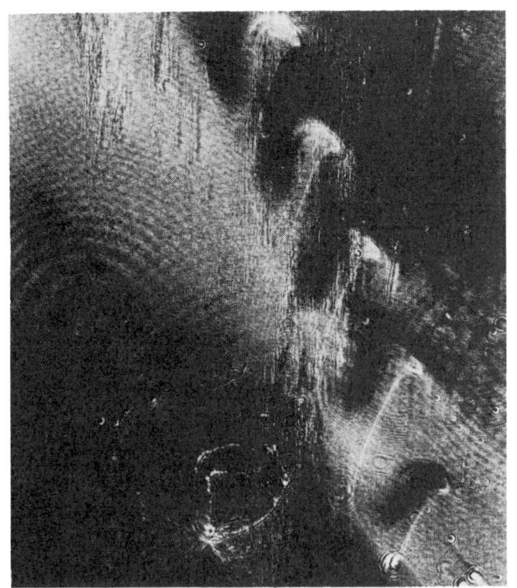

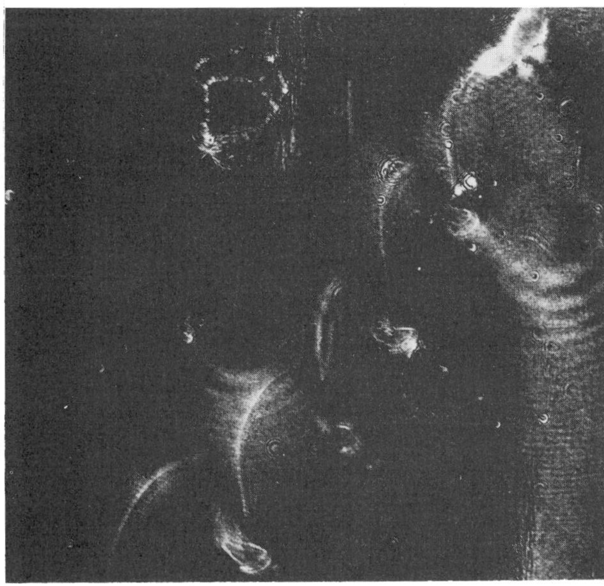

FIGURE 27.3
Hop-and-sink swimming behavior of a copepod.
Two views of *Cyclops scutifer* swimming by time-lapse Schlieren microphotography described by Strickler (1977). (From Kerfoot et al., 1980.)

animal generating turbulent flow (Kerfoot, 1980). Doubtless predators of similar habit are suspended in all the mixed layers of the world oceans as well, ready for a deadly turbulent rush on organisms dropping quietly through treacley, laminar flow.

VERTICAL MOVEMENT THROUGH THE UPPER OCEAN LAYERS

For the animal life of the open sea the horizontal layering of the ecosystem is of paramount importance. Light and food are at the surface of a relatively warm mixed layer. Darkness, cold, and short commons are below. In this milieu, vertical travel is to be expected, if only for the feeding foray on the surface, then back to the depths to hide, as do trout in similarly stratified cold mountain lakes.

Yet a more universal pattern of vertical movement is found in animals of all sizes, in both oceans and lakes. The animals swim up and down on a diurnal rhythm, appearing near the surface at night then swimming down by day. The phenomenon is known as "**vertical diurnal migration.**"

Limnologists have long remarked this habit by zooplankton in lakes. When the small size of the animals is considered, some of the distances covered are remarkable—a vertical journey of about 10 meters for an animal 1 mm or so long, for instance (Table 27.3). Most of the larger limnic zooplankters undertake these daily journeys. Smaller ones like the rotifers migrate also, but not with such consistency, and they may be out of phase with the large plankton (Wetzel, 1975).

Vertical journeys in the sea can be much

TABLE 27.3
Vertical Migration of Zooplankton
Data are maxima that have been measured. The larger amplitudes and velocities are for the larger animals, there being a general correlation between range of migration and size. (Approximated from data compiled from various sources by Wetzel, 1975.)

	Maximum Amplitude (m)	Maximum Velocity (m/hr)	
		Ascent	Descent
Cladocera			
Daphnia retrocurva	24	10.6	5.0
Daphnia schoedleri	2	1.4	1.4
Bosmina longirostris	20	19.0	12.0
Leptodora kindtii	9	4.2	2.0
Copepoda			
Limnocalanus macrus	24	18.0	9.8
Amphipoda			
Pontoporeia affinis	40	11.7	13.9

greater. Oceanic copepods can climb and drop a hundred meters in their day–night cycles, and the larger animals much further still. The really big movements show up on a ship's sonar. Massed reflections in the 500–800-meter depth range are known to oceanographers as the "**deep scattering layer**." But the deep scattering layer can be found only by day. In the evening it dissolves until it is gone for the dark hours. The layer represents massed fish, crustaceans, and other motile animals that pass their days just below the mixed layer but come to the surface at night.

A diurnal cycle obviously seems to be linked to light, and this has been well documented in lakes (Hutchinson, 1967). Light is the trigger for the movements, the animals rising in darkness but letting themselves sink in proportion to rising light intensity. But vertical travel through the surface layers of stratified ocean or lake has more immediate effects on the well-being of an animal than might come from differential illumination alone. Thus, although light is the trigger for the behavior, the selective advantages might be various. Three general hypotheses offer different selective advantages.

Hypothesis 1: Predator Avoidance

The advantages of not being suspended in a brightly lighted fluid in a world inhabited by predators that hunt by sight is self-evident. Zooplankton can feed in algal-rich waters of the surface at night with less danger from fish predation. Hunting zooplankters can feed at night since they do not depend on light for prey detection. The night-feeding argument may be less potent for the fish that hunt by sight, yet they would find a dense concentration of prey in the surface waters at night. The dark depths offer safety to both groups by day.

Hypothesis 2: Metabolism at Low Temperature

This hypothesis rests on the undoubted fact that there are advantages in "feeding warm and digesting cool." The hunting efficiency of ectotherms increases with temperature as they become more active (see Chapter 6). On the other hand, their metabolic costs go down with temperature, so it is an advantage to remain in colder water when not actively feeding. Thus the hypothesis states that vertical migration takes advantage of the fact that mixed layer and epilimnion are both warmer than the dark water beneath (McLaren, 1963). This hypothesis can be coupled with the suggestion that algae are more nutritious at night, since this is when they synthesize proteins. So: eat quality food at night at the warm surface, and digest by day in the cool of the depths.

Hypothesis 3: Niche Separation and the Food Search

A pattern of regular movement should be useful in the food search, and it may also serve in separation of niches (see discussion of the paradox of the plankton, Chapter 26). In a transparent mass of water, ordination cues are rare, the most obvious being light from above and gravity from below. Gravity may be less useful in orientation for low-density animals than light. Hence many regular movements will be constrained to the vertical, using light intensity as the ordinator.

These hypotheses have been tested largely in lakes, because lakes are more tractable systems for experiment than the great oceans. Other studies have taken advantage of the 30-meter-high experimental marine tank at the Dalhousie marine station in Canada in which layered systems can be established.

In Lake Gatun in Panama a common planktivorous fish can be shown to forage voraciously on the copepod *Diaptomus* in the laboratory, but the stomachs of wild-caught fish hold few *Diaptomus*. It was shown that vertical migration of *Diaptomus* kept it away from the foraging fish in the lake, thus demonstrating the selective advantage directly (Zaret and Suffern, 1976).

In another large tropical lake, Lanao in the Philippines, *Chaoborus* larvae are restricted to parts of the lake where deep water lets them descend more than 30 meters to where fish cannot get them in daytime (Lewis, 1979).

That predation cannot be the sole answer, however, is shown by data for water mites in a lake in Quebec. The mites go through a pronounced vertical migration, although it is known that they are distasteful to fish and thus are almost never eaten. Both energetic efficiency and pursuit of food seem to be plausible advantages to these water mites (Riessen, 1981).

One very elegant demonstration used the Dalhousie tank to show how the habit can serve as a food search: a population of marine copepods went through the classic diurnal migrations until a layer of algae was injected into the water, after which they tended to remain around the algal layer (Bohrer, 1980).

Finally, studies in Toolik Lake of northern Alaska, where there is permanent daylight in the summer months, found no vertical migration, but the animals positioned themselves according to temperature and environmental cues (Buchanan and Haney, 1980).

It therefore seems certain that the selective advantages of vertical migration can be various. The hunt and be hunted equations of predation are probably the most all-pervading selective pressures, particularly in the larger marine vertical migrations, but the other explanations must all operate to varying degrees (Neill, 1992). The habit essentially is a property of the system in which the animals live—transparent, thermally stratified, and intermittently lighted from above.

LIFE INSHORE

The oceans have remarkable littoral zones. On cold coasts in temperate regions, a few meters at the edges of the sea support rooted, or rather anchored plants. With the sole exception of the sea grasses, these belong to ancient lineages long antedating the evolution of flowering plants on land. Brown or red pigments augment chlorophyll in scavenging all useful wavelengths of the attenuated light (see Chapter 3).

On the warmer shores of the tropics very different ecosystems develop. Where cool water washes hot tropical land, rocky shores look barren because they lack the forests of large algae characteristic of colder latitudes. Elegant examples are the coasts of the Galapagos Islands, where black rocks emerge from the sea covered by nothing more obvious than marine iguanas and red crabs. Ruinous herbivory, or exposure to the tropical sun at low tide, are possible hypotheses to explain the barrenness of such coasts.

But over much of the tropics, where warm water abuts hot land, the classic coastal community is the coral reef. These are carbonate-depositing systems, their ubiquity in warm tropical waters undoubtedly due to the temperature-dependent solubility of calcium carbonate.

Coral animals, which are colonies of coelenterates, deposit calcium carbonate in so conspicuous a way that we talk of "coral reefs" (Veron, 1986). But other animals do so too, like the nearly microscopic foraminifera whose accumulated calcareous skeletons are one of the important components of the reef mass, and heavy-shelled mollusks, like the giant clams *Tridacna* of coral reefs. More important to the economy of the oceans, however, is deposition of calcium carbonate by photosynthetic algae.

Early in the geologic record the great reefs taking the place of modern coral reefs were manifestly built by photosynthetic primary producers; geologists call their fossils "**stromatolites**" (Schopf, 1975). The parent blue-green algae were microscopic and colonial, living at the surface of the growing carbonate mass. In modern carbonate reefs, microscopic photosynthetic organisms are probably nearly as abundant as they ever were, and they still provide most of the energy input to the reef structure. But they live as symbionts (called "**zooxanthellae**") of the coelenterate colonies, so that the framework of a modern carbonate reef may be likened to a veneer of microscopic primary producers supported by the carbonate structures of the coral animals (Muscatine and Porter, 1977; Rowan and Powers, 1991).

But the coral reef system also supports large free-living plants, most of which deposit calcium carbonate just as the coral animals do. Coralline red algae form the rock-like reef ridge that takes the direct onslaught of the waves, and in the shelter of which the coral animals themselves live. But probably of more importance to the economy of the whole reef system are green calcareous algae, especially of the genus *Halimeda* (Hillis, 1980).

Coral reef systems have shallow areas of active coral growth that are well-lighted to support the zooxanthellae and are in the turbulence of the waves (Figure 27.4). But the larger area of the reef surface, immensely larger in atolls, is in darker, deeper water where most of the reef-building corals do not grow. These are prime habitats for *Halimeda* and for such relatives as the merman's shaving brush (*Penicillus*) of the Caribbean.

The green, segmented, calcareous thalli of rock-dwelling halimedas live draped on the dead parts of the reef matrix down to depths of more than 100 meters. More im-

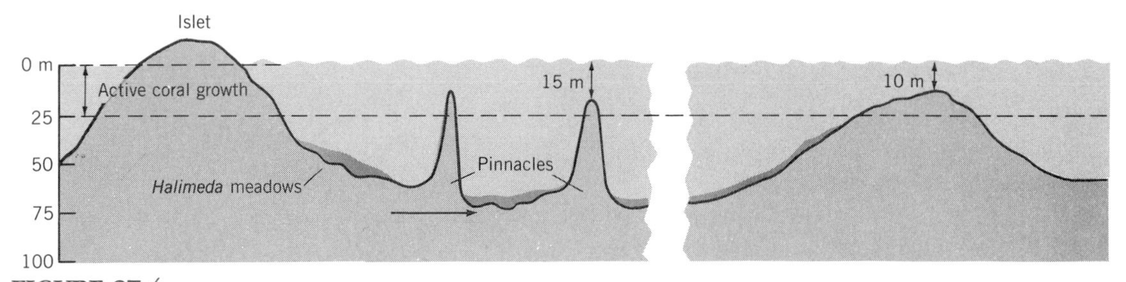

FIGURE 27.4

Stylized section through a coral reef system.

Active growth of the coelenterate coral communities is limited to the top 15–25 meters. Growth by the calcareous green alga *Halimeda* in the larger areas of deeper water probably contributes more carbonate mass to the reef.

portantly, the broad sandy floors of lagoons at 25 meters or so can be meadows of upright halimedas, growing from bulbous holdfasts in the sand and reproducing by creeping rhizoids that serve them like the underground stems of pasture grasses (Figure 27.5). Throughout life, both rock- and sand-dwelling halimedas shed segments, whose aragonite skeletons add to the calcareous sand of the lagoonal bottom.

Because the total area of reef represented by *Halimeda* habitat is far larger than the ridges of living corals, it is reasonable to expect that the contribution of carbonate by the green algae to the reef system might be crucially important. This is confirmed by sediment studies in the Great Barrier Reef and elsewhere (Orme, 1985), and by the discovery of shoal deposits built by halimedas on portions of the Great Barrier Reef of Australia that receive nutrient-rich upwelled water, where the population densities of *Halimeda* are so high that the shoals have been called "**bioherms**" (Searle and Flood, 1988).

The huge solar-powered industry of massed ranks of *Halimeda* may well contribute the largest overall fraction of the reef mass as fallen segments, later to be cemented into place. Coral animals make the framework, but in association with their symbiont producers, the zooxanthellae. Red algae form the rock-like masses that endure the breaking waves. It is thus reasonable for a botanist to cry foul at the naming of reefs for the coralline coelenterates, saying that they should be called algal reefs like their predecessors of long ago, the stromatolites (Hillis, 1986).

Like the seaweed communities of cold northern waters, algal and coral reefs are anchored structures made possible where water is so shallow that the seabed is lighted. Large photosynthetic structures can be maintained because of this attachment, whether kelp plants 100 meters long (*Macrocystis*), the more widespread brown, red, and green algae less than 1 meter high of temperate coasts, *Halimeda* and sea grasses of tropical bottoms, or coral structures themselves with their symbiont zooxanthellae. These photosynthetic structures build biomass in a low-nutrient medium the way a tropical rain forest does, by accretion, storage, and recycling (Smith and Kinsey, 1988).

Once established, the large plants of shallows at the edge of the sea become the energetic and structural bases of food chains and consumer trophic levels, whose components function in ways directly comparable to their equivalents in terrestrial ecosystems.

FIGURE 27.5
***Halimeda* building reefs: the Enewetak lagoon 21 meters deep.**
The halimeda meadow (top) has eight species of *Halimeda*, the commonest being *H. incrassata* (bottom). The sand is mostly made from *Halimeda* segments and piled into mounds by the ghost shrimp *Callianassa*.

THE SODIUM CYCLE AND THE AGE OF THE OCEANS

It is an old idea that the sea is salt because rivers eternally bring salt down from the land. The carrying water evaporates to return to the land as rain and the salt remains in the ocean. Setting aside for the moment the obvious fact that the composition of salt in the oceans is very different from typical river solutes, this simplistic view of oceans' getting steadily saltier since the dawn of time suggested a method of aging the oceans.

As late as the 1940s, the conventional method of aging the oceans was to estimate the annual tonnage of sodium delivered to the sea by the world's rivers and to divide this into an estimated tonnage of sodium held by the oceans as sodium chloride. The answer given in textbooks of the period was about 100 million years. But then radiometric dating of rocks was invented and strata holding undoubted marine fossils were shown to be 500 million years old. Thus the oceans of half a million years ago were already salty, and the conventional calculation was shown to be wildly wrong.

To investigate the cause of the calculational error, Livingstone (1963) reversed the logic, first calculating how much sodium should have been added to the world oceans in the 500 million years that have elapsed since Cambrian time:

> Sodium carried by rivers to the sea during 500 million years of post-Cambrian time: 119.4×10^{15} tons

This figure is vastly more than all the sodium known to be in solution in the world oceans at present:

> Sodium dissolved in world oceans: 14.1×10^{15} tons

So where did the rest of the sodium go? Livingstone estimated all possible depositories for sodium that had once been in the oceans, sediments, rocks, salt domes, etc., finding that little of the missing sodium could be accounted for in this way:

> Sodium in deep-sea sediments: 5.1×10^{15} tons
>
> Sodium in other suboceanic sediments: 5.4×10^{15} tons
>
> Sodium in old sedimentary rocks now raised out of the oceans: 2.6×10^{15} tons
>
> Known reserves of rock salt: 0.4×10^{15} tons

Adding in the tonnage now in the oceans, this leaves us with:

> Total sodium accounted for: 27.6×10^{15} tons

Thus, even with these generous assumptions, most of the sodium known to have been carried to the sea in rivers over the last half billion years cannot be accounted for. Where is it? The generally accepted answer is that no missing deposit remains to be found; rather, a smaller mass of sodium is recycled between land and sea. The proposed return mechanism is in two parts: return through the air of sodium picked up in spray to be carried by wind, and sodium in sediments returned to the land as sedimentary rock (Figure 27.6). If as much sodium were to be returned to the land each year in these ways as travels to the sea in rivers, the oceanic concentration of sodium would be maintained as a steady state.

The geochemical sodium cycle is complicated by the fact that fresh sodium, like many other elements, is constantly added to the system by volcanoes, even as so-

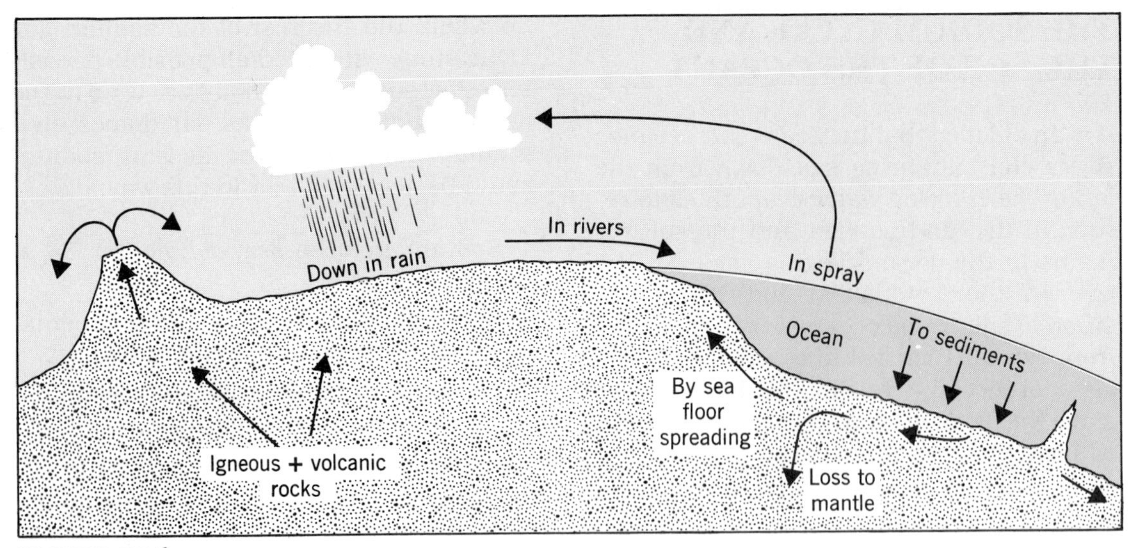

FIGURE 27.6
The geochemical cycle.
Minerals are cycled between the oceans, the land, and the crust of the earth. Rivers carry soluble minerals to the sea, but wind and rain carry some back; most of the rest are probably returned by crustal movements. Minerals in sediments are moved across the ocean bottoms by seafloor spreading from the mid-ocean ridges (right of figure) to the continents. Loss to the mantle in the deep ocean trenches may be made good by the extrusion of fresh rocks.

dium-rich sediments are lost by subduction in deep ocean trenches. Whether sodium in the world oceans really is in steady state is thus hard to compute from contemporary data alone, despite the evidence of fossils suggesting the existence of a saline ocean since Cambrian times.

THE HYPOTHESIS OF OCEANS WITH STEADY-STATE CHEMISTRY

The obvious way to test the hypothesis that ocean chemistry and salinity have been constant since, say, Cambrian times at the start of the fossil record is to look for chemical data in sedimentary rocks. These are often hard to interpret because of the changes that take place between sea, sediments, and rock, and because of what can happen to a rock in its long history on the land. The best that can be done is to compare concentrations of some elements in young and old rocks of similar type. These data are still somewhat equivocal, though the preponderance of the evidence, as the lawyers say, is consistent with an ocean of near constant chemistry over at least the last 500 million years (Garrels and Mackenzie, 1971).

One of the nicest sets of data comes from analyses of the boron content of the clay mineral illite in similar rocks spanning not just the rocks of the fossil record but the better part of 3000 million years (Figure 27.7). The boron content of the illite is so constant that there must be a strong presumption that each illite sample collected its boron from solutions of comparable chemistry.

Thus the concentration of boron ap-

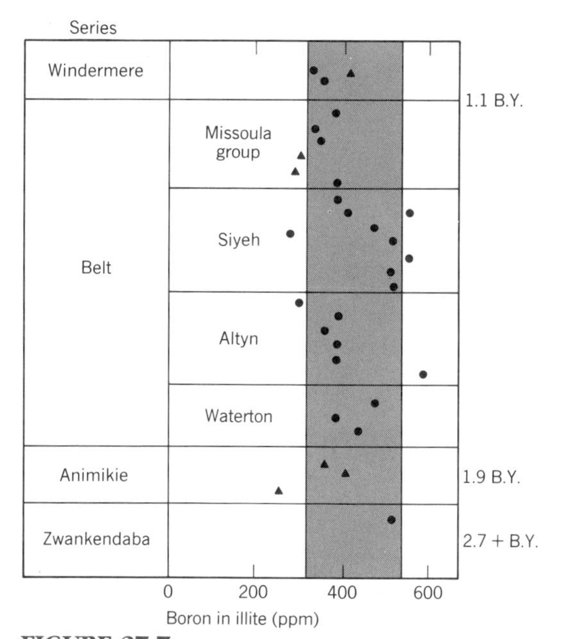

FIGURE 27.7
Evidence of boron for constant ocean chemistry. Data are parts per million (ppm) of boron in the clay illite within carbonate rocks and shales. The diagram covers only pre-Cambrian time, but analyses of similar rocks of the last 500 million years are all in the range shown by the shading. The data suggest that illites of all ages have collected boron from similar solutions. (From Reynolds, 1965, and Garrels and Mackenzie, 1971.)

pears to have been constant throughout the 2000 million years covered by the data, far longer than the half billion years of the marine fossil record that followed. Such constancy at least encourages the idea that the geochemical cycle can result in steady-state concentration of all solutes in the oceans.

Biologists have one special datum of their own which suggests significant antiquity for the present concentration of salt in the oceans. It is that vertebrate land animals have blood with an ionic concentration close to that of seawater. The best explanation for this is that their marine ancestors had blood that was isotonic with the seawater in which they lived, suggesting that something close to a steady-state concentration of sea salt has been maintained for at least the last 300 million years for which vertebrates have lived on land.

WHY THE SEA IS SALT

The oceans cannot be merely evaporated river water because the mix of ions in seawater is quite different from that found in rivers (Table 27.4). What happens if river water is evaporated in a fairly simple system is shown by the alkali lakes of deserts, like the waters of the Great Salt Lake in Utah; instead of sodium chloride, it contains a brew of carbonates and sulfates. Special processes clearly must exist in the

TABLE 27.4
Sea and River Water Compared
The table shows the more abundant elements in oceans and typical river water. It is evident that simple concentration of river water does not produce seawater. (From data in Garrels et al., *1975.)*

	ppm (mg/l)	
	Sea	Rivers
Chlorine	19,000	7.8
Sodium	10,500	6.3
Magnesium	1,300	4.1
Sulfur	904	5.6
Potassium	380	2.3
Calcium	400	15.0
Silicon	2.9	6.1
Bromine	65	0.02
Strontium	8	0.07
Aluminum	0.001	0.4
Boron	4.5	0.01
Fluorine	1.3	0.1
Phosphorus	0.09	0.02
Iron	0.003	0.67
Manganese	0.002	0.007
Zinc	0.002	0.02

oceans to conserve sodium chloride and rid the system of most of the other ions that come down yearly in the rivers.

River water is the product of dissolving crustal rocks in the dilute carbonic acid of rain. Important minerals of the rocks are feldspars like potassium feldspar ($KAlSi_3O_8$), anorthite ($CaAl_2Si_2O_8$), and sodium feldspar ($NaAlSi_3O_8$).

Solutions of these minerals are rich in potassium, sodium, and calcium but poor in chlorine. Such is the typical composition of river water (Table 27.4). For this solution to be changed to the composition of seawater, potassium and calcium must be removed and chlorine must be added. In general, it may be said that potassium is removed on mud, calcium is precipitated as carbonate, and chlorine is supplied from the mid-ocean ridges.

Potassium Removal

Clay minerals of mud are supplied to rivers and the oceans as a by-product of the solution of feldspars. A clay mineral like kaolinite [$Al_2Si_2O_5(OH)_4$] is made from the insoluble residue of feldspar and is carried to the sea in suspension. It is a flat, platelike mineral, charged on all surfaces. To thrust minerals like this from the dilute waters of rivers into the concentration of ions found in seawater is to cause complex rearrangements of their structures. It is of particular importance that kaolinite reforms into the potassium-based clay illite. This chemistry is subtle, slow, and poorly understood. As MacIntyre (1970) has it, "Graduate students who study these reactions invariably leave before the reactions are complete." But there is little doubt that the reactions take care of the potassium donated to the sea. The resulting illite takes potassium to the bottom and holds it in the mud.

Calcium Removal

The fate of calcium is possibly more satisfying to an ecologist because the biota are involved. Calcium is deposited in shallow oceans as calcium carbonate, particularly in the shells or skeletons of plants and animals, and most spectacularly in coral reefs. The emphasis is on the word *shallow*, because calcium carbonate redissolves under high pressure, effectively in water below 4000 meters. Calcium carbonate, therefore, does not collect in muds of the abyssal depths of the really deep sea, and any settling skeletons that penetrate there return to solution.

Like everything about ocean chemistry, the chemistry of precipitating calcium carbonate is not simple (Milliman, 1974). Inorganic precipitation is possible, as when cold water saturated with calcium carbonate rises from the depths and is heated at the surface. This is a well-known phenomenon in parts of the Caribbean where the surface of the sea turns milky white from the small crystals of the carbonate mineral aragonite. Some of the mass of carbonate in coral atolls may be inorganic aragonite also, probably deposited alongside the plant skeletons in the lagoons, behind the framework set by coral animals (Hillis, 1980). But the great mass of calcium carbonate deposition seems to be by animals or plants, and has been so since the start of the fossil record. Life disposes of calcium, therefore, leaving sodium in lonely state as *the* alkali metal of the oceans.

Chlorine Enrichment

Two sources of chlorine are known—volcanoes and mid-ocean ridges. Chlorine appears in the form of hydrochloric acid (HCl) both from terrestrial volcanoes and in juvenile waters from the submarine volcanoes of the mid-ocean ridges. Either way, chlorine enters the ocean system as aqueous hydrochloric acid. It then reacts with bicarbonate ions to produce water, carbon dioxide, and chloride ions. The carbon dioxide returns to the atmosphere leaving the chlorine to take the place of the bicarbonate as a negative valence, balancing sodium.

Requiring chlorine to enter the sea from volcanic vents almost certainly implies a lesser annual flux of chlorine than there is of sodium, potassium, and calcium. For the sea to remain rich in sodium chloride, therefore, it seems certain that chlorine must remain in the sea even as sodium and the other elements are removed in the geochemical cycle. This is, indeed, what seems to happen. The reactions that deposit clays or carbonates do not deposit chlorides, and the minerals of sediments are chloride-poor (Holland, 1984). It may be proper to say that the ocean conserves chlorine even as sodium is passed backward and forward between land and sea. But it is also advisable to say that these are still early days in our understanding of the chemical stability of the oceans.

SCAVENGING OTHER ELEMENTS FROM THE SEA: MUD AND THE BIOTA

Not only is the sea salty, but the ocean solution is almost nothing but salt, since the rest of the periodic table is present in only small amounts. Since rivers carry most of the periodic table continuously to the oceans, mechanisms must exist to sweep the ocean clean again.

Geochemists have looked for the places where the sweeping occurs by using uranium as a tracer. Uranium and its decay products (the "uranium daughter series") are particularly useful for this, since the decay of the parent into the daughters at known rates provides clocks that time many of the processes (Figure 27.8). Following the uranium tracers has shown that the process of sweeping goes on at every stage in the journey of minerals from land to sea: in soils, streams, rivers, estuaries, and in the deep sea itself (Turekian, 1977).

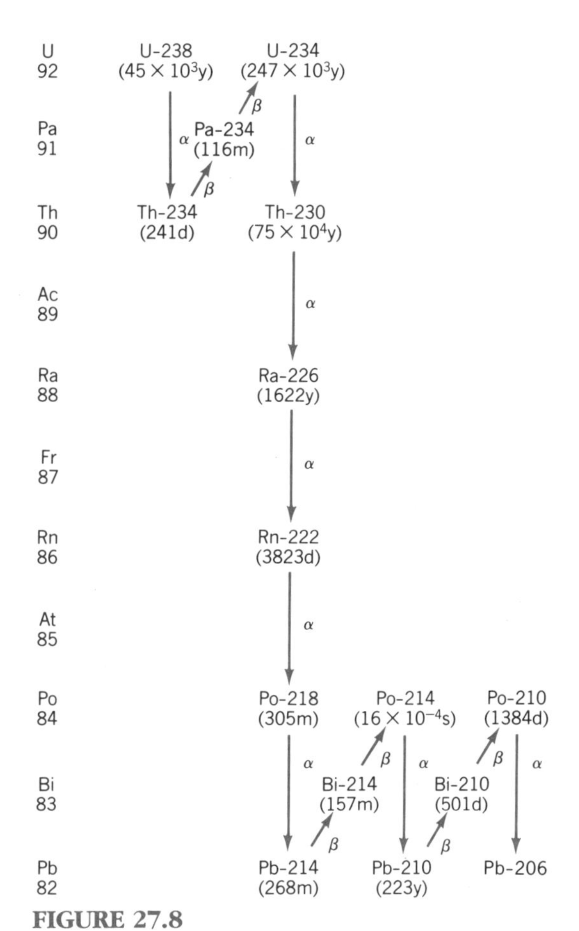

FIGURE 27.8

Uranium decay chain.

Elements of the uranium daughter series serve oceanographers both as tracers representative of heavy metals and as clocks. The half-life of each element is given in days, months, or years. (From Turekian, 1977.)

First among the sweeping mechanisms is attachment to clay minerals in the same way that potassium is attached to illite. Other elements are attached to the long-chain organic molecules in solution called humic acids. The resulting clay and humic complexes are then precipitated when entering the high-electrolyte medium of the oceans.

Additional elements may be removed from solution as bacterial flocs or by algae that extract phosphorus, copper, and other minor elements. Bacterial flocs and algal corpses then join the rain of clay and humic complexes to the seafloor, leaving behind an ocean containing little more than sodium chloride.

The living parts of ecosystems, particularly in estuaries, probably serve to refine this process of sweeping elements from the sea. This is because respiration controls the redox potential of mud in the high BOD circumstances of worm burrows, just as it does below the hypolimnion in eutrophic lakes (see Chapter 26). Burrowing animals pump solutes in and out of their surroundings of reduced bottom mud. If the open water holds oxygen, then the pumping by burrowing animals can run a chemical plant in which reduced solutes are pumped from mud to open water, where oxidation precipitates them back to the surface mud.

AN ECOLOGIST'S VIEW OF THE OCEANS

The oceans are basically a blue void of nutrient-poor saline fluid. The color of the deep sea is the blue remnant of white light from which all other colors have been extracted by clear water. The blue color itself is evidence of the desert properties of oceans (see Chapter 24), for if productive they would be green or brown with plants. Over the small areas of upwellings and shallows the sea is, indeed, green, the waters opaque to light.

No plants have evolved that can maintain large floating biomass in which nutrients can be hoarded, probably because the cost to individual fitness of loss by drifting would be too great. Instead, selection pressures for smallness have driven plant size down to the microscopic. The desert state of the oceans is thus maintained.

Because the ocean is fluid, fine particles of debris (**detritus**) as well as the tiny plants can only be seized at low Reynolds number under conditions of laminar flow. But algae and detritus together make up the larger part of the flux of reduced carbon in the oceans. Not only the zooplankton, therefore, but also many of the larger animals of the sea live by filter feeding. Apparently the entire ocean floor, from coastline to abyss, supports populations of sessile filter feeders, from bivalve mollusks to worms.

Rocky shores, upwellings, and even the filter feeders of the ocean bottoms all support food chains ending in the motile predators of the open sea. Yet the vastness of the open water ensures that the larger part of the animal support system of the seas depends on primary production in the euphotic zone. The oceans for an ecologist are thus largely the lighted part of the mixed layer.

Life in the lighted mixed layer is either the low Reynolds number world of the tiny drifting plankton at the base of the ocean food chains or it is the life of vertical travel, of hunting forays at the surface and periodic descents to darker water for safety or physiological need. This whole way of life is predicated on the absence of vegetation in the ocean. And that seems to be a property of the fluidity of the medium.

Preview to Chapter 28

White cliffs of Dover: carbon sink from the Cretaceous.

AN atmosphere of oxygen, nitrogen, and minimal carbon dioxide is bizarrely unlikely without the intervention of life processes. Strangely, carbon dioxide concentrations may be least under living control, the main control being by solution in the oceans. Carbon dioxide concentrations are now rising because human activities release carbon from terrestrial stores faster than the oceans can take up the surplus. Reduction of carbon dioxide during the last ice age was probably a result of increased absorption by the oceans. The oddest property of terrestrial air, the presence of free oxygen, has now been conclusively shown to have resulted from photosynthesis in the early oceans. The crucial process was rapid burial of sufficient reduced carbon to balance the free oxygen produced. This carbon is present at low concentrations in crustal rocks, forever in escrow. Oxygen is now apparently regulated within broad limits, between 13 and 35% by volume, but the method of regulation is not understood. Sulfur bacteria, or burial of phosphates, are leading candidates. The reservoir of nitrogen is the air itself, a reservoir that is maintained principally by bacteria of swamps that use nitrates as hydrogen acceptors. Were it not for life processes in anoxic wetlands, the atmosphere of the planet probably could not be maintained.

Chapter 28

The Origin and Maintenance of the Air

Terrestrial air is an improbable brew, quite without equal in the solar system. Most startling is free oxygen, despite the fact that the surface rocks of the crust of the earth are all chemically reduced, except where they come into actual contact with the air, and that volcanoes continue to put fresh, partly reduced gases into the atmosphere (Figure 28.1). Every other planet examined so far has an atmosphere of reducing gases; only earth has an oxidizing air. A cynic observing us from orbit around a distant star might well suggest that we ought spontaneously to combust. At times, of course, some of our ecosystems do.

Apart from water vapor, the air is made up of nitrogen (79%), oxygen (21%), and carbon dioxide (0.03%), with other gases present in trace amounts only. Nitrogen, oxygen, and carbon dioxide all are manipulated by individual organisms in ecosystems. Over geologic time, it seems likely that the concentration of all three in the air has either been determined or at least drastically altered by life.

It seems very likely that the mixture of gases in the air has been roughly constant at least since the first large plants and animals appear in the fossil record. Ten percent more oxygen, and forest fires would be so frequent as to banish all trees. Much more carbon dioxide would be toxic. Life on earth seems organized to cope with the present mixture, suggesting that something close to our present air has persisted since at least the Cambrian epoch, 500 million years ago.

REGULATION OF ATMOSPHERIC CARBON DIOXIDE

In the 1990s nobody can think of atmospheric CO_2 without hearing the haunting

FIGURE 28.1
Raw material for the atmosphere.
Mayon volcano in the Philippines spews gases into the atmosphere. Volcanic gases are the major source of carbon, nitrogen, and sulfur for the atmosphere over geologic time.

sounds of environmental debate. The dangers of burning coal, the greenhouse effect, air pollution; these are the phrases that come to mind. And yet atmospheric carbon dioxide ought to be as well buffered as any part of a terrestrial system we can easily imagine.

The air is in direct contact with the sea, and there is free exchange of CO_2 at the air–sea interface. The oceans hold CO_2 in solution in the various oxide species equivalent to about 50 atmospheres' worth. The oceans, therefore, should act as an enormous shock absorber for the air against which humanity's puny efforts at burning coal and oil should have little effect.

Moreover, the oceans are themselves shock-absorbed by the mechanism, largely under biological control, that deposits excess dissolved CO_2 as carbonate rock. A rough estimate of the carbonate rocks of the earth suggests that they represent about 40,000 atmospheres' worth of CO_2. So we have a potentially enormously powerful buffering system of oceans with carbonate sediments behind the scanty mass of CO_2 held in the air (Figure 28.2).

These facts were pointed out by G. E. Hutchinson in 1949 in a seminal paper analyzing, among other things, the possibility of serious perturbation of atmospheric CO_2 by industrial activity. Hutchinson's conclusion was that it was unlikely that significant changes in atmospheric CO_2 could result from industrial activity because the rate of release of CO_2 from industrial chimneys and vehicle exhausts would be too

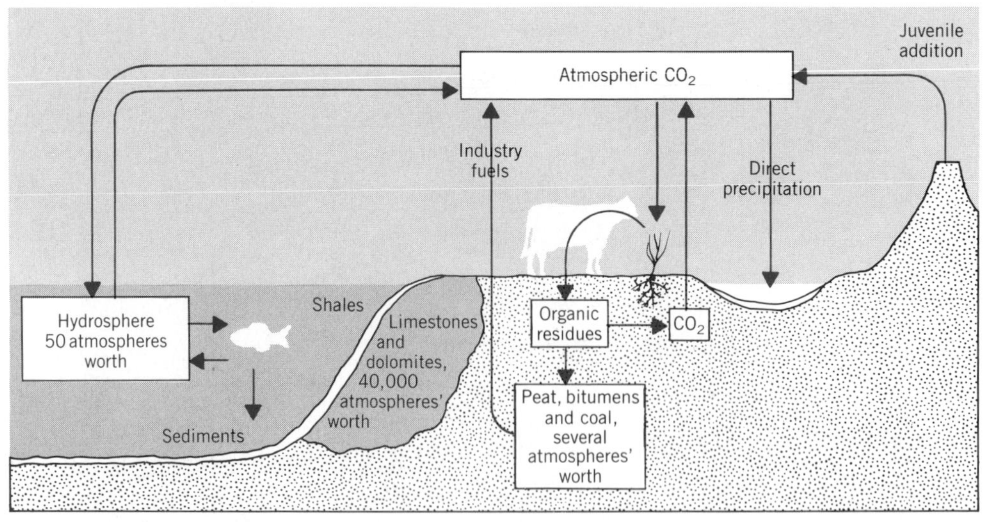

FIGURE 28.2
The carbon system.
There is apparently a steady-state supply of carbon to the biosphere, the input being from volcanos and the sink being carbonate rocks. Within this steady state there are three orders of cycling: a cycle between land and sea, through crustal folding and solution of limestones and dolomites, which has a mean time span of geologic epochs; a cycle involving reduced organic deposits, which is only now being closed by human activity; and the short-term cycles between animals and plants.

low to swamp so massive a shock absorber. Interestingly, he suggested that if changes in the CO_2 of the air were to be detected, they would probably result from the cutting down of forests and the draining of bogs rather than from industrial activity. This was because a more massive and sudden release of CO_2 might be effected by these agricultural means than by mining and burning coal.

These conclusions of Hutchinson (1949) follow naturally from a first look at the geochemistry of carbon, for it is hard to imagine any system more thoroughly buffered. And yet we now know that the conclusion was partly wrong and that there is in fact a sustained and significant secular rise in the CO_2 of the air.

The most spectacular data come from air chemistry measurements at the observatory at Mauna Loa in Hawaii. Hawaii is well-sited for taking these measurements because the islands are in the middle of the Pacific amid well-mixed oceanic air. There are no continental industries nearby whose chimneys could produce purely local effects. The data show that the CO_2 concentration in the air over Hawaii has been rising, and at an increasing rate, ever since the measurements began (Figure 28.3). Less complete data from Antarctica show a comparable rise, and other studies confirm that an increase in the atmospheric concentration of CO_2 has been proceeding for about a century (Bacastow and Keeling, 1973; Stuiver, 1978).

Despite these rather alarming data showing that the concentration of CO_2 in our air is rising, the original arguments about the potential buffer of oceans and rocks remain sound. In the long term, the CO_2 of the air is kept a rare gas by the buf-

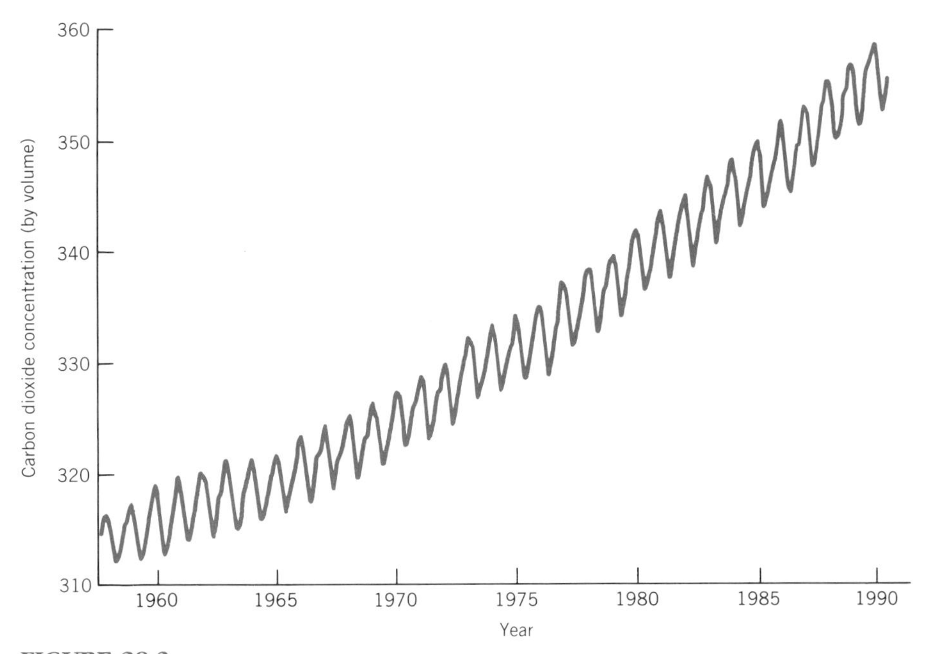

FIGURE 28.3
Secular changes in carbon dioxide: Mauna Loa.
The data are concentrations of CO_2 in the air at the Mauna Loa Observatory, Hawaii. Peaks in the curve reflect the respiration of northern ecosystems during winter. Troughs reflect the excess of photosynthesis over respiration of northern ecosystems in summer. The upward trend of the curve is a direct demonstration that the CO_2 concentration of the air is rising. (Data from L. Machta, NOAA Air Resources Laboratories, 1992.)

fering system. The long-term fate of carbon is a one-way journey from volcano to limestone, with the chance of some future wanderings in solution as part of the geochemical cycle (Figure 28.4).

But in the short term, as we now know, CO_2 can vary. Perhaps the aspect of the Mauna Loa data most revealing of this is the annual fluctuation (Figure 28.3). Carbon dioxide is pumped in and out of the atmosphere with a yearly rhythm, one inhalation and one exhalation every year. This is because most of the land surface is in the northern hemisphere and most plant production is on the land. In the northern summer, vegetation takes in more CO_2 than it respires; in the northern winter, vegetation respires more than it takes in. The concentration of the gas in the entire atmosphere changes in synchrony with the production and respiration of the ecosystems of northern lands (Figure 28.5).

Seasonal and longer-term changes in CO_2 are noticeable only because the gas is at such a low concentration in the atmosphere. The actual masses of gas involved are small compared with the mass of the atmosphere as a whole. Similar mass changes in the oxygen or nitrogen of the air would not be detectable by our instruments. Add a gram to a gram and the result is noticeable, but add a gram to a ton and nobody knows the difference.

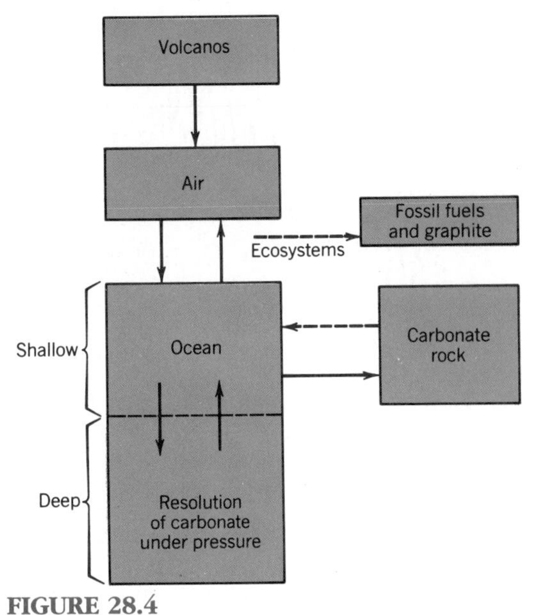

FIGURE 28.4
Carbon system in geologic time.
The main feature is continuous production by volcanos that is disposed of by incremental storage as carbonate rocks. Because carbonate minerals dissolve in the deep sea under pressure, there probably is little return to the mantle in subduction trenches. The surface of the earth now stores the CO_2 production of all geologic time. The CO_2 stored as organic compounds is trivial compared with that stored as carbonate.

We have a short and convincing answer as to why the carbon buffer system is less than instantaneous in its action. This is that the world oceans stir only slowly. It is one thing to dissolve excess CO_2 in waves, but quite another to carry the solute to the depths for equilibrium of the entire ocean with the air to be achieved. Direct evidence that slow mixing of the oceans is the cause comes from radiocarbon dating of water from the deep sea, for often this water turns out to have last been exposed to fresh carbon more than 100 years ago, and some samples of bottom water are 1000 years old.

The Global Carbon Budget

Since we now know that the concentration of CO_2 in the air is rising, it is natural to suspect that industrial use of fossil fuels is responsible. Yet it is equally plausible that clearing forests or bogs is to blame. But for that matter, why not an increase in the emission of volcanoes, or some disturbance in the oceans? When the Mauna Loa observations at last gave us incontrovertible evidence that CO_2 really was rising, all these possibilities and more had to be explored.

Considering all possible causes of the CO_2 rise requires drawing up a budget of all possible sources and sinks for CO_2. From the point of view of the atmosphere, a **source** is *any reservoir that may contribute carbon dioxide to the air* and a **sink** is any *reservoir into which high atmospheric CO_2 may drain.* Sources and sinks of atmospheric CO_2 are listed in Table 28.1 and Figure 28.6.

Over geologic time the important source is volcanoes, as we have seen, and the output of volcanoes is taken up by carbon-

TABLE 28.1
Sources and Sinks of Atmospheric CO_2
Reservoirs in parentheses are probably of little quantitative importance.

Sources	Sinks
Volcanos	Carbonate rock
(Synthesis of ^{14}C from ^{14}N)	(Decay of ^{14}C to ^{14}N)
	(Descent to mantle)
	Graphite
Oceans	Oceans
Fossil fuels	Fossil fuels
The biota (living plants and animals)	The biota (living plants and animals)
Detritus	Detritus

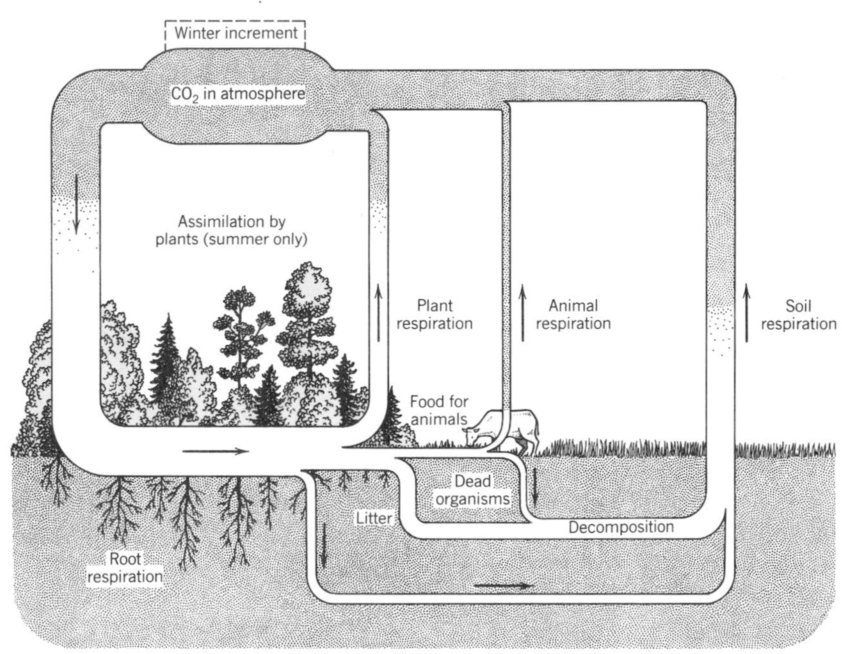

FIGURE 28.5
Ecosystem cycling of CO_2.
Notice that plants assimilate only in summer but that respiration proceeds year-round. The "pulse" in the Mauna Loa data of Figure 28.3 reflects the seasons in the northern hemisphere.

ates. Any change in volcanism should have short-term effects on CO_2 in the atmosphere, but direct observation of volcanoes makes it certain that volcanic events have not been important in the present CO_2 rise. Our data are good enough to declare that volcanoes have not emitted significant extra CO_2 in the recent past, but there could have been big changes in CO_2 in the air due to volcanism in the more distant past.

We are left with fossil fuels, the organic parts of ecosystems, both living and dead, and the oceans as possible culprits (Table 28.1, Figure 28.6). These three possibilities will be reviewed briefly, and then the isotope evidence that we can use to choose among them will be described.

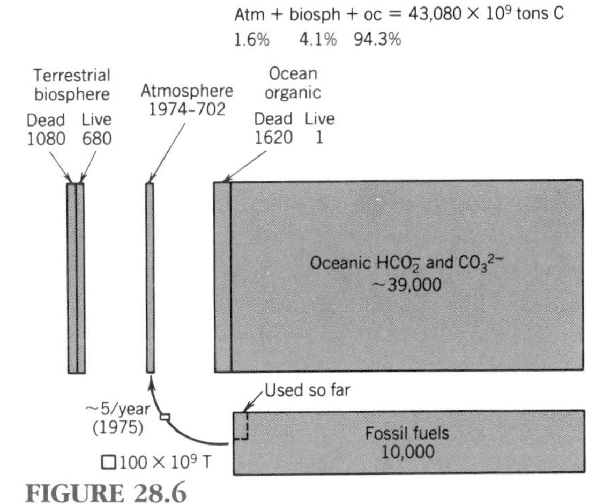

FIGURE 28.6
Major reservoirs of the contemporary carbon cycle.
Figures are in units of 10^9 tons of carbon. Reservoirs are drawn to scale. (From Stuiver, 1978, based on data of C. D. Keeling and R. M. Rotty.)

Fossil Fuels as a Source

The total reservoir of fossil fuel carbon at the start of the industrial revolution was about one quarter of the reservoir in the oceans (Figure 28.6). Of a total reservoir of about 10,000 x 10^9 tons of carbon in petroleum, peat, coal, and shale deposits, we have released about 180 x 10^9 tons in the last century (Stuiver, 1978). This compares with a present atmospheric reservoir of about 650 x 10^9 tons. The amount released from fossil fuels so far, therefore, is equal to 28% of the present atmospheric reserve, certainly sufficient to account for the increase shown by the Mauna Loa data even if the other sources had not changed (Figure 28.7).

It is worth noting that most of the 10,000 x 10^9 tons of fossil fuel in the global reservoir is out of our reach as fuel, being in the more diffuse or inaccessible shale and coal deposits. The most that we could release is perhaps several times what we have burned so far, possibly as much as three or four atmospheres' worth. So far, however, we have burned only about a quarter of one atmosphere's worth.

FIGURE 28.7
Carbon released from an ancient ecosystem sink.
A coal-fired power station in Arizona makes its contribution to the enrichment of the contemporary atmosphere with CO_2. The carbon released will find its way to the reservoir of carbonate rocks via solution in the oceans within a few centuries.

The Biota: Carbon Sink or Source?

Clearing forests, burning peat, draining marshes, plowing prairies: all these release CO_2 that has been stored in ecosystems. Usually CO_2 released in these ways is balanced by CO_2 captured by regrowth so that the carbon cycle is closed (Figure 28.5). But, by suddenly releasing carbon stores that were slowly built by ecosystems, modern humanity has been releasing organic carbon faster than it can be replaced.

Yet it is also true that photosynthesis tends to be carbon-limited (see Chapter 3). Increasing the CO_2 concentration of the air should increase the rate of photosynthesis in all of the more productive ecosystems. As the CO_2 of the air rises, therefore, we should expect the biota to fix carbon more rapidly, acting, like the oceans, as a CO_2 shock absorber. This is not just a theoretical construct or one known only on a laboratory scale, because field crops have been shown to increase their productivity when grown under plastic tents supplied with extra CO_2.

Furthermore, the fields and forests of a humanized world are all young fields and forests. We cut down the climax and promote secondary succession (see Chapter 20). The resulting young plant associations are those of most rapid growth, making it certain that the flux of CO_2 into the biota has been maximized. Thus the prime ecological view of CO_2-enriched air is that it should sponsor plant growth thus tending to return the air to normal.

An ingenious use of the Mauna Loa data suggests that extra uptake of CO_2 by photosynthesis is negligible, however (Hall et al., 1975). If there had actually been a net return of carbon to the biota in the last few decades, then this should be reflected in the Mauna Loa data (Figure 28.3) as the northern ecosystems took in more carbon with each successive summer. Rigorous analysis of the data reveals no systematic change in the amplitude of the fluctuations, showing that any tendency of the biota to act as a sink for fossil fuel carbon through increased productivity was slight. This is consistent with recent empirical findings that increased photosynthesis of wild vegetation exposed to high CO_2 is short-lived as the ecosystem adjusts to the new regime (Shaver et al., 1992).

The other obvious way to decide whether the biota are now sink or source is to make a budget of organic gains and losses from the world ecosystems. This is a difficult undertaking, one fraught with assumptions that may be perilous. A cost accounting by Woodwell et al. (1978) used the net productivity estimates of world ecosystems by Whittaker and Likens (1973), which are described in Chapter 24. Starting with regional estimates for net productivity, they then made the following assumptions:

1. World forests are being cleared at a rate of about 1% per year.
2. The response of photosynthesis to increased concentrations of CO_2 is linear.

Both these assumptions are reasonable. The scanty data we have from the Amazon region, where the largest remaining forests are, suggest that a 1% reduction is conservative, and the exploitation of other tropical countries with growing populations is at least as great. Only in the northern temperate regions are forests coming back (in New England, for instance). A linear response by photosynthesis to CO_2 concentrations is probably generous.

In these calculations the carbon released by cutting and clearing exceeds the carbon fixed by contemporary photosynthesis (Table 28.2). Thus the biota turn out to be a source. Woodwell et al. (1978) claimed to be conservative when they concluded that the biota now contribute between 2×10^{15} and 8×10^{15} grams of carbon to the atmosphere each year. This range spans the present contribution of fossil fuel carbon from industry (5×10^{15} g C/yr). This conclusion, however, is vulnerable to attacks on the assumptions. If the world forests are being cleared at a rate other than 1% a year the conclusions fail; if the regrowth of forests has been underestimated the conclusions again fail (Bolin, 1970).

The Oceans: Carbon Sink or Source?

We naturally tend to think of the oceans as a carbon sink. The oceans are the great shock absorber, and over the course of geologic time they do usually serve as a sink. But the fact remains that 50 atmospheres' worth of CO_2 are held in the oceans in intimate proximity to the air. Anything that changed the solution chemistry at the ocean surface might easily convert the oceans temporarily from sink to source.

The CO_2 regimen in the oceans as a whole is extremely complicated, such that no detailed simulation of the system is

TABLE 28.2
Carbon Budget for the Biosphere
These are the results calculated by Woodwell et al. *(1978). See text for assumptions of the calculations.*

		Carbon Budget ($\times 10^{15}$ g C/yr)			
	Area (10^6 km)	Total Net Primary Production	Forest Clearing and Harvest	Increase in Net Ecosystem Production	Net Release (total effect of human activity)
Tropical forests	24.5	22.2	4.4	0.9	3.4
Temperate forests	12.0	6.7	1.7	0.3	1.4
Boreal forests	12.0	4.3	1.0	0.2	0.8
Other vegetation (including agriculture)	98.5	19.2	0.3	0.1	0.2
Detritus and humus	147	—	—	—	2.0

possible yet (Broeker, 1971; Garrels and Mackenzie, 1971). But it is clear that the concentration of carbon oxide species in the surface waters is influenced by the concentration of CO_2 in the air, by temperature, and by ocean chemistry, as this is directed by biological processes.

Of particular importance to solution equilibria is the concentration of the bicarbonate ion, HCO_3^-. This ion may be formed directly by solution of CO_2, thus

$$H_2O + CO_2 + CO_3^{--} \rightleftharpoons 2HCO_3^-$$

But a second important source of bicarbonate is rivers that bring the ion down to the sea in solution. Bicarbonate is formed as a result of the aqueous weathering of crustal rocks in the presence of gaseous CO_2 in various reactions of which the following, for magnesium silicate minerals, is an example:

$$2CO_2 + 3H_2O + MgSiO_3 \rightarrow Mg^{++} + H_4SiO_4 + 2HCO_3^-$$

This flux of riverain bicarbonate into the ocean appears to be the main reason that seawater is alkaline. The pH of seawater is weakly buffered around pH 8 by the carbon oxide solution system

$$CO_2 \rightleftharpoons HCO_3^- \rightleftharpoons CO_3^{--}$$

This system must be sensitive to demands on particular carbon oxide species within the ocean. Photosynthesis, for instance, removes bicarbonate so that increased productivity should draw CO_2 into the sea and reduced productivity should release CO_2. Fixing of the electrically neutral form of silica known as OPAL by diatoms and other organisms also influences the reaction as more bicarbonate is needed to replace the valence of the lost silicate ion. Through reactions like these, therefore, biological activity in the sea can affect the solution equilibria of carbon oxides. Since the full complexity of this system is not known, it remains possible that the relative state of the oceans as sink or source can change as productivity changes (Broeker, 1971, 1973).

Ocean temperature can also be important to the retention of CO_2 because the solubility of CO_2 is strongly temperature-dependent. Solubility falls as water warms. A rise in mean water temperature of 0.1°C will increase atmospheric CO_2 concentration by 10 parts per million (Machta, 1973). This means that the oceans become a net source of CO_2 if they warm while other parameters remain the same. If the earth is warming slightly, from the result of human activities or from natural climatic change, then the oceans may be contributing CO_2 to the air. It is possible that warmer geological periods always have higher concentrations of CO_2 in the air in this way.

The Ice Age Air

Brilliance has conspired with good fortune to yield a carbon analysis of air from the last ice age. The air samples come from bubbles in the Vostok core of glacial ice from Antarctica (Figure 28.8). The core, taken by a Russian drilling crew and analyzed by a French research team, is more than 2000 meters long. Dating of inclusions in the upper section, evidence of seasonal banding, and ice flow equations provide a chronology good enough to identify ice from throughout the last glacial period back to about 150,000 years ago (Neftel et al., 1982). And in the ice are bubbles of air large enough to permit accurate analysis.

Air in the last ice age was dramatically carbon-poor, only about 0.02% by volume, or about two thirds the concentration at the start of the industrial revolution. The change to modern air was abrupt with the onset of the Holocene 10,000 years ago.

The carbon-poor air of the ice age might have been associated with carbon-poor land also, because calculations of the likely vegetation and bog cover of the ice age earth by pooling all known paleoecological data suggest only about half the mass of stored organic matter present on the modern earth (Adams et al., 1991). This calculation is disputed by alternative measures of ice age plant cover based on the contribution of terrestrial biota to marine deposits (Crowley, 1991). The controversy is probably only about the degree of vegetation depression, because some reduction in plant cover seems inevitable, given a colder earth and a drier Africa. Whether the reduced atmospheric CO_2 contributed to reduced vegetation cover through lowering the rate of photosynthesis remains to be determined.

The best guess for the cause of low ice age CO_2 blames the oceans, as the largest carbon sink. Known cooling of the ocean surface, however, is not sufficient to allow solution of so much atmospheric CO_2 into the mixed layer at the top of the ocean column, and a mechanism to draw carbon

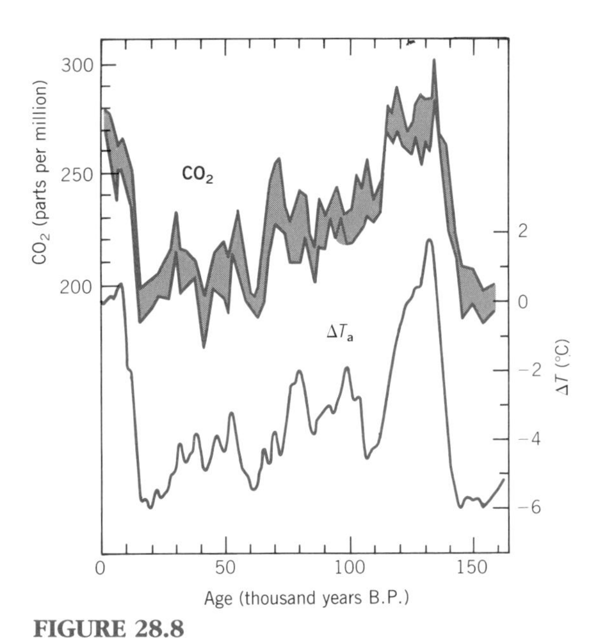

FIGURE 28.8
Carbon dioxide record from the Vostok ice core. The concentration of CO_2 in the atmosphere of the ice age earth has been measured from air bubbles trapped in a core of Antarctic ice. Lowered CO_2 in an ice age is now thought more likely to have been a consequence of cooling than a cause. Temperature is determined from oxygen isotope ratios of the ice. (From Lorius et al., 1990.)

rapidly down from the surface is required. Wholly satisfactory models for this have yet to be made (Broeker, 1991).

Because normality is an ice age, as emphasized repeatedly in this book, it follows that lowered CO_2 is the normal ambience for most of the animals and plants of the contemporary earth. Productivity of the ice-age Earth might have been below that of the Holocene from reduced CO_2 as well as from low temperature.

Isotope and Tree-Ring Data to Identify Carbon Sources

The most persuasive evidence for the cause of the secular rise in CO_2 comes from the studies of carbon isotopes in air, water, fuels, and the biota. Ratios of the different isotopes tell us the origin of any carbon sample, whether fossil fuel, the contemporary air, or organic matter. Moreover, we can take samples of air from different years over the past century or two by cutting out rings of trees and using the carbon isotopes in the resulting dated slivers of wood as vouchers for the past atmosphere. In this way it has been possible to trace in outline the course of the present CO_2 enrichment. A general conclusion is that both fossil fuels and cleared forests have contributed to the CO_2 rise, but that the oceans probably are still acting as a sink.

Carbon occurs naturally as three isotopes: ^{12}C, ^{13}C, and ^{14}C. Both ^{12}C and ^{13}C are stable, permanent parts of the earth's elemental species, the one abundant, the other rare. But ^{14}C is produced in the upper atmosphere by cosmic bombardment of nitrogen, and it decays back to nitrogen with a half-life of about 5700 years. Fossil fuels, being organic matter that has been sequestered for millions of years, contain no ^{14}C; the atmosphere does contain ^{14}C.

One of the consequences of our burning fossil fuel these last 100 years has been a dilution of atmospheric ^{14}C with extra ^{12}C, because none of the CO_2 coming from fossil fuels contains ^{14}C. This dilution is known as the **Suess effect**, after its discoverer, Hans Suess. We can measure the actual reduction in ^{14}C of the contemporary air from the Suess effect by taking samples of wood from before the industrial revolution and measuring the $^{12}C/^{14}C$ ratio. The actual ^{14}C reduction is between 1.7 and 2.3% (Broeker, 1971). This reduc-

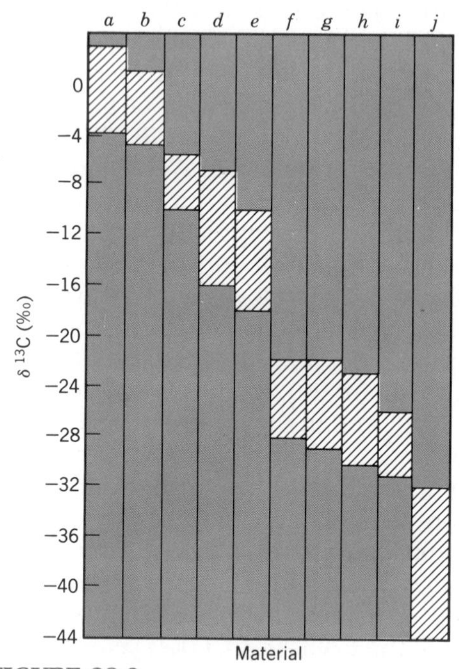

FIGURE 28.9

Excess ^{13}C ($\delta\ ^{13}C$) over ^{12}C of various carbonaceous materials.

Because plants discriminate between the isotopes in photosynthesis, and because fossil fuels have characteristic isotope ratios, it is possible to identify the source of CO_2 used in photosynthesis in the past from pieces of wood. *a*, Carbonates; *b*, volcanic CO_2; *c*, marine animals; *f*, coal; *g*, land plants; *h*, liquid hydrocarbons; *i*, organic sediments; *j*, natural gas. (From Farmer and Baxter, 1974.)

tion lets us calculate that between one third and one half of fossil fuel carbon so far burned is still in the air. The rest, if not in the biota, must be in the world oceans.

During photosynthesis, plant enzymes discriminate against the unusual carbon isotopes, both ^{14}C and ^{13}C. These are taken up and synthesized into organic compounds but proportionately less easily than is ^{12}C. The ratio of ^{13}C to ^{12}C in a plant, therefore, is different from the ratio found in the air on which that plant draws. Chemical processing of organic matter into fossil fuels over geological time also skews the isotope ratios. The result of this is that fossil fuels, and different contemporary organic materials, have different and characteristic ratios of the stable isotopes ^{12}C and ^{13}C as well as of ^{14}C (Figure 28.9).

Yet the actual ratio of isotopes fixed by a plant in photosynthesis must be a function of the ratio in the air from which the plants were drawing their supplies. Several investigators have now taken trees of known age, extracted samples of tree rings, and measured the carbon isotope ratios of each. With all three isotopes known for a time series of samples we can proceed as follows, following the analysis of Stuiver (1978):

1. Measure actual lowering of ^{14}C content for each tree-ring increment in the time series and calculate the actual contribution of fossil fuel carbon to each increment.
2. Calculate from the results of (1) the lowering of ^{13}C for each increment due to fossil fuels.
3. Adjust each ^{13}C measure for the contribution of fossil fuels, then any trend remaining is due to the flux of carbon from the biota to the atmosphere.

Figure 28.10 gives the results of Stuiver's calculations. There has been a more or less progressive fall in the ^{13}C concentration over the last century, with a particularly sharp decline between 1880 and 1930. This decline is due to the dilution of the atmospheric carbon pool with ^{13}C-deficient carbon released from the biota.

Stuiver concludes that about half the increment in CO_2 floating about in our air

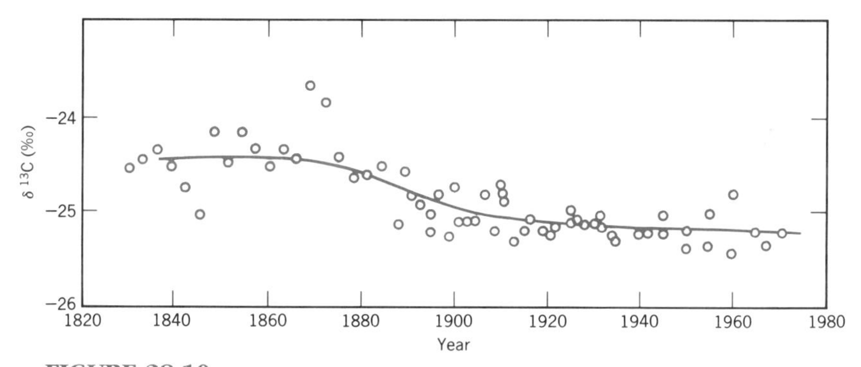

FIGURE 28.10

Reduction in ^{13}C in atmosphere resulting from reduction in biota.

Data are residual reduction in ^{13}C after allowing for the effect of fossil fuels. The modern reduction in atmospheric ^{13}C shows that the biota are acting as a source of CO_2 rather than a sink. The curve is a visual aid drawn by hand. (From Stuiver, 1978.)

comes from fossil fuels and the other half comes from the biota. On this analysis, therefore, the biota are at present a *source* of CO_2, and the old belief that plants would compensate for our burning of fossil fuels with extra photosynthesis is found to be wrong (Figure 28.11).

The following tentative conclusions seem warranted:

1. The mass of CO_2 in the atmosphere is rising.
2. Human activities are the cause of this rise.
3. Both burning fossil fuels and the clearing of forests are contributing to this rise.
4. The rise is possible because the rate at which the oceans can take up CO_2 is kept low by the slow mixing of the oceans.
5. Unless ocean temperatures change, the extra CO_2 eventually will be absorbed by the oceans, allowing the atmospheric concentration to return to what it was before the industrial revolution.
6. Changes in the mass of CO_2 in the atmosphere comparable to that now being caused by human activity certainly occurred in the past with global changes of temperature or displacement of continents.

FIGURE 28.11
Carbon dioxide source for the greenhouse.
A carbon reservoir in a British Columbian forest is released into the atmosphere. Both carbon isotope fractionation and model calculations suggest that forestry and agriculture now release carbon dioxide more rapidly than plants can return it to the reservoir.

HOW THE AIR GOT OXYGEN

The primeval atmosphere was made of reduced gases from the original "outgasing" of the earth's beginning. Chemical reduction was the universal norm, and the chances of a molecule of oxygen remaining free over this original earth were those of the proverbial snowflake in a very hot place.

For free oxygen to be put into the earth's atmosphere, all the primeval oxygen sinks had to be swamped. Surface rocks were all chemically reduced, as indeed they still are beneath the thin film of oxidized material that has since been draped over the earth. Two kinds of mechanism are known that pumped oxygen into this reducing world. One is purely physical, the dissociation of water vapor in the upper atmosphere, with the loss of hydrogen to outer space. The other is photosynthesis, now shown to be by far the most important.

Dissociation of water is the more straightforward mechanism. The upper atmosphere is exposed to ionizing radiation from the sun, and this radiation provides the energy to split water into molecular oxygen and molecular hydrogen. Hydrogen, being the lighter gas, is then lost to outer space. This mechanism is in place still, though working effectively only above the ozone shield.

The beauty of the dissociation mechanism from the geochemist's point of view is that the hydrogen is removed from the earth system entirely; the mechanism is a real whole-earth oxidizer. But a most serious objection is that calculations of the rate of net oxygen production by the process are ruinously slow.

Photosynthesis in its various forms also uses solar energy to release oxygen, typically from water. The scale of production is huge, so much larger in fact than any other imagined source that most geochemists have long assumed that photosynthesis by green plants was the primary source of atmospheric oxygen. But the oxygen released in photosynthesis is less final an act because the other product of the reaction is still around. Photosynthesis balances the oxygen released with reduced carbon compounds that are the object of the synthesis. Photosynthesis, therefore, can yield free oxygen only by burying carbon.

If all the oxygen used to oxidize the primeval gases and the exposed parts of the earth's crust, together with all the oxygen in the air, came from photosynthesis, then there must be an enormous mass of reduced carbon buried in the crust of the earth. The carbon required is far larger than all the known reserves of fossil fuels, including oil, coal, natural gas, and oil shale. These combined cannot even balance the mass of oxygen in the air, let alone the far larger mass that must have been used to oxidize minerals before any oxygen got into the air at all. For photosynthesis to have been the principal cause of the oxygenation of the atmosphere, therefore, requires that there must exist in the crust of the earth a reservoir of reduced carbon so massive that the oil reserves of the Middle East or the coal of Poland become trivial by comparison. So where is it?

Thus the dissociation hypothesis was unconvincing because the rate of production seemed far too slow and the photosynthesis hypothesis had a missing carbon problem.

The Missing Carbon Problem Solved

When the missing carbon problem became evident in the 1980s, two solutions seemed plausible: giant hidden reserves of natural gas or tiny fragments of carbon scattered through all the more ancient sedimentary rocks, unmeasurable and unrecoverable but adding up to an immense tonnage. The natural gas hypothesis probably appealed to oil company executives, though perhaps less to environmentalists worried about the greenhouse effect. But we now know that the scattered fragments hypothesis is correct.

The definitive study was done by Andrew Knoll of Harvard and his colleagues (Knoll et al., 1986). The trick was to measure the effect that the burial of organic carbon would have had on the isotopic composition of the dissolved carbon in ancient seas. Because plants discriminate against the heavy carbon isotope ^{13}C, they are relatively richer in the common light carbon isotope ^{12}C than the air or water from which they extract their CO_2 for photosynthesis. When organic matter sinks to be buried on an ocean floor, therefore, it inevitably leaves behind ocean water enriched with heavy carbon. For the modern oceans we can measure the ratios of heavy carbon to light, which we know describes an ocean burying organic carbon at the modern, known rate. If at any time in the past organic carbon was buried more rapidly, then the ancient ocean should have had a "heavier" mix of carbon. And we can measure the carbon mix of ancient oceans from samples of both carbonate and organic matter preserved in ancient sedimentary rocks.

Figure 28.12 shows the Knoll team's carbon isotope measurements from drill cores spanning a sedimentary rock sequence about 6000 meters thick and covering more than 300 million years of the time known, from the presence of animal and plant fossils, to have been critical for the evolution of oxygen. The two curves describe the increase or decrease of the heavy isotope ^{13}C relative to the light isotope ^{12}C in organic carbon and carbonate fossils throughout the record.

The organic curve of Figure 28.10 shows that organic carbon was always about 28% lighter than carbonate carbon, the expected result of plant discrimination between the isotopes. But the curves are closely parallel, showing that both plants and precipitating carbonate were sampling the same ocean. That ocean, for more than 100 million years in the middle of the record, had a carbon mix much heavier than the baseline of normal oceans. To account for an isotope mix that heavy, the rate of burial of light organic carbon must have been truly massive. Here then is direct evidence that organic carbon was indeed being buried in unprecedented quantities in the years of oxygen buildup.

An intuitive feel for the vastness of the burial comes from the size of the deposits. The old sediments, now rock, are more than 6000 meters thick, and samples used were taken from as far as 600 kilometers apart. Clearly we are dealing with deposits as large as ocean basins, say at least three quarters of the surface of the globe, and of a thickness measurable in miles. Typical organic carbon contents of the rocks are only 1 to 3%, but a tiny percentage of deposits so enormous adds up to a vast accumulation.

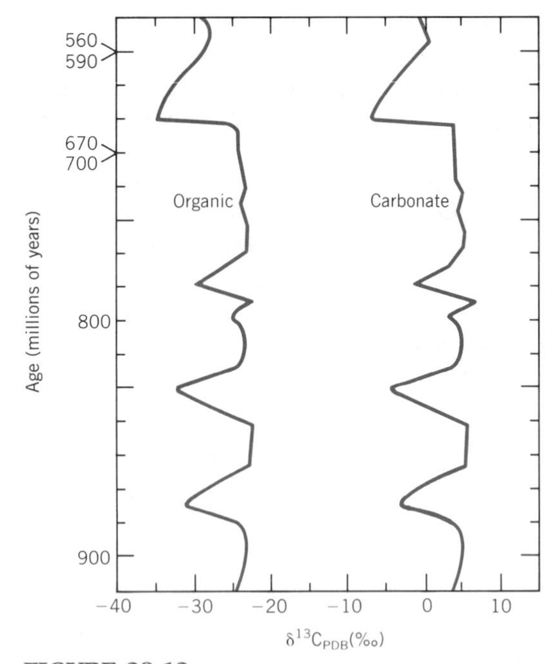

FIGURE 28.12
The missing carbon problem solved.
Data are carbon isotope ratios from the organic and carbonate components of drill cores through a 6000-meter-thick sedimentary sequence laid down while the atmosphere acquired free oxygen. Because the ancient ocean was "carbon-heavy" (see text) carbon must have been accumulating in the sediments. (From Knoll et al., 1986.)

Knoll's team used their excess ^{13}C measurement to estimate actual rates of carbon burial through comparisons with modern oceans, finding that in the critical 100 million years the oxygen equivalent of the reduced carbon burial was equal to ten atmospheres of oxygen at present concentrations. The missing carbon problem, therefore, is solved. The carbon is present in old sedimentary rocks, finely divided, forever beyond the reach of energy engineers and natural processes alike.

The Evidence of Banded Iron

Fossil evidence that oxygen was being released in vast quantities in these early days is at its most spectacular with the banded iron formations (Figure 28.13). These are deposits of ferric oxides of a kind most certainly not deposited anywhere on the contemporary earth. They imply oceans, or at least water, with high concentrations of dissolved ferrous compounds that are supplied with a source of oxygen. Ferric compounds would then be precipitated. The iron itself was thus a very large sink for oxygen, but in the reducing world in which ferrous oceans were possible, there would have been many other mineral sinks for oxygen also. There was no oxygen in the air, yet oxygen was being produced rapidly.

Fossils show that life, probably photosynthetic life, was already well established when the banded iron deposits were laid down. In particular, stromatolites, the carbonate relics of ancient algal reefs, leave little doubt that algae rather like the present blue-greens were building reefs of carbonates in shallow seas of the time, rather as other algae and coral animals are building reefs in present-day oceans. Photosynthesis, therefore, is demonstrated and, being present, becomes a parsimonious hypothesis for the release of oxygen demonstrated by banded iron (Cloud, 1976; Schopf, 1975).

The bands in the banded iron formation are the second prop of the photosynthesis hypothesis, because they look like seasonal bands. Certainly whatever caused the banding had its influence over large pieces of real estate at the same time, making seasonal change a likely candidate. Presumably the algal reefs produced less oxygen in winter than in summer so that the bands are annual. The photosynthesis hypothesis, therefore, explains the complete set of phenomena of those early days, at least in a qualitative way.

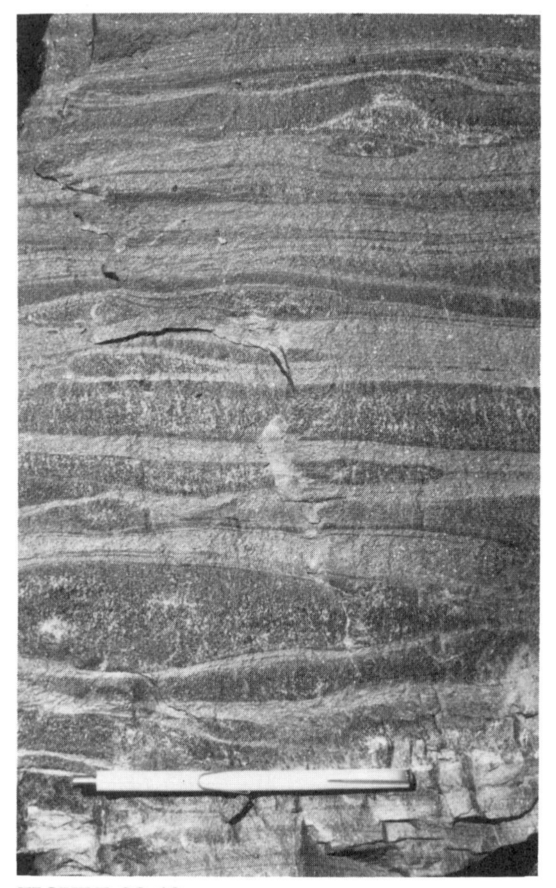

FIGURE 28.13
Banded iron formation: a trace of ancient photosynthesis.
The finely layered deposit could not have been emplaced by any process continuing on the present earth, but required oceans saturated with ferrous iron solutes and subject to repeated infusions of oxygen at short intervals. Photosynthesis in the surface water could have produced the necessary oxygen. Scale object about 18 cm long.

Advocates of the dissociation hypothesis suggested that bands in the iron were simply a result of seasonal mixing; for winter and summer, read stormy and quiet. In the stormy season, oxygen got down to the bottom and dumped ferric iron. The hypothesis saw oxygen being produced in the upper atmosphere at a steady rate as hydrogen is lost, but being pumped into the world ocean only with seasonal events of the global climate (Towe, 1978).

But the demonstration of immense burial of carbon now leaves the photosynthesis hypothesis as by far the most credible explanation of the banded iron formations. Burial of this carbon is, in itself, prima facie evidence for photosynthesis in the oceans on a massive scale. No one now seriously doubts that the banded iron formations represent massive seasonal photosynthesis at the surface of world oceans that were nearly saturated with ferrous iron.

Two Giga (G = 10^9 or 1 billion) years ago banded iron deposition ceased and the geological record begins to supply widespread evidence of an atmosphere that holds some free oxygen, notably the evidence of "red beds", the color of iron and aluminum oxides. Between 2 G and 1 G years ago this atmosphere with free oxygen acquired more and more of the gas until concentrations in the range of 20% by volume were reached, almost certainly by the time of the splendid array of larger organisms that appeared in the Cambrian 0.5 G years ago.

IS THE OXYGEN SUPPLY CONSTANT?

Free oxygen has been present in the air in sufficient concentrations to support air-breathing life since the Cambrian, but this does not mean constancy of supply. The actual limits we know about are rather crude, and take us back only to the coal-measure deposits of the Carboniferous, about 350 million years ago. The evidence is that of old forest fires, showing that these occurred with only modest frequency.

Coal deposits include fossils of trees, showing that forest fires were not as ruinously frequent as they would have been if oxygen concentration had been above about 35%. Yet charcoal, known as "fusinite" in the trade, is also present in the coal deposits. The only known source of fusinite in ecosystems is incomplete combustion associated with forest fires, showing that Carboniferous forests did burn occasionally. Spontaneous forest fires are probably impossible with less than about 13% free oxygen. Thus oxygen has remained between 13 and 35% by volume for the last 0.35 G years (Holland, 1990). A less than satisfactory degree of precision, this.

The hunt is on for actual data. Best of all would be entrapped samples of air far more ancient than the bubbles preserved in even the oldest ice. The exciting candidate of the moment is bubbles in amber dating from Cretaceous and early Tertiary times, say before and after the great extinction of 60 million years ago (Brener and Landis, 1988). If the bubbles truly represent ancient air (disputable), then oxygen fell from 35% to near its modern 21% across this great transition.

For all we know, free oxygen might have continued to oscillate within these limits with the slow passage of secular time, and may even now be changing. But the changes are slow even by the standards of paleoecology. Confinement to within a range of 13 to 35% still requires a controlling mechanism with a negative feedback. We do not know what this is, although likely candidates are available.

Sources and Sinks for Atmospheric Oxygen

Regulation of atmospheric oxygen requires sources and sinks comparable to those that maintain the CO_2 concentration. There is an important difference between these two systems, however, in that the reservoir of oxygen (21%) in the atmosphere is very much larger than that of CO_2 (0.03%). All the oxidation and reduction reactions of the carbon cycle are sinks and sources for atmospheric oxygen and, to this extent, the fate of oxygen is linked to that of carbon. But the carbon–oxygen cycle is only a small part of the oxygen system of the biosphere (Figure 28.14).

Important sources and sinks of oxygen are listed in Table 28.3. Photosynthesis as a source is very nearly balanced by respiration, and the dissociation mechanism is thought now to be trivial. The ocean surface generally is saturated with oxygen. The slight excess production of oxygen from dissociation and from burial of organic matter in contemporary ecosystems may just be sufficient to balance the sink of oxygen used in what is left of the process

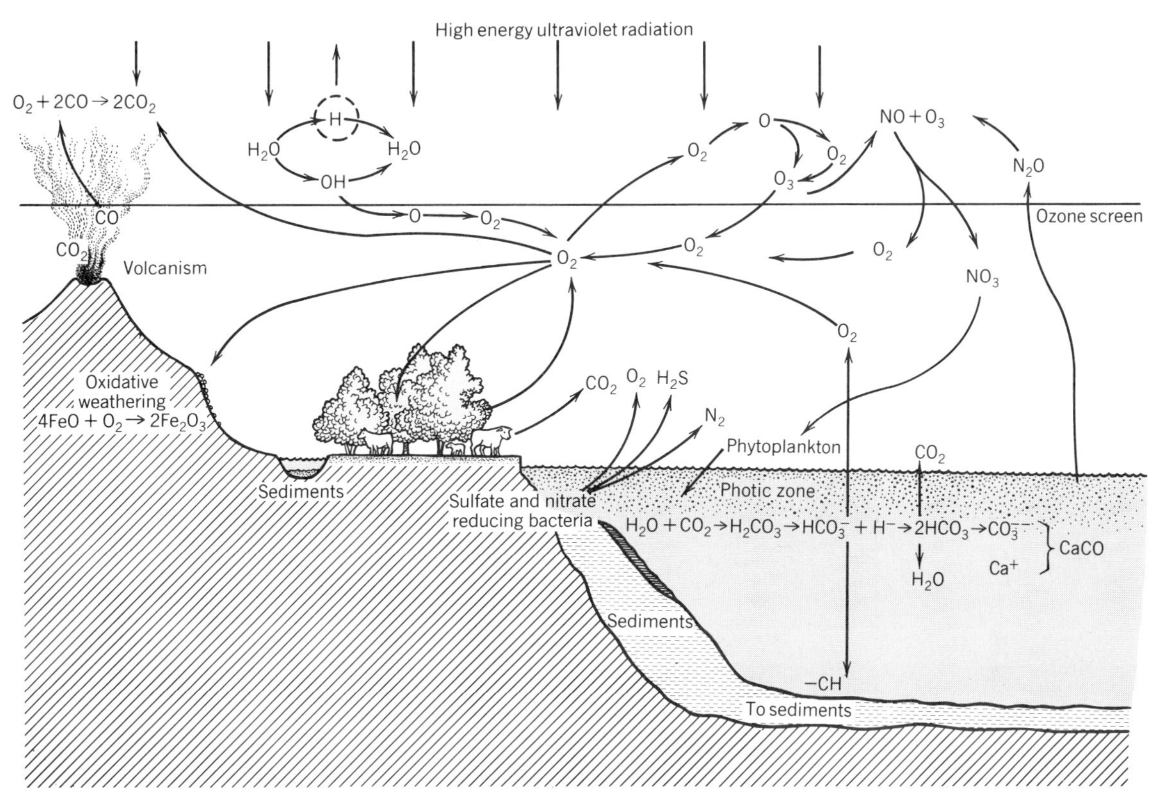

FIGURE 28.14

The oxygen cycle.

The diagram identifies important pathways in the oxygen cycle. It is important to remember, however, that the flux of all the processes illustrated is small compared with the volume of O_2 present in the air. (Based on a figure of Cloud and Gibor, 1970.)

TABLE 28.3
Sources and Sinks of Atmospheric Oxygen
Fossil fuels are ignored as a sink (when burned) because they are probably of negligible importance. Fossil fuels and other organic deposits as a sink are allowed for when listing photosynthesis as a source.

Sources	Sinks
Escape of hydrogen from water to space	Oxidation of crustal minerals
Sulfate-reducing bacteria	Sulfur bacteria
Nitrate-reducing bacteria	Nitrogen-fixing organisms
	Volcanic gases
Oceans	Oceans
Photosynthesis	Respiration

of oxidizing iron minerals of crustal rocks and volcanic gases, and in the production of marine carbonates (Garrels et al., 1975).

One promising mechanism for oxygen control has just come into what looks like fatal conflict with data. This hypothesis was that organic carbon was deposited in the deep sea as a function of oxygen concentration in ocean bottom water. With more oxygen in the water carbon is oxidized; with less oxygen carbon is buried. The feedback is obvious, and with the oceans occupying three quarters of the globe, the scale should have been adequate. But recent data show that burial of carbon seems independent of oxygen in the bottom water across a wide range of tensions (Figure 28.15).

In publishing these data for carbon burial independent of oxygen, Holland (1990) suggests that burial of phosphate in the oceans might serve as the regulator in-

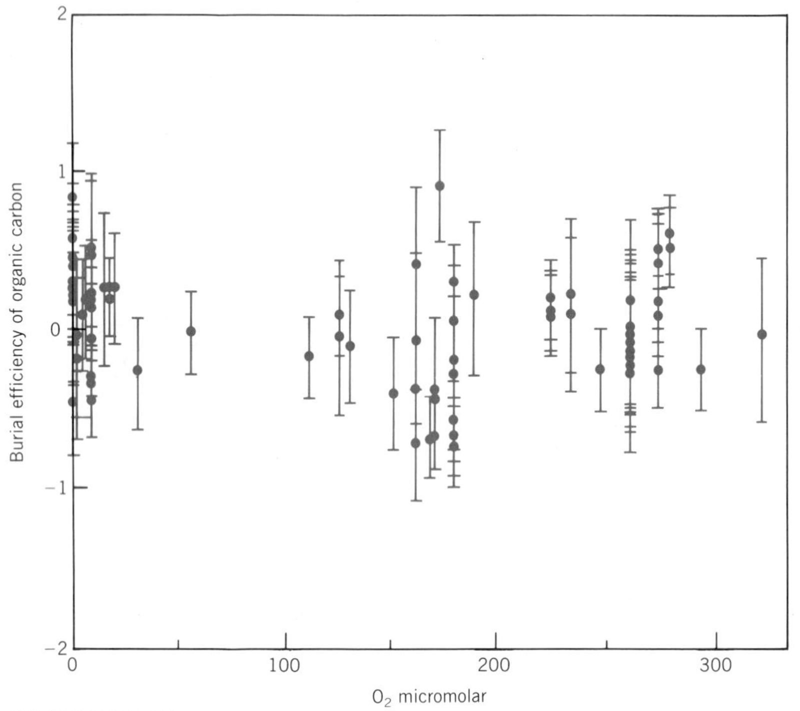

FIGURE 28.15
Carbon deposition in ancient oceans was independent of oxygen tension.
These data indicate that the oxygen concentration of the atmosphere cannot be regulated by the reduced carbon flux. (From Holland, 1990.)

stead of carbon, because phosphorous combines with oxygen in a family of oxides. Meanwhile, it is known that large sinks and sources of oxygen are regulated by biological processes. These result from the oxidation and reduction of sulfur and nitrogen (Table 28.3).

Sulfate-Reducing Bacteria

Spontaneous oxidation of sulfide minerals and SO_2 from volcanoes is potentially a large sink of both oxygen and sulfur. The sulfates produced are stable and soluble ions. They are cycled in ecosystems, where they are treated by living things as resources to be conserved. Inevitably they leak from the ecosystem, into the drainage channels and thence to the sea to be dropped into its mud. This process might be expected to result in accumulations of sulfate rock just as carbonate rock accumulates—sulfur stone instead of limestone. The reason that this does not happen is that specialized bacteria take oxygen from sulfate in anoxic mud.

Sulfate-reducing bacteria like *Desulfovibrio* and *Desulfotomaculo* live in places without free oxygen, typically reduced mud in bogs, lakes, or estuaries. These are places where energy in the form of dead organic matter abounds but oxygen for respiration is absent. Sulfate-reducing bacteria respire by using the oxygen of sulfate ions as the required electron and hydrogen acceptor:

$$H_2SO_4^{--} + 2C \rightarrow 2CO_2 + H_2S$$

This self-interested undertaking of the bacteria releases hydrogen sulfide gas and CO_2 into the air or to the overlying water column. Marshes smell of sulfur, and oxygen starts its journey back to the air.

The CO_2 released by the sulfate-reducing bacteria into the deep sea has a special significance in that its release is accompanied by a flux of nutrients leaving the mud when made soluble by their chemical reduction. When these nutrients and CO_2 are transported together as a package into lighted water, the CO_2 can be expected to be reduced in photosynthesis and free oxygen released. The return to the air of oxygen once used to make sulfate out of sulfide is now complete (Deevey, 1970; Kellog et al., 1972).

No other ways are known in which oxygen can be released from sulfates. It is hard to escape the conclusion that the oxygen concentration in the atmosphere would be less than its steady-state figure of 21% were it not for these bacteria. The sulfate-reducing process is possible only in anoxic mud. Ecosystems of swamps and fertile waters, therefore, are essential to the maintenance of the biosphere.

Other bacteria use sulfur in their daily lives, though they do not have consequences for the oxygen budget of the earth like those of the sulfate-reducing bacteria. Some even reverse the work of the sulfate reducers by oxidizing sulfide back to sulfate. An outline of the global sulfur cycle is given in Figure 28.16.

Sulfur Bacteria as Possible Oxygen Regulators

It is known that a large flux of sulfur comes out of seawater back to land. The data supporting this conclusion are measures of sulfate entering the sea in rivers showing this flux to be many times larger than can be accounted for by weathering. If these river sulfates have not been weathered out of rock, they must come from the air. The air in turn must receive its sulfur from volcanoes or the sea. Since the contribution of volcanoes is trivial, it follows that a large annual flux of sulfur goes from the sea to the air, thence into rivers as sulfate and once more back to the sea.

It can be shown that most of the sulfur

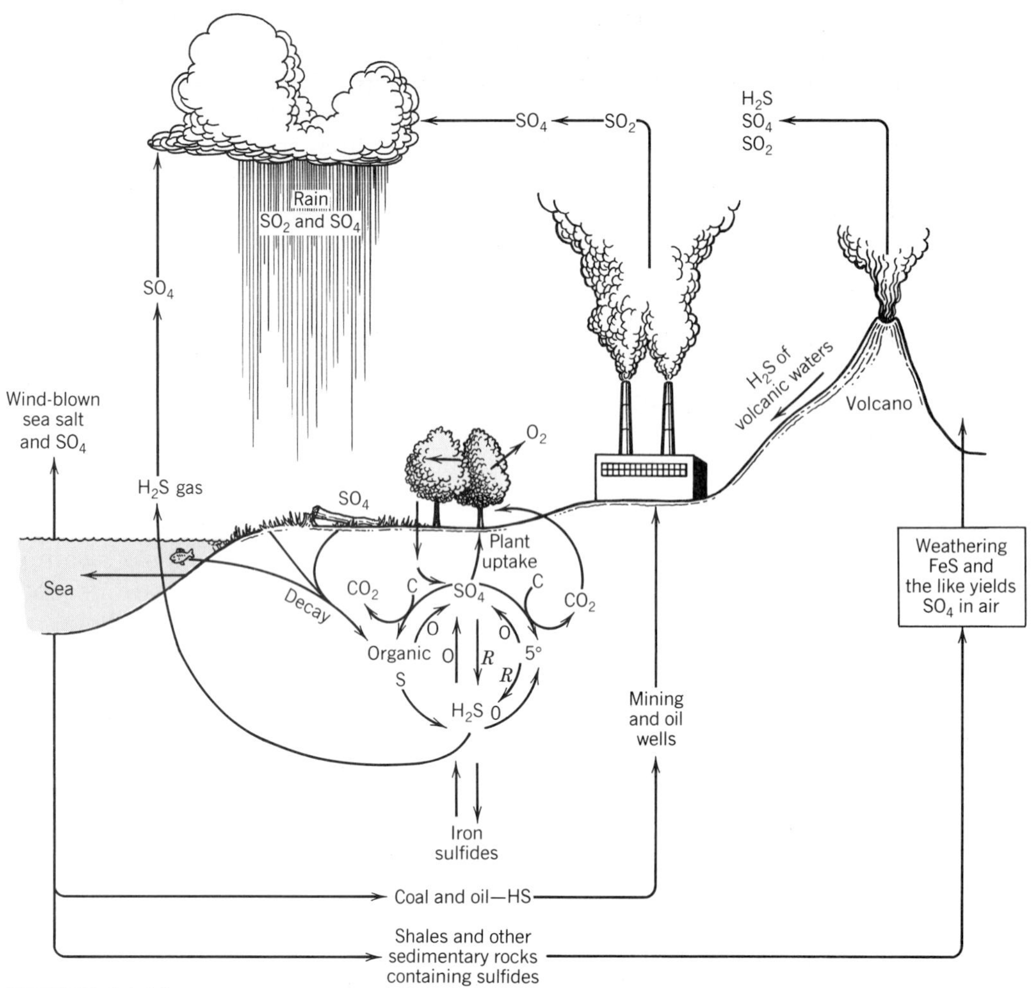

FIGURE 28.16

The sulfur cycle.

Ecosystems are provided with sulfates, which they cycle, from the oxidation of the products of weathering. In marine and freshwater muds, sulfur is cycled between sulfate-reducing bacteria, ***Desulfovibrio*** and *Desulfotomaculo*, which reduce sulfates to S and H_2S, and sulfide-oxidizing bacteria, such as the photosynthetic green or purple sulfur bacteria and the chemosynthetic colorless bacteria (Leucobacteriaceae), which oxidize sulfides. H_2S returned to the atmosphere is spontaneously oxidized and delivered by rain. Sulfides incorporated in fossil fuels and sedimentary rocks are eventually oxidized following human combustion or crustal movements and weathering. Primary production accounts for the incorporation of SO_4 into organic matter, and anaerobic and aerobic heterotrophic microorganisms transform organic S to H_2S and SO_4, respectively.

reaching the air from the sea does so via biogenic processes, a conclusion based on the isotopic composition of atmospheric sulfur (Erikson, 1960). This makes it virtually certain that the prime source of this sulfur is sulfate-reducing bacteria in oceanic mud.

The complete ocean–atmosphere–land

sulfur cycle, therefore, can be deduced (Figure 28.17). Sulfate concentration in the mud regulates the rate at which sulfate-reducing bacteria vent H_2S to the water column and the atmosphere. This flux of H_2S regulates the concentration of sulfate in rainwater, which determines the concentration in rivers. Finally, the concentration in rivers regulates the concentration in ocean mud, completing the feedback system.

This speculative scheme has a place in an ecology text because it shows the extent to which life might be involved in regulating oxygen supplies. With the triumph of the photosynthesis hypothesis in explaining the source of oxygen, it would be agreeable to biologists to find that life also regulates oxygen. But the truth is that we still know far too little about the global dynamics of oxygen. Authorities in the field now brood most about the possible role of phosphates (Holland, 1990).

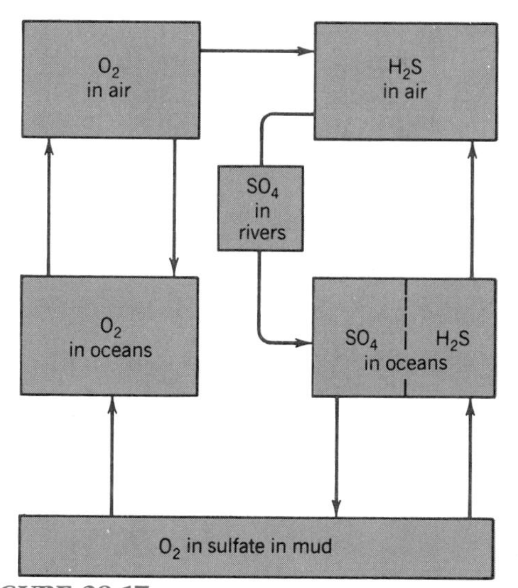

FIGURE 28.17
Oxygen regulation: the sulfur hypothesis.
Oxygen leaves the air to oxidize sulfides emitted from the oceans. Oxygen is released from these sulfates by sulfate-reducing bacteria in mud. This process may be a prime regulator of oxygen concentration of the oceans, and hence of the air. The sulfur system is itself regulated by release of hydrogen sulfide back to the air.

THE REGULATION OF NITROGEN IN THE AIR

The almost inert gas nitrogen makes up 79% of the atmosphere by volume. The prime source seems to be volcanoes that have been emitting small quantities of nitrogen throughout geologic time. Since the gas has a low solubility in water, most has collected in the air itself. The main sinks and sources are given in Table 28.4.

The mass of nitrogen held in the oceans is trivial compared to the mass in the air (Table 28.5). This is quite different from the circumstance of carbon dioxide and oxygen, where the oceans are principal reservoirs. For nitrogen the atmosphere is the principal reservoir (though the total mass in the biota may be large also), so the oceans are unimportant as a buffer. The annual input of volcanoes is small, though large enough to suggest that the atmosphere might continually be enriched with nitrogen throughout geologic time without corresponding sinks (Table 28.5). But there are two very significant sinks, both directing nitrogen into the stable oxide ni-

TABLE 28.4
Sources and Sinks of Atmospheric Nitrogen
The biological sinks and sources are by far the most significant.

Sources	Sinks
Volcanos	Synthesis of nitrates in electric storms
Reduction of nitrates by bacteria	Biological nitrogen fixation
Oceans	Oceans
Decay of ^{14}C	Formation of ^{14}C

TABLE 28.5
Nitrogen in Air, Oceans, and Contributed by Volcanos
(Data of Delwiche, 1970.)

N_2 in atmosphere	3.8×10^{15} tons
N_2 in oceans	2.0×10^{13} tons
Annual input of N_2 from volcanos	2.0×10^{5} tons
Annual output of N_2 from oxidation in electric storms	4.9×10^{9} tons

trate. One is nitrate synthesis in electrical storms and the other is nitrogen fixation in ecosystems.

The manufacture of nitrates by storms and organisms represents a sink for atmospheric nitrogen that is potentially bottomless (Figure 28.18). Nitrogen could be removed from the air continuously until both nitrogen and oxygen ended up as nitrate rock, destroying the atmosphere as we know it and ending life on earth. In the face of these cumulative drains it appears that the nitrogen concentration of the air can be maintained only by the nitrogen-return functions of ecosystems. These nitrogen-return functions arise because some bacteria use nitrate ions as hydrogen acceptors in their oxidations of organic matter, just as other bacteria use sulfates for the same purpose. Nitrogen in the air, therefore, is maintained and moderated by living systems.

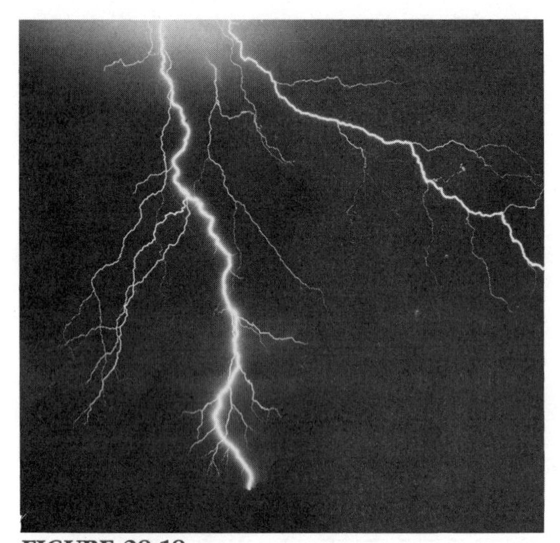

FIGURE 28.18
High energy nitrogen loss.
Lightning storms release sufficient free energy for the oxidation of nitrogen gas without benefit of a catalyst. This process would denude the air of free nitrogen were nitrogen oxides not subsequently reduced in ecosystems.

Nitrogen Fixation: The Traditionalist or Farmer's View

Farming increases the leak of combined nitrogen from a habitat. Clearing land and letting soil humus rot always increases the leak of combined nitrogen, as it does for the other soluble nutrients. In the wet tropics this loss of nitrogen is part of the catastrophic cost of cutting down a rain forest. Even in a temperate habitat the loss can be dramatic, as the Hubbard Brook clear-cutting experiment showed (see Chapter 23), because the loss of nitrate two years after clear-cutting was an order of magnitude greater than the loss of any other ion. Regular farming increases the leak still further by carrying protein-rich crops from the land, passing them through the digestive tract of various bipedal and quadripedal animals, and burying the nitrogenous remains in privy, lake, or sea.

Old-style farming was spared paying the crippling cost for this prodigality by the activity of nitrogen-fixing bacteria during years when land was left fallow. New-style agriculture is impatient of the delays inherent in waiting on the bacteria and employs an industrial synthesis to the same end to such an extent that the flux of nitrogen from the air to soluble ion in land and sea from industrial process is at least as large as that worked by the nitrogen-

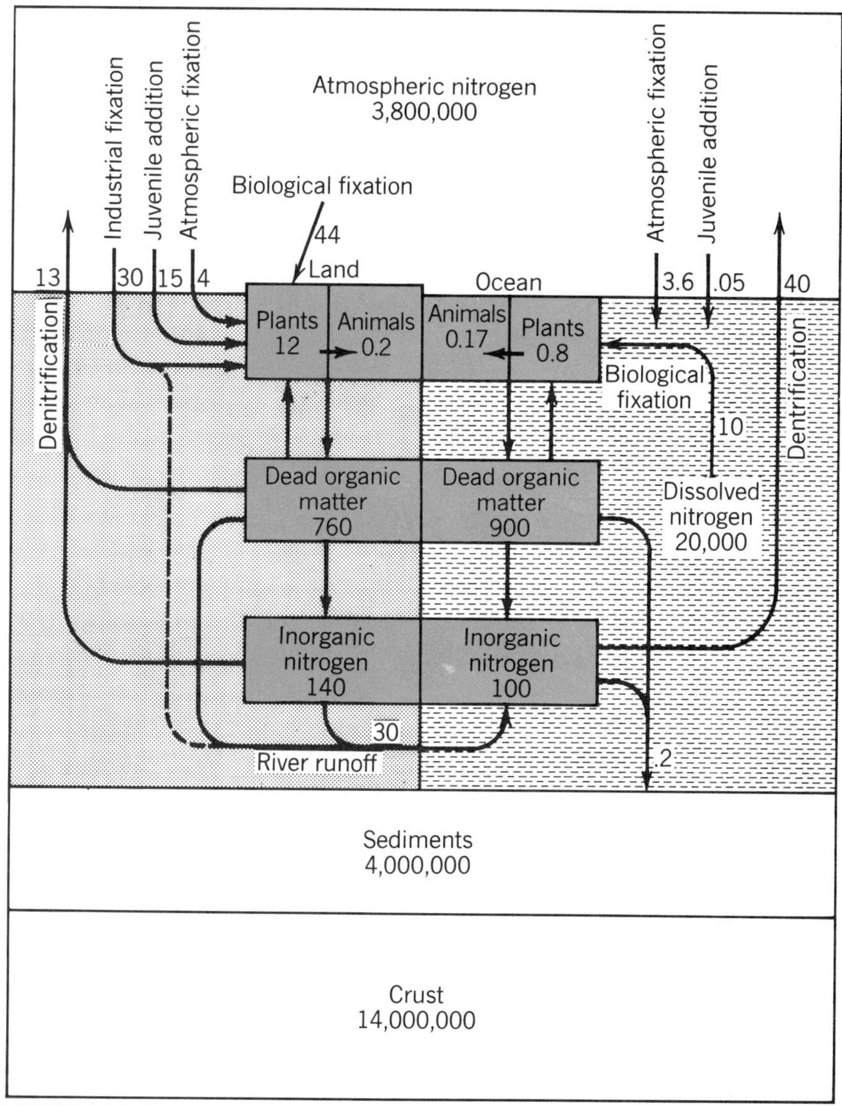

FIGURE 28.19

Nitrogen in the biosphere: Sinks, sources, and transfers.

Figures in reservoirs are $\times 10^9$ metric tons. Figures for transfer rates (arrows) are $\times 10^6$ metric tons per year. There are large uncertainties in the estimates of transfer rates, some of which may be off by as much as a factor of 10. The very large size of the atmospheric reservoir compared to the transfer rates is, however, apparent. (From Delwiche, 1970.)

fixing bacteria of the whole world put together (Figure 28.19).

The traditionalist view has it that nitrogen-fixing organisms are friends of humankind. It is known that their work is continually being undone by other organisms that break down nitrogenous fertilizer and void nitrogen back to the air. It must follow that these other organisms are enemies of humankind for eating into our fertilizer supplies. And yet the nitrogen-fixing organisms can also be seen as being in the business of destroying the atmosphere and the nitrogen releasers as grimly opposing them to restore the air. The potential major reservoirs for nitrogen on the surface of the earth, therefore, turn out to be fertilizer (combined nitrogen in soil) and air. Various bacteria arrange how much nitrogen will be in each.

The Mechanism of Nitrogen Fixation

What is called the activation of nitrogen

$$N_2 \rightarrow 2N$$

requires much free energy. In the atmosphere it appears to result only from lightning discharges, when a series of reactions takes place that ends up with nitrogen oxides and, in the presence of water, with nitric acid (Ferguson and Libby, 1971). The nitric acid comes down with rain, reacts with basic minerals, and becomes a nitrate input to the ecosystems below. As a practical source of nitrates to ecosystems it is generally of small importance (Figure 28.16), but it may provide a significant part of ecosystem nitrate inputs in exceptional, barren, stormy places like Antarctica.

In biological nitrogen fixation the energy for the activation of nitrogen is supplied by the oxidation of organic matter, and the process is facilitated by an enzyme, nitrogenase. Once N_2 has been activated to 2N at a heavy cost in energy, it is combined with hydrogen to make ammonia, an exothermic reaction that pays back a small portion of the original energy cost. Ammonia, dissolved in water as the ammonium ion $NH_4^{-}{}^{-}$, is then available for bacterial and plant metabolism and has entered the biosphere.

If Plants Could Fix Their Own Nitrogen

Nitrogen-fixing organisms are all single-celled procaryotes—various bacteria and blue-green algae. The nitrogen-fixing ability is not possessed by any multicellular plants or animals. It is, therefore, possible to find the very odd circumstance of large organisms dying for want of nitrogen even though they are bathed in a fluid that is 79% nitrogen gas. Perhaps procaryotes inherited the nitrogenase mechanism from ancient days before the air acquired oxygen and when different chemical pathways were required for life. But, if so, why has it not been possible to evolve the mechanism since? We have no idea.

Root nodules in which bacteria live are apparently the best stratagem that natural selection has been able to provide for large plants. The plants pay a ransom in energy supplied to the bacteria and thus get combined nitrogen back at a greater cost than they would have paid if they had made it themselves. This cost explains why legume crops take their nitrogen from the soil rather than from their bacteria when the soil is fertilized (Child, 1976). A similar stratagem for obtaining nitrogen has apparently been undertaken for termites, which have nitrogen-fixing bacteria in their guts along with their cellulose-digesting protozoa (Benemann, 1973).

It is likely that the failure of all large plants to fix their own nitrogen has important consequences for the atmosphere. Were all plants able to do so cheaply it is easy to imagine the ecosystems of the world adopting the sort of attitude to utilizing resources that people have: take it and throw it away. Nitrogen from the air might be cheaper than nitrogen from the soil, and the consequence of this would certainly be a different steady-state mixture of gases in the air as both nitrogen and oxygen were withdrawn to the nitrate reservoir in "nitrostone."

Ecosystem Nitrogen Cycles

A complex series of modifications of combined nitrogen as it cycles around ecosystems depends on the fact that each stage in the reaction series—amino acid, ammonia, nitrite, nitrate—yields free energy. Groups of bacteria specialize in sponsoring each reaction in this series, using the free energy won as a prime energy source. The bacteria are amino acid, ammonia, or nitrite "feeders," nicely illustrating the ecological principle that if free energy is available through a chemical reaction, some organism has probably evolved to use it.

Figure 28.20 illustrates the principal reactions in this series of bacterial oxidations and reductions of various forms of combined nitrogen and compares the free energy of the various undertakings with that released in respiration or required for nitrogen fixation. Notice the regrettable names given to many of the reactions: "nitrification," "denitrification," and the like. These names were given to processes discovered piecemeal in the biological long-ago. They are both illogical and confusing, but they keep turning up in the literature and need to be handy for reference. What is important is not the names, neither of process nor bacteria, but the fact that series of bacteria use nitrogen compounds as their prime energy sources.

In wet anaerobic mud, nitrates can be used by bacteria as hydrogen acceptors, just as sulfates can be used. These bacteria complete the nitrogen cycle that moves nitrogen to and from the air. We know of no other mechanism to complete the cycle, which is thus essential to maintenance of the present steady-state balance of gases in the air. Figures 28.19 and 28.21 summarize the whole nitrogen system.

CATASTROPHE HYPOTHESES

The most lurid of the scares put forward by "environmentalists" of the late 1960s were those that suggested we could destroy the atmosphere as a life-support system. In its most spectacular form, the catastrophe hypothesis suggested that we might so interfere with oxygen sources that the air would be critically depleted of this essential gas. The argument, at first sight, had a certain plausibility because of the growing awareness by geochemists that oxygen had been put into the atmosphere by life processes in the first place and that it is maintained by life processes now. With this plausible start the argument went as follows:

1. Oxygen in the air is maintained principally by photosynthesis.
2. Since the oceans cover more than two thirds of the globe it is reasonable to conclude that it is ocean photosynthesis that maintains our oxygen.
3. The world oceans are being polluted already. A major spill of a tanker conveying herbicides like 2,4-D and 2,4,5-T or picloram was inherently likely (perhaps on its way to Viet Nam). A few such spills would kill most of the plants in the sea.
4. With the plants of the sea dead the air's oxygen begins to run down without being replaced.

This hypothesis was always absurd on the grounds of one unjustified inference and two massive errors of fact.

Reaction	Energy Yield (kcal)
DENITRIFICATION	
1 $C_6H_{12}O_6 + 6KNO_3 \rightarrow 6CO_2 + 3H_2O + 6KOH + 3N_2O$ Glucose, Potassium Nitrate, Potassium Hydroxide, Nitrous Oxide	545
2 $5C_6H_{12}O_6 + 24KNO_3 \rightarrow 30CO_2 + 18H_2O + 24KOH + 12N_2$ Nitrogen	570 (per mole of glucose)
3 $5S + 6KNO_3 + 2CaCO_3 \rightarrow 3K_2SO_4 + 2CaSO_4 + 2CO_2 + 3N_2$ Sulfur, Potassium Sulfate, Calcium Sulfate	132 (per mole of sulfur)
AMMONIFICATION	
4 $CH_2NH_2COOH + 1\frac{1}{2}O_2 \rightarrow 2CO_2 + H_2O + NH_3$ Glycine, Oxygen, Ammonia	176
NITRIFICATION	
5 $NH_3 + 1\frac{1}{2}O_2 \rightarrow HNO_2 + H_2O$ Nitrous Acid	66
6 $KNO_2 + \frac{1}{2}O_2 \rightarrow KNO_3$ Potassium Nitrite	17.5
NITROGEN FIXATION	
7 $N_2 \rightarrow 2N$ "Activation" of nitrogen	−160
8 $2N + 3H_2 \rightarrow 2NH_3$	12.8
RESPIRATION	
9 $C_6H_{12}O_6 + 6O_2 \rightarrow 6CO_2 + 6H_2O$ Carbon Dioxide, Water	686

FIGURE 28.20
Energetics of the nitrogen cycle.
Steps 1 through 6 yield free energy and are undertaken by the various bacteria for that reason. The production of N_2O in the dentrification of step 1 may release the gas to the atmosphere with consequences for the ozone cycle of the upper atmosphere. The costs of all the bacterial operations of steps 1 through 7 have to be met eventually by green plants, either by metabolizing nitrates or by "feeding" nitrogen-fixing bacteria. Nitrogen fixation (steps 7 to 8) requires free energy. The energy yield of respiration (9) is given for comparison. (Modified from Delwiche, 1970.)

The unjustified inference was that it would be possible to kill all the plants of the sea with a herbicide spill or two. Ocean plants are tiny phytoplankton, the populations of which are so large as to defy easy comprehension. Rapidly reproducing populations of enormous size include very great genetic diversity such that strains resistant to spilt chemicals are certain. In other words, natural selection should cope with herbicides spread on phytoplankton even if the virtually impossible goal of spreading and maintaining a concentration of the stuff in the world's oceans was achieved.

But the two massive errors of fact are

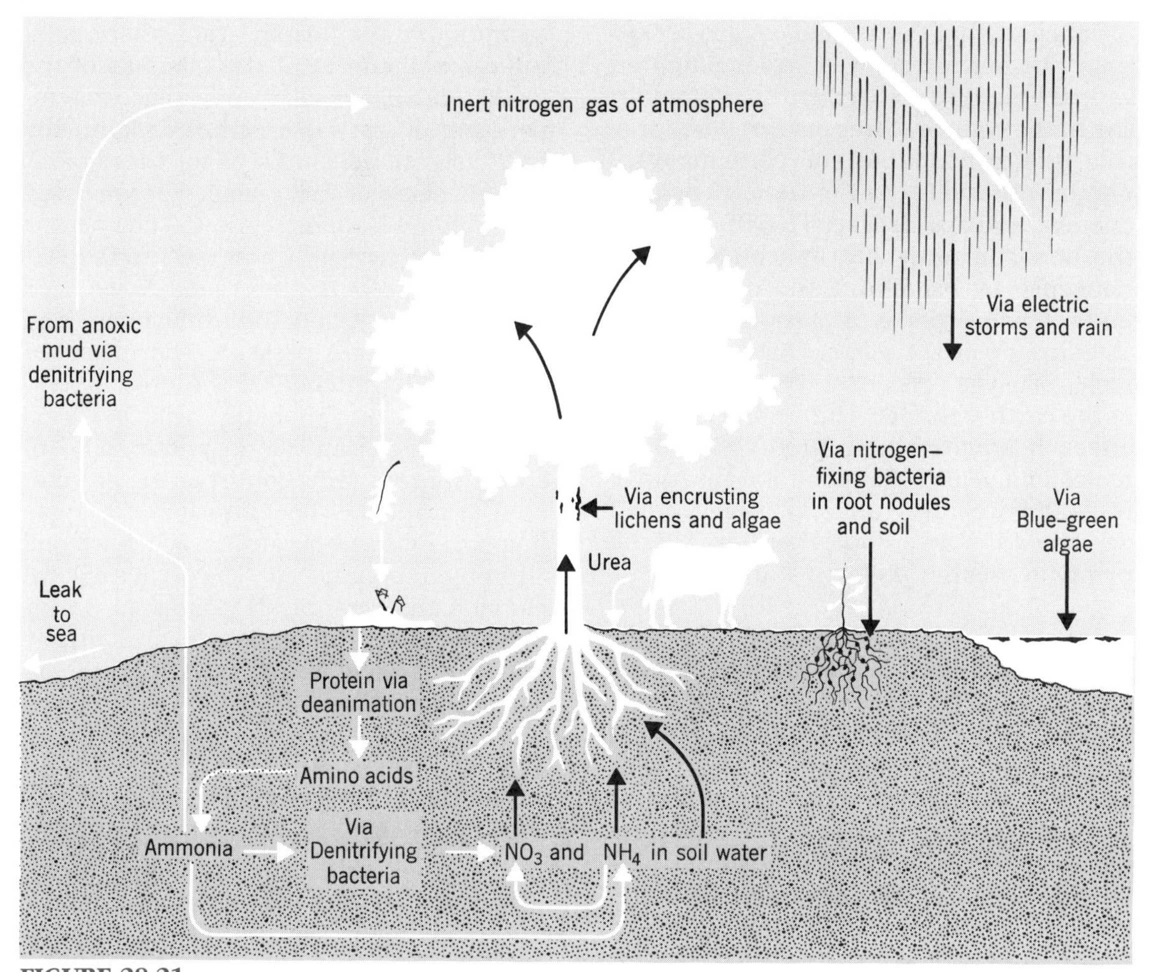

FIGURE 28.21

The nitrogen cycle.

Nitrogen is cycled through the ecosystem in combined form either as oxides or reduced as in ammonia, amino acids, or proteins, and the like. There is a continuous leak from the system in flowing waters that must be made good by fresh supplies from the atmospheric reservoir. Many kinds of microorganisms fix nitrogen, principally by reducing it to ammonium compounds, and there may be additional supplies (as oxides of nitrogen) in rain water that have been synthesized in electric storms. Nitrogen is returned to the atmospheric reservoir by dentrifying bacteria living in reduced muds of fertile waters and bogs. Black arrows are paths of nitrogen synthesis and white arrows are paths of nitrogen release from more complex molecules.

more immediately fatal to the oxygen catastrophe hypothesis. The first of these is to think that the oceans are the principal source of photosynthetic oxygen. They are not. The oceans are unproductive deserts (see Chapter 24) that, although covering much of the globe, contribute only a quarter of the total plant production. Killing all ocean plants would interfere with no more than a quarter of the photosynthetic oxygen return.

The even more massive error of fact was

to conclude that the annual oxygen return from photosynthesis was large compared with the amount of oxygen in the air. In truth, the volume of free oxygen is so large that the contribution of contemporary photosynthesis is close to trivial. The geochemist Wallace Broeker (1970) deserves the honor of most clearly scotching this nonsense by examining the oxygen flux and reserves over a typical square meter of the surface of the globe. Sixty thousand moles of oxygen rest on every square meter of the earth's surface. But photosynthesis on each square meter produces only 8 moles in a year; 8 out of a total reserve of 60,000! If we killed all the plants on earth we should starve, but there would be plenty to breathe while we starved.

THE GREENHOUSE EFFECT

To ecologists generally it has long been known that the atmospheric gas with which we can and do tinker in significant ways is not oxygen but CO_2. The Mauna Loa data (Figure 28.3) offer the most spectacular evidence. The reason human activities are significant to the level of CO_2 in the air is, of course, that the gas is present in such low concentration. It is likely, however, that the scale of the changes we are now causing is within the range of similar events in the past. These changes will definitely have consequences for life on earth, ours included with the rest, but how serious these changes will be we do not yet know.

The effects of rising CO_2 will depend on how concentrated the gas becomes finally. To predict what this final concentration will be we need data on fossil fuel consumption in the future, atmospheric mixing rates, changes in the biomass of the world, and, above all, trends in ocean temperature. Models are being made on the basis of various estimates for these parameters, some of them suggesting perhaps a doubling of atmospheric CO_2 sometime in the next century (from 0.03 to 0.06%). To choose between these models we need many more data on the atmosphere and oceans, past and present. The most important consequence of the enrichment, whatever it is, probably will be local changes in climate. Atmospheric CO_2 absorbs infrared radiation from the ground, so that a higher concentration of the gas increases air temperatures, what we call "the greenhouse effect." The details of local climatic change from CO_2-induced warming are a subject of debate and uncertainty, but such change is expected to be considerable.

LIFE IN THE MAINTENANCE OF THE AIR

In the creation and maintenance of the air the role of life is certain. It seems safe to say that free oxygen is present in the atmosphere as a result of photosynthesis by green plants. Moreover, the best explanations we now have for the regulation of the mix of nitrogen and oxygen in the air depend on the activities of bacteria that use sulfate and nitrate as hydrogen acceptors in reduced mud. The flux of oxygen, nitrogen, sulfur, and carbon into the oceans from the activities of these organisms maintains the air that the green plants build.

Preview to Chapter 29

Masai adolescent with wealth: modest impact through herding.

THE great peculiarity of humans is that they learn their niches. Other animals learn, but humans learn far more, including most of the traits important to their survival and reproduction. This trait of learning let our early ancestors accommodate to virtually every kind of terrestrial habitat without speciating. By 30,000 years ago we had already achieved a global distribution, as our subpopulations had taken the basic habits of using fire, clothing, and weapons to most parts of the earth. As hunters, we exterminated larger animals in at least several countries, with the inevitable cascading consequences on local food webs. Later, in the years after the close of the last ice age, we learned to herd and cultivate what before we had hunted or gathered. Dense populations in subsequent millennia resulted in local extinctions and habitat changes. Finally, divisions of labor in dense communities and systems of training outside the home group allowed people to be raised in different niches. We can live anywhere, using the resources of everywhere. Our long juvenile periods and longer lives have always acted to constrain rates of changing habit, essentially helping maintain a fixed niche. In early humans these traits served to prevent infants learning from other than the appropriate role models with which they lived. The offsetting advantages of intelligence may well have been that it increased fitness by allowing the efficient training of young. Now this ancient apparatus is beginning to serve us well as generations are trained to the aspirations of new role models. Female liberation can end population growth, even as educated people of broad-niche living add an appreciation of wilderness to their perceived necessities of life.

Chapter 29

Postscript: The Human Impact

Natural selection eventually permitted intelligence. We have no evidence that this happened more than once, and plenty of reason for thinking that we were the first species to be so equipped. In the tiny span of evolutionary time for which our species has existed we have made such a mark on the earth's surface that it is impossible to think that any comparable animals have gone before without leaving highly visible traces in the fossil and sedimentary records. The emergence of *Homo sapiens* sometime within the last two glacial cycles, therefore, was the first injection of intelligent life into the biosphere.

Of first ecological importance among our unique intellectual properties is that we must learn how to live. Other animals learn; some of them learn much; but the learning required of a human infant before it can be an effective breeding adult is uniquely large. We learn the important parameters of our niches, having an almost absolute reliance on learning for the acquisition of skills for survival, foraging, and reproduction. A one-line definition of the human species could well be "*the animal that learns its niche.*"

Learning as an ecological process serves to refine niches in ways that are impracticable when behavior is rigidly constrained by genetic instruction. Many birds, for instance, learn to sing by copying the songs they hear when they are very young. They learn local dialects of song from their parents, effectively fixing their niches more narrowly than would be possible by genetic instruction alone. Because songs are recognition signals, the birds thenceforth belong to local populations of common training; effectively to a common, narrowly defined niche (Marler and Mundinger, 1971).

Learning in nonhuman primates appears to serve the same function of fitting individuals to local resources that it does for birds. Typically, primates are foragers of highly diverse ecosystems like the tropical rain forest. Patterns of resource are in constant flux, driven by seasonality of rains, months of flowering or fruiting, and the monthly changes in insect numbers. An adequate amount of food must be found on any particular day. For fashioning foraging niches in these ecosystems, learning is ideal. The tactic is to learn the monthly patterns of fruiting in the local forest during the juvenile years when the young range with their knowledgeable parents. When they themselves are mature and ready to breed, the young individuals have niches perfectly attuned to the unique local forest in which they must raise their own young in turn. In this way, gibbons in the Malaysian rain forests are able routinely to use at least 260 species of plant food (Gittins and Raemaekers, 1980).

Human learning let the process of replacing niche parameters fixed by genes by those acquired through learning go even further, particularly by the advent of language, through which experiences of adults could be passed to the young in extraordinary detail. When accompanied by the important property of intelligence that can be called curiosity, the human niche-learning system let individuals be equipped with a niche suited to almost any sufficiently productive ecosystem. When the use of fire, weapons, and clothing was learned, local populations could learn the habits needed for life in all parts of the earth.

The first great human impact came from our occupation of every major ecosystem, from arctic tundras to tropical rain forests. This happened long before the invention of agriculture or any of our later traits of modifying habitats. The technical skills needed were no more than to clothe ourselves, to heat food, and to kill with sticks and stones (Figure 29.1).

No other species of animal or plant has ever achieved the range that preagricultural humans reached, more than 30,000 years ago in the midst of the last ice age. We have radiocarbon-dated the traces of our presence at this remote period from Australia, through Aşia and Africa, to northern Europe close to the ice sheets. Only the Americas were immune to our presence, cut off by open oceans from the Old World lands of our origin.

This was the triumph of a fully learned niche. All other hereditary lines could put individuals into new habitats only by speciating, which is to say by finding genetic combinations that should adapt individuals to the new conditions. Humans took the same animal and trained it differently. In some habitats the arrival of this cannily equipped, alien forager had significant consequences for the food webs to which it was added.

THE OVERKILL HYPOTHESIS

People arrived late in the New World, most probably from central Asia via the Bering Strait region and Alaska between 14,000 and 12,000 years ago (West, 1981). If there were people on the continent earlier than this, which is considered unlikely by most students of the subject, they have left so few traces that they must have been rare (Dillehay and Meltzer, 1991). Thus the first great expansion of humans into the Americas was a recent happening, occurring in the period for which we have the best paleoecological evidence. Our records show that the human arrival coincided with the extinction of the North American megafauna.

Ice-age North America was home to elephants of two species (mammoths and mastodons), giant bison, giant ground-sloths, large wolves and bears, great cats, and others, all of which are now extinct. For the soul of a zoologist, their loss was agonizingly recent, the latest radiocarbon

FIGURE 29.1
Human armament need be only sticks and stones: a fine mind makes a great hunter.
Gathering and hunting peoples equipped with simple clothing and weapons occupied every habitat in the world outside the Americas thirty thousand years ago. Lasting impacts were the extinction of the more vulnerable prey.

dates from their remains being around the start of the Holocene, about 10,000 years ago. People lived alongside them. That people and megafauna had fatal meetings has been demonstrated by finding stone lance points in mammoth ribs at ancient kill sites.

The megafauna overkill hypothesis argues that the sudden debouchment out of Siberia and Alaska to the western American plains of a population of big game hunters 12,000 years ago had fatal consequences for mammoths and the rest (Martin, 1973). American megafauna were without experience of human hunting. The human hunters had lived previously on the arctic plains of the old Bering land bridge (see Chapter 17) where big game hunting was of necessity the primary way of life. On the prairies the killing was easy.

That humans killed off the American megafauna is without proof; we have only the tantalizing coincidence of the human arrival and the disappearance of the animals. Elsewhere, though, the exterminating effect of stone age hunting in pristine ecosystems has been demonstrated well enough. New Zealand and Hawaii were some of the last lands to be reached, just a few thousand years before the modern period. A guild of huge flightless birds vanished in New Zealand, their passing recorded by cracked bones scattered in the hearths of the Maori settlers. Recent finds have shown that perhaps half of all the species of birds in Hawaii perished within a few centuries of the human arrival, including all the large ones (Olson and James, 1982).

Those who argue that the extinction of the American megafauna at the time of human arrival must be more than coincidence can make comparisons with the records from New Zealand and Hawaii to show that the sudden arrival of stone age hunters can be devastating. Stronger support still is emerging from Australia, as the remains of a vanished megafauna suggest extinction 30,000 or more years ago, roughly coincident with the first human arrivals from the large ice age islands that spanned what is now the Indonesian archipelago. Overkill of the continental Australian megafauna provides a better model for exterminating mammoths and mastodons in America than the taking out of flightless birds on islands.

Even the African record, where a megafauna with elephants and the rest still survives, is consistent with the overkill hypothesis. Human origins were in Africa, so that the impact of evolving humans was felt only gradually. Nevertheless, there were megafaunal extinctions in Pleistocene Africa, more selective extinctions of just part of the fauna, and they occurred early, in the range of 100,000 years ago at about the time that *H. sapiens* first appeared.

Critics of the overkill hypothesis for the American megafauna point out the enormity of the task of exterminating huge animals all across a continent by people who walked and threw spears, or they put forward the equally plausible hypothesis of death by climatic change, as the ice age waned and every American habitat was altered (Martin and Kline, 1984).

To show that overkill was theoretically possible, defenders of the hypothesis have modeled possible scenarios (Figure 29.2; Martin, 1973; Mosimaun and Martin, 1975). The numbers seem reasonable enough. The people had 2000 years to do the job, long enough to breed the necessary population and travel the necessary distance.

The death by climate hypothesis draws strength from the fact that the critical millennia were in the late-glacial period, when climatic change was more rapid and extreme than at any other time in the last

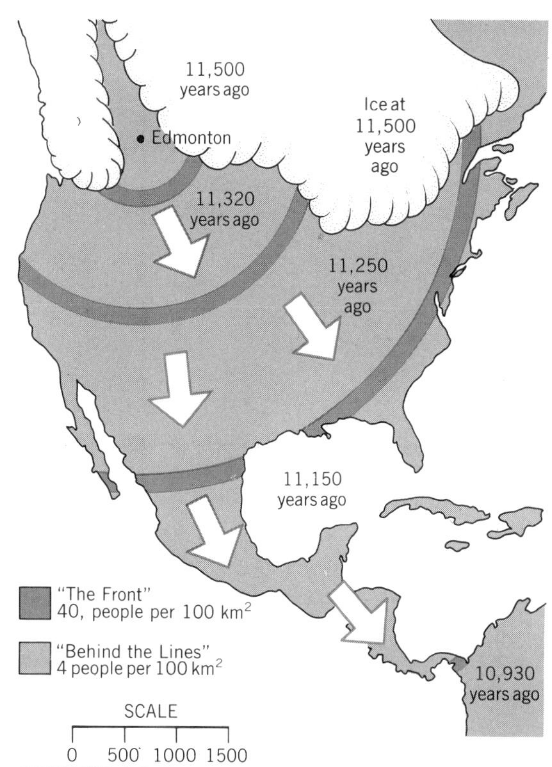

FIGURE 29.2

Overkill hypothesis: computer simulation of an advance by hunting peoples into North America. One hundred hunting people, fresh from life on the old Bering land bridge, start from Edmonton 11,500 years ago. Game animals of extinct species are assumed to have lived at the density of the modern game herds on the East African plains. Hunting is easy, so the one in four humans assumed to do the hunting can kill for their families an average of 13 animals per person per year. Population doubles every 20 years. People are assumed to concentrate their population on a moving front where game is still to be had. In just 350 years the first people will have reached the gulfs of Mexico and California. The human population is 590,000, and most of the Pleistocene megafauna is extinct. (From Martin, 1973; Mosimann and Martin, 1975.)

glacial cycle (see Chapter 17). The pollen record shows that habitats changed radically, as when the whole of the Midwest changed from the spruce parkland and tundra ranges of Pleistocene elephants to its present condition of prairie to the west and mixed forest to the east. Thus it is entirely plausible that the American megafauna would have been lost through climate change even if the first Americans had not been allowed by those same climatic changes to advance southward from their ancient homelands around Bering Strait. This argument can be answered, however, by noting that the earlier megafaunal extinctions in Australia and Africa did not occur at times of extreme climatic change but during the long ice age normality. For the Australian and African extinctions, the only obvious correlate is the arrival of populations of weapon-wielding humans.

Proponents of overkill and of habitat loss through climatic change are not persuaded by each other's arguments, and books continue to be published setting out the controversy. Many on the sidelines tend to think that both processes were involved, seeing human predation as particularly devastating to animal populations already in decline through loss of habitat.

The preponderance of the evidence thus shows that migrating human populations of the Pleistocene frequently did hunt large animals to extinction. In New Zealand they certainly did. In the continents they were probably a principal factor. This was the first great impact of humans on the biosphere.

AGRICULTURE AND NUMBERS: THE FATEFUL ALGORITHM

The second great impact came when humans learned to manipulate their environment through farming. Inextricably associated with farming was an increase in human numbers, made possible by more certain food supplies, then made necessary to provide labor for the farms. The extinctions and food web changes that followed made the earlier results of hunting overkill seem trivial by comparison.

Farming was an invention, slowly assembled then firmly learned. It was surely invented several times over—certainly at least twice, because no connection that isn't fantasy can be found between the agriculturalists of the Americas and the Middle East. The paleoecological record everywhere is of gradual addition of herded animals or cultivated plants to human diets as generations and centuries passed. This is quite in keeping with the properties of the animal that learned its niche while possessing curiosity. People added parameters of farming to the niche by learning from successful role models, just as expanding populations of hunter–gatherers had earlier learned the habits suited to life on their parents' frontier.

Around the world, farming began in the early postglacial period. The earliest traces are variously dated in the 11,000- to 7000-year time range. Allowing for the uncertainty of the record and the possibility that we may not yet have found traces of the earliest cultivation, this is a very narrow interval of time compared to the tens of thousands of preceding years in which hunting and gathering peoples had lived over much of the earth. The end of the ice age must surely have had something to do with this postglacial invention of agriculture, but what?

It cannot be that climate was suddenly suited to agriculture in the warm postglacial because large parts of the earth had always had comparable climate throughout the ice age. One school of thought suggests that postglacial warming favored food supplies and successful child rearing so that human populations tended to crowd, thus forcing people to be inventive. A variant is to suggest droughts or other climatic shocks, again forcing invention. I prefer an entirely opposite hypothesis, one of populations so reduced by the climatic shocks of the late glacial changes that densities were very low. This would produce uncrowded peoples in the returned favorable climates of the Holocene, thus removing the pressure of brute necessity for young people to learn exactly what their elders did. The young should then have had time to experiment with such untraditional things as keeping pets or growing pot.

All we know with certainty is that herding and agriculture were invented in late-glacial or early postglacial times, whatever the motivation of the innovators might have been. The ecological consequences for an animal that relied on learning to define its niche were then inevitable. Farming provided more certain food supplies for the raising of young, so that those who learned it left more surviving offspring. Cultural selection then ensured that communities of those who farmed well came to have the larger populations.

Cultural selection should not be confused with natural selection. Natural selection is of phenotypes constrained by genes. Natural selection always works to promote traits that maximize the number of surviving offspring. Cultural selection, however, is of habits acquired by learning. Success in cultural selection merely means that the culture is adopted by future generations. This could be through breeding success, or it could be through suppression of other cultures, or from simple emulation.

When hunter–gatherers learned to supplement their diets by farming, however, the expected result is the simplest. Numbers should grow as the combination of traditional breeding habits and more food led to larger, successful families. Farming cultures would populate the land more densely than would neighboring hunters or gatherers.

With the benefit of historical hindsight, we can also see that traditional breeding habits might have been relearned at the same time. Subsistence farmers in recent history strive to raise large families for the express purpose of providing labor for the family farm. We have no way of knowing when such breeding attitudes were first acquired, but it is a good guess that it would not take many generations of farming for young couples to see the benefit of having lots of willing hands.

It is a matter of archeological record that agricultural societies did indeed develop dense populations (Figure 29.3). These multitudes of farmers had impacts on the ecosystems of their times by subverting whole habitats to their crops and beasts. At the same time, people maintained many of the learned habits of the past, like foraging for firewood or hunting to supplement diets, habits that became environmentally important when shared by so many people.

Several thousand years of agriculture and expanding numbers preceded the classical civilizations of Greece and Rome. By then, the accumulated effects of cutting trees for firewood and building materials had stripped whole landscapes of trees, even as continuous digging let the soils erode. Bare red subsoils were exposed on Mediterranean lands, and Greek writers complained of countrysides left as nothing but skin and bones. Ionic and Corinthian pillars for Greek temples had to be built out of stone, because there were none of the big trees left out of whose trunks the classic columns had originally been designed (Figure 29.4).

FIGURE 29.3
Temple of Abu Simbel: 3250 years ago agriculture in the Nile delta allowed dense populations. People lived in many niches including those of nonproducing crafts like statue making. Human impacts were severe but local.

Similar havoc undoubtedly went with all the denser agricultural settlements around the world. Large or predatory animals were driven to local extinction, forests were driven back to early succession stages or replaced by burned scrublands. Food webs changed profoundly as the effects cascaded through them.

Yet the impact of agricultural humans was probably less than devastating, except on purely local scales, even after several thousand years. The really dense settlements were always patchily distributed, with large areas of less intensively used land nearby. The constant restarting of secondary successions must actually have introduced diversity into landscapes, changing food webs undoubtedly, yet leaving generous arrays of habitats. For impacts that would change forever the balance of life on earth, further niche changes were necessary, changes that not only allowed still denser populations but that also increased the impact per capita.

THE INVENTION OF MANY HUMAN NICHES

The human niche had really altered very little in the few thousand years it took to invent and consolidate the habits of subsistence farming. We had gone only from being a clothe-wearing gatherer using sticks and stones as weapons to a rather similar animal that had learned to herd and cultivate what it hunted and gathered. This had led to huge population increases with minimal changes in human habit. Young humans continued to learn what resources were good and how to extract them. They learned about the locality in which they lived and how to make or use the simple tools of foraging or cultivating. They learned the social duties of their group, certainly including the necessary rituals of reproduction and child rearing.

All individuals continued to learn their local niches from adult role models around them, as songbirds learn their dialects from neighbors. As for earlier peoples, the local niche, once learned, served the adult well for gaining a livelihood while raising children. Before the rise of civilizations in the last few millennia, therefore, the learned human niche still served most people in the same way that learning the resources of a rain forest serves the individual gibbon.

The greater change came when different people learned to live in utterly different ways. This was an inevitable consequence

FIGURE 29.4
Temple of Apollo at Corinth: 2600 years ago wood was short in Greece.
Greek civilization had so cleared and plowed once forested land that architects copied the old wooden columns of temples in stone. Intense human impact on a whole countryside was a foretaste of what would result from the more widespread civilizations of modern times.

of the organizations that came with dense and settled living. People needed to be trained for specialized functions required by the divisions of labor inseparable from any organized society. In ecological language, different people would be trained to different niches. This process of inventing different niches has proceeded ever since the first construction of cities. Priests, builders, artisans, and bureaucrats occupied different niches from the farmers and gatherers out of whose ranks they were selected and trained. The eventual industrial revolution added still more niches: those of technologists and other specialists to run the industrial enterprises. Each new niche has its own unique suite of resources and needs.

Division of labor among humans is the ultimate consequence of our total reliance on training to acquire the parameters of niche. It lets individual human lives be so different from each other that they might have been the lives of different species. We all remain, in fact, members of a single species population, whatever part of the world we come from and despite whatever odd genes for stature or color marking we might carry, and yet we can and do occupy an extraordinary array of different niches, each with its own hypervolume of necessary resources.

Thus a consequence of the trait of acquiring the ancestral niche entirely by learning is that the ancestral niche could be replaced. Over tens of thousands of years of human history, the process of replacement was very slow, testimony to the constraints on learning inherent in a system of learning only from parents and neighbors. Across the spans of generations, people could learn to forage in the different circumstances of another country, or they could add herding to hunting, and farming to gathering, but the invention of new ways of life was inevitably of glacial slowness. The great change came when professional teachers replaced parents for much of the instruction of the young. The professional instructors were aided by new technologies to assist teaching, like writing as well as language, and books instead of memory. After this, new generations could be trained in niches quite unlike those of their parents.

Modern societies and their powerful technologies certainly depend on our properties of abstract reasoning and curiosity for the necessary inventions. Yet our primary ecological trait of niche learning perhaps explains more about the conditions of our societies and our collective environmental impacts. We are trained in the uses of technology. Innumerable individuals learn the uses of each new thing, so that whole subpopulations are trained to the use of new resources in new niches.[1]

This process of mass acquisition of new resource-demanding niches is the ecological explanation of the increased environmental impact per capita now so evident in richer nations. The impact is massively compounded by the high densities at which people in niche-differentiated societies can live.

In a few instances, mass demands for new resources are so large that they can affect the physical systems of the whole earth. Potential disruption of the ozone shield by halogenated hydrocarbons is an elegant example of an ecological process leading to bizarre effects. People have been trained to live lives that require refrigeration, for without it they cannot live densely in cities while maintaining an appetite for fresh food. Inserting the parameter called refrigeration into the niches of crowded city folk is the underlying ecological process. A side effect, or consequence, is that escaped halogenated hydrocarbons break the ozone shield.

Greenhouse warming from the rapid release of carbon stores from fossil fuels is another such consequence (see Chapter 28). Again, the ecological process is the training of massed human populations to need fluxes of energy that can be met only

---

[1]The concept of diverse human niches fixed by learning can be used to study human social events. I have argued elsewhere that aggressive wars and the establishment of caste systems result, at least in part, as rising numbers of the richer subpopulations seek to maintain adequate resources for their expansive ways of life (Colinvaux, P.A. *The Fates of Nations*, Simon and Schuster, N.Y. 1982 [Penguin Edition, London 1982, Hebrew edition Massada, Tel-Aviv 1984]).

by burning fossil fuels. The summed carbon release from us all is so great that the composition of the air will be altered for centuries, with interesting consequences for not a few of the earth's inhabitants.

Generally, however, the physical systems of the biosphere are so massive that even the massed demands of technological humans can have little effect on them. Significantly altering the oxygen content of the air, for instance, seems beyond the capabilities of human folly, despite scares once raised in the environmental literature (see Chapter 28). What high densities of niche-differentiated humans can achieve is extinction and elimination of diversity on a scale not seen in 100 million years.

Dense populations in cities have huge demands on the primary productivity of the countryside. In an ecological sense, city people cannot be said to live in their cities alone but rather in all the land needed to supply their wants. They need food from the land, land to run about in (when their niches are broad enough to include some intervals of running free), and land for raw materials. Demand for food and running land rises perhaps little more than linearly with numbers, but demand for other resources from the land probably rises exponentially as increasing numbers have rising expectations of life. The land must be taken from wild ecosystems and put to human use.

An urgent additional consequence of our technological triumphs is that extinctions and forest destruction of the kind that were once purely local have taken on continental scales and spread to the last places of wilderness. This is cruel paradox. The lands of the Amazon basin and elsewhere are not needed to feed the human population, neither can the poor people who clear the forests expect to achieve modern high standards of living by doing so. Production from fertile lands already farmed around the world can easily feed the present world population. It is not from ecological need that the last wild places are being subdued but because the means have been invented to subdue them.

Clearing the Amazon forest, like cutting down the last of the old-growth trees in the western United States, is a government make-work project. The work of clearing is, for the participants, much better than going on welfare (Figure 29.5). The cost to society in lost discretionary income to pay for the work (tax breaks, standing timber sold at less than free-market cost) is less than public assistance would be, and the real cost of lost diversity and lost amenity will be paid by later generations.

The plow and the chain saw are far more potent as agents of extinction than the hunting rifle or almost any imaginable pollutant. Their use has now escaped the constraints of wisdom. This third level of human impact on the biosphere followed our learning to live in the different niches of technological society.

ON THE ORIGIN OF ADVANCED LEARNING

It is tempting to think the selective advantage of intelligence to be self-evident. Cleverness pays, whether in a computer programmer or a stone age hunter. And yet ecological logic suggests that cleverness, particularly half-cleverness on the way up, could be dangerous to fitness.

Ecological logic requires that niches be fixed and that individuals whose habits deviate significantly from others of their kind be winnowed by natural selection (see Chapter 9). The dawn of cleverness and advanced learning obviously raises the possibility of deviance from the parental niche, with the consequent penalty of loss

FIGURE 29.5
Favela in Rio de Janeiro: as the numbers crowd.
Rapidly growing modern populations can be temporarily released from favela misery by welfare or make-work projects like the cutting down of remaining forests.

of fitness. Thus, while it is easy to see how better powers of learning might let our remotest ancestors get more food or survive hazard, it is also necessary to recognize that there could be large costs connected to the habit of learning.

Learning in the earliest humans should have given selective advantage only if it were so constrained that dangerously deviant behavior was suppressed. It is reassuring to an ecologist to find properties of humans that could provide the necessary constraints. We have extremely long juvenile periods of more or less helpless dependency on adults, from whom we learn how to live. In our youthful years we are easiest to train, we pick up languages without apparent effort, and our tastes in food or even politics become so firmly entrenched that changing them in later life is difficult. It is highly plausible that these traits served our first ancestors well by fixing us to suitable role models (parents and their friends) while we did our learning. Once we had become breeding adults ourselves, our tendencies to learn were reduced, which is to say that our niches, once learned, were then fixed.

Yet the independence possible with the levels of intelligence achieved by humans should surely have made some individuals act differently from the rest despite the constraints. Perhaps some region of ice age Africa had so unsettled an environment that this deviance actually gave an advantage in fitness analogous to the colonist strategy of gambling on a few of varied genotypes being successful in an unknowable future (see Chapter 12). Otherwise a selection pressure against the nonconformity resulting from intelligence should have been strong.

The primary selective advantage of intelligence might, however, have come from its use in the rearing of children rather than for its value to foraging or survival behavior. If a human female could train her young better than her neighbors could, then her offsprings' chances of being successful parents in the next generation would be much improved. On this model, intelligence is firmly linked with language as part of a device to increase female fitness through increasing the survival, and eventual effective fecundity, of her young. I suggest, therefore, that intelligence may have arisen in humans primarily in the service of the breeding strategy and not, as often supposed, to make a better forager.[2]

Whatever the original selective advantage of intelligence may have been, the fates of all later human populations have been influenced not only by the use of intelligence itself, but also by the ways in which we learn. We still have the constraints of a long, trainable juvenile period followed by an even longer time of being relatively fixed in our ways. New ideas come from the young. Training the young is the method we use to promote change.

[2]This model is probably controversial and has only recently been set out in the formal literature. (Colinvaux, P.A. "A model for the selective advantage of intelligence to breeding females." *Evolutionary Theory* 10: 15–32, January, 1991).

THE IMPACT CONTROLLED

Younger generations in the wealthier societies already choose to regard the preservation of diverse nature as one of the goals for living. Appreciation of wilderness is thus inserted into the learned human niche, inducing people to act in ways that might conserve wilderness. Of much larger importance, though, is the way that life in broad technological niches can reduce the rate of reproduction.

Demographers talk of the "**demographic transition**," by which they mean the observation that average family size in the modern world has generally fallen as nations have grown rich. To the extent that this is true, there is a simple ecological explanation. Educating children to niches of affluence is expensive, in monetary commitment. People who want their children to grow up to live broad and varied lives, and who want to live similar lives themselves, choose to have small families. To convert this almost self-evident statement to ecological jargon, we can say broad niches require so many resources that the resources available to any breeding couple will only divide for the support of few offspring.

The process most likely to achieve lasting control of human reproduction, however, is learning to emancipate women. If I am right in my conjecture that a crucial selective advantage of intelligence to those first Pleistocene human females was that it increased their effective fecundity by making their child rearing more efficient, then we now have the happy paradox that learning has begun to serve us well by letting women spend their energies on tasks other than raising children.

Events of the next century will probably decide the diversity of the biosphere for the next 5 or 10 million years. Whether high diversity survives must largely depend on the rate at which new cultural and reproductive attitudes are learned by contemporary societies. Pessimists must note that niche-learning changes work only with the tempo of human generations—the young must learn the new thing to influence events more than 20 years later. If the tropical forests are cleared in the next 20 years, as some forecasts suggest, then the pending cultural changes will be too late (Terborgh, 1992). Pessimists must also note what seems to be strong resistance in so many cultures to the idea of freeing women from the duties of child rearing.

Optimists, however, can take heart from the fact that, because all important niche parameters of humans are learned, our behavior, and its impact, does change. People are valuing wilderness. Women are being liberated. By any ecological measure of time, these fundamental changes in human attributes are happening with extreme rapidity. The world population will be stabilized. If it happens quickly enough, then conservationists can plan the management of future reserves at their leisure.

Glossary

adiabatic lapse rate: cooling of rising air without external source of energy for expansion; 10°C (dry), 6°C (wet) per kilometer

albedo: reflectivity of the earth's surface to sunlight

alfisol: gray-brown podzolic soil of temperate latitudes

allelochemics: study of use of chemical agents by organisms to influence other organisms, most particularly as defenses or as lures

alkaloid: toxic compound used in plant qualitative defense

allelopathy: use of secondary compounds by plants for herbivore defense

allometric growth: principle that growth in total size involves disproportionate growth in different body parts

allometry: relative changes of parts as a structure is enlarged

allopatric: living in regions that are clearly separated

alluvium: water-borne deposit

alpha diversity: within-habitat

altricial: has naked or helpless offspring

aluvium: deposit of water-borne material

anemophily: wind pollinated

annual: plant completing life-history from germination to death in one year

ant-guard: symbiosis of plants providing resources for ants and ants providing defense for plants

apomixis: (apomictic) asexual reproduction, especially when resembling sexual reproduction but in which the egg and sperm do not fuse

arena: alternative term for lek

aridisol: desert soil; A-C soil

artiodactyl: 2-toed ungulate (e.g., cow)

assimilation: for animals, absorption of food energy from the gut; for plants the uptake of carbon dioxide

assimilation efficiency: ratio of food absorbed to food ingested

association: plant community unit; variously, often subjectively, defined

autecology: ecology of single species populations

Batesian mimicry: in which a vulnerable mimic resembles a defended model

benthos: organisms living on the bottom of sea or lake (adjective: benthic)

beta diversity: between-habitat

biennial: plant completing life-history in two years, typically with the main reproductive episode in the second year

biocoenose: (biocoenosis) animal and plant communities combined to one unit, roughly equivalent to "ecosystem," though a version more closely comparable to ecosystem is biogeocoenosis, as widely used in Russian literature

bioherm: living reef structure in which the plant contribution is manifestly more important than the contribution of coralline animals: particularly mounds built from *Halimeda* remains

biological accommodation: fitting life-styles more to the presence of other organisms than to physical factors

biomass: mass of living organisms, originally expressed as a mass-density (e.g., grams per square meter) but now sometimes expressed as calories per unit area

biome: ecosystem of a large geographic area in which plants are of one formation and for which climate sets the limits

biotype: genotype reproduced asexually

Blytt-Sernander: sequence of Holocene climatic events in northern Europe

BOD: biological oxygen demand

boundary layer: thin layer alongside leaf or other object in which air or other fluid is motionless

brown earth: soil of temperate forest; alfisol

caliche: carbonate deposit at the line of maximum water penetration in mollisol or chernozem, or term may refer to the carbonate-layered soil itself

CAM: crassulacean acid metabolism, use of C4 photosynthesis and storage of carbon dioxide collected at night as C4 acids for photosynthesis by day behind closed stomates

cascade hypothesis: in food web studies any hypothesis of processes impacting a trophic level influencing lower trophic levels

catena: related set of soil series differentiated by drainage or topography

character displacement: selection for characters that avoid interspecific competition

character species: indicator of qualitatively defined community type

chernozem: black earth of prairies through which percolation is incomplete; mollisol

climax: plant community resulting after an ecological succession and in which further change is slow

clone: asexually produced offspring of a common ancestor

cohort: standard-size population used in construction of life tables

colluvium: deposit of material slowly worked down a slope

competitive exclusion: principle that strongly competing species cannot coexist indefinitely

constant species: one present in at least 80% of communities sampled (Uppsala school of phytosociology)

consumer: herbivores and carnivores that consume energy originally transformed into producer biomass

continuum: community unit describing changing species composition along a gradient

croppers: browsers or grazers but not animals that eat seeds or fruits that must be sought out in a manner analogous to hunting

cropping principle: predation that keeps prey populations low (works to allow increased diversity)

cyclomorphosis: changing of shape with succeeding generations

dark reaction: synthesis of reduced carbon compounds following light energy transformation in photosynthesis

decomposer: organism that receives energy by oxidizing dead organic matter (most are bacteria or fungi)

deep scattering layer: plankton at depths in the oceans by day as recorded by sonar

deme: panmictic population with general and random mating

dendrochronology: dating by tree rings

density: number of individuals per unit area

detritivore: consumer of detritus

detritus: fragmented dead plant matter

diapause: process of surviving hostile season with resting stage of reduced metabolic rate

dominant: (ecological dominant) more abundant than allowed for in a random assortment, or competitively superior; sometimes also used to refer merely to the most common species

ecdysis: molting of exoskeleton during growth of arthropods

ecocline: gradient of changing species composition

ecological time: duration of interactive process among existing species

ecological (Lindeman) efficiency: ratio of gross production between trophic levels
ecological equivalents: organisms of comparable function or niche
ecological succession: sequential appearance of species or communities
ecotone: region of rapidly changing species composition at an environmental disjunction
ecotope: term intended to define niche or niche space as a function of habitat
ectocrine: chemical secreted into the environment that influences other organisms
ectothermy: use of environmental heat (principally the sun) to regulate body temperature
Eltonian pyramid: the pyramid of numbers
eluvial horizon: soil layer from which matter has been removed
endorheic: interior drainage (used of closed basin lakes)
endothermy: use of metabolic heat to regulate body temperature
energy flow: concept that food transfer between prey and predator is calorie transfer with energy release, hence a unidirectional flow
enthalpy: state of order in a thermodynamic system
entomophilous: insect pollinated
entropy: physicist's term describing state of maximum disorder or randomness toward which natural systems spontaneously move
epideictic display: sexual display postulated to serve the purpose of a density measuring device (a view not generally held by ecologists)
epilimnion: warm surface water of a lake
equilibrium species: species adapted to persist; *K*-strategist
equitability: component of relative abundance in measures of diversity
eukaryote: has nucleus in cells
euphotic: zone of surface water (about 100 m in sea) where light is sufficient for photosynthesis to exceed respiration
eury- : broadly tolerant, as in eurythermal or euryhaline for organisms tolerating wide ranges in temperature or salt
eusociality: insect social systems with divisions of labor among morphs, many of which do not reproduce themselves
eustacy: changing sea level resulting from changes in ice or ocean volumes
eutrophic: fertile
exclusion: see *competitive exclusion*
exploitation efficiency: ratio of food ingested to net production of food species

fecundity: rate at which female produces offspring
fertility: (demography studies) population reproductive rate
floristic: refers to species composition of vegetation food chain: predation series linking animals to ultimate plant food
food cycle: old concept now replaced with concepts of nutrient cycles (nutrients being the "food") and energy flow
food web: concept of intersecting food chains
formation: vegetation of a large climatic region recognized by a characteristic shape or lifeform
frass: solid insect excrement
frequency: proportion (%) of samples in which a species occurs
frugivore: seed or fruit eater
fugitive species: opportunist or *r*-strategist adapted to disperse away from competitors
functional response: high prey density leads to changes in predator behavior
fundamental niche: niche in the absence of interspecific competition

gamma diversity: regional over geographic clines
GAS: general adaptation syndrome—hormonal response to stress in vertebrates
graben: basin formed by geological faulting
gross production: used in ecological energetics, term means energy input, but used differently in related disciplines
guild: a set of coexisting species that share a common resource

halocline: gradient of increased salinity (and hence density) with depth in the ocean surface
halflife: time taken for half of radioactivity to decay

hemimetabolous: insect life history with no distinctive larval stage
hemolymph: body fluid of arthropods
heterogonic: life cycle in which sexual and asexual reproductive epidodes alternate
Holocene: time since end of last ice age (about 10,000 years)
holometabolous: insect life histories with separate larval and pupal stages
homeostasis: maintenance of constancy or near-uniformity in organism, community or other entity
homeotherm: animal that maintains a core temperature by release of metabolic heat
home range: wandering or feeding area of an animal
hydrarch: succession on wetlands
hydrarch succession: sequential colonization of marshy habitats
hypolimnion: cold bottom water of a lake

illuvial horizon: soil layer to which matter has been added from above
inclusive fitness: includes genes carried by offspring and genes carried by relatives into the next generation
infauna: animals living within sediments of sea floor
ingestion: taking of food into the gut
instar: insect form between molts
interspecific: between members of different species
intraspecific: among members of a species population
intrinsic rate of increase: exponential growth rate of a popuiation with a stable age distribution
inversion: stable system of warm air floating over cold air
isohyet: line on a map connecting places of equal precipitation
isopleth: line of constant precipitation–evaporation ratio
isostacy: changing sea level due to crustal movements under ice masses
iteroparity: breeding repeatedly

karst: lake occupying a solution basin
kin selection: selection favoring individuals by preserving genes carried in relatives
kranze syndrome: leaf morphology required for C4 photosynthesis in which chloroplasts are in cells below the epidermis
***K*-strategy:** life-style in which fecundity is reduced to divert resources to persistence

landnam: hypothetical land clearing by early European agriculturalists and recorded by elm pollen decline
langley: (ly) flux of solar energy measured as gram-calories per unit area in unit time
large young strategy: concentrates resources into a few young, either by making young large or through parental care
laterite: tropical red earth, oxisol or latosol
lek: communal breeding ground where males display and females come to be mated
light reaction: stage of photosynthesis in which light energy is transformed
limiting factors: concept that distribution or abundance may be set by a critically limiting resource
Lincoln index: mark and recapture index
loess: wind-blown deposits of characteristic grain size
luxury consumption: uptake of nutrients by algae in excess of immediate needs

maar: volcanic explosion crater
margallitic soil: tropical mollisol important to tropical agriculture
meromictic lake: permanently stratified lake
mesic: environment midway between extremes; especially habitats supporting dense, diverse plant communities
mesotrophic: moderately fertile
metalimnion: volume of lake water across which a thermocline runs
microcosm: once used as synonymous with "ecosystem," now widely taken to mean a small experimental ecosystem
minimum area: sum of area of quadrats to be sampled to include all but chance species of a habitat
mixolimnion: upper water of a chemically stratified lake
mollisol: black and brown soils of prairies through which percolation is incomplete
monimolimnion: bottom water of a chemically stratified lake

monolayer: postulated leaf arrangement for shade trees in which leaf shape and pattern achieves maximum light occlusion in a narrow vertical distance
mortality: rate of death
Müllerian mimicry: in which defended animals share warning signals
multilayer: postulated leaf arrangement for trees of bright sunlight that allows deep diffusion of light
mutualism: living together of species in which both partners gain and both incur cost
mycorrhizae: fungal root symbionts

nannoplankton: the very smallest plankton that passes through most collecting nets, includes protozoa as well as small phytoplankters
natality: rate of offspring production
net production: energy input less respiration (but has other meanings in related disciplines)
niche: role, function, or place of organism
niche-packing: conceptual process of community building with final species number set by maximum possible interspecific competition
niche space: environmental parameters defining a niche, or the resource flux required for an individual to survive and reproduce
nidicolous: young fed in nest
nidifugous: young leave nest as soon as hatched
numerical response: high prey density results in high reproduction of predators

oligotrophic: infertile
opportunist: colonist or *r*-strategist
optimal defense: theory that secondary compounds are allocated to defense in a way that minimizes cost and maximizes inclusive fitness
optimal foraging: theory that animals should behave so as to maximize energy intake for the time spent foraging
orogeny: mountain building
overrepresentation: taxa too numerous in a pollen diagram because of large local source of pollen
oxisol: tropical red soil depleted of silicate minerals

paleoecology: use of fossils to test ecological hypotheses (but see alternative meanings in text)
palynology: study of pollen and spores
paradox: statement apparently conflicting with common sense but actually well founded or reasonable
parapatric: living in adjacent regions
parthenogenesis: reproduction with unfertilized eggs
pedalfer: soil through which water percolates to a drainage system
pedocal: soil through which water does not percolate completely
pedon: smallest describable patch of a soil series
peneplain: flat dissected surface from erosion by an ancient river system
perennial: plant living many years
periphyton: organisms attached to aquatic plants
perissodactyl: 1-toed ungulate (e.g., horse)
permafrost: permanently frozen ground, only a thin surface layer of which thaws in summer
phenology: properly the study of phenomena with constrained timing but now loosely used in ecology to refer to the process of constraining biological process like flowering into particular times of year or day as part of a system of coevolution
pheromone: chemical used as a social cue
phoresy: dispersal of communal group by host organism
photorespiration: respiration of carbon dioxide in light resulting from properties of the enzyme RuBP carboxyllase
philopatry: returning to the place of birth to breed
phycobilisomes: colored pigments of algae that are adaptations to increase light absorption in dim light
physiognomic: refers to shape, structure, or form
phytolith: silica concretion within a plant cell
phytoplankter: individual plant of the plankton

phytoplankton: plant plankton
phytosociology: study of plant communities
plankter: organism of the plankton
plankton: drifting organisms of open water, mostly small to microscopic
pleuston: organisms suspended from water surface
pluvial: time of increased rainfall
podzol: soil with bleached A horizon, typical of boreal forest; spodosol
poikilotherm: animal that does not rely on metabolic heat to maintain a core temperature
point diversity: in very small samples
pollen influx: pollen grains deposited per unit area in unit time
pollen spectrum: histogram of pollen percentages
pollen zone: section of pollen diagram in which spectra remain similar and which can be distinguished from neighboring zones
precocial: has fully formed young able to feed themselves
primary succession: ecological succession on ground that has not hitherto supported vegetation
proclimax: community maintained by repeated disturbance
producer: primary energy transformer of an ecosystem, most are green plants but chemosynthetic bacteria are producers in a few ecosystems
production: used variously in ecology as energy input, energy stored as biomass, or biomass sequestered for reproduction
production efficiency: ratio of growth and reproduction to assimilation
profligate reproduction: relies on large numbers of very small eggs, seeds, etc.
prokaryote: no nucleus in cells; principally bacteria and some algae
propagule: unit of propagation, whether live young, egg, seed, or asexually cloned individual or dispersal unit
protandrous flower: flower that changes from male to female with age
prudential reproduction: relies on large young or parental care
pubescent: hairy
pycnocline: gradient of increasing density with depth in surface ocean waters
pyramid of numbers: observation that food chains run in parallel with animals at any level having comparable size and relative abundance, the animals high on the food chains being both large and rare

quadrat: quantitative sample; particularly number per unit area
qualitative defense: allelopathy employing specific toxins (e.g., alkaloids)
quantitative defense: allelopathy employing universal toxins (e.g., tannin)

***r* and *K* selection:** selection within a population for traits tending towards *r* or *K* strategies
R-C-S continuum: postulated replacement series of plants of ruderal, competitive, and stress-tolerant strategies
reactive distance: effective targeting range of a predator
realized niche: niche in the presence of interspecific competition
realm: (biogeographic) latitudinal expanse in which organisms have comparable adaptations
red queen: hypothesis that evolution results as selection tracks persistent environmental change
region: (biogeographic) continental expanse, the organisms of which are taxonomically related
regolith: top layer of unconsolidated material covering crustal rocks
relaxation: in island biogeography means loss of species following decreased access or reduction in area
respiration: energy represented by carbon dioxide given off from oxidation of reduced carbon compounds to do work
Reynold's number: relationship between length, speed and viscosity that defines motion through fluids
richness: total species number
***r*-strategy:** colonist life-style of high fecundity, low persistence
ruderal: plant with colonizer strategy

saprobe: synonym for decomposer—hence "saprobe chain"
saprophage: synonym for decomposer
secondary compounds: plant physiological term for chemicals with no obvious physiological function and which ecologists generally consider to have coercive functions on other organisms
secondary plant metabolites: toxic compounds manufactured by plants, frequently tending to be autotoxic
secondary succession: ecological succession on ground from which previous vegetation was removed
seed bank: accumulation in soil of seeds with delayed germination
seedling bank: accumulation of seedlings maintaining themselves in dim light on a forest floor
semelparity: breeding only once in a lifetime
seral stage: community identifiable as part of a successional sequence
sere: a particular successional sequence
sesquioxides: generic term for a mixture of iron and aluminum oxides
sexual dimorphism: physical differences between sexes
sigmoid: growth or response curve that is s-shaped
signal receiver: term for dupe in a mimicry system
small egg strategy: relies on many small young
sociation: plant community unit defined by measured minimum area
sociobiology: the study of how selection for individual fitness leads to group or social phenomena
soil horizon: soil layer resulting from weathering processes
soil series: local soil type dependent on parent material and drainage
solar constant: flux of solar energy reaching the upper atmosphere (about 2 langleys)
species richness: total number of species present
spodosol: podzol
standing crop: biomass present at time of sampling
steno- : narrowly restricted, as in stenothermal or stenohaline for organisms with narrow tolerances to temperature or salt
stomate: gas-exchange aperture through plant epidermis that can be controlled by turgor in surrounding cells
stress-tolerator: plant adapted to allocate resources between competition and physical stress resulting from competition
stromatolite: carbonate structure made by marine algae
structured deme: system of isolated populations without general and continuous mixing of genes throughout the population
subduction: descent of leading edge of a continental plate at line of collision
succession: allogenic = responding to habitat changes imposed by outside physical forces; autogenic = from biological interactions within the ecosystem; primary = on land never before vegetated; secondary = on old habitat that has been devegetated
Suess effect: dilution of atmospheric ^{14}C by ^{12}C from burning fossil fuels
survivorship: cohort survivors of stipulated age
symbiosis: living together of species in which both partners benefit
sympatric: living in the same region
synecology: ecology of communities, study of community properties

taxon cycle: selection series on islands as *K*-strategists evolve from colonists, followed by *r*-selection among island endemics
tannins: compounds serving plants as quantitative defenses that make tissues indigestible
teleology: the doctrine that developments are due to the purpose that is served by them
tetrapod: 4-legged animal
thermocline: middle depths of a stratified lake where temperature changes rapidly
thermokarst: lake occupying a melted depression in permafrost
till: unsorted mineral deposit left by a retreating glacier (also called boulder clay)
time-stability: hypothesis that places with stable and permanent physical environments

collect many species because extinction is unlikely

tramp species: extreme opportunists, particularly species dispersed by human commerce

transpiration: transfer of water from soil to atmosphere by plants, the motive force of which is evaporation from the leaves

trichome: fine barbed hair, a mechanical defense

trophic level: common feeding at the same link on food chains—level in a pyramid of numbers

trophogenic: upper zone of lake or ocean where light allows photosynthesis

VAM: vesicular arbuscular mycorrhizae, fungal symbionts important to nutrient uptake

varve: couplet of bands in sediment deposited in a single year

Wallace's line: separates Australia from Indonesia on biogeographic evidence

xerarch succession: sequential colonization of dry or desert habitats

yield: may mean net production or any harvestable portion of net production, according to context

younger dryas: warm interval of <1000 years in lateglacial time, detected in northern hemisphere and arguably of global importance

zooplankter: an animal of the zooplankton

zooplankton: animal plankton

zooxanthellae: filamentous algae symbiotic with reef-building corals

References

Adams, J. M., 1989, Species diversity and productivity of trees. *Plants today*, November–December, p. 183–187.

Adams, J. M., and F. I. Woodward, 1989, Patterns in tree species richness as a test of the glacial extinction hypothesis. *Nature*, v. 339, p. 699–701.

Adams, J. M., H. Faure, L. Faure-Denard, J. M. MacGlade, and F. I. Woodward, 1991, Increases in terrestrial carbon storage from the Last Glacial. *Nature*, v. 348, p. 711–714.

Alcaraz, M., G. A. Paffenhofer, and J. R. Strickler, 1980, Catching the algae: A first account of visual observations on filterfeeding calanoids, p. 241–248 *in* Kerfoot, W. C., *ed.*, *Evolution and Ecology of Zooplankton Communities*. Hanover, N. H., Univ. Press of New England.

Allen, J. A., 1871, Mammals and winter birds of east Florida, and a sketch of the bird fauna of eastern North America. *Bull. Museum of Comp. Zool.*, v. 2, p. 161–450, pt. V, p. 375–450.

Alvarez, L. W., W. Alvarez, F. Asaro, and H. V. Michel, 1980, Extraterrestrial cause for the Cretaceous-Tertiary Extinction. *Science*, v. 208, p. 1095–1108.

Andrewartha, H. G., and L. C. Birch, 1954, *The Distribution and Abundance of Animals*, Chicago, Univ. of Chicago Press.

Armstrong, J., 1960, The Dynamics of *Dugesia tigrina* populations and of *Daphnia pulex* populations as modified by immigration, Ph.D. dissertation, Univ. of Michigan.

Ayala, F. J., 1968, Genotype, environment, and population numbers. *Science*, v. 162, p. 1453–1459.

Bacastow, R., and C. D. Keeling, 1973, Atmospheric carbon dioxide and radiocarbon in the natural carbon cycle. II. Changes from A.D. 1700 to 2070 as deduced from a geochemical model, p. 86–135 *in* Woodwell, G. M., and E. V. Pecan, *eds.*, *Carbon and the Biosphere.* Washington, D.C.

Baker, H. 1937, Alluvial meadows: A comparative study of grazed and mown meadows. *Ecol.*, v. 25, p. 408–420.

Baker, H. G., and I. Baker, 1975, Studies of nectar constitution and pollinatorplant coevolution, p. 100–140 *in* Gilbert, L. E., and P. R. Raven, *eds.*, Coevolution of animals and plants. Austin Univ. of Texas Press.

Barnosky, C. W., E. C. Grimm, and H. E. Wright Jr, 1987. Towards a postglacial history of the northern Great Plains: A review of the paleoecologic problems. *Ann. of the Carnegie Museum*, v. 56, p. 259–273.

Baskerville, G. L., 1965, Estimation of dry weight of tree components and total standing crop in conifer stands. *Ecol.*, v. 46, p. 867–869.

Bazzaz, F. A., and S. T. A. Pickett. 1980, Physiological ecology of tropical succession: a comparative review. *Annu. Rev. Ecol. Syst.*, vol. 11, p. 287–310.

Bell, G., 1982, *The Masterpiece of Nature: The Evolution and Genetics of Sexuality.* Univ. of California Press, Berkeley.

Benecke, V., 1972, Wachstum, CO_2-gaswechsel und Pigmentgehalt einiger Baumarten nach Ausbringung in verochiedene Hohenlagen. *Angew. Bot.*, v. 46, p. 117–135.

Benemann, J. R., 1973, Nitrogen fixation in termites. *Science*, v. 18, p. 164–166.

Bennet, F. B., and J. A. Ruben, 1979, Endothermy and activity in vertebrates. *Science*, v. 206, p. 649–654.

Benninghoff, W. S., 1966, The releve method of describing vegetation. *Michigan Botanist*, v. 5, p. 109–114.

Berger, A., J. Imbrie, J. Hays, G. Kukla, B. Saltzman, eds., 1984, *Milankovitch and Climate.* Dordrecht Reidel.

Berger, W. H., J. S. Killingley, and E. Vincent, 1985, Timing of deglaciation from an oxygen isotope curve for Atlantic deep-sea sediments. *Nature*, v. 314, p. 156–158.

Berner, R. A., 1990, atmospheric carbon dioxide levels over phanerozoic time. *Science*, v. 249, p. 1382–1386.

Berner, R. A., and G. P. Landis, 1988, Gas bubbles in fossil amber as possible indicators of the major gas composition of ancient air. *Science*, v. 239, p. 1406–1409.

Bertram, B. C. R., 1978, Living in groups: Predators and prey, p. 64–96 *in* Krebs, J. R., and N. B. Davies, *eds., Behavioural Ecology*, Sunderland, Mass., Sinauer Associates.

Billings, W. D., 1938, The structure and development of old field shortleaf pine stands and certain associated physical properties of the soil. *Ecol. Monog.*, v. 8, p. 437–499.

Birks, H. J. B., and H. H. Birks, 1980, *Quaternary Paleoecology.* Baltimore, Univ. Park Press.

Bischoff, J. L., and R. J. Rosenbauer, 1981, Uranium series dating of human skeletal remains from the Del Mar and Sunnyvale sites, California. *Science*, v. 213, p. 1003–1005.

Björkman, O., and J. Berry, 1973, High efficiency photosynthesis. *Scientific Am.*, v. 229, p. 80–93.

Blum, U., and E. L. Rice, 1969, Inhibition of symbiotic nitrogenfixation by gallic and tannic acid, and possible roles in old field succession. *Bull. Torrey Bot. Club*, v. 96, p. 531–544.

Blumenstock, D. I., and S. W. Thornthwaite, 1941, *Climate and the world pattern: Climate and man (USDA Yearbook)*, Washington, D.C., U.S. Government Printing Office, p. 98–127.

Bode, H. R., 1958, Beitrage zur Kenntnis allelоysathischer Erscheinungen beieinigen Juglandaceen. *Planta*, v. 51, p. 440–480.

Boersma, P. D., 1977, An ecological and behavioral study of the Galapagos penguin. *The Living Bird*, v. 15, p. 43–93.

Bohrer, R. N., 1980, Experimental studies of diel vertical migration, p. 111–121 *in* W. C. Kerfoot, *ed., Evolution and Ecology of Zooplankton Communities.* Hanover, N.H., Univ. Press of New England.

Bolin, B., 1970, The carbon cycle. *Scientific Am.*, v. 223, p. 124–132.

Bonnefille, R., J. C. Roeland, and J. Guiot, 1990, Temperature and rainfall estimates for the past 40,000 years in ecuatorial Africa. *Nature*, v. 346, p. 347–349.

Bonner, J., 1962, The upper limit of crop yield. *Science*, v. 137, p. 11–15.

Bormann, F. H., 1976, An inseparable linkage: Conservation of natural ecosystems and the conservation of fossil energy. *BioSci.*, v. 26, p. 754–760.

Bormann, F. H., and G. E. Likens, 1967, Nutrient cycling. *Science*, v. 155, p. 424–429.

Bormann, F. H., G. E. Likens, T. G. Siccama, R. S. Pierce, and J. S. Eaton, 1974, The export of nutrients and recovery of stable conditions following deforestation at Hubbard Brook. *Ecol. Monog.*, v. 44, p. 255–277.

Bormann, F. H., G. E. Likens, and J. S. Eaton, 1969, Biotic regulation of particulate and solution losses from a forest ecosystem. *BioSci.* v. 19, p. 600–610.

Botkin, D. B., J. F. Janak, and J. R. Wallis, 1972, Some ecological consequences of a computer model of forest growth. *Ecol.*, v. 60, p. 849–872.

Bowman, R. I., 1961, Morphological differentiation and adaptation in the Galapagos finches. *University of California Publications in Zoology*, v. 58, p. 1–302.

Boyce, J. B. 1946, The influence of fecundity and egg mortality on the population growth of *Tribolium confusom* Duval. *Ecol.*, v. 27, p. 290–302.

Boyce, M. S., 1984, Restitution of r^- and K^- selection as a model of density-dependent natural selection. *Ann Rev. Ecol. Syst.*, v. 15, p. 427–447.

Braun, E. L., 1955, The phytogeography of unglaciated Eastern United States and its interpretation. *Botan. Rev.*, v. 21, p. 297–375.

Braun-Blanquet, J., 1932, *Plant Sociology: The Study of Plant Communities* (trans., rev., and ed. by Fuller, G. D., and H. S. Conard). New York, McGraw-Hill.

Briand, F., and J. E. Cohen, 1987. environmental correlates of food chain length. *Science*, v. 238, p. 956–959.

Bridges, E. M., 1978, *World Soils*, 2nd. ed. Cambridge, Cambridge Univ. Press.

Broecker, W. S., 1970, Man's oxygen reserves, *Science*, v. 168, p. 1537–1538.

Broecker, W. S., 1971, A kinetic model for the chemical composition of sea water. *Quat. Res.*, v. 1, p. 188–207.

Brooks, J. L., 1946. Cyclomorphosis in *Daphnia. Ecol. Monogr.* v. 16, p. 409–447.

Brooks, J. L., 1965, Predation and relative helmet size in cyclomorphic *Daphnia. Proc. Nat. Acad. Sci.*, v. 53, p. 119–126.

Brooks, J. L., and S. I. Dodson, 1965, Predation, body size, and composition of plankton. *Science*, v. 150, p. 28–35.

Brower, L. P., and J. V. Z. Brower, 1964, Birds, butterflies and plant poisons: A study in ecological chemistry. *Zoologica*, v. 49, p. 137–159.

Brown, J. H., 1971, Mammals on mountaintops: Nonequilibrium insular biogeography. *Am. Nat.*, v. 105, p. 467–478.

Brown, J. H., and R. C. Lasiewski, 1972, Metabolism of weasels: The cost of being long and thin. *Ecol.*, v. 53, p. 939–943.

Brown, J. L., 1969, Territorial behavior and population regulation on birds. *Willson Bull.*, v. 81, p. 293–329.

Brown, K. S., Jr., 1982. Paleoecology and regional patterns of evolution in neotropical forest butterflies, p. 255–308 in G. T. Prance, ed. *Biological Diversification in the Tropics.* New York, Columbia Univ. Press.

Brown, K. S., Jr., J. R. Trigo, R. B. Francini, A. B. Barros de Morais, and P. C. Motta. 1991. Aposematic insects on toxic host plants: coevolution, colonization and chemical emancipation, p. 375–402. *in* P. W. Price, T. M. Lewinsohn, G. W. Fernandes, and W. W. Benson eds. *Plant-Animal Interactions.* New York, John Wiley.

Brown, L. L., and E. O. Wilson, 1956, Character displacement. *Systematic Zool.*, v. 5, p. 49–64.

Brugam, R. B., 1978, Pollen indicators of land–use change in southern Connecticut. *Quat. Res.*, v. 9, p. 349–362.

Bryson, R. A., 1966, Air masses, streamlines, and the boreal forest. *Geog. Bull.*, v. 8, p. 228–269.

Buckman, H. O., and N. C. Brady, 1969, *The Nature and Properties of Soils*, 7th ed. Toronto, Collier–Macmillan.

Bunt, J. S., 1975. Primary productivity of marine ecosystems, p. 169–215 in Leith, H. and R. H. Whittaker, *eds. Primary Productivity of the Biosphere*, New York, Springer-Verlag.

Burris, R. H., and C. C. Black, *eds.*, 1976, *CO_2 Metabolism and Plant Productivity.* Baltimore, Univ. Park Press.

Burton, T. H., and G. E. Likens, 1973, The effect of strip–cutting on stream temperatures in the Hubbard Brook experimental forest, New Hampshire. *BioSci.*, v. 23, p. 433–435.

Bush, M. B., 1991, Modern pollen-rain data from South and Central America: A test of the feasibility of fine-resolution lowland tropical palynology. *The Holocene*, v. 1(2), p. 162–167.

Bush, M. B., D. R. Piperno, and P. A. Colinvaux, 1989, A 6000 year old history of Amazonian maize cultivation. *Nature*, v. 340, p. 302–303.

Bush, M. B., P. A. Colinvaux, M. C. Wiemann, D. R. Piperno, and K-b Liu, 1990, Late Pleistocene temperature depression and vegetation change in Ecuadorian Amazonia. *Quat. Res.*, v. 34, p. 330–345.

Bush, M. B., and P. A. Colinvaux, 1990, A pollen record of a complete glacial cycle from lowland Panama. *J. of Vegetation Science* v. 1, p. 105–118.

Bush, M. B., and R. J. Whittaker, 1991, Krakatau: colonization patterns and hierarchies. *J. of Biogeog.*, v. 18, p. 341–356.

Calhoun, J. B., 1962, Population density and social pathology. *Scientific Am.*, v. 206, p. 139–148.

Carpenter, S. R., J. F. Kitchell, and J. R. Hodgson. 1985, Cascading trophic interactions and lake productivity *BioSci.*, v. 35, p. 634–639.

Chapman, R. N., 1928, The quantitative analysis of environmental factors. *Ecol.*, v. 9, p. 111–122.

Chapman, R. N., 1931, *Animal Ecology with Especial Reference to Insects.* New York, McGraw-Hill.

Charnovo, E. L., 1982, *The Theory of Sex Allocation.* Princeton, Princeton Univ. Press.

Chesson, P. L. 1985, Environmental variation and the coexistence of species. p. 240–256 in J. M. Diamond and T. J. Case, eds. *Community ecology.* New York, Harper and Row.

Chew, R. M., and A. E. Chew, 1970, Energy relationships of the mammals of a desert shrub (*Larrea tridentata*) community. *Ecol. Monog.*, v. 40, p. 1–21.

Child, J. J., 1976, New developments in nitrogen fixation research. *BioSci.*, v. 26, p. 614–617.

Chisholm, S. W., and F. M. M. Morel, 1991, ed. What controls phytoplankton production in nutrient-rich areas of the open sea. *Limnol.* and *Oceanogr.*, v. 36, 1970 p.

Christian, J. J., 1950, The adreno-pituitary system and population cycles in mammals. *J. Mammol.*, v. 31, p. 248–259.

Christian, J. J., V. Flyger, and D. E. Davis, 1960, Factors in the mass mortality of a herd of Sika deer, *Cervus nippon. Chesapeake Science*, v. 1, p. 79–95.

Clapperton, C. M., 1987, Glacial geomorphology, Quaternary glacial sequence and paleoclimatic inferences in the Ecuadorian Andes, *in* V. Gardiner, *ed.*, Part 2, "International Geomorphology", p. 843–870. Chichester, John Wiley.

Clements, F. E. 1916, Plant succession: An Analysis of the Development of Vegetation. *Carnegie Inst. of Wash. Publ.* 242. Facsimile reprint by Haffner, New York.

CLIMAP, 1976, The Surface of the Iceage Earth. *Science*, v. 191, p. 1131–1144.

Cloud, P., 1976, Beginnings of biospheric evolution and their biochemical consequences. *Paleobiology*, v. 2, p. 351–387.

Cloud, P. and A. Gibon, 1970. The oxygen cycle. *Scientific Am.*, v. 223, p. 110–123.

Clutton-Brock, T. H., 1991, *The Evolution of Parental Care.* Princeton, Princeton Univ. Press.

Cody, M. L., 1966, A general theory of clutch size. *Evol.*, v. 20, p. 174–184.

Cody, M. L., 1974, *Competition of the Bird Communities.* Princeton, Princeton Univ. Press.

Cody, M. L., 1975, Towards a theory of continental species diversity, p. 214–257 *in* Cody, M. L., and J. M. Diamond, *eds.*, *Ecology and Evolution of Communities.* Cambridge, Harvard Univ. Press.

Cohen, J. E., 1978, *Food Webs and Niche Space.* Princeton, Princeton Univ. Press.

Cohen, J. E., F. Briand, and C. M. Newman, 1990, *Community Food Webs: Data and Theory.* New York, Springer-Verlag.

COHMAP, 1988, Climatic changes of the last 18,000 years: observations and model simulations. *Science*, v. 241, p. 1043–1052.

Cole, L. C., 1954, Some features of random population cycles. *J. Wildlife Management*, v. 18, p. 1–24.

Colinvaux, P. A., 1968, Reconnaissance and chemistry of the lakes and bogs of the Galapagos Islands. *Nature*, v. 219, p. 590–594.

Colinvaux, P. A., 1972, Climate and the Galapagos Islands. *Nature*, v. 240, p. 17–20.

Colinvaux, P. A., 1973, *Introduction to Ecology.* New York, John Wiley.

Colinvaux, P. A., 1978, On the use of the word "absolute" in pollen statistics. *Quat. Res.*, v. 9, p. 132–133.

Colinvaux, P. A., 1982, Towards a theory of history: Fitness, niche and clutch of *Homo sapiens. J. Ecol.*, v. 70, p. 393–412.

Colinvaux, P. A., 1987, Amazon diversity in light of the paleoecological record. *Quat. Sci. Rev.*, v. 6, p. 93–114.

Colinvaux, P. A., 1993, Pleistocene biogeography and diversity in Tropical forests of South America, *in* P. Goldblatt, *ed.*, *Biological Relationships between Africa and South America.* New Haven, Yale Univ. Press.

Colinvaux, P. A., and B. D. Barnett, 1979, Lindeman and the ecological efficiency of wolves. *Am. Nat.*, v. 114, p. 707–718.

Colinvaux, P. A., and E. K. Schofield. 1976, Historical ecology in the Galapagos Islands: I. A Holocene pollen record from el Junco Lake, Isla San Cristobal. *J. Ecol.*, v. 64, p. 989–1012.

Colinvaux, P. A., and M. Steinitz-Kannan, 1980, Species richness and area in Galapagos and Andean lakes: Equilibrium phytoplankton communities and a paradox of the zooplankton communities, p. 697–712 *in* Kerfoot, W. C., *ed.*, *Evolution and Ecology of Zooplankton.* Hanover, N.H., Univ. Press of New England.

Connell, J. H., 1961, The influence of interspecific competition and other factors on the distribution of the barnacle *Chthamalus stellatus. Ecol.*, v. 42, p. 710–723.

Connell, J. H., 1978, Diversity in tropical rain forests and coral reefs, *Science*, v. 199, p. 1302–1310.

Connell, J. H., and E. Orias, 1964, The ecological regulation of species diversity. *Am. Nat.*, v. 98, p. 387–414.

Connell, J. H., and R. O. Slatyer, 1977, Mechanisms of succession in natural communities and their role in community stability and organization. *Am. Nat.*, v. 111, p. 1119–1144.

Connor, E. F., and D. Simberlof, 1978, Species number and compositional similarity of the Galapagos flora and avifauna. *Ecol. Monog.*, v. 48, p. 219–248.

Cooke, G. D., 1967, The pattern of autotrophic succession in laboratory microecosystems. *BioSci.*, v. 17, p. 717–721.

Coope, G. R., 1959, A Late Pleistocene insect fauna from Chelford, Cheshire. *Proc. R. Soc. Lond. Ser. B.*, v. 151, p. 70–86.

Coope, G. R., 1977, Fossil coleopteran assemblages as sensitive indicators of climatic changes during the Devensian (last) cold stage. *Philos. Trans. of R. Soc. Lond. Ser. B.* v. 280, p. 313–340.

Coope, G. R., 1979, Late Cenozoic fossil Coleoptera: Evolution, biogeography, and ecology: *An. Rev. Ecol. and Systematics*, v. 10, p. 247–267.

Cowles, H. C., 1899, The ecological relations of the vegetation on a structural basis. *Ecol.*, v. 32, p. 172–229.

Cox, C. P., and P. D. Moore, 1980, *Biogeography: An Ecological and Evolutionary Approach*, Oxford, Blackwell.

Crocker, R. L., and J. Major, 1955, Soil development in relation to vegetation and surface age at Glacier Bay, Alaska. *Ecol.*, v. 43, p. 427–428.

Crombie, A. C., 1946, Further experiments on insect competition. *Proc. R. Soc. of Lond. Ser. B.* v. 133, p. 76–109.

Crowley, T. J., 1991, Ice age carbon. *Nature*, v. 352, p. 575–576.

Cullen, J. J., 1991, Hypotheses to explain high-nutrient conditions in the open sea. *Limnol. and Oceanogr.*, v. 36, p. 1578–1599.

Daly, M., 1978, The cost of mating. *Am. Nat.*, v. 112, p. 771–774.

Darlington, P. J., 1959, *Zoogeography: The Geographical Distribution of Animals.* New York, John Wiley.

Davies, N. B., 1978, Ecological questions about territorial behavior, p. 317–350 *in* Krebs, J. R., and N. B. Davies, *eds.*, *Behavioural Ecology: An Evolutionary Approach.* Sunderland, Mass., Sinauer Associates.

Davis, M. B., 1965, Phytogeography and Palynology of northeastern United States *in* Wright, H. E., and D. G. Frey, *eds.*, *Quaternary of the United States.* Princeton, Princeton Univ. Press.

Davis, M. B., 1969, Climatic changes in southern Connecticut recorded by pollen deposition at Rogers Lake. *Ecol.*, v. 50, p. 409–422.

Davis, M. B., 1982. Holocene vegetational history of the Eastern United States, p. 166–181 in Wright, H. E., Jr., *ed. Late Quaternary Environments of the United States.* Minneapolis, Univ. of Minn. Press.

Davis, M. B., *ed.*, 1986, Vegetation-climate equilibrium. *Vegetatio*, v. 67, p. 1–141.

Davis, R. B., 1967, Pollen studies on nearsurface sediments in Maine lakes, p. 143–173 *in* Cushing, E. J., and H. E. Wright, *eds.*, *Quaternary Paleoecology.* New Haven, Yale Univ. Press.

Davis, R. B., 1987. Paleolimnological diatom studies of acidification of lakes by acid rain:

an application of Quaternary science. *Quat. Sci. Rev.*, v. 6, p. 147–163.

DeBach, P., *ed.*, 1964, *Biological Control of Insect Pests and Weeds.* New York, Reinhold.

deCandolle, A. P. A., 1874, Constitution dans le règne vègètal de groupes physiologiques applicables à la géographie ancienne et moderne. *Archives des Sciences Physiques et Naturelles*, Geneva, Switzerland.

deCandolle, A. P. A., and C. deCandolle, 1824–1873, *Prodomus systematis naturalis regni vegetabilis*, 17 vols. Paris.

Deevey, E. S., 1947, Life tables for natural populations of animals. *Quart. Rev. of Biol.*, v. 22, p. 283–314.

Deevey, E. S., 1949, Biogeography of the Pleistocene. *Bull. Geol. Soc. Am.*, v. 60(9), p. 1315–1416.

Deevey, E. S., 1955, The obliteration of the hypolimnion. *Mem. 1st Ital. Idrobiol., Suppl.*, v. 8, p. 9–38.

Deevey, E. S., 1969, Coaxing history to conduct experiments. *BioSci.*, v. 19, p. 40–43.

Deevey, E. S., 1970, In defense of mud. *Bull. Ecol. Soc. Am.*, v. 51, p. 5–8.

Delwiche, C. C., 1970. The nitrogen cycle. *Scientific Am.* v. 223, p. 136–146.

Diamond, J. M., 1969, Avifaunal equilibria and species turnover rates on the Channel Islands of California. *Proc. Natl. Acad. Sci.*, v. 64, p. 57–63.

Diamond, J. M., 1973, Distributional ecology of New Guinea birds. *Science*, v. 179, p. 759–769.

Dillehay, T. D., and David J. Meltzer. 1991, *eds. The first Americans: Search and Research.* Boca Raton, CRC Press.

Dodd, A. P., 1940, *The Biological Campaign against Prickly Pear.* Brisbane, Australia.

Dowding, P., F. S. Chapin, III, F. E. Wielgolaski, and P. Kilfeather, 1981, Nutrients in tundra ecosystems, p. 647–683 *in* Bliss, L. C., O. W. Heal, and J. J. Moore, eds., *Tundra Ecosystems: a Comparative Analysis.* Cambridge, Cambridge Univ. Press.

Downhower, J. F., 1971, Darwin's finches and the evolution of sexual dimorphism in body size. *Nature*, v. 263, p. 558–563.

Downhower, J. F., and K. B. Armitage, 1971, the yellow-bellied marmot and the evolution of polygamy. *Am. Nat.*, v. 105, p. 355–370.

Drury, W. H., and I. C. T. Nisbet, 1973, *Succession.* J. Arnold Arboretum, v. 54, p. 331–368.

DuRietz, G. E., 1930, Classification and nomenclature of vegetation. *Svensk Botanisk Tidskrift*, v. 24, p. 489–503.

Duvigneaud, P., and S. Denaeyer-de Smet, 1970, Biological cycling of minerals in temperate deciduous forests, p. 199–225 *in* Reichle, D. E., *ed.*, *Analysis of Temperate Forest Ecosystems.* New York, Springer-Verlag.

Edmondson, W. T., 1991, Sedimentary record of changes in the condition of Lake Washington. Seattle, Univ. Washington Press.

Edmunds, G. F., Jr., and D. N. Alstad, 1978, Coevolution in insect herbivores and conifers. *Science*, v. 199, p. 941–945.

Elias, T. S., and H. Gelband, 1975, Nectar: Its production and functions in trumpet creeper. *Science*, v. 189, p. 289–291.

Elner, R. W., and R. N. Hughes, 1978, Energy maximation in the diet of the shore crab, *Carcinus naenas* (L.). *J. of Animal Ecol.*, v. 47, p. 103–116.

Elton, C. S., 1927, *Animal Ecology.* New York, Macmillan.

Elton, C. S., 1942, *Voles, Mice, and Lemmings.* Oxford, Oxford Univ. Press.

Elton, C. S., 1966, *The Pattern of Animal Communities.* London, Methuen.

Emiliani, C., 1966, Isotopic paleotemperatures. *Science*, v. 154, p. 851–856.

Emiliani, C., 1992, Pleistocene paleotemperatures. *Science*, v. 257, p. 1462.

Eriksson, E., 1960, The yearly circulation of chloride and sulfur in nature: meteorological, geochemical and pedological implications, Part 2. *Tellus*, v. 12, p. 63–109.

Errington, P. L., 1963, *Muskrat Populations.* Ames, Iowa State Univ. Press.

Erwin, T. L., 1982, Tropical forests: their richness in Coleoptera and other arthropod species. *Coleoptera Bulletin* v. 36, p. 74–75.

Faegri, K., and J. Iversen, 1989, *Textbook of Pollen Analysis*, 4th ed. New York, John Wiley.

Farmer, J. G., and M. S. Baxter, 1974, Atmospheric carbon dioxide levels as indicated by the stable isotope record in wood. *Nature*, v. 247, p. 273–275.

Ferguson, E. E., and W. F. Libby, 171, Mechanism for the fixation of nitrogen by lightning. *Nature*, v. 229, p. 37.

Finerty, J. P., 1980, *The Population Ecology of Cycles and Small Mammals.* New Haven, Yale Univ. Press.

Fischer, A. G., 1960, Latitudinal variations in organic diversity. *Evol.*, v. 14, p. 64–81.

Fisher, R. A., 1930, *The genetical theory of Natural Selection.* Oxford, Clarendon Press.

Fitch, H. S., 1965, An ecological study of the garter snake, *Thamnophis sirtalis. Univ. of Kansas Mus. of Nat. Hist.*, v. 15, p. 493–564.

Flenley, J. R., 1979. *A geological history of tropical rainforest.* London, Butterworths.

Flint, R. F., 1971, *Glacial and Quaternary Geology.* New York, John Wiley.

Flohn, H., 1969. Climate and weather. New York. McGraw-Hill.

Forbes, S A., 1887, The lake as a microcosm. *Bull. of Peoria (Illinois) Sci. Assoc.*, Reprinted in *Ill. Nat. Hist. Surv.*, v. 15, p. 537–550, 1925.

Fortey, R., 1991, *Fossils: The key to the past.* London British Museum (Natural History).

Fretwell, S. D., 1972, *Populations in a seasonal environment.* Princeton, Princeton Univ. Press.

Fridriksson, S., 1975, *Surtsey.* New York, John Wiley.

Fritts, H. C., 1976, *Tree Rings and Climate.* New York, Academic Press.

Frost, B. W., 1991, The role of grazing in nutrient-rich areas of the open sea. *Limnol. and Oceanogr.*, v. 36, p. 1616–1630.

Gaarder, T., and H. H. Gran, 1927, Investigations of the production of plankton in the Oslo Fjord. *Rapp. et Proc. Verb., Cons. Int. Explor. Mer.*, v. 42, p. 1–48.

Gaastra, P., 1958, Light energy conversion in field crops in comparison with photosynthetic efficiency under laboratory conditions. *Medel., Landbouwhogeschool Wageningin,* v. 58, p. 1–12.

Gantt, E. 1975. Phycobilisomes: Light harvesting pigment complexes. *BioSci.*, v. 25, p. 781–788.

Garrels, R. M., F. T. Mackenzie, and C. Hunt, 1975, *Chemical Cycles and the Global Environment.* Los Altos, Ca., Kaufmann.

Garrels, R. M., and F. T. Mackenzie, 1971, *Evolution of Sedimentary Rocks.* New York, Norton.

Gates, D. M., 1968*a*, Energy exchange between organisms and environment. *Australian J. Sci.*, v. 31, p. 67–74.

Gates, D. M., 1968*b*, Energy exchange between organisms and environment, p. 122 *in* Lowry, W. P., *ed.*, *Biometeor. Proc. of 28th An. Biol. Colloquium 1967.* Corvallis, Oregon State Univ. Press.

Gates, D. M., 1965, Energy, plants and ecology. *Ecol.*, v. 46(1,2), p. 1–13.

Gause, G. F., 1934, *The Struggle for Existence.* Baltimore, Williams & Wilkins.

Gause, G. F., 1936, The principles of biocenology. *Quart. Rev. of Biol.*, v. 11, p. 320–396.

Gentry, A. H., 1988, Tree species richness of upper Amazonian forests. *Proc. Natl. Acad. Sci.*, v. 85, p. 156–159.

Georgi, J., 1934, *Mid-Ice* (translated by Lyon, F. H.) London, Kegan Paul.

Gilbert, J. J., 1980, Feeding in the rotifer *Asplanchna:* Behavior, cannibalism, selectivity, prey defenses, and impact on rotifer communities, p. 158–172 *in* Kerfoot, W. C., *ed.*, *Evolution and Ecology of Zooplankton Communities.* Hanover, N. H., Univ. Press of New England.

Gilbert, L. E., 1971, Butterfly-plant coevolution: Has *Passiflora adenopoda* won the selectional race with Heliconiine butterflies? *Science*, v. 172, p. 585–586.

Gilbert, L. E., and M. C. Singer, 1975, Butterfly ecology. *An. Rev. Ecol. and Systematics*, v. 6, p. 365–397.

Gilpin, M. E., 1975, *Group Selection in Predator–Prey Communities.* Princeton, Princeton Univ. Press.

Gleason, H. A., 1926, The individualistic concept of the plant association. *Bull. Torrey Bot. Club*, v. 53, p. 7–26.

Glesener, R. R., and D. Tilman, 1978, Sexuality and the components of environmental uncertainty: Clues from geographic parthenogenesis in terrestrial animals. *Am. Nat.*, v. 112, p. 659–673.

Godwin, H., 1956, *The History of the British Flora, a Factual Basis for Phytogeography.* Cambridge, Cambridge Univ. Press.

Goodman, D., 1974, Natural selection and a cost-ceiling on reproductive effort. *Am. Nat.*, v. 113, p. 735–748.

Grant, P. R., 1975, The classical case of character displacement. *Evol. Biol.*, v. 8, p. 237–337.

Grant, P. R., 1986, *Ecology and Evolution of*

Darwin's Finches. Princeton, Princeton University Press.

Gray, J., and P. Thompson, 1977, Climatic information from $^{18}O/^{16}O$ analysis of cellulose, lignin and whole wood from tree rings. *Nature*, v. 270, p. 708–709.

Grigg, R. W., and J. E. Maragos, 1974, Recolonization of hermatypic corals on submerged lava flows in Hawaii. *Ecol.*, v. 55, p. 387–395.

Grime, J. P., 1979, *Plant Strategies and Vegetation Processes.* New York, John Wiley.

Grinnel, J., 1904, The origin and distribution of the chestnut-backed chicadee. *Auk*, v. 21, p. 364–382.

Haffer, J., 1969, Speciation in Amazonian forest birds. *Science*, v. 165, p. 131–136.

Hairston, N. G., F. E. Smith, and L. B. Slobodkin, 1960, Community structure, population control, and competition. *Am. Nat.*, v. 94, p. 421–425.

Hairston, N. G., D. W. Tinkle, and H. M. Wilbur, 1970, Natural selection and the parameters of population growth. *J. Wildl. Mgmt.*, v. 34, p. 681–690.

Hall, C. A. S., and R. M. Moll, 1985. Methods of assessing aquatic primary productivity, p. 19–53, in Lieth, H. and R. H. Whittaker, eds. *Primary Productivity of the Biosphere.* New York. Springer–Verlag.

Hall, C. A. S., A. Eckdahl, and D. E. Wartenberg, 1975, A fifteen-year record of biotic metabolism in the northern hemisphere. *Nature*, v. 225, p. 136–138.

Halvorson, H. O., and A. Monroy, 1985, *The Origin and Evolution of Sex.* New York, A. R. Liss.

Hamilton, W. D., 1964, The genetical evolution of social behaviour. *J. Theoret. Biol.* v. 7, p. 1–52.

Hardin, G., 1960, The competitive exclusion principle. *Science*, v. 131, p. 1292–1297.

Hardy, A. C., 1956, *The Open Sea and the World of Plankton.* London, Collins.

Hrdy, S. B., 1981, *The Woman That Never Evolved.* Cambridge, Harvard Univ. Press.

Harrington, F. H., and L. D. Mech, 1983, Wolf pack spacing: Howling as a territory-independent spacing mechanism in a territorial population. *Behav. Ecol. Sociobiol.*, v. 12, p. 161–168.

Harrington, F. H., and L. D. Mech, 1979, Wolf howling and its role in territory maintenance. *Behaviour*, v. 68, p. 207–249.

Harris, W. F., R. A. Goldstein, and G. S. Henderson, 1973, Analysis of forest biomass pools, annual primary production and turnover of biomass for a mixed deciduous forest watershed. p. 41–64, *in* Young, H. E., *ed.*, *IUFRO Biomass Studies: International Union of Forest Research Organization Papers.* Orono, Maine.

Harvey, H. W., 1950, On the production of living matter in the sea off Plymouth. *J. Mar. Biol. Assoc. of U.K.*, v. 29, p. 97–137.

Hassell, M. P., 1985, Insect natural enemies as regulating factors. *J. of Animal Ecol.* v. 54, p. 323–334.

Hastenrath, S., and J. Kutzbach, 1985, Late Pleistocene climate and water budget of the South American Altiplano. *Quat. Res.*, v. 24, p. 249–256.

Hatch, M. D., and C. R. Slack, 1970, Photosynthetic CO_2-fixation pathways. *Ann. Rev. Plant Physiol.* v. 21, p. 141–162.

Hays, J. D., J. Imbrie, and N. J. Shackleton, 1976, Variations in the earth's orbit: Pacemaker of the ice ages. *Science*, v. 194, p. 1121–1132.

Heatwole, H., and R. Levins, 1973, Biogeography of the Puerto Rican bank: Species-turnover on a small cay, Cayo Ahogado. *Ecol.*, v. 54, p. 1042–1055.

Heinselman, M. L., 1963, Forest sites, bog processes, and peatland types in the glacial Lake Agassiz region, Minnesota. *Ecol. Monog.*, v. 33, p. 327–374.

Heinselman, M. L., 1975, Boreal peatlands in relation to environment, p. 93–103 *in* Hasler, A. D., *ed.*, *Coupling of Land and Water Systems, Ecological Studies 10.* New York, Springer-Verlag.

Hensley, M. M., and J. R. Cope, 1951, Further data on removal and repopulation of the breeding birds in a spruce-fir forest community. *Auk*, v. 68, p. 483–493.

Hillis, L. (as Hillis-Colinvaux), 1966, Distribution of marine algae in the Bay of Fundy, New Brunswick, Canada. *Proc. of 5th Int. Seaweed Symp.*, Halifax, N.S., p. 92–98.

Hillis, L. (as Hillis–Colinvaux), 1980, The genus

Halimeda: Primary producer in coral reefs. *Adv. in Mar. Biol.*, v. 17, p. 1–327.

Hillis, L., 1986, Historical perspectives of algae and reefs: have reefs been misnamed? *Oceanus*, v. 29, p. 433–448.

Holland, H. D., 1984, *The Chemical Evolution of the Atmosphere and Oceans*. Princeton, Princeton Univ. Press.

Holland, H. D., 1990, Origins of breathable air. *Nature*, v. 347, p. 17.

Holloway, J. K., 1964, Projects in biological control of weeds, p. 650–670 *in* DeBach, P., *ed.*, *Biological Control of Insect Pests and Weeds*. New York, Reinhold.

Hooghiemstra, H. 1989, Quaternary and upper-pliocene glaciations and forest development in the tropical Andes: evidence from a long high-resolution pollen record from the sedimentary basin of Bogota, Colombia. *Paleogeography, Palaeoclimatology, Palaeoecology*, v. 75, p. 11–26.

Horn, H. S., 1971, *The Adaptive Geometry of Trees*. Princeton, Princeton Univ. Press.

Horn, H. S., 1974, The ecology of secondary succession. *An. Rev. Ecol. and Systematics*, v. 5, p. 25–37.

Horn, H. S., 1975, Markovian properties of forest succession, p. 196–211 *in* Cody, M. L., and J. M. Diamond, *eds.*, *Ecology and Evolution of Communities*. Cambridge, Harvard Univ. Press.

Horn, H. S., 1976, Succession, p. 187–204 *in* May, R. M., *ed.*, *Theoretical Ecology, Principles and Applications*. Philadelphia, Saunders.

Horn, H. S., 1978, Optimal tactics of reproduction and life history, Chapt. 14 *in* J. R. Krebs and N. B. Davies, *eds.*, *Behavioral Ecology*, Sunderland, Mass., Sinauer Associates.

Horner, J. R., 1982, Evidence of colonial nesting and 'site fidelity' among ornithischian dinosaurs. *Nature*, v. 297, p. 675–676.

Hornocker, M. G., 1969, Winter territoriality in mountain lions. *J. Wildlife Management*, v. 33, p. 457–464.

Huckel, W., 1951, *Structural Chemistry of Inorganic Compounds* (transl. by Long, L. H.) v. 11. New York, Elsevier.

Huntley, B., and H. J. B. Birks, 1983, *An atlas of past and present pollen maps for Europe 0–13,000 years ago*. Cambridge University Press, Cambridge.

Hurlbert, S. H., 1971, The nonconcept of species diversity: A critique and alternative parameters. *Ecol.*, v. 52, p. 577–586.

Hutchinson, G. E., 1949, Circular causal systems in ecology. *An. of N.Y. Acad. of Sci.*, v. 51, p. 221–246.

Hutchinson, G. E., 1951, Copepodology for the ornithologist. *Ecol.*, v. 32, p. 571–577.

Hutchinson, G. E., 1957, Concluding remarks. *Cold Spring Harbor Symp. on Quantitative Biol.* 22, *Population Studies: Animal Ecology and Demography*. Cold Spring Harbor, Biol. Lab., p. 415–427.

Hutchinson, G. E., 1959, Homage to Santa Rosalia or why are there so many kinds of animals? *Am. Nat.*, v. 93, p. 145–159.

Hutchinson, G. E., 1961, The paradox of the plankton. *Am. Nat.*, v. 95, p. 137–145.

Hutchinson, G. E., 1967, *A Treatise on Limnology: Vol. 2, Introduction to Lake Biology and the Limnoplankton*, New York, Wiley-Interscience.

Hutchinson, G. E., 1970, Ianula: An account of the history and development of the Lago di Monterosi, Latium, Italy. *Trans. Am. Philo. Soc.*, v. 60, p. 1–170.

Hutchinson, G. E., 1978, *An Introduction to Popular Ecology*. New Haven, Yale Univ. Press.

Huxley, J. S., 1932, *Problems of Relative Growth*. New York, Dial.

Imbrie, J., and N. G. Kipp, 1971, A new macropaleontological method for quantitative paleoclimatology: Application to a late Pleistocene Caribbean core, p. 71–181 *in* Turekian, K. K., *ed.*, *Late Cenozoic Glacial Ages*. New Haven, Yale Univ. Press.

Irion, G., 1984, Sedimentation and sediments of Amazonian rivers and evolution of the Amazonian landscape since Pliocene times. p. 201–204, *in* J. Siolo, *ed.*, *The Amazon. Limnology and Landscape of a Mighty Tropical River and its Basin*. The Netherlands, Dordrecht, Dr. W. Junk.

Jablonski, D., 1986, Evolutionary consequences of mass extinctions, *in* D. M. Raup and D. Jablonski, *eds.*, *Pattern and Process in the History of Life*. New York, Springer-Verlag.

Jablonski, D., 1989. The biology of mass extinction: a paleontological view. *Phil. Trans. Roy. Soc. London* B v. 325, p. 357–368.

Jablonski, D., 1991. Extinctions: A paleontological perspective. *Science*, v. 253, p. 754–757.

Janzen, D. H., 1968, Host plants as islands in evolutionary and contemporary time. *Am. Nat.*, v. 102, p. 592–594.

Janzen, D. H., 1970, Herbivores and the number of tree species in tropical forests. *Am. Nat.*, v. 104, p. 501–528.

Janzen, D. H., 1971, Escape of juvenile *Dioclea megacarp* (Leguminosae) vines from predators in a deciduous tropical forest. *Am. Nat.*, v. 105, p. 97–112.

Jerlov, N. G., 1951, Optical studies of ocean waters. *Rep. Swedish Deep Sea Expedition*, v. 3, p. 1–59.

Johnsen, S. J., H. B. Clausen, W. Dansgaard, K. Fuhrer, N. Gundestrup, C. U. Hammer, P. Iversen, J. Jouzel, B. Stauffer, and J. P. Steffensen, 1992, irregular glacial interstadials recorded in a new Greenland Ice Core. *Nature*, v. 359, p. 311–313.

Johnson, M. P., and P. H. Raven, 1973, Species number and endemism: The Galapagos archipelago revisited. *Science*, v. 179, p. 893–895.

Juday, C., 1940, The annual energy budget of an inland lake. *Ecol.*, v. 21, p. 438–450.

Kauppi, P. E., K. Mielikainen, and K. Kuusela, 1992, Biomass and carbon budget of European forests, 1971 to 1990. *Science*, v. 256, p. 70–79.

Kellog, W. W., R. D. Cadle, E. R. Allen, A. L. Lazrus, and E. A. Martell, 1972, The sulfur cycle. *Science*, v. 175, p. 587–596.

Kemeny, J. G., and J. L. Snell, 1960, *Finite Markov Chains*. New York, Van Nostrand.

Kemp, T. S., 1982, Mammal-like reptiles and the origin of mammals. Academic Press, London.

Kendall, R. L., 1969, An ecological history of the Lake Victoria basin. *Ecol. Monog.* v. 39, p. 121–176.

Kenward, R. E., 1978, Hawks and doves: Attack success and selection in goshawk flights at woodpigeons. *J. of Animal Ecol.*, v. 47, p. 449–460.

Kerfoot, W. C., D. L. Kellogg, and J. R. Strickler, 1980, Visual observations of live zooplankters: Evasion, escape, and chemical defenses, p. 10–27 *in* Kerfoot, W. C., *ed.*, *Evolution and Ecology of Zooplankton Communities*. Hanover, N.H., Univ. Press of New England.

Kira, T., and T. Shidei, 1967, Primary production and turnover of organic matter in different forest ecosystems of the western Pacific. *Japanese J. Ecol.*, v. 17, p. 70–87.

Kleiber, M., 1961, *The Fire of Life. An Introduction to Animal Energetics*. New York, John Wiley.

Klopfer, P. H., 1969, *Habitats and Territories*. New York, Basic Books.

Knoll, A. H., J. M. Hayes, A. J. Kaufman, K. Swett, and I. B. Lambert, 1986, Secular variation in carbon isotope ratios from Upper Proterozoic successions of Svalbard and East Greenland. *Nature*, v. 321, p. 832–838.

Koblentz-Mishke, O. J., V. V. Volkovinsky, and J. G. Kabanova, 1970, Plankton primary production of the world ocean, *in* Wooster, W. S., *ed.*, *Scientific Exploration of the South Pacific*. Wash., D.C., Nat. Acad. Sci. p. 183–193.

Köppen, W., 1884, Die Warmezonen der Erde, nach der Dauer der Heissen, Gemassigten und Kalten Zeit, und nach der Wirkung der Warme auf die Organische Welt betrachter. *Meteorologische Zeitschrift*, v. 1, p. 215–226.

Krebs, J. R., 1971, Territory and breeding density in the great tit, *Parus major* L. *Ecol.*, v. 52, p. 2–22.

Krebs, J. R., 1978, Optimal foraging: Decision rules for predator, p. 23–63 *in* Krebs, J. R., and N. B. Davies, *eds.*, *Behavioural Ecology*. Sunderland, Mass., Sinauer.

Krebs, J. R., 1979, Coevolution of bees and flowers. *Nature*, v. 278, p. 689.

Krebs, J. R., and N. B. Davies, 1981, *An introduction to behavioural ecology*. Sunderland, Mass., Sinauer.

Kruuk, H., 1972, *The Spotted Hyena*. Chicago, Univ. of Chicago Press.

Kullenberg, B., 1955, Deep sea coring. *Rept. Swedish Deep Sea Expedition*, v. 4(2), p. 35–96.

Kutzbach, J. E., and P. J. Guetter, 1986, The influence of changing orbital parameters and surface boundary conditions on climate simulations for the past 18,000 yrs. *J. Atmospheric Sciences*, v. 43, 1726–1759.

Lack, D. L., 1945, Ecology of closely related

species with special reference to cormorant (*Phalacrocorax carbo*) and shag (*P. aristolelis*). *J. of Animal Ecol.*, v. 14, p. 12–16.

Lack, D. L., 1947, *Darwin's Finches*. Cambridge, Cambridge Univ. Press.

Lack, D. L., 1954, *The Natural Regulation of Animal Numbers*. New York, Oxford Univ. Press.

Lack, D. L., 1968, *Ecological Adaptations for Breeding in Birds*. London, Methuen.

LaMarche, V. C., 1974, Paleoclimatic influences from long tree-ring records. *Science*, v. 183, p. 1043–1048.

LaMarche, V. C., D. A. Graybille, H. C. Fritts, and M. R. Rose, 1984, Increasing atmospheric carbon dioxide: Tree-ring evidence for growth enhancement in natural vegetation. *Science*, v. 225, p. 1019–1021.

Lamb, H. H., 1972, *Climate: Present, Past and Future, Vol. 1*. London, Methuen.

Langmuir, I., 1938, Surface motion of water induced by wind. *Science*, v. 87, p. 119–123.

Lawton, J. H., and R. M. May, 1983, The birds of Selborne. *Nature*, v. 306, p. 732–733.

Lawton, J. H., 1989. Food webs. p. 43–78 in J. M. Cherrett, *ed. Ecological concepts*. Oxford, Blackwell.

Lawton, J. H., 1992. Feeble links in food webs. *Nature*, v. 355, p. 19–20.

Leuenberger, M., U. Siegenthaler, and C. C. Langway, 1992, Carbon isotope composition of atmospheric CO2 during the last ice age from an Antarctic ice core. *Nature*, v. 357, p. 488–490.

Lewis, W. M., 1979, *Zooplankton Community Analysis*. New York, Springer-Verlag.

Libby, L. M., L. J. Pandolfi, P. H. Payton, J. Marshall, B. Becker, and V. Giertz-Sienbenlist, 1976, Isotopic tree thermometers. *Nature*, v. 261, p. 284–288.

Lieth, H., 1975, Primary productivity of the major vegetation units of the world, p. 203–216 *in* Lieth, H., and R. H. Whittaker, *eds.*, *Primary Productivity of the Biosphere*. New York, Springer–Verlag.

Likens, G. E., F. H. Bormann, N. M. Johnson, D. W. Fisher, and R. S. Pierce, 1970, Effects of forest cutting and herbicide treatment on nutrient budgets in the Hubbard Brook watershed-ecosystem. *Ecol. Monog.*, v. 40, p. 23–47.

Likens, G. E., F. H. Bormann, R. S. Pierce, J. S. Eaton, and N. M. Johnson, 1977, *The Biogeochemistry of a Forested Ecosystem*. New York, Springer–Verlag.

Lindeman, R. L., 1941, Seasonal food-cycle dynamics in a senescent lake. *Am. Midland Nat.*, v. 26, p. 636–673.

Lindeman, R. L., 1942, The trophic dynamic aspect of ecology. *Ecol.*, v. 23, p. 399–418.

Liu, K-b, M. L. Fearn, and X. Li, 1991, A 5000-year history of hurricane climate and vegetation change along the Gulf Coast of Alabama. *Abstracts, Assoc. of American Geographers*, p. 117.

Liu, K-b, and P. Colinvaux, 1985, Forest changes in the Amazon Basin during the last glacial maximum. *Nature*, v. 318, p. 556–557.

Livingstone, D. A., 1955, A lightweight piston sampler for lake sediments. *Ecol.*, v. 36, p. 137–139.

Livingstone, D. A., 1963, The sodium cycle and the age of the ocean. *Geoch. et Cosmoch. Acta*, v. 27, p. 1055–1069.

Livingstone, D. A., and W. D. Clayton, 1980, An altitudinal cline in tropical African grass floras and its paleoecological significance. *Quat. Res.*, v. 13, p. 392–402.

Livingstone, D. A., 1993, *Evolution of the African Climate in* P. Goldblatt, *ed.*, *Biological Relationships between Africa and South America*. New Haven, Yale Univ. Press.

Lloyd, J. E., 1975, Aggressive mimicry in *Photuris* fireflies: signal repertoires by femme fatales. *Science*, v. 187, p. 452–453.

Lloyd, M., and R. J. Ghelardi, 1964, A table for calculating the equitability component of species diversity. *J. of Animal Ecol.*, v. 33, p. 217–225.

Lomnicki, A., 1988, *Population ecology of individuals*. Princeton, Princeton University Press.

Long, S. P., L. D. Incoll, H. W. Woolhouse, 1975, C_4 photosynthesis in plants from cool temperate regions, with particular reference to *Spartina townsendii*. *Nature*, v. 257, p. 622–624.

Longwell, C. R., R. F. Flint, and J. E. Sanders, 1969, Physical geology. New York, John Wiley.

Loomis, R. S., W. A. Williams, and A. E. Hall, 1971, Agricultural productivity. *An. Rev. Plant Physiol.*, v. 22, p. 431–468.

Lorius, C., J. Jouzel, D. Raynaud, J. Hansen, and H. Le Treut. 1990, The ice-core record: Climate sensitivity and future greenhouse warning. *Nature*, v. 347, p. 139–145.

Lotka, A. J., 1925, *Elements of Physical Biology*. Baltimore, Williams & Wilkins. Reprinted as *Elements of Mathematical Biology*. New York, Dover Press, 1956.

Lund, J. W. G., 1965, The ecology of the freshwater phytoplankton. *Biol. Rev.*, v. 40, p. 231–293.

MacArthur, R. H., 1958, Population ecology of some warblers of northeastern coniferous forests. *Ecol.*, v. 39, p. 599–619.

MacArthur, R. H., 1965, Patterns of species diversity. *Biol. Rev.*, v. 40, p. 510–533.

MacArthur, R. H., 1972, *Geographical Ecology*. New York, Harper & Row.

MacArthur, R. H., and E. R. Pianka, 1966, On optimal use of a patchy environment. *Am. Natur.*, v. 100, p. 603–609.

MacArthur, R. H., and E. O. Wilson, 1967, *The Theory of Island Biogeography*. Princeton, Princeton Univ. Press.

MacArthur, R. H., and E. O. Wilson, 1963, An equilibrium theory of insular zoogeography. *Evol.* v. 17, p. 373–387.

MacArthur, R. H., and J. H. Connell. 1966, *The Biology of Populations*. New York, John Wiley.

MacArthur, R. H., and R. Levins, 1967, The limiting similarity, convergence and divergence of coexisting species. *Am. Nat.*, v. 101, p. 377–385.

MacFayden, A., 1957, *Animal Ecology: Aims and Methods*. London, Pitman.

Machta, L., 1973, Prediction of CO_2 in the atmosphere, p. 21–31 *in* Woodwell, G. M., and E. V. Pecan, *eds.*, *Carbon and the Biosphere*. Washington, D.C.

MacIntyre, F., 1970, Why the sea is salty. *Scientific Am.*, v. 223, p. 104–115.

Maguire, B., Jr., 1963, The passive dispersal of small aquatic organisms and their colonization of isolated bodies of water. *Ecol. Monog.*, v. 33, p. 161–185.

Maguire, B., Jr., 1971, Phytotelmata: Biota and community structure determination in plant-held waters. *An. Rev. Ecol.* and *Systematics*, v. 2, p. 439–464.

Maher, L. J., 1972, Absolute pollen diagram of Redrock Lake, Boulder County, Colorado. *Quat. Res.*, v. 2, p. 531–553.

Maley, J., 1991, The African Forest Vegetation and Paleoenvironments during Late Quaternary. *Climatic Change* v. 19, p. 19–79.

Margalef, D. R., 1963, On certain unifying principles in ecology. *Am. Nat.*, v. 97, p. 374.

Margulis, L., and D. Sagan, 1984, *Origins of Sex*. New Haven, Yale Univ. Press.

Marks, P. L., and F. H. Bormann, 1972, Revegetation following forest cutting: Mechanisms for return to steady-state nutrient cycling. *Science*, v. 176, p. 914–915.

Marler, P., and P. Mundinger, 1971, Vocal learning in birds, in Moltz, H. ed. *Ontogeny of Vertebrate behavior*, New York, Academic Press.

Marshall, L. G., 1974, Why kangaroos hop. *Nature*, v. 248, p. 174–175.

Martin, P. S., 1958, Pleistocene ecology and biogeography of North America, p. 375–420 *in* Hubbs, C. L., *ed.*, *Zoogeography: Publ. 51*, Amer. Assoc. Adv. Sci., Washington, D.C.

Martin, P. S., 1973, The Discovery of America. *Science*, v. 179, p. 969–974.

Martin, P. S., and R. G. Klein, 1984, *Quaternary Extinctions*. Tucson, Univ. of Arizona Press.

Martin, J. H., R. M. Gordon, and S. E. Fittzwater, 1991, The case for iron. *Limnol. and Oceanogr.*, v. 36, p. 1793–1802.

Maxwell, D. C., 1974, *Marine primary productivity in the Galapagos Islands:* Ph.D. dissertation, The Ohio State Univ.

May, R. M., 1976, Models for two interacting populations, p. 49–70 *in* May, R. M., *ed.*, *Theoretical Ecology, Principles and Applications*. Philadephia, Saunders.

May, R. M., 1986*a*, The search for patterns in the balance of nature: Advances and retreats. *Ecol.*, v. 67, p. 1115–1126.

May, R. M., 1986*b*, When two and two do not make four: nonlinear phenomena in ecology. *Proceedings of the Royal Society of London, B.*, v. 228, p. 241–266.

May, R. M., and G. F. Oster, 1976, Bifurcations and dynamic complexity in simple ecological models. *Amer. Nat.*, v. 110, p. 573–599.

Maynard Smith, J., 1978, *The evolution of sex*. Cambridge, Cambridge Univ. Press.

Maynard Smith, J., 1987, Sexual selection—A classification of models. p. 9–20 *in* Bradbury,

J. W., and M. B. Andersson, *eds.*, *Sexual selection: testing the alternatives.* New York, John Wiley.

McLaren, I. A., 1963, Effects of temperature on growth of zooplankton and the adaptive value of vertical migration. *Jour. of Fisheries Res. Board of Canada*, v. 20, p. 685, 727.

McNab, B. K., 1963, Bioenergetics and the determination of home range size. *Am. Nat.*, v. 97, p. 133–140.

McQueen, D. J., M. R. S. Johannes, J. R. Post, T. J. Stewart, and D. R. S. Lean, 1989, Bottom-up and top-down impacts on freshwater pelagic community structure. *Ecol. Mono.*, v. 59, p. 289–309.

Mech, L. D., 1966, *The Wolves of Isle Royale: Fauna of Nat. Parks of U.S. Fauna Series 7*, Washington, D.C.

Mertz, D. B., 1971, The mathematical demography of the California condor population. *Am. Nat.*, v. 105, p. 437–453.

Michod, R. E. and B. R. Levin, *eds.*, 1988, *The evolution of sex.* Sunderland, Mass., Sinauer.

Milliman, J. D., 1974, *Recent Sedimentary Carbonates Part I: Marine Carbonates.* New York, Springer–Verlag.

Möbius, K., 1877, *Die Auster und die Austernwirtscraft.* Berlin. (Transl. 1880. *The oyster and oyster culture.* Rep. US Fish Comm. p. 683–751.)

Moore, P. D., J. A. Webb, and M. E. Collinson, 1991, *Pollen Analysis*, 2nd ed., Oxford, Blackwell.

Moran, P. A., 1950, Some remarks on animal population dynamics. *Biometrics*, v. 6, p. 123.

Morgan, A., 1973, Late Pleistocene environmental changes indicated by fossil insect faunas of the English midlands. *Boreas*, v. 2, p. 173–212.

Mosimann, J. E., and P. S. Martin, 1975, Simulating Overkill by Paleoindians. *Am. Sci.*, v. 63, p. 304–313.

Moulton, M. P., and S. L. Pimm, 1983, The introduced Hawaiian avifauna: biogeographical evidence for competition. *Am. Nat.*, v. 121, p. 669–690.

Mueller, L. D., Pingzhong Guo, and F. J. Ayala, 1991, Density-dependent natural selection and trade-offs in life history traits. *Science*, v. 253, p. 433–435.

Mueller-Dombois, D., and H. Ellenberg, 1974, *Aims and Methods of Vegetation Ecology.* New York, John Wiley.

Muir, A., 1961, The Podzol and Podzolic soils. *Adv. Agron.*, v. 13, p. 1–56.

Murie, A., 1944, *The Wolves of Mount McKinley: Fauna of Nat. Parks of U.S. Fauna Series 5.* Washington, D.C.

Muscatine, L., and Porter, J. W., 1977, Reef corals: mutualistic symbioses adapted to nutrient-poor environments. *BioSci.*, v. 27, p. 454–460.

Myers, N., 1991, *Tropical Forests and Climate.* Boston, Kluwer.

Nagy, K. A., 1972, Water and electrolyte budgets of a free-living desert lizard *Sauvomalus obesus*. *J. Comp. Physiol.*, v. 79, p. 39–62.

Neftel, A., H. Oeschger, J. Schwander, B. Stauffer, and R. Zumbrunn, 1982, Ice core sample measurements give atmospheric CO_2 content during the past 40,000 years. *Nature*, v. 295, p. 220–223.

Neil, W. E., 1992, Population variation in the ontogeny of predator-induced vertical migration of copepods. *Nature*, v. 356, p. 54–57.

Nelson, B., 1968, *Galapagos: Islands of Birds.* New York, Morrow.

Nelson, M. E., and L. D. Mech, 1981, Deer social organization and wolf predation in northeastern Minnesota. *Wildl. Monogr.*, v. 77, p. 1–53.

Nelson, M. E., and L. D. Mech, 1985, Observation of a wolf killed by a deer. *J. Mamm.*, v. 66, p. 187–188.

Nicholson, A. J., 1954, An outline of the dynamics of animal populations. *Aust. J. Zool.*, v. 2, p. 9–65.

Nicholson, A. J., and V. A. Bailey, 1935, The balance of animal populations. *Proc. Zool. Soc. London*, v. 1935, p. 551–603.

O'Brien, W. J., 1979, The predator-prey interaction of planktivorous fish and zooplankton. *Am. Sci.*, v. 67, p. 572–581.

O'Brien, W. J., D. Kettle, and H. P. Riessen, 1979, Helmets and invisible armor: Structures reducing predation from tactile and visual planktivores. *Ecol.*, v. 60, p. 287–294.

Odum, E. P., 1969, The strategy of ecosystem development. *Science*, v. 164, p. 262–270.

Odum, E. P., 1971, *Fundamentals of Ecology*, 3rd ed., Philadelphia, Saunders.

Odum, H. T., 1956, Efficiencies, size of organisms, and community structure. *Ecol.*, v. 37, p. 592–597.

Odum, H. T. 1957, Trophic structure and productivity of Silver Springs, Florida. *Ecol. Monog.*, v. 27, p. 55–112.

Odum, H. T., *ed.*, 1970, *A Tropical Rain Forest: A Study of Irradiation and Ecology at El Verde, Puerto Rico.* Washington, D.C.

Odum, H. T., and E. P. Odum, 1955, Trophic structure and productivity of a windward coral reef community on Eniwetok Atoll. *Ecol. Monog.*, v. 25, p. 291–320.

Ollason, J. C., and G. M. Dunnet, 1988, Variation in breeding success in Fulmars, *in* T. H. Clutton-Broock, ed., *Reproductive success: Studies of individual variation in contrasting breeding systems.* Chicago, Univ. of Chicago Press.

Olson, J. S., 1958, Rates of succession and soil changes on southern Lake Michigan sand dunes. *Bot. Gaz.*, v. 199, p. 125–170.

Olson, S. L., and F. H. James, 1982, Fossil Birds from the Hawaiian Islands: Evidence for wholesale extinction by man before western contact. *Science*, v. 217, p. 633–635.

Orians, G., 1969, On the evolution of mating systems in birds and mammals. *Am. Nat.*, v. 103, p. 589–603.

Orians, G. H., 1980, *Some Adaptations of Marsh-nesting Blackbirds.* Princeton, Princeton Univ. Press.

Orme, G. R., 1985, The sedimentological importance of *Halimeda* in the development of back reef lithofacies, northern Great Barrier Reef (Australia). *Proceedings of the Fifth International Coral Reef Symposium*, v. 5, p. 31–37.

Osman, R. W., and R. B. Whitlach, 1978, Patterns of species diversity: fact or artifact? *Paleobiology*, v. 4, p. 41–54.

Oster, G. F., and E. O. Wilson, 1978, *Caste and Ecology in the Social Insects.* Princeton, Princeton Univ. Press.

Owen-Smith, N., 1971, Territoriality in the white rhinocerus (*Ceratotherium simun*) Burchell. *Nature*, v. 231, p. 294–296.

Paine, R. T., 1966, Food web complexity and species diversity. *Am. Nat.*, v. 100, p. 65–75.

Paine, R. T., 1992, Food-web analysis through field measurement of per capita interaction strength. *Nature*, v. 355, p. 73–75.

Park, T. D., B. Mertz, W. Grodzinski, and T. Prus, 1965, Cannibalistic predation in populations of flour beetles. *Physiol. Zool.*, v. 38, p. 289–321.

Pasteur, G., 1982, A classificatory review of mimicry systems. *An. Rev. Ecol. Syst.*, v. 13, p. 169–199.

Pennington, W., R. S. Cambray, J. D. Eakins, and D. D. Harkness, 1976, Radionuclide dating of the recent sediments at Blelham Tarn. *Freshwater Biol.*, v. 6, p. 317–331.

Peters, R. P., and L. D. Mech, 1975, Scent-marking in wolves. *Am. Sci.*, v. 63, p. 628–637.

Petersen, R., 1975, The paradox of the plankton: An equilibrium hypothesis. *Am. Nat.*, v. 109, p. 35–49.

Peterson, R. O., R. E. Page, and K. M. Dodge, 1984, Wolves, moose and the allometry of population cycles. *Science*, v. 224, p. 1350–1352.

Pianka, E. R., 1970, On r and K selection. *Am. Nat.*, v. 104, p. 592–597.

Pielou, E. C., 1975, *Ecological Diversity.* New York, John Wiley.

Pielou, E. C., 1979, *Biogeography.* New York, John Wiley.

Pielou, E. C., 1991, *After the Ice Age: The Return of Life to Glaciated North America.* Chicago, Univ. of Chicago Press.

Pilcher, J. R., M. G. L. Baillie, B. Schmidt, and B. Becker, 1984, A 7,272-year tree-ring chronology for western Europe. *Nature*, v. 312 p. 150–152.

Pimm, S. L., 1982, *Food Webs.* London, Chapman and Hall.

Pimm, S. L., J. H. Lawton, and J. E. Cohen, 1991, Food web patterns and their consequences. *Nature*, v. 350, p. 669–674.

Pimm, S. T., 1991, The balance of nature?: ecological issues in the conservation of species and communities. Chicago, Univ. of Chicago Press.

Pomiankowski, A., 1990, How to find the top male. *Nature*, v. 347, p. 616.

Pough, F. H., 1980, The advantages of ectothermy for tetrapods. *Am. Nat.*, v. 115, p. 92–112.

Pough, F. H., 1983, Amphibians and Reptiles as Low-Energy Systems. p. 141–188, in W. P. Aspey and S. I. Lustick, ed. *Behavioral Energetics: the cost of survival in vertebrates.* Columbus, Ohio State Univ. Press.

Pratt, T. K., and E. W. Stiles, 1985, The influence of fruit size and structure on composition of frugivore assemblages in New Guinea. *Biotropica*, v. 17, p. 314–321.

Prentice, I. C., P. J. Bartlein, and T. Webb, 1991, Vegetation and climate change in eastern North America since the last glacial maximum. *Ecol.*, v. 72, p. 2038–2056.

Price, P. W., 1975, Insect Ecology. New York, John Wiley

Price, P. W., T. M. Lewinsohn, G. W. Fernandes, and W. W. Benson, *eds.* 1991, *Plant-Animal interactions: evolutionary ecology in Tropical and Temperate Regions.* New York, John Wiley.

Pruett-Jones, S. G., and M. A. Pruett-Jones, 1990, Sexual selection through female choice in Lawes' Parotia, a Lek-Mating bird of Paradise. *Evol.*, v. 44, p. 486–501.

Pulliam, H. R., 1983, Ecological community theory and the coexistence of sparrows. *Ecol.*, v. 64, p. 45–52.

Pyke, G. H., 1978, Optimal foraging in bumble bees in coevolution with their plants. *Oecologia*, v. 36, p. 281.

Quinn, W. H., 1971, Late Quaternary meteorological and oceanographic developments in the equatorial Pacific. *Nature*, v. 229, p. 330–331.

Ralls, K., J. D. Ballou, and A. Templeton, 1988, Estimates of the cost of inbreeding in mammals. *Conserv. Biol.*, v. 2, p. 185–193.

Ralph, E. K., and J. Klein, 1979, Composite computer plots of ^{14}C dates for tree-ring-dated bristlecone pines and sequoias, p. 545–553 *in* Berger, R. and H. E. Suess, *ed. Radiocarbon Dating.* Berkeley, Univ. California Press.

Ramus, J., 1981, The capture and transduction of light energy, p. 458–492 *in* C. S. Lohban and M. J. Wynne, *eds.*, *Biology of Seaweeds* Oxford, Blackwell.

Ramus, J., S. K. Beale, D. Mauzerall, 1976, Changes in pigment content with photosynthetic capacity of seaweeds as a function of water depth. *Marine Biology*, v. 37, p. 231–238.

Rankama, K., and T. G. Sahama, 1950, *Geochemistry.* Chicago, Univ. of Chicago Press.

Rastetter, E. B., and R. A. Houghton, 1992, Carbon budget estimates. *Science*, v. 258, p. 382.

Rathcke, B. J., and R. W. Pool, 1975, Coevolutionary race continues: Butterfly larval adaptation to plant trichomes. *Science*, v. 187, p. 175–176.

Raup, D. M., 1986, Biological extinction in earth history. *Science*, v. 231, p. 1528–1533.

Raup, D. M., and G. Boyajian, 1988, Patterns of generic extinction in the fossil record. *Paleobiology*, v. 14, p. 109–125.

Raup, D. M., and J. J. Sepkoski, Jr., 1986, Periodic extinction of families and genera. *Science*, v. 231, p. 833–836.

Rechav, Y., R. A. I. Norval, J. Tannock, and J. Colborne, 1978, Attraction of the tick *Ixodes neitzi* to twigs marked by the klipspringer antelope. *Nature*, v. 275, p. 310–311.

Reiners, W. A., 1973, Terrestrial detritus and the carbon cycle, p. 303–327 *in* Woodwell, G. M., and E. V. Pecan, *eds.*, *Carbon and the Biosphere.* Washington, D.C.

Reynolds, R. C., 1965, The concentration of boron in Precambrian seas. *Geoch et Cosmoch. Acta*, v. 29, p. 1–16.

Rhoades, D. F., 1979, Evolution of plant chemical defense against herbivores, p. 3–54. *in* Rosenthal, G. A., and D. H. Janzen, *eds.*, *Herbivores. Their Interaction with Secondary Plant Metabolites.* New York, Academic Press.

Richards, F. A., 1968, Chemical and biological factors in the marine environment, p. 259–303 *in* Brahtz, J. F., *ed.*, *Ocean Engineering.* John Wiley, New York.

Richards, P. W., 1973, Africa, the "odd man out," p. 21–26 *in* Meggers, B. J., E. S. Ayensu, and W. D. Duckworth, *eds.*, *Tropical Forest Ecosystems in Africa and South America.* New York, Random House.

Richman, S., 1958, The transformation of energy by *Daphnia pulex. Ecol. Monog.*, v. 28, p. 273–291.

Ricker, W. E., 1954, Stock and recruitment. *J. Fish. Res. Bd. Canad.*, v. 11, p. 559–623.

Riley, G. A., 1963, Marine biology, I., p. 69–70 *in* Riley, G. A., *ed.*, *Proceedings of the First International Interdisciplinary Conference.* Washington, D.C., Am. Inst. Bio. Sci.

Rind, D., and D. Peteet, 1985, Terrestrial conditions at the last glacial maximum and CLIMAP sea-surface temperature estimates: Are they consistent? *Quat. Res.*, v. 24, p. 1–22.

Rind, D., C. Rosenzweig, and R. Goldberg, 1992, Modelling the hydrological cycle in assessments of climate change. *Nature*, v. 358, p. 119–122.

Ritland, D. B., and L. P. Brower, 1991, The viceroy butterfly is not a batesian mimic. *Nature*, v. 350, p. 497–498.

Root, J. B., 1967, The niche exploitation pattern of the blue-grey gnat catcher. *Ecol. Monog.*, v. 37, p. 317–350.

Rosenthal, G. A., and D. H. Janzen, *eds.*, 1979, *Herbivores. Their Interaction with Secondary Plant Metabolites.* New York, Academic Press.

Rowan, R., and D. A. Powers, 1991, A Molecular Genetic Classification of zoozanthellae and the evolution of animal-algal symbioses. *J. of Phycology*, v. 26, p. 490–494.

Rubinoff, I., 1968, Central American sea-level canal: possible biological effects. *Science*, v. 161, p. 857–861.

Rubenstein, P. I., and R. W. Wrangham, 1986, eds. *Ecological aspects of social evolution.* Princeton, New Jersey.

Ruttner, F., 1963, *Fundamentals of Limnology, 3rd. ed.* (transl. by Frey, D. G., and F. E. J. Fry). Toronto, Univ. Toronto Press.

Ryther, J. H., 1959, Potential productivity of the sea. *Science*, v. 130, p. 602–608.

Ryther, J. H., 1969, Photosynthesis and fish production in the sea. *Science*, v. 166, p. 72–76.

Salo, J. 1987, Pleistocene forest refuges in the Amazon: evaluation of the biostratigraphical, lithostratigraphical and geomorphological data. *Annales Zoologici Fennici*, v. 24, p. 203–211.

Sanders, H. L., 1968, Marine benthic diversity: A comparative study. *Am. Nat.*, v. 102, p. 243–282.

Sanders, H. L., and R. R. Hessler, 1969, Ecology of the deep-sea benthos. *Science*, v. 163, p. 1419–1424.

Scheutzow, M., 1992, The epidemiology and environmental impact of organophosphate pesticide use in Ecuador with emphasis on parathion. Ph.D. Thesis, The Ohio State University.

Schindler, D. W., 1978, Factors regulating phytoplankton production and standing crop in the world's lakes. *Limnol. Oceanogr.* v. 23, p. 478–486.

Schluter, D., T. D. Price, and P. R. Grant, 1985, Ecological character displacement in Darwin's finches. *Science*, v. 227, p. 1056–1059.

Schmidt-Nielsen, K., 1972*a*, *How Animals Work.* Cambridge, Cambridge Univ. Press.

Schmidt-Nielsen, K., 1972*b*, Locomotion: Energy costs of swimming, flying, and running. *Science*, v. 177, p. 222–228.

Schmidt-Nielsen, K., 1975, *Animal Physiology. Adaptation and Environment.* Cambridge, Cambridge Univ. Press.

Schoener, T. W., 1974, Resource partitioning in ecological communities. *Science*, v. 185, p. 27–39.

Schofield, E. K., and P. A. Colinvaux, 1969, Fossil *Azolla* from the Galapagos Islands. *Bull. Torrey Bot. Club*, v. 96, p. 623–628.

Schopf, J. W., 1975, Precambrian paleobiology: Problems and perspectives: *An. Rev. Earth and Planet, Science*, v. 3, p. 213–249.

Schultz, A. M., 1969, A study of an ecosystem: The arctic tundra, p. 77–93 *in* VanDyne, G. M., *ed.*, *The Ecosystem Concept in Natural Resource Management.* New York, Academic Press.

Searle, D. E., and Flood, P. G., 1988, *Halimeda* bioherms of the Swain Reefs-Southern Great Barrier Reef. *Proceedings of the Sixth International Coral Reef Symposium*, v. 3, p. 139–144.

Selye, H., 1950, *The Physiology and Pathology of Exposure to Stress; a Treatise Based on the Concepts of the General-Adaptation-Syndrome and the Diseases of Adaptation.* Montreal, Acta.

Sernander, R., 1908, On the evidences of postglacial changes of climate furnished by peat mosses of northern Europe. *Geol. Foreningens Forhandlingar*, v. 30, p. 465–473.

Shackleton, N., 1968, Depth of pelagic Foraminifera and isotopic changes in Pleistocene oceans. *Nature*, v. 218, p. 79–80.

Shaver, G. R., W. D. Billings, F. S. Chapin, A. E. Giblin, K. J. Nadelhoffer, W. C. Oechel, and E. B. Rastetter, 1992, Global Change and the Carbon Balance of Arctic Ecosystems. *BioSci.*, v. 42, p. 433–441.

Sheail, J., 1971, *Rabbits and Their History.* Newton Abbot, David and Charles.

Shields, W. M., 1982, *Philopatry, Inbreeding, and the Evolution of Sex.* New York, SUNY Press.

Simberloff, D., 1974, Equilibrium theory of biogeography and ecology. *An. Rev. Ecol. and Systematics*, v. 5, p. 161–182.

Simberloff, D. S., and E. O. Wilson, 1970, Experimental zoogeography of islands. A two-year record of colonization. *Ecol.*, v. 51, p. 934–937.

Skutch, A. F., 1949, Do tropical birds rear as many young as they can nourish? *Ibis*, v. 91, p. 430–455.

Slobodkin, L. B., 1962, Energy and animal ecology. *Adv. Ecol.*, v. 4, p. 69–101.

Smith, S. V., and Kinsey, F. J. R., 1988, Why don't budgets of energy, nutrients, and carbonates always balance at the level of organisms, reefs, and tropical oceans? An overview. *Proceedings of the Sixth International Coral Reef Symposium*, v. 1, p. 115–121.

Soholt, L. F., 1973, Consumption of primary production by a population of kangaroo rats (*Dipodomys merriami*) in the Mojave Desert. *Ecol. Monog.*, v. 43, p. 357–397.

Sollbrig, O. T., and B. B. Simpson, 1974, Components of regulation of a population of dandelions in Michigan. *Ecol.*, v. 62, p. 473–486.

Stanley, S. M., 1987, *Extinction.* New York, Scientific American Library.

Steinitz-Kannan, M., P. A. Colinvaux, and R. Kannan, 1983, Limnological studies in Ecuador: I A survey of chemical and physical properties of Ecuadorian lakes. *Archiv. Fur. Hydrobiol.*, v. 1, p. 61–105.

Stehli, F. R., R. G. Douglas, and N. D. Newell, 1969, Generation and maintenance of gradients in taxonomic diversity. *Science*, v. 164, p. 947–949.

Stenseth, N. C., 1985, Why mathematical models in evolutionary ecology? p. 239–287, in J. H. Cooley and F. B. Golley, *ed., Trends in Ecological Research for the 1980s.* New York: Plenum.

Stevens, G., and R. Belling, *eds.*, 1988, *The Evolution of Sex.* San Francisco, Harper and Row.

Stewart, R. E., and J. W. Aldrich, 1951, Removal and population of breeding birds in a spruce-fir forest community. *Auk*, v. 68, p. 471–482.

Stiles, F. G., 1977, Coadapted competitors: the flowering seasons of hummingbird-pollinated plants in a tropical forest. *Science*, v. 198, p. 1177–1178.

Strahler, A. N., 1969, Physical geography. 3rd ed. New York, John Wiley.

Street, F. A., and A. T. Grove, 1979, Global maps of lake level fluctuations since 30,000 yr. B.P. *Quat. Res.*, v. 12, p. 83–118.

Strickland, J. D. H., 1960, Measuring the Production of marine phytoplankton, *Fisheries Research Board of Canada.* Bulletin No. 122, p. 172.

Strickland, J. D. H., and T. R. Parsons, 1972, A practical handbook of seawater analysis. *J. of Fisheries Res. Board of Canada*, v. 167, p. 1–311.

Strickler, J. R., 1977, Observation of swimming performances of planktonic copepods. *Limnol. and Oceanog.*, v. 22, p. 165–170.

Strong, D. R., Jr., 1974, Rapid asymptotic species accumulation in phytophagous insect communities: the pests of cacao. *Science*. v. 185, p. 1064–1065.

Strong, D. R., E. D. McCoy, and J. R. Rey, 1977, Time and the number of herbivore species: the pests of sugarcane. *Ecol.*, v. 58, p. 167–175.

Struhsaker, T., 1977, Infanticide and social organization in the redtail monkey (*Cercopithecus ascanius schmidti*) in the Kibale Forest, Uganda. *Zeitschrift fur Tierpsych.*, v. 45, p. 75–84.

Struhsaker, T. T., and M. Leakey, 1990, Prey selectivity by crowned hawk-eagles on monkeys in the Kibale forest, Uganda. *Behav. Ecol. Soc.*, v. 26, p. 435–443.

Stuijts, I., J. C. Newsome, and J. R. Flenley, Jr., 1988, Evidence for late quaternary vegetational change in the Sumatran and Javan highlands. *Review of Palaeobotany* Palynology, v. 55, p. 207–216.

Stuiver, M., 1978, Atmospheric carbon dioxide and carbon reservoir changes. *Science*, v. 199, p. 253–258.

Summerhayes, E. S., and C. S. Elton, 1923, Contributions to the ecology of Spitsbergen and Bear Island. *J of Ecol.*, v. 11, p. 214–286.

Talbot, L. M., and M. H. Talbot, 1963, The wildebeest in Western Masailand, East Africa. *Wildlife Monog.*, v. 12, p. 88.

Tansley, A. G., 1935, The use and abuse of vegetational concepts and terms. *Ecol.*, v. 16, p. 284–307.

Tansley, A. G., and R. S. Adamson, 1925, Studies of the vegetation of the English chalk. III. The chalk grasslands of the Hampshire-Sussex border. *J. Ecol.*, v. 13, p. 177–223.

Taylor, C. R., and V. J. Rowntree, 1973, Running on two or on four legs: Which consumes more energy? *Science*, v. 179, p. 186–187.

Tedrow, J. C. F., 1965, Arctic soils. *Proc. First Permafrost Int. Conf.*, p. 50–55.

Tedrow, J. C. F., and D. E. Hill, 1955, Arctic brown soil. *Soil Sci.*, v. 80, p. 265–275.

Temple, S. A., 1977, Plant animal mutualism: Coevolution with dodo leads to near extinction of plant. *Science*, v. 197, p. 885–886.

Terborgh, J., 1973, Chance, habitat, and dispersal in the distribution of birds in the West Indies. *Evolution*, v. 27, p. 338–349.

Terborgh, J., 1974, Preservation of natural diversity: The problem of extinction prone species. *BioSci.*, v. 24, p. 715–722.

Terborgh, J., 1988, The big things that run the world—a sequel to E. O. Wilson. *Conser. Biol.*, vol. 2, p. 402–403.

Terborgh, J. 1992, Diversity and the Tropical Rainforest. *Scientific American Library*, New York, W. H. Freeman.

Thomas, M. F., 1974, *Tropical Geomorphology*, New York, Macmillan.

Thompson, D. W., 1942, *On Growth and Form.* Cambridge, Cambridge Univ. Press.

Till, A. R., 1979, Nutrient cycling, p. 277–285, in R. T. Coupland, *ed. Grassland Ecosystems of the World: Analysis of Grasslands and their uses.* Cambridge, Cambridge Univ. Press.

Towe, K. M., 1978, Early Precambrian oxygen: A case against photosynthesis. *Nature*, v. 274, p. 657–661.

Tranquillini, W., 1979, *Physiological Ecology of the Alpine Timberline.* New York, Springer-Verlag.

Transeau, E. N., 1926, The accumulation of energy by plants. *Ohio J. of Sci.*, v. 26, p. 1–10.

Tucker, V. A., 1975, The energetic cost of moving about. *Am. Sci.*, v. 63, p. 413–418.

Turekian, K. K., 1977, The fate of metals in the oceans. *Geoch.* et *Cosmoch. Acta*, v. 41, p. 1139–1144.

Urey, H. C., H. A. Lowenstam, S. Epstein, and C. R. McKinney, 1951, Measurement of paleotemperatures of the Upper Cretaceous of England, Denmark, and the southeastern United States. *Bull. Geol. Soc. of Am.*, v. 62, p. 399–416.

Valentine, J. W., and C. A. Campbell, 1975, Genetic regulation and the fossil record. *Am. Sci.*, v. 63, p. 673–680.

Van Balgooy, M. M. J., 1971, Plant geography of the Pacific as based on a census of phanerogam genera. *Blumea Sup.* v. 6, p. 1–222.

Van Valen, L., 1973, A new evolutionary law. *Evol. Theory*, v. 1, p. 1–30.

Vane-Wright, R. I., 1991, A case of self-deception. *Nature*, v. 350, p. 460–461.

Varley, G. C., G. R. Gradwell, and M. P. Hassell, 1973, *Insect Population Ecology.* Oxford, Blackwell.

Vaurie, C., 1951, Adaptive differences between two sympatric species of nuthatches (*Sitta*): *Proc. Int. Ornithol. Cong.*, v. 19, p. 163–166.

Vernon, J. E. N., 1986, Corals of Australia and the Indo-Pacific. Angus and Robertson, Townsville, Australia.

Vigilant, L., M. Stoneking, H. Harpending, K. Hawkes, and A. C. Wilson, 1991, African populations and evolution of human mitochondrial DNA. *Science*, v. 253, p. 1503–1507.

Vollrath, F., and G. A. Parker, 1992, Sexual dimorphism and distorted sex ratios in spiders. *Nature*, v. 360, p. 156–159.

von Post, Lennart, 1916, Om skogstradspollen; sydsvenska torfmosselagerfoljder (foredragsreferat). Geolgiska Foereningen i stockholm. *Foerhandlingar*, v. 38, p. 334–384. (Forest tree pollen in South Swedish Peat Bog deposits)

Walker, D., 1970, Direction and rate in some British postglacial hydroseres, p. 117–139 *in* Walker, D., and R. West, *eds.*, *The Vegetational History of the British Isles.* Cambridge, Cambridge Univ. Press.

Wallace, A. R., 1876, *The Geographic Distribution of Animals.* 2 vols. London, Macmillan.

Wallace, A. R., 1878, *Tropical Nature and Other Essays*. London, Macmillan.
Walter, H., 1973, *Vegetation of the Earth*. New York, Springer-Verlag.
Wardle, P., 1968, Englemann spruce (*Picea engelmannii* Engel) at its upper limits on the Front Range, Colorado. *Ecol.*, v. 49, p. 483-495.
Wassink, E. D., 1959, Efficiency of light energy conversion in plant growth. *Plant Physiol.*, v. 34, p. 356-361.
Webb, T., and R. A. Bryson, 1971, Late and postglacial climatic change in the northern Midwest, USA: Quantitative estimates derived from fossil pollen spectra by multivariate statistical analysis. *Quat. Res.*, v. 2, p. 70-115.
Webb, T., 1986, Is vegetation in equilibrium with climate? How to interpret late Quaternary pollen data. *Vegetatio*, v. 67, p. 75-91.
Webb, T., 1988, Eastern North America, pp. 385-414 *in* B. Huntley and T. Webb III, *eds., Vegetation History*. Kluwer, Dordrecht, The Netherlands.
Wegener, A., 1966, *The Origin of Continents and Oceans* (trans. by Biram, J.). New York, Dover.
Went, F. W., and N. Stark, 1968, Mycorrhiza. *BioSci.*, v. 18, p. 1035-1039.
Wetzel, R. G., 1975, *Limnology*. Philadelphia, Saunders.
White, T. C. R., 1969, An index to measure weather-induced stress of trees associated with outbreaks of Psyllids in Australia. *Ecol.*, v. 50, p. 905-909.
Whitham, T. G., 1977, Coevolution of foraging in *Bombus* and nectar dispensing in *Chilopsis:* A last dreg theory. *Science*, v. 197, p. 593-596.
Whitmore, T. C., and G. T. Prance, 1987, *Biogeography and Quaternary History in Tropical America*. Oxford, Oxford Univ. Press.
Whittaker, R. H., 1956, Vegetation of the Great Smoky Mountains. *Ecol. Monog.*, v. 26, p. 1-80.
Whittaker, R. H., 1960, Vegetation of the Siskiyou Mountains, Oregon and California. *Ecol. Monog.*, v. 30, p. 279-338.
Whittaker, R. H., 1962, Classification of natural communities. *Botan. Rev.*, v. 28, p. 1-239.
Whittaker, R. H., 1975, *Communities and Ecosystems*. 2nd ed. New York, Macmillan.
Whittaker, R. H., 1977, Evolution of species diversity in land communities. *Evol. Biol.*, v. 10, p. 1-67.
Whittaker, R. H., F. H. Bormann, G. E. Likens, and T. G. Siccama, 1974, The hubbard brook ecosystem study: Forest biomass and production. *Ecol. Monog.*, v. 44, p. 233-254.
Whittaker, R. H., and P. P. Feeny, 1971, Allelochemics: Chemical interactions between species. *Science*, v. 171, p. 757-770.
Whittaker, R. H., and G. E. Likens, 1973, Carbon in the biota, p. 281-302 *in* Woodwell, G. M., and E. V. Pecan, *eds., Carbon and the Biosphere*. Washington, D.C.
Whittaker, R. H., and P. L. Marks, 1975, Methods of assessing terrestrial productivity, p. 55-118 *in* Lieth, H., and R. H. Whittaker, *eds., Primary Productivity of the Biosphere*. New York, Springer-Verlag.
Whittaker, R. H., and G. M. Woodwell, 1968, Dimension and production relations of trees and shrubs in the Brookhaven Forest, New York. *Ecol.*, v. 56, p. 1-25.
Whittaker, R. J., M. B. Bush, and K. Richards, 1989, Plant recolonization and vegetation succession on the Krakatau Islands, Indonesia. *Ecol. Monog.*, v. 59, p. 59-123.
Wicklow, D. T., and G. C. Carroll, 1981, *The Fungal Community*. New York, Dekker.
Wiens, J. A., 1966, On group selection and Wynne-Edwards' hypothesis. *Amer. Sci.*, p. 273-287.
Wiley, R. H., 1973, Territoriality and nonrandom mating in sage grouse, *Centrocercus urophasianus*. *An. Behav. Monog.*, v. 6, p. 85-169.
Williams, E. C., Jr., 1941, An ecological study of the floor fauna of the Panama rain forest. *Bull. Chicago Acad. Sci.*, v. 6, p. 63-124.
Williams, G. C., 1975, *Sex and Evolution*. Princeton, Princeton Univ. Press.
Wilson, D. S., 1980, *The Natural Selection of Populations and Communities*. Menlo Park, Calif., Benjamin Cummings.
Wilson, E. O., 1969, The species equilibrium, p. 38-47 *in* Woodwell, G. E., and H. H. Smith, *eds. Diversity and Stability in Ecological Systems*. Brookhaven Symp. in Biol., v. 22, p. 38-47.
Wilson, E. O., 1971, *The Insect Societies*. Cambridge, Harvard Univ. Press.

Wilson, E. O., 1975, *Sociobiology*. Cambridge, Harvard Univ. Press.

Wilson, E. O., and D. S. Simberloff, 1969, Experimental zoogeography of islands. Defaunation and monitoring techniques. *Ecol.*, v. 50, p. 267–278.

Wilson, E. O., *ed.*, 1988, *Biodiversity*. Washington, D.C., National Academy Press.

Wilson, A. T., and M. J. Grinstead, 1975, Palaeotemperatures from tree rings and the D/H ratio of cellulose as a biochemical thermometer. *Nature*, v. 257, p. 387–388.

Winter, K., 1985, Crassulacean acid metabolism, *in* J. Barber and N. R. Baker, *eds.*, *Photosynthetic Mechanisms and the Environment*. New York, Elsevier.

Winter, K., and H. Ziegler, 1992, Induction of Crassulacean acid metabolism in *Mesembryanthemum crystallinum* increases reproductive success under conditions of drought and salinity stress. *Oecologia* (in press).

Wolda, H., 1992, Trends in abundance of tropical forest insects. *Oecologia*, v. 89, p. 47–52.

Wolf, L. L., and F. R. Hainsworth, 1982, Economics of foraging strategies in sunbirds and hummingbirds, p. 223–259 *in* Aspey, W. P., and S. Lustick, *eds.*, *Behavioral Energetics: Vertebrate Costs of Survival*. Columbus, Ohio State Univ. Press.

Wood, B. J., 1971, Development of integrated control programmes for pests of tropical perennial crops in Malaysia, p. 422–457 *in* Huffaker, C. B., *ed.*, *AAAS Symposium on Biological Control, Boston*. New York, Plenum Press.

Woodwell, G. M., and D. B. Botkin, 1970, Metabolism of terrestrial ecosystems by gas exchange techniques: The Brookhaven approach, p. 73–86 *in* Reichle, D. E., *ed.*, *Analysis of Temperate Forest Ecosystems*. New York, Springer-Verlag.

Woodwell, G. M., and W. R. Dykeman, 1966, Respiration of a forest measured by CO_2 accumulation during temperature inversions. *Science*, v. 154, p. 1031–1034.

Woodwell, G. M., R. H. Whittaker, W. A. Reiners, G. E. Likens, C. C. Delwiche, and D. B. Botkin, 1978, The biota and the world carbon budget. *Science*, v. 199, p. 141–146.

Wu, L. S-Y, and D. B. Botkin, 1980, Of elephants and men: A discrete stochastic model for long-lived species with complex life histories. *Am. Nat.*, v. 116, p. 831–849.

Wynne-Edwards, V. C., 1962, *Animal Dispersion in Relation to Social Behavior*. New York, Hafner.

Yadara, P. S., and J. S. Singh, 1977, *Grassland Vegetation, its Structure, Function, Utilization and Management*. New Delhi, Indo-Am.

Yang, S. Y., and J. L. Patton, 1981, Genic variability and differentiation in Galapagos finches. *Auk*, v. 97, p. 230–242.

Yoda, K., 1967, A preliminary survey of the forest vegetation of eastern Nepal. II. General description, structure and floristic composition of the sample plots chosen from different vegetation zones. *J. Col. Arts* and *Sci.*, Chiba Univ., Nat. Sci. Ser., v. 5, p. 99–140.

Zajic, J. E., 1969, *Microbial Biogeochemistry*. New York, Academic Press.

Zaret, T. M., 1980, *Predation and Freshwater Communities*. New Haven, Yale Univ. Press.

Zaret, T. M., and R. T. Paine, 1973, Species introduction in a tropical lake. *Science*, v. 182, p. 449–455.

Zaret, T. M., and J. S. Suffern, 1976, Vertical migration in zooplankton as a predator avoidance mechanism. *Limnol. and Oceano.*, v. 21, p. 804–813.

Photo Credits

Chapter 1

Figure 1.1: Henry M. Mayer/Photo Researchers. Figure 1.3: Science Vu/Visuals Unlimited.

Chapter 2

Preview: Omikron/Photo Researchers. Figure 2.1: The New Yorker. Figure 2.8: Ray Bishop/Stock, Boston.

Chapter 3

Preview: Stephen Dalton/NHPA. Figure 3.1: Joe Schuyl/Stock, Boston. Figure 3.9: Margarite Bradley/Stock, Boston. Figure 3.17: L. Hillis. Figure 3.18 (top): Paul Colinvaux/Photo Researchers. Figure 3.18 (bottom): Runk/Schoenberger/Grant Heilman Photography. Figure 3.19 : Paul A. Colinvaux.

Chapter 4

Preview: Scott Barry/The Image Works. Figure 4.2: M. Cognac/National Film Board of Canada, Phototeque. Figure 4.3: Steven Fuller/Peter Arnold. Figure 4.4: Tom McHugh/Photo Researchers.

Chapter 5

Preview: Julie O'Neil/Stock, Boston. Figure 5.6: Julie Bartlett/Photo Researchers. Figure 5.9 (top): Verna Johnston/Photo Researchers. Figure 5.9 (bottom): YLLA/Photo Researchers. Figure 5.11(a): Verna R. Johnston/Photo Researchers. Figure 5.11(b): Robert Lamb/Photo Researchers. Figure 5.11(c): Jim Yoakum. Figure 5.12: Paul A. Colinvaux. Figure 5.13: Peter Zwerger/Forest Research Institute.

Chapter 6

Preview: George H. Harrison/Grant Heilman Photography. Figure 6.1: Anthony Mercieca/Photo Researchers. Figure 6.3: Jack Dermid/National Audubon Society/Photo Researchers. Figure 6.5: Jen & Des Bartlett/Photo Researchers. Figure 6.6: Ron Austing/Photo Researchers. Figure 6.7 (left): Hal Harrison/Grant Heilman Photography. Figure 6.7 (bottom): A. W. Ambler, National Audubon Society/Photo Researchers. Figure 6.8: R. S. Virdee/Grant Heilman Photography. Figure 6.9: Peter Menzel/Stock, Boston. Figure 6.10: Stephen Dalton/Photo Researchers. Figure 6.12: A. W. Ambler/Photo Researchers. Figure 6.13: Joyce Photographics/Photo Researchers.

Chapter 7

Preview: Courtesy NOAA. Figure 7.4: Max & Bea Hunn/Visuals Unlimited. Figure 7.7: Klans D. Francke/Peter Arnold. Figure 7.10: Courtesy NASA. Figure 7.11: Eric Grave/Photo Researchers. Figure 7.13: E. K. Schofield.

Chapter 8

Preview: Manfred Danegger/NHPA.

Chapter 9

Preview: Hubertus Kanus/Rapho/Photo Researchers. Figure 9.1 (left): A. W. Ambler/Photo Researchers. Figure 9.1 (right): Eric Hosking/Photo Researchers. Figure 9.2: Frank Schleicher/VIREO. Figure 9.5: Alvin E. Staffen/Photo Researchers. Figure 9.8 (left): Miguel Castro/Photo Researchers. Figure 9.8 (right): D. Cavagnaro/Peter Arnold.

Chapter 10

Preview: Jen & Des Bartlett/Photo Researchers. Figure 10.17: Courtesy Department of Lands, Queensland, Australia. Figure 10.23: Kelvin Aitken/Peter Arnold.

Chapter 11

Preview: Jim Balog/Photo Researchers. Figure 11.7: Leonard Lee Rue, III/National Audubon Society/Photo Researchers. Figure 11.8: Denver Bryan/Comstock. Figure 11.11: George Holton/Photo Researchers. Figure 11.13: Renee Lynn/Photo Researchers. Figure 11.15: D. Overcash/Bruce Coleman. Figure 11.16: Karl H. & Stephen Maslowski/Photo Researchers. Figure 11.17: L. Hillis.

Chapter 12

Preview: Comstock. Figure 12.4: Allan Morgan/Peter Arnold. Figure 12.7: Photofile.

Chapter 13

Preview: Roger Wilmhurst/Bruce Coleman. Figure 13.1: Bob & Ira Spring. Figure 13.5: Toni Angermayer/Photo Researchers. Figure 13.6: Leonard Lee Rue, III. Figure 13.8: Scott Barry/Image Works. Figure 13.10: C. H. Greenewalt/Academy of Natural Sciences. Figure 13.11: Leonard Lee Rue, III/Photo Researchers. Figure 13.13: S. Nagendra/Photo Researchers. Figure 13.14: Department of Photography, Ohio State University. Figure 13.18: Fred Breunner. Figure 13.19: Courtesy National Marine Fisheries Service, Honolulu Laboratory.

Chapter 14

Preview: Tom McHugh/Photo Researchers. Figure 14.1: L. David Mech. Figure 14.2: Stephen J. Krasemann/Photo Researchers. Figure 14.5: Robert Caputo/Photo Researchers. Figure 14.6: YLLA/Rapho/Photo Researchers. Figure 14.7: Bob & Ira Spring.

Chapter 15

Preview: Hans Pfletschinger/Peter Arnold. Figure 15.1: Toni Angermayer/Photo Researchers. Figure 15.2: Leonard Lee Rue, Jr./Photo Researchers. Figure 15.3 (top): Courtesy Lawrence E. Gilbert, Department of Zoology, University of Texas, Austin. Figure 15.4: Pierre Berger/Photo Researchers. Figure 15.6: Thomas S. Elias/Rancho Santa Ana Botanic Garden. Figure 15.7: Courtesy Carl W. Rettenmeyer, University of Connecticut. Figure 15.8: B. Miller/Biological Photo Service. Figure 15.9: Z. Lesczczynski/Animals Animals.

Chapter 16

Preview: Paul Sterry/Nature Photographers. Figure 16.2: Russ Kinne/Photo Researchers.

Chapter 17

Preview: Courtesy Scott W. Rogers. Figure 17.2: Nicholas B. Carter. Figure 17.12: Owen Franken/Stock, Boston. Figure 17.16: Dr. G. R. Gope, Department of Geological Sciences, University of Birmingham, England.

Chapter 18

Preview: Luiz C. Marigo/Peter Arnold. Figure 18.4: Steve McCutcheon/Alaska Pictorial Service. Figure 18.5: Paul A. Colinvaux. Figure 18.6: Courtesy U.S. Forest Service. Figure 18.7: Grant Heilman/Grant Heilman Photography. Figure 18.8: Pierre Berger/Photo Researchers. Figure 18.9: Hans Pfletschinger/Peter Arnold.

Figure 18.10: Paul A. Colinvaux. Figure 18.11: Nicholas B. Carter. Figure 18.12: James R. Simon/Photo Researchers. Figure 18.13 (left): L & D Klein/Photo Researchers. Figure 18.13 (center): Nicholas B. Carter. Figure 18.13 (right): Tom McHugh/Photo Researchers. Figure 18.14: Mark N. Boultan/National Audubon Society/Photo Researchers. Figure 18.15: Howard Sochurgk/Woodfin Camp & Associates. Figure 18.16: Russ Kinne/Photo Researchers. Figure 18.17: Leonard Lee Rue, III/ Photo Researchers. Figure 18.21: R. E. Wallace/Courtesy USGS.

Chapter 19

Preview: Ian Adams Photography. Figure 19.10: Grant Heilman Photography.

Chapter 20

Preview: Tom & Pat Leeson/Photo Researchers. Figure 20.9: Paul A Colinvaux. Figure 20.10: Frank J. Staub/The Picture Cube.

Chapter 21

Preview: Georg Gerster/Comstock. Figure 21.3: Peter Menzel/Stock, Boston. 21.4: George Harrison/Grant Heilman Photography.

Chapter 22

Preview: Ross E. Hutchins/Photo Researchers.

Chapter 23

Preview: Tom Bean/The Stock Market. Figure 23.8: Nicholas B. Carter.

Chapter 24

Preview: Jen and Des Bartlett/Photo Researchers. Figure 24.7: Paul A. Colinvaux. Figure 24.8: George W. Gardner/Stock, Boston. Figure 24.16: Gregory Ochocki/Photo Researchers. Figure 24.17: Runk/Schoenberger/Grant Heilman Photography.

Chapter 25

Preview: Gilles Guitard/The Image Bank.

Chapter 26

Preview: Sinclair Stammers/Science Photo Library/Photo Researchers. Figure 26.9: Paul A. Colinvaux.

Chapter 27

Preview: James T. Spencer/Photo Researchers. Figure 27.3: Rudi Strickler. Figure 27.5: Courtesy Dr. P. Colin.

Chapter 28

Preview: John Heseltine/Science Photo Library/Photo Researchers. Figure 28.1: Fritz Henle/Photo Researchers. Figure 28.7: Lisa Law/The Image Works. Figure 28.11: Karen Preuss/The Image Works. Figure 28.18: H. Armstrong Roberts.

Chapter 29

Preview: Lynn McLaren/Photo Researchers. Figure 29.1: Nancy Tucker/Photo Researchers. Figure 29.3: George Holton/Photo Researchers. Figure 29.4: Allan Cash/Rapho/Photo Researchers. Figure 29.5 : Nicholas B. Carter.

Index